Sustainable Treatment and Reuse of Municipal Wastewater

Sustainable Treatment and Reuse of Municipal Wastewater

For Decision Makers and Practicing Engineers

Menahem Libhaber and Álvaro Orozco-Jaramillo

IWA Publishing
London • New York

Published by **IWA Publishing**
Alliance House
12 Caxton Street
London SW1H 0QS, UK
Telephone: +44 (0)20 7654 5500
Fax: +44 (0)20 7654 5555
Email: publications@iwap.co.uk
Web: www.iwapublishing.com

First published 2012

British Library Cataloguing in Publication Data
A CIP catalogue record for this book is available from the British Library

ISBN 9781780400167 (Hardback)
ISBN 9781780400631 (eBook)

MIX
Paper from responsible sources
FSC® C007785

Printed by Bell & Bain Ltd., Glasgow

Contents

About the Authors

Menahem Libhaber, PhD, Consulting Engineer. Dr. Libhaber received an MSc in Chemical Engineering and a PhD in Water Resources and Environmental Engineering from the Technion, Israel Institute of Technology, Haifa, Israel. Prior to joining the World Bank in 1991, he worked for 18 years for the consulting firm Tahal Consulting Engineers as a water and sanitation engineer in Israel and many other countries including Brazil, Costa Rica, Peru, El Salvador, Chile, Mexico, Honduras, Turkey, Spain, Yugoslavia, and Nigeria. He served for three years as a consultant to UNEP - United Nations Environmental Program, the Mediterranean Action Plan. He joined the World Bank in 1991 as a Lead Water and Sanitation Engineer in the Latin America Region. He has served as task manager of water and sewerage projects in Colombia, Bolivia, Argentina, Costa Rica, the Dominican Republic, Peru, and Trinidad and Tobago. He has also worked on other projects in Brazil, Paraguay, St. Lucia, Guyana, Venezuela, Jamaica, Honduras, Panama, Haiti, India, and China. Many of these projects involved design and construction of wastewater treatment plants. Dr. Libhaber is Coauthor of the Book "Marine Wastewater Outfalls and Treatment Systems" (2010). He also published numerous articles and reports and delivered numerous presentations in international conferences and seminars. Dr. Libhaber retired from the World Bank in July 2009. He is presently a private consulting engineer to the World Bank and other institutions. E-mail: mlibhaber@gmail.com

Alvaro Orozco-Jaramillo, MSc, Consulting Engineer. Civil Engineer of the School of Mining, National University of Colombia (1971) and MSc in Sanitary Engineering of the Pennsylvania State University (1976). Worked in the fields of environmental engineering, water treatment, solid wastes management and global warming. Winner of the Award Diódoro Sánchez of the Colombian Society of Engineers in 1986 and Mention of Honor of the same award in 1981. Was also granted the Merit Award "World Water Year" in 2003 from the Commission of

Sanitary Engineering of the Colombian Society of Engineers, Category Academic-Investigator. Works as an International Consultant for the World Bank, the Inter-American Development Bank, Biwater and other institutions in various countries, including Brazil, Argentina, Uruguay, Bolivia, Central America, the Caribbean, Peru and Colombia. Was a professor in the Los Andes University in Bogota, Colombia, for 20 years and in the University of Antioquia, Colombia, in various opportunities. Author of the Textooks "Bioingeniería de Aguas Residuales" (2005) (Wastewater Bioengineering), "Desechos Sólidos" (1980) (Solid Wastes) and Coauthor of the book "Tratamiento Biológico de las Aguas Residuales" (1985) (Biological Wastewater Treatment). Also published numerous articles and delivered international presentations in conferences and seminars. E-mail: alvaro.orozco.jaramillo@hotmail.com

Acknowledgements

Many people contributed to the writing of this book. The authors are particularly indebted to SAGUAPAC, the outstanding water and sanitation utility of the city of Santa Cruz, Bolivia, especially to its Senior Mangers, the General Manager Fernando Ibanez Cuellar and the Director of Projects and Works Fernando Yavari Mendez, as well as to the Engineers Tito Calvimonte and Fernando Trigo for involving the authors in selecting the processes of upgrading the wastewater treatment plants of the city, for authorizing the publication of performance data of these treatment plants and for supplying and authorizing the publication of photographs and illustrations of the treatment plants.

The authors are grateful to Klaus Dieter Neder, Adviser to the President of CAESB, the water and sanitation utility of the city of Brasilia, Brazil, for his support, important advice, hospitality in presenting the wastewater treatment plants of Brasilia during the authors' visit to the city and for authorizing the publication of photographs of these treatment plants. Similarly, the authors are grateful to Cleverson Vitorio Andreoli, Research Engineer in SANEPAR, the water and sanitation utility of the State of Parana, Brazil, and Professor in the Federal University of Parana, for his hospitality in presenting the wastewater treatment plants of SANEPAR during the authors' visit in the State of Parana.

Thanks are due to Yechiel Menuchin, CEO of Environmental Protection Technologies for his advice and ideas, to Cicero Onofre de Andrade Neto, Professor at the Federal University of Rio Grande do Norte, Natal, Brazil, for his contribution on Anaerobic Filters and to Marcio Schittini and Luiz Pereira, of ACESA Bioengeneria in Rio de Janeiro, Brazil, for their advice and contribution on utilization of biogas for generation of energy. The authors are grateful to Albert Gordon, President of the National Water Commission (NWC) of Jamaica, and to Vernon Barrett, Vice-president of NWC, for providing information on the Wastewater Treatment Plant of Kingston, Jamaica, and authorizing the publication of this information.

Several colleagues contributed, each in his own way, to the writing of this book. They include, among others: Haim Weinberg, Fernando Troyano, Miguel Vargas-Ramirez, Luz Maria Gonzalez, David Sislen, Franz Drees-Gross, Paula Dias Pini and Renan Alberto Poveda. Thanks are due to each of them.

The Authors are thankful to Marcelo Juanico, CEO of Environmental Consultants Ltd. for authorizing the publication of illustrations from his articles and book. In addition, one of the Authors (Orozco) extends his thanks to the School of Mining's former Professor Alvaro Pérez Arango, who was the first person to teach him the importance of developing a design algorithm or Orderly Design Method based on process fundamentals, and not on trial and error methods so common in engineering design.

The Authors are thankful and acknowledge the contribution and authorization of Huber Technology and its CEO, George Huber, to publish photographs and illustrations of its products, especially related to equipment of Rotating Micro Screens, and the contribution and authorization of Gurney Environmental and its CEO, John Gillett, to publish photographs and illustrations of the products of this company, especially mixers for lagoons.

Dedication

This book is dedicated to:

- The memory of my beloved parents Shifra and Jacob Libhaber
- My son Barak Libhaber
- My sister Klara Glesinger, her husband David and her Children Ronen, Iris and Merav
- Paula Dias Pini
- And to the memory of my dear friend, mentor and teacher Dr. Emanuel Idelvitch, who was taken from us before his time

Menahem Libhaber

- To my loving wife, Beatriz Munera, for a life of companionship and patience
- My daughters Lina and Fernanda and her husband, Rainer Viertel
- Last but not least, to my adored grandsons Friedrich and Martin Viertel-Orozco

Alvaro Orozco-Jaramillo

Preface

The uncontrolled disposal to the environment of municipal, industrial and agricultural liquid, solid and gaseous wastes constitutes one of the most serious threats to the sustainability of the human race by contaminating water sources, land and air, and by its potential contribution to global warming. With increasing population and economic growth, treatment and safe disposal of wastewater is essential to preserve public health and reduce intolerable levels of environmental degradation. In addition, adequate wastewater management is also required for preventing contamination of water bodies for the purpose of preserving the sources of clean water.

Effective wastewater management is well established in developed countries, but is still limited in developing countries. In most developing countries many people are lacking access to water and sanitation services. Collection and conveyance of wastewater out of urban neighborhoods is not yet a service provided to all the population and adequate treatment is provided only to a small portion of the collected wastewater, in most cases covering less than 10% of the municipal wastewater generated. In slums and peri-urban areas it is not rare to see raw wastewater flowing in the streets. The inadequate water and sanitation service is the main cause of diseases in developing countries.

In 2011 the population of the planet was 7 billion. Population growth forecasts indicate a rapid global population growth which will reach 9 billion in 2030. The forecasts also indicate that: (i) most of the population growth will occur in developing countries while the population of developed countries will remain constant at about 1 billion; and (ii) a strong migration from rural to urban areas will take place. Considering the expected population growth and the order of priorities in the development of the water and sanitation sector in developing countries (water supply and sewerage first and only then wastewater treatment), as well as the financial difficulties in these countries, it cannot be assumed that the current low percentage of the coverage of wastewater treatment in these countries will increase in the future, unless a new strategy is adopted and innovative, affordable wastewater treatment options are used. Application of appropriate wastewater treatment technologies, which are effective, low cost (in investment and especially in operation and maintenance), simple to operate, proven technologies, should be a key component in any strategy aimed at increasing the coverage of wastewater treatment. Appropriate technology processes are also more environment-friendly since they consume less energy and have therefore a positive impact on mitigating climate change effects. Also, with modern design, appropriate technology processes cause less environmental nuisance than conventional processes, for example they produce lower amounts of excess sludge and their odor problems can be effectively controlled.

Unfortunately the need to adopt appropriate technology processes is in many cases not understood to decision makers in developing countries. There is a tendency to apply cutting edge technologies consisting of highly mechanized, complex treatment plants which are of high investment costs and of high operation and maintenance costs. Investment financing for complex treatment plants can sometimes be mobilized in developing countries in the form of grants and/or soft loans; however, it is almost impossible to obtain grants or subsidies for operation and maintenance of such plants. Usually, the authorities (municipalities or water and sanitation utilities) do not have the capacity to finance high operation and maintenance costs of complex treatment plants from their internal cash generation, and so this type of treatment plants tend to deteriorate rapidly due to insufficient budget for operation and maintenance, and many of them are abandoned a short time after being commissioned. This indicates that complex plants are not sustainable in developing countries and points to the need for the employment of plants based on alternative, simpler and low cost appropriate technology processes.

A variety of unit process of appropriate technology with a proven track record are known and in operation for many years, each yielding a different effluent quality. Some provide low quality effluents and some, effluents of good quality. When an effluent quality higher than what a single unit process of appropriate technology can produce is required, a treatment plant consisting of a series of appropriate technology unit processes can be used (2, 3 or more), in which the effluent of the first unit process is fed into the second process for polishing and the effluent of the second process is fed to the third and so on, if necessary. This approach can produce practically any final effluent quality required. *The idea of the ability to combine unit processes to create a treatment plant based on a series of appropriate technology processes which jointly can generate any required effluent quality is the main message of this book.* A plant based on a combination in series of appropriate technology unit processes is still easy to operate and is usually of lower costs than conventional processes in terms of investments and certainly in operation and maintenance. So in essence, this book present the concept of sustainable appropriate technology processes and the basic engineering design procedures to obtain high quality effluents by treatment plants based on simple, low cost and easy to operate processes.

The concepts of appropriate technology for wastewater treatment and issues of strategy and policy for increasing wastewater treatment coverage are presented in the first part of the book. In the second part each chapter is dedicated to a selected unit process of appropriate technology and provides the scientific basis, the equations and the parameters required to design the unit processes, with some design and process innovations developed by the authors. The book also presents some chapters on design procedures for selected combined processes which are in use in developing countries. Once the fundamentals of each unit and combined process have been established, the book proposes in each chapter an innovative Orderly Design Method (ODM), easy to be followed by practicing engineers, using the equations and formulas developed in the first section of each chapter. At the end of each chapter, a numeric example for the basic design of each selected appropriate technology process is solved for a city with a population of 20,000 using the ODM and an Excel program which is provided to the readers for download from an online web site (http://www.iwawaterwiki.org/xwiki/bin/view/Articles/Software+Developed+for+Sustainable+Treatment+and+Reuse+of+Municipal+Wastewater). The book also presents ideas of many additional combinations of unit processes of appropriate technology, classified according to their adequacy for functioning in different temperature zones and in accordance with the size of land area occupied by the wastewater treatment plant. Finally, the book contains a chapter on climate change and the potential impact of wastewater treatment on climate change.

The book title contains the concept of sustainability of wastewater treatment. It is intuitively clear that the use of appropriate technology wastewater treatment plants can significantly enhance their sustainability. They are simple to operate and their operation and maintenance costs are low so there are no financial and technical difficulties to keep them adequately operating over an extended period of many years and no reason to abandon them a short time after their commissioning. However the sustainability aspects of appropriate technology treatment plants have a much wider scope. First they contribute to improving the overall environmental sustainability since the use of appropriate technology enables the expansion of the coverage of wastewater treatment in developing countries. In addition, appropriate technology processes can contribute to enhancing the sustainability of utilities in several ways: (i) by enhancing the financial sustainability of the utilities due to reduced investment as well as operation and maintenance costs; (ii) by enhancing the technical and operational sustainability of the treatment plants through the employment of simple to operate and maintain processes based on simple, mostly locally manufactured equipment; and (iii) by enhancing the institutional sustainability of the utilities since due to the limited financial demand and technical efforts, they do not present meaningful problems to the utilities' managements, do not impose additional managerial efforts, reduce the institutional burden and challenges of the water and sanitation utilities and thereby contribute to enhancing institutional sustainability. In fact, the use of appropriate technologies in wastewater treatment helps to alleviate the main problems of the water and sanitation sector in developing countries, which are: financial weakness, low technical capacity and institutional weakness, thereby contributing to improving the sustainability of the sector as a whole.

The inclusion in the book title of the concept of reuse (which refers to reuse of effluents for irrigation) requires an explanation. Seemingly the book contains only one chapter on reuse, chapter 7, which presents the concept of stabilization reservoirs as an important component of any reuse system, and applies an innovative algorithmic design approach. The proposed reuse concept provides that the general scheme of a reuse system consists of preliminary treatment followed by a stabilization reservoir. The preliminary treatment system can be any installation able to reduce the organic matter content of the wastewater to a level which prevents development of anaerobic conditions in the reservoir. If the pretreatment system is based on any one of the appropriate technology processes presented in the other chapters of the book, then the entire reuse system is an appropriate technology system. So in fact the entire book applies to wastewater reuse for irrigation. However, the focus of reuse in the book is on the technical aspects and design of reuse systems and practical implementation of reuse projects, and it does not analyze other aspects of reuse, which can be found in the professional literature.

Part 1 of the book (theory and concepts) is directed to policy and decision makers, utilities managers and staff, as well as to practitioners and scholars interested in concepts but not in design. The objective of Part 1 is to explain that there are alternatives to mechanized technologies which can be as effective in terms of effluent quality and advantageous from other perspectives. Part 2 of the book is directed to water and sanitation engineers, consulting firms, staff of water and sanitation utilities, project managers, water and sanitation practitioners, technicians and other professionals dealing with water and environmental issues, academic scholars, professors, teachers and students, providing them with an innovative tool which employs for each process an algorithmic Orderly Design Method applied through an Excel program to perform the calculations once the input information has been introduced.

Although the focus of the book is the resolution of wastewater treatment and disposal problems in developing countries, the concepts presented are valid and applicable anywhere and plants based on combined unit processes of appropriate technology can be used also in developed countries and provide to them the benefits described in the book.

The authors hope that the book provides information that will be of value to all who are involved in any way with wastewater treatment and disposal, including those involved in the decision-making process, those involved in the design of treatment plants, and those concerned with their environmental impacts. We especially hope that the book will contribute to rational choices of wastewater treatment and disposal schemes and to sound wastewater management, especially in developing countries.

Contents: The first part of the book presents the concepts of appropriate technology and of combining unit processes to achieve higher quality effluents, as well as issues of strategy and policy for expanding the coverage of wastewater treatment. The second part deals with the fundamentals of wastewater treatment, process design and design examples including: Decomposition Processes of Organic Matter, Calculation of Municipal Wastewater Flow and BOD Load, Rotating Micro Screens, Treatment in Stabilization Lagoons, Anaerobic Treatment (Upflow Anaerobic Sludge Blanket Reactor-UASB, Anaerobic Filter, Piston Anaerobic Reactor), Stabilization Reservoirs, Horizontal Flow Constructed Wetland, Chemically Enhanced Primary Treatment (CEPT), Other complementary processes like Sand Filtration, Dissolved Air Flotation (DAF) and UV Disinfection, as well as Combinations of appropriate Technology Processes: (i) Rotating Micro Screens Followed by UASB followed by Anaerobic Filter, (ii) Rotating Micro Screens Followed by UASB followed by Facultative Lagoons, (iii) Rotating Micro Screens Followed by UASB followed by Sand Filtration, (iv) Rotating Micro Screens Followed by CEPT followed by Sand Filtration, and (v) Rotating Micro Screens Followed by UASB followed by Anaerobic Filter followed by DAF followed by Membrane Filtration, and Global Warming and the impact of Wastewater Treatment on Climate Change.

Menahem Libhaber
Alvaro Orozco-Jaramillo

Nomenclature

a	Net area, m^2/m^3
a	Pipeline orifice area
a_p	Passing area in the UASB separator
a_g	Gas exit area of the UASB separator
A_J	Surface area of process J (i.e des: grit channel, s: settling basing, M: maturation lagoon, etc.)
A_p	Main pipeline area
AAL	Aerated Aerobic Lagoons
ABR	Anaerobic Baffle Reactor
ACF	Altitude Correction Factor (masl)
AD	Anaerobic digestion
AF	Anaerobic Filter
AMet	Methanogenic Activity
AOR	Actual Oxygen Requirement
A/S	Air to solids ratio
A/V	Area to volume ratio
A_a	Afferent area of infiltration, ha
A_c	Crop area in SR, ha
As	Surface area
A_T	Transversal area
A_{UASB}	Surface area of the UASB, m^2
α	Earth Albedo, approximately 0,30
α	Transfer correction factor of O_2, tap water/WW
B	Fecal Coliforms, MPN/100mL
B	Wetland width
BOD	Biochemical Oxygen Demand
BOD_5	Biochemical Oxygen Demand at day five
BOD_u	Ultimate Biochemical Oxygen Demand

BODT	Total BOD_5
β	Transfer correction factor of O_2 for salinity
c	Concentration, mg/L
C_0	Concentration of O_2 at operating conditions
C_s	Saturation concentration of O_2 at Standard Conditions
C_{sT}	Saturation concentration of O_2 at temperature T
CAESB	The water and sanitation utility of the Federal District of Brasilia, Brazil
Cal	Calories
CBOD	Carbonaceous BOD
CEMS	Chemically Enhanced Rotating Microscreening
CDM	Clean Development Mechanism
CEL	Cost-Effective Level, $kgBOD_{5r}/USD_i$
CEL	Cost-effective Level, $USD\$/kgCOD_r$
CEPT	Chemically Enhanced Primary Treatment
CER	Carbon Emission Reduction
CMAS	Completely Mixed Activated Sludge
CMI	Mean Investment Cost
CMLP	Mean Long-term Cost
CMO	Mean Operating Cost
Coli Fecal	FC, NMP/100 mL
COD	Chemical Oxygen Demand
COPASA	The water and sanitation utility of the State of Minas Gerais, Brazil
CO_2	Carbon Dioxide
CRE	The Power Utility of Santa Cruz, Bolivia
CRT	Cell retention time, θ_c (Sludge Age)
CWW	Combined waste waters
d	Particle effective size
d	Dispersion factor
D	Dose UV, $W \cdot s/m^2$ or J/m^2.
D	Axial dispersion coefficient, m^2/h
DAF	Dissolved air flotation (or diffused)
DNA	Deoxyribonucleic Acid
DO	Dissolved Oxygen, mg/L
DOM	Degradable Organic Matter
DTC	Developing and Transition Countries
DWW	Domestic wastewater
DWWT	Domestic Wastewater Treatment
D_{10}	Sand effective size
D_T	Drum diameter of a MS, m
D_w	Total monthly demand for agriculture water in a SR, $m^3/ha \cdot mes$
ΔCH_4	Methane produced, mg/L CH_4
$\Delta G^{\circ\prime}$	Standard free energy, kJ/reaction
$\Delta G'$	Real free energy, kJ/reaction
Δh	Hydraulic head loss
ΔO_2	Oxygen uptake, mg/L of O_2

ΔS	Substrate Removal, mg/L BOD_u or COD
ΔX	Biomass Production, mg/L SSVLM
EF	Emission Factor
EHSA	Extremely High Sludge Ages
ET	Real evapotranspiration, mm/month
ET_0	Evapotranspiration Potential, mm/month
EU	European Union
EW	Equivalent Weight, eq/L
ε_0	Bed porosity
f	Factor of proportionality in photosynthetic lagoons, 0.5 m/d
FC	Fecal Coliforms
FLC	Food limiting conditions
F_{O2}	Oxygenation factor
FDS	Fixed Dissolved Solids
FSS	Fixed Suspended Solids
F/M	Organic Load, kg COD/kg MLVSS · d
φ	Granule form factor; 1 if spherical
G	Hydraulic Gradient, s^{-1}
GCM	Global Climate Model
GHG	Greenhouse Gases
GSLS	Gas-Solid-Liquid Separator, in a UASB and PAR
GSLS-SM	Standard Model of the Gas-Solid-Liquid Separator, in a UASB and PAR
γ	Water specific weight, N/m^3
σ	Boltzmann's Constant, $5{,}6697 \times 10^{-8}$ W/m^2K^4
h	Head loss, depth, m
h	Depth
h_f	Head loss, m
ha	Hectare
H	Depth
H_J	Depth of process J (i.e UASB, lagoon, etc.)
H_G	UASB's GSLS Depth
H_L	UASB's Sludge depth
H_T	UASB's Total depth
HCR	Hydrograph Controlled Release
HDT	Hydraulic detention time, t_d
HDPE	High Density Poly Ethylene
HP	Horse power
I	UV Ray intensity, W/m^2
IAT	Innovative Appropriate Technology
IO	Inverse Osmosis
IPCC	Intergovernmental Panel on Climate Change
IWW	Industrial wastewater
IWWT	Industrial Wastewater Treatment
k	Eckenfelder's equation constant
k	Anaerobic metabolic change rate

k	Bottle constant of the CBOD base e
K	Bottle constant of the CBOD base 10
K	Screens Coefficient
K_B	FC removal constant
K_H	Henry's constant
K_{Hi}	Wetland constant of first order for $i = BOD_5$, TKN, NO_3 and FC, d^{-1}
K_p	First order area constant for P. K_p is 0,0273 m/d, in SSFCW
K_h	Proportional constant of Percolator Filter
K_{La}	Aeration Coefficient
K_O	Orozco's constant (depends on θ_c)
K_w	Ion product constant, $[H+][OH-] = 1 \times 10^{-14}$
k_0	Net maximum rate of substrate removal
k_c	Contois saturation constant
k_c	Coefficient of each crop (ET real on ET_0)
k_e	Endogenous Coefficient, d^{-1}
k_L	McKinney's equation constant
k_m	Monod's saturation constant
k_s	Hydraulic Conductivity, m/d
KWH	Kilo Watt Hour
L	Remnant CBOD, mg/L
L	Length, m
L	Liter
l	Liter
LAC	Latin America and Caribbean Region
LAS	Low Power Level mixers
Lps	Liters per second
L_{BOD}	DBO load, kg/d
L_{OD}	Dissolved Oxygen Load, kg/d
L_q	Air load in the biofilter
L_s	Surface Organic Load, $kgBOD_5/ha \cdot d$
L_v	Volumetric Organic Load, $kg/m^3 \cdot d$ BOD_5 or sCOD
L/w	Length to width ratio
λ	Substrate maximum biodegradability, %
m	$-K_h/q_a n$ in percolating filter
MBR	Membrane Biological Reactor
MCF	Methane correction factor
MF	Micro Filtration
ML	Mixed Liquor
MLC	Mass limiting conditions
MLSS	Mixed Liquor Suspended Solids
MLVSS	Mixed Liquor Volatile Suspended Solids
MPN	Most Probable Number, E-Coli per 100 mL
MPS	Method of Process Selection
MS	Micro Screens
MW	Molecular Weight, g/mole

masl	Meters above sea level
mole	Gram molecular weight
n	Potential constant of Percolating Filter
n	Manning's rugosity coefficient
n	Filter media Porosity, %
NF	Nanofiltration
n	Undefined number
η	Methane Concentration in biogas
NBOD	Nitrogenous Biochemical Oxygen Demand
NH_3	Ammonia
NO_3	Nitrate
N_0	O_2 Transfer of mixing aerator, kg/h · HP
O_2	Oxygen, mg/L
ODM	Orderly Design Method
OECD	High Income Countries (members of the Organization of Economic Cooperation and Development)
OM	Organic Matter
O&G	Oil and Grease, mg/L
O&M	Operation and Maintenance
p	Barometric pressure, kPa
psi	Pounds per square inch
P	Population of design, hab
P	Power, in HP or kW
P	Pressure, atm
PAR	Plug-flow anaerobic reactor or Piston anaerobic reactor
PF	Peak Factor
PFE_i	Percentage of "fresh" effluent during the last "i" days
PL	Power Level, $kW/1000\ m^3$ or $HP/1000\ ft^3$
PT	Primary treatment
P_x	Sludge Production in the reactor, kg/d
Q	Flow
Q_D	Design Average Flow of a WWTP
Q_{DH}	Design Hydraulic Flow of a WWTP
Q_{dom}	Flow of domestic WW
q	Per capita flow, L/hab · d
q_a	Hydraulic Load, Lps/m^2
q_{BOD_5}	Per capita BOD load, kg BOD_5/capita · d
q_{dom}	Domestic per capita flow, L/hab · d
q_F	Average Filtration Rate ($m^3/m^2.\ d \equiv m/d$)
q_{H_2O}	Hydraulic load in trickling filter, $m^3/m^2 \cdot h$
q_I	Infiltration, Lps/ha
Q_{DWW}	Flow of domestic WW
Q_I	Infiltration Flow, Lps
Q_{maxd}	Maximum daily flow, k_1Q_D
Q_{maxh}	Maximum hourly flow, $k_1k_2Q_D$

Q_r	Return flow
Q_s	Solids Load, $kgMLSS/m^2 \cdot d$
Q_S	BOD_5 Load, kg/d
Q_w	Excess sludge discharge
r_a	Removal rate of anaerobic substrate
r_s	Net rate of substrate removal, dS/Xdt
R	Recirculation Ratio, Q/Qr
R	dO_2/dt
R	Pressurized Recirculation at DAF
R	Hydraulic Radius, m
RAFA	Reactor Anaerobio de Flujo Ascendente (UASB in Spanish and Portuguese)
RMS	Rotating Micro Screens
RNA	Ribonucleic Acid
RNG	Renewable Natural Gas
RO	Reverse Osmosis
RW	Rain Water
ρ_p	Particle Density
S	Solar Radiation $(cal/cm^2 \cdot d)$
S_a	MLSS at DAF
s	Hydraulic Slope
s_a	Air Solubility, at DAF
S_a	Influent suspended solids, mg/L
S	Substrate, mg/L BOD_5 o COD
SAGUAPAC	The water and sanitation utility of the city of Santa Cruz, Bolivia
SANEPAR	The water and sanitation utility of the State of Parana, Brazil
SAT	Soil Aquifer Treatment
SBOD	Sulfur Biochemical Oxygen Demand
sBOD	Soluble BOD
SBR	Sequencing Batch Reactor
sCOD	Soluble COD
SEPA	Sidestream Elevated Pool Aeration
SF	Sand Filtration
SFCW	Surface Flow Constructed Wetland
SL	Stabilization Lagoon
SM-GSLS	Standard model of GSLS
SP	Stabilization Ponds
SR	Stabilization Reservoir
s	Slope, decimal
SOR	Surface Overflow Rate, Q/A_s
SOTR	Standard Oxygen Transfer Rate
SS	Suspended Solids
SSC	Steady State Conditions
SSFCW	Sub-Surface Flow Constructed Wetland
SST	Total Suspended Solids
ST	Septic Tank

SVI	Sludge Volume Index
SW	Solid Wastes
S/X	Substrate to biomass ratio, Orozco's equation
TKN	Total Kjehldahl Nitrogen
TSS	Total Suspended Solids
t	Contact time, s
T	Temperature
T_a	Atmosphere Temperature, °K, °C
T_g	Temperature of earth's surface, ° K
T_i	Temperature of influent DWW, °C
T_L	Lagoon temperature, °C
T_{aire}	Filter air draft, mm H_2O
TOC	Total Organic Carbon
t_d	Detention time, V/Q
t_{dmj}	Average detention time at SR
θ	Temperature correction coefficient
θ_c	Sludge Age or CRT
θ_{cmin}	Minimum Sludge Age
U	dS/Xdt
UASB	Upflow Anaerobic Sludge Blanket Reactor
UC	Uniformity Coefficient in Sand Filters (D_{60}/D_{10})
UF	Ultra Filtration
USA	United States of America
US$	United States Dollar
USD_i	US $ invested
USEPA	United States Environmental Protection Agency
UUFCCC	United Nations Framework Convention on Climate Change
UV	Disinfection with ultraviolet rays
μ	Biomass net growth rate
μ	Dynamic viscosity, $N \cdot s/m^2$
μ_m	Maximum biomass net growth rate
μF	Micro Filtration
μS	Micro sieves
V	Volume
V_J	Volume of process J (i.e. A: anaerobic lagoon, F: facultative lagoon, etc.)
V_a	Settled Volume in ½ hour, mL
V_{biogas}	Volume of produced biogas
V_F	Final volume (or definitive)
V_g	Gas exit load in a UASB, $m^3/m^2 \cdot h$
v	Velocity, m/s, m/h, m/d
v_b	Velocity of WW in the RAP's baffles
v_L	Sludge accumulation rate, $m^3/hab \cdot year$
v_r	Ascent Velocity in the UASB, m/h
v_g	Gas exit velocity in a UASB, $m^3/m^2 \cdot h$
vp	Pass velocity through units of the GSLS, m/h

v_s	OFR in the Settling basin
VFA	Volatile Fatty Acids
VDS	Volatile Dissolved Solids
VNG	Vehicular Natural Gas
VSS	Volatile Suspended Solids
υ	Water kinematics viscosity
W	*Parshall* flume throat width, inches
W	Width, m
w	Width, m
WHO	World Health Organization
WSP	Water and Sanitation Program of the World Bank
WW	Wastewater
WWT	Wastewater Treatment
WWTP	Wastewater Treatment Plant
WWTS	Wastewater Treatment System
w_b	Width of tilted baffles of the GSLS at UASB
X	Biomass, mg/L MLVSS
X_e	Effluent SS secondary settling basin
X_{ib}	Influents biodegradable SS
X_{ii}	Influents inorganic SS
$X_{\inf}$	Influents SS
X_T	Total Mass of the ML, TMLSS
Y	Yield Coefficient, g VSS/g COD removed
$\mathbf{Y_{obs}}$	Observed Yield Coefficient
Y_{O_2}	Production of O_2 (kg O_2/ha · d)
$\mathbf{Y_a}$	Yield Coefficient of Acidogenic Bacteria
$\mathbf{Y_m}$	Yield Coefficient of Methanogenic Bacteria
γ_a	Viable Fraction, X/XT

Part 1

Concepts

Chapter 1

Appropriate technologies for treatment of municipal wastewater

1.1 INTRODUCTION

1.1.1 Wastewater treatment issues in developing countries

The uncontrolled disposal to the environment of municipal, industrial and agricultural liquid, solid and gaseous wastes constitutes one of the most serious threats to the sustainability of the human race by contaminating water sources, land and air, and by its potential contribution to global warming.

It is commonly accepted that adequate collection, treatment and disposal of municipal wastewaters is required in order to prevent public health risks and environmental degradation. A lot of attention is currently being directed to the global water crisis and to water scarcity issues. Wastewater management needs also to be evaluated in the context of the global water crisis. A short discussion of this crisis is presented in Section 1.18, from which it is concluded that in addition to protection of public health and of the environment, adequate wastewater management is also required for preventing contamination of water bodies for the purpose of maintaining additional sources of clean water. The water crisis discussion also indicates the importance of municipal effluents reuse to irrigate farmland as both means for generating an additional source of water for irrigation and as a method for totally eliminating discharge of effluents to clean surface water bodies. Similarly it highlights the need for adopting low cost and simple to operate treatment processes in developing countries as means to alleviate economic problems by lowering investments and O&M expenses, and especially as means to alleviate institutional problems by applying processes that are easily manageable.

On July 28, 2010 the UN General Assembly declared that "Safe drinking water and sanitation is a human right essential to the full enjoyment of life and all other human rights". Worth noting is the fact that not only safe drinking water but also sanitation services are considered by the UN as a human right. We profoundly agree that water and sanitation services are a Human Right. These are basic rights essential for sustaining human life. Our interpretation regarding the UN declaration is that all governments have the obligation to supply potable water in sufficient quantities and sanitation services to all their citizens at an affordable cost. This is not being done by most governments in developing countries, in which part of the population does not have access to water and sanitation services. Although water and sanitation are human rights, they are also economic goods. There is a cost for collection, conveyance, treatment, storage and distribution of water, and a cost for collection of the wastewater, its treatment and safe disposal. In principle, the beneficiaries of the services (known as service users) should pay the full cost of these services. However, governments, national or locals, should provide subsidies to poor users who cannot pay for the full cost of the services.

The text of the UN General Assembly resolution of July 28, 2010 also expresses deep concern that an estimated 0.9 billion people lack access to safe drinking water and a total of more than 2.6 billion people do not have access to basic sanitation. The resolution does not specifically indicate that the water and sanitation problem is mainly located in developing countries, but the 192-memner assembly called on the UN member states and international organizations to offer funding, technology and other resources to help poorer countries scale up their efforts to provide clean, accessible and affordable drinking water and sanitation for everyone. The fact is that most of the people lacking access to water and sanitation services reside in developing countries and the water crisis is, for the moment, mainly concentrated in developing countries and for the most part affects the poor population in those countries. Unfortunately, in developing countries collection and conveyance of wastewater out of urban neighbourhoods is not yet a service provided to all the population and adequate treatment is provided only to a small portion of the collected wastewater, usually covering less than 10% of the municipal wastewater generated. In slums and peri-urban areas, in which a large portion of the population of developing countries resides (an average of about 38% of the total urban population in 2005 with significant differences among geographical region), it is not rare to see raw wastewater flowing in the streets. The inadequate water and sanitation service is the main cause of diseases in developing countries. About 80% of the diseases in these countries result from the effects of contaminated water or lack of water. Water Borne diseases are responsible for more deaths than any other cause. About 3.3 million people die annually from water borne diseases, 1.5 million of them children under the age of 5. In developed countries, water and sewerage coverage are higher than in developing countries, however, wastewater treatment coverage, although much higher than in developing countries, is still far from reaching universal coverage. In spite of the low coverage of wastewater treatment in developing countries, given the prevailing water and sanitation sector problems, governments consider, and rightfully so, that provision of safe drinking water and expanding sewerage coverage are of a higher priority than wastewater treatment.

In 2011 the population of the planet was approaching 7 billion. Forecasts indicated a rapid global population growth which will reach 9 billion in 2030. Given this rapid growth, the challenges in the water sector are enormous. The ensuing global warming and the adaptation actions which need to be undertaken magnify the challenges. The world's Population growth forecast (World Bank, 2003) indicates that: (i) most of the population growth will occur in developing countries while the population of developed countries will remain constant at about 1 billion; and (ii) a strong migration from rural to urban areas will take place, mostly in developing countries. As a result of this urbanization process, about 60% of the population will live in urban areas in 2030 (up from 45% in 2010).

The implications of the trends of population growth and urbanization are: (i) increased urban water demand; (ii) increased generation of municipal wastewater; and (iii) increased demand of water for agriculture to increase irrigated farmland for the purpose of generating sufficient food for the growing population. Future water resources management strategy will have to be based on: (i) efficient irrigation; (ii) water conservation (reduction of water losses, efficient use of water, water demand management); (iii) better use of all available resources including small scale solutions and use of marginal water such as saline sources; (iv) reuse of municipal effluents for irrigation; (v) effluent recycling for industrial and non potable municipal reuse; (vi) application of adequate tariff policies and regulation to reduce consumption; (vii) desalination and (viii) additional measures. A more detailed discussion on the global water crisis and a strategy to mitigate it is presented in Section 1.18.

Considering the expected population growth and the order of priorities in the development of the water and sanitation sector in developing countries (water supply and sewerage first and only then wastewater treatment), as well as the financial difficulties in these countries, it is difficult to expect that the current low percentage of the coverage of wastewater treatment in developing countries will increase in the future, unless a new strategy is adopted and innovative, affordable wastewater treatment options are used. Application of appropriate wastewater treatment technologies, which are effective, low cost technologies as detailed in the following sections, is a key component in any strategy aimed at increasing the coverage of wastewater treatment in developing countries. Appropriate low cost solutions are more adequate for such countries than highly mechanized and complex conventional solutions such as activated sludge, which have high operation and maintenance cost due to their high energy consumption and the production of large quantities of excess biomass which needs to be treated and properly disposed, and which usually also constitutes an environmental nuisance. The modification of the approach towards wastewater treatment and adoption of innovative, sustainable appropriate technology treatment processes forms part of a more general strategy for facing the global water crisis, as discussed in Section 1.18.

There are several reasons to the sanitation problems in developing countries. Those include governance weakness at the central and local governments' levels, institutional capacity weakness at the utilities level including low technical level, financial weakness and assigning low priority to sanitation problems. The main problem and most difficult to resolve is the institutional weakness of the utilities. Institutional weakness is a problem not only in the water and sanitation sector and is, in general, the main problem of developing countries. The concept outlined in this book mainly provides advice regarding means for relief of the technical and financial wastewater treatment problems of water and sanitation utilities. But in addition, it also provides advice on relief of the institutional problems of the utilities, since adoption of simple, low cost treatment processes alleviates not only the financial problems, but also the institutional burden.

1.1.2 Effluent quality standards

Defining the level of wastewater treatment and selecting the treatment processes depends mainly on the effluent quality standards prescribed by the Law, so these standards have an important impact on the development of the sector. Developed countries have usually stringent standards which are becoming more stringent with time and do not take into account the assimilation capacity of the receiving water bodies but rather require almost complete elimination of pollutants prior to the discharge of effluents to the environment. There is a continuous endeavour in these countries to achieve higher and higher effluent quality levels, through the development of more sophisticated treatment technologies. The problem is that the investment costs and especially the operation and maintenance costs of processes based on such technologies are high. This approach may be reasonable for developed countries, in which users have the capacity to pay for the high level of treatment and in which the governments usually provide significant subsidies for environmental projects like grants for construction of wastewater treatment plants. Even in developed countries standards have evolved during many years and were not very stringent at the outset, since the principle of gradual development was well understood in these countries.

In most developing countries, wastewater treatment and reuse standards are inspired by standards of developed countries, mainly the USA and the EU, without taking into account economic aspects and financing limitations related to such high standards. These standards usually require effluents of such quality that secondary or higher treatment levels are required to produce it, irrespective of the assimilation capacity of the receiving water body (i.e., the capacity of the receiving body is not taken into account in the standard setting process but rather the best available treatment technology). Since the costs of complying with these standards are high, sometimes beyond reach in developing countries, the overly stringent effluent standards induce a strategy of "No Action" with devastating public health and environmental impacts. In most cities in developing countries raw wastewater is discharged to receiving bodies, usually causing pollution problems. In many cases, raw wastewater is discharged to the streets due to lack of sewerage systems. It is common to find that when local authorities take the first step in the sanitation program and request an environmental license to discharge to a receiving body partially treated effluent, the environmental regulation agency denies the request, insisting on achievement of the stringent effluent quality standards right from the start. Since achieving high effluent quality at the first stage of a sanitation program is usually not financially feasible, the planned first stage of partial treatment cannot be implemented and raw wastewater keeps flowing into the receiving body. This is an irrational behavior of regulation agencies; which accepts discharge of raw wastewater to receiving bodies but do not accept discharge of partially treated effluents. Nevertheless, most of such agencies in developing countries consider that this is the correct course of action.

As an example, there is no rationale in requiring secondary treatment prior to discharge of effluent to the ocean through an effective submarine outfall. Partial treatment of the wastewater is sufficient in this case, as explained by Roberts J.W.P. *et al.* (2010). However, standards of many developing countries require secondary treatment prior to ocean discharge. Similarly, stringent standards of effluent reuse for irrigation prevent in most cases implementation of adequate effluents reuse projects due to the high costs required for reaching these standards, thereby inducing reuse of raw wastewater for irrigation, which is an inadequate outcome.

It is important for authorities in developing countries to understand that not in all cases is the highest effluent quality required. The assimilation capacity of receiving bodies needs to be taken into account, at least at the first stage of project development. As an example, the wastewater of a small city located on the bank of a large river does not need to undergo a high level of treatment. And even the wastewater of a large city, if it is located on the bank of a very large river (like the Parana, De la Plata, Magdalena, Orinoco or the Amazon, to mention some such rivers in Latin America) does not need to undergo a high level of treatment. Also, wastewater contains different types of pollutants, whose relative impact in damaging the environment depends on the type of the receiving body into which the effluent is discharged. Consequently, the type of treatment needs to be tailored to the type of the receiving body. For instance, if effluent is discharged to a river, organic matter is the main pollutant of importance due to its capacity to deplete the oxygen of the river water, if the effluent is discharged to a lake, nutrients are pollutants of importance due to their capacity for causing eutrophication, whereas if the effluent is used for irrigation of food crops or is discharged to the sea, pathogens are the pollutants of concern due to the risk they impose on the health of consumers of the irrigated crops or on the health to people bathing in the sea water.

It is impractical to expect that cities in developing countries that never treated their wastewater can achieve in one stage the highest effluent quality level. Effluent standards in developing countries should therefore take into account the need for staged development of treatment systems. Wastewater management schemes should be developed in stages in accordance with availability of financial resources and capacity, while taking into account in the first stage the assimilation capacity of the receiving bodies and the systems' construction time, so that the stringent effluent standard are achieved after time, at the ultimate project stage, while maintaining reasonable environmental standards at the first stages so that the systems are developed in such a manner that no environmental nuisance is caused by them even at the first stage. This can usually be achieved by taking into account the assimilation capacity of the receiving water bodies.

1.2 WASTEWATER TREATMENT PRINCIPLES

1.2.1 Introduction

The selection and design of an adequate wastewater treatment processes depends to a large extent on the final disposal option of the effluent. The selected treatment process must be tailored to fit the proposed disposal method, which may include: discharge to an ocean or sea, to a river or stream, to a lake, infiltration to groundwater, reuse for irrigation, or recycling for other purposes. The size and design of wastewater treatment plants is based on the forecasted raw wastewater flow rate (in m^3/day) and on the contaminants loads (in ton/day) contained in the raw wastewater.

Reliable forecasts of wastewater flows and contaminants loads are important since treatment plants are designed to handle conditions typically 10 to 20 years in the future. Plants are frequently constructed in a modular form, with new modules added as the raw wastewater flow increases. Forecasts of flows and contaminants loads are based on forecasts of population growth, changes of water consumption, contribution of contaminants per capita, and evolution of the sewerage coverage area. Flow forecasts should be prepared with care, since the flow may not always grow much with time, especially in developing countries. Changes of tariff policy may induce more efficient water use and moderate the evolution of water consumption and generation of wastewater. An example is Bogotá, Colombia, where water consumption has been constant over the past 15 years in spite of significant population growth. This resulted from: (i) periods of failures of the water supply infrastructure, with consequent severe rationing that brought people to consume less water, and (ii) significant tariff increases which resulted in reduced consumption.

1.2.2 Key pollutants in municipal wastewater

Typical municipal wastewater consists of about 99.9% water and 0.1% pollutants. About 60 to 80% of the pollutants are dissolved and the rest are suspended matter. The pollutants include mineral and organic matter, suspended solids, oil and grease, detergents, nitrogen (in various forms such as ammonia, nitrate, and organic nitrogen), phosphorous in various forms, sulphur, phenols, and heavy metals. The organic matter consists of many compounds, and since it is practically impossible to identify and quantify all of them, it is usually represented by the Biochemical Oxygen Demand (BOD), which is the amount of oxygen required to biologically decompose the organic matter, or by Chemical Oxygen Demand (COD), which is the amount of oxygen required to chemically decompose the organic matter. Municipal wastewaters also contain large amounts of bacteria and viruses, some of them pathogenic. Total bacteria counts in raw wastewater are typically about 10^7–10^8 MPN/100 ml. As with organic matter, since it is difficult to identify and quantify all the organisms present in a sample of wastewater, organisms are usually represented by an indicator organism. The most commonly used indicator for pathogenic organisms is fecal coliforms (although others are now being used, such as intestinal enterococci). Fecal coliform counts in raw municipal wastewater are typically in the range 10^6–10^7 MPN/100 ml.

1.2.3 Treatment processes and sequencing of treatment units

1.2.3.1 Treatment units sequencing and processes

The principles of wastewater treatment are shown in Figure 1.1. Treatment is conventionally divided into a sequence of preliminary, primary, secondary, and tertiary processes. Preliminary and primary treatments are based on physical processes; secondary treatment is based on biological processes, and tertiary usually on physicochemical processes, although it sometimes includes biological processes. For each of these four treatment units there are several alternative processes that can achieve similar results, so the total number of combinations is large and that means that the total number of possible overall treatment process schemes is large. Many treatment processes also generate sludge as a by-product. There are also several alternatives for sludge treatment and disposal so the number of combinations of treatment methods is myriad; one can find treatment plants with completely different unit processes that achieve similar results with widely differing costs.

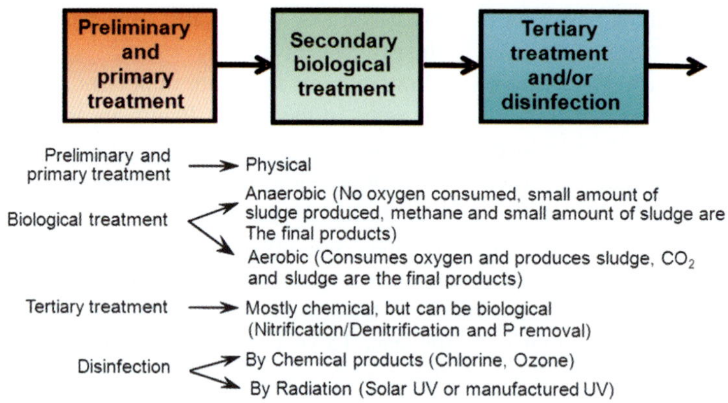

Figure 1.1 The principles of wastewater treatment

Following are some details on each treatment stage:

Preliminary treatment. All treatment processes, with the exception of septic tanks and household systems, require some sort of preliminary or screening process to remove large and floatable objects. Most widely used are screening and grit removal, which also removes floatable material and rags.

Primary treatment. Primary treatment often precedes biological treatment in conventional secondary wastewater treatment facilities. Its main purpose is to reduce the portion of the suspended solids and BOD, which can be reduced by gravity settling. Sedimentation tanks are the most common form of primary treatment of domestic sewage. In sedimentation tanks the suspended solids settle on the bottom, are scraped to a central point and then removed by gravity or by a sludge pump. Scum, primarily oil and grease, floats to the surface where it is collected by a mechanical arm and periodically drawn off. A dissolved air flotation (DAF) process can also be used as a primary treatment unit and can remove oil and grease in less space. In the DAF process the wastewater and air are pressurized and released into a tank open to the atmosphere. Small bubbles are formed in the tank and float to the top. On their way the bubbles become enmeshed with the light solids and oil particles and bring them to the surface where a skimmer collects them. The clarified liquid continues to downstream processes. DAF is more commonly used for industrial wastewater. A conventional sedimentation tank removes 25–40% of BOD, 40–60% of total suspended solids, and about 50% of the bacterial load. The main residuals collected are settled solids, scum, and oil. The cost of primary treatment is about half that of full secondary treatment. Although primary treatment processes are relatively simple mechanical processes, they require some routine maintenance. This mostly consists of upkeep of pumps, sludge scrapers, scum collectors, and motors. Most of the Operation and Maintenance (O&M) costs of primary treatment are due to the treatment and disposal of the sludge. The capital cost of the sludge treatment unit is a significant part of the cost of primary treatment.

Secondary treatment. The main purpose of secondary treatment is removal of colloidal and dissolved organic matter (BOD). This is achieved through biological processes carried out by aerobic or anaerobic bacteria which feed on organic matter, transforming the BOD to bacterial mass and to by products of BOD decomposed to generate the life maintenance energy of the bacteria. Aerobic processes are by far the most common for large treatment facilities. Aerobic bacteria need sufficient oxygen to metabolize the organic material. The two main types of biological reactors are suspended growth reactors and attached growth reactors. In suspended-growth systems oxygen is provided by aeration through mechanical aerators, a diffused air system, or some other process. The bacteria must be subsequently separated from the wastewater. Most secondary processes have a separate secondary sedimentation tank to settle the flocculated cell mass. The effluent, which is the overflow stream of the sedimentation tank, continues to downstream treatment processes or is discharged to a receiving water body for final disposal. The sludge is returned from the sedimentation tank to the aeration tank, which maintains a viable concentration of bacteria to metabolize the incoming organic material. This is called return activated sludge. A portion of the sludge must be removed and not returned to the aeration tank in order to prevent sludge concentrations from increasing to levels that would be too high to sustain sedimentation. The removed sludge is called excess activated sludge or waste activated sludge. Attached growth systems are similar in principle to suspended growth systems but the reactor is filled with a fixed media of stones or plastic media on which the bacteria grow. This type of reactor is not flooded, the primary effluent is sprayed on the media and the bacteria extract the oxygen directly from the air. Sometimes assisted oxygen supply by aeration is required. There are many variants of the aerobic treatment processes. Lagoons are often used in developing countries. They are natural systems that can provide secondary treatment. In aerobic lagoons the oxygen is provided by algae that develop in the lagoon and generate oxygen through the photosynthesis process. In anaerobic lagoons oxygen is not required (nor is it present) since the metabolism of the organic matter is done by anaerobic bacteria that do not consume oxygen.

Tertiary treatment for nutrient removal. Conventional secondary treatment removes some nitrogen and phosphorus; however, specialized processes are needed to remove more. These include nitrification and denitrification for nitrogen removal, partial phosphorus removal by biological processes, and chemical precipitation for complete phosphorus removal. Nutrient removal processes are more complex and expensive than secondary treatment. They are rarely required in developing countries, except for discharges into enclosed water bodies where eutrophication may occur such as lakes, estuaries, or bays. Sometimes nitrification is required also prior to discharge to rivers, to avoid nitrogenous oxygen consumption in the river after the discharges of the effluent to it. Capital costs for nutrients removal are higher and include construction of additional tanks, pipes, and recirculation pumps. O&M costs generally increase when nutrient removal is included, for maintaining the additional installations. Chemical costs for phosphorus precipitation also increase O&M expenses. The processes of nutrients removal are more complex and require skilled labour for efficient operation.

Disinfection removes pathogens and other bacteria from treated wastewater effluent. Common processes include chlorination, ultraviolet (UV) radiation, ozonation, and disinfection in lagoons (by natural UV radiation). Chlorine and ozone are strong oxidizing agents, which oxidize organic and inorganic matter and quickly kill all the pathogens they contact. Chlorine can be added to wastewater in gas, liquid, or tablet form; ozone is added as a gas only. UV radiation sterilizes pathogens and is applied through low-pressure mercury lamps. Disinfection in lagoons is a natural process that occurs in successive stabilization ponds due to natural visible light and ultraviolet radiation from the sun, sedimentation, and natural die-off. Wastewater effluent discharged into the ground by infiltration basins (not injection) generally undergoes significant pathogen and bacteria removal as it travels through the soil, so a properly designed system of infiltration of effluent to groundwater can also achieve disinfection. Typical maintenance of disinfection installations includes maintaining the supply of chemicals, adjusting feed rates, and maintaining mechanical components. Most chlorine systems are designed for minimal maintenance; UV radiation requires little maintenance other than regular cleaning and lamp replacement. Ozone needs to be generated on site since it is not a stable compound. The ozone generating and feeder equipment uses large amounts of electricity and is complicated to operate. The USEPA estimates that 8 to 10 KWH are used for each pound of ozone generated. Chlorination is the most widely used disinfection process around the world. UV radiation performs less well with effluents that are high in turbidity or suspended solids so sand filtration prior to radiation is common. Ozonation disinfects more powerfully than chlorine, with no harmful by-products. It is usually used to disinfect highly treated secondary or filtered effluents. Disinfection in lagoons is a simple technology and maintenance-free but requires a large land area. Chlorination produces contain many undesirable organic compounds that are toxic to humans and aquatic life. For this reason, it is desirable to remove as much of the organic material as possible before adding chlorine. Dechlorination may be needed to lower the residual chlorine concentration in the effluent so as to prevent damage in the receiving body. Chlorine gas is very hazardous and robust safety features must be employed in its storage site. Because of these safety concerns, liquid sodium hypochlorite is commonly used instead of chlorine.

1.2.3.2 Conventional secondary treatment processes

Commonly used secondary treatment processes

The most commonly used secondary wastewater treatment processes, especially in developed countries, are aerobic processes. Anaerobic processes are not commonly used in developed countries as secondary treatment for treating liquid wastewater, especially not in large cities. They are mainly used for treating excess sludge generated as a by product in the aerobic processes. The activated sludge processes, based on aerobic decomposition of the organic matter contained in the wastewater, is the most frequently used process for secondary treatment. There are other secondary treatment processes which are commonly used but most of them are variants of the activated sludge process. Since this type of conventional processes is not the focus of this book, they are not discussed here in detail. Some typical processes are presented briefly below. For more details, many specialized textbooks can be consulted, for example Orozco (2005) and Tchobanoglous *et al.* (2003). Most of the secondary treatment processes are not unit processes but rather a combination of several unit processes, so it is not just one unit that produces the final effluent. For example, the activated sludge process is composed of a primary sedimentation unit, a biological reactor, a secondary clarifier, sludge treatment installations and more. This is important to understand when discussing appropriate technology processes in the following chapters of this book. Some commonly used secondary conventional treatment processes are the following:

The *activated sludge* process (Figure 1.2), where wastewater is brought into an aeration tank usually after primary sedimentation, air is bubbled into the wastewater mixture (mixed liquor) or supplied by mechanical aerators of various types, and aerobic bacteria which naturally develop in the reactor metabolize the dissolved and suspended organic material. From the aeration tank, the effluent flows into a secondary sedimentation tank, where the bacterial cells mass settles out and the treated effluent overflows from the tank. The activated sludge process consists of a combination of unit processes including preliminary treatment units (bar screens and grit chambers), a primary sedimentation unit, a biological aerated reactor, a secondary clarifier, sludge recirculation equipment from the secondary clarifier to the biological reactor, sludge treatment installations and more. If properly operated, an activated sludge plant yield a high quality effluent but it is quite a complex factory which is of high investment cost and requires high O&M costs.

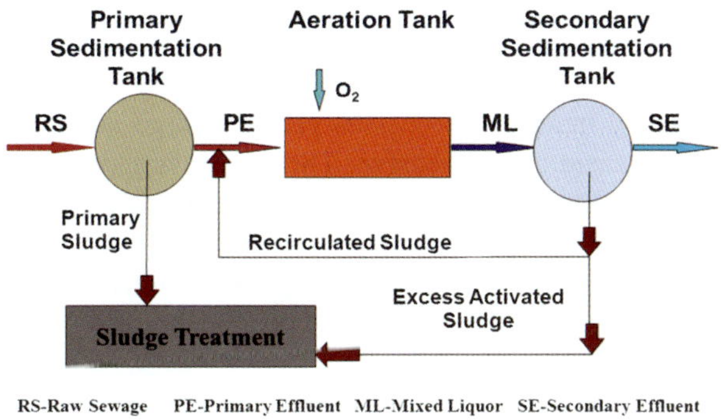

Figure 1.2 The conventional activated sludge schematic process diagram

The *oxidation ditch* process, which is an activated sludge consisting of a ring-shaped channel as the biological reactor. Oxygen is supplied by horizontal surface aerators and is not evenly mixed throughout the ditch as in a conventional activated sludge process. This provides zones of varying reaction, allowing more operational control. Cell mass is settled out in a secondary sedimentation tank and recycled back to the oxidation ditch and the treated effluent overflows from the sedimentation tank.

The *trickling filter*, in which primary effluent is evenly distributed over an aerobic attached growth reactor composed of a circular bed of fist-sized stones or plastic filling 900 to 1800 mm deep. Bacteria, fungi, and algae grow on the rock surface. As wastewater flows between the rocks, aerobic bacteria metabolize the organic material in the wastewater. As the biomass grows, the influent wastewater flow sloughs off the excess, which settles out in a secondary sedimentation tank. Settled sludge is usually recirculated to the trickling filter reactor to improve its performance. The trickling filter fixed bed reactor is not flooded with wastewater so the bacteria can absorb oxygen from the air that surrounds them. Sometimes, air is blown through the fixed media to provide more oxygen and thus improve the process performance.

The *Sequencing Batch Reactor (SBR)*, which is variant of activated sludge in which all the treatment steps take place in a single completely mixed tank to which influent is intermittently directed. The treatment consists of discrete, timed processes: fill, mix/aerate, settle, withdraw effluent, and withdraw sludge. Historically, SBRs have been used for only small treatment facilities but this is changing. There has been a recent resurgence of interest in the SBR process because it eliminates the need for secondary sedimentation and pumps.

The *Membrane Biological Reactor (MBR)*, which is the most important development in water and wastewater treatment in recent years. In the wastewater treatment industry microfiltration, ultrafiltration and nanofiltration membranes are used for replacing secondary sedimentation in activated sludge treatment. The process is basically an activated sludge process in which the solids in the mixed liquor are separated not by sedimentation but rather by membrane filtration. This yields an extremely high quality effluent, basically void of suspended solids and of very low organic matter content and pathogenic organisms count.

These processes can produce good quality effluent if operated and maintained properly, but very poor effluent if operated improperly. They are of high cost and high skill levels are required for their operation and maintenance. Energy costs can be substantial and a standby generator must be provided, especially in developing countries, to provide a solution for oxygen supply in cases of power failures lasting longer than a few hours. In trickling filters flies can be a serious nuisance as they live and breed within the filter medium.

Sludge treatment and disposal

The mentioned aerobic secondary treatment processes generate large quantities of excess sludge whose treatment and disposal is complicated and expensive. The sludge is generally high in volatile solids and can become septic quickly, producing offensive odours and becoming a severe nuisance if not treated or disposed immediately. Sludge treatment installations form part of the wastewater treatment plant and should not be omitted at the design or construction stages (as sometimes happens). Sludge treatment consists of a succession of processes including thickening, stabilization, dewatering and drying. It can also be treated by cold digestion/drying lagoons. These processes are briefly described below. There are many variants of the mentioned processes and also other processes of sludge handling and disposal. Final disposal methods of sludge include conveyance and disposal in landfills, along with municipal solid wastes, land application by spraying on agricultural farmlands as an organic fertilizer and soil conditioner, or in large scale projects also incineration. Recent land application regulations require sterilization of the sludge before its application as a fertilizer. Sludge handling is one of the main problems of municipal wastewater treatment. A brief description of the main sludge treatment processes is described below. For more details, many specialized textbooks can be consulted, for example Andreoli *et al.* (2001).

Sludge thickening removes water from sludge to reduce the cost of subsequent treatment or sludge disposal as a concentrated liquid. Typical processes include thickening by gravity, lagoon, gravity belt, and centrifuge. Gravity thickening feeds liquid sludge to a settling tank. The effluent is discharged over a weir and the thickened sludge is pumped from the tank bottom to be digested or disposed as liquid sludge. Lagoon thickening is gravity thickening in an earthen basin. Gravity belt thickening uses the gravity zone of a belt filter press for sludge thickening. Centrifuge thickening involves pumping the sludge to a solid bowl centrifuge rotating at up to 3,000 rpm. Lagoon thickening is appropriate for many applications in low to medium population density communities; it is simple and inexpensive but requires a large land area. Gravity thickening uses less land area, but requires more operator attention and equipment maintenance. Gravity belt thickening and centrifuge thickening are used in high population density communities and in industrial uses. Gravity belt thickening requires higher operator's attention and regular maintenance by qualified technicians. Centrifuge thickening has high power requirements. Maintenance can require highly skilled technicians and expensive shipment of spare parts from outside the country of use.

Sludge stabilization is performed on thickened sludge to reduce its volatile solids and pathogen content so it can be safely disposed or used for land application. It also reduces the solids volume. Typical processes include: anaerobic digestion; aerobic digestion; composting; lime stabilization and air drying. Anaerobic digestion is decomposition in the absence of free oxygen in closed tanks; aerobic digestion is oxidation in aerobic conditions. The stabilized sludge is drawn off the bottom or from the mixing tank. Composting is a process where aerobic organisms degrade and disinfect already thickened sludge. The sludge is mixed with bulking material, such as wood chips, to provide porosity for aeration and then laid over a network of porous piping and aerated. The composted sludge can be used as a fertilizer. Lime stabilization is the addition of alkaline compounds to raise the pH of the sludge mixture. Equipment and

O&M costs can be high in lime stabilization, and trained operators are needed for proper operation. Air drying beds are shallow paved or earthen basins where thickened waste sludge is allowed to naturally dry in the sun. Composting and air drying require large land areas and large quantities of organic materials such as wood chips or waste plant material as a bulking agent. Air drying is easiest to operate, but may not be suited in rainy areas.

Sludge dewatering removes water from sludge to reduce the cost of subsequent treatment processes or prior to sludge disposal as concentrated solids. The processes are similar to thickening processes, but higher solids concentrations are achieved. Typical processes include belt filter press dewatering, centrifuge dewatering, screw press dewatering, and plate and frame dewatering. In Belt filters press the sludge is dewatered by belts that apply pressure and squeeze out the liquid from the sludge. In centrifuge dewatering, sludge is pumped to a solid rotating bowl. In screw press dewatering the sludge is pumped into a perforated cylinder surrounding a rotating screw. The screw forces the sludge toward the end of the container and progressively dewaters it by the pressure of the screw against the sludge. Plate and frame presses are an established, but high maintenance and high cost dewatering processes. Belt filter press, centrifuge, and screw pump operations require close operator attention for control of the loading rate. The equipment requires regular maintenance and may require periodic import of maintenance parts from outside the country of use.

Cold digestion/drying (CDD) lagoons are a low-technology alternative that incorporates thickening, stabilization, dewatering, and storage in a series of earthen basins. Digestion and stabilization takes place at ambient temperatures. Two lagoons are needed, one for fill and one for maturation. At the end of the one year filling period the first lagoon is isolated and allowed to dry for up to one year and sludge fill is directed to the second lagoon. Rooted aquatic plants grow on the surface during the maturation period and assist in sludge drying by evapotranspiration. This sludge handling method is limited to hot climates with a prolonged dry season. CDD lagoons require a large land area but little operation or maintenance during filling.

Advantages and disadvantages of the conventional processes

The conventional aerobic secondary treatment processes mentioned above, which are basically variants of the activated sludge process, are well known processes with a long trajectory of successful service, serving cities of all sizes. They are reliable processes which yield good quality effluents in terms of low organic matter and suspended solids content. These effluents can be reliably disinfected and the processes can be adjusted to also remove nutrients when necessary. These processes can also be designed to perform well at low temperatures and have a small footprint.

In spite of these advantages they also have some disadvantages: their investment costs are high. They consume significant amounts of energy for the supply of oxygen and for operation of all the mechanical equipment which forms part of this type of plants, and they generate large amounts of excess sludge. Consequently their O&M costs are high. They are complex processes which require highly skilled personnel for their operation, and that further increases their O&M costs. The excess sludge produced by these processes generates an environmental problem. All these disadvantages constitute a special problem for the use of these processes in developing countries, where financial difficulties persist, especially for providing O&M costs, and where the complexity of the processes presents difficulties.

Investment financing for complex treatment plants can sometimes be mobilized in developing countries in the form of grants and/or soft loans; however, it is almost impossible to obtain grants or subsidies for operation and maintenance of such plants. Usually, the local or regional authorities (municipalities or water and sanitation utilities) do not have the capacity to finance high operation and maintenance costs of complex treatment plants from their internal cash generation, and so this type of treatment plants tend to deteriorate rapidly due to insufficient budget for O&M and many of them are abandoned a short time after being commissioned. This indicates that complex plants are not sustainable in developing countries and points to the need for the employment of plants based on alternative, simpler and lower cost processes, that is, on appropriate technology.

1.3 THE APPROPRIATE TECHNOLOGY CONCEPT

Developing countries cannot afford expensive and difficult to operate wastewater treatment installation. Experience shows that treatment plants based on complex aerobic mechanical biological processes are usually out of operation in developing countries a short time after their commissioning due to high operation costs or lack of capacity to operate them adequately. It is therefore necessary to use in developing countries simple, low cost treatment processes, otherwise known as processes based on appropriate technologies.

Appropriate technologies for wastewater treatment means simple treatment processes of proven technology, of low investment costs and especially of low Operation and Maintenance costs (much less costly than conventional processes), simple to operate and with the capacity of yielding any required effluent quality. Such processes do exist and are especially fitted for countries of warm climates, as most developing countries are, since biological processes perform better at higher temperatures. Appropriate technology processes can also be fitted to perform in cold climates. Appropriate technology processes are often ignored, certainly in developed countries, and to a large extent also in developing countries, due to: (i) lack of understanding or interest of environmental authorities, professionals and politicians; (ii) fashion: the drive to install cutting edge technologies even when not necessary and not affordable; or (iii) combination of the above. Also, appropriate technology processes do not contain sophisticated equipment and in most cases do not include equipment at all; just local construction materials, so there are no equipment manufacturers to push these processes, while a lot of manufacturers and consultants promote complex process based on complicated equipment and convince decision makers to stay away from appropriate technologies.

Not all the appropriate technology processes achieve the same effluent quality, that is, each process yields an effluent of a different quality, and some do not achieve a high effluent quality, however, this does not present any impediment for the use of these processes since: (i) as explained, not always is a high effluent quality required (cities located on the banks of large rivers or coastal cities may not need to produce effluents of high quality); and (ii) high effluent quality can be achieved by combining various appropriate technology processes into one treatment plant in which they function in series to yield higher quality effluents.

Appropriate technology processes can be divided into two types: (i) unit processes, each of which is one basic process of a proven track record, usually the unit processes do not produce a very high quality effluent; and (ii) combined processes which are various combination of unit processes which can produce high quality effluents. Some of the combinations of appropriate unit processes are in use in various plants in the world and others have not yet been used and have the potential to be used in the future. The fact that high effluent quality can be achieved by combining various appropriate technology unit processes into one treatment plant is a very important and often a misunderstood point. In many cases, since a certain unit process does not achieve a required effluent quality, it is rejected without taking into account that its effluent can be subjected to an additional treatment step of a different appropriate technology treatment process. An analysis reveals that most of the conventional aerobic secondary treatment processes discussed above consist of a combination of unit processes. For example, the activated sludge process is composed of a combination of unit processes including preliminary treatment units (bar screens and grit chambers), a primary sedimentation unit, a biological aerated reactor, a secondary clarifier, sludge recirculation equipment from the secondary clarifier to the biological reactor, sludge treatment installations and more. No one would expect to obtain a high quality effluent from an activated sludge process that lacks a secondary clarifier. The same concept should be applied to appropriate technology processes: high effluent quality can be achieved by combining unit processes. A combined treatment plant composed of two or more processes is more costly than a unit process plant, however, it can still be simpler to operate and less costly than conventional processes.

Appropriate technology unit processes include (but are not limited to) the following: (i) Preliminary Treatment by Rotating Micro Screens; (ii) Vortex Grit Chambers; (iii) Lagoons Treatment (Anaerobic, Facultative and Polishing), including recent developments in improving lagoons performance; (iv) Anaerobic Treatment processes of various types, mainly, Anaerobic Lagoons, Upflow Anaerobic Sludge Blanket (UASB) Reactors and Anaerobic Filters; (v) Physicochemical processes of various types like Chemically Enhanced Primary Treatment (CEPT); (vi) Constructed Wetlands; (vii) Stabilization Reservoirs for wastewater reuse and other purposes; (viii) Overland Flow; (ix) Infiltration-Percolation; (x) Septic Tanks; (xi) Submarine and Large Rivers Outfalls and more. Of these processes various combinations can be set up. Combinations can also include some other simple processes like Sand Filtration and Dissolved Air Floatation (DAF) which are not considered appropriate processes per se, but they are in fact appropriate processes. One interesting combined process is the generation of effluents suited for reuse in irrigation based on pretreatment by one of the mentioned unit processes followed by a stabilization reservoir. Aerated lagoons is not strictly an appropriate technology process but can be combined with appropriate technology processes. It is a kind of hybrid processes since it is of low investment cost and simple to operate, but when it serves as the main treatment step it is expensive in O&M because of the high energy consumption required to operate the aeration systems that supplies oxygen for decomposing the bulk of the organic matter. However, when an aerated lagoon serves as a polishing step of the effluent of an upstream process which removes the main part of the organic matter from the wastewater, the oxygen demand (and energy consumption) of the aerated lagoons is not high and under such a configuration, the process as a whole can be considered an appropriate technology process.

Detailed explanations on the appropriate technology processes and the design procedures of part of the mentioned appropriate technology unit processes are presented in the following chapters of this book. Detailed explanations and design procedures are also presented for some selected combinations of unit processes which can be used to achieve high quality effluents. Ideas regarding additional combination of unit processes are presented as well, albeit without design procedures.

Treatment in lagoons is most common for small and medium size towns and not commonly used for large cities. However, in special cases lagoons can be used for large cities too, for instance, in case that several smaller plants can be constructed in various parts of a large city, instead of one large plant. The most widely used anaerobic process for wastewater is anaerobic lagoons. The UASB Reactor process became popular in Latin America in recent years. Originally it was used mainly for industrial wastes treatment applications and later for small cities, but now it is being used also for medium size and large cities. Reuse of effluents for irrigation using Stabilization Reservoirs can be applied for small and large cities; however, the concept requires a high level of inter-institutional coordination between the municipalities and the farmers' organizations. Physicochemical treatment requires a certain level of operating capacity. Constructed Wetlands and Overland Flow are mostly suitable for small cities because these processes require large extensions of land. Preliminary Treatment by Rotating Micro Screens can be applied for cities of any size. It can be applied as the sole treatment when a preliminary level of treatment is sufficient, or as the first step of any combination of processes used to achieve a higher effluent quality.

Since most plants based on appropriate technologies do not contain complex equipment and use only local materials and labour, they can be readily implemented in developing countries and this is one of their great advantages.

A comparison of the performance of the appropriate technology processes (unit processes as well as some frequently used combined processes) to that of conventional activated sludge process in terms of effluent quality, as well as in terms of investment costs and O&M costs shows that the investment costs in unit appropriate technology plants fall in the range of 20–50% of investments in activated sludge and O&M costs are mostly in the range of 5–25% of the O&M costs of activated sludge, except for CEPT which is in the range of 25–40%. Costs of combined appropriate technology processes need to be calculated for each specific case and they are more costly than the unit processes, however, they are still of lower investment cost than conventional processes and certainly have lower O&M costs, which is an important feature of these processes. This is an important fact which can be of benefit for the efforts to close the gap and increase the coverage of wastewater treatment in the world, especially when considering the forecasted population growth.

In terms of effluent quality, most of the processes (except the rotating Micros Screen process) achieve results that are not very different from those achieved by the activated sludge process. This is especially true for the combined treatment processes, whose effluents are of similar quality to that of an activated sludge effluent. If the dilution in the rivers to which the effluents are discharged is significant, than there is practically no difference between the effluents of the combined appropriate technology processes and the effluent of an activated sludge process, in terms of their impact on the oxygen content of the river.

The general conclusion that can be drawn is that by using appropriate technology treatment processes or a combination of such processes: (i) a significant amount can be saved in investment costs; (ii) a lot of O&M expenses can be saved; (iii) the operation and maintenance of the plants is much simpler than that of plants based on conventional process; and (iv) almost nothing is sacrificed in terms of effluent quality.

The appropriate technology processes are not new processes. All of them are known for many years and have a good track record. The newest one is the UASB process. This process has been in use for decades in industrial wastewater treatment plants. Its application to treatment of municipal wastewater is relatively new, but even for that application it has been in use for about 20 years. *The message of this book refers to the innovation of the use of combinations of appropriate technology unit processes.* When an effluent quality higher than what a single unit process of appropriate technology can produce is required, a treatment plant consisting of a series of appropriate technology unit process functioning in series can be used (2, 3 or more), in which the effluent of the first unit process is fed into the second process, the effluent of the second process is fed to the third and so on, or in other words, to improve the quality of the effluent of a certain unit process, this effluent can be subjected to one or more polishing steps, each of which consists of an additional appropriate technology unit process. This approach can produce practically any final effluent quality required. *The idea that it is possible to combine unit processes to create treatment plants each based on a series of appropriate technology processes and that it is possible to combine unit process which jointly can generate any required effluent quality, is the main message of this book.* A plant based on a combination in series of appropriate technology unit processes is still easy to operate and is usually of lower costs than conventional processes in terms of investments and certainly in operation and maintenance.

The combinations of various unit processes yield interesting process diagrams for wastewater treatment plants, some of them already being applied in existing treatment plants in many parts of the world and others are only ideas and proposals, which have not yet been tested but there is no reason that they will not function well, and they might be advantageous for specific local condition in various parts of the world. *So in essence, this book shows the way to obtain high quality effluent by treatment plants based on simple, low cost and easy to operate processes.*

As mentioned, one of the main problems of using in development countries conventional technology based on aerobic processes is the high cost of energy required for supply of oxygen to such processes and the difficulty in ensuring proper operation of the oxygen supply equipment (the aeration equipment) over a long time horizon. It is therefore good practice to avoid in developing countries the use of aerobic processes and base wastewater treatment, to the extent possible, on anaerobic processes. Aerobic processes which utilize mechanical air supply for decomposing the bulk of the organic matter contained in the wastewater cannot be considered appropriate technology processes.

The appropriate technology concept can contribute to satisfying the demand for increasing the wastewater treatment coverage by lowering investment and especially O&M costs. The simplicity of operation and maintenance of appropriate technology plants and the low O&M costs reduce the institutional burden and challenges of the water and sanitation utilities. In this way, the use of appropriate technologies for wastewater treatment helps to confront the main problems of the water and sanitation sector in developing countries, which are financial weakness, low technical capacity and institutional weakness.

It is worth mentioning that the concept of appropriate technology for wastewater treatment is in wide application in Brazil, where treatment plants of this type are used for all sizes of cities and where the application of this concept is rapidly expanding. The utilities which are pioneers in this respect are SANEPAR, the water and sanitation utility of the State of Parana, CAESB, the water and sanitation utility of the Federal District of Brasilia and COPASA, the water and sanitation utility of the State of Minas Gerais.

Finally, it is clarified that, although a lot of the discussion was dedicated to the situation in developing countries, it was not meant to imply that appropriate technology treatment processes should be used only in this type of countries. Appropriate technology processes can be used anywhere, on condition that they are well designed and take into account the temperature conditions at the site of installation. Developed countries will be well advised to use more often wastewater treatment plants based on appropriate technology processes if they want to save money and provide more sustainable solutions. Appropriate technology processes are also more environment friendly, since these processes consume less energy and this fact has a positive impact on mitigating climate change effects. Also, with modern design appropriate technology processes cause less environmental nuisance than conventional processes, for example they produce lower amounts of excess sludge and their odor problems can be effectively controlled.

1.4 SUSTAINABILITY ASPECTS OF APPROPRIATE TECHNOLOGY PROCESSES

One of the main problems in developing country is ensuring the sustainability of the investments. In this case we refer to sustainability of investments in wastewater treatment. Experience shows that in many cases, the existing treatment plants in developing countries are based on non-sustainable processes and many of them are abandoned shortly after their commissioning. Investment financing for complex treatment plants can sometimes be mobilized in developing countries in the form of grants and soft loans; however, it is almost impossible to obtain grants or subsidies for operation and maintenance of the treatment plants after their construction has been completed. Usually, the local or regional authorities (municipalities or water and sanitation utilities) do not have the capacity to finance from their internal cash generation the high operation and maintenance costs of complex treatment plants, and so this type of treatment plants tend to deteriorate rapidly due to insufficient budget for O&M. This is the main reason that many of the treatment plants are abandoned a short time after being commissioned and it indicates that complex plants are not sustainable in developing countries.

The objective of achieving sustainability of wastewater treatment and of expanding the current low coverage of wastewater treatment in developing countries is an important and worthy objective. In order to attain this objective, it is necessary to change the paradigm of constructing complex treatment plants and move towards the employment of plants based on alternative, simpler and lower cost processes. Those are basically plants based on appropriate technology processes.

It is intuitively clear that the use of appropriate technology wastewater treatment plants will significantly enhance their sustainability. They are simple to operate and their O&M costs are low so there will be no reason to abandon them a short time after their commissioning. However the sustainability aspects of appropriate technology treatment plants have a much wider scope.

First there is the aspect of improving environmental sustainability through the expansion of the coverage of wastewater treatment in developing countries. With the current low coverage of wastewater treatment in these countries and the forecasts of rapid population growth, the perspective of reaching a more sustainable environmental management by expanding wastewater treatment coverage through the use of conventional processes is slim, because of the high investment and O&M costs of these processes. However, given the lower costs of the appropriate technology processes, especially their lower O&M cost, they can be the basis of an innovative approach in supporting the efforts to close the gap in wastewater treatment services and in increasing the coverage of wastewater treatment in developing countries. Increase in the coverage of wastewater treatment will contribute to the sustainability of water resource and environmental quality by eliminating or reducing the sources of contamination of water bodies.

In addition, appropriate technology processes can contribute to enhancing the sustainability of utilities in several ways:

- *by enhancing the financial sustainability of the utilities*: as mentioned, investment costs of unit appropriate technology treatment plants are in the range 20–50% of investments in conventional processes, and more important, O&M costs are in the range 5–25% of the O&M costs of conventional processes. These figures make a huge difference from the financial standpoint. Utilities which may collapse as a result of employing conventional processes can safely cope with the use of appropriate technology processes.
- *by enhancing the technical and operational sustainability of the treatment plants*: the employment of simple to operate and maintain processes based on simple, mostly locally manufactured equipment, can ensure ease in overcoming technical difficulties and keeping the treatment plants functioning properly as long as required without the risk of failing and being abandoned, ensuring thereby long terms sustainability in operating the plants.
- *by enhancing the institutional sustainability of the utilities*: given that the appropriate technology treatment processes are reliable and keep functioning with limited financial and technical efforts, they do no present meaningful problems to the utilities' managements, do not impose additional managerial efforts, reduce the institutional burden and challenges of the water and sanitation utilities and thereby contribute to institutional sustainability. Appropriate technology treatment plants can also generate income and further enhance institutional and financial sustainability as discussed below.

Municipal wastewater contains some resources which can be recovered and are of value. Those include the water contained in the wastewater, nutrients (nitrogen, phosphorous and some potassium) which can be used as fertilizers and organic matter which can generate some energy.

Properly treated municipal wastewater can generate effluents adequate for agricultural or industrial reuse. When effluent is used for irrigation, the nutrients contained in it function as fertilizers. Municipal and industrial wastewaters contain organic matter measured as COD. This organic matter contains a certain amount of energy, which can be recovered in anaerobic processes. The end products of anaerobic decomposition of wastewater are biogas, which contains about 65–85% methane, and small amounts of excess sludge. If the first treatment unit in a treatment plant is an anaerobic process unit like covered anaerobic lagoons, UASB or anaerobic filters, the biogas can be captured, cleaned and used as an energy source. A more detailed discussion on options for energy generation is presented in Section 1.13.

The sale of effluent for irrigation or of the energy which can be produced in anaerobic processes can generate additional financial income to utilities which use appropriate technologies for wastewater treatment. The reduction of emission of methane, a strong greenhouse gas, can in certain cases generate income from the sale of Carbon Emission Reduction. All this additional income can further improve the financial sustainability of such utilities. In certain cases it can also improve the institutional sustainability. In many cities wastewater is not treated at all but rather discharged directly to receiving bodies, and utilities are institutionally weak to even manage appropriate technology treatment plants. If the discharge of raw wastewater causes severe water contamination problems (which is the situation in many cases), an appropriate technology treatment plant can be constructed (usually with the support of government grants) to resolve the problem. Such plant can include installations for energy recovery or effluent reuse. The income generated from the recovery of the resources can be used to finance the hiring of a private operator to operate the treatment plant. This can improve the overall institutional sustainability of the utility by ensuring the sustainability of the operation of the treatment plant and by enhancing the sustainability of the environmental management through the reduction in the contamination of natural water bodies.

In summary, appropriate technology wastewater treatment processes are more sustainable since they are of low cost and simple to operate. Moreover, the use of appropriate technologies for wastewater treatment helps to alleviate the main problems of the water and sanitation sector in developing countries, which are: financial weakness, low technical capacity and institutional weakness, thereby contributing to improving the sustainability of the sector as a whole.

1.5 PROPOSED STRATEGY FOR WASTEWATER MANAGEMENT IN DEVELOPING COUNTRIES

1.5.1 The government's perspective

Expanding the coverage of wastewater treatment services is a common objective of the central government, local governments and water and sanitation utilities. Experience shows that a government support program is required to stimulate the expansion of sanitation services. Such programs were executed in developed countries like USA, Japan, the EU countries and others, and should certainly be implemented in developing countries because if cities in developed countries needed support, cities in developing countries need it even more. In essence such programs provide grants or soft loans to water and sanitation utilities to partially finance the investment costs of the sanitation infrastructure. The programs should not provide financing of O&M costs because financing O&M expenses is not a sustainable approach. O&M cost is a continuous everlasting expense and while the government that starts the support program can ensure provision of investment funds during its term in office, it cannot ensure provision of O&M by future governments. To ensure sustainability, O&M costs need to be financed by user tariffs. Government investment subsidies are well justified as they provide support to improve the quality of life of the poor who have difficulties to participate in financing the sanitation infrastructure, and support environmental protection.

Governments need to ensure that reasonable effluent quality standards are adopted and realize that the most stringent standards cannot be achieved within a short period; right at the first stage of implementing a sanitation project, and that more time is required to reach them. The standards need to be flexible in terms of quality and time so that they take into account the stages of project development and also the assimilation capacity of the receiving bodies. It is not an easy task for a government to convince ministries of environment to relax standards but if a reasonable approach is not adopted, costs will be too high and the goal of expansion of the sanitation services will not be achieved.

In most cases, governments in developing countries contract concessionary loans from multilateral financing institutions to finance the grants for investments in wastewater treatment infrastructure. Since governments have limited funding for that purpose, they need to allocate the investment grants in such a way that from the national perspective the method of allocation would maximize the benefits for the investments, or in other words, with a fixed amount of grants the maximum benefits needs to be achieved. To achieve that goal, the proposed eligibility criteria of a city to obtain access to sanitation investment grants are the following:

- Grants should be provided only to priority cities whose untreated wastewater discharge generates a real public health or environmental problem, such as contamination of critical water bodies or severe contamination of water resources which are used downstream as a water supply source, for example, a large city discharging its raw wastewater to a small river, or a coastal tourism city discharging its raw wastewater on beaches used for bathing and recreation. Priority cities need to be identified upfront. Cities whose raw wastewater discharge does not cause severe problems (for instance cities which discharge to receiving bodies with high assimilation capacity) are of secondary priority. In this way, the government subsidies will support the implementation of the priority wastewater management program of the country;
- Grants should be provided only to cities which agree to use one of the appropriate technology unit processes or combination of processes prescribed by the government. A list of acceptable appropriate technology treatment processes need to be identified upfront and the treatment process selected for each participating city needs to be one of the processes included in this list. In this way the government ensures that the investment costs are kept reasonable while not sacrificing the effluent quality;
- Grants should be provided only to cities which accept to undertake a strict institutional strengthening program to ensure sustainability in operating the constructed wastewater treatment installations. Private sector participation should be considered as one of the institutional strengthening options.

The level of the grant for each city should be based on the financial analysis of the utility, with view of minimizing the subsidy to each city so that the government will be able to support as many cities as possible. Also, it is advisable to develop for each case financial models that would define a differential set of tariffs to ensure that the government subsidies benefit only the poor (i.e., that the poor pay a lower tariffs), while users that have financial capacity pay the full cost of treatment.

1.5.2 The utility's perspective

Following are some guidelines for good practices of wastewater management in developing countries from the utility perspective:

- The order of priority of investments in the water and sanitation sector should be: (i) first water supply; (ii) then sewerage; and (iii) finally wastewater treatment;
- The highest health benefit to users stems from the evacuation of the wastewater from their neighbourhood, not from treatment. Therefore wastewater should be first conveyed downstream of the city limits and treated there. Construction of multiple wastewater treatment plants within the city limits, with discharge of effluents to water courses that cross the city, might be a lower cost

solution (depending on the specific conditions of each case) but it bears risks because it is difficult to ensure in developing countries the production, on a continuous basis, of pathogens free effluents, especially in multiple plants. So adopting such decentralized solution will most probably render all the water courses crossing the city contaminated with pathogens (although the water might have a better appearance than that of raw sewage) and that poses risks to people that come in contact with the water flowing in the water course.

- Wastewater contains many types of pollutants whose relative importance in generating public health risks and harm to the environment depends, to a large extent, on the type of the receiving body into which the treated effluent is discharged. The treatment process needs to be tailored in each specific case to the type of the receiving body;
- Appropriate technology processes should be preferably used for treatment of wastewater. Such processes are based on simple technologies which are less costly than conventional processes in terms of investments and Operation and Maintenance, simple to operate and have the capacity of yielding any required effluent quality;
- Wastewater management schemes should be developed in stages in accordance with financing availability, while taking into account in the first stage the assimilation capacity of the receiving body (using water quality simulation models) and the system's construction time, so that the stringent effluent standards are achieved after time, at the ultimate stage, while maintaining reasonable environmental standards even during the first stages of the project. Projects can be developed in such a manner that no environmental nuisance is caused by them even at the first stage. This can usually be achieved by taking into account the assimilation capacity of the receiving body;
- It is necessary to apply a comprehensive water quality monitoring program prior to and following the construction of the treatment installations. If the monitoring results demonstrate that no environmental problems are caused after the first stage, delaying or abolishing construction of subsequent stages should be considered, so as to prevent unnecessary investments. If the monitoring results demonstrate that environmental nuisance persists after the first stage, implementing the subsequent management stages needs to be continued;
- It is advisable to involve all the stakeholder through participatory processes right from the initiation of project planning, so as to ensure support and mitigate opposition;
- Tunnelling and trenchless technology can be used for trans-boundary transfers of raw sewage between basins and for crossing municipal areas in which open cut trenches would cause severe environmental and social interruptions. Tunnelling may be more expensive but would be more convenient from the social and environmental perspectives;
- Contrary to the position held by many, poor users in developing countries cannot pay the full cost of wastewater treatment, so for achieving the goal of expanding wastewater treatment services, it is necessary to secure subsidies for this type of users. Such subsidies are well justified as they improve the quality of life of the poor and support environmental protection.

1.5.3 The strategy pillars

In summary, the proposed conceptual approach for wastewater management in developing countries is the following:

– First and foremost, appropriate technology processes, unit or combined processes, should be used in the wastewater treatment plants. Application of appropriate wastewater treatment technologies is the key for success in any strategy aimed at increasing the coverage of wastewater treatment. Appropriate low cost solutions are more adequate for developing countries than highly mechanized and complex conventional solutions as activated sludge and its variants, which consume large amounts of energy and produce large quantities of excess biomass which needs to be treated and properly disposed. The appropriate technology approach provides the benefits of requiring low investment costs as well as low O&M costs and providing simplicity of operation and thus allowing overcoming weak institutional capacity issues which are often a problem in these countries.

– When warranted, it is proposed to advance in stages, for instance, applying in the first stage the first unit of a treatment processes composed of a combination of unit processes. Advancing in stages simplifies the handling of the first stage and operation results may indicate that the first stage is sufficient in producing the required results.
– the assimilation capacity of the water body into which the effluent is discharged should be taken into account and, when justified, the stringent effluent quality standards should be relaxed, at least temporarily for the first years of the treatment plant's operation.
– the government should provide investment subsidies for treatment plants in priority towns. Poor users in small towns cannot pay the full cost of wastewater treatment, so for achieving the goal of treatment, it is necessary to secure subsidies for this type of users. Such subsidies are well justified both from the standpoint of environmental protection and the standpoint of poverty alleviation.

1.6 ANAEROBIC AND AEROBIC PROCESSES OF DECOMPOSITION OF ORGANIC MATTER

In view of the importance assigned within the appropriate technology concept to anaerobic processes of decomposition of organic matter and in view of the innovation in using anaerobic processes as the main unit processes for removing the bulk of organic matter from the raw wastewater, it is important at this point to provide information on the difference between anaerobic and aerobic processes of wastewater treatment.

The most commonly used method for decomposing the organic matter contained in wastewater is aerobic decomposition, which is a biochemical pathway that requires supply of oxygen for the breakdown of organic matter, usually by injection of air, and presence of aerobic bacteria. The final products of the decomposed organic matter are CO_2 and water, as well as sludge composed of new cells of aerobic bacteria, also known as biosolids. About 50–60% of the organic matter contained in the raw wastewater is transformed under the aerobic pathway to sludge and that means that aerobic decomposition of wastewater generates large quantities of sludge and creates a new significant environmental problem which is the treatment, handling and disposal of the large quantities of sludge produced in this processes. Another biochemical pathway of decomposition of the organic matter contained in wastewater is anaerobic decomposition which takes place in an oxygen void environment in presence of anaerobic bacteria. The anaerobic biochemical pathway is different than the aerobic pathway and its final products are: (i) a biogas which contains a large portion of methane (65%–85%), in addition to lower quantities of CO_2 (10–20%) as well as nitrogen and some other impurities as water and hydrogen sulphide; and (ii) sludge composed of new cells of anaerobic bacteria. The sludge quantities produces in the anaerobic process are much smaller than the sludge quantity formed while decomposing the same amount of organic matter under an aerobic pathway. This is a result of the fact that only about 5–15% of the organic carbon is converted to biomass during anaerobic decomposition of organic matter, while in aerobic decomposition; the equivalent number is about 50–60%.

Anaerobic decomposition is a two-stage process, each stage being carried out by a different group of bacteria, with the second stage bacteria being more sensitive than the first to environmental conditions such as pH. In the first stage acid-forming or nonmethanogenic bacteria convert the organic matter present in the sewage to organic acids, whereas in the second stage, methane forming-bacteria or methanogens convert the organic acids to methane gas and carbon dioxide. For efficient performance, the methanogens require a pH in the range 6.5–7.5, and they cannot develop at all below a pH of 6.2. However, if too many acids are produced by the acid-forming bacteria, which develop and multiply easily, the result is a low pH which may impede the production of methanogens. In the presence of high concentrations of sulphates, the methanogens also have to compete with sulphur-reducing bacteria. The main consequences of such a situation are the appearance of unpleasant odours and the reduction in the efficiency of the anaerobic process. Although the problem can be resolved by adding lime or other chemicals to raise the pH, it is preferable to prevent it by controlling the pH and the volatile acids concentration.

The anaerobic process is sensitive to low temperatures and its rate declines strongly at temperatures lower than 12°C (in the liquid, not in the air), however, it is well functioning at higher temperatures and therefore appropriate for developing countries which are usually located in hot climate zones. In spite of the common perception that anaerobic processes do not function at temperatures lower than 12°C, there is evidence from the highlands of Bolivia and Peru that anaerobic lagoons function perfectly well in air temperatures well below zero.

In developed countries, the anaerobic process is mainly used for digestion of excess sludge in activated sludge plants. All over the world anaerobic decomposition is considered a slow rate process adequate mainly for digestion of sludge. Only in recent years the anaerobic process started to be considered a process suitable also for municipal wastewater treatment, not only for sludge treatment. The capacity and rate of decomposition of organic matter in raw wastewater under anaerobic conditions at temperatures higher than 12°C are similar to those of aerobic processes, as evidenced by the fact that the hydraulic detention times in anaerobic reactors treating municipal wastewater are similar to the detention times in activated sludge aerobic reactors. The sludge retention time in the anaerobic reactors is, however, much higher than that of aerobic reactors since the portion of the raw wastewater organic matter that is transformed to sludge is much smaller in the anaerobic process. So an anaerobic process can perform well as the main treatment process of wastewater, especially in hot climates.

An anaerobic treatment process has some advantages over an aerobic process: (i) it does not require oxygen thus it does not consume energy; (ii) excess sludge quantities are much smaller than those generated in aerobic process, thus expenses for sludge treatment and disposal are much smaller and so are negative environmental effects generated by the sludge; (iii) the gas it produces, mainly methane, can be collected and used to generate energy or be burned, reducing the emission of greenhouse gases, and carbon fund credit can be obtained for the emission reduction, increasing the financial benefits of the process; (iv) smaller amounts of nutrients are required in anaerobic processes because less biomass is produced; (v) the occupied area of an anaerobic reactor is small, except in the case of anaerobic lagoons, which require a somewhat larger area; (vi) electromechanical equipment is not required for these processes; (vii) construction is simple and uses mostly local materials; (viii) operation is quite simple; and (ix) both investment cost and operation and maintenance costs are way smaller than those of aerobic processes. Anaerobic treatment also has some disadvantages: (i) it needs longer start-up time to develop the necessary amount of anaerobic biomass in the reactor; (ii) it is more sensitive to lower temperatures; (iii) it is more sensitive to toxic substances; (iv) biological nitrogen and phosphorus removal cannot be achieved in an anaerobic process; (v) it may require alkalinity addition; (vi) it may produce odours and corrosive gases, however those can be controlled; and (vii) the effluent quality of a one stage anaerobic treatment process may not meet discharge standards and, depending on the specific project conditions, a subsequent treatment stage might be necessary.

The fact that the rate of anaerobic decomposition of organic matter contained in raw wastewater is similar to the decomposition rate under aerobic conditions is still not known to many professionals, who consider that because anaerobic decomposition of excess sludge is a slow rate process, then all anaerobic processes are slow rate processes. The fact is that excess sludge, which is basically excess bacterial cells, is a very difficult matter to decompose due to the need to break the cell membranes, and even aerobic digestion of excess sludge is an equally slow process. So in anaerobic digestion of sludge the problem is the type of substrate (which in this case is the excess bacterial cells) and not the anaerobic process itself.

Considering the advantages of anaerobic processes, it is surprising that their use is not more widespread. There are several reasons for that fact. First, they are not promoted by professionals due to lack of familiarity and experience. Second, they do not contain complex equipment and in fact contain minimal amounts of equipment and are mainly based on local materials of construction. As such they are not attractive to equipment manufacturers, who usually discourage their use. Lack of marketing and advocating the use of anaerobic process by professionals and equipment manufacturers to politician and decision makers is probably the main reason for the limited use of these processes. However, this is starting to change. With the increase of the microbiological and biochemical knowledge of anaerobic processes, an improvement has occurred in the optimization and efficiency of anaerobic reactors and anaerobic processes are becoming more frequently used.

Anaerobic processes achieve organic matter removal in the range 40%–85%, depending on the type of reactor used, and as such they are most valuable for developing country since they have the potential to achieve such a meaningful removal at low cost and simple to operate installation. If a higher level of organic matter removal is required, anaerobic processes can be the first stage of treatment followed by a variety of polishing processes, such as lagoons of various types, constructed wetlands, sand filtration, dissolved air floatation and others. It is also possible to use one anaerobic process as the first stage treatment and another as a following polishing process, for instance, a UASB followed by an anaerobic filter, or even anaerobic lagoons followed by UASB followed by an anaerobic filter. In South America, polishing of an effluent of an anaerobic first treatment stage is done in many cases by an activated sludge treatment stage or another aerobic polishing process such as trickling filters or aerated lagoons. This combination of Anaerobic-Aerobic processes works well and produces a high quality effluent, but such a process as a whole cannot be considered an appropriate treatment process since it includes an aerobic stage that consumes energy, increases costs of investment and operation and complicates operation and maintenance. Aerated lagoons is an exception, since when used as a polishing step, it can be considered an appropriate technology unit due to its simplicity of operation and relatively low cost.

There is a large variety of types of anaerobic reactors for treatment of wastewater including: anaerobic digesters of excess sludge (of low loading, high loading and two stages), septic tanks, anaerobic lagoons, high loading reactors including fixed bed reactors also known as anaerobic filters (of ascending or descending flows), piston anaerobic reactor, rotating bed reactor, expanded bed reactor, fluidized bed reactor, two stage anaerobic reactor, up-flow Anaerobic sludge blanket reactor (UASB), expanded bed granular reactor, internal recirculation anaerobic reactor and various combinations of these reactors. Detailed information on all these types of reactor is presented by Chernicharo (2007) and Campos (2009). Most of these reactors are used to treat industrial wastewater. The three anaerobic reactors (or anaerobic unit processes) which are used most frequently for treatment of municipal wastewater are: Anaerobic Lagoons, UASB reactors and Anaerobic Filters. Anaerobic filters achieve the highest organic matter removal, in the range 70–80% of BOD_5 removal, followed by UASB with BOD_5 removal levels in the range 60–75%. Anaerobic lagoons achieve a lower BOD_5 removal level, in the range 40–70%. The reason for the different removal level of organic matter in each of these reactors is that the mixing mechanisms and the intensity of the contact of the wastewater with the bacterial biomass are different in each type of reactor. In anaerobic lagoons the mixing in the reactor is not effective and the contact of the organic matter dissolved in the raw sewage flowing into the lagoon with the biomass which develops on the bottom of the lagoon is not intense, so only partial removal of dissolved organic matter is achieved. The removal of the suspended solids contained in the raw sewage, which settle in the lagoon, is high, but as a whole, the organic matter removal in an anaerobic lagoon is low in comparison to other anaerobic processes. In a UASB reactor the granular biomass particles float in the liquid due to the flow pattern and the action of the biogas released in the reactor. Consequently, the contact of the liquid with biomass is more intense than in anaerobic lagoons and as a result, the organic matter removal is higher. However, there is a limit to the capacity of the UASB reactor to remove organic matter since a higher concentration of biomass is required for achieving higher removal efficiencies, but it is impossible to achieve higher sludge concentrations since when the concentration increases, more sludge is carried away with the effluent, so the organic matter removal in UASB is still not very high. In an anaerobic filter, where the biomass is attached to the filter media and the reactor is flooded with the wastewater, the biomass concentration levels in this reactor are high and the contact of the liquid that flows into the reactor with the biomass is more effective, and as a result, the level of organic matter removal is higher than in UASB and certainly higher than in anaerobic lagoons. Detailed discussion and design procedures of the unit processes of Anaerobic Lagoons, UASB, Anaerobic Filters and Piston Anaerobic Reactors, are presented in the following chapters of the book.

In regard to selecting the anaerobic unit process to be used, some experts hold the position that UASB and anaerobic filters do not have an advantage over anaerobic lagoons, especially covered lagoons. With detention time in anaerobic lagoons of 24 hours or less and depth of more than 4 meters, the area occupied by lagoons is not much larger than that occupied by UASB or anaerobic filters, while the operation of lagoons is easier. Analysing the attributes of each process, the following can be observed: (i) Anaerobic lagoons are the simplest to operate, are of lowest cost in terms of investment and O&M and covered lagoons are free of nuisance, on the other hand lagoons achieve the lowest removal level of organic matter and occupy more area than the other processes; (ii) UASB is of higher cost in terms of investment and O&M and is more complicated to operate, but occupies less area and achieves a better removal of organic matter; and (iii) Anaerobic filter is of similar investment and O&M costs to those of UASB, is comparable to UASB in terms of complexity of operation and occupied area, but achieves a better removal of organic matter. The selection of the most adequate anaerobic unit is project specific and depends on the level of relative importance assigned to specific plant's characteristics: ease of operation, effluent quality and minimum land occupation.

1.7 UNIT PROCESSES OF APPROPRIATE TECHNOLOGY FOR TREATMENT OF MUNICIPAL WASTEWATER

1.7.1 Introduction

A series of commonly used unit process of appropriate technology for wastewater treatment is presented and briefly discussed in the following paragraphs. Part of them is discussed in more detail in the following chapters of the book and design procedures for some of them are also presented there. As mentioned, each of these unit processes alone does not produce a very high quality effluent, although there are differences between the processes presented and some produce quite good quality effluents. As explained, depending on the receiving bodies, even effluents of unit processes which do not produce a very high effluent quality might be sufficient for discharge without downstream polishing treatment. For example, if an effluent is discharged into a large river or to the sea, a high effluent quality is not required and a single unit process might be sufficient as the sole unit in the treatment plant. The discussion on the unit processes is divided into two sections: (i) main unit processes; and (ii) additional unit processes. The division is somewhat arbitrary and is not based on the importance of the processes but rather on the level of details provided in the book on each process. For the unit processes included in group (i) the design procedures are provided in the book while for the unit processes included in group (ii), for the most part, the design procedures are not included in the book. The unit processes contained in each group are presented in Table 1.1.

Table 1.1 Commonly used unit process of appropriate technology of wastewater treatment.

Unit process name	Performance (% Removal of BODT, SST and Fecal Coliforms)
Main Unit Processes	
1. Rotating Micro Screens	BODT (0–30%) SST (0–30%) FC: Practically no Removal
2. Vortex Grit Chamber	BODT: Practically no Removal SST: Practically no Removal FC: Practically no Removal
3. Anaerobic Lagoons (including covered lagoons)	BODT: (40–70%) SST: (40–60%) FC: Removal very low, if necessary, effluent can be disinfected
4. Upflow Anaerobic Sludge Blanket Reactor (UASB)	BODT (60–75%) SST (60–70%) FC: Removal very low, if necessary, effluent can be disinfected
5. Piston Anaerobic Reactor (PAR): Baffled Anaerobic Reactor -BAR- for temperatures below 20°C.	BODT (60–70%) SST (70–80%) FC: Removal very low, if necessary, effluent can be disinfected
6. Anaerobic Filters	BODT (70–80%) SST (70–80%) FC: Removal very low, if necessary, effluent can be disinfected
7. Lagoon Systems (facultative and maturation lagoons including lagoons assisted by Mixers)	BODT (80–95%) SST (70–90%) FC: in effluent (30–10^5) MPN/100 ml, depending on the design of the lagoons. if necessary, effluent can be disinfected
8. Stabilization Reservoirs	BODT (75–95%) SST (70–90%) FC: in effluent (10–10^3) MPN/100 ml, depending on the design of the reservoir. if necessary, effluent can be disinfected
9. Constructed Wetlands	BODT (80–90%) SST (80–90%) FC: Removal of up to 3 logarithmic units which is low, if necessary, effluent can be disinfected

(*Continued*)

Table 1.1 Commonly used unit process of appropriate technology of wastewater treatment (*Continued*).

Unit process name	Performance (% Removal of BODT, SST and Fecal Coliforms)
10. Chemically Enhanced Primary Treatment (CEPT)	BODT (70–75%) SST (80–90%) FC: Removal low, if necessary, effluent can be disinfected
Additional Unit Processes	
11. Sand Filtration	Used for polishing of effluents of other units, performance data not given separately for this unit but are present for combined processes in following sections
12. Dissolved Air Flotation	Used for polishing of effluents of other units, performance data not given separately for this unit but are present for combined processes in following sections
13. Overland Flow	BODT (70–80%) SST (70–80%) FC: Removal low, if necessary, effluent can be disinfected
14. Infiltration-Percolation	BODT: Almost complete removal SST: Complete removal FC: Complete removal
15. Septic Tanks	BODT (20–80%) SST (20–90%) FC: Removal low, if necessary, effluent can be disinfected
16. Submarine and Large River Outfalls	BODT 99.9% SST 99.9% FC: Practically complete removal

1.7.2 Main unit processes of appropriate technology

1.7.2.1 Rotating micro screens (RMS)

The first treatment unit in a wastewater treatment plant is usually the preliminary treatment unit. Preliminary treatment is physical treatment usually consisting of bar screens, which remove from the liquid wastewater large floating material, and a grit chamber which removes coarse solids and oil/grease. This type of materials needs to be removed mostly to protect equipment in subsequent treatment units of the treatment plant. Preliminary treatment is usually not the sole treatment unit in a plant but rather the first in a train of units; however, in certain cases, like discharge of effluent of a coastal city to the sea, it might be sufficient to have only preliminary treatment. As mentioned, conventional preliminary treatment usually consists of coarse screens followed by a grit chamber, in most cases an aerated grit chamber. This type of unit requires considerable manual work for ensuring proper operation, and is one of the problematic units in a conventional activated sludge treatment plant. In the past decades, innovative, automatic, efficient and reliable equipment for preliminary treatment was developed. The coarse screens have been replaced by rotating micro screens with a sieve opening in the range of 0.2–6 mm, according to the requirements in each specific case. With such a fine screen, the grit chamber unit is not always required. Various configurations of rotating screens are used, but the rotating drum type seems to be the most efficient and easy to operate. As an example, a schematic presentation of a rotating drum of micro screens manufactured by "Huber Technology" is presented in Figure 1.3. The raw sewage enters the rotating drum from the bottom to its inside and passes to the outside while floating and suspended material is retained on the internal side of the micro screen. The drum is cleaned with a water or effluent jet spray. The drum does not rotate continuously but rather starts rotating when the screen becomes clogged. As the drum rotates, a jet of high pressure water or effluent is sprayed on its upper part causing the material clogging the screen to fall to the center of the drum, onto a screw conveyor which transports it upwards. The material retained on the conveyor is then dropped into a compactor that compresses it. The compressed solids are then transported out by a conveyer screw or belt conveyor into a central hopper and hauled for final disposal in a solid wastes disposal site such as a sanitary landfill. After rotating several times the screen is cleaned and the drums stops rotating until new material clogs the internal side of the screen. At that point the drum starts rotating again and so on. Drums are available in diameters ranging from 600 to 2600 mm. The units are made completely of stainless steel. The wastewater flow which can be handled by one screening drum unit of a certain diameter depends on the screen opening and the drum's diameter. For example, a 2600 mm diameter rotating screen with an opening of 0.5 mm can handle a wastewater flow of about 0.5 m^3/s, while the same diameter drum with a screen opening of 6 mm can handle a wastewater flow of about 1.8 m^3/s. Larger flows are accommodated by arranging several drum units in parallel.

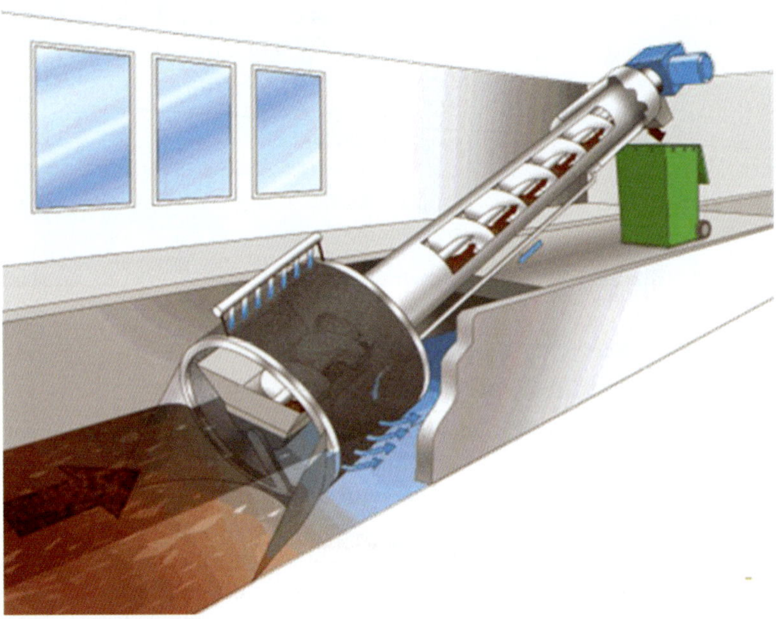

Figure 1.3 Diagram of a rotating micro screen for preliminary treatment (Courtesy Huber Technology)

Removal levels of suspended solids and organic matter depends on the screen opening. According to information from the Huber Group, a manufacturer of this type of rotating screens, for a 1 mm screen opening, removal efficiencies of up to 20% of BOD, up to 90% of floatable solids, and up to 70% of floatable oil and grease can be achieved. Koppl and Frommann (2004) report removal of 20–30% BOD and of 30–60% filterable solids (suspended solids) in rotating screens with screen opening of 0.2 to 0.3 mm. These high removals of suspended solids result in large quantities of sludge, similar to that of primary treatment, and would require the provision of sludge treatment installations.

Preliminary treatment by rotating micro screen is now a well-proven technology with robust design. It has reduced investment cost, is reliable and simple to operate and can be adjusted to particular hydraulic inflow conditions. Land and maintenance requirements are low, and package plants are available. It is a unit process of appropriate technology because it is less costly than conventional preliminary treatment that uses bar screens and aerated grit chambers, more reliable, requires less land and maintenance and removes more solids. It includes only small electrical motors which operate intermittently and therefore has small energy consumption.

A preliminary treatment unit consisting of at least rotating micro screens should be the first unit in any type of wastewater treatment plant. It removes coarse and floating materials from the raw wastewater in a reliable and cost effective manner, and it is an exemplary appropriate technology treatment unit.

Investment costs of preliminary treatment based on a rotating micro screens unit are in the range 3–10 US $/Capita, depending on the unit size and level of plant finishing (outdoor plants are of lower cost than indoor plants with elaborated site development). Operation and maintenance costs are low, estimated at 0.1–0.15 US$/Yr/Capita.

1.7.2.2 Vortex grit chambers

If a grit chamber is required as part of the preliminary treatment, a vortex type grit chamber is a better choice as an appropriate technology unit, although a conventional aerated grit chamber is also quite a simple unit. A grit chamber, including vortex grit chamber, is a physical treatment unit. In a vortex type unit of grit removal, a free vortex is generated by the flow that enters tangentially at the top of the cylindrical tank unit. Effluent exits the upper part of the unit cylinder. Grit settles by gravity at the bottom, where it is removed by a belt conveyor or lift station, and organic solids exit with the effluent. Another type of vortex grit chamber uses a rotating turbine to maintain constant angular flow velocity and to promote separation of organic matter from the grit (see Figure 1.4). The grit settles by gravity into a hopper, from which it is removed by a grit pump or lift pump. Vortex grit chambers are more compact than conventional grit chambers, and are efficient and reliable. Because of their smaller size they are also less expensive.

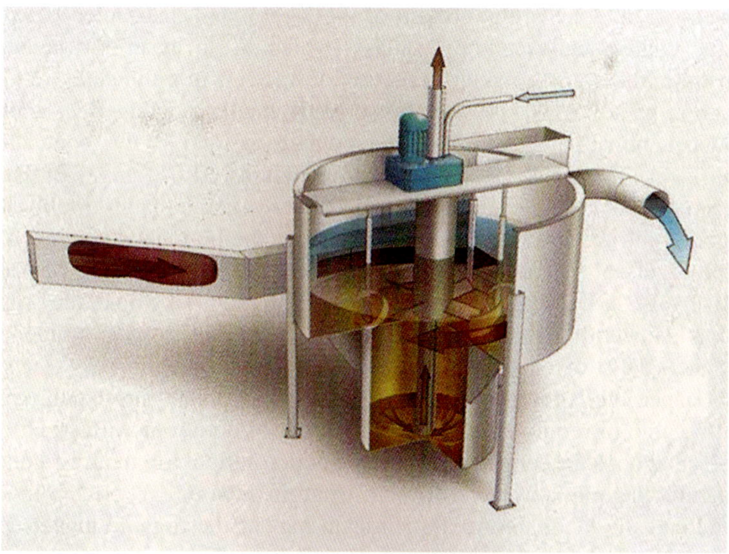

Figure 1.4 Vortex grit chamber (Courtesy Huber Technology)

1.7.2.3 Anaerobic lagoons

Anaerobic lagoons technology is well known and widely spread all over the world. Organic matter is removed in anaerobic lagoons through an anaerobic biological process; the anaerobic lagoon is in fact an anaerobic reactor. It is usually the first stage in lagoon treatment plants, followed by other types of lagoons. Anaerobic lagoons are designed to receive such a high organic loading that even if algae would have developed in these lagoons they would not be able to produce sufficient oxygen to keep the lagoon aerobic, so this type of lagoons are completely devoid of oxygen and under such condition they favour the development of anaerobic bacteria, which decompose the organic matter contained in the wastewater flowing into the lagoons through the anaerobic biochemical pathway. A typical anaerobic lagoon is basically a "hole in the ground"; usually with no equipment in it, except for inlet and outlet hydraulic structures. Commonly, the depth of anaerobic lagoons is about 4 meters, but there is no specific limit to the lagoons depth. An anaerobic lagoon is in fact a low rate anaerobic reactor. The concept "low rate" should not be interpreted to mean a second grade or low performing system, but rather that larger reactor volumes, in relation to other higher rate anaerobic processes, are required to achieve high removal of organic matter. Low rate systems can be designed to provide organic matter removals equal to those achieved in high rate systems. The volumetric loading rate of anaerobic lagoons is between 0.1 and 0.5 kg BOD_5/day/m^3. The low range loading is used for areas with a profound cold season of about 10°C, and the high range loading is used in areas where there are uniform annual warm temperatures in the range 25–30°C. These loading rates are equivalent to area BOD_5 loading rates of between 4,000 and 16,000 kg BOD/day/ha assuming an anaerobic lagoon depth of 4 meters. The depth of anaerobic lagoons is usually in the range of 3–5 meters, 4 meters being a common figure. Detention times in anaerobic lagoon range from 2 to 6 days. There appears to be little treatment advantage in extending detention time to more than 2 days and recently there is even a tendency to reduce the detention time in an anaerobic lagoon to 1 day. Mara (2004) provides data on experience with anaerobic lagoons from northeast Brazil which shows that at 25°C little variation in BOD removal was found at detention times in the range 0.8–6.8 days. To reduce the detention time and still have an efficient performance, the inflow to the anaerobic lagoon needs to be done in the bottom of the lagoon and distributed between several feeding points. This way, the contact between the biomass (which in an anaerobic lagoon develops on the bottom) and the substrate (the organic matter in the raw sewage) is more intensive and effective.

An anaerobic lagoon removes 40–70% of the organic matter contained in municipal wastewater. Considering that typical municipal raw wastewater contains a total BOD_5 concentration in the range 200–300 mg/l, an effluent of anaerobic lagoons usually contains 60–180 mg/l BOD_5. The mixing in an anaerobic lagoon reactor is not effective and the contact of the organic matter dissolved in the raw sewage flowing into the lagoon with the biomass which develops in the bottom of the lagoon is not intense, so only partial removal of dissolved organic matter is achieved. The removal of the suspended solids contained in the raw sewage, which settle in the lagoon, is high, but as a whole, the organic matter removal in an anaerobic lagoon is lower than in other anaerobic processes. The mixing regime in anaerobic lagoons is also the reason that the use of anaerobic lagoons in series is not effective when treating municipal wastewater. The first anaerobic lagoon will remove the organic matter it is capable of removing and the second one will not be able to remove a significant amount of additional dissolved organic matter. The approach of anaerobic lagoons in series might be a good solution for treating concentrated industrial wastewater. An anaerobic lagoon is a powerful unit, small in size, of low cost in investment and O&M, simple to operate and capable of removing a significant part of the organic matter contained in the raw wastewater. If additional removal of the remaining organic matter is necessary, the removal effort of the residual organic matter is smaller since the major part was removed by the anaerobic lagoon.

The flow regime in an anaerobic lagoon is such that the mixing in the lagoon is not effective. Conventional anaerobic lagoons do not contain mechanical mixers. The solids contained in the raw sewage flowing into the lagoon settle to its bottom and new anaerobic biomass develops mainly in the bottom of the lagoon. Under the flow regime of the lagoon, the liquid does not come into effective contact with the bottom sediments, that is, with the anaerobic bacteria, so the removal of the dissolved organic matter is not very efficient. The mechanism by which the liquid phase comes in contact with the bottom sediments is through the assistance of gas bubbles, which develop at the bottom, grow and carry sludge particles upwards through the liquid column to the liquid surface, where they collapse and slowly settle back to the sludge bed at the bottom. Sludge particles go through many up and down cycles during their residence time in the lagoon. Settled solids stay a long time in the bottom of the lagoon and get well digested, therefore, the excess sludge quantities generated in an anaerobic lagoon are much lower than in the other anaerobic process. However, due to the ineffective contact between the liquid and the sludge, a long detention time is required to remove even part of the dissolved organic matter in the liquid, and this is the reason that the detention time in anaerobic lagoons is much higher than that of other anaerobic processes and even so the organic matter removal in an anaerobic lagoon is low in comparison to other anaerobic processes. Due to the flow regime in anaerobic lagoons they are most advantageously used to pre-treat wastewaters which have a high solid content. The solids settle to the bottom where they are digested anaerobically. The successful operation of anaerobic lagoons depends on the delicate balance between the acid forming bacteria and the methanogenic bacteria, thus temperature higher than 10°C in the lagoon liquid is required and the pH in the lagoon must be higher than 6.2. Under such conditions, sludge accumulation is minimal, and desludging, which is required when the lagoon is half full of sludge, is necessary only once every 5–10 years. Although it is commonly accepted that anaerobic lagoons do not function at temperatures lower than 10°C, evidence from the highlands of Bolivia and Peru (with double the solar radiation than at sea level) show that anaerobic lagoons perform well even in zones with prolonged low air temperatures of up to −7°C. The temperature in the lagoon liquid is of course not that low, but it is certainly lower than 10°C. It seems that anaerobic bacteria can adjust to low temperatures if they are kept without severe fluctuations during prolonged periods of several months.

Lagoons plants are, in general, suitable for serving small and medium size cities, of up to 400,000 inhabitants in one plant. That applies also to anaerobic lagoons, which usually form part of various configurations of lagoons systems. A lagoon plant serving larger populations requires a large land area which is usually not available in the vicinity of a large city or becomes prohibitively expensive. However, if several smaller plants can be constructed at different locations, lagoon systems can serve cities with populations much larger than 400,000.

The advantages of anaerobic lagoons are low cost, simplicity of operation; effectiveness in handling highly concentrated sewage, the ability to cope with high concentrations of suspended solids (TSS), the ability to handle shock loads and success in treating a variety of biodegradable industrial wastes. The main disadvantage of anaerobic lagoons is the potential for generation of odours, caused by emission of H_2S when they are submitted to a too low or too high organic loading. A low loading may occur during the early years of operation and a high loading, during late years. Another disadvantage associated with the odour problem is that once designed and constructed, there are no operational means of controlling the process and resolving the odour problems when they appear. The odour problem associated with anaerobic lagoons is the reason for the resistance to use this process or when using it, locating the plant at a long distance from urban areas, which increases the cost of conveyance of the raw sewage. The relationship between odour development and organic loading is now reasonably well understood so that problems can usually be overcome by proper design. Recently a method for completely eliminating the risk of odours was developed by covering the anaerobic lagoons with a geomembrane, capturing the biogas produced in the process and either burning it or using it for energy generation. Covering the lagoons provides a measure of operational control on the performance of anaerobic lagoons which totally eliminates the odour problem and is a response to the disadvantage of lack of means of operational control in uncovered anaerobic lagoons. The idea of covering lagoons is not new and is in common use in industrial wastewater treatment plants. However, the cost of geomembranes was prohibitive for municipal treatment plants, in which the area of the anaerobic lagoons is larger than in industrial plants. In the recent years the cost of an HPDE geomembrane in Latin America dropped to 12 US$ per square meter, which is a reasonably low unit cost which facilitates its use in municipal treatment plants. The practice of covering anaerobic lagoons in municipal wastewater treatment plants is just now initiating to be implemented and will probably change the negative attitude towards the utilization of anaerobic lagoons in municipal wastewater treatment plants. Covering of anaerobic lagoons provides many benefits: (i) capturing the odorous gas for removal, flaring or treatment, instead of its release to the environment at the vicinity of the plant, thus eliminating odour problems; (ii) enabling to flare or use the captured biogas which contains strong greenhouse gases, thus preventing their emission to the atmosphere and contributing to the worldwide efforts of reducing the emission of greenhouse gases; (iii) enabling the mobilization of additional financial resources for the utility through the sale of "greenhouse gases emission reductions" to industrialized countries, (iv) reducing process heat loss; (v) reducing water evaporation, (vi) blocking sunlight, thus inhibiting algae growth in the anaerobic lagoons; and (vii) if the energy policy of the country in which the lagoons are located is favourable, generating energy from the biogas, thus mobilizing additional financial resources for the utility.

Methane, which is an important component of the biogas produced in anaerobic lagoons (the biogas is typically a mixture of methane and other gases such as carbon dioxide ammonia, nitrogen, hydrogen and other gases, in which the methane content is in the range 65–85%), is also a powerful greenhouse gas, 22 times as potent as carbon dioxide. Covering anaerobic lagoons, capturing the biogas and selling the "emission reductions" of the greenhouse gases, can generate an additional revenue stream to a project and make it more financially viable, thus contributing to economic development while also contributing to the mitigation of climate change.

The "Bolivia Urban Infrastructure Project" financed by the World Bank included the expansion of sewerage coverage in poor areas of the city of Santa Cruz, which has a population of 1.3 million, as well as increasing the capacity of the four city's lagoons wastewater treatment plants so as to enable them to process the increased volume of wastewater. As part of increasing the capacity, the anaerobic lagoons of the four plants were covered with HDPE geomembranes and the captured biogas is being flared, while the feasibility of generating energy from it is being studied. After covering the anaerobic lagoons the odour problems of the four plants were completely eliminated, as was the problem of proliferation of mosquitoes in the vicinity of the plants. SAGUAPAC, the city's water and sanitation utility has negotiated a contract for sale of the biogas emission reduction. This is the first case in developing countries of an emission reduction contract based on capturing of anaerobic lagoons biogas.

Off-gas removal piping connects directly to the membrane cover system. The membrane material is chosen to be chemically resistant throughout the project life to the biogas and to the wastewater treated, under rigorous weather conditions. Features of the floating covers can include automatic rainwater removal, baffles and sampling ports. The floating covers are durable, UV resistant thus protected against sunlight, can be installed quickly without disrupting plant operation and are easy to maintain while in service. Floating covers are strong enough to support foot traffic, rainwater loads, snow loads and light vehicles. They can be designed to automatically drain rainwater or snowmelt. The membrane cover is securely anchored to the lagoon perimeter and suspended across the lagoon surface. Rainwater drains into the lagoon through a weight pipe and a drain holes system. Support columns can be added when necessary and provide easy access to internal components. The membrane cover is gas tight, but can be quickly detached and easily rolled up. This gives operators access to inspect and maintain internal components. Reattaching the membrane cover is quick and easy. Optional hatches allow accesses by plant operators without retracting the cover. A photo of a covered anaerobic lagoon in one of the treatment plants of Santa Cruz is presented in Figure 1.5.

Figure 1.5 Covered anaerobic lagoons in a wastewater treatment plant in Santa Cruz, Bolivia

The anaerobic lagoon is an important unit process of appropriate technology. It is a low cost unit in terms of investment and O&M, simple to operate, effective in handling highly concentrated sewage, able to cope with high concentrations of suspended solids and to handle shock loads and successful in treating a variety of biodegradable industrial waste. It requires relatively little extension of land and therefore has a small footprint, and removes a significant part of the organic matter from the wastewater at a low cost, without energy input, while producing only very small amounts of excess sludge. The recent approach of covering anaerobic lagoons eliminates environmental problems of odours and vectors generated by uncovered anaerobic lagoons, rendering it a very attractive appropriate technology unit process. The biogas generated in the covered anaerobic lagoons can be captured and used for energy generation and this is an additional advantage of this system.

Usually, anaerobic lagoons are not constructed as a sole treatment unit, but rather as a part of several lagoon units. Based on lagoons plants cost it is estimated that the investment cost of only anaerobic lagoons would be about 5–10 US$/Capita. Operation and maintenance cost is extremely low. The cost of covering lagoons in Latin America amounts to 12 US$/m^2. Additional information on Anaerobic Lagoons can be found in the books by Von Sperling (2002), Campos (2009), Mara (2004) and Arthur (1983).

1.7.2.4 Upflow anaerobic sludge blanket reactor (UASB)

An Upflow Anaerobic Sludge Blanket Reactor is a high rate anaerobic biological reactor for treatment of wastewater based on anaerobic granular sludge bed technology. UASB technology is being increasingly considered for municipal wastewater treatment applications in warm-weather locations given its low cost and simple operations. Although this technology can not by itself produce an effluent of the quality of a conventional secondary process like activated sludge, it can still achieve significant organic matter removal rates in the range of 60–75% of BOD_5 at a fraction of the construction and O&M costs of activated sludge. Experience to date in Latin America has shown that a UASB-based plant is simpler, less reliant on mechanical components and can easily achieve double the organic matter removal rates of conventional primary treatment. In addition, the UASB process generates substantially lower quantities of sludge than aerobic processes, reducing therefore the associated sludge disposal costs. Given the anaerobic nature of the process, a potential for H_2S generation (thus odour and corrosion problems) exists in municipal UASB applications when the sulphate concentration in the raw sewage is high. Appropriate selection of construction materials can reduce corrosion problems.

UASB reactors were developed in the 1970s for the treatment of highly concentrated industrial wastewater. Following the initial reactor designs for the treatment of sugar industry wastes, the benefits of the system, such as low sludge production, small footprint, low energy requirements and valuable biogas production, made the UASB reactors an attractive and hence widely applicable treatment alternative for highly concentrated industrial wastewaters at mesophilic temperatures. These advantages encouraged investigating the application of the UASB process to the treatment of domestic wastewater for municipal applications, where the low BOD concentrations coupled with high particulate BOD fractions results in insufficient methane production for heating the reactor to mesophilic temperatures. However, since even moderate ambient temperatures favour satisfactory removal rates within a reasonably sized reactor, the use of UASB reactors for domestic wastewater is becoming widespread in tropical countries such as Brazil, Colombia and others. In applications requiring higher effluent quality levels than those which can be achieved by UASB, then UASB reactors are usually followed by an additional treatment unit for polishing the UASB effluent. UASB technology followed by a polishing unit is a treatment option being increasingly implemented in warm-weather countries with minimum wastewater temperatures of over 10°C.

A schematic diagram of a UASB reactor is shown in Figure 1.6. From a hardware perspective, a UASB reactor appears to be nothing more than an empty tank with installations of three phases separation (liquid, solids and gas) on its upper part, thus extremely simple and inexpensive. UASB reactors are constructed as tanks with a rectangular or circular cross section. Wastewater is fed into the tank in the bottom, through appropriately spaced inlets. The wastewater passes upwards through an anaerobic sludge bed where anaerobic bacteria in the sludge come into contact with the wastewater (the substrate). The sludge is composed of microorganisms that naturally form granules (pellets) of 0.5 to 2 mm in diameter, which have a high sedimentation velocity and thus resist washout from the system even at high hydraulic loads. The anaerobic decomposition process resulting from the contact between the granular sludge and the wastewater generates biogas as in other anaerobic processes. The upward motion of released gas bubbles causes hydraulic turbulence which provides effective reactor mixing without any mechanical equipment. At the top of the reactor, the liquid phase is separated from the sludge solids and gas in a three-phase separator (also known as the gas-liquid-solids separator). This separator is usually a gas cap with a settler situated above it. Baffles are used below the opening of the gas cap to deflect gas to the gas-cap opening.

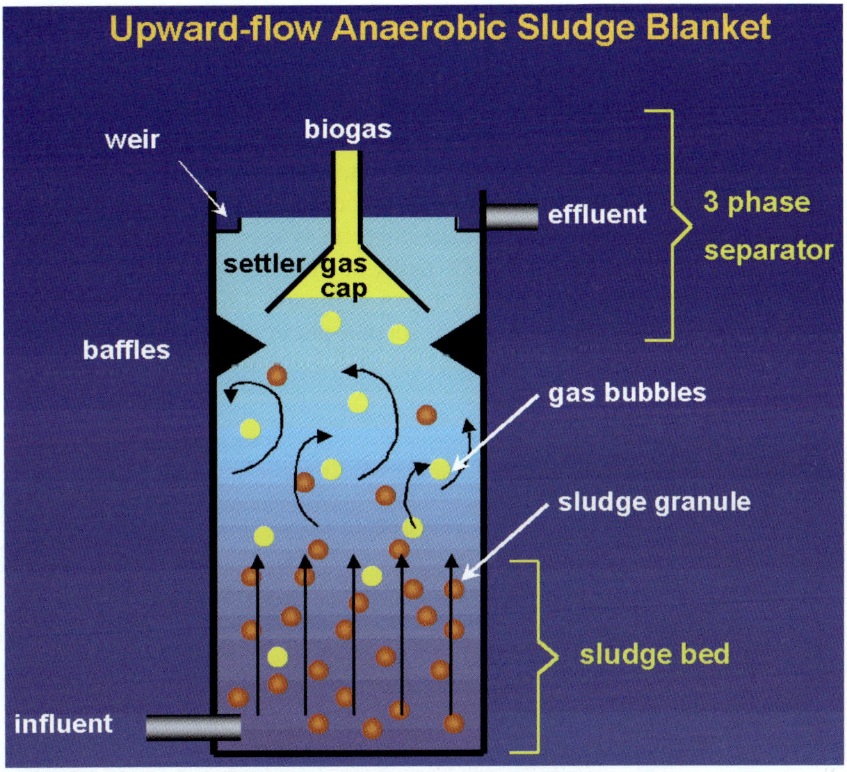

Figure 1.6 Schematic diagram of a UASB reactor (*Source*: http://www.uasb.org/discover/agsb.htm)

The level of removal of organic matter in a UASB reactor is limited by the concentration of the sludge in the reactor. Higher sludge concentrations would have resulted in higher organic matter removal, but would have also result in higher escape of sludge in the effluent, which then controls the maximum achievable sludge concentration.

To achieve higher sludge concentrations and higher removal of organic matter, an Expanded Granular Sludge Bed (EGSB) reactor, which is a variant of UASB, can be used. The distinguishing feature of EGSB is that it includes effluent recycle, so sludge concentration can reach higher values and a faster rate of upward flow velocity is designed for the wastewater passing through the bed. The increased flux permits partial expansion (fluidization) of the granular sludge bed, improving wastewater-sludge contact. The increased flow velocity is accomplished by a combination of utilizing tall reactors and effluents recycle. Organic matter removal of up to 90% and even higher can be achieved with the EGSB reactor. This type of reactor is used for treatment of industrial wastewaters but is not yet under widespread use for treatment of municipal wastewater.

UASB reactors treating municipal wastes yields an effluent that usually contains 35–100 mg/l Total BOD_5, 30–40 mg/l Soluble BOD_5 and 50–130 mg/l TSS (removal efficiencies are 60–75% for total BOD and 65–80% for TSS).

Typical hydraulic detention time values in a UASB reactor is in the range 4–5 hr and can be somewhat larger. Maximum up-flow velocity is 1.0 m/hr and can be lower, water depth in the reactor is about 5 meters and the recommended maximum flow per reactor is 70 l/s. Additional information on USAB reactors can be found in the books by Orozco (2005), Campos (2009), Chernicharo (2007) and Mara (2004).

Several large UASB plants have been constructed in Brasilia and Belo Horizonte in Brazil, each serving a population of about 1 million. Many plants serving smaller population are in operation in Latin America. The UASB technology is therefore suitable for serving all range of cities size, including large cities. Most of the UASB tanks are constructed as over the ground level concrete tanks, as seen in Figure 1.7, which shows the Onca UASB treatment plant in the city Belo Horizonte, Brazil, serving a population of 1 million. This type of tanks is relatively expensive. A less costly method of constructing UASB tanks is the use of tanks buried in the ground. In this case concrete is used only for lining the tanks, not for structural purposes. Concrete lined buried UASB tanks of the treatment plant of the city of Uberaba, Brazil, serving a population of about 300,000 are shown in Figures 1.8 and 1.9. Figure 1.8 shows one of the eight UASB tanks of the plant during the construction phase and Figure 1.9 shows the same tank after the plant has been commissioned.

Figure 1.7 The Onca UASB treatment plant in the City Belo Horizonte, Brazil

UASB plants treating industrial wastes are usually constructed of stainless steel or plastic materials and are designed to capture the biogas, and in most cases use it for generation of energy. The existing municipal UASB plants were not design to capture the biogas and its seems that with concrete tanks and without proper design attention to biogas capture, the biogas escapes the system, so currently there are no municipal UASB plants that manage to capture and utilize the biogas. However, it is expected that with proper attention to this issue and adequate design, the biogas in UASB plants can be captured and, if economically viable, utilized for energy generation.

Figure 1.8 A buried UASB tank under construction in the treatment plant of the City of Uberaba, Brazil

Figure 1.9 (a) A buried UASB tank under operation in the treatment plant of the City of Uberaba, Brazil. (b) Baffled reactor in Zambia (*Source*: http://www.germantoilet.org/en/projects/zambia/menu-sambia/progress.html)

With total plant construction costs (including sludge handling facilities) similar to, and in many cases lower than, a conventional primary treatment plant, a UASB-based plant is simple, less reliant on mechanical components than conventional secondary treatment, does not consume energy and can achieve significant removal levels of organic matter. In addition, the UASB process generates substantially lower quantities of sludge than conventional primary treatment and certainly lower than conventional secondary treatment, reducing thereby the associated sludge disposal costs. These characteristics make it a member of the appropriate technology unit treatment processes. UASB is also a unit process with a small footprint, so it becomes an important process when the land available for the municipal wastewater treatment plant is scarce. The biogas generated in the UASB reactor may be captured and used to drive a downstream aerobic polishing processes (if such processes are included in the treatment plant), leading to energy-autonomous treatment plants. Alternatively, the energy produced from the biogas can be used for other purposes.

Based on the recent construction of several medium to large size UASB facilities (serving equivalent populations of over 200.000) in Curitíba, Brazil, and some years ago in Bucaramanga, Colombia, construction costs in the range 20–40 US$/Capita have been documented for the UASB technology. Even including additional treatment steps for polishing the UASB effluent, these costs compare favourably with the cost of over 100 US$/Capita considered typical for a conventional activated sludge treatment plant. Operation and maintenance costs of UASB are in the range of 1.0–1.5 US$/Yr/Capita.

A different type of blanket reactor is the Baffled Anaerobic Reactor (BAR) initially proposed by Perry McCarty in the early 1980s. The piston (or plug-flow) anaerobic reactor (PAR) is a variant of baffled reactors which uses a Gas-Solid-Liquid-Separator (GSLS) in the final chamber to improve the entrapment of solid particles so as to prevent their washout with the effluent and maintain thereby the sludge age at a level as high as possible. The PAR was developed at the Los Andes university in Bogota, Colombia, (Orozco, 1988, 1997) for temperatures below 20°C, taking advantage of the higher efficiency of substrate removal obtained with the more perfect plug flow regime with baffles. Full scale plants based on RAP reactors were constructed and operated achieving COD removal efficiencies of 70% at temperatures between 13 and 17 °C (See Metcalf & Eddy, 2003, pp. 1005, 1018). These efficiencies can be improved to the extent that the retention of solids becomes more effective, because the effluent contains a high level of suspended solids which represent most of the COD in it. The PAR reactor in the city of Tunja (Colombia) was the first baffled reactor constructed in the world in a full scale plant (Metcalf & Eddy, 2003, pp. 1005). Moreover, the use of PAR reactors for operation at sub-optimal temperatures (lower than 15°C) has influenced the development of other baffled reactors that have been built at an accelerated pace in several countries, especially in China (Zonglian She *et al.* 2006), New Zealand (Foxon *et al.* 2003) and Bolivia (Medina *et al.* 1993). Figures 1.9b and 6.7 present photos of PAR reactors.

1.7.2.5 Anaerobic filter

An Anaerobic filter is a biological fixed bed anaerobic reactor which performs as an attached growth reactor, in which anaerobic bacteria grow on and are attached to a media of rocks or plastic filling of large surface area and the wastewater flows through the media. The reactor works under flooded conditions, which means that the media is flooded by the incoming wastewater so air, and consequently oxygen does not have contact with the biomass attached to the media and the biomass which grows in this reactor is strictly anaerobic. The bacteria are in continuous contact with the wastewater and consume the organic matter contained in it. An anaerobic filter can be described as the anaerobic equivalent of a trickling filter. Both reactors are based on attached growth and look the same: a tank filled with a bacterial support media. However, the trickling filter is not flooded, the wastewater trickles through the media and the bacteria growing on the media have access to oxygen so they are aerobic bacteria, while the anaerobic filter works when it is flooded, the bacteria growing on the media do not have access to oxygen and therefore they are anaerobic bacteria. Attached growth bacterial reactors can be divided to fixed bed, rotating bed (rotating discs) and expanding bed reactors. They also have a variety of shapes and structures. This section deals with fixed bed anaerobic reactors, also known as anaerobic filters, which are the most commonly used anaerobic reactors. Two main types of fixed bed reactors are ascending and descending flow reactors. The anaerobic fixed bed attached growth reactors consist of a tank, which can be of circular or rectangular cross section, containing the support media, rock particles or various types of plastic media. The bacteria develop in an attached form on the media and the consumption of substrate (i.e., the decomposition of the organic matter dissolved in the wastewater) as well as the growth of the anaerobic microorganism takes place on the external biofilm of each media particle. Anaerobic microorganisms can also develop in the form of flakes or granules trapped between the media particles. With increased accumulation of the biomass, parts of it detach sporadically from the media particles to which it is attached and flow out with the liquid effluent. In spite of this process, the average residence time of the biomass in the anaerobic filter is very long due to its attachment to the support media. The average residence time of biomass in this type of reactor is more than 20 days, resulting in an effective performance of the treatment process. Ascending flow filters are more commonly used than descending flow filters. In descending flow filters, recirculation of effluent to the filter inflow is commonly used. The elevated residence time of the biomass in the filter which results in high microorganisms concentration, associated with short hydraulic detention time of the liquid in the filter and the effective contact between the liquid and the biomass result in high organic matter removal (higher than in anaerobic lagoons and UASB reactors) and provide the anaerobic filter with great potential for its use for treatment of non-concentrated wastewater.

To prevent clogging of the filter, effective removal of suspended solids from the raw sewage is required prior to its inflow to the filter. Such removal of suspended solids can be achieved by rotating micro screens. For small treatment plants, decanter-digesters (otherwise known as large septic tanks or Imhoff tanks) are used to remove suspended solids prior to an anaerobic filter. This is a common practice in Brazil. In addition to the use of anaerobic filters as the main treatment unit, they can be used as polishing unites for improvement of the quality of the effluent of a preceding unit. They are in fact more suitable to perform as polishing units since when fed with treated effluent of a preceding unit they are less bound to clog by suspended solids and have to treat a lower content dissolved organic matter. In Brazil anaerobic filters are used as the main treatment unit mainly for small and medium size municipalities, while for polishing application anaerobic filters are used in the entire range of cities size, including in treatment plants of large cities.

With time, the anaerobic filter becomes clogged because of the growth of the biomass. When the pressure drop on the filter becomes too high, it needs to be washed. Similarly to sand filters for water treatment, the anaerobic filter is washed by a stream of water or treated effluent which is passed through it in a counter wise direction to its operation flow. The wash water, which contains the excess sludge, is directed to the sludge handling facilities, which are usually just drying beds. Since the process is anaerobic, the excess sludge quantities are small and the washing, which is specific to each systems, is required infrequently, once every several days to once every few months.

Chernicharo (2000) reports Total BOD removal efficiencies of anaerobic filters in the range of 70–80% when the anaerobic filter is the main treatment unit and 75–95% when it used as a polishing units. Andrade Neto (1997, 2000) reports results of a research on an anaerobic filter system treating municipal wastewater which was in operation for six years. This system, consisting of a septic tank followed by an up-flow anaerobic filter followed by a down-flow anaerobic filter, achieved a COD removal of up to 90%. The effluent of this anaerobic filter system contained less than 30 mg/l of BOD and less than 20 mg/l total suspended solids. Sludge withdrawal from the filter was undertaken once every six months.

A photo of a circular cross section ascending flow anaerobic filter in the treatment plant of the city of Tibagi in the state of Parana, Brazil, is presented in Figure 1.10. The filter is similar to a circular sedimentation tank. It is filled with a crushed rock media. The filter washing is done by a descending water stream. The wheels in the front of the filter operate the valves which change the flow direction to initiate the filter washing cycle. Anaerobic filter media composed of plastic material shapes with a large surface area is presented in Figure 1.11. To reduce the media cost, instead of buying factory manufactured plastic media, rock media can be used or plastic media can be prepared on site by using portions of electrical wires plastic conduits or by cutting small diameter plastic pipes to small sections and use them as the filter media.

Figure 1.10 An anaerobic filter in the wastewater treatment plant of the City of Tibagi in Parana, Brazil

Figure 1.11 Anaerobic filter media composed of plastic material shapes with a large surface area

Detention time of the liquid in an anaerobic filter is in the range of 4–12 hours. The organic load in an anaerobic filter treating municipal wastewater is in the range 1.0–1.6 kg COD/d/m^3 (considering total filter volume). The effective depth of the filter media is 1.2–1.8 meters. Based on Brazilian experience, construction costs for anaerobic reactors as the main treatment unit are in the range 10–25 US$/Capita. Operation and maintenance costs are about 0.8–1 US$/Yr/Capita.

1.7.2.6 Lagoons systems

Conventional technology

Along with the impressive advances and developments in man-made wastewater treatment processes in the past decades, some natural, old treatment systems are still being used successfully and should be considered as appropriate technology alternatives. However, most of these natural systems require large extensions of land, which may limit their applicability to small and medium-size cities, with populations of up to 400,000. Besides soil absorption, which is the natural process used in on-site disposal systems (cesspools and septic tanks); there are three major groups of natural wastewater treatment systems: (i) stabilization lagoons (also known as oxidation pond or stabilization ponds); (ii) land treatment systems (overland flow, infiltration into the soil); and (iii) aquatic systems (natural wetlands and constructed wetlands). Stabilization ponds are used extensively in developing countries and present one of the most cost effective reliable and easy to operate processes for treating domestic and industrial liquid wastes. They can produce high quality effluents, maintenance requirement are very simple and no energy is needed to make them function, other than solar energy.

There are three types of natural (non-aerated) stabilization lagoons: anaerobic, facultative and maturation lagoons. Anaerobic lagoons were discusses above and are used for reducing the major portion of the organic matter in the raw wastewater by the activity of anaerobic bacteria. Facultative lagoons also remove organic matter, however, the removal mechanism is aerobic and the organic matter in such lagoons is decomposed by aerobic bacteria. Maturation ponds (or polishing ponds) are designed to remove pathogenic organisms but they also remove residues of organic matter which reach these lagoons, trough the activity of aerobic bacteria. Some removal of fecal bacteria and of most of the helminth eggs occurs in anaerobic and facultative lagoons. Significant removal of fecal bacteria can be achieved, through proper design, in facultative and maturation lagoons. Facultative and maturation lagoons are aerobic phothosynthetic ponds. The organic matter reaching these ponds with the influent flow is decomposed by aerobic bacteria. The oxygen needed by the aerobic bacteria to oxidize the organic matter is mainly supplied by algae that develop naturally in lagoons in which the hydraulic detention times is over 2 days and the organic load is not excessive so that aerobic conditions can prevail in such lagoons. The oxygen is released by the algae as a product of the photosynthesis process. The carbon dioxide needed by the algae in addition to sun light for the photosynthesis process is provided by the aerobic bacteria in the lagoon, and is the end product of the decomposition of the organic matter by these bacteria. The algae and the bacteria in the lagoon perform in coordination to decompose the organic matter. The typical green colour of facultative and maturation lagoons is a result of the presence of algae in these lagoons. Facultative and maturation lagoons are large water bodies in which the hydraulic detention time (in each lagoon) is 5–10 days, so naturally the mixing in such lagoons is not very effective and the composition of the liquid in a lagoon is not uniform. Sections of a conventional lagoon which does not contain mixing equipment remain anaerobic in spite of the oxygen supplied by the algae. The algae tend to develop in the upper layers of the lagoon, up to the depth of penetration of the sun light, and in this part of the lagoon oxygen is available, but not in lower layers. Typically the upper layer of a facultative lagoon is aerobic but the lower layers are anaerobic, and the meaning of the term facultative lagoon is that it contains aerobic and anaerobic portions. Anaerobic decomposition of organic matter takes place in the lower, anaerobic layers of a lagoon. Suspended organic matter, dead algae and excess sludge settle to the bottom of the lagoon and undergo there anaerobic decomposition. The detention time of the sludge in the bottom of the lagoon is very long and as a result, the decomposition of the sludge is very effective. This is the reason that lagoon systems do not generate a lot of excess sludge. The mineral portion of the digested sludge accumulates in the bottom of the lagoons and so once every several years the lagoons need to be cleaned. Due to the large size of facultative lagoons, hydraulic short circuits can occur in this type of lagoons if the hydraulic design is not adequate, and in such cases, even portions of upper layers of facultative lagoons remain anaerobic and in fact do not conform part of the aerobic reactor. This is a common problem in facultative lagoons, which results in decreased efficiency of treating wastewater in the lagoons.

Maturation ponds have similar dimensions to those of facultative lagoons and their main purpose is to eliminate pathogenic organisms, bacteria, viruses and helminth eggs. Bacteria are removed by two mechanisms: (i) sedimentation while absorbed to settleable matter or contained within flocks of settling matter, predation by protozoa and starvation; and (ii) elimination by high sunlight intensity and high oxygen concentration. Long detention times in lagoons and high temperatures increase the removal of bacteria. It is believed that viruses are removed by absorption to settleable solids. Helminth eggs and protozoa cysts are removed by sedimentation and their removal occurs mostly in the first lagoon in the series, the anaerobic or facultative lagoon. The removal of pathogenic organisms in a lagoon systems can be very effective, achieving less than 100 fecal coliforms per 100 ml in the effluent, however to achieve such a high removal level, a series of lagoons with sufficient total detention time needs to be provided. A common problem in design of lagoon systems is the belief that effective removal of pathogenic organisms can be achieved within short retention times.

Usually, a lagoon plant is not constructed as one single lagoon but rather as a series of lagoons. For maximum efficiency, especially in warm climates, lagoons need to be operating in series: an anaerobic lagoon followed by one or two facultative lagoons and one or more maturation (or polishing) lagoons, each one treating the effluent of the previous lagoon. The inclusion of an anaerobic lagoon in a lagoons treatment plant is very important due to its high efficiency in removing a significant portion of the BOD (about 60%) occupying a relatively small area. Lagoons in series can produce excellent effluents with a microbial quality suitable for unrestricted irrigation. A typical lagoon system, the East treatment plant (Plant E) in the city of Santa Cruz, Bolivia is presented in Figure 1.12. The plant consists of three parallel trains, each one including one anaerobic lagoon (the small lagoon) followed by two facultative lagoons followed by one maturation pond. The effluent is discharged to a local creek that flows into the Pirai River. Figure 1.12 includes in fact two treatment plants, the other one, located at the left hand side of the figure, is the industrial park plant treating the wastewater of a nearby industrial estate. It is also a lagoons plant, much smaller in the area it occupies, has the same process flow diagram as the municipal plant and treats an organic load not much smaller than that treated by municipal Plant E. However, the flow of the industrial wastewater is much smaller than that of the municipal wastewater so the dimensions of the industrial plant are smaller. As seen in Figure 1.12 plant E is surrounded by residential areas which basically reach the fence of the plant on all sides. The residential areas are especially close to the anaerobic lagoons. This is a common phenomenon. Usually the treatment plant is constructed outside the city limits but with time the city grows and surrounds the plant, and then the residents start complaining about nuisance generated by the plant. This was also the case with the four treatment plants of Santa Cruz. The main complains are against the odours generated by the anaerobic lagoons. The Santa Cruz plants were recently upgraded to resolve the odour problems, as discussed below. Figure 1.12 presents plant E as it was before its upgrading. At that time the picture was taken Plant E served a population of 240,000 persons and treated a wastewater flow of 36,000 m^3/d. To resolve the odour problems, SAGUAPAC, the water and sanitation utility of Santa Cruz, covered the anaerobic lagoons in its four lagoons treatment plants. That completely resolved the odour problem as well as that of mosquitoes proliferation. Figure 1.13 shows the lagoons Plant E after the anaerobic lagoons have been covered. The biogas collected from the anaerobic lagoons is being flared and alternative for its utilization for generating energy are under study. Upgrading of the facultative lagoons will be undertaken by introducing mixers in these lagoons, under the principles discussed in the following section.

Figure 1.12 Air photo of a conventional anaerobic-facultative-maturation lagoons plant for municipal wastewater treatment, Plant E, one of three in the City of Santa Cruz, Bolivia

Figure 1.13 Air photo of Plant E after covering the anaerobic lagoons

Many lagoon systems do not include an anaerobic lagoon in the flow diagram and they are based only on a facultative lagoon followed by maturation ponds. Such a flow diagram is usually adopted for the purpose of preventing odour problems which might be generated by the anaerobic lagoons. The disadvantage of such an approach is that a plant without an anaerobic lagoon occupies a much larger area than a plant with an anaerobic lagoon as the first stage of treatment. As mentioned, the odour problem can be resolved by a covered anaerobic lagoon, so there is no reason to omit the anaerobic lagoon and lose the benefits provide by such an important unit process. In a plant without an anaerobic lagoon, recirculation of the effluent of a second facultative lagoon to the entrance to the first facultative lagoon, at a recirculation ratio of 2:1 in relation to the raw wastewater flow, permits increasing the organic load on the first lagoon and reducing the area required for the plant. But recirculation is costly and energy consuming.

Total BOD and TSS removal in conventional lagoons systems is in the range 70–90% for each of these contaminants, depending on the configuration and design of the lagoon system. Effluent BOD and TSS can be somewhat high but most of it is due to the algae in the lagoons effluent. In BOD Analysis tests Algae appears as an oxygen consumer but in a receiving watercourse algae is an oxygen contributor. In this respect, algal BOD and SS are different from non-algal BOD and SS. Effluent standards in many countries (for instance in the European Community) require that lagoons BOD effluent quality be less than 25 mg/l on a filtered (non-algal) basis. This should also be the approach in all developing countries. However, the effluent quality is related also to the receiving water body. If the effluent is discharged to a lake a lagoon effluent may generate problems because of its nutrient content. Discharge of lagoons effluent to the sea without an effective outfall is also a problematic practice. Algae removal from lagoons effluent is not an easy task. Many processes are available for this purpose, but they significantly increase the treatment cost. Recently, a new process using ultra-sound waves was developed and seems to be cost effective (see www.lgsonic.com).

One of the main drawbacks of lagoons is the large extension of land they require, which makes a single lagoon plant suitable for small and medium-size cities since for larger populations availability of area for a single plant usually becomes a problem. However, a city with a population larger than 400,000, can be served by several lagoon plants. As an example, the city of Santa Cruz, Bolivia, with a population of approximately 1.3 million is served by three lagoon plant for treatment of municipal wastewater (and one plant for treatment of industrial wastes). A forth lagoon treatment plant for municipal wastewater is under construction, so all the wastewater of a city of this size is treated in lagoon plants, but not in one single plant.

An important advantage of lagoon systems over other treatment processes is their ability to remove pathogenic organisms without the need for chlorination or other type of disinfection (Ozonation, UV, etc.), if the detention time of the effluent in the ponds is sufficiently long. Lagoon systems are also extremely efficient in removing organic matter and suspended solids, and if well designed, can easily achieve removal rates of BOD and suspended solids of higher than 90%. Lagoon systems are very robust due to their long hydraulic detention time, so they can handle hydraulic shock loads better than other processes. They are also resistant to heavy metal and toxic materials, so they can cope well with industrial effluents. Other advantages of lagoon systems include the low capital investment and the simple construction, using only local materials and no equipment. Operation and Maintenance costs are also very low since a lagoon system does not consume energy. The operation and maintenance is also quite simple and does not require operators of high expertise. Lagoon systems generate very small quantities of excess sludge (in practice they require bottom cleaning for sludge removal just once every several years). As such; lagoon treatment can be classified as an excellent appropriate technology process, ideal for use in developing countries and elsewhere. The biogas generated in the anaerobic lagoons can be captured and used for energy generation and this is an additional advantage of this system. Lagoons systems have two perceived disadvantages. The first one is the potential generation of odours. To avoid odours lagoons (both anaerobic and facultative) must be adequately designed with correct loading rates and overloading of the lagoons with time must be avoided since overloading is the main reason for odour generation. In addition, as explained previously, anaerobic lagoons can be covered and this totally eliminates odours and vectors in such lagoons. Regarding odours generation by facultative lagoons, as explained in the following section, wind operated mixers can be installed in facultative lagoons and this practically eliminates odours and vectors in this type of lagoons. The second disadvantage is the large footprint of lagoons systems. The extension of area of lagoons plants can be significantly decreased by inclusion of anaerobic lagoons in the flow diagram. It can be further decreased by installing mixers in the facultative lagoons, as is explained in the following section. However, even with all this measures, the area required for a lagoon plant is larger than that of other types of wastewater treatment plants. If land at the vicinity of the city is not available or is expensive, the wastewater can be conveyed further downstream to rural areas where land is available and cheaper. The conveyance system will increase the investment cost, but even if the lagoons system investment cost becomes comparable with the cost of a conventional treatment plant located closer to the city, it would still be more reasonable to adopt the lagoon system solution because the O&M costs are much lower and a lagoon system is a more sustainable solution, especially in developing countries. The choice is between paying for conveyance and land on one hand and paying for complex equipment and high O&M costs, especially energy costs, on the other hand. Complex equipment and high O&M cost is not a good solution for developing countries. In addition, investment in energy costs is lost forever while investment in land is not lost.

The depth of facultative lagoons is usually in the range of 1.5–2 meters and that of maturation ponds is similar or somewhat smaller. The organic load on facultative lagoons is generally in range of 100 to 250 kg BOD/day/ha, where the low value is applied during the low temperature season and the high value during the high temperature season. It is obvious that a treatment plant needs to be designed in such manner that it is able to cope also with the conditions of the low temperature season, which are the critical conditions. GTZ reports that the uncovered anaerobic lagoons of Puchukollo, El Alto in the highlands of Bolivia, are operating with efficiencies of more than 50% with detention times of 6 days and reactor temperatures as low 8°C. However, the solar radiation at that altitude is about double than at sea level (Juanico & Weinberg, 1995) and that may be the reason that the anaerobic lagoons maintain higher temperatures than those with lower solar radiation, that is, at lower altitudes.

It is difficult to provide information on cots of stabilization lagoons system since the cost depends on the configuration on the lagoons, on the size of population served and on the cost of land, however, representative values can be mentioned. Information based on Arthur (1983), Yanez (1993), Rolim (2000), Mara (2004) and on experience of the authors indicates that construction costs of a lagoons system is in the range of 10–40 US$/Capita, depending on the size of population served (the lower range refers to larger size plants). Operation and Maintenance costs are in the range of 0.2–0.4 US $/Yr/Capita. The sources mentioned along with the book of Von Sperling (2002) are also good sources of information on design, performance, operation and maintenance of lagoon systems.

Aerated lagoons, a process developed to reduce the extension of land required by lagoons treatment is based on installing aeration equipment in the lagoons so it is not a natural process of organic matter metabolism in a lagoon but rather a process based on mechanical supply of oxygen to the lagoon for supporting the decomposition of the organic matter. When aerated lagoons are the main process of treating the raw wastewater they should not be considered an appropriate technology process. Although their construction costs may be lower than those of other lagoon types, with the aeration equipment their investment cost becomes higher. But the main problem of aerated lagoons is that their energy consumption is high since in this process the oxygen required for the treatment process is supplied by aeration equipment operated by electric energy. Usually in developing countries, and especially in small and medium size cities, there is no capacity to finance the electricity cost for operating the aeration equipment of the aerated lagoons and there is no technical capacity to keep the equipment running for extended periods of time, so frequently an aerated lagoons plant functions as a raw sewage reservoir in which non-functioning aerators are floating, or are deposited on the lagoons dykes. However, aerated lagoons can be used as a polishing process of effluent of another process (for example anaerobic lagoons, UASB or an anaerobic filter). In such a case, they can be considered an appropriate technology treatment process, since the amount of organic matter which needs to be removed in such polishing processes is not large, the aeration equipment is not intensive and the electricity cost is moderate. If the process preceding the aerated polishing lagoons is an anaerobic process, the biogas released in this process can be used to generate the electricity required for operating the polishing aerated lagoons.

Facultative and maturation lagoons equipped with mixers, an innovative technology

Interesting innovative processes have been developed during the recent years to improve the performance of existing facultative and maturation stabilization lagoons and to improve the design of new ones. Common problems in facultative and maturation lagoons are: (i) short circuiting which drastically reduces the treatment efficiency; (ii) odours at night due to disappearance of sunlight which stops the oxygen production by algae; (iii) stratification which disturbs or prevents treatment processes in lower layers; (iv) over concentration of algae in the upper layer, preventing sunlight penetration and reducing process efficiency; (v) oil film build-up in the water surface, disturbing the treatment process; (vi) continuous odour problems resulting from overloading of the lagoons, and (vii) in areas with very low temperatures in the cold season, development of a snow or ice layer on the water surface, disturbing the treatment process by preventing penetration of sunlight. One of the disadvantages of the lagoons treatment process is that once designed and constructed, there are no operational means of controlling the process and resolving problems when they appear. Two types of solution have been developed to resolve the facultative and maturation lagoons problems. Both are based on the principle of installing in the lagoons equipment which does not change the nature of the process in the lagoons but rather help it to perform better. Several companies manufacture such equipment. Information on the equipment and the processes can be found for example on the web site of one of these companies, Gurney Environmental (see www.GurneyEnvironmental.com).

The first instrument (denominated by Gurney environmental as the Accel-O-Fac Process) is installing in the lagoon gentle mixers which pump liquid from the depth of the lagoon, without breaking the facultative boundary layer or disturbing the anaerobic layer at the bottom of the lagoon, and discharge the liquid near the surface. The mixers prevent short-circuits, bring to the upper layer organic matter from deeper layers for processing, distribute to the entire lagoon volume the oxygen generated by the algae in the upper layer thus creating uniform oxygen profiles from the top of the lagoon to the facultative layer, prevent stratification, break oil films, break snow and ice layers, and actually resolves all the mentioned problems of facultative lagoons and thereby improve the performance of such lagoons. The mixers used for this purpose are not aggressive aerators which consume large amounts of energy, but rather gentle mixers, each operated by a small motor of 0.75 HP that consumes little energy and can be operated by wind when the wind velocity in the treatment plant site is sufficiently high (over 5–6 km per hour). The mixer switches automatically to wind operation when the wind velocity passes a selected threshold. Each mixer has an influent range of 0.65–3 hectare, depending on the type of lagoon. In higher loaded lagoons the influence area of each mixer is smaller, between 0.65 to 1 hectare and in maturation ponds the influence area of each mixer is between 0.75 to 3 hectares, so higher loaded lagoons require more mixers. When designing a new lagoons plant with mixers, the depth of the lagoons should be 2–3 meters and even up to 4 meters, that is, higher than lagoons without mixers. The installation of the mixers in a facultative lagoon can significantly increase its treatment capacity. Organic loads on mixers equipped lagoons can reach values of 450 Kg BOD/d/ha in the high temperatures season of 25–30°C and even higher, with lower loading values at lower temperatures.

The second instrument (denominated by Gurney environmental as the Aero-Fac Process) is injecting air into the lagoon through blowers (not compressed air) to augment the oxygen quantity produced by the algae without changing the flow regime, that is, without transforming the lagoon to an aerated lagoon. For that purpose, a combination of blowers with air distributors and gentle mixers need to be installed in the lagoon. The addition of air to the lagoons increases its treatment capacity to a required level.

The use of mixers in lagoons provides some measure of operational control on the performance of the lagoon system and is a partial response to the disadvantage of lack of means of control in lagoons that are not equipped with mixers.

Mixers can be used as part of a design of a new lagoon plant. In such case, the new plant will occupy a smaller area than a plant without mixers, since it can handle higher organic matter loads. Mixers can also be installed in existing lagoon plants to upgrade their capacity. In many cases lagoon plants become overloaded but there is no area available to increase the capacity of the plant by adding lagoon units because the city expanded and occupied the land areas around the plant. Moving the plant to another site is expensive and abandoning the lagoons process to replace it with a conventional mechanical treatment process is very expensive. In this case mixers can be installed in the facultative lagoons without the employment of blowers, and they will increase the capacity of the plant without increasing the lagoons area, and at a very small cost. If further capacity increase is required without increasing the lagoon area, then blowers can be added. An excellent method to upgrade an existing conventional lagoons plant consisting of anaerobic lagoons followed by facultative lagoons followed by polishing ponds is covering the anaerobic lagoon and installing mixers in the facultative lagoons. This was done in the municipal wastewater treatment plants of Santa Cruz Bolivia and is discussed in a following chapter.

The mixing equipment principles of operation are presented in Figures 1.14 and 1.15. The mechanism by which the mixers improve the performance of facultative lagoons and resolves the above mentioned lagoons problems are presented in Figure 1.16. A photo of several mixers in a facultative lagoon (in Plant North 2 in Santa Cruz, Bolivia) is presented in Figure 1.17.

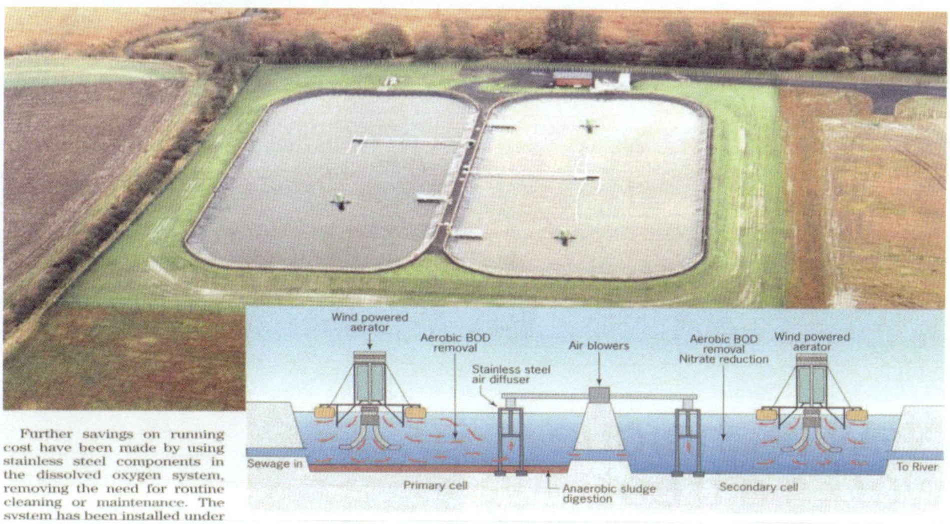

Figure 1.14 Mixers and air injectors in facultative lagoons (Courtesy Gurney Environmental)

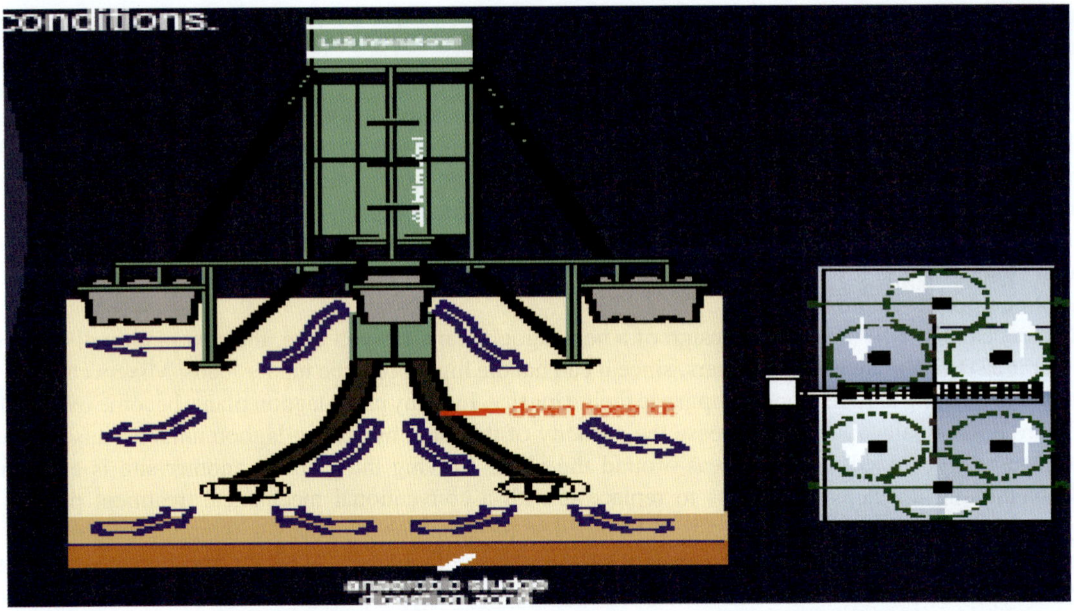

Figure 1.15 The principle of the lagoons mixer function (Courtesy Gurney Environmental)

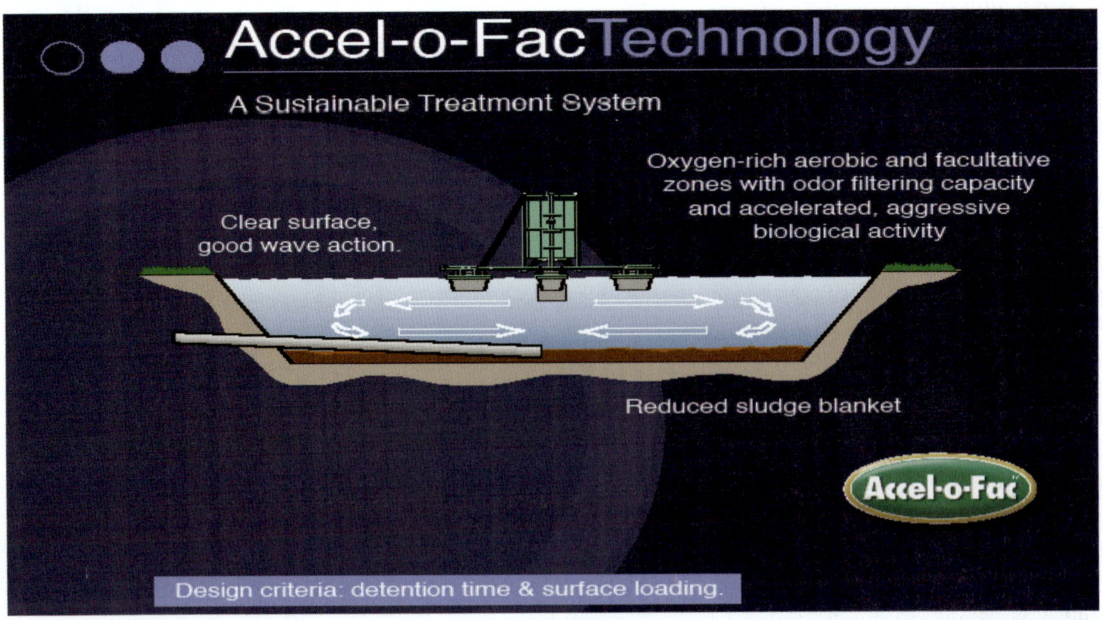

Figure 1.16 The mechanism of improvement in facultative lagoons performance by the use of mixers (Courtesy Gurney Environmental)

Figure 1.17 Mixers in a facultative lagoon in the N2 plant in Santa Cruz, Bolivia

Upgrading of existing plants can be done in stages, first mixers are installed in shallow lagoons, after time, lagoon depth is increased to 3–4 meters, since the mixers are more effective at such depths, and finally, air blowers and injection equipment can be added. This is an effective method to increase lagoons plants capacity when additional area is not available. For design of new lagoons, it is proposed to provide a plant consisting of a basic unit of a covered anaerobic lagoon followed one or two facultative lagoons in series, each one 3–4 meters deep, equipped with mixers, followed by a shallower maturation pond which might also be equipped with mixers. An ultrasonic installation for reducing algae content in the effluent can be installed in the maturation pond. This yields effective treatment and good quality effluent occupying less area. With mixing equipment aided lagoons it is possible to achieve effluents of higher quality than those of conventional lagoons systems and reach Total BOD removals of 90–95% and TSS removals of 80–90%.

A treatment plant of a small town in England, Sutton St. James, based on facultative lagoons (without anaerobic lagoons in the flow diagram) with mixers and air injection equipment is presented in Figure 1.18. The six box shaped structures in the right side lagoons are modules of plastic media on which nitrifying bacteria grow and convert the ammonia to nitrate by the nitrification process, so the effluent of this plant does not contain nitrogenous BOD (in addition to the elimination of the carbonaceous BOD).

Facultative lagoons, as any aerobic process, can function at low air temperatures, below zero, as long as the wastewater and the liquid in the lagoons do not freeze, that is, the liquid temperature does not fall to less than 1°C. Such conditions require an adequate design, which leads to longer retention times but by constructing deeper lagoons and adding mixers it is possible to retain cost effectiveness. A lagoon equipped with mixers functioning under severe winter conditions is presented in Figure 1.19.

Figure 1.18 A facultative lagoons plant of Sutton St. James, England, with mixers and air injection systems (Courtesy Gurney Environmental)

Figure 1.19 A facultative lagoon equipped with mixers functioning under severe winter conditions

A facultative lagoon equipped with mixers is still an appropriate technology wastewater treatment unit process. It is simple to operate and is of low investment and O&M costs. The mixers add cost but increase capacity thus decrease the area occupied by the lagoons and decrease cost, so overall, the investment cost remains low. Air injection to a lagoon is not strictly an appropriate technology process because of the energy it consumes. However, if an upgrade of an existing plant is needed, it might be the cheapest alternative, so it can be considered a hybrid appropriate technology process.

It is difficult to provide data on costs of upgrading existing plants because such costs are specific to each plant, depending on local conditions. The cost of one mixer for facultative lagoons manufactured by Gurney is approximately US$ 35,000. Rough estimates would suggest an additional investment cost range in mixers of 5–10 US$/Perosn for upgrading an existing lagoons treatment plant. Incremental operation and maintenance costs are very small. The investment cost of newly constructed mixers aided lagoons treatment plants is similar to that of conventional lagoons systems, because although an additional investment in mixers is required, the population served is larger since the plant capacity increases when using mixers, so the cost per Capita does not increase. The same applies for operation and maintenance costs. The investment cost of a newly constructed lagoons treatment plant with covered anaerobic lagoons and mixers aided facultative lagoons is 20–50 US$/Capita and the O&M costs are similar to those of a conventional lagoons system.

1.7.2.7 Stabilization reservoirs for effluent reuse for irrigation and for intermittent discharge of effluents to rivers

Introduction

A stabilization reservoir is an appropriate technology unit process developed in Israel in the early 1970s. It is a seasonal storage reservoir for effluents which in addition to storage provides also polishing treatment through biological processes. A stabilization reservoir must be preceded by a pre-treatment unit which provides a higher level of treatment than preliminary. The pre-treatment process can be of various types and can be based on most of the unit processes mentioned in this book and also on other conventional treatment processes as activated sludge and others. The purpose of the pre-treatment is to reduce the organic matter load which is diverted to the reservoir. The pre-treatment followed by the reservoir is already a combined process. In this section we discuss the reservoir unit process itself and in following sections combined processes which include reservoirs are discussed. The stabilization reservoir concept was developed as an element in projects of wastewater reuse for irrigation. Stabilization reservoirs are large volume deep earth reservoirs constructed for seasonal storage of effluents during periods which last several months. These reservoirs are aimed to store effluents during the wet season so that they can be used for irrigation during the dry season. Stabilization reservoirs range in volume from 0.02 to 12 million cubic meters and can be even larger. The depths of stabilization reservoir are in the range of 8 to 12 meters. A stabilization reservoir is an appropriate technology unit because it is very similar to a lagoon: an earth structure with no equipment (although mixers of the type installed in facultative lagoons can also be installed in reservoirs to improve their performance). The stabilization reservoir unit process is not a well known process, although it has been in widespread use in Israel during the past 35 years, with over 350 systems under successful operation. Detailed information on the concept of the stabilization reservoir, theory, design and performance is presented in a book by Juanico and Dor (1999). Stabilization reservoirs have two main uses: (i) an important element in projects of wastewater reuse for irrigation; and (ii) storage of effluent for intermittent discharge to rivers when the flow of the rivers is high and the dilution is favourable. These two used are discussed in the following sections. Most of the details on the stabilization reservoirs concept are presented in the following section.

Stabilization reservoirs for effluent reuse in irrigation

Wastewater reuse for irrigation is not a treatment method but rather a concept of combined treatment and disposal. It can offer an additional final effluent disposal alternative on top of conventional ones. This additional disposal alternative is disposal on the fields, not to a receiving water body. The reuse alternative can also generate agricultural economic benefits. Considering the water crisis in many countries, especially in developing countries located in water scarce zones, and considering the population growth forecasts, regulated wastewater reuse for irrigation will gain more and more importance with time (non regulated reuse is already been practices in most countries under unsafe conditions). Wastewater reuse for irrigation is a wide topic that cannot be fully discussed in this book; however it has a close relation to appropriate treatment technologies, which needs to be clarified. Reuse of wastewater for agriculture is usually associated in people's minds with highly sophisticated wastewater treatment technologies aimed at generating an effluent of very high quality, close to that of potable water. The fact is that an effluent adequate even for unrestricted irrigation of crops eaten raw can be produced using simple treatment processes, and entire reuse systems can be constructed in such a way that they fall within the definition of appropriate technology.

Implementation of wastewater reuse projects is not a simple matter. On top of technical matters as treatment technologies, efficiency and reliability, it entails resolution of a series of other complicated issues as legal and regulatory framework issues, public health risk aspects (risk assessment, reuse standards, monitoring), public acceptance, financial and economic issues, and especially institutional issues of coordination between multiple stakeholders such as the municipality, the water and sanitation utility, farmers and farmers' organizations, health and environmental authorities, regulators, NGOs and the public. Information on the concept and details of effluents reuse for irrigation is abundant. Two publications are referenced as including much more information about the subject: Asano *et al.* (2007) and Scheierling *et al.* (2010). We find that in spite of the availability of a lot of information on the concept of reuse and on all the problems associated with the reuse concept, there is not much on practical implementation of reuse project, and certainly not on the importance and use of stabilization reservoirs in reuse projects. Stabilization reservoirs are the subject of this section.

As mentioned, in areas of limited natural water resources, wastewater represents an unconventional water source which may be utilized for agriculture, industry, recreation and recharge of aquifers. The use of domestic wastewater for irrigation of agricultural crops is one of the most attractive and popular reuse options in many regions of the world, for the following reasons:

- Irrigation is usually needed in areas where water tends to be scarce and wastewater reuse is necessary for supplementing available fresh water sources. In terms of availability it is also a safe water source, independent on climatic conditions.
- Agriculture requires large amounts of water, which are used only once, since irrigation is a consumptive use. Thus, there is no danger of gradual accumulation in the water of undesirable substances by continuous recycling.
- Agriculture can use not only the water, but also additional resources found in wastewater, such as organic matter or nutrients (nitrogen and phosphorous), which are thus converted from an environmental nuisance to an asset.
- Irrigation is relatively flexible with respect to water quality requirements. Some crops can be irrigated with low quality water, and some water quality problems can be overcome by suitable agricultural practices.
- Use of wastewater for irrigation prevents its discharge to receiving water bodies and pollution of water resources.

When considering irrigation with treated wastewater, several related factors and terms should be mentioned, some of which are presented below:

- *Restricted and unrestricted irrigation*: "restricted irrigation1" refers to the use of low quality effluent in specific areas, where only certain crops should be cultivated. The restrictions imposed usually refer to the type of crops to be cultivated, the irrigation methods, the harvesting method, fertilizer application rates, distance of irrigated fields from residential areas and paved roads and distance between non-potable and potable water supply mains. "Unrestricted Irrigation" refers to the use of high quality effluents for irrigation of all crops on any soil type, without any adverse environmental effect or the soil, crops, animals, persons involved in the production process or persons consuming agricultural products.
- *Effluent quality*: the effluent quality is the main factor that dictates the type of reuse (i.e., restricted or unrestricted irrigation), types of crops and irrigation methods. The effluent quality should be considered regarding two aspects: public health and agronomics. Most existing standards for wastewater reuse deal only with public health aspects and prescribe the treatment processes or quality parameters that the effluent must meet before it can be used to irrigate a certain category of crops. These include mostly bacteria, viruses and other pathogens and sometimes also organic matter (BOD, COD, TSS), dissolved oxygen, pH and residual chlorine. However, agronomic aspects related to crops and soils must also be taken into account. These include salinity, sodium absorption ratio of the water, nitrogen, phosphorus, chloride, bicarbonate, heavy metals, boron and other trace elements, pH and synthetic organics (including pesticides). Considering the agronomic aspect of effluent quality, water quality standards established for crop irrigation with fresh water are, at present, also the best available criteria for effluent reuse. However, there are additional constituents in wastewater, usually absent from or unimportant in fresh water. For such constituents, specific reuse standards will have to be developed in the future. At present, only preliminary guidelines can be established based an available knowledge.
- *Irrigation methods*: the effluent quality has an effect on the irrigation method to be used. Surface irrigation methods can utilize low quality effluents, whereas sprinkler and mainly drip irrigation require higher-quality effluents, because of the danger of clogging the orifices of the irrigation equipment by suspended solids. This problem can be overcome by utilizing filters at the head of the irrigation system.
- *Seasonal storage*: while municipal wastewater is available at a relatively constant flow, irrigation water is required mainly during the dry season. In order to compensate between the constant supply and variable demand, seasonal storage is required.
- *Economic aspects*: From the point of view of overall economic evaluation, the wastewater treatment system and the effluent reuse system are regarded as an integrated system for which a combined cost-benefit analysis should be carried out in order to assess the feasibility of the project. Similarly to other irrigation projects, the economy of a reuse projects depends on many factors, most of which are specific to each project and include soil type climatic conditions and irrigation methods.

Several factors have to be taken into account when aiming at a feasible reuse project:

- *Efficient water use*: maximum efficiency of water use must be attempted in order to attain the maximum possible irrigated areas and thus maximize crop production. The efficiency of water use is interrelated with selection of the irrigation method, and high efficiency irrigation methods are preferable.
- *Provision of seasonal storage*: the inclusion of Seasonal Storage in a reuse scheme is a prerequisite. Provision of seasonal storage is basically part of the efficient water use. Without such storage, an

alternative disposal method of the effluent needs to be provided during the seasons in which irrigation is not necessary. This has two drawbacks: (i) decreased irrigated areas and thus lower income from agriculture; and (ii) increased environmental contamination.

- *Distance of the reuse area from the wastewater source*: the closer the reuse area to the sewage source, the more profitable is the project because of a lower investment in the conveyance system.
- *Deduction of alternative disposal costs from the reuse project cost*: a reuse project can be economically feasible only if alternative disposal costs are deducted from the project costs. Such a deduction is justified because sewage disposal should be provided in any case.
- Existence of consumption of the effluent for irrigation, preferably the existence of farmers with agricultural infrastructure, who need the effluent.
- Shortage of alternative water sources, because if alternative fresh water sources are available, farmers would usually prefer to use the alternative sources.

The public health risks represent the main concern related to the reuse of wastewater for irrigation. Inadequate reuse can expose to risk the consumers of crops irrigated with wastewater, the farmers and the residents of areas in the vicinity of the irrigated fields. Pathogenic organisms (bacteria, viruses, parasites and intestinal nematode or helminths) are the parameters of significance when considering health effects.

Previous public health orthodoxy which held that every excreted pathogen than can persist in the environment, in soil or on crops irrigated with wastewater is a potential cause of serious disease in humans, is basically overly conservative because human infection is not dependent only on the presence of pathogens in soils and crops. It depends also on the minimum dose of a pathogen necessary to cause infection (which varies greatly for various types of organisms), on the persistence time of the pathogens in the environment and on the level of immunity of the population to endemic diseases. Combination of the above factors led to the conclusion that the amount of excess infection and diseases, caused by various classes of pathogens when irrigating with untreated wastewater is in the following order of descending magnitude:

- Intestinal nematode infections (Ascaris, Trichuris and the hookworms)
- Excreted bacterial infection (bacterial diarrheas and typhoid)
- Excreted viral infections (rotavirus diarrhea and hepatitis A)

Effluent quality standards for various types of irrigation (restricted and non restricted) are important and change and evolve with time, as more knowledge becomes available. Norms of developed countries for unrestricted irrigation are very stringent and require advanced treatment processes which yield very high quality effluents. These processes are certainly not appropriate technology processes. The reuse standards proposed by the WHO are less stringent than those of developed countries and they also evolve and change with time. We do not provide here information on specific standards, but it is emphasized that the health risk imposed on the public by irrigation with effluent is caused by pathogenic organisms contained in the effluent. The elimination of such organisms from the effluent reduces or eliminates this risk, even if the effluent still contains other contaminants such as organic matter and suspended solids.

Based on the recent findings a series of practical alternative remedial measures for controlling negative health effects and for protection of public health may be applied as follows:

- Imposing restrictions on type of crops irrigated so as to prevent consumers from being exposed directly to infection through contaminated vegetables or salad crops eaten raw. Not all crops need to be irrigated with high quality effluents. Industrial crops such as cotton can be irrigated with low quality effluents while crops eaten raw need to be irrigated with pathogens free effluents.

- Selecting irrigation techniques and procedures that minimize direct contact between wastewater and crops. For example, drip irrigation of fruit trees does not impose high risk of contamination of the fruit.
- Using wastewater treatment and/or storage practice aimed at effectively reducing the concentrations of priority pathogens to low level for which the incidence of excess infection is essentially controlled.

In addition, improved occupational health and hygiene conditions are necessary.

Stabilization reservoirs can be designed to provide low quality effluents adequate for restricted irrigation or high quality effluent, void of pathogens, adequate for unrestricted irrigation. Simple filtration units of various types (gravel filters, disk filters and others) eliminate suspended matter from reservoirs' effluents and make such effluents adequate for being applies to the fields by drip irrigation systems.

The widespread practice of inappropriate wastewater reuse for irrigation in developing countries can be remedied by adopting a rational approach to wastewater reuse, through the attractive concept of stabilization reservoirs. It is not a treatment method but rather a reuse concept, which combines effluent polishing treatment with storage aimed at highly increasing the efficiency of reuse and agricultural productivity. But it can also be applied as a wastewater treatment method, and one with good resistivity to low temperatures.

The key element of any wastewater reuse system is the storage installation which compensates between the relatively constant discharge of wastewater throughout the year and the variable seasonal water demand for irrigation, which is required mainly during the dry season.

There are two possibilities for storage alternatives: surface storage and underground storage. Surface storage can be applied in any project, while infiltration into the ground, especially for large cities, depends on the soil type and on hydro-geological condition of each specific site. In both cases seasonal storage installations act also as efficient treatment processes which may provide, in addition to storage, either polishing of highly treated effluents or serve as the main wastewater treatment step.

The concept of appropriate technology refers mainly to reuse projects with seasonal surface storage in stabilization reservoirs. This type of reservoirs was originally built in Israel by farmers in the 1970s to store effluents and sometimes raw sewage, to be reused for cotton irrigation during the three-month peak summer season. It was soon observed that the quality of the effluent after several months of storage was significantly better than the quality of the influent to the reservoir, mainly with respect to organic matter content and content of pathogenic organisms. Since then, the stabilization reservoir treatment has been developed as an innovative treatment and management method and has been successfully applied in small, medium size and large irrigation reuse projects in Israel. The reservoir is full at the beginning of the irrigation season and empty at the end of this season, when it's filling restarts. The patterns of inflow to and outflow from the reservoir along the year, as well as the water level patterns in the reservoir are presented in Figure 1.20. Inspection of the inflow pattern, which is the constant raw wastewater flow along the year, and the outflow pattern which is the variable pattern of consumption for irrigation, explains the need for the seasonal storage.

Figure 1.21 illustrates the common practice of water resources management in zones in which irrigation of crops is practiced. In such zones there is usually no relation between wastewater management and irrigation water management. Figure 1.22 presents the wastewater reuse concept through the utilization of stabilization reservoirs.

The effluents of all the stabilization reservoirs in Israel are used for irrigation in productive agriculture. However, the effluent of a reservoir can be used for non-productive irrigation such as irrigation of forests and woods. In this case the idea is to avoid the discharge of effluents to any receiving water body and the effluent is disposed of by evapotranspiration and infiltration. The reservoir serves in this case as a regulating vessel which adds flexibility in managing the effluent disposal system.

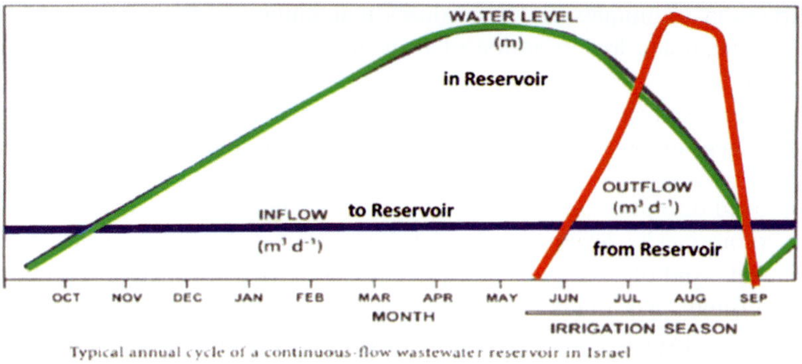

Figure 1.20 The need for effluent seasonal storage: a typical annual cycle of a stabilization reservoir

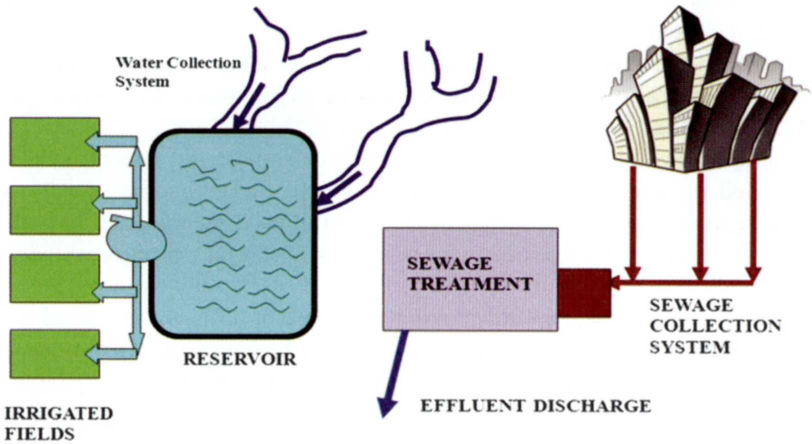

Figure 1.21 The common practice – no relation between irrigation and wastewater disposal

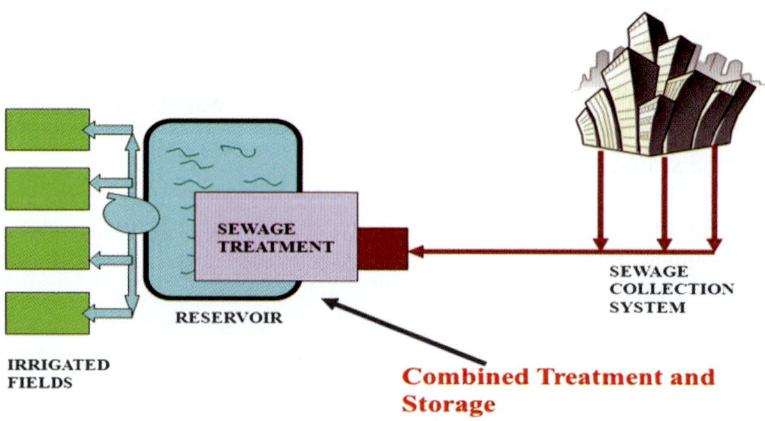

Figure 1.22 Wastewater reuse utilizing a stabilization reservoir which combines storage and treatment

A stabilization reservoir is in fact a deep facultative lagoon with a variable water level which fluctuates throughout the year. The purification processes which occur in a stabilization reservoir are biological processes. During the detention period of effluent in the reservoir, which varies between two to several months, the processes which take place in it include sedimentation, organic matter decomposition, biomass growth, and some nitrogen transformation through nitrification, denitrification and ammonia stripping. Biological populations which develop in the reservoir ecosystem include bacteria, algae and zooplankton. Water purification is effected by activity of aerobic bacteria and algae in the upper layer as well as by anaerobic bacteria in the bottom of the reservoir. Most of the time the reservoir is stratified, with most of its volume acting as an anaerobic reactor and only the upper layer is acting as an aerobic zone from which the final effluent is extracted. The reservoir is totally mixed only during winter or transition seasons.

The key to proper functioning of the reservoir is prevention of anaerobic incidents at the water surface. Existence of aerobic conditions, at least in the upper layer, prevents generation of odours and ascertains improvement of water quality. Decomposition of organic matter is an oxygen-consuming oxidation process. In the event that the quantity of oxygen supplied by algal photosynthesis and re-aeration is smaller than consumption for bacterial respiration, anaerobic conditions may develop. Such conditions can be prevented by control of the organic loads diverted to the reservoir.

The concept of stabilization reservoirs reuse scheme is presented in Figure 1.23. A reservoir system includes a wastewater pre-treatment unit whose task is to reduce organic matter to levels which will prevent development of anaerobic conditions on the surface layer of the reservoir when the effluent of the pre treatment unit flows into the reservoir. Pre-treatment processes and the combined system of Pre-treatment and Reservoir are discussed in one of the following section. Oxygen balance models of reservoirs, as well as experience, established that the maximum permissible organic loads which still prevent generation of anaerobic conditions in the upper layer of stabilization reservoirs are 30–50 kg BOD/day/ha during the winter low temperature season and 60–100 kg BOD/day/ha during autumn and high temperature summer season. Mixers (identical to the mixers for facultative lagoons discussed above) can be installed in odour generating reservoirs to improve their performance and increase the maximum permissible organic load on a reservoir. Preliminary results indicate that an increase in the organic load of up to 250 kg BOD/day/ha is possible for a reservoir equipped with mixers. Additional information on stabilization reservoirs can be found in publications by Libhaber *et al.* (1987), Libhaber (1987), Libhaber (1996) and Juanico and Dor (1999). A typical reservoir almost filled with stored effluent is presented in Figure 1.24. An empty reservoir at the end of the irrigation season is shown in Figure 1.25. In this case the reservoir is lined with a geomembrane to prevent water loss by infiltration. This is necessary in areas of permeable soils but not necessary in areas of impermeable soils.

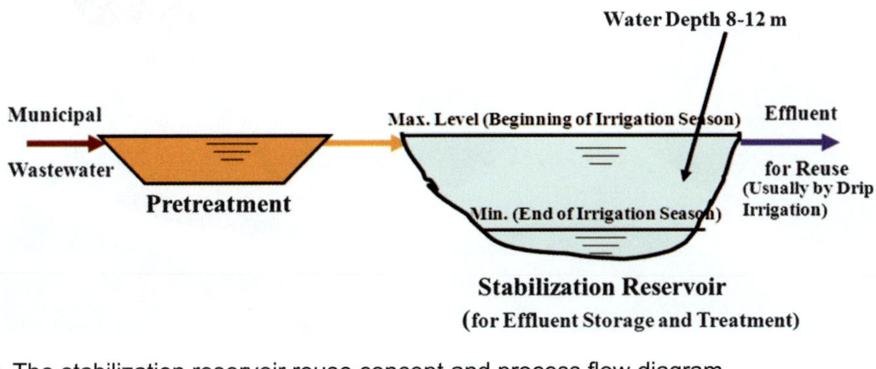

Figure 1.23 The stabilization reservoir reuse concept and process flow diagram

Figure 1.24 A typical reservoir almost filled with effluent towards the beginning of the irrigation season

Figure 1.25 An empty lined reservoir at the end of the irrigation season

The expected effluent quality of a well designed pre-treatment system and single cell reservoir at the beginning and during a large portion of the irrigation season is: Filtered BOD 5–10 mg/l, suspended solids 5–20 mg/l and fecal coliforms 10^2–10^3 MPN/100 ml. The system can also achieve 25 to 50% ammonia removal. The quality of the effluent of a single cell reservoir tends to deteriorate towards the end of the irrigation season, when the water level and water volume of the reservoir drop. The reason is that the inflow stream with high concentrations of colifirms continues to reach the reservoir, the mixing of the reservoir water by the wind and the relatively small water volume enable fresh coliforms that just came into the reservoir reach the effluent withdrawal structure. The effluent quality gets worse as the water volume in the reservoir decreases. This is not a problem if the reservoir effluent is used for irrigating industrial crops or crops which do not come in contact with the effluent (for instance fruit trees irrigated by drip irrigation). However, if the effluent is to be used for unrestricted irrigation, pathogen reduction needs to be high and secure up to the end of the irrigation season. Higher reduction levels of pathogens may be achieved up to the complete withdrawal of the effluent from the reservoir if the influent stream to the reservoir is stopped before the initiation of the irrigation period and is diverted to and stored in a secondary reservoir. Reduction of enteric bacteria and viruses may be explained by the influence of radiation and predation by other organisms. In terms of water quality for irrigation, the critical quality parameter is the concentration of pathogenic organisms. Reservoirs completely remove Helminthes eggs and nematodes with no difficulty. The level of removal of other pathogens (bacteria and viruses) depends on the mode of operation of the reservoir. To obtain a fecal coliform (indicator organism for pathogens) free effluent, a system consisting of at least two reservoirs is required, in which during the irrigation season the influent is stored in the secondary reservoir, not in the main one. An additional reservoir renders the system more expensive, but also produces a higher quality effluent, which can be used for unrestricted irrigation and can generate more profits by irrigating higher value crops. Alternative schemes for operating reservoir systems to obtain higher level effluent qualities are presented in Figure 1.26. These are just two alternatives (schemes B and C) presented to demonstrate the principles, while in fact there are many more alternatives to operate two reservoirs within one reuse system, as presented in Juanico (1995). The variation in the concentration of fecal coliforms in a single reservoir system operated as scheme A in Figure 1.26 is presented in Figure 1.27 (adapted from Juanico 1995), in which the dotted red line is the fecal coliforms count in the effluent of the reservoir, demonstrating that when the inflow to the reservoir is continuous throughout the year including during the irrigation season, the fecal coliforms counts in the effluent starts to increase by the middle of the irrigation period, rendering the drip irrigated effluent of a single cell reservoir to be adequate only for restricted irrigation (industrial crops as cotton, sugar beet etc., cereals, dry fodder, seeds, green fodder, olives, peanuts, citrus, bananas, almonds, nuts, dates, orchards, etc.). If the influent to the principal reservoir is discontinued about a month prior to initiating the irrigation and is diverted to the secondary reservoir, the effluent of the principal reservoir can be maintained with very low pathogens concentrations during the entire irrigation season, as demonstrated in Figure 1.28 (adapted from Juanico 1995), which corresponds to operation under scheme C in Figure 1.26 and in which the dotted red line is the fecal coliforms count in the effluent of the principal reservoir. In this case, there is no entrance of fresh pathogens to the principal reservoir during the irrigation season, and the pathogens introduced prior to the cut off of inflow to the principal reservoir die off due to the effect of sunlight, so the effluent is pathogens free during the entire irrigation season and can be used for unrestricted irrigation of all crops. Figures 1.27 and 1.28 present results of theoretical simulations, however this theory has been proved to be correct in full scale operating reservoir systems with data presented by Juanico (1995).

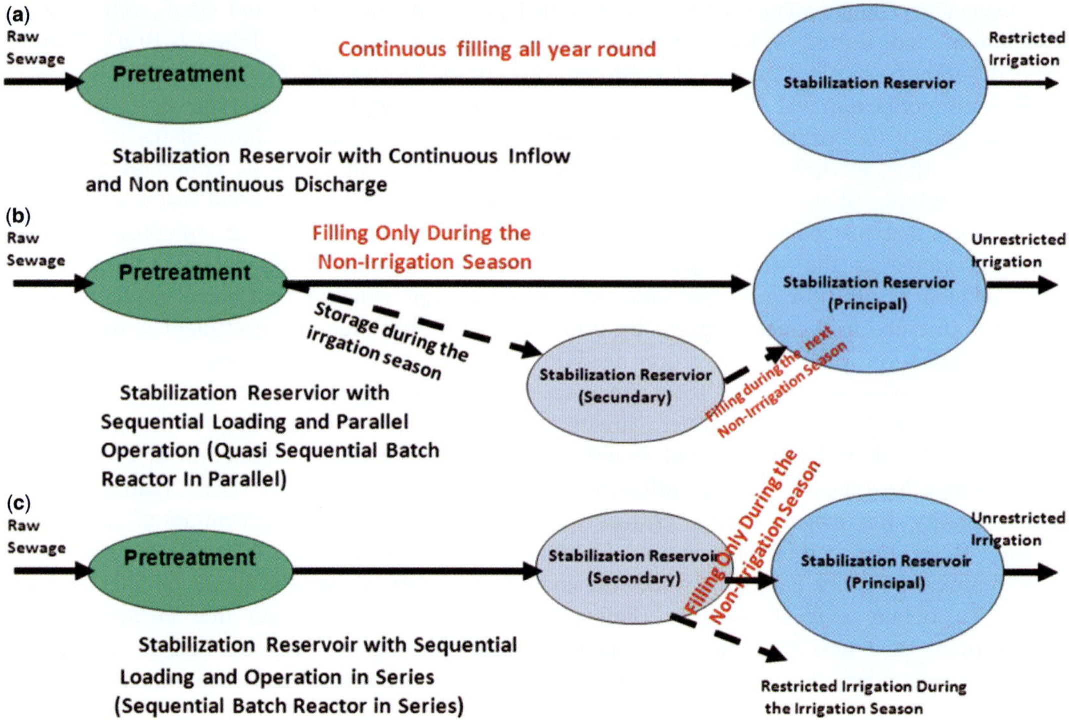

Figure 1.26 Alternative schemes for operating stabilization reservoirs

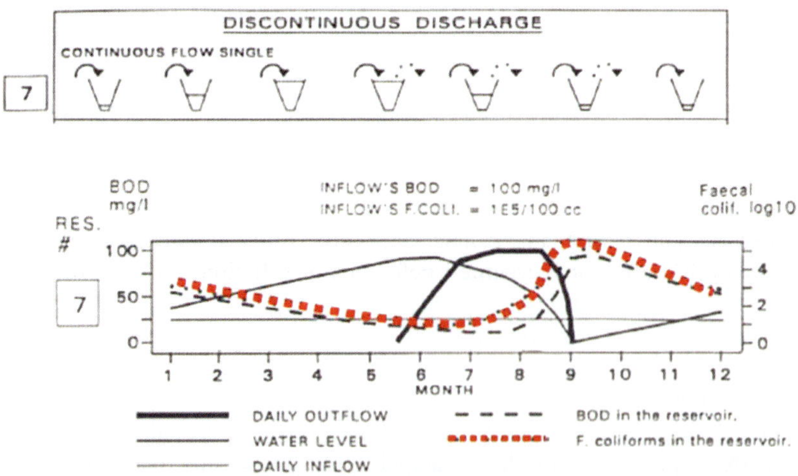

Figure 1.27 Annual effluent quality variations in a single reservoir with discontinuous discharge (adapted from Juanico, 1995)

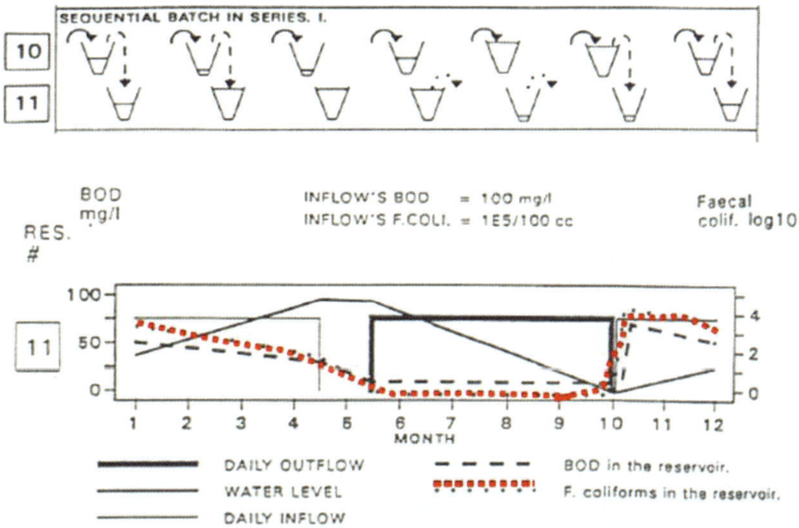

Schematic representation of four selected operational alternatives and their effect on the removal of BOD (a low rate removal pollutant) and Faecal coliforms (a high rate removal pollutant). Reservoir numbers (framed) correspond to those of Fig. 1.

Figure 1.28 Annual effluent quality variation in a system of two sequential batch reservoir reactors in series (adapted from Juanico, 1995)

A photo of a system of two reservoirs in series (Maale Hakishon reservoirs System) is shown in Figure 1.29. This is the largest reservoir system in Israel storing the wastewater of the city of Haifa, with a population of over 700,000. The volume of each of the two reservoirs is about 6.5 million m^3, with a total volume of about 13 million m^3. This system is in operation for over 30 years yielding a high quality effluent with less than 100 MPN/100ml fecal coliforms, which is used for unrestricted irrigation. A photo of the effluent of the Maale Hakishon reservoir is shown in Figure 1.30.

Another important element in the reuse scheme is the irrigation system. In most of the reuse projects in Israel the drip irrigation method is successfully applied. In addition to its advantage of efficient water use, this method is very effective in diminishing public health risk by preventing air dispersion of bacteria and viruses through aerosols and minimizing contact between effluent and irrigated crops. Some additional treatment of the reservoir's effluent is required before its distribution to the fields by drip irrigation. A complete system includes a primary straining stage (40 to 60 mesh), rapid gravel filtration (at rates of 40 to 70 m/hr) or other type of filtration (such as the use of disk filters) and a secondary finer straining stage (80 to 150 mesh), which is typically done in each plot. A typical reservoir effluent disk filtration system (of the Kfar Menahem reservoir) is shown in Figure 1.31. Another element of the reservoir effluent treatment is chlorination, which is usually done by injection of a hypochlorite solution at the entrance to each plot. The chlorine ensures that residues of algae do not clog the drippers and also complete the disinfection of the effluent.

As mentioned, mixers identical to the mixers for facultative lagoons discussed above can also be installed in reservoirs to improve their performance. This can be done if a reservoir generates odours. Odour generation mean that the organic load discharged to the reservoir is too high. One way to resolve the problem is to augment the capacity of the pre-treatment unit so that it will remove more organic matter and thereby the organic load on the reservoir will be reduced. Another way is to install mixers in the reservoir. The mixers enable a reservoir to handle much higher organic loads of up to 250 kg BOD/day/ha. The installation of a mixer in a reservoir (the Og reservoir) is shown in Figure 1.32. Figure 1.33 shows the Og reservoir equipped with the mixers installed in it.

Figure 1.29 Photo of a two cells in series reservoir (Maale Hakishon reservoir)

Figure 1.30 Effluent sample of Maale Hakishon reservoir

Figure 1.31 A typical disk filtration system for reservoir effluent filtration at the Kfar Menahem reservoir

Figure 1.32 Installation of mixers in a stabilization reservoir (the Og reservoir in Israel)

Figure 1.33 Mixers installed in the Og stabilization reservoir

The cost of a complete stabilization reservoir system, including pretreatment, depends to a large extent on the pretreatment system. The additional investment cost of the reservoir is about 20–30 US$/Capita. When the pretreatment is based on appropriate technology, the investment cost of a complete reservoir system, including pretreatment, is about 30–50 US$/Capita. When the reservoir is in the same site of the pretreatment installations and both are operated by the same institution the additional operation and maintenance cost of the reservoir is negligible. In many cases, the pretreatment and the reservoir are not in the same site and are operated by different entities, the pretreatment by the municipality or the water utility of the city and the reservoir by a farmers association. In this case, the operation and maintenance costs of the reservoir are similar to those of operating a stabilization lagoons system, that is, in the range of 0.2–0.4 US$/Yr/Capita.

The combined pretreatment and stabilization reservoir reuse system is a simple, economic wastewater reuse (and disposal) method which does not require high skills for operation and maintenance. The required pretreatment of the wastewater before it is stored in the reservoir can be achieved using a variety of treatment process, among them all the appropriate technology unit treatment processes described above. If simple pretreatment processes are selected, the entire pretreatment, reservoir and irrigation system can be considered an appropriate technology system.

Finally, when considering the application of wastewater reuse for irrigation in developing country, several facts needs to be taken into account:

- Seasonal storage is an essential component in any reuse scheme.
- Irrigation is relatively flexible with respect to water quality requirements. Some crops can be irrigated with low quality effluents, while others require high quality effluents.
- Surface reservoirs serve, on top of storing effluents, as treatment units. Depending on the operation regimes, the combined system of pretreatment and reservoir can produce low quality effluents (in

terms of pathogenic organism content) for restricted irrigation, or high quality effluents for unrestricted irrigation.

- The basic condition for application of a wastewater reuse project is the existence of consumers of the effluent and a commitment on their part for long term consumption of effluent, that is, the existence of infrastructure for agricultural activity based on consumption of the effluent for irrigation. The agricultural activity can be based on farms owned by the wastewater producing municipality, such as in the case of Melbourne, Australia and Muskegon County, USA, or by private farmers or agricultural communities as is the case in all wastewater reuse projects in Israel.
- Wastewater reuse projects are integrated projects which require a high level of inter-institutional cooperation for their successful implementation and operation. The operation of the treatment installation needs to be done in close cooperation with and participation of the farmers (in many small to medium size reuse schemes in Israel the farmers are in charge of operation of the treatment installations). This is a sensitive point for developing countries, where institutions are weak and achieving cooperation between different institutions is difficult.

Stabilization reservoirs for intermittent discharge of effluents to rivers

Many of the municipal treated effluents are discharged to rivers. Wastewater treatment plants which discharge their effluent to a river are constructed to comply with the stringent discharge standards even under the most critical conditions, which usually occur during a short part of the year in the low flow season in the river, when the dilution of the effluent in the river water is the smallest. At critically low dilution of the effluent in the natural water of the river, the effluent quality needs to be very high and that means that the treatment plant needs to be constructed based on a process which can produce a very high quality effluent. But in fact its high capacity is utilized only during a short part of the year, when the river water flow is low, while during the rest of the year, the effluent quality could be lower.

A different strategy would be: (i) to construct a treatment plant that produces a lower effluent quality, which is still adequate for discharge to the river during most of the year; (b) to construct a deep storage reservoir (basically a stabilization reservoir) to store the effluent during the low flow season in the river, which might last up to several months; and (iii) the mode of operation would be the following: during the part of the year which is not low flow season, the plant effluent will be discharged directly into the river whereas during the low flow season, the effluent will be stored in the stabilization reservoir and no river discharge will take place during this season. At the end of the low flow season, the effluent of the treatment plant will again be discharged to the river, and also the effluent stored in the reservoir will be discharged to the river, at a flow adequate for the conditions of the specific river, so that the discharge of the treatment plant effluent plus the flow of effluent from the reservoir will still comply with the quality standards permitted for the river. Usually the low flow season coincides with the high temperature season so the treatment of the effluent in the reservoir during the storage period will be effective.

If quality consideration require, the effluent of the treatment plant can be discharged all year round to the reservoir and from there to the river. In that case the reservoir acts also as a polishing unit of the treatment plant effluent. During the low flow season the treatment plant effluent fills the reservoir and effluent is not discharged from the reservoir to the river. During the rest of the year the treatment plant effluent is discharged to the reservoir and an effluent stream of a flow larger than the incoming flow is discharged to the river. The reservoir is never totally emptied but rather remains with a minimum water depth of about 2 metres, so the reservoir act as a polishing lagoon (maturation pond) for the treatment plant effluent.

One of the key design aspects of such a system is the proper sizing of the reservoir relative to the flow and water quality characteristics of the effluent-receiving river or stream. The design of the reservoir is based on preparing a water balance, with the first stage being the determination of the amount of effluent which may be discharged to the river as a function of the river flow. An analysis of water quality changes should be performed considering the flow characteristics of the effluent receiving river. A time span between discharge events is then determined and the reservoir storage volume is calculated.

This strategy is much better for developing countries because it makes the maximum use of the river's assimilative capacity by discharging to the river an effluent flow proportional to the river flow, thereby allowing the use of appropriate, low cost and easy to operate processes instead of complex equipment for achieving higher levels of treatment which might otherwise be required. It ensures the benefits of sustainable, low cost and simple to operate appropriate technology installations while still attaining the required environmental standards. According to Martin *et al.* (1991) over 18 systems of this type are in operation in the USA (e.g., in Linder, AL, in Heidlberg and Canton, MS and in West Monroe, LA). These systems are denominated Hydrograph Controlled Release (HCR) systems. They include a river flow monitoring system and an automatic effluent discharge structure from the reservoir, which can be motor operated valves, motor operated sluice gates, floating weirs or variable speed pumps. The discharge of the reservoir effluent to the river is automatically controlled by the river flow monitoring system. According to the mentioned source, there have been no major operational problems related to the HCR system.

1.7.2.8 Constructed wetlands

The use of constructed wetlands for wastewater treatment has increased dramatically in the past decade, particularly for small scale applications such as individual homes and small villages. Interest in constructed wetlands has been driven by a variety of factors including enhanced enforcement of environmental standards, limitation of regional sewerage networks to cost effectively serve remote areas in developing countries and the fact that because they use local materials and labour, constructed wetlands can readily be implemented in developing countries.

A constructed wetland bio-filter is a gravel or volcanic rock aerobic biological filter sown with marsh plants, through which preliminary treated wastewater is flowing horizontally or vertically. A film of aerobic bacteria is formed on the filter bed and consumes organic matter dissolved in the wastewater. To avoid clogging of the filter bed, large suspended solids need to be removed from the inflowing wastewater before it reaches the wetland. This can be achieved by preliminary treatment, preferably by rotating micro screens, or by the use of settling units such as septic tanks or Imhoff tanks or settling tanks. Plant roots allow airflow from the atmosphere to the subsoil thus maintaining the aerobic conditions in the filter bed and supplying the oxygen consumed by the aerobic bacteria. The pre-treated wastewater is uniformly distributed over the whole filter bed surface and flows through the bed to the effluent outlet structure. Feeding intervals must be long enough to allow good wastewater distribution and air filling into the empty bed spaces. Once installed and properly operating, the constructed wetland plant can have a long life span due to the balance between plant growth and death cycles and biomass production. The plants to be used in the constructed wetland are selected to fit the type of contaminants that need to be removed from the wastewater. *Plantanillo, zacate taiwan, tule* and reed were found to be effective.

Advantages of constructed wetlands are: (i) operating without energy consumption; (ii) an easy to operate system since it does not employ equipment; (iii) integrates well in rural areas; and (iv) produces vegetal biomass (50–70 tons of dry matter/hectare/year). Disadvantage include: (i) requires large areas; and (ii) solid wastes and sludge are generated in the pre-treatment unit.

Additional information and design details regarding constructed wetlands and a review of the constructed wetland process is presented in the publication WSP (2008). A section of a horizontal flow constructed wetland is presented in Figure 1.34. A photo of a constructed wetland treatment plant in Copacabana, Bolivia, consisting of an Imhoff tank followed by three constructed wetland bio-filter units in series, is presented in Figure 1.35.

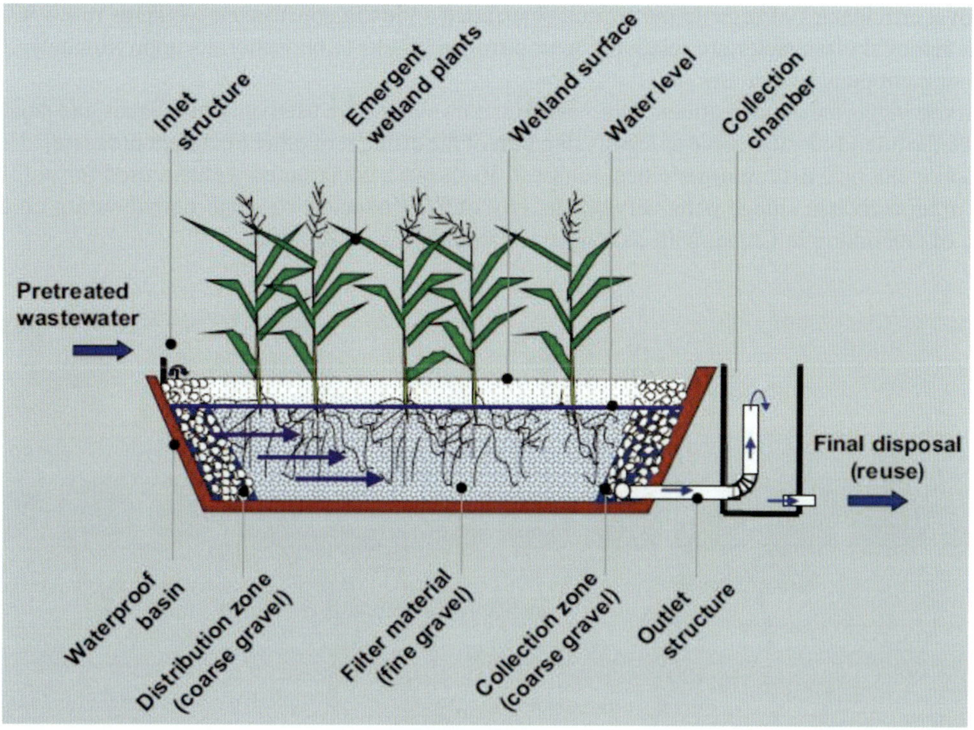

Figure 1.34 Section of a horizontal flow constructed wetland (adapted from WSP, 2008)

Figure 1.35 Photo of a constructed wetland plant in Copacabana, Bolivia

Removal efficiencies of organic matter and suspended solids in constructed wetlands reach 80 to 90%. Nutrients removal efficiencies are relatively low, nitrogen 30–40%, phosphorous approximately 20%, and removal of pathogens is also low.

When serving as the main treatment unit, constructed wetlands are used only in villages and small towns. Constructed wetland can also serve as a polishing unit of the effluent of other treatment processes. Under this application it can be used for larger cities. Figure 1.36 shows a constructed wetland used for polishing the effluent of an activated sludge plant serving the city of Zhuliao and two small neighbouring cities in the province of Guandong in China, with a total served population of 120,000.

Figure 1.36 A constructed wetland unit used for polishing the effluent of an activated sludge plant serving a population of 120,000 in the province of Guandong in China

WSP (2008) reports an investment cost in constructed wetlands in the range of 20–40 US$/Capita and operation and maintenance costs in the range of 1–1.5 US$/Yr/Capita. Constructed wetlands are more expensive than facultative lagoons because for the same function they require a larger land area and because the gravel filling is plentiful and costly.

1.7.2.9 Chemically Enhanced Primary Treatment (CEPT)

The first process which comes to mind when considering physicochemical treatment is Chemically Enhanced Primary Treatment (CEPT), a well known process which has been in use for over one hundred years, yet is not in widespread use as would be expected on the basis of its performance. Its main uses are: (i) treatment prior to ocean discharge; (ii) phosphorus reduction; and (iii) as a compact treatment unit. CEPT is a process by which chemicals, typically metal salts as ferric chloride or aluminium sulphate, are added to primary settling tanks. Coagulation and flocculation cause particles to cling together and then settle faster, thereby achieving much higher removal efficiencies of TSS and BOD than those experienced in conventional primary treatment, while also obtaining higher performance efficiencies measured by working at double to triple the surface overflow rate in relation to the one achieved in conventional primary tanks. The performance of CEPT approaches that of conventional secondary biological treatment in terms of removal of BOD and TSS, achieving 70–75% Total BOD removal and 80–90% TSS removal in CEPT, compared to only about 20–35% Total BOD and 40–50% TSS removal in conventional primary sedimentation tanks. The increased performance efficiency of the CEPT allows for the design of smaller settling tanks in relation to those of conventional primary treatment, for the same wastewater flow. That means that for the same wastewater flow, a CEPT settling tank would require about half the surface area of a conventional settling tank (i.e., would be half the size of the conventional settling tank and therefore of much smaller investment) and would yield a much better effluent quality.

A schematic diagram of the CEPT process is presented in Figure 1.37. Figure 1.38 shows a photo of a CEPT effluent sample aside a regular primary effluents sample of bench scale experiments, demonstrating the superior quality of the CEPT effluent.

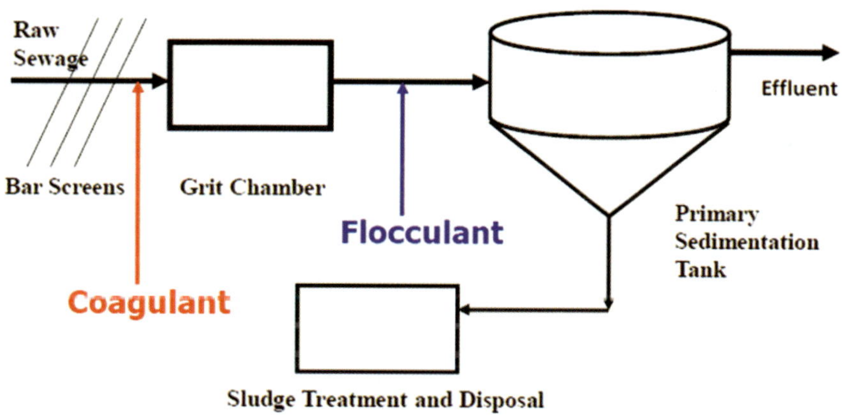

Figure 1.37 Schematic diagram of the CEPT process

To upgrade a conventional primary treatment facility to a CEPT facility, all that is needed is the addition of a chemical coagulant and optionally a flocculant. An upgraded facility has double the capacity of the original conventional settling system so it is able to serve future flows without an additional upgrade. CEPT relies on small doses of metal salts, usually in the range of 10–50 mg/l and the chemicals themselves make only a slight contribution to the total sludge production. However, CEPT produces more sludge than conventional primary treatment because it is through the production of sludge that the organic matter is removed in this process. The greatest portion of the increased sludge production in CEPT is the result of the increased solids (and therefore also organic matter) removal in the settling tank through the physicochemical (not biological) process, which is precisely the CEPT's goal.

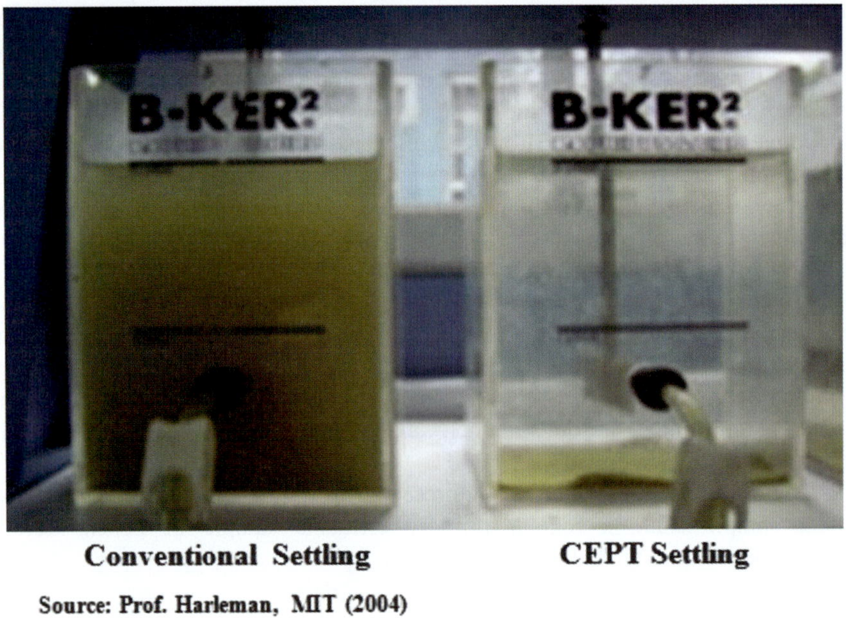

Figure 1.38 Bench-scale CEPT and conventional primary treatment effluents

Since CEPT is a physicochemical process based primarily on chemical reaction, it is not dependent on the dynamics of bacterial cultures and occurs rapidly, so the detention time of the wastewater in CEPT tanks is much shorter than in biological reactors and is about 1 hour. For the same reasons it also responds much better to wastewater flow variations. Another of its important features is that it is less sensitive to temperature variations than biological processes. The flocculation process loses some efficiency at low temperatures but the efficiency can be recovered by adjusting the doses of coagulant and flocculant. Consequently, the CEPT process can perform well in all the temperature range of wastewater and therefore in all type of conditions and zones, from tropical to sub-polar. The CEPT process destroys H_2S gas present in the wastewater preventing thereby generation of unpleasant odours and corrosion of pipes and structures. It does not generate aerosols and occupies a small area, therefore a CEPT plants can be located within the limits of urban areas, if necessary.

The main options for sludge treatment, which needs to be part of the CEPT installations, are: (i) lime stabilization and use of drying beds to dry the sludge; (ii) anaerobic digestion; and (iii) composting. The lime treated sludge is stabilized and disinfected by the lime and can be disposed by distributing on farmland as an organic fertilizer and soil conditioner. The same goes for composted sludge. Digested sludge can be disposed in sanitary landfills, spraying on farmland (in some countries) or reused for other purposes.

The quality of a CEPT effluent approaches that of a biological secondary effluent in terms of removal efficiencies of TSS and BOD. The CEPT effluent can be effectively disinfected, and that is an important feature for developing countries, where due to the high level of morbidity originating in water borne diseases, disinfection is a priority. Additional advantages of CEPT are the high overflow rate that it can handle, which is 2 to 3 times higher than that of conventional settling, the high phosphorous removal that it achieves and the fact that it reduces the size of subsequent treatment units if polishing the CEPT effluent is required.

According to Harleman *et al.* (1998, 2001) and Murcott (1996) the ratio of investment cost in conventional activated sludge to investment in CEPT is about 4 to 1. Construction costs of CEPT system of various sized are in the range 20–40 US$/Capita. Operation and maintenance costs are in the range of 1.5–2 US$/Yr/Capita.

Removal of 70% BOD by CEPT can in many cases be sufficient prior to discharge of effluent to rivers, when the rivers' assimilation capacity is taken into account. When removal of phosphorus is required, for example before discharge of effluents to lakes, it can be achieved through the use of CEPT. And since all that is achieved at a reduced cost, CEPT can be attractive for developing countries. An additional advantage of CEPT is that its effluent can undergo efficient disinfection at low doses of disinfectant (due to the low suspended solids concentration in the effluent) which is sometimes an important characteristic.

CEPT treatment does not preclude subsequent additional treatment of its effluent. It makes any subsequent treatment unit smaller and less costly due to the CEPT increased efficiency. CEPT is a relatively simple technology, which provides low cost, easy to implement and effective treatment. The CEPT process can be applied to medium size and large cities. Small cities might have difficulties to operate a CEPT plant since its operation required a certain technical level.

1.7.3 Additional unit processes of appropriate technology

1.7.3.1 Sand filters

Rapid sand filtration is a physicochemical process for elimination of turbidity and suspended solids from water and wastewater. It is one of the oldest unit processes used in the treatment of potable water and is frequently used for the filtration of effluents of wastewater treatment plants, mainly conventional plants. Sand filtration is therefore a polishing unit of the effluent of upstream treatment units, and cannot be the forst treatment unit of municipal wastewater. In addition to providing removal of suspended solids, rapid sand filtration also removes the BOD contained in the suspended particles. Filtration is a process in which the effluent passes through single, dual or multilayer filtration materials. The solids are retained on the filter media because of their size and the liquid passes through the media and leaves the filter free of solids. With time the filter clogs because of the solids it retains and needs to undergo backwashing to release the retained solids. Flocculation of the effluent before its arrival to the filter improves the filtration process. Sedimentation before the filtration is usually not required (so the filtration of effluents is in fact direct filtration or contact filtration). Typically, sand or anthracite is used as the filtering material in single bed filters. Dual media filters usually consist of a layer of anthracite over a layer of sand. Multi-media filters typically consist of a layer of anthracite over a layer of sand over a layer of garnet or of resin filter media. Activated carbon is also used sometimes as a filtering media. There is a number of different type of filters depending on depth (shallow, conventional and deep bed), the type of filtering medium used (single, dual and multi-media), the type of operation (downflow or upflow) and the working pressure (gravity filters or pressure filters).

In wastewater treatment, filtration is an effluent polishing processes. It is widely used as tertiary treatment of conventional secondary treatment effluents and can be also used to polish effluents of appropriate technology treatment processes. Filtration is a well known process by water and sanitation utilities of developing countries since it is used for treatment of water. It is a simple process, does not consume a lot of energy and is not expensive in operation and maintenance. It requires chemicals for flocculation and compressed air for backwashing. It can be considered an appropriate technology unit process. Figure 1.39 shows a sand filtration unit installed as a polishing unit of the effluent of an activated sludge plant in the Guandong province in China. When taking the picture, the plant was new and the sand filtration unit was not yet in continuous use since the quality of the activated sludge effluent was sufficiently high and the filtration unit was used only as an emergency unit. The investment cost in a sand filtration system is in the range 10–15 US$/Capita.

Figure 1.39 A sand filtration unit used for polishing an effluent of an activated sludge plant in the province of Guandong in China

1.7.3.2 Dissolved air flotation (DAF)

Dissolved Air Flotation (DAF) is a proven and effective physicochemical process for removing suspended solids from effluent and from raw sewage. DAF removes also particulate organic matter and therefore in addition to removing SST it reduces also the BOD and COD of the effluent it treats. It is also effective in reducing oil and grease. These contaminants are removed through the use of a dissolved air-in-water solution produced by injecting air under pressure into a recycle stream of clarified DAF effluent. This recycle stream is then combined and mixed with the incoming streamflow to the DAF unit in a contact chamber where the dissolved air comes out of solution in the form of very fine bubbles that attach to the suspended solids. The attachment of the air bubbles to the solids decreases the apparent density of the solids and the bubbles with the attached suspended solids rise to the surface of the flotation tank and form a floating bed of material, which is basically sludge that is removed by a surface skimmer into a hopper for further handling. The detention time in the flotation chamber is 20–60 minutes. The attraction between the bubbles and the particles is largely the result of surface charges on the particles. A process flow diagram of a DAF unit is presented in Figure 1.40.

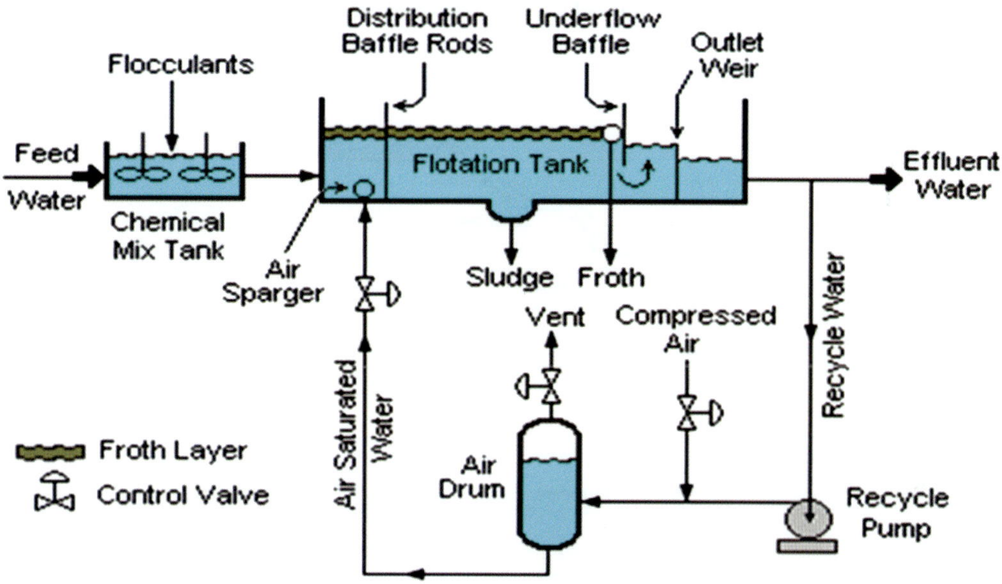

Figure 1.40 Flow diagram of a dissolved air flotation unit (DAF)

The DAF process is not recommended for use in villages and very small cities because its operation requires a certain technical level, however it is not a complicated process, not expensive in operation and maintenance and can be very effective as a polishing unit of effluents of other appropriate technology processes. Therefore it is considered an appropriate technology unit process. The investment cost in a DAF system is in the range 10–12 US$/Capita.

1.7.3.3 Overland flow

Overland flow is a land treatment process in which wastewater or treated effluent is applied at the upper end of sloped vegetated terraces and allowed to flow down the terraces in a sheet flow to a series of runoff collection ditches or pipes. The terraces are planted with a mixture of grasses. The overland flow processes has two objectives: (i) reducing the flow of the effluent that it handles, and (ii) improving the quality of the effluent flowing out of the overland flow unit. The reduction of the flow is achieved by losing a significant part of the wastewater through evapotanspiration and infiltration. The improved quality is achieved by a combination of physical, chemical and biological processes which take place before the liquid reaches the toe of the terrace where it is collected in runoff channels and discharged, further treated or reused. Overland flow can be used as a secondary treatment unit or as a polishing unit for other secondary treatment units. In some cases, overland flow is basically irrigation with effluent as a disposal method. The pre-treatment of wastewater before land application is minimal and consists of either septic tanks or anaerobic ponds. In certain cases oxidation ponds are utilized as pre-treatment.

The effluent application system is a gravity or low pressure surface distribution system. The effluent is distributed by 50 mm diameter perforated polyethylene pipes with fixed openings of 5 mm at intervals ranging from 0.5 to 1.5 meters. These pipes must be levelled to achieve uniform distribution. Details on the distribution system are summarized schematically in Figure 1.41. The effluent distribution systems do not tend to clog and operation and maintenance of the combined treatment-distribution system is quite simple. In general it is considered as a successful disposal solution for small systems in mountain zones.

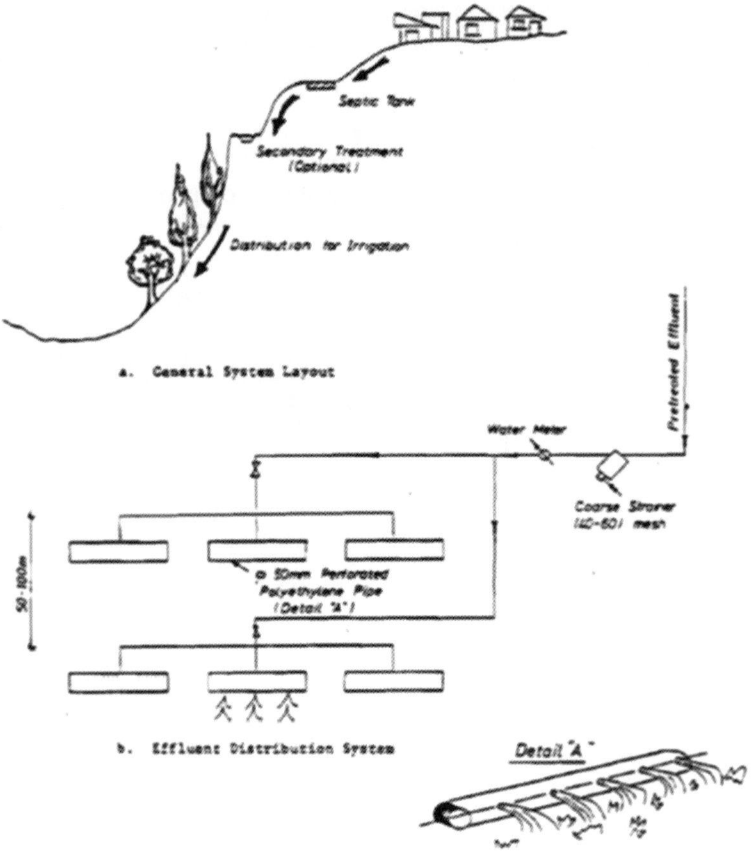

Figure 1.41 Schematic diagram of the overland flow system and details of the effluent distribution system

The vegetation on the terraces is very important since proper selection will allow the desired smooth sheet flow down the terraces. Vegetation also protects the terraces from erosion. To achieve these two goals, a water tolerant, tuft grass is preferred. The most commonly used grasses in warm climates have been Bermudagrass, Reed canarygrass and mixtures of several types of grasses. The Capin overland flow system in the city of Uberaba, Brazil, serving a neighbourhood of the city with a population of 2,000, is presented in Figure 1.42. A photo of one of the wastewater distribution pipelines of the Capin plant is presented in Figure 1.43.

Overland flow is applicable only to small localities of up to several thousand people. For larger localities, very large extensions of land will be required and since the irrigated areas do not yield valuable agricultural produce, large extensions of land become prohibitive. Design information on overland flow systems is presented by Campos (2009). Investment cost in an Overland Flow system used as the main treatment step is in the range 15–30 US$/Capita and O&M cost is in the range 1–1.5 US$/Yr/Capita.

Figure 1.42 The capin overland flow system in the city of Uberaba, Brazil, serving a neighbourhood of the city with a population of 2,000

Figure 1.43 One of the wastewater distribution pipelines of the capin overland flow plant in the city of Uberaba, Brazil

Overland flow is a low cost simple to operate unit process. It can be used as a secondary treatment unit or as a polishing unit for other secondary treatment units, including appropriate technology treatment units. As such it is an appropriate technology unit process.

1.7.3.4 Infiltration-percolation

In the infiltration-percolation process, the applied effluent percolates through the soil and eventually reaches groundwater. So it is in fact an effluent disposal method. In many cases it is also a land treatment technology for indirect water reuse since the groundwater is pumped out for a variety of uses. The process requires highly permeable soils such as sandy and loamy soils so that the effluent will be able to infiltrate through the use of infiltration basins (known also as spreading basins or recharge basins). On its way to the water table, the effluent passes through the unsaturated layer of the soil and undergoes there a series of treatment processes including biological, physical and chemical processes, as well as disinfection through die off of microorganisms due to inadequate survival condition in the soil. The processes which take place during effluent infiltration and retention in the groundwater are known as soil aquifer treatment – SAT. Additional information on this process is given by Idelovitch *et al.* (1984) and Libhaber *et al.* (1987).

Infiltration may be a disposal solution in cases where there is no water receiving body in the vicinity of a city, or there is a strong opposition to discharge any type of effluent to a river. The infiltration process can form part of projects of any size, even very large ones. The information presented by Idelovitch (1984) refers to the wastewater project of the Tel Aviv metropolitan area (the Dan Region) which to date serves a population of over 2 million and in which about 5 m^3/s of effluent is infiltrated (and then re-pumped for reuse in unrestricted irrigation). However, large projects require large extensions of permeable soil for infiltration, which are not always available, and very good knowledge and management capacity of the groundwater in the vicinity of the infiltration basins. Large Projects of this type are not exactly appropriate technology operations. But infiltration projects of small towns do not have the complications of large projects, especially in zones where groundwater is not in use, and an infiltration project of a small town can be considered as based on an appropriate technology process.

The level of treatment of the wastewater prior to infiltration does not have to be very high. Effluents of anaerobic reactors of various types and of CEPT can be infiltrated without additional polishing. Infiltration of facultative lagoons effluents may be a problem because of clogging of the infiltration basins by algae. The design of a SAT system depends to a large extent on the soil type and the infiltration rates that can be achieved in that type of soils.

Figure 1.44 shows one of the infiltration basins of the Dan Region project in Israel and from which the infiltration process can be appreciated. Each infiltration site contains four basins. The first basin on the right is already recovered and ready to receive a new batch of effluent for infiltration. The second basin from the right is already inundated with a batch of effluent and has initiated the infiltration process. The third basin from the right is already in the final stage of infiltrating the batch of effluent that was diverted to it (see details in Figure 1.45) and the first basin on the left is already in the process of recovering after its batch of effluent has been infiltrated. From time to time the infiltration capacity of the basins needs to be recovered by a tractor or a similar type of mechanical equipment which turn up the soil, as shown in Figure 1.46. After over 30 years of operating the Dan region project it is known that the average infiltration rate in this project is in the range of 0.7–1.0 cm/hr. The infiltration rate is specific to each project and depends on the soil type in the project area. The soil at the Dan Region project is basically sand dunes which have a high infiltration rates. The infiltration rate in other project may be lower.

Figure 1.44 Satellite photo of one of the effluent infiltration basins of the Dan Region wastewater reuse project in Israel, serving the metropolitan area of Tel Aviv

Figure 1.45 Photo of an infiltration basin after its inundation with effluent, initiating the infiltration process

Figure 1.46 Photo of cleaning of an infiltration basin with mechanical equipment, to recover its infiltration capacity

1.7.3.5 Septic tanks

Treatment of domestic wastewater in a septic tank is a very old technology developed in 1860. A septic tank performs various functions including sedimentation and removal of floating material. In addition it functions as a low load non-mixed and non-heated anaerobic digester (decanter-digester). A septic tank followed by a soil absorption bed is the traditional on-site system for the treatment and disposal of domestic wastewater of individual household or establishments in areas not served by a public sewerage network. Septic tanks are still in widespread use all over the world. A section view of a typical septic tank is presented in Figure 1.47. The suspended solids present in the inflow stream settle in the tank and form a sludge layer. The oil, grease and other light materials present in the inflow stream float to the top of the liquid in the tank and form a scum layer. The outflow stream leaves the tank without the materials that settled and floated, and is subjected to additional treatment stages or to final disposal. The settled solids undergo anaerobic decomposition and are converted to more stable materials such as carbon dioxide, methane, hydrogen sulphide and water. Although hydrogen sulphide is formed in septic tanks, odour problems are not common because the sulphide combines with metals accumulated in the sludge, forming insoluble metal sulphates. The anaerobic decomposition of the settled material reduces its quantity; however, after months of operation, sludge accumulates and needs to be removed from the tank, along with the scum. The removed material is usually hauled to the municipal wastewater treatment plant where it is combined with the raw wastewater and undergoes treatment in the plant. The investment cost in a Septic Tank System is in the range of 12–20 US$/Capita and the O&M cost is in the range 0.5–1.0 US$/Yr/Capita.

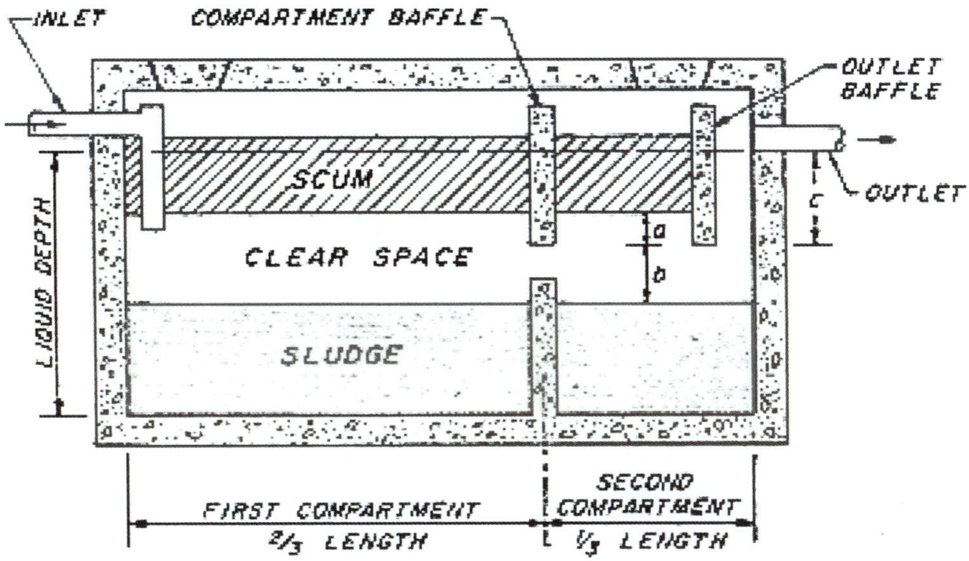

Figure 1.47 A section view of a typical septic tank

Septic tanks can also be used as pre-treatment units in a treatment plant of small communities of up to several thousand inhabitants, as shown in Figure 1.35, where in the city of Copacabana, Bolivia, a septic tank serves as a pre-treatment unit in a constructed wetlands plant. In performing such a function, the septic tank can be an effective pre-treatment unit.

Septic tanks are easy to operate and simple to construct and a septic tank is certainly an appropriate technology unit process. The performance efficiency of a septic tank varies greatly with its configuration, hydraulic loading and organic loading, and with temperature. The minimum recommended hydraulic detention time in a septic tank is two days and the tank needs to have multiple compartments and baffles. The recommended organic loading is less than 0.1 kg BOD/day/m^3. Such a loading enables achieving a removal efficiency of more than 50% of BOD at the temperature range of 10–20°C, and 80% at higher temperatures. The removal efficiency of SST is in the range 20–90% and of oil and grease is in the range 70–90%. The tank's volume should be sufficient keep the solids at a depth of 1 to 3 meters. Such a configuration allows the tank to function two to three years without cleaning.

Septic tanks are used as a pre-treatment of anaerobic filters and constructed wetlands in small treatment plants serving communities of up to several thousands of people.

1.7.3.6 Submarine and large rivers outfalls

About 50% of the world's population resides in or near coastal areas. The most probable receiving body of sewage and treated effluents generated by this population are the oceans and seas. Ocean disposal of sewage and its treatment prior to discharge are therefore issues of major importance. Wastewater treatment is often seen as the only way to cope with water quality problems; although effective outfalls combined with lower pre-treatment levels may substantially reduce costs while still reaching the same environmental objectives as higher treatment levels. For developing countries, in which resources for wastewater treatment and disposal are scarce, low cost solutions are of major importance and a system of simple pre-treatment followed by an effective submarine outfall is an appropriate technology unit process, since it is usually a low investment cost system, simple to operate and of low operation and maintenance costs. The length of the sea outfall and therefore its investment costs depend on the specific bathymetry and sea currents and the outfall site, but usually the investment is lower than that of other solutions. Detailed information on the design of pretreatment systems and submarine outfall is an ample subject and is not treated in this book. The reader is directed to the recently published book of Roberts and al (2010) which is dedicated to all aspects of submarine outfalls for treatment and disposal of wastewater of coastal cities. This chapter only presents general information on submarine outfalls and emphasises the fact that the combination of pretreatment and a submarine outfall is an appropriate technology unit process. Similarly, a combination of pretreatment and an outfall to a large river is an appropriate technology unit process for discharge of effluents of cities located on banks of large rivers. The pretreatment followed by an effective outfall are treatment units which do not need to be combined with additional treatment units, provided that the outfalls are adequately designed and constitute effective outfalls.

The World Health Organization (WHO) has published in October 2003 guidelines for recreational water quality protection (WHO, 2003). Table 1.2 below presents the WHO findings on relative health risks to human health by exposure to effluent discharged from submarine outfalls, classified according to various schemes of combination of sewage pretreatment and submarine outfall length.

Table 1.2 Relative risk potential to human health by exposure to sewage through submarine outfalls (from WHO, 2003).

Treatment	Discharge type		
	Directly on beach	**Short outfall**	**Effective outfall**
None	Very high	High	NA
Preliminary	Very high	High	Low
Primary (including septic tanks)	Very high	High	Low
Secondary	High	High	Low
Secondary plus disinfection	–	–	–
Tertiary	Moderate	Moderate	Very low

The discharge methods of sewage or effluent into oceans can be classified into three principal types: (i) discharge directly onto the beach, which also includes discharge into rivers streams or canals, a short distance from the coastline; (ii) "short" outfalls, where sewage or effluent is likely to contaminate recreational waters; and (iii) "effective" outfalls, designed so that the sewage or effluent is efficiently diluted and dispersed so as to ensure that it does not pollute recreational water areas. While the terms "short" and "long" are often used, outfall length is generally less important than proper location and effective dilution. An effective outfall is assumed to be properly designed, with sufficient length and diffuser discharge depth to ensure that the sewage does not reach the recreational area.

According to the WHO, if an effective outfall is provided for discharge of effluent to the sea, the level of pretreatment of domestic sewage practically does not have a bearing on the risk to human health of the discharged effluent. The risk to human health of secondary effluent discharged through an effective outfall is classified as low. The same low risk is assigned to primary effluent discharged through an effective outfall, and even the risk of preliminary treated effluent discharged through an effective outfall is classified as low. The WHO findings lead to a conclusion that if an effective outfall is provided, preliminary treatment is sufficient prior to discharge of domestic sewage to the sea. And on the other hand, if a short (ineffective) outfall is provided, even secondary treatment is not sufficient to reduce the high risk to human health of effluent discharged trough that type of outfall. These findings of the WHO are of high importance for developing countries since they imply that only preliminary treatment is required prior to discharge of effluent to the sea through an effective outfall. Preliminary installations as pretreatment prior to discharge of effluents to the sea can save a lot of money in developing countries, in which financial resources are usually scarce. The investment costs, as well as the operation and maintenance costs of preliminary treatment are about 1/10 of the cost of secondary treatment, while the use of preliminary treatment yields results comparable to those of secondary treatment. The mentioned conclusions refer to typical municipal sewage which does not contain excessive levels of toxic wastes originating from industrial wastewater.

For coastal cities in developing countries, the strategy of domestic wastewater disposal by preliminary treatment followed by an effective submarine outfall is an affordable, effective, and reliable solution. It is simple to operate and free of negative health and environmental impacts, and falls into the definition of appropriate technology. Many outfalls of this type are successfully functioning and they have a proven track record in many coastal cities all over the world.

A submarine outfall system, which includes the outfall itself and the nearfield zone, should in fact be considered a treatment plant. This treatment plant provides a high level of treatment; highly superior to that which any conventional land based plant can reach. Land based plants can reach, in extreme cases, removal levels of BOD and TSS of up to 95%, and if the effluent is not disinfected, they remove 50–80% of pathogenic organisms, which is a very low removal level, leaving the effluent with practically the same level of risk to human health as that of raw sewage. On the other hand, a well designed outfall system removes over 99% of all contaminants, including pathogenic organisms. Removal of 99% of pathogenic organisms might not be sufficient; however, the mentioned value refers to physical dilution. Taking into account also the biological decay of pathogenic organisms in the marine environment, a well designed outfall can ensure their overall removal to levels lower than that of bearing risk to human health. The setup and dilution mechanism of a submarine outfall is presented in Figures 1.48 and 1.49.

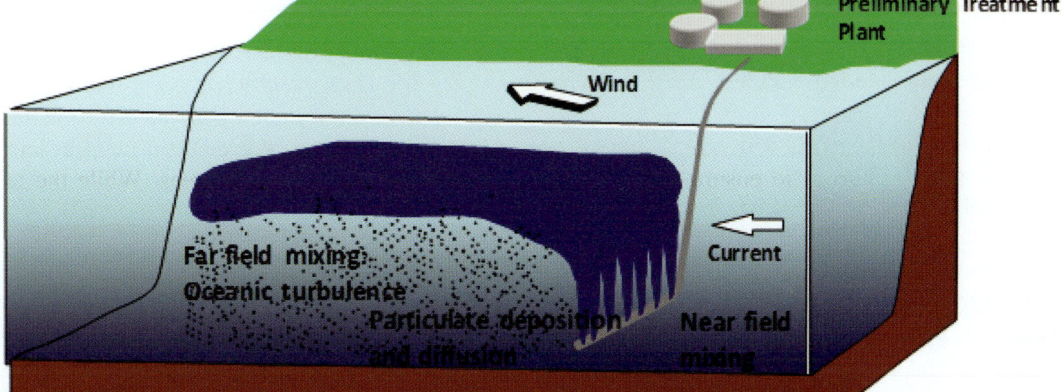

Figure 1.48 Schematic description of a coastal outfall dispersion process

Figure 1.49 Setup and dilution mechanism of an effective submarine outfall

As an example, operating results of the submarine outfall of the city of Tome, in the Concepción region, Chile, consisting of preliminary treatment followed by an effective outfall, are presented in Table 1.3, based on information provided by Leppe and Padilla (1999). The results are average results of five years of measurements near the outfall discharge point.

Table 1.3 Marine water quality near the tome outfall, concepción, chile (average of 5 years monitoring).

Parameter	Effluent discharged into outfall	Maximum according to local standards	At discharge point	At 100 m from the discharge point	Background values
pH	7.3	5.5–9.0	7.6	7.7	7.8
Temperature, °C	17.8	–	13.1	14.1	14.2
Oil and Grease, mg/l	46.3	150	4.6	9.2	2.4
TSS (Total Suspended Solid), mg/l	206	300	7.6	1.8	3.0
BOD_5 (Biochemical Oxygen Demand), mg/l	348	–	2.3	2.2	2.4
TOC (Total Organic Carbon), mg/l	203	–	2.1	3.4	2.0
O.D. (Dissolved Oxygen), mg/l	0.5	–	4.8	7.5	7.2
Detergents, mg/l	9.2	15	0.07	0.06	0.05
Nitrogen Kjeldal, mg/l	72	–	0.52	0.41	0.57
Nitrites, mg/l	0.01	–	0.02	0.02	0.02

(*Continued*)

Table 1.3 Marine water quality near the tome outfall, concepción, chile (average of 5 years monitoring) (*Continued*).

Parameter	Effluent discharged into outfall	Maximum according to local standards	At discharge point	At 100 m from the discharge point	Background values
Nitrates, mg/l	0.17	–	0.25	0.28	0.18
Phosphorus Total, mg/l	12.8	–	0.20	0.21	0.20
Phosphate, mg/l	38	–	1.00	1.00	1.00
Sulfur, mg/l	1.0	5	0.20	0.15	0.24
Phenol	<0.002	1	<0.002	<0.002	<0.002
Fecal Coliforms, MPN/100 ml	3.5×10^7	–	2546	30	11
Total Coliforms, MPN/100 ml	5×10^7	–	3450	41	18

The data in Table 1.3 show that for all quality parameters, expect fecal and total coliforms, the concentration values at a distance of 100 m from the outfalls discharge point are identical to the background sea concentrations, and that demonstrates the high treatment capacity of the outfall system. As for coliforms, their concentrations in the raw sewage and preliminary treated effluents are extremely high. Even so, their concentrations are markedly reduced at a distance of 100 m from the discharge point of the outfalls, to levels that meet the most stringent standards even though they are higher than the background level. Nonetheless, with the biological decay, the concentration of the coliforms is reduced to the background level after an additional short distance from the outfalls discharge point. In the same paper, Leppe and Padilla (1999) report on another outfall, of the city of Penco, which demonstrated similar results.

Based on the above it is concluded that only preliminary treatment needs to be provided prior to ocean effluent discharge through an effective submarine outfall. This is extremely important for developing countries and emphasizes the importance of preliminary treatment as a pre-treatment process. It demonstrates that even preliminary treatment can sometimes be the main treatment process prior to final disposal of the effluent.

The most appropriate preliminary treatment process prior to disposal of wastewater through a submarine outfall is rotating micro screens followed by a vortex grit chamber. This is the preliminary treatment selected for the city of Cartagena, Colombia (Population about 1 million) prior to discharge of the effluent to the Caribbean Sea (Libhaber & Roberts, 2002).

The preliminary treatment followed by a submarine outfall is a scheme adequate for cities of all sizes: small, medium and large cities. The cost of a preliminary treatment plus outfall system depends on the length of the effective outfall, which is site specific. In Santa Marta, Colombia, a resort city on the Caribbean coast with a fixed population of about 400,000 and an additional floating population, the sea bed is very steep and only a 400 meter long outfall was required. The preliminary treatment consists of only bar screens. The investment cost in Santa Marta was 3 US$/Capita. In Cartagena, a larger resort city on the Colombian Caribbean coast, the preliminary treatment includes rotating micro screens followed by a vortex grit chamber, and the submarine outfall is 4.2 km long. The investment cost in

Cartagena amounted to about 28 US$/Capita. A range of 3–30 US$/Capita would be indicative of the investment cost of an outfall system, including preliminary treatment. Operation and maintenance costs would be similar to those of the preliminary treatment system itself.

1.8 COMMONLY USED COMBINED UNIT PROCESSES OF APPROPRIATE TECHNOLOGY

1.8.1 Introduction

A series of unit processes of appropriate technology for wastewater treatment was presented in the previous sections, each yielding a different effluent quality. The presented series of processes might not be the complete set of existing appropriate technology unit processes and perhaps there are others which can be considered as part of this group. When an effluent quality higher than what a single unit process of appropriate technology can produce is required, a treatment plant consisting of a series of appropriate technology unit process can be used (2, 3 or more), in which the effluent of the first unit process is fed into the second unit process, the effluent of the second is fed to the third and so on. Another way to look at it is that to improve the quality of the effluent of a certain unit process, this effluent can be subjected to one or more polishing steps, each of which consisting of an additional unit process. This approach can produce practically any final effluent quality required. *The main message of this book is the idea that it is possible to combine unit processes to create various treatment plants, each based on a series of appropriate technology processes, and that it is possible to combine unit process in such a manner that jointly they can generate any required effluent quality.* A plant based on a combination in series of appropriate technology unit processes is still easy to operate and is usually of lower costs than conventional processes in terms of investments and certainly in operation and maintenance. However, the use of a complex process as a polishing step of an effluent of an appropriate technology process should be avoided, because it transforms the entire treatment plant to a complex and costly treatment plant. There are cases of plants consisting of a UASB unit followed by an activated sludge unit or a trickling filter unit. Such plants are not appropriate technology plants.

In this section we present several combined processes which are already in use in developing countries. Many other additional combinations of unit processes of appropriate technology can be formulated. They are not currently in common use but may be adequate for specific project conditions and can be used in the future. Such combined processes are discussed in following sections of the book and it may well be that treatment plants based on some of them already exist somewhere and are in operation. In all the combined processes, Rotating Micros Screens (RMS) is proposed as the first unit of the preliminary treatment section of any treatment plant, since we consider that RMS is the preferred, most reliable and most effective screening process. If grit removal is required, a Vortex Grit Chamber is recommended as the preferred unit process of the preliminary treatment, complementing the RMS. If the raw wastewater contains large amounts of oil and grease which need to be removed, then an aerated grit chamber should be used as part of the preliminary treatment installation, complementing the RMS. In the schematic flow diagrams of the treatment processes which are presented in the book, the scheme of a Rotating Micro Screen is used to represent the entire preliminary treatment unit. Also, when discussing combined treatment processes, the mention of preliminary treatment is sometimes omitted, but it is clarified that preliminary treatment forms part of every treatment process.

The focus of the discussion in Chapter 1 of this book is the resolution of wastewater disposal problems in developing countries. However, the concepts presented are applicable anywhere and plants based on unit processes or on combined unit processes of appropriate technology can be used also in developed countries and provide the same benefits described in this book also to such countries.

Several commonly used combined unit processes of appropriate technology for wastewater treatment are presented in Table 1.4. Details on each process are presented in the following sections and design procedures of these combined processes are presented in following chapters.

Table 1.4 Commonly used combined processes of appropriate technology for wastewater treatment.

Combined process structure	Performance (% Removal of BODT, SST and fecal coliforms)
1. A series of Conventional Stabilization Lagoons: Anaerobic Lagoons followed by Facultative Lagoons followed by Maturation Lagoons	BODT (70–95%) SST (70–90%) FC in effluent (30–10^5) MPN/100 ml, depending on the design of the lagoons. If necessary, effluent can be disinfected
2. A Series of Upgraded Stabilization Lagoons: Rotating Micro Screens (RMS) followed by Covered Anaerobic Lagoons followed by Facultative Lagoons with Mixers followed by Maturation Lagoons	BODT (80–95%) SST (70–90%) FC in effluent (30–10^5) MPN/100 ml, depending on the design of the lagoons. If necessary, effluent can be disinfected
3. UASB followed by Facultative Lagoons	BODT (80–90%) SST (70–80%) FC in effluent (30–10^5) MPN/100 ml, depending on the design of the lagoons. If necessary, effluent can be disinfected
4. UASB followed by Anaerobic Filter	BODT (80–90%) SST (80–90%) FC Removal very low, if necessary, effluent can be disinfected
5. UASB followed by Dissolved Air Flotation (DAF)	BODT (70–90%) SST (70–80%) FC Removal very low, if necessary, effluent can be disinfected
6. Chemically Enhance Primary Treatment (CEPT) followed by Sand Filtration	BODT (80–90%) SST (80–90%) FC Removal very low, if necessary, effluent can be disinfected
7. Pre-Treatment of Various Types followed by a Stabilization Reservoir	BODT (75–95%) SST (70–90%) FC in effluent (10–10^3) MPN/100 ml, depending on the design of the reservoir. If necessary, effluent can be disinfected
8. UASB followed by Anaerobic Filter followed by Dissolved Air Flotation followed by Membrane Filtration	BODT Practically 100% SST Practically 100% FC Practically 100%

The suspended solids content in the effluents of all the combined processes mentioned in the table is sufficiently low to allow for performing effective disinfection of these effluents by any existing disinfection method. Since the final unit process in the combined processes 1, 2, 3 and 6 in the table above are lagoons or a stabilization reservoir, they can be designed to achieve an effluent of low pathogens content even without disinfection.

Some experts do not approve of using an anaerobic process for polishing the effluent of a preceding anaerobic process, arguing that the final effluent is strongly anaerobic. They promote the use of an aerobic process to polish the effluent of an anaerobic process. For example, in Brazil, the use of activated sludge or trickling filter is promoted for polishing of UASB reactors effluents. This is a somewhat overly cautious approach. Effluents of activated sludge and trickling filters are also anaerobic or close to anaerobic after an effluent detention time of 1–2 hours in secondary clarifiers. The level of oxygen in the effluent needs to be considered taking into account the dilution of the effluent in the receiving body. If the dilution is high, an oxygen void effluent does not present a problem. If the dilution is low, the effluent can be hydraulically aerated before its discharge to the receiving body (treatment plants are usually topographically located several meters higher than the receiving body and cascade aeration can be provided before the effluent discharge to the receiving body). If the natural river flow is really small and most of its flow downstream the discharge point is a result of the effluent discharge, for most types of treatment processes the oxygen in the river will be depleted a short distance after the discharge point, irrespective of the oxygen level in the effluent. In such cases the river water needs to be aerated at several points along its trajectory. An efficient and environmentally friendly aeration method is the SEPA (Sidestream Elevated Pool Aeration) method (Butts 1988).

As already mentioned, the concept of appropriate technology for wastewater treatment is in wide application in Brazil, where treatment plants of this type are used for all sizes of cities and where the application of this concept is rapidly expanding. Many of the examples of combined appropriate technology processes mentioned in Chapter 1 of the book are from Brazil, mainly from three utilities which are pioneers in the use of this type of processes: (i) CAESB, the water and sanitation utility of the Federal District of Brasilia; (ii) SANEPAR, the water and sanitation utility of the State of Parana; and (iii) COPASA, the water and sanitation utility of the State of Minas Gerais. The information on treatment plants located in Brasilia is based on personal communication with Klaus Dieter Neder, Chief, Special Projects, CAESB, on visit of the authors to Brasilia, and on the CAESB Report (2006). The information on treatment plants located in the State of Parana is based on a visit of one of the authors to Parana and on the MSc dissertation by Prado (2010). The information on treatment plants located in Minas Gerais is based on visits of one of the authors to the state of Minas Gerais and on papers related to specific projects, as referenced with the description of the projects.

In Brasilia, out of 17 treatment plants, two large plants and one small plant are activated sludge plants while the rest are based on appropriate technology combined processes, twelve with a UASB reactor as the first unit, one with a CEPT unit as the first unit and one with lagoons as the first unit. In Parana, out of 23 treatment plants in the Alto Iguacu river basin, which includes also the State Capital, Curitiba, only one old plant is an activated sludge plant. All the rest are based on appropriate technology combined processes and in all of them, the first unit is an anaerobic unit. In 21 of the plants the first unit is a UASB reactor and in one plant the first unit is an anaerobic lagoon.

1.8.2 A series of conventional stabilization lagoons

The most common and widespread combined appropriate technology process is a system of a series of conventional lagoons composed of anaerobic lagoons followed by facultative lagoons followed by maturation lagoons. A schematic flow diagram of such a system is presented in Figure 1.50. A photo of such a typical lagoons system is presented in Figure 1.51, which depicts Plant N2 in the city Santa Cruz, Bolivia, composed of two parallel units each containing an anaerobic lagoon followed by two facultative lagoons followed by a maturation lagoon. The arrows show the flow pattern in the plant, red arrows in one unit and blew arrows in the other. The N2 plant was recently upgraded. The figure presents Plant N2 before its upgrade. The N2 plant produced a good effluent (30–40 mg/l BOD, 50–90 mg/l TSS and 1.10^3–8.10^3 MPN/100 ml Fecal Coliforms) which complies with the local standard and is discharged to the Pirai River. Downstream of the city this river flows through a very low density populated region and the effluent discharged to the river causes no nuisance or problems.

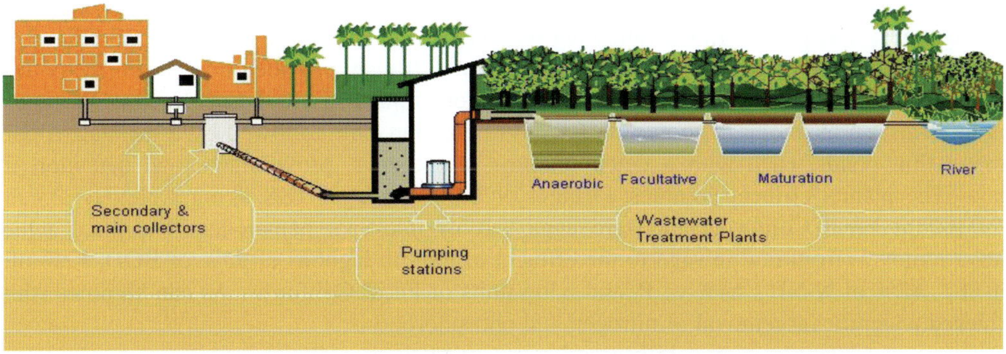

Figure 1.50 Schematic flow diagram of a series of conventional stabilization lagoons

Figure 1.51 Air photo of a conventional anaerobic-facultative-maturation lagoons plant for municipal wastewater treatment, plant N2, one of three in the city of Santa Cruz, Bolivia

This type of conventional stabilization lagoons plant is in widespread use all over the world, both in developing and developed countries. In many cases the system consists only of facultative lagoons followed by maturation lagoons and the anaerobic lagoons are omitted for the purpose of avoiding generation of odours. Lagoon plants are usually constructed outside the limits of the city which generates the wastewater treated by them, but with time the city spreads and surrounds the plant. At that stage people start complaining about odours and mosquitoes. With time the plant becomes overloaded and if it includes anaerobic lagoons, the odour problem becomes intensive.

Inclusion of an anaerobic lagoon as the first unit in a stabilization lagoons plant is of great importance since this lagoon removes about 60% of BOD from the raw wastewater while occupying a small area, thus significantly reducing the total area of the plant in relation to an area of a lagoons plant composed of only facultative and maturation ponds. But it suffers from the drawback of generation of odours when the plant becomes overloaded. Fortunately, today there is a way to prevent odours generation by anaerobic lagoons, as explained in the following section.

A stabilization lagoons plant can produce a high quality effluent, provided that it is correctly designed. Mara (1987) and Mara (2004, PP 137) reports on a lagoons plant in Northeast Brazil consisting of five lagoons in series: an anaerobic lagoon with 6.8 days detention time followed by a facultative pond and three maturation ponds, each with a detention time of 5.5 days, providing a total detention time of 28.8 days. The plant's effluent contains 17 mg/l of BOD, 45 mg/l TSS, 8 mg/l N-NH_3, 2.4 mg/l P, 30 MPN/100 ml Fecal Coliforms and zero eggs/liter of intestinal nematodes. The plant achieved a cumulative removal (from raw sewage to the effluent of the 5th lagoon) of 93% of BOD, 82% of COD, 85% of SST, 82% of N-NH_3, 64% of P, and 6 log units of Fecal Coliforms. As can be noted, this plant also achieved a significant removal of nitrogen and phosphorous.

One important advantage of a stabilization lagoons plant is its ability to achieve natural disinfection, without the use of chemicals or energy, provided that the plant is correctly designed to achieve such a target, as is the mentioned plant in Northeast Brazil. The term correctly designed means well designed maturation ponds and a sufficiently long overall detention time of about 30 days or more. One common disappointment of laymen with lagoons systems is that they are expected to reduce pathogens but are usually designed with a detention time too short to achieve an effective pathogens reduction, so the effluent does not comply with the required quality standard and sometimes presents a public health risk.

Because of the large extension of land occupied by a stabilization lagoons plant, most of this type of plants is designed to serve medium and small size cities but not large cities. However, if a large city has several sewage drainage basins, each basin can be served by a lagoons plant treating the wastewater of the population residing in this basin. In this manner, even a large city can be completely served by lagoon plants. This is the case in Santa Cruz, Bolivia, a city with a population of 1.3 million served by three lagoon plants for treatment of municipal wastewater, with an additional one under construction in 2011. Several large stabilization lagoons plants do exist and are in operation, such as the Dandora Plant in Nairobi, Kenya, currently serving a population of about 1.2 million and is to be expanded to serve about 2 million people.

In summary, conventional stabilization lagoons systems are used extensively for wastewater treatment in moderate and tropical climates, and present one of the most cost-effective, easy to operate and reliable appropriate technology processes for treating domestic and industrial wastewater. The simplicity and low cost of this system made it an attractive solution in both developing and developed countries, but effective performance depends on correct design. Investment cost in a conventional lagoons system is in the range 10 40 US$/Capita and the O&M cost is in the range 0.2–0.4 US$/Yr/Capita. Recent process developments have improved the stabilization lagoons technology, as presented in the following section.

1.8.3 A series of improved stabilization lagoons

A most common and wide spread combined process is a lagoon system composed of anaerobic lagoons followed by facultative lagoons followed by maturation lagoons, as presented in Figure 1.50. As explained, inclusion of an anaerobic lagoon as the first unit in this plant is of great importance since it significantly reduces the total area of the plant in relation to an area of a lagoons plant composed of only facultative and maturation ponds. The drawback of the inclusion of an anaerobic lagoon is that when it becomes overloaded it generates odours.

Another innovative lagoons based process is RMS followed by **Covered** Anaerobic Lagoons followed by Facultative Lagoons with **Mixers** installed in them followed by Maturation Lagoons. The Schematic flow diagram of this process is presented in Figure 1.52.

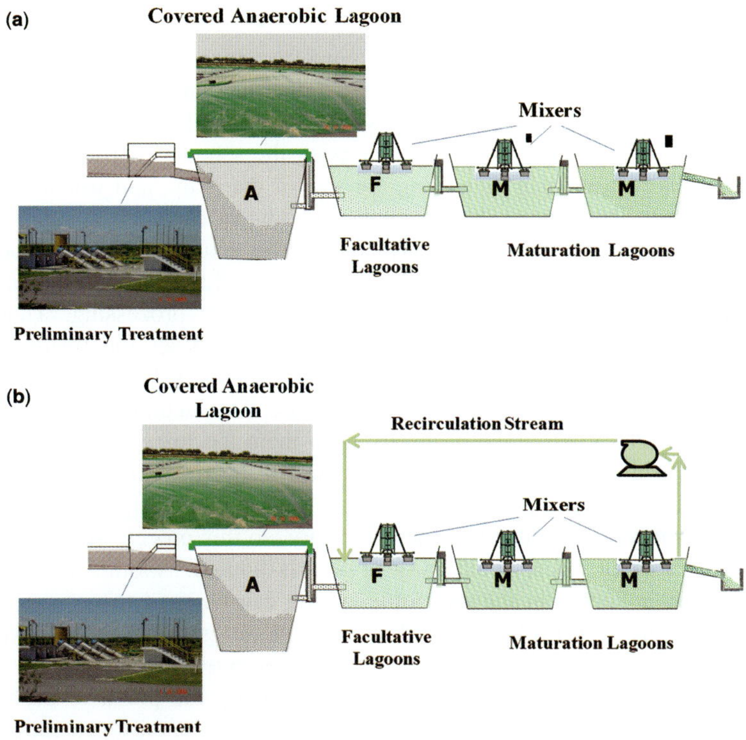

Figure 1.52 Schematic process flow diagram of RMS followed by covered anaerobic lagoons followed by facultative lagoons with mixers followed by maturation lagoons, (a) Process flow diagram without recirculation; (b) Process flow diagram with recirculation from the final lagoon to the first facultative lagoon

A covered anaerobic lagoon is presented in Figure 1.5 and a facultative lagoon with mixers is presented in Figure 1.17. The benefits of covering anaerobic lagoons and of inserting mixers in facultative lagoons are discussed in Sections 1.7.2.3 and 1.7.2.6 respectively.

Figure 1.52 presents two variants of the lagoons improved process. In Figure 1.52a the first facultative lagoon is only equipped with mixers. As mentioned in Section 1.7.2.6, organic loads on mixers equipped lagoons can reach values of 450 Kg BOD/d/ha in the high temperatures season of 25–30°C, with lower loading values at lower temperatures. The critical point in the process presented in Figure 1.52a is the first facultative lagoon, which received the highest organic loading. To avoid problems of anoxic conditions in the entrance zone of this facultative lagoon and allow increasing organic load on this lagoon, that is the flow of the raw wastewater to the system, even more, it is possible to operate the system using the process presented in Figure 1.52b. In this process, the effluent of the final lagoon is recirculated to the inflow of the first facultative lagoon at a ratio of about 2:1 (recirculation stream to raw wastewater). The recirculated flow dilutes the inflow to the first facultative lagoon and brings in a stream rich in oxygen and in algae which immediately start to generate more oxygen. This prevents development of anoxic condition in the first facultative lagoon and enables increasing the organic loading on this lagoon, that is, increasing the flow to the plant. The recirculation requires energy and increases the O&M cost of the process in Figure 1.52b but it also increases the capacity of the plant, so the cost of reciculation is the cost of increasing the capacity, which is lower than the cost of increasing the capacity using other methods.

Each process (of Figure 1.52a or 1.52b) can be designed from inception to function under its respective flow diagram or can be used to upgrade existing conventional lagoons plants to increase their treatment capacity without increasing their area. Most of the existing lagoons plants are conventional plants. When they become overloaded, the option to upgrade them using the scheme proposed in Figure 1.52 is very attractive since it is much cheaper than increasing the size of the plant, and usually there is no room to increase the existing plant. The process has several important advantages in relation to conventional lagoons plants. First, it eliminates the odours which may be generated in the anaerobic lagoons by preventing emission of gases from the lagoon to the external environment. This is achieved by collecting the biogas generated in the lagoon and subjecting it to further processing. That is an important environmental improvement of the lagoons plant. Second, it increases the treatment capacity of the lagoons units, without increasing their area. The anaerobic units can be subjected to increased loads because the limiting factor of odour generation was eliminated. The facultative and maturation ponds can be subjected to higher organic loads since by inserting in them the mixers they are turned into a more efficient reactors, as explained in Section 1.7.2.6, and using the recirculation concept allows loading them even more. Overall, a lagoons plant based on this flow diagram requires less area than that of a conventional lagoons plant with the same treatment capacity, so this process saves area and this fact can be an important factor in cases where availability of area is limited. Another important benefit of this process is that the biogas collected in the anaerobic lagoon can be used to generate energy. There are several ways to generate energy from biogas, and all of them can generate income which can cover the entire O&M expenses of the plant and still leave significant revenue to the plant's owner. The feasibility for generating energy depends on the size of the town served by the plant. In the case of very small plants it might be unfeasible to generate energy. When not used for energy generation, the biogas must be flared. Flaring the biogas or generating energy has one more benefit (on top of preventing odours), it prevents the emission to the atmosphere of methane, a strong Green House Gas (GHG), thus supporting the effort of Carbon Emission Reduction (CER) and contributing to the alleviation of global warming. It is also possible to obtain CER credits for flaring the biogas or using it for generation of energy.

After years of operation lagoons plants, as any other plant, reach the stage of operating at full capacity and in many cases they are even being overloaded. Often there is no area available in the plant site to expand their capacity under the principles of the conventional lagoons treatment process, that is, there is no room to construct additional lagoons. The process presented in Figure 1.52 can be used to extend the treatment capacity of such plants without extending their area. The city of Santa Cruz, Bolivia, has 4 conventional lagoons plants and all of them are being upgraded using the process of Figure 1.52a. The upgrade of Plant N2 in Santa Cruz has already been completed and a satellite photo of the upgraded plant is presented in Figure 1.53. As mentioned, this plant consists of 2 parallel lines, each composed of an anaerobic lagoon followed by 2 facultative lagoons followed by one maturation lagoon. The photo of the upgrade plant, Figure 1.53, can be compared to the photo of the plant before the upgrade, which is presented in Figure 1.51. After the upgrade, the anaerobic lagoons are covered, as seen in Figures 1.53

and 1.54 and the facultative lagoons contain mixers. Because of their small size, it is impossible to see the mixers clearly in Figure 1.53, but they can be seen in Figure 1.55, and part of them is also presented in Figure 1.17. The RMS unit is located outside the N2 plant site, in the site of the pumping station that conveys the raw wastewater to the plant. A photo of the RMS unit is presented in Figure 1.56.

Figure 1.53 A satellite photo of the upgraded N2 lagoons plant in Santa Cruz, Bolivia

Figure 1.54 The covered anaerobic lagoons in the upgraded N2 lagoons plant in Santa Cruz, Bolivia

Figure 1.55 Mixers in the facultative lagoon of plant N2 after its upgrade

Figure 1.56 The rotating micro screens (RMS) unit serving plants N2 and N1

The concept of the biogas collection system in the covered anaerobic lagoons is presented in Figure 1.57. The collected biogas is being flared in the flaring system shown in Figure 1.58 (which is not seen on Figure 1.53 because of its small size). A study of alternatives for energy generation from the biogas is being conducted for all the wastewater treatment plants in Santa Cruz and installations will be constructed once the best alternative is selected. A request for financial credits for CER has also been submitted.

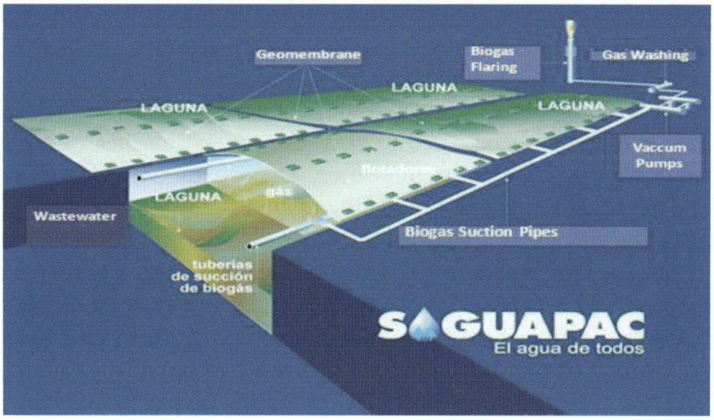

Figure 1.57 The biogas collection system in the covered anaerobic lagoons

Figure 1.58 The biogas flaring system serving plants N2 and N1

As seen in Figure 1.53 residential areas are very close to the anaerobic lagoons of plant N2 and before the upgrade residents complained about odour problems. After the upgrade, the odour problems were completely eliminated. The upgrade method of a conventional existing lagoons system which was executed in plant N2 in Santa Cruz can serve as an example to many other cases of lagoon plants which become overloaded, need to be upgraded, but area for constructing additional lagoons is unavailable. The upgrade method selected in Santa Cruz is the lowest cost method of upgrade within the existing available area, The plant remained a simple to operate lagoons plant, odour problems were completely eliminated, the emission of Green House Gases was stopped and an opportunity for generation energy from the captured biogas has been created.

Before the upgrade the N2 plant was serving a population of 170,000 and after the upgrade it is serving a population of 270,000, that is, the plant's capacity was increased by about 60%, without an increase of the plant's area, and at a very low investment cost.

The quality of the effluent that can be achieved in a plant of an upgrade lagoons system is similar to the quality which can be achieved in a conventional lagoons plant and depends to a large extent on the design of the plant. A high quality effluent can be achieved with an adequate design.

Based on the Santa Cruz experience, the investment for upgrading the N2 lagoon plant is about 15 US$/Capita, which is a low value compared to cost of other approaches of transforming the existing plants to more complex plants that do not need more area. The operation and maintenance cost is similar to that of conventional lagoons systems and practically did not increase as a result of the upgrade. If the process of Figure 1.52b is selected, the O&M cost will increase because of the recirculation, however, this process is more reliable and the treatment plant would have a greater capacity. Investment cost in a Newly Constructed improved lagoons system is in the range 20–50 US$/Capita and the O&M cost of this system is in the range 0.2–0.4 US$/Yr/Capita. The mentioned costs do not take into account the potential income which may be generated from the sale of emission reduction of Green House Gases or potential income from production of energy from the biogas. These two income streams may further reduce the cost of this process.

1.8.4 UASB followed by facultative lagoons

Under this scheme, the effluent of a UASB plant is polished by facultative lagoons. This is a very logical scheme and an exemplary appropriate technology process. Preliminary treatment is achieved by rotating micro screens and, if needed, a vortex grit chamber, which is the best available preliminary treatment scheme. The removal of the main portion of organic matter is achieved in this scheme by the USAB process, which is one of the unit processes of appropriate technology and is a simple, effective low cost process. The facultative lagoons system used for polishing the UASB effluent is a simple reliable system. Removal of pathogenic organisms can be achieved in this process naturally (without disinfection), if sufficient detention time is provided in the lagoons system. This process includes also the disposal of the screened material and the handling of the UASB excess sludge, mainly by drying beds, without mechanical equipment. The main disadvantage of this scheme is that the lagoons require large extensions of land. If area is scarce, deeper lagoons with mixers can be used. An ultrasound device can be used to decrease algae content in the final effluent. Because of the use of lagoons as part of this process, it is suitable for small and medium size cities, but not for large cities. As a whole this is a good and recommended process. A schematic flow diagram of this process is presented in Figure 1.59.

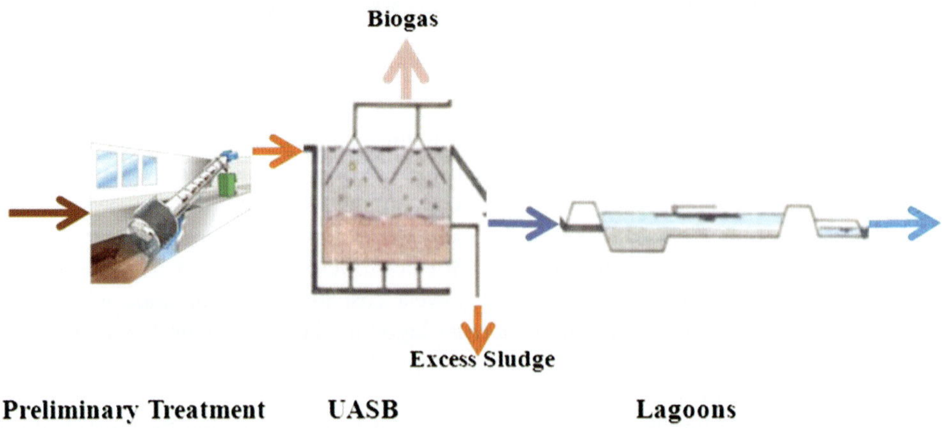

Figure 1.59 A schematic flow diagram of the process of preliminary treatment followed by UASB followed by facultative lagoons

Frassinetti and van Haandel (1996) describe the advantages of the UASB followed by lagoons process. A way to look at this process is replacing the anaerobic lagoon in a traditionally waste stabilization ponds system by a UASB unit. The UASB Reactor is smaller than the anaerobic lagoon and the biogas produced by the UASB reactor can be captured, so the odour problem can be eliminated. According to authors, the UASB reactor is neighbour friendly and can be used in densely populated areas. The UASB reactor is more efficient than the anaerobic lagoon so the residual organic load in its effluent is small and the subsequent pond configuration can be designed with the specific purpose of eliminating pathogens. By adopting a plug flow regime the required area of the lagoons can be greatly reduced. Also, since the UASB effluent has a good transparency, photosynthesis in the lagoons is intense and the pH increases due to biological carbon dioxide consumption, accelerating the death rate of pathogens and opening the possibility for reducing nutrients: nitrogen by evaporation of dissolved gaseous ammonia (whose portion in the lagoons increases with the increase of the pH) and phosphorous by phosphate precipitation, which also intensifies with the increase of the pH. Since the operation of the UASB reactor and of the lagoons is simple, Frassinetti and van Haandel propose to construct in a city, for the purpose of saving significant investments in the sewerage network, several small treatment plants based on this process instead of one large central treatment plant.

This process is being applied in several plants in Brazil and in Colombia, using conventional preliminary treatment, not RMS. The largest plant of this type in Colombia is the Rio Frio plant in Bucaramanga. The process flow diagram and an air photo of the Rio Frio plant are presented in Figure 1.60 and in Figure 1.61 respectively. This plant, commissioned in 1991, is one of the oldest large scale UASB-based facilities treating municipal wastewater in the world. The plant's original design flow was 740 l/s serving a population of 240.000. The UASB reactors are located at the right hand side of the bottom of Figure 1.61 and next to them, to the left are the sludge drying beds. The UASB reactors are equipped with aluminium covers and the collected biogas is flared. Pretreatment for the UASB reactors is provided by 6 mm screens followed by grit chambers. The reactors are followed by two facultative lagoons. Removal efficiency of BOD is about 80% in the USAB plant and 33% in the facultative lagoons, achieving a total BOD removal in the range of 85–90%. Construction costs for the existing facilities were approximately 15 US$/Capita. Operating costs are 0.79 US$/Year/Capita or 0.015 US$/m^3 treated (Osorio, 1996).

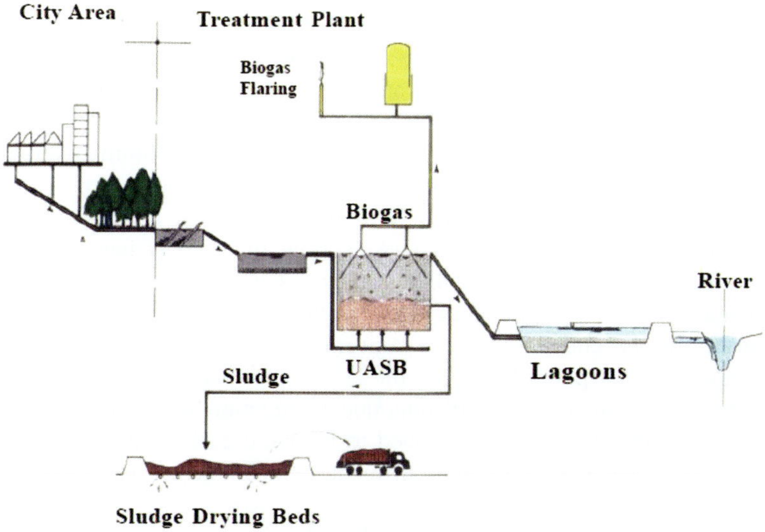

Figure 1.60 Process flow diagram of the Rio Frio treatment plant in Bucaramanga, Colombia

Figure 1.61 Air photo of the Rio Frio treatment plant in Bucaramanga, Colombia

The plant has been able to achieve essentially secondary treatment levels compatible with that of conventional processes at a very low capital and O&M costs, thereby demonstrating the potential of application of this process in developing countries. The Rio Frio plant has two problems: (i) the detention time in the facultative lagoons is way too low (about 2 days) and this limits their performance; and (ii) the potable water in Bucaramanga has elevated sulphate levels, and most of the sulphates present in the raw wastewater are effectively reduced to H_2S in the anaerobic reactors, generating odour around the plant and promoting corrosion of plant equipment. Addition of aluminium covers and the collection and flaring of the biogas has helped to reduce the intensity of the odours and the corrosion of the equipment.

With time the population served by the Rio Frio plant increased and its capacity needed to be augmented. The plant is now being upgraded and its flow diagram is being modified. The UASB reactors section is being expanded but the facultative lagoons have been eliminated since there was no area available to increase them. The lagoons are being replaced by an activate sludge unit. As previously mentioned, this transforms the entire plant to a conventional treatment plant, and it ceases to be an appropriate technology treatment plant. However, the UASB unit performed well during a period of 20 years and has not been eliminated, so it will continue to function in the foreseeable future, and the configuration of UASB followed by lagoons proved to be a successful configuration which performed well and provided during 20 years a high quality effluent, comparable to that of conventional treatment, at a fraction of the cost.

SANEPAR, the water utility of the state of Parana in Brazil, is implementing the project of environmental sanitation of Parana (Projeto de Saneamento Ambiental do Paraná) which consists, in addition to a sewerage component, of expansion and upgrading of existing wastewater treatment facilities, as well as construction of new treatment plants, a large part of them in the state capital, Curitiba and its surroundings. SANEPAR has adopted the principle of using combinations of simple treatment unit processes and most of its plants are based on this type of processes. It incorporated in most of its plants anaerobic treatment unit processes and has developed in-house a UASB reactor design that has been widely used throughout the state of Parana. Several of the treatment plants in Parana are based on the flow diagram of UASB followed by facultative lagoons.

Figure 1.62 shows a photo of a UASB plant followed by a facultative lagoon in Ronda, a small town, part of the city of Ponta Grossa in the state of Parana. In this plant the two UASB reactors are circular. The plant includes also a conventional preliminary treatment unit and sludge drying beds, which are not shown in the figure. Photos of the effluent at various stages of the treatment process (Raw Wastewater, UASB effluent and Lagoon effluent) are presented at the bottom of the figure and the improvement of the effluent quality at each stage is notable.

Figure 1.62 UASB reactors followed by a facultative lagoon in Ronda, Ponta Grossa, Parana, Brazil

Another plant in Parana based on the flow diagram of UASB followed by facultative lagoons is the Padilha Sul Plant located in the city of Curitiba, the capital of Parana. A photo of this plant is presented in Figure 1.63. The first stage of this plant was designed for an average wastewater flow of 440 l/s and is serving an equivalent population of 318,000. It will be expanded to handle an average flow of 600 l/s. The UASB reactors occupy the small area at the left hand side of the bottom of the figure. As can be seen, the UASB reactors unit occupies only a small portion of the total plant area. Also seen in the photo is that the UASB reactors are located at a close vicinity to residential areas. That means that the UASB units do not present an environmental nuisance to the nearby residential areas.

Figure 1.63 The Padilha Sul Plant in Curitiba, Parana, Brazil, based on UASB followed by facultative lagoons

It has been noted that in most cases the detention time in the lagoons that serve as the polishing unit of the UASB reactors is too short. This was the case in Rio Frio, where the detention time in the lagoons was only 2 days. For some reason the designers of this process consider that a short detention time would be sufficient, but in fact lagoons with such a short time are not able to do much. They do not become facultative lagoons and remain some sort of settling ponds. The lagoons need to be adequately designed as facultative lagoons. A plant based on adequate design of the lagoons following a UASB reactor will produce a better effluent than that reported in existing plans, however the effluent quality presented in Table 1.4 are based on results achieved in existing plants. The investment cost in a UASB reactor followed by lagoons system is in the range 30–50 US$/Capita and the O&M cost is in the range 1.0–1.5 US$/Yr/Capita.

UASB reactors followed by a polishing stage consisting of facultative lagoons, is gaining popularity in warm-weather countries with minimum wastewater temperatures of 10°C.

1.8.5 UASB followed by anaerobic filter

In addition to the use of anaerobic filters as the main treatment unit, they can be used as polishing units for improvement of the quality of the effluent of a preceding unit. They are in fact more suitable to perform as polishing units since when fed with treated effluent of a preceding unit they are less bound to be clogged by inorganic suspended solids.

A process composed of Rotating Micro Screens as preliminary treatment, followed by UASB as the unit for removal of the main portion of organic matter, then followed by an anaerobic filter for polishing of the UASB effluent, and finally, if necessary, followed by a disinfection unit (chlorination or UV disinfection) for removal of pathogens is a reasonable schemes and an effective appropriate technology process. It is very similar to the process described in the previous section and is basically the replacement of the lagoons by an anaerobic filter followed by a disinfection unit. The advantages of this process are that it yields a high quality effluent and is compact, that is, does not occupy a large land area. This process can be applied in cases where area for locating a treatment plant is scarce and insufficient for installation of lagoons. The process includes also the disposal of the screened material of the primary treatment and the handling of the UASB and anaerobic filter excess sludge, mainly by drying beds, without mechanical equipment. The process is adequate for application in small, medium and large cities.

This process has been applied in several pilot plants and full scale plants in Brazil, using conventional preliminary treatment or an Imhoff tank as the preliminary treatment stage. In pilot plant experiments reported by Chernicharo (2000) a final effluent containing an average of 22 mg/l total BOD (soluble and suspended) and 15 mg/l of TSS was achieved. Such an effluent quality is similar to that of an activated sludge process effluent. Total BOD and total COD removal efficiencies achieved were 90%, soluble COD removal achieved was 81% and TSS removal achieved was 95%. Additional investigations reported by Chernicharo (2007) demonstrated that plants based on UASB followed by anaerobic filters are capable to maintain in the final effluents concentrations of COD, BOD and SST lower than 120 mg/l, 60 mg/l and 30 mg/l respectively.

Andrade Neto (2006) reports that several plants based on the UASB followed by an anaerobic filter process are under operation since 1996 in the state of Parana, Brazil, serving populations in the range of 1,500 to 50,000 persons. Effluent quality achieved in these plants is total BOD lower than 60 mg/l and TSS lower than 20 mg/l. The same type of plants is under operation since 1997 in the state of Minas Gerais, Brazil, serving populations in the range 2,000 to 15,000 persons. One plant in Minas Gerais, at Ipatinga, serves a population of about 200,000.

Andrade Neto (2006) also reports that the investments costs of the UASB followed by anaerobic filter systems in Parana were in the range of 5 to 30 US$/Capita. This process is of special importance since it is able to achieve essentially secondary treatment effluent levels at a low capital and O&M costs, occupying a small similar are and even a smaller area than that of activated sludge, thereby demonstrating the potential of applying this process in developing countries as a favourable alternative to activated sludge.

A schematic flow diagram of the process of UAB followed by Anaerobic filter is presented in Figure 1.64 and a photo of such a plant in the small town of Tibagi, in the State of Parana, Brazil, serving a population of about 20,000 is presented in Figure 1.65. This plant, operated by SANEPAR, the water and sanitation utility of Parana, achieves an effluent quality of BOD 5–25 mg/l, COD 40–100 mg/l and TSS 4–10 mg/l. A photo of the Anaerobic Filter of the Tibagi plant is presented in Figure 1.66 and a photo of the effluent of this plant is presented in Figure 1.67. The anaerobic filter is presented in a separate photo since in the general photo of the plant (Figure 1.65) the filter is not seen in detail because it is the last unit of the plant. The anaerobic filter in the Tibagi plant is a circular cross section filter of an ascending flow. This filter is similar to a circular sedimentation tank. It is filled with a crushed rock media on which the anaerobic bacteria grow, attached to the media. The filter washing is done by a descending stream of the incoming flow from the UASB reactor, which is basically a backwashing of the filter. The backwashing is done once every couple of days. The incoming flow direction is changed and the stream used for washing of the rock media is directed, after the washing operation, to the drying beds.

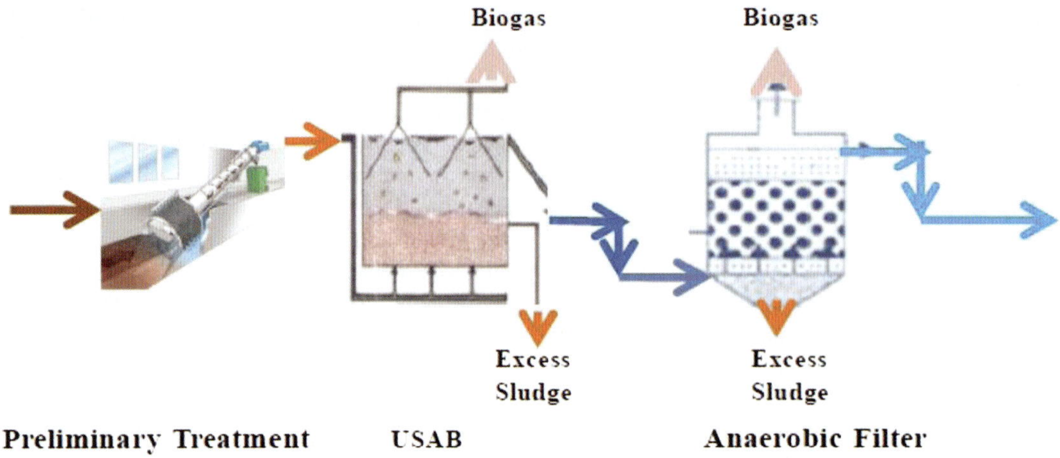

Figure 1.64 A schematic flow diagram of a UASB reactor followed by an anaerobic filter

Figure 1.65 UASB reactor followed by anaerobic filter in Tibagi, Parana, Brazil

During the visit of the Authors in the Tibagi plant a glass flask for collection of a final effluent sample was not available since the visit was conducted on a Sunday and the office and laboratory were closed. So an empty soft drink bottle was used to collect a sample of the effluent of the trickling filter, which is the final effluent of the plant, and is shown in Figure 1.67. The effluent is of high quality and has a nice appearance.

Figure 1.66 The circular anaerobic filter at the Tibagi wastewater treatment plant

Figure1.67 The effluent of the Tibagi treatment plant, Parana, Brazil (Effluent Quality: BOD 5–25 mg/l, COD 40–100 mg/l, TSS 4–10 mg/l)

COPASA, the water and sanitation utility of the State of Minas Gerais in Brazil operates several plants based on the process of UASB followed by anaerobic filters. The treatment plants of Arenal, Bela Vista, Horto, and Ipanema located in the municipality of Ipatinga are some of these plants. The performance of the Ipanema plant was evaluated during the period July 2005–September 2006 (Greco, 2007). The wastewater flow to this plant was 180 l/s and it serves a population of about 200,000. The plant achieved an average BOD removal efficiency of 84%, an average COD removal efficiency of 87% and an average SST removal efficiency of 80%. A photo of the wastewater treatment plant in Ipatinga, Minas Gerais, which is a UASB followed by anaerobic filters plant serving a population of about 200,000 is presented in Figure 1.68. The UASB units and the Anaerobic Filters units in this plant are rectangular and that enables to achieve a compact configuration of the plants. As seen in the figure, the plant also includes drying bed of the excess sludge, located in the upper part of the photo. The plant contains 10 UASB units located in two rows (at the bottom of the photo) and two anaerobic reactors located in a raw between the UASB reactors and the drying beds.

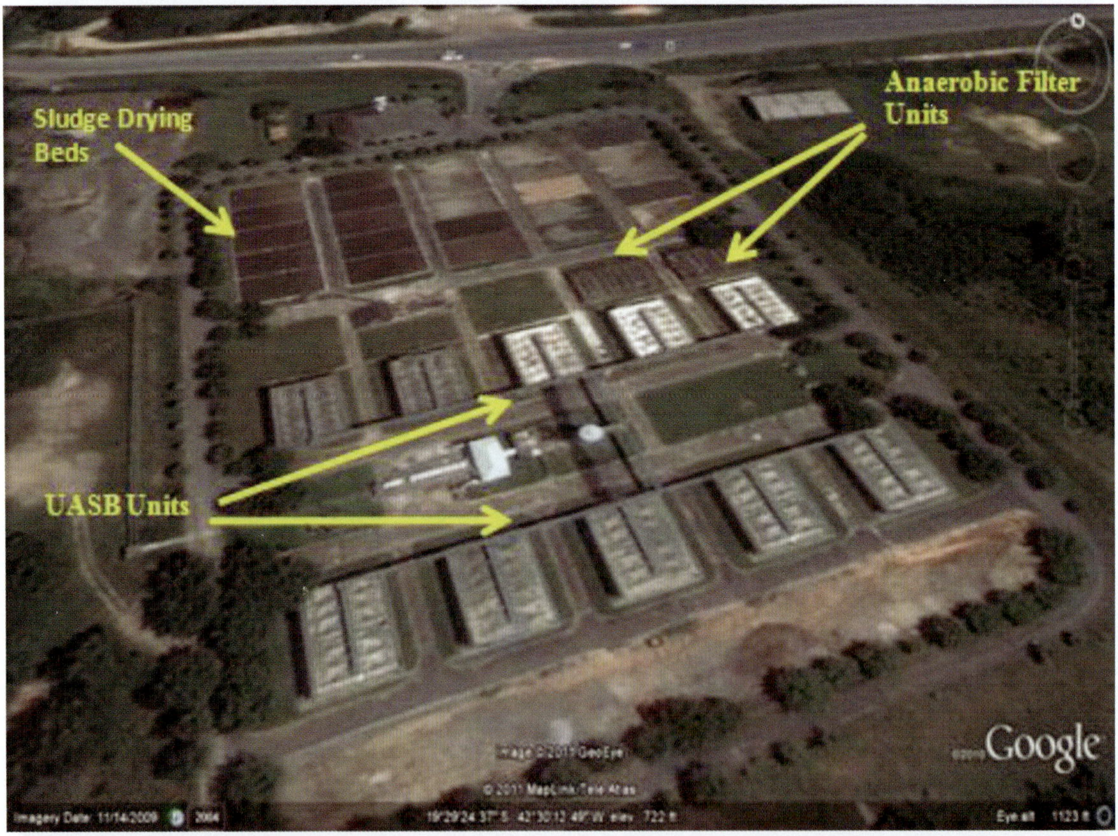

Figure 1.68 The treatment plant of Ipatinga, State of Minas Gerais, Brazil, based on UASB reactors followed by anaerobic filters

The investment cost in the system UASB reactor followed by Anaerobic Filter is in the range 20–40 US$/Capita and the O&M cost is in the range 1.0–1.5 US$/Yr/Capita.

1.8.6 UASB followed by dissolved air flotation

An additional interesting process is a process composed of preliminary treatment followed by UASB as the unit for removal of the bulk portion of organic matter, then followed by a Dissolved Air Flotation (DAF) unit as a polishing unit for removal of the suspended solids from the UASB effluent and by that also removing part of the BOD of the effluent. Finally, if necessary, a disinfection unit (chlorination or UV disinfection) can be added for removal of pathogens. This process is similar to the processes of the previous two sections and is basically a process of polishing a UASB effluent by an additional polishing unit, in this case by DAF. The advantages of this process are that it can yield a reasonably high quality effluent in terms of removal of BOD, suspended solids and phosphorous (although the removal of nitrogen and coliforms is low), and is compact, that is, it occupies a small land area. The removal of phosphorous is a result of the use of flocculants in the flotation process. This process can be applied in cases where area for locating a treatment plant is scarce and insufficient for installation of lagoons. The process includes also the disposal of the screened material of the primary treatment and the handling of the UASB and DAF unit excess sludge, mainly by drying beds, without mechanical equipment. The process is adequate for application in small, medium and large cities. A schematic flow diagram of the process is presented in Figure 1.69. According to Chernicharo (2007) this process can achieve 83–93% removal of BOD, 90–97% removal of TSS, 75–88% removal of total phosphorous, but only 1–2 logarithmic units removal of Fecal Coliforms. The effluent contains 20–50 mg/l of BOD, 10–30 mg/l TSS, 1–2 mg/l phosphorous and 10^6–10^7 MPN/100 ml Fecal Coliforms.

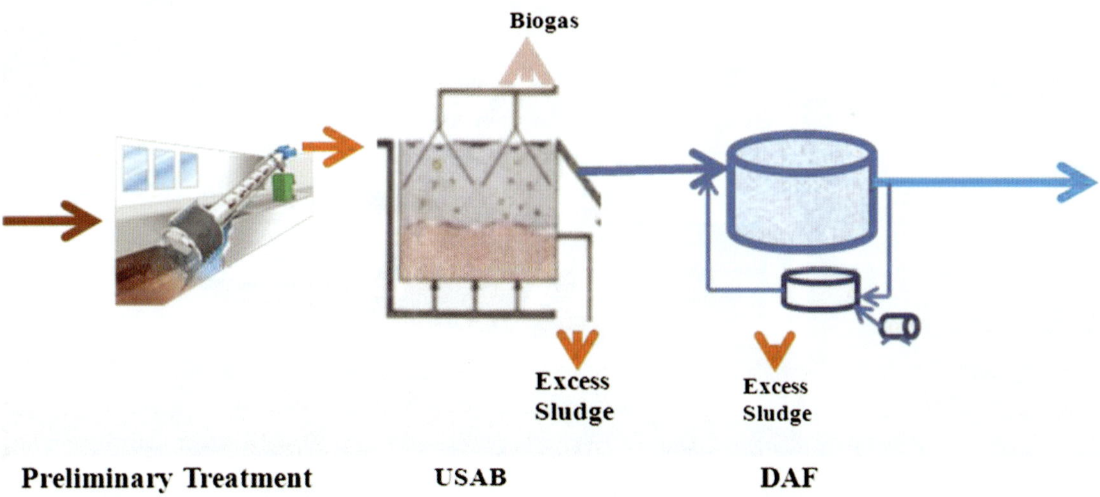

Figure 1.69 A schematic flow diagram of a UASB reactor followed by a dissolved air flotation unit

Several treatment plants based on this process are in operation in Brazil. The Atuba Sul treatment plant in Curitiba, State of Parana, Brazil, based on this process, is a large plant which handles a raw wastewater flow of 1.12 m^3/s and serves a population of 544,000 (Prado, 2010). A photo of the plant is presented in Figure 1.70. The UASB reactors were constructed in 1998 and the DAF units were added in 2008. The UASB section contains 16 reactors, each handling a flow of 70 l/s (totalling 1.12 m^3/s). The DAF section consists of 4 units, each handling 280 l/s. The flocculant used in the flotation units is Ferric

Chloride Hexahydrate. The effluent is discharged to the Atuba River. In 2009 the plant achieved about 80% removal of BOD and 70% removal of TSS, values lower than expected. About 15 l/s of the effluent undergo a higher level of treatment denominated water recovery and its effluent is not discharged to the river but rather being reused. The water recovery process consists of ozonation of the DAF effluent, sedimentation of solids in a CEPT unit, dual media filtration by sand and anthracite layers, chlorination by hypochlorite and finally absorption of residual organic matter by an activated carbon filter. This is a complex process.

Figure 1.70 The treatment plant Atuba Sul in Curitiba, State of Parana, Brazil, based on UASB followed by dissolved air flotation

Another treatment plant based on UASB followed by DAF is the Sao Jorge plant treating a flow of 70 l/s and serving part of the city of Almirante Tamandare in the State of Parana, Brazil. The effluent of this plant is disinfected using the UV method. A photo of the Sao Jorge plant is presented in Figure 1.71.

The investment cost in the system UASB reactor followed by DAF is in the range 30–40 US$/Capita and the O&M cost is in the range 1.0–1.5 US$/Yr/Capita. The process is simple to operate and is an appropriate technology process.

Figure 1.71 The treatment plant Sao Jorge in the City of almirante tamandare in the state of Parana, Brazil, based on UASB followed by dissolved air flotation followed by UV disinfection

1.8.7 Chemically Enhanced Primary Treatment (CEPT) followed by Sand Filtration

CEPT is an important appropriate technology unit process because of its capacity to function at the entire range of temperatures, including very low temperatures. If a higher effluent quality than that which can be produced by the CEPT unit process is required, sand filtration can be used to polish the CEPT effluent, so a combination of CEPT followed by sand filtration and, if required, UV disinfection can provide a good quality effluent, similar to that of conventional treatment. The filtration can be a single or multimedia sand filtration. The schematic process flow diagram is presented in Figure. 1.72. The CEPT followed by Sand Filtration process is based on physicochemical and physical processes, with no biological processes involved, so it is not very sensitive to temperature and will function well at very low temperatures, and as long as the wastewater does not freeze the process will perform even if the air temperature in the plant site is much lower than zero.

According to Cooper-Smith (2001) an extensive pilot scale investigation showed that the combination of CEPT followed by sand filtration and UV disinfection proved capable of producing a high quality effluent (averaging 20 mg/1 BOD, 15 mg/l TSS) using a Tetra Deep Bed Filter at a filtration rate of 5 m/h and a coagulant (Kemira PAX XL60, a polyaluminium silicate) dose of 40 mg/l. UV disinfection of the filtered effluent complied with the required Bathing Water Directive standard of the EU and Net Present Value calculations showed this option to be considerably cheaper than a secondary biological treatment option. Sludge treatment is identical to that of CEPT sludge.

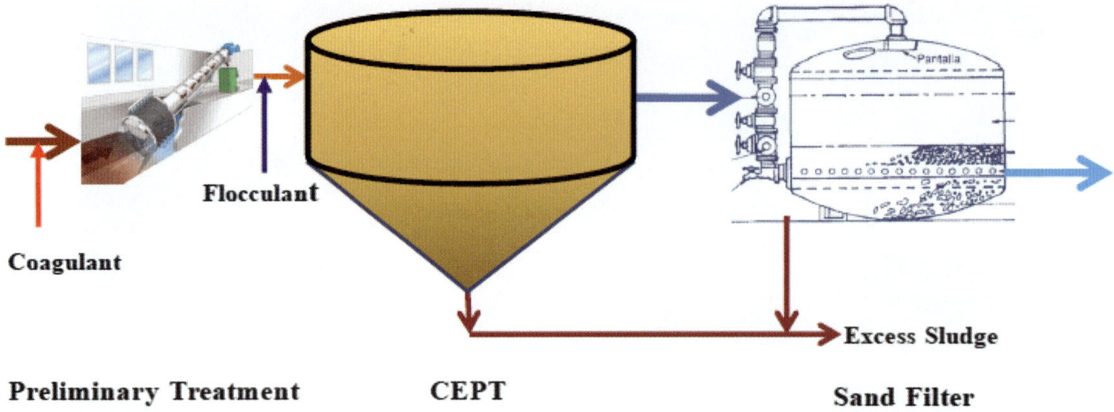

Figure 1.72 A schematic process flow diagram of CEPT followed by Sand Filtration

CEPT followed by filtration can be used to produce an almost secondary quality effluent which would represent a good option for developing countries, as its attributes include: (i) the ability to provide a secondary effluent quality at lower cost than conventional processes; (ii) reduced power needs; (iii) easy operation; and (iv) occupies a small land area, smaller than that of conventional processes. This process is suitable for medium and large cities. Small cities may find it difficult to operate.

CEPT followed sand filtration is an appropriate technology process since it is a relatively simple technology, which provides low cost, easy to implement and effective treatment. The CEPT unit process is an easy to operate process since it is basically consisting of a settling unit and a sand filter. Sand filtration is a process that most water utilities know how to operate since they operate filtration plants in the water supply system, and it is basically an easy process to operate. The only complication might be the handling of the sludge, but as a whole, it is an appropriate technology process.

It is estimated that the ratio of investment cost in conventional activated sludge to investment in CEPT followed by multimedia sand filtration is about 2 to 1. The investment cost of the system CEPT followed by a Sand Filter is in the range 40–50 US$/Capita and the O&M cost is in the range 1.5–2.0 US$/Yr/Capita.

1.8.8 Pre-treatment of various types followed by a stabilization reservoir (Wastewater reuse for irrigation, the stabilization reservoirs concept)

The stabilization reservoir process was presented and discusses as one of the unit processes of appropriate technology. As part of this discussion, it was clarified that before discharging the wastewater to the reservoir for storage, it needs to be treated to remove a part of the organic matter contained in it; so as to prevent development of anaerobic conditions in the upper layer of the reservoir. Prevention of such conditions is important since anaerobic condition may generate odours. So in fact, the reservoir stores effluent and not raw wastewater. The required pre-treatment of the wastewater before it is stored in the reservoir is not a specific treatment processes. It can be achieved using a variety of treatment process, among them; all the appropriate technology processes or combination of processes described in this book, as long as they remove the required load of organic matter and ensure that the surface organic loading on the reservoir does not exceed the permitted limit. The combination of the pretreatment unit and the reservoir form as a whole an appropriate technology process, if the pretreatment unit is based on appropriate technology.

The schematic process flow diagram of pretreatment followed by a stabilization reservoir is presented in Figure 1.73. This figure is not the flow diagram of just one combined process but rather of a large series of combined processes, since the pretreatment unit is not just one specific unit, but can be one of a variety of unit processes or combination of unit processes. For example, Figure 1.74 shows a photo, taken in the 1980s, of a typical wastewater pretreatment and stabilization reservoir scheme of a medium size city in Israel. In this case it is the treatment system of the city of Nazareth, in the north of Israel, and the pretreatment is based on anaerobic lagoons followed by aerated lagoons. This pretreatment process is not strictly an appropriate technology process because it includes aerated lagoons which are based on mechanical aeration and consume energy, but it is close to being an appropriate technology process because the anaerobic lagoons which precede the aerated lagoons decompose the bulk of the organic matter and the aerated lagoons just serve for polishing and do not consume a lot of energy.

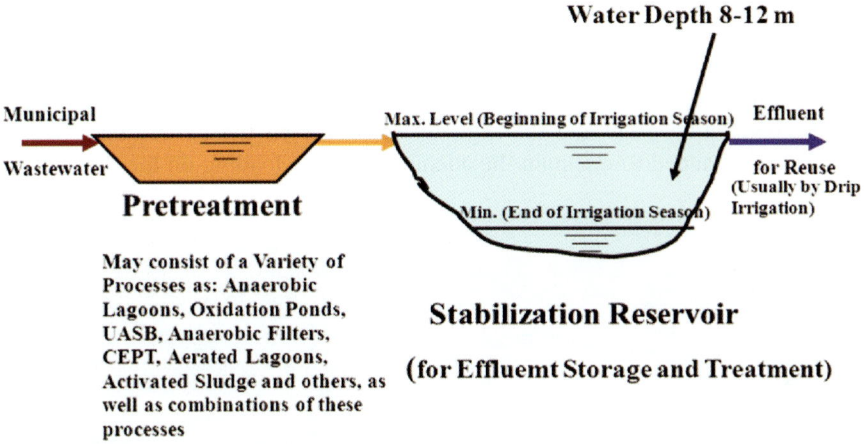

Figure 1.73 The stabilization reservoir reuse concept and schematic process flow diagram

Figure 1.74 The wastewater treatment system and stabilization reservoir of the City of Nazareth–Photo taken in the 1980s

In the system of Nazareth, the anaerobic lagoons can be covered to prevent odours and capture the biogas, which can then be used for generating the energy to operate the aeration system in the aerated lagoons. Preliminary estimates indicate that the biogas would be more than sufficient to supply all the energy required for operating the aerated lagoons, however, the economic feasibility of such a system needs to be analyzed.

In Israel, the pretreatment units preceding reservoirs are mostly of two types: (i) a unit based on various combinations of lagoons; and (ii) a unit based on activated sludge. When the pretreatment is based on lagoons, the entire scheme is an appropriate technology process, however, when the pretreatment is based on activated sludge, it is not an appropriate technology process. There are about 350 reservoirs systems in Israel, each unique in its pretreatment process, in the volume of the reservoir designed to fit the size of the town it serves, and in the shape of the reservoir, adjusted to the local topography in the reservoir site. The volumes of the reservoirs vary between the smallest of about 20,000 m^3 serving a very small town, to 12 million m^3 serving the city of Haifa with a population of about 700,000.

A recent satellite photo of the treatment system of Nazareth, with the city on the horizon, is shown in Figure 1.75, and a close-up satellite photo of the treatment plant is shown in Figure 1.76. In these two figures it is possible to note that the number of lagoons in the pretreatment unit has grown since the 1980s, because the population grew and the organic load received by the treatment plant grew, so the pretreatment had to be expanded to maintain the adequate organic loading on the reservoir.

Figure 1.75 The wastewater treatment system and stabilization reservoir of the city of Nazareth – recent satellite photo

Another combined reservoir treatment system, the Naan Reservoir, is shown in Figure 1.77. This is one of the first reservoir systems constructed in Israel and is in operation since the 1970s. The pretreatment unit in this case is very simple and consists of only two anaerobic lagoons. The configuration of the pretreatment installations depends on the surface area of the reservoir, which in turn depends on its design, that is, the relation of volume to depth and the availability of land.

Figure 1.76 A close-up satellite photo of the treatment system of the city of Nazareth

Figure 1.77 The Naan wastewater treatment system and stabilization reservoir

Reservoirs which start generating odour problems after years of satisfactory operation, because of the natural increase with time of the raw wastewater flow, can be equipped with mixers and that would improve the their performance as biological reactors and make them capable of receiving increased organic surface loadings without becoming anaerobic and without generation odours. Figure 1.78 shows a satellite photo of the Og reservoir which receives wastewater of the eastern basin of Jerusalem. The pretreatment unit is based on a combination of anaerobic and aerated lagoons, which are incapable of reducing the organic load to an adequate level, so mixers were installed in the reservoir and made it capable of processing higher loads without becoming anaerobic. The mixers are clearly seen in Figure 1.78 and the process of their installation is shown in Figures 1.32 and 1.33.

Figure 1.78 A satellite photo of the Og reservoir equipped with mixers which enable to increase the organic matter surface loading it is capable of handling

In Israel, pretretment units in reservoir plants do not include UASB reactors, anaerobic filters, CEPT or other appropriate technology processes or their combinations, but these unit processes can be used as pretreatment unit in future projects. The stabilization reservoir system is a concept that can be structured of combinations of various pretreatment processes. An important consideration in the design of a reservoir system is the location of the pretreatment unit. The reservoir is always located in the vicinity of the irrigated farmland, since the area it occupied is large and its use for treatment and supply of irrigation water can only be done in locations outside city limits, where the value of land is lower. The pretreatment unit can be located either inside the city limits or next to the reservoir. The location of the pretreatment unit is related to institutional issues and cooperation between the city and the farmer's organization. When located inside the city, the pretreatment unit is usually operated by the municipal utility and when located next to the reservoir it may be operated by the utility or, in many cases, by the farmers' organization. When the pretreatment unit is located inside the city, it is usually necessary to base it on a process that occupies a small size of land. In this respect unit processes such as UASB, Anaerobic Filters and CEPT are advantageous since they occupy small land areas. In the reservoir systems of Nazareth, Naan and Og, the pretreatment units are located next to the reservoirs, where availability of land is not restrictive and therefore all of them are lagoons based systems. But there are many reservoir systems in Israel in which the pretreatment units are located inside the cities, at a large distance from the reservoirs. The systems of the city of Haifa and of the western basin of Jerusalem can be mentioned as such examples. In each of these cities the pretreatment is an activated sludge plant located inside the respective city limits and operated by the municipal water and sanitation utility. The effluents of these pretreatment units are conveyed to the reservoirs located outside the cities, next to the farmlands.

The combination of CEPT as a pretreatment unit and a stabilization reservoir can be advantageous from the standpoint of resistance to low temperatures and can be used in regions with very low temperatures during the cold season. The CEPT unit performs well in low temperatures, and the CEPT effluent can be stored in a reservoir without causing nuisance until temperatures rise to the level that enables resumption of bacterial activity.

The effluent of stabilization reservoirs is usually used for irrigation, and in Israel, the irrigation method used is, for the most part, drip irrigation. But any other irrigation method can of course be used. The combined Pretreatment – Stabilization Reservoir – irrigation reuse system is a simple, economic wastewater reuse (and disposal) method which does not require high skills for operation and maintenance. The required pretreatment of the wastewater before it is stored in the reservoir can be achieved using a variety of treatment process, among them, all the appropriate technology processes. If simple pretreatment processes are selected, the entire pretreatment, reservoir and irrigation system can be considered an appropriate technology system.

The quality of the effluent of a reservoir system depends on the system's design. If the effluent is used for restricted irrigation, the effluent quality does not need to be very high and the system is designed accordingly. Usually a single continuous flow reservoir is used to produce effluent for restricted irrigation. The annual mean Total BOD removal percentage of such a system is 70–85% and the annual mean TSS removal is 40–80%. The total coliforms removal in continuous flow reservoirs is only one order of magnitude. If the effluent is used for unrestricted irrigation it needs to be of high quality and, most importantly, contain a low count of pathogenic organism. This can be achieved by the use of two or more reservoirs system in which the reservoirs are operated under a sequential batch mode. Such systems can achieve effluents of a very high quality with BOD, COD and TSS removals of 85–95% each, continuously during the irrigation season. They produce an effluent with a BOD_5 concentration of 5–10 mg/l and a TSS concentration of 5–20 mg/l. Moreover, the faecal coliforms removals in such systems are very high, producing effluents with faecal coliforms levels of 0–100 MPN/100 ml. This means that sequential batch reservoir systems perform disinfection without the use of chemicals. In addition, the stabilization reservoirs completely remove helminth eggs and nematodes without difficulty. The effluents of sequential batch reactors are adequate for unrestricted irrigation.

When appropriate technology processes are used for pretreatment, the investment cost in the system of pretreatment followed by a Stabilization Resrevior is in the range of 30–50 US$/Capita and the O&M cost is in the range 0.2–0.4 US$/Yr/Capita.

1.8.9 UASB followed by anaerobic filter followed by dissolved air flotation followed by membrane filtration

All the combined processes which were presented above are processes used in full scale plants or processes which have been studies in pilot scale plants. The process discussed in this section, RMS followed by

UASB followed by Anaerobic Filter followed by Dissolved Air Flotation followed by Membrane Filtration, has most probably not yet been applied in a full scale plant. It cannot be considered an appropriate technology process because the membrane filtration units are not based on an appropriate technology process; however the entire section of the plant preceding the membrane filtration is an appropriate technology section. This process is presented here to show that it is possible to combine appropriate technology units with the cutting edge existing technology to obtain an effluent of the highest possible quality. By using combined appropriate technology units it is possible to produce an effluent adequate to be fed to a membrane filtration unit. A schematic process flow diagram is presented in Figure 1.79. The effluent of this process would be of very low content of BOD and SST and practically void of Fecal Coliforms. An effluent of such quality can be reused in industrial plants. The section of the membrane filtration may in fact be installed in an industrial plant and operated by the industry itself.

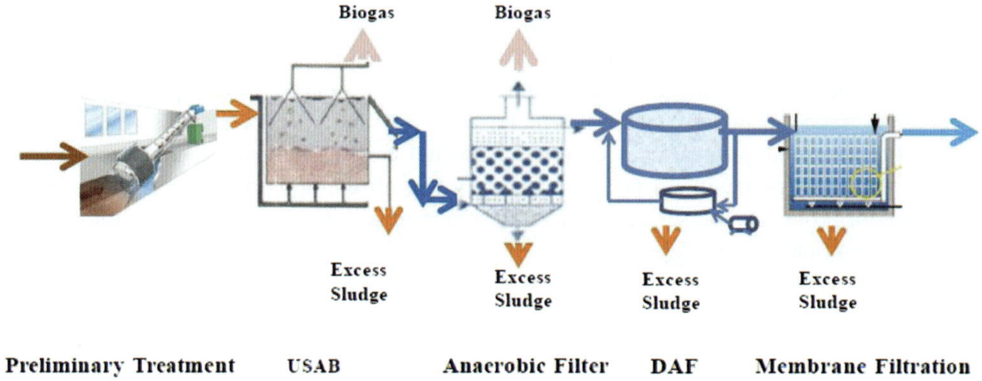

Fig 1.79 A schematic process flow diagram of UASB followed by anaerobic filter followed by DAF followed by membrane filtration

The section preceding the membrane filtration unit would not be costly in investment and in O&M, nor would it be difficult to operate. The membrane filtration unit would add cost and its setup will depend on the required effluent quality, which will determine if one or a combination of Micro, Ultra or Nano filtration units would need to be used. The membrane filtration units are more complex to operate, but not too complex. The process is not adequate for small cities but utilities of medium and large cities would have no problem to operate it. Information regarding cost data and operation problems is unavailable because this process has not yet been in use.

1.9 ADDITIONAL POTENTIAL COMBINED PROCESSES OF APPROPRIATE TECHNOLOGY

1.9.1 Introduction

Several combined unit processes which are in use for treatment of municipal wastewater were presented in Section 1.8. In this section, additional potential combined unit processes of appropriate technology which can be used for treatment of municipal wastes are presented. The number of possible combined processes is large and not all of them are presented here, only some which seem to have the potential to be applied in practice, that is, may be adequate for specific conditions of certain projects. Part of them are already existing and functioning as full scale plants, and the others are presented as ideas and proposals for use. The processes presented are not discusses here in detail but rather presented for the purpose of providing ideas and information regarding the range of combined processes which may be applied. Rotating Micro Screens (RMS) is proposed as part of the pretreatment unit of all the processes.

Before discussing the additional combined appropriate technology processes we would like to mention some combined processes which are in use and cannot be considered appropriate technology processes. Several combined processes which are considered less costly and simpler than conventional processes are being used in Latin America, mostly in Brazil but also in other countries such as Colombia and Bolivia. These processes are a combination of an appropriate technology unit process followed by a conventional treatment process for polishing the effluent of the appropriate technology process. The use of a complex process as a polishing step of an effluent of an appropriate technology process transforms the entire treatment plant to a complex plant and it would be advisable to avoid such an approach. For example, many plants in Brazil use activated sludge units or trickling filters units as polishing units of a UASB reactor effluent. The Rio Frio plant in Bucaramanga, Colombia, which was originally a UASB followed by lagoons plant is also being refurbished to replace the lagoons by an activated sludge unit, so the process in this plant will be UASB followed by Activated sludge (replacing the UASB followed by lagoons plant). The reason for doing this change in the Rio Frio plant is that the raw sewage flow to the plant needs to be increased but there is no area available to increase the stabilization lagoons unit. In El alto, Bolivia, a conventional stabilization lagoons plant, consisting of anaerobic lagoons followed by facultative lagoons followed by maturation lagoons, was refurbished so that its process is now anaerobic lagoons followed by trickling filters. Such combinations do not form appropriate technology processes. The conventional treatment units which are used for polishing (activated sludge and trickling filters) are very close in size and investments to activated sludge and trickling filters which would have been constructed as the sole treatment unit since the design of these polishing units is based to a large extent on the flow of the liquid stream they have to handle (the hydraulic design is based on the flow and defines to a large extent the size of the structures), and the UASB effluent flow that is fed to them (for polishing) is identical to the flow of the raw wastewater.

A plant based on UASB followed by activated sludge is perhaps a little less costly in investment than activated sludge alone, and the activated sludge unit of the combined plant would consume less energy for aeration than the only activated sludge plant, but the difference is not large. On the other hand, running a plant composed of UASB followed by activated sludge is as complicated as running an activated sludge plant or perhaps even more. The same arguments apply for the trickling filter as a polishing unit of a UASB unit.

A satellite photo of the Gama plant in Brasilia, Brazil, which is a UASB followed by Activated Sludge is shown in Figure 1.80. It treats a flow of 328 l/s and serves a population of about 180,000. It is basically an activated sludge plant in which the primary treatment has been replaced by a UASB process consisting of four units, but in essence it is still an activated sludge plant. The process is not an appropriate technology process and is not recommended as the treatment process of choice if sustainability, simplicity and low cost are required features of the treatment plant.

An additional process in use is UASB followed by aerated lagoons. An example of this process is the main treatment plant of the city of Uberaba, in the state of Minas Gerais, Brazil, which has a population of about 300,000. A satellite view of the Uberaba treatment plant is presented in Figure 1.81. Six UASB units operating in parallel are located at the lower left part of the figure.

The effluent of the UASB units flows into a battery of six aerated lagoons in series; in which fine bubble aeration is the source of oxygen. It can be noted in the figure that the aeration intensity decreases along the lagoons, as the organic matter is consumed. In this case, although the polishing process is aerated lagoons, that is, an aerobic process which consumes energy, the combined process is very close to being an appropriate technology process. It is, as a whole, a simple process to operate since the operation of an aerated lagoons plant is not complicated. The polishing aerated lagoons do not consume a lot of energy since the bulk of the organic matter is removed in the UASB unit and the aerated lagoons need to remove only part of the residual organic matter contained in the UASB effluent. In addition, it is possible to capture the biogas released in the UASB process and use it to generate the energy required to operate the aeration system of the aerated lagoons, transforming thereby the entire process to an energy self-sufficient process. This has not yet been done in a full scale plant but we believe it can be done. Another plant whose process is based on UASB followed by aerated lagoons is the Recanto Das Emas plant in Brasilia. It was designed to handle a flow of 246 l/s and serves a population of about 125,500. It achieves a Total BOD removal of 82% and SST removal of about 67%.

Figure 1.80 The Gama plant in Brasilia, Brazil, based on UASB followed by activated sludge, which does not constitute an appropriate technology process

The difference in the size of the UASB units and the aerated lagoons should be noted. The UASB section of the treatment plant occupies only a small size of the entire plant area, while the aerated lagoons occupy most of the plant area. The hydraulic detention time in a UASB reactor is 4–8 hr while in aerated lagoons it is 2–4 days, and that is the reason for the difference in the area occupied by these tow processes. The issue of processes adequate for circumstances of limited available area for a wastewater treatment plant is discusses in one of the following sections.

Figure 1.81 The treatment plant of Uberaba consisting of UASB followed by aerated lagoons

Another process used, although not frequently, is anaerobic lagoons followed by aerated lagoons. A plant of this type is in operation in the city of Nazareth, in Israel (as pre-treatment to a stabilization reservoir) and is presented in Figure 1.82. In this case, as in the case of UASB followed by aerated lagoons, although the polishing process is an aerobic process (aerated lagoons), the combined process is very close to being an appropriate technology process. It is, as a whole, a simple process to operate since the operation of an aerated lagoon plant is not complicated. The aerated lagoons acting as a polishing unit do not consume a lot of energy since the bulk of the organic matter is removed in the anaerobic lagoons unit. The process can be upgraded by covering the anaerobic lagoons and capturing the biogas. This can ensure that the plant will not generate odour problems and enables utilizing the biogas released in the anaerobic lagoons to generate the energy required to operate the aeration system of the aerated lagoons, transforming thereby the entire process to an energy self-sufficient process. Capturing the biogas generated by covered anaerobic lagoons has already been done successfully in Santa Cruz Bolivia, so there is no doubt that it can be used for generating electricity to operate the aeration system of the aerated lagoons. Preliminary estimates indicate that the biogas captured in the covered anaerobic lagoons would be more than sufficient to supply all the energy required for operating the aerated lagoons, however, the economic feasibility of such a system needs to be studied on a case by case basis and depends on energy cost and energy policy, which are country specific.

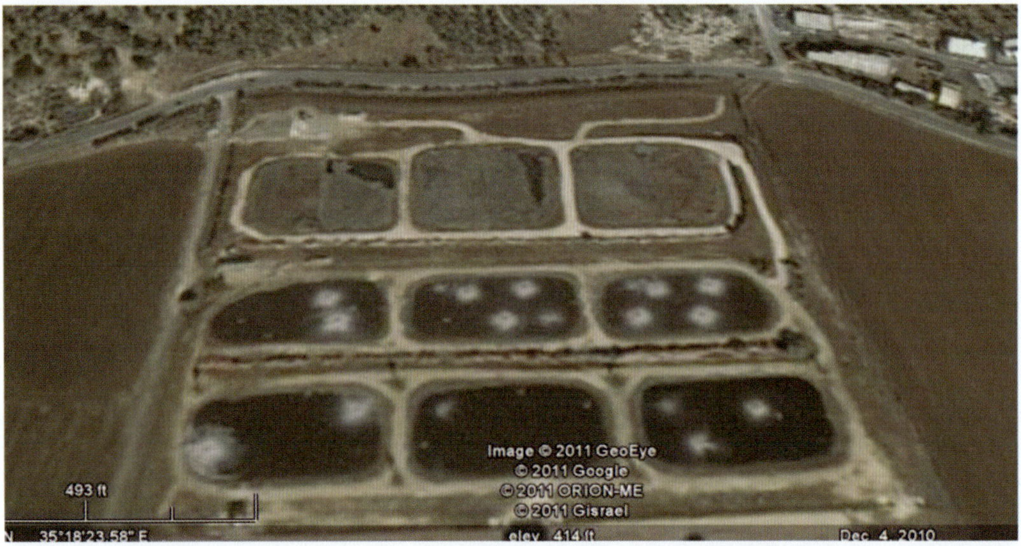

Figure 1.82 The pre-treatment unit of the wastewater treatment plant of Nazareth consisting of anaerobic lagoons followed by aerated lagoons

The system of anaerobic lagoons followed by aerated lagoons is quite efficient in removing organic matter. A full scale plant of this type serving a population of about 250,000 was operated in Santiago, Chile, and was used for investigation. This was the Santiago Poniente plant (which does not exist anymore since the site was used to construct the La Farfana Activated sludge treatment plant which serves most of Santiago). The Santiago poniente plant had a system of anaerobic lagoons followed by aerated lagoons whose effluent was discharged to a stabilization reservoir. The combined process of anaerobic lagoon followed by aerated lagoons achieved 90% removal of soluble BOD and 60% removal of Total BOD. That means that most of the BOD was removed from the liquid phase. If the effluent of such an aerated lagoon is deposited in a settling pond or a series of settling ponds (similar to what is done in Uberaba) the suspended matter, which is basically the bacterial biomass developed in the first aerated lagoon, would have settled out and undergo anaerobic decomposition. The system as a whole would achieve a Total BOD removal of about 90% and the effluent of the final settling pond would be a high quality effluent.

1.9.2 Additional potential combined processes

Additional combined appropriate technology processes for treatment of municipal wastewater include:

- *Preliminary Treatment followed by UASB followed by Sand Filtration* and if required, Disinfection. The schematic flow diagram of this process is presented in Figure 1.83. This is a simple process which occupies a small area. The sand filter will capture and remove most of the solids contained in the UASB effluent, so the final effluent will be clear. The removal of total of BOD will be medium to high and the removal of total Suspended Solids will be high. Coagulants and/or Flocculants need to be injected to the UASB effluent to achieve an efficient filtration. The biogas generated in the UASB reactor can be collected and put to use, if economically feasible.

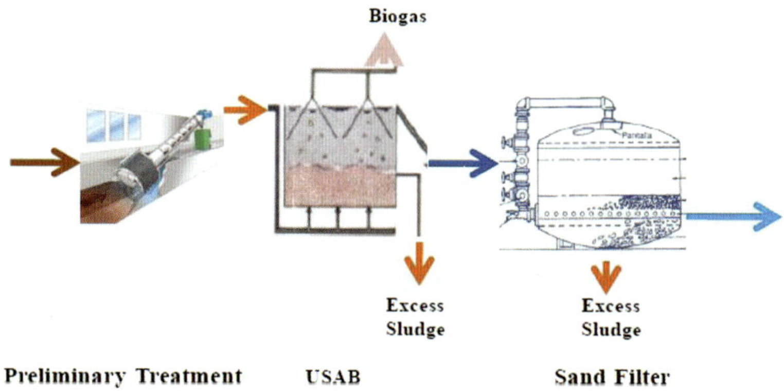

Figure 1.83 Schematic flow diagram of UASB followed by sand filtration

– *Preliminary Treatment followed by UASB followed by Overland Flow* and if required, Disinfection. The schematic flow diagram of this process is presented in Figure 1.84. This is a simple process which occupies a large area because of the overland flow unit. The removal of total BOD and total suspended solids is expected to be high in this process and its effluent quality will be good. The Sao Sebastiao treatment plant in Brasilia is based on this process flow diagram, treating 226 l/s and serving a population of about 80,000. The Overland Flow effluent in the Sao Sebastiao plant undergoes an additional polishing step by a maturation lagoon as shown below, so this process is only part of the Sao Sebastiao plant process, however, the fact is that a full scale plant based on UASB followed by Overland Flow is already under operation. The biogas generated in the UASB reactor can be collected and put to use, if economically feasible.

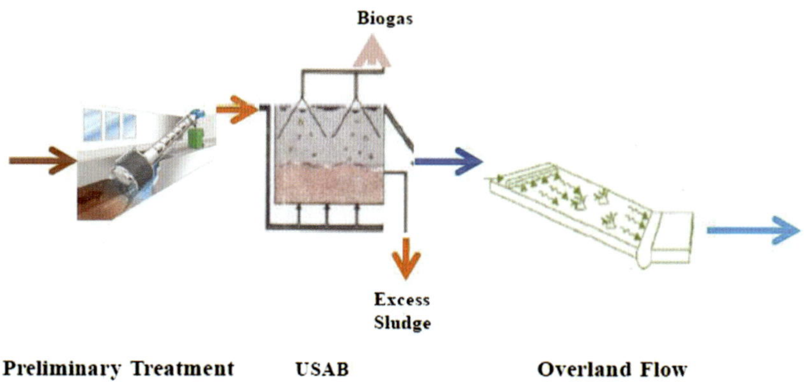

Figure 1.84 Schematic flow diagram of UASB followed by overland flow

– *Preliminary Treatment followed by Anaerobic Lagoons followed by Overland Flow* and if required, Disinfection. The schematic flow diagram of this process is presented in Figure 1.85. This is a simple process which occupies a large area. It does not contain mechanical equipment except simple equipment in the preliminary treatment unit, and it practically does not generate excess sludge. The removal of total BOD and total suspended solids may not be high and the effluent quality may not be the highest, but the effluent will be of decent quality and if using a very simple, low cost and easy to operate process is the objective, this process is a good candidate. The anaerobic lagoon can be a covered lagoon so the process will not cause nuisance, and the collected biogas can be put to use, if economically feasible.

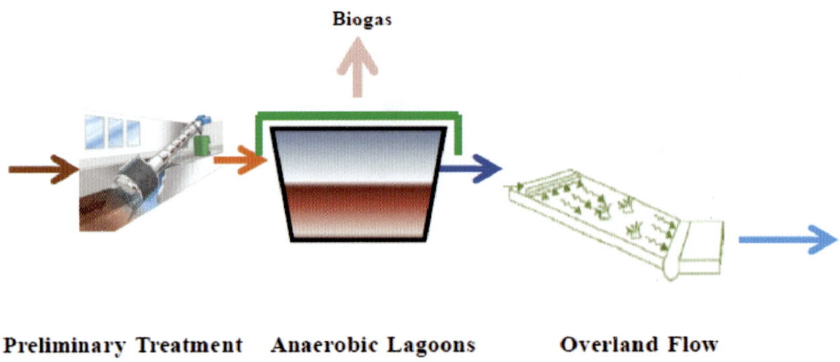

Figure 1.85 Schematic flow diagram of covered anaerobic lagoons followed by overland flow

– *Preliminary Treatment followed by Anaerobic Filter followed by Overland Flow* and if required, Disinfection. The schematic flow diagram of this process is presented in Figure 1.86. This is a simple process which occupies a large area. It does not contain mechanical equipment except simple equipment in the preliminary treatment unit. The preliminary treatment often used prior to an anaerobic filter is a septic tank, and this can be a good solution for treatment plants of small communities. The removal of total BOD and total suspended solids in this process will be reasonably high and the effluent quality will be high. The biogas generated in the anaerobic filter can be collected and put to use, if economically feasible.

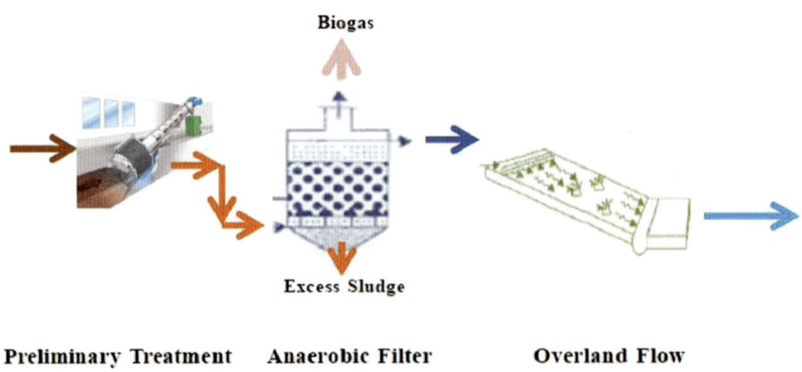

Figure 1.86 Schematic flow diagram of anaerobic filter followed by overland flow

– *Preliminary Treatment followed by UASB followed by Constructed Wetlands* and if required, Disinfection. The schematic flow diagram of this process is presented in Figure 1.87. This is a simple process which occupies a large area due to the constructed wetlands unit. It does not contain mechanical equipment except simple equipment in the preliminary treatment unit. According to

Chernicharo (2007) this process achieves medium level removals of total COD and total suspended solids and achieves good removal of phosphorous and partial removal of nitrogen. The biogas generated in the UASB reactor can be collected and put to use, if economically feasible.

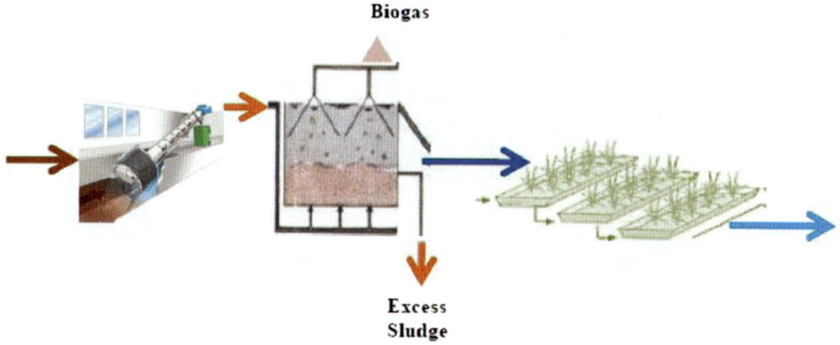

Figure 1.87 Schematic flow diagram of UASB followed by constructed wetlands

– *Preliminary Treatment followed by UASB followed by Lagoons followed by Overland Flow* and if required, Disinfection. The schematic flow diagram of this process is presented in Figure 1.88. This is a simple process which occupies a very large area because of the lagoons and the overland flow units. It does not contain mechanical equipment except simple equipment in the preliminary treatment unit. This process is used in two plants in Brasilia, Brazil: the Alagado plant and the Santa Maria plant. Each of them is designed to handle an average flow of 154 l/s and serve a population of 85,000. These plants achieve a Total BOD removal of about 85% and a total suspended solids removal of about 80%. The biogas generated in the UASB reactor can be collected and put to use, if economically feasible. Mixers can be installed in the lagoons to improve performance and reduce their area. A satellite photo of the Alagado plant is presented in Figure 1.89. In addition to the preliminary treatment installations the Alagado plant consists of 4 UASB reactors located at the bottom of the photo. The effluent of the UASB reactors flows into a series of 13 high rate lagoons located at the left hand side of the photo and the effluent of the lagoons flows into the overland flow area at the right hand side of the photo. The Santa Maria plant is similar in its layout to the Alagado plant.

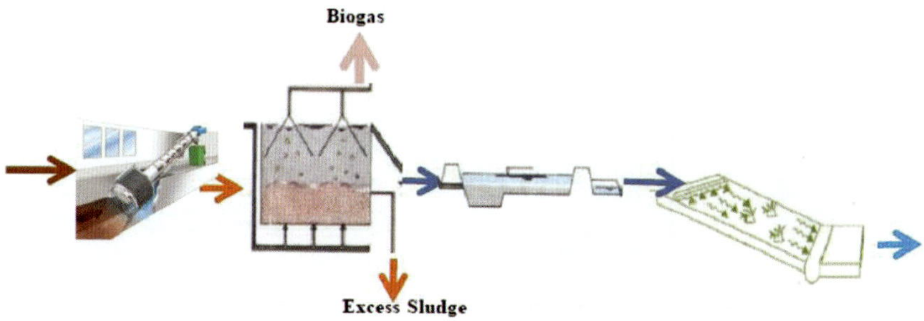

Figure 1.88 Schematic flow diagram of UASB followed by lagoons followed by overland flow

Figure 1.89 A satellite photo of the alagado treatment plant in Brasilia, Brazil, consisting of UASB followed by lagoons followed by overland flow

– *Preliminary Treatment followed by UASB followed by Overland Flow followed by Lagoons* and if required, Disinfection. The schematic flow diagram of this process is presented in Figure 1.90. This is a simple process, similar to the previous processes with the difference that the overland flow unit comes before the lagoons unit. This process occupies a very large area because of the lagoons and the overland flow units. It does not contain mechanical equipment except simple equipment in the preliminary treatment unit. This process is used in the Sao Sebastiao plant in Brasilia, Brazil designed to handle an average flow of 226 l/s and serve a population of about 80,000. This plant achieves a Total BOD removal of about 90% and a total suspended solids removal of about 85%, yielding a good quality effluent. The biogas generated in the UASB reactor can be collected and put to use, if economically feasible. Mixers can be installed in the lagoons to improve performance and reduce their area. A satellite photo of the Sao Sebastiao plant is presented in Figure 1.91. In addition to the preliminary treatment installations the plant consists of 4 UASB reactors whose effluent flows into the overland flow area from which it is collected and flows into two lagoons.

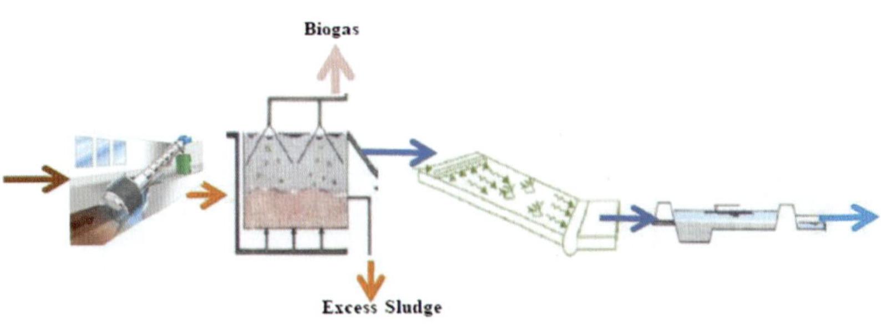

Figure 1.90 Schematic flow diagram of UASB followed by overland flow followed by lagoons

Figure 1.91 Satellite photo of the sao sebastiao treatment plant in Brasilia, Brazil, consisting of UASB followed by overland flow followed by lagoons

- *Septic Tank followed by Anaerobic Filter* and if required, Disinfection. The schematic flow diagram of this process is presented in Figure 1.92. A plant consisting of a decanter-digester (the septic tank) followed by an anaerobic filter constitutes an appropriate plant for treatment of domestic sewage. It combines two reactors in series, the first resistant to variations in the influent and the second efficient in removing dissolved organic matter. Its operation is very simple and does not require continuous presence of operators neither does it require specialized operators. It performs well since startup, is resistant to shock loads and toxic materials and it does not lose efficiency with time. It does not contain mechanical equipment. It occupies a very small area. Since the septic tank is fitted to serve only small communities, this process is basically applicable to small communities. Plants of this type are in operation in Brazil. At temperatures in the range 10–20°C the septic tank achieves 50% removal level of total BOD and even more, so it plays an important role in the removal of organic matter in this process. At higher temperature it can achieve up to 80% total BOD removal. The combined system of a septic tank followed by an anaerobic filter achieves over 80% total BOD removal and over 80% TSS removal. Information on this process is presented by Andrade Neto (1997).
- *Preliminary Treatment followed by Anaerobic Filter followed by Lagoons.* The schematic flow diagram of this process is presented in Figure 1.93. A lagoons system (facultative and maturation lagoons, or just maturation lagoons) can be used as a polishing system of an anaerobic filter effluent. The lagoons can further reduce the BOD and TSS content of the anaerobic filter effluent, and if necessary, the lagoons system can be designed to reduce the content of pathogenic organisms in the effluent. The lagoons system can be equipped with mixers to improve its efficiency. For systems serving small communities, the preliminary treatment can consist of a septic tank. The anaerobic filter followed by lagoons is a simple process. It does not contain mechanical equipment except simple equipment in the preliminary treatment unit. This process occupies a large area because of the lagoons unit. The biogas generated in the anaerobic filter can be collected and put to use, if economically feasible.

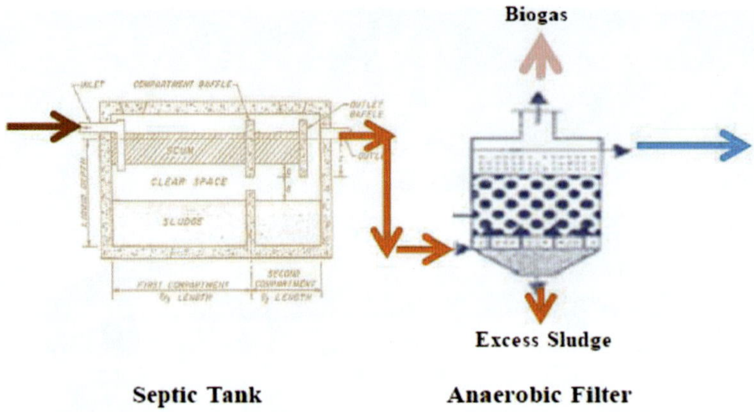

Figure 1.92 Schematic flow diagram of septic tank followed by anaerobic filter

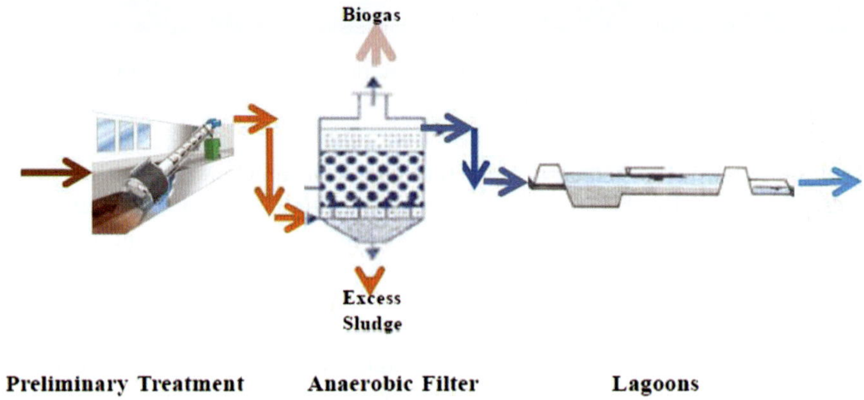

Figure 1.93 Schematic flow diagram of anaerobic filter followed by lagoons

- *Preliminary Treatment followed by Anaerobic Filter followed by Sand Filter* and if required, Disinfection. The flow diagram of this process is presented in Figure 1.94. A Sand filter can be used to polish an anaerobic filter effluent, basically, to reduce suspended solids and with that to reduce also the BOD content of the effluent. This process occupies a very small area. For systems serving small communities, the preliminary treatment can consist of a septic tank. It is a simple process and it can produce a good quality effluent. The biogas generated in the anaerobic Filter can be collected and put to use, if economically feasible.
- *Preliminary Treatment followed by Anaerobic Filter followed by DAF* and if required, Disinfection. The schematic flow diagram of this process is presented in Figure 1.95. A Dissolved Air Flotation (DAF) unit can be used to polish an anaerobic filter effluent, basically, to reduce suspended solids and with that to reduce also the BOD content of the effluent. This process occupies a very small area. For systems serving small communities, the preliminary treatment can consist of a septic tank. It is a simple process and it can produce a good quality effluent. The biogas generated in the anaerobic filter can be collected and put to use, if economically feasible.

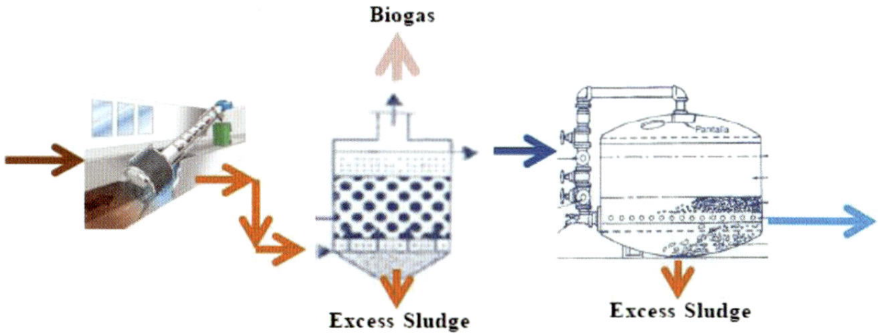

Figure 1.94 A schematic flow diagram of anaerobic filter followed by sand filter

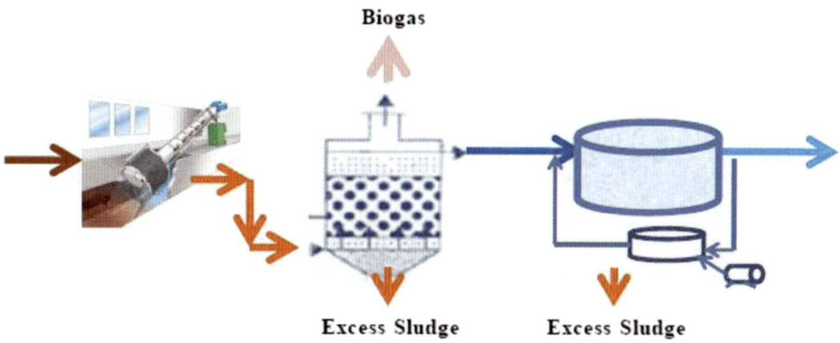

Figure 1.95 Schematic flow diagram of anaerobic filter followed by DAF

- *Preliminary Treatment followed by Anaerobic Filter followed by Constructed Wetlands* and if required, Disinfection. The schematic flow diagram of this process is presented in Figure 1.96. A constructed wetlands unit can be used to polish an anaerobic filter effluent to reduce suspended solids and with that to reduce also the BOD content of the effluent. This process occupies a large area because of the constructed wetlands. For systems serving small communities, the preliminary treatment can consist of a septic tank. It is a simple process and it can produce a good quality effluent. The biogas generated in the anaerobic filter can be collected and put to use, if economically feasible.
- *Septic Tank followed by Constructed Wetlands* and if required, Disinfection. This process can be used only in small communities because of the septic tank which is fitted to serve only such communities. This is the most frequently used process configuration involving constructed wetlands for small communities. The septic tank is an anaerobic reactor which removes a significant part of the organic matter from the raw sewage and the constructed wetlands remove an additional part. The process is simple to operate. It does not contain mechanical equipment and yield a good quality

effluent. A photo of a plant of this type is presented in Figure 1.35. The WSP (2008) publication provides ample information on this process.

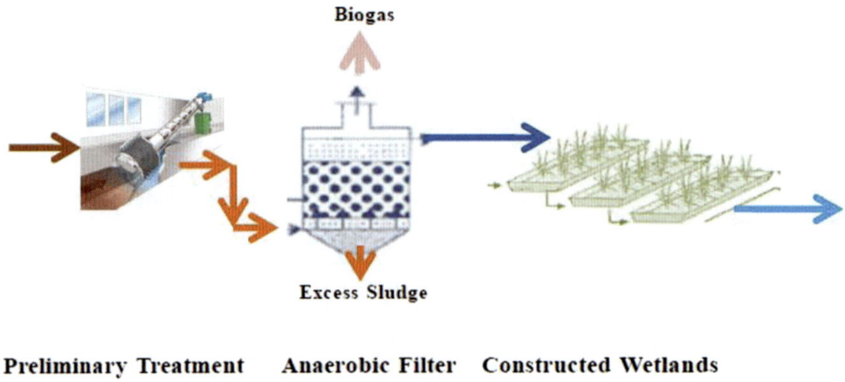

Figure 1.96 Schematic flow diagram of anaerobic filter followed by constructed wetlands

- *Preliminary Treatment followed by Anaerobic Lagoons followed by Anaerobic Filter* and if required, Disinfection. The schematic flow diagram of this process is presented in Figure 1.97. An anaerobic filter unit is used in this process to polish an anaerobic lagoons effluent to remove additional dissolved and suspended organic matter as well as suspended solids. This process occupies a relatively small area. It is a simple process, does not contain mechanical equipment except for a small amount in the preliminary treatment unit, and it can produce a good quality effluent. The anaerobic lagoons can be covered to reduce environmental nuisance and collect the biogas. The biogas generated in the anaerobic lagoons and in the anaerobic filter can be collected and put to use, if economically feasible.

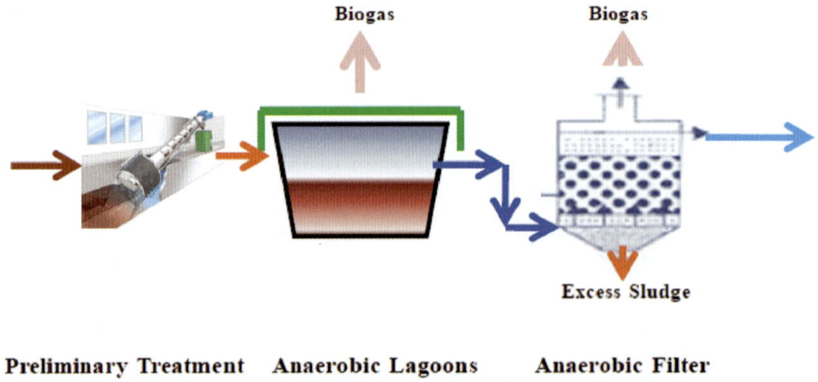

Figure 1.97 Schematic flow diagram of covered anaerobic lagoons followed by anaerobic filter

- *Preliminary Treatment followed by Anaerobic Lagoons followed by Anaerobic Filter followed by Sand Filter* and if required, Disinfection. The schematic flow diagram of this process is presented in Figure 1.98. This is a process similar to the previous one, with a sand filter unit added to improve the quality of the effluent of the anaerobic lagoon followed by an anaerobic filter system. It is still a simple process, does not contain mechanical equipment except for a small amount in the preliminary treatment unit, and it can produce a high quality effluent. The area occupied by this process is similar to that of the previous process and is still small. The anaerobic lagoons can be covered to reduce environmental nuisance and collect the biogas. The biogas generated in the anaerobic lagoons and in the anaerobic filter can be collected and put to use, if economically feasible.

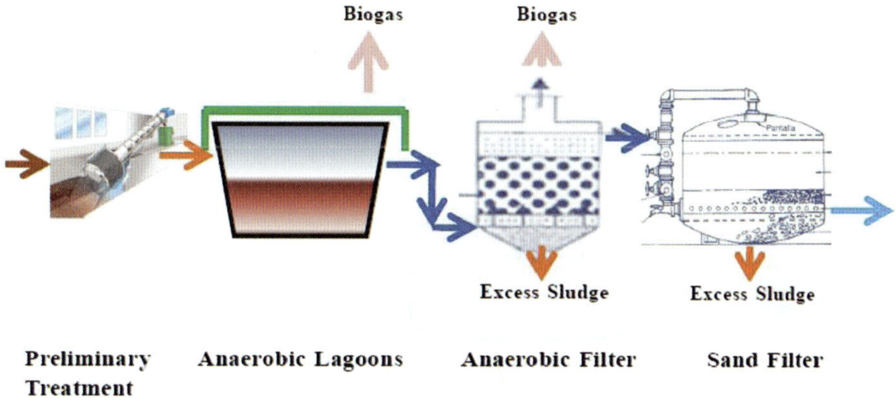

Figure 1.98 Schematic flow diagram of covered anaerobic lagoons followed by anaerobic filter followed by sand filter

- *Preliminary Treatment followed by CEPT followed by DAF* and if required, Disinfection. The schematic flow diagram of this process is presented in Figure 1.99. This is a physicochemical process which, relative to other process presented, contains some mechanical equipment like dosing and flocculation equipment in the CEPT and DAF units and some equipment in the preliminary treatment unit. But as a whole, it is not a complicated process to operate. It occupies a very small area and its great advantage is that as a physicochemical process it is not strongly dependent on variations of temperature and it functions well even in very low temperatures. It can produce a good quality effluent. It generates significant quantities of sludge since the removal mechanism of organic matter in this process is the transformation of the organic matter to settling solids. The sludge needs to be treated and adequately disposed.

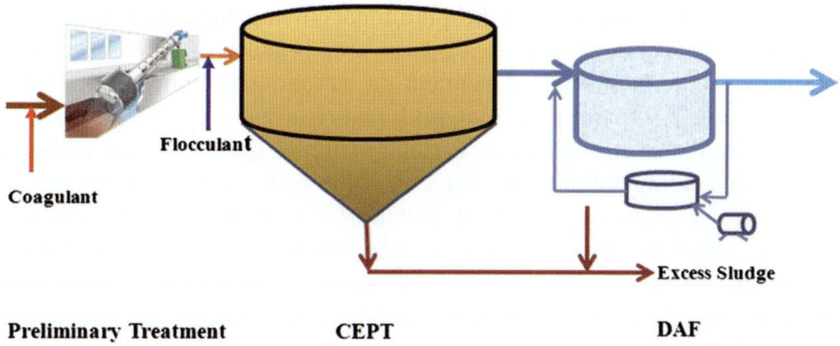

Figure 1.99 Schematic flow diagram of CEPT followed by dissolved air flotation (DAF)

– *Preliminary Treatment followed by CEPT followed by Overland Flow* and if required, Disinfection. The schematic flow diagram of this process is presented in Figure 1.100. In this process the Overland Flow unit is used polishing unit of the CEPT effluent. This process is not recommended for small communities because such communities do not have the capacity to operate a CEPT unit. This is a relatively simple process. It contains some mechanical equipment like dosing and flocculation equipment in the CEPT unit and some equipment in the preliminary treatment unit. But as a whole, it is not a complicated process to operate. It occupies a large area because of the overland flow unit. Its advantage is that it is not strongly dependent on variations of temperature and it functions well even in low temperatures. It can produce a good quality effluent. It generates significant quantities of sludge since the removal mechanism of organic matter in the CEPT unit is the transformation of the organic matter to settling solids. The sludge needs to be treated and adequately disposed.

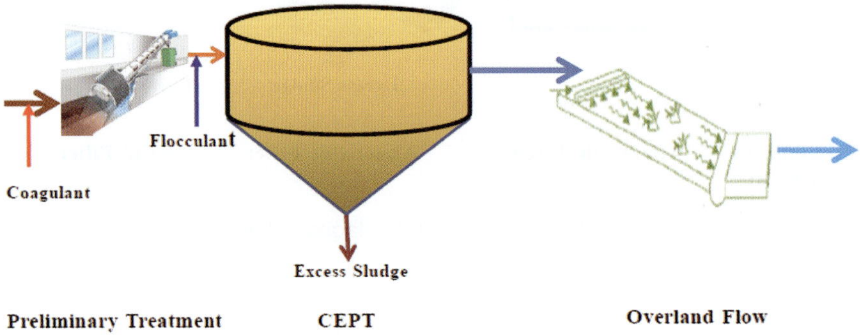

Figure 1.100 Schematic flow diagram of CEPT followed by overland flow

– *Preliminary Treatment followed by CEPT followed by Constructed Wetlands* and if required, Disinfection. The schematic flow diagram of this process is presented in Figure 1.101. In this process the Constructed wetlands unit is used as a polishing unit of the CEPT effluent. This process is not recommended for small communities because such communities do not have the capacity to operate a CEPT unit. However, in larger plants, where the main treatment unit is CEPT, the Constructed Wetlands may serve as a polishing unit of the CEPT effluent. This is a relatively simple process. It contains some mechanical equipment like dosing and flocculation equipment in the CEPT unit and some equipment in the preliminary treatment unit, but as a whole, it is not a complicated process to operate. It occupies a large area because of the Constructed Wetlands unit. Its advantage is that it is not strongly dependent on variations of temperature and it functions well even in low temperatures. It can produce a good quality effluent. It generates significant quantities of sludge since the removal mechanism of organic matter in the CEPT unit is the transformation of the organic matter to settling solids. The sludge needs to be treated and adequately disposed.
– *Preliminary Treatment followed by CEPT followed by Stabilization Lagoons* and if required, Disinfection. The schematic flow diagram of this process is presented in Figure 1.102. In this process the stabilization lagoons unit is used for polishing the CEPT effluent. This is a relatively simple process. It contains some mechanical equipment like dosing and flocculation equipment in the CEPT unit and some equipment in the preliminary treatment unit. But as a whole, it is not a complicated process to operate. It occupies a large area because of the lagoons unit. This unit can be

a complete lagoons system consisting of anaerobic lagoons followed by facultative lagoons followed by maturation lagoons, or a partial lagoons system consisting of facultative and maturation lagoons. The facultative lagoons may be equipped with mixers to improve their performance and reduce their size. It can produce a good quality effluent. It generates significant quantities of sludge since the removal mechanism of organic matter in the CEPT unit is the transformation of the organic matter to settling solids. The sludge needs to be treated and adequately disposed.

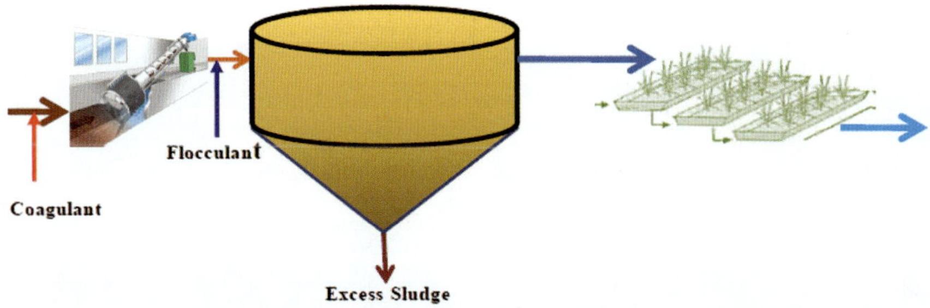

Figure 1.101 Schematic flow diagram of CEPT followed by constructed wetlands

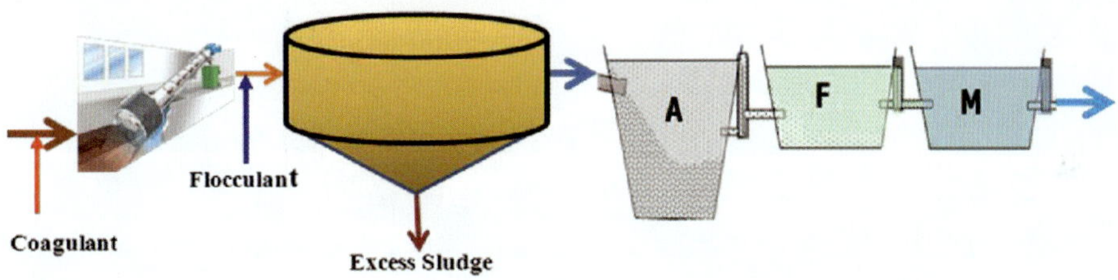

Figure 1.102 Schematic flow diagram of CEPT followed by stabilization lagoons

The CEPT followed by Lagoons process is used in the treatment plant of the city Riviera Sao Lourenco (Tsukamoto, 2002) which is a vacation resort on the Atlantic coast of the state of Sao Paulo, Brazil, not far from the state's capital Sao Paulo. Originally, the wastewater treatment system of the city consisted of a conventional lagoon system with a capacity of handling the wastewater of 32,000 persons. The system has a pretreatment unit, an anaerobic lagoon of 7,200 m^2, three parallel facultative lagoons (12,000 m^2, 17,000 m^2, and 18,000 m^2) and an effluent chlorination system. During the summer season the population of Riviera Sao Lourenco increases significantly because of tourism. That results in increased wastewater flows and organic matter loads. All the lagoons became strongly anaerobic and generated odours which reached the city, 3 km away. The problems were resolved by adding two CEPT units between the preliminary treatment and the anaerobic lagoon. The sedimentation tank of each unit has an area of 180 m^2 and a capacity for treating the wastewater of 40,000 persons, increasing thereby the treatment capacity by 250%. The upgraded plant functioned successfully when the population increased from 7,000 to 65,000 during a period of

days, removing 80%–95% of the total suspended solids, 50%–66% of BOD and 82%–87% Phosphorous. The CEPT unit is operated during the summer season which is the tourism season, and is bypassed during winter, when the population residing in the city is small and the CEPT process is not needed. The advantages of the CEPT unit is that it takes care of high shock loads in an immediate manner since the CEPT process is a physicochemical process which does not require the development and acclimatization of bacterial cultures. The CEPT is also a low investment cost processes. A photo of the Riviera Sao Lourenco wastewater treatment plant is presented in Figure 1.103.

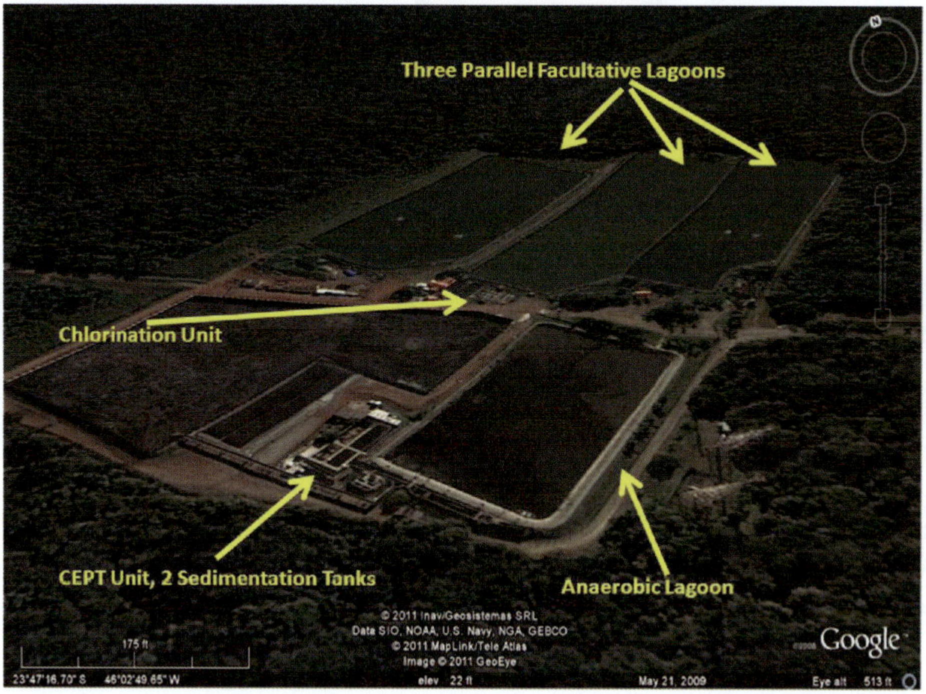

Figure 1.103 The treatment plant of Riviera Sao Lourenco, Sao Paulo State, Brazil, based on CEPT followed by stabilization lagoons

– *Wastewater Treatment in Cities with Large Seasonal Variation in Wastewater Loads: CEPT followed by Additional Treatment Units*. The example of Riviera Sao Lourenco raises the issue of wastewater treatment for cities which experience large variations in wastewater flows and contamination loads, such as cities which are summer vacation resorts, or cities with a seasonally active industrial sector (for instance during a fruit harvesting and processing season). The solution for such cases is a treatment plant consisting of a CEPT unit followed by additional treatment units. The additional treatment units can consist of one or more of the unit processes discussed previously such as one of the anaerobic processes, lagoons, reservoirs, overland flow, constructed wetlands and others. During the low load season, the CEPT unit is bypassed and the additional treatment units have the capacity to handle the loads of the low load season. During the high load season the CEPT unit is also put on stream, and the plant (CEPT followed by the additional units) has the capacity to handle the loads of the high load season. The concept is presented in Figure 1.104. The reason that the CEPT followed by additional units is the adequate process is that in vacation resort cities, the population can grow tenfold within a short period, during the vacation season and holydays. The increase in population can occur during a short time causing sharp peaks in wastewater flows. Biological treatment processes are incapable of handling sharp oscillations in wastewater loads because they needs to develop the biomass capable of handling the peaks and that takes time. The CEPT, which is a physicochemical process, has the capacity to handle the increased loads within a short time. The application of the concept of Figure 1.104 to the case of Riviera Sao Lourenco yielded the flow diagram of the Treatment plant of this city presented in Figure 1.105.

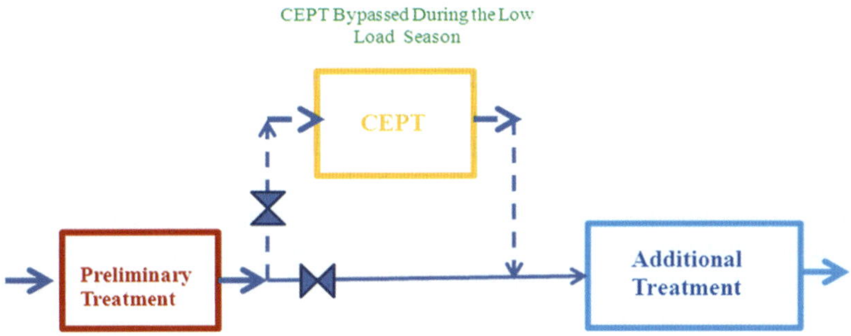

Figure 1.104 A treatment process concept adequate for cities with sharp seasonal variation of wastewater flows or organic matter loads

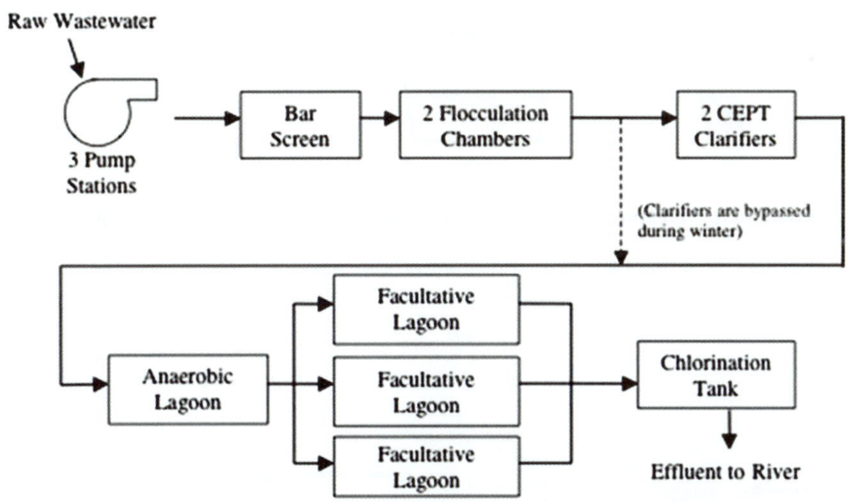

Figure 1.105 The treatment process of Riviera Sao Lourenco (*Source*: Prof. Harleman, MIT, 2004)

The concept presented in Figure 1.104 can also be used in treatment plants located in zones with a very cold low temperature season (and in this case, without sharp seasonal variations of wastewater loads). During the high and medium temperature seasons the CEPT unit is bypassed and the additional treatment units, which are basically biological treatment units, have the capacity to treat the wastewater. During the low temperature season the capacity of the biological treatment units drops and then the CEPT unit, which functions well even at low temperatures, is put on stream providing an additional capacity.

– *Preliminary Treatment followed by one of the Anaerobic Units (Anaerobic Lagoons or UASB or Anaerobic Filter) followed by Infiltration.* The schematic flow diagrams of these processes are presented in the following figures: Figure 1.106 shows the flow diagram of Anaerobic Lagoons followed by infiltration basins, Figure 1.107 shows the flow diagram of UASB followed by infiltration basins and Figure 1.108 shows the flow diagram of an Anaerobic Filter followed by infiltration basins. These three processes are simple processes. The area occupied by each of them is large because of the infiltration basins that require a large area. However, the actual size of the infiltration basins depends on the soil type and may vary in accordance with the location of the plant. If soil conditions are adequate for a rapid infiltration rate, this process can be applied also to large plants. Application of these processes requires thorough familiarity with the groundwater conditions and they can be implemented only when they do not present a risk of contaminating groundwater used for domestic supply. These three processes do not generate effluent on the surface so they do not contaminate surface water bodies. The anaerobic lagoons can be covered to reduce environmental nuisance and collect the biogas. The biogas generated in any of the anaerobic processes can be collected and put to use, if economically feasible. The Torto plant in Brasilia, Brazil, is an example to the process of UASB followed by infiltration. It serves a population of 2,500 and has a capacity of 6 l/s.

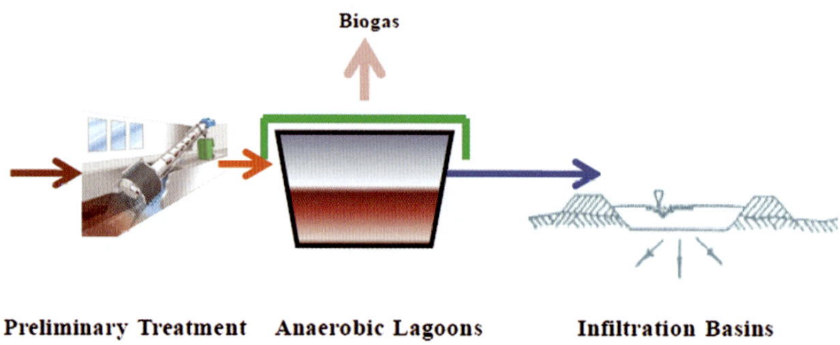

Figure 1.106 Schematic flow diagram of covered anaerobic lagoons followed by infiltration basins

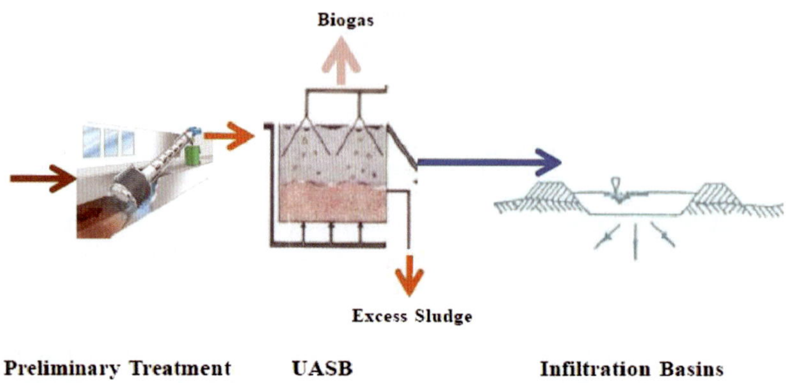

Figure 1.107 Schematic flow diagram of UASB followed by infiltration basins

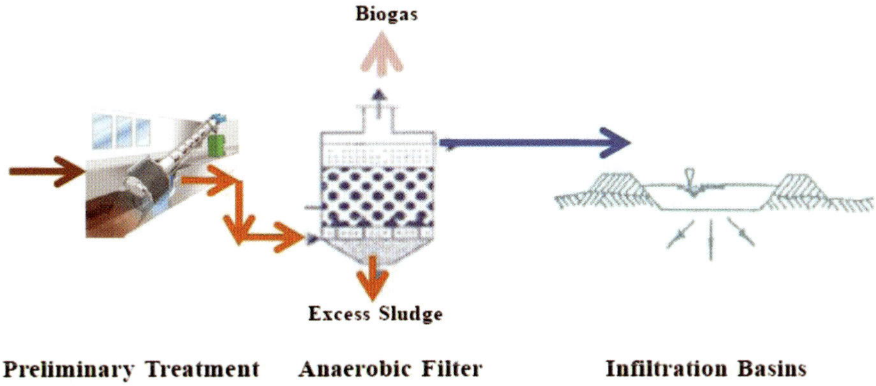

Figure 1.108 Schematic flow diagram of anaerobic filter followed by infiltration basins

– *Preliminary Treatment followed by Anaerobic Lagoons followed by UASB* and if required, Disinfection. The schematic flow diagram of this process is presented in Figure 1.109. A UASB unit is used in this process to polish an anaerobic lagoons effluent in order to remove some more dissolved and suspended organic matter as well as suspended solids. The use of UASB reactor to polish an effluent of a preceding unit has already been done, as referenced in the following paragraph. This process occupies a relatively small area. It is a simple process, does not contain mechanical equipment except for a small amount in the preliminary treatment unit, and it can produce a fair quality effluent. The anaerobic lagoons can be covered to reduce environmental nuisance and collect the biogas. The biogas generated in the anaerobic lagoons and in the UASB reactor can be collected and put to use, if economically feasible.

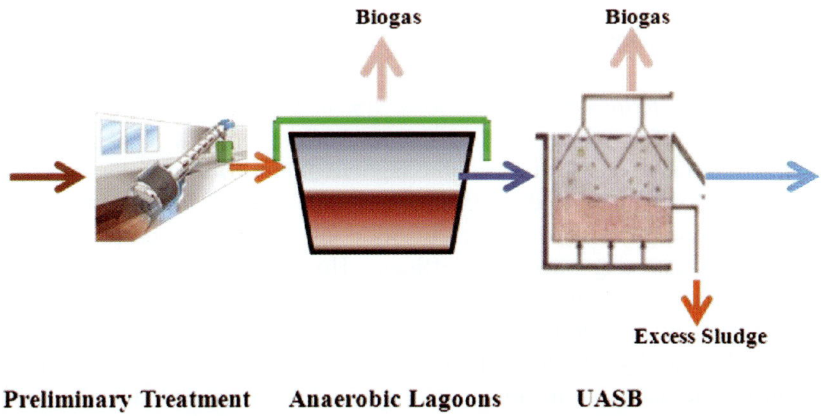

Figure 1.109 Schematic flow diagram of covered anaerobic lagoons followed by USAB

– *Preliminary Treatment followed by CEPT followed by UASB* and if required, Disinfection. The schematic flow diagram of this process is presented in Figure 1.110. A UASB unit is used in this case to polish the effluent of a CEPT unit to remove residual dissolved and suspended organic matter as well as suspended solids. This process occupies a very small area. It is a simple process, does not contain a lot of mechanical equipment except for a small amount in the preliminary treatment unit and in the CEPT unit, and it can produce a good quality effluent. The biogas generated in the UASB reactor can be collected and put to use, if economically feasible. Aiyuk *et al.* (2004) conducted a laboratory investigation of this process. They used an additional unit of a zeolite ion exchange column to remove ammonia from the UASB effluent. They used $FeCl_3$ as coagulant and an anionic organic flocculant. The CEPT unit removed 73% of Total COD and 85% of Total suspended solids. The integrated system removed over 90% of Total COD and 88% of Total suspended solids, as well as 99% of nitrogen (by the zeolite column) and 94% of phosphorous. The investigators concluded that the simple design and relatively low operating costs, due to low costs of added chemicals and low energy input, combined with the excellent treatment performance, means that this system can be used as a novel domestic wastewater treatment system for developing countries. Therefore, they called the system a Low Investment Sewage Treatment (LIST) system.

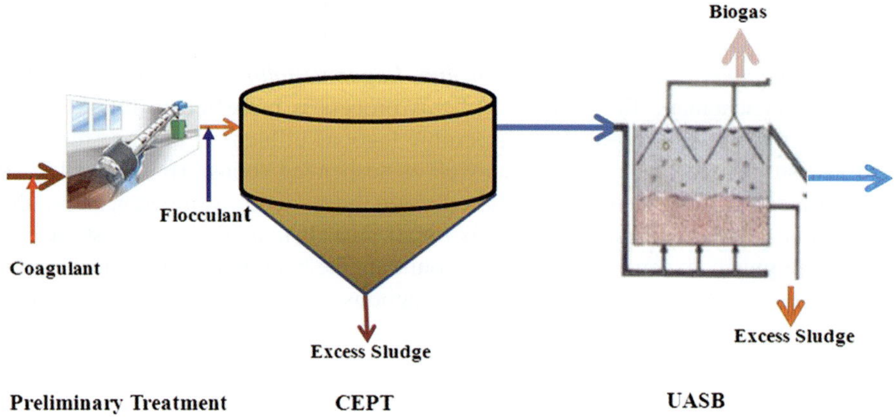

Figure 1.110 Schematic flow diagram of CEPT followed by USAB

- *Preliminary Treatment followed by a Lagoons System followed by Constructed Wetlands* and if required, Disinfection. The schematic flow diagram of this process is presented in Figure 1.111. The lagoons system may consist of anaerobic lagoons followed by facultative lagoons, or only of facultative lagoons. The constructed wetlands act as a polishing unit replacing maturation lagoons. This process is very simple to operate, practically does not include equipment, does not consume energy and can generate a good quality effluent. However, it occupies a very large area. The wastewater treatment plant of the town of Crowley, southern Louisiana, is shown in Figure 1.112. It is basically a facultative lagoons plant followed by constructed wetlands. Details about this plant are provided by Tillman Lyle (1994). The population of Crowley is 18,000. When the mechanical wastewater treatment plant of the town, which had functioned sporadically and unreliably for several years, needed an overhaul, the mayor decided to take a different approach and build a new plant based on natural processes. The facultative lagoons are equipped with mixers that improve their performance and the final effluent undergoes UV disinfection before it's discharged into a nearby small river (bayou). The detention time in the facultative lagoons is 50 to 60 days and in the constructed wetlands system, a few days more. The Crowley wastewater treatment plant is described as a beautiful tranquil place, a rich landscape that recalls the nearby natural marshes. There are no odours in the plant site other than those of a productive natural marsh.

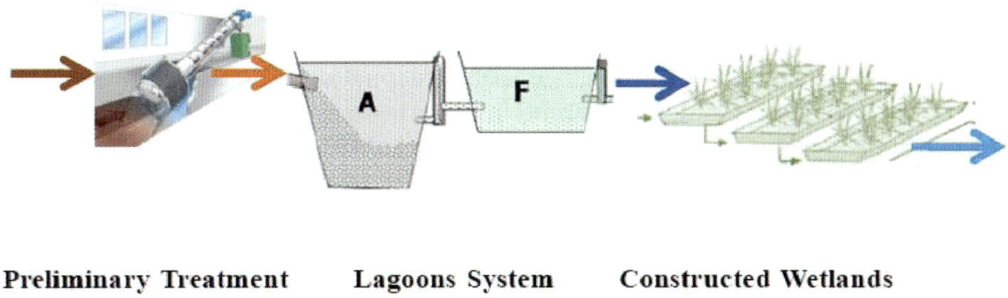

Figure 1.111 Schematic flow diagram of lagoons system followed by constructed wetlands

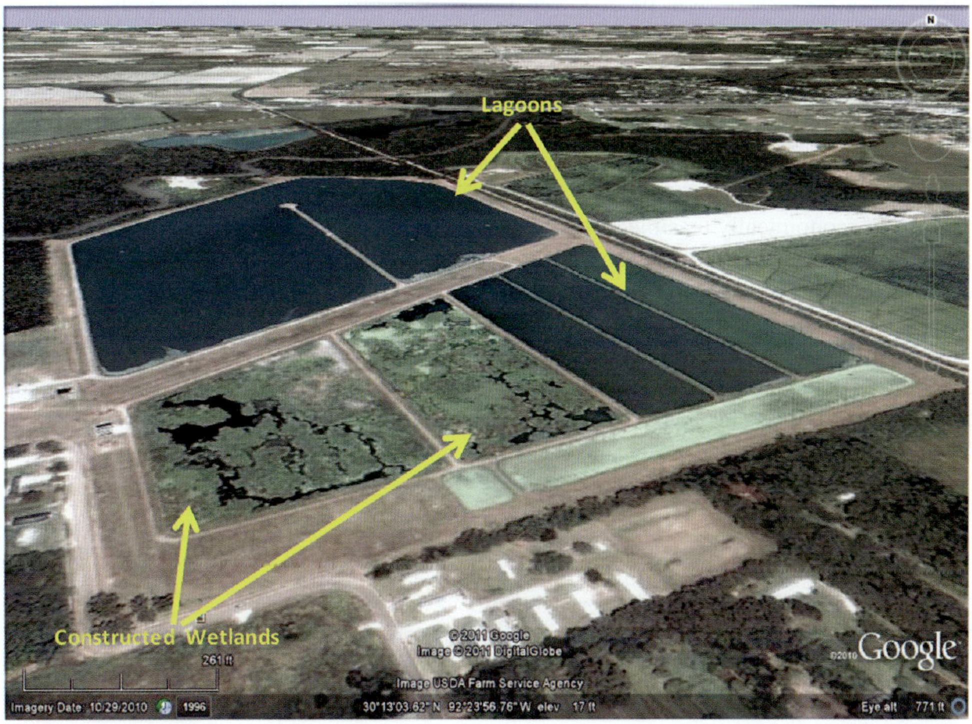

Figure 1.112 The treatment plant of Crowley, Louisiana, based on facultative lagoons followed by constructed wetlands

– *Preliminary Treatment followed by one of the Anaerobic Units (Anaerobic Lagoons or UASB or Anaerobic Filter) followed by Aerated Lagoons,* and if required, by Disinfection. The schematic flow diagrams of these processes are presented in the following figures: Figure 1.113 shows the flow diagram of Anaerobic Lagoons followed by Aerated Lagoons; Figure 1.114 shows the flow diagram of UASB followed by Aerated Lagoons and Figure 1.115 shows the flow diagram of an Anaerobic Filter followed by Aerated Lagoons. As explained in a previous section, when an aerated lagoon is used for polishing of the effluent of a preceding unit process, it can be considered as an appropriate treatment technology unit process since it does not consume a lot of energy and its operation is simple. The three processes contain some equipment which is the aeration equipment in the aerated lagoons, but this is simple equipment. The area occupied by each one of the three processes is of medium size. The anaerobic unit of each process occupied a small area, but the aerated lagoons occupy a larger area, since the detention time in these lagoons is 2–4 days. In the flow diagrams a series of two aerated lagoons is presented, the first one with more intensive aeration and the second one with partial aeration. The use of two aerated lagoons in series is a sound approach. In the first one, the residual organic matter is decomposed and in the second one, which is a facultative lagoon, the biomass created in the aerated lagoons settles down to the bottom of the lagoon and undergoes an anaerobic decomposition process, so the aerated lagoons do not produce a lot of sludge and do not need to be cleaned frequently. Each of the three processes can produce a good quality effluent. Although the use of two aerated lagoons in series is beneficial, the second lagoon can be omitted, but in such a case the effluent will be of lower quality. It is also recommended to install in the aerated lagoons mixers since the combination of mixers and aerators produces better results. In each of the flow diagram figures of the three processes the aeration system is portrayed as consisting of floating surface aerators, but it can also be a fine bubble aeration system. Today there are very efficient fine bubble aeration systems which can be employed in the aerated lagoons. In the process which includes anaerobic lagoons, these lagoons can be covered to reduce environmental nuisance and collect the biogas. The biogas generated in each of the three anaerobic processes can be collected and put to use, if economically feasible. Preliminary estimates indicate that the biogas captured in each of the anaerobic processes would be more than sufficient to supply all the energy required for operating the aerated lagoons, however, the economic feasibility of such a system needs to be studied on a case by case basis and depends on local energy cost and energy policy, which are country specific. If the captured biogas can be used in an economically feasible fashion to generate the energy required to operate the aeration system of the aerated lagoons, the entire process becomes an energy self-sufficient process. This has not yet been done in a full scale plant but we believe it can be done.

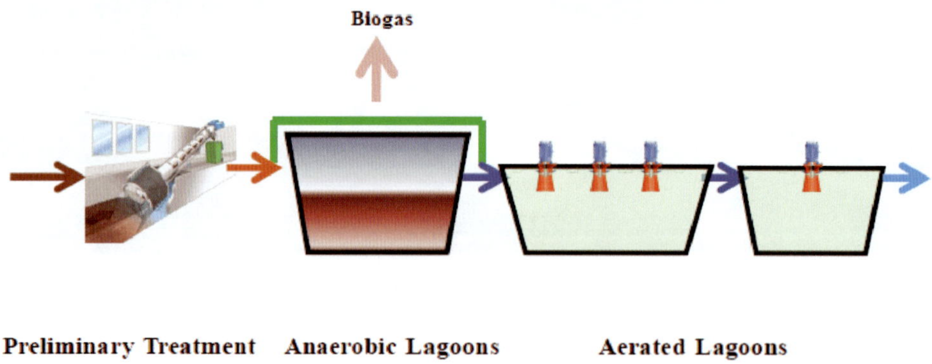

Figure 1.113 Schematic flow diagram of covered anaerobic lagoons followed by aerated lagoons

The process of anaerobic lagoons followed by aerated lagoons is not a new process. The photo of such a treatment plant is presented in Figure 1.82 which shows part of the treatment plant of the city of Nazareth. As seen in the figure, the system consists of two sets of aerated lagoons, the first one with high aeration intensity and the second with lower aeration intensity. The aeration method used is floating surface aerators. Figure 1.116a presents a photo of the Carapongo treatment plant located near Lima, Peru, which consists of covered anaerobic lagoons followed by aerated lagoons. The interesting fact in this plant is that the anaerobic lagoons are covered to prevent generation of unpleasant odours in the neighbouring industrial and residential areas. The flow diagram of the Carapongo plant is exactly the diagram presented in Figure 1.113. The process of UASB followed by aerated lagoons is applied in several plants in Brazil, in the state of Minas Gerais, in the state of Parana and in the capital, Brasilia. A photo of such a treatment plant is presented in Figure 1.81 which shows the treatment plant of the city of Uberaba, Minas Gerais. This system also consists of two sets of aerated lagoons, the first one with high aeration intensity and the second with lower aeration intensity. The aeration system used in Uberaba is a diffused air system. Another plant whose process is based on UASB followed by aerated lagoons is the Recanto Das Emas plant in Brasilia, the capital of Brazil. It was designed to handle a flow of 246 l/s to serve a population of about 125,500. The aeration is done by mechanical aerators. The plant includes two sets of aerated lagoons: the first is completely mixed with intensive aeration, followed by partially mixed facultative lagoons. The treatment plant Vale do Amanhecer in Brasilia, is also based on the same process. It has a capacity of 35 l/s and serves a population of 15,000. The treatment plant Fazenda Rio Grande In the state of Parana, Brazil, is also based on UASB followed by an aerated lagoon followed by a facultative lagoon in which the sedimentation of the suspended matter occurs. The aeration in this plant is achieved by floating surface aerators. A photo of this plant is presented in Figure 1.116b.

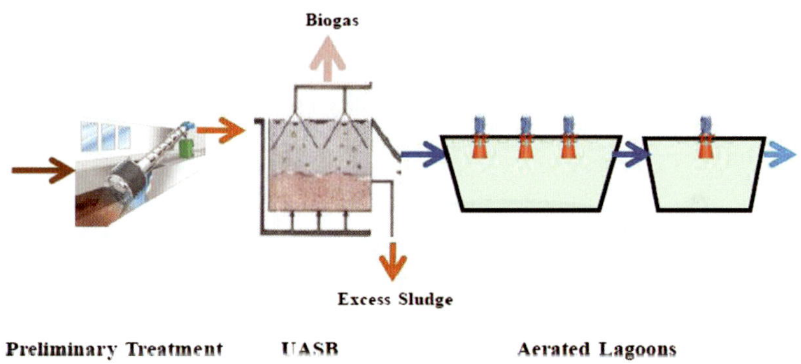

Figure 1.114 Schematic flow diagram of UASB followed by aerated lagoons

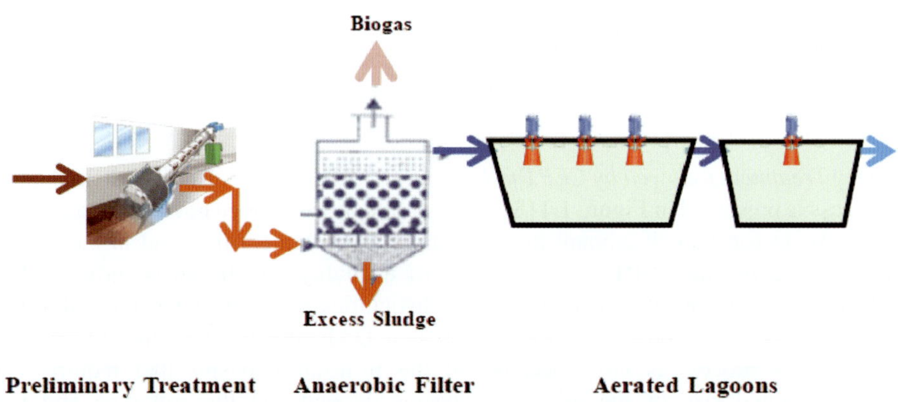

Figure 1.115 Schematic flow diagram of anaerobic filter followed by aerated lagoons

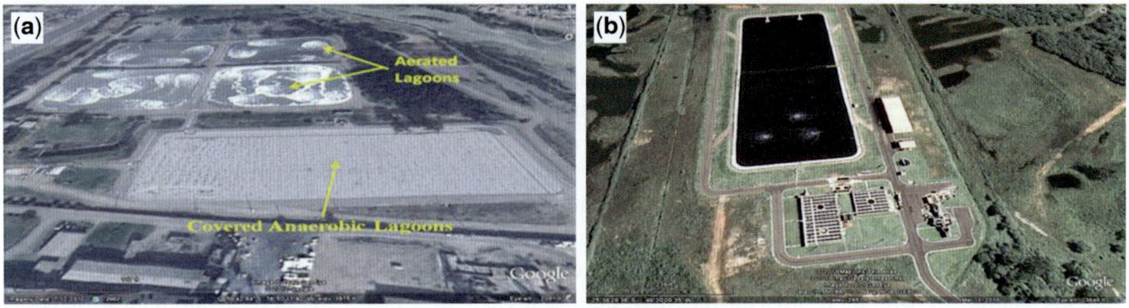

Figure 1.116 (a) Carapongo Treatment Plant near Lima, Peru, based on Covered Anaerobic Lagoons followed by Aerated Lagoon; (b) The Fazenda Rio Grande Treatment Plant in Parana, Brazil, based on UASB followed by an Aerated Lagoon followed by a Facultative Lagoon

- *Preliminary Treatment followed by CEPT followed by Aerated Lagoons,* and if required, by Disinfection. The schematic flow diagram of this process is presented in the Figure 1.117. Aerated lagoons are used in this case to polish the effluent of a CEPT unit to remove residual dissolved and suspended organic matter as well as suspended solids. This process occupies a medium size area. It contains quite a lot of mechanical equipment, but it is simple equipment and the process is simple to operate. An important feature of this process is that it can function well at very low temperatures. This process can produce a good quality effluent.

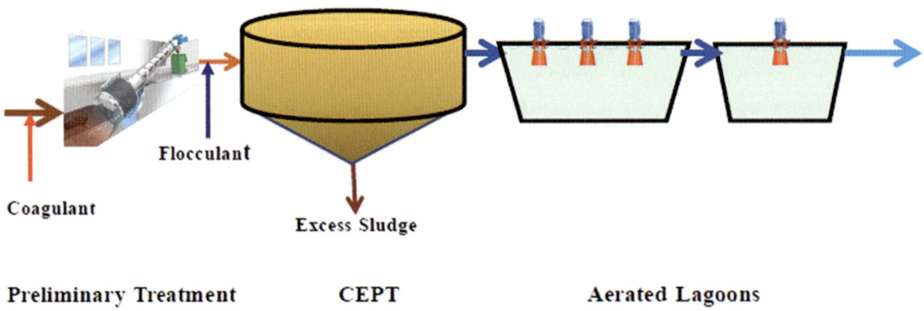

Figure 1.117 Schematic flow diagram of CEPT followed by aerated lagoons

- *Preliminary Treatment followed by CEPT followed by Infiltration Basins*. The schematic flow diagram of this process is presented in Figure 1.118. It is a simple process, does not contain a lot of mechanical equipment except for a small amount in the preliminary treatment unit and in the CEPT unit. The effluent produced by the CEPT unit is usually of a quality which can be infiltrated in recharge basins. In cases where infiltration may present difficulties because of the type of soil, the CEPT effluent can be polished by a Sand Filter unit or a DAF unit prior to the infiltration. The area occupied by this process is large because of the infiltration basins that require a large area. However, the actual size of the infiltration basins depends on the soil type and may vary in accordance with the location of the plant. If soil conditions are adequate for a rapid infiltration rates, this process can be applied also to large plants. Application of this process requires thorough familiarity with the groundwater conditions and it can be implemented only when it does not present a risk of contaminating groundwater used for domestic supply. The process does not generate effluent on the surface so it does not contaminate surface water bodies. An important feature of this process is that it can function well at very low temperatures.

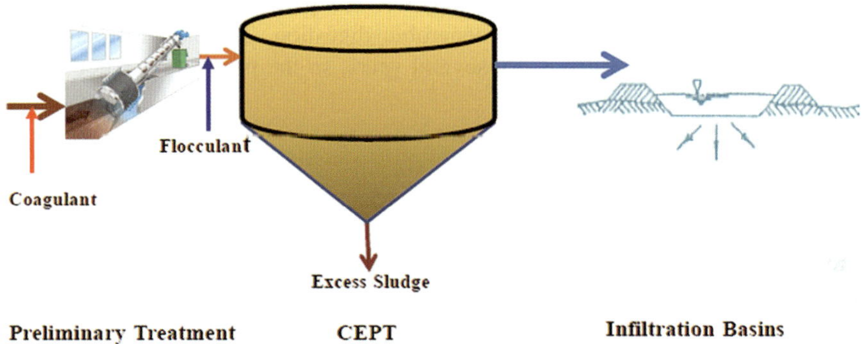

Figure 1.118 Schematic flow diagram of CEPT followed by infiltration basins

– *Preliminary Treatment followed by Lagoons followed by DAF followed by Sand Filtration*. The schematic flow diagram of this process is presented in Figure 1.119. This is an interesting process since it is a process that removes the suspended solids, mostly algae, from a lagoons system effluent. The lagoon system may consist of any common type combination of lagoons such as Anaerobic-Facultative-Maturation lagoons or Facultative-Maturation lagoons. Effluents of such lagoons system contain relatively high concentrations of solids composed mainly of algae which give these effluents a green colour. If the discharge standards require a higher quality effluent, the suspended solids, the organic matter content and the green colour can be removed by submitting the lagoons effluent to a treatment by a combination of units DAF followed by sand filtration. These two units can effectively remove high levels of suspended solids and significantly improve the effluent quality. These two units can in fact improve the quality of any effluent of high suspended solids content, not just lagoons effluents. In all the previously presented processes in which the final unit is DAF, if the effluent quality needs to be upgraded, a sand filtration unit can be added. In all the previously presented processes in which the final unit is sand filtration and it is suspected that the TSS in the effluent is too high so that frequent filter washing cycles will be required, a DAF unit can be added in front of the sand filtration unit. The quality of the effluent of the lagoons followed by DAF followed by sand filtration is very high, even higher than an activated sludge effluent, as presented in the following data of such plant. The process effectively removes Suspended Solids, BOD and Phosphorous (the letter is precipitated by the polyelectrolyte used in the DAF process and then filtered out by the sand filter). However, it does not fully remove nitrogen, since lagoons do not effectively remove nitrogen. The process is relatively simple, however it contains some mechanical equipment in the DAF and sand filtration units and a certain level of technical capacity is required to operate this process. This process is not recommended for small communities but medium size and large communities can cope with the operation of the DAF and sand filter units since similar installations are used in the water supply system. The area occupied by this process is large because the lagoons require a large area (the DAF and Sand filtration units occupy a very small area). This process can be applied to medium and large communities and also to small communities which have some technical capacity. If the lagoons system does not include anaerobic lagoons, the process can function well at low temperatures. If the lagoons system includes anaerobic lagoons, the process cannot function in low temperatures. An advantage of this process is that the solids removed in the DAF and sand filter units can be recirculated to the lagoons (to the anaerobic lagoons if such lagoons are part of the process or to the facultative lagoons) where they settle and undergo anaerobic digestion. The mineral matter which remains after the digestion is removed once every several years, so the process does not generate sludge on a continuous daily basis.

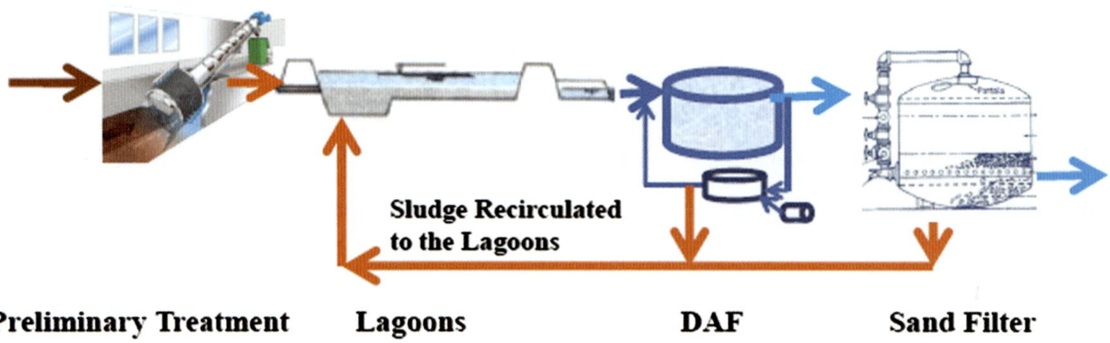

Figure 1.119 Schematic flow diagram of lagoons followed by DAF followed by sand filtration

An example of a full scale plant based on the process of lagoons followed by DAF followed by sand filtration is the Soapberry wastewater treatment plant serving the city of Kingston, Jamaica. A photo of this plant is presented in Figure 1.120. This plant is the first phase of the project, with a capacity to serve a population of about 250,000. Additional modules will be added in the future so that the plant will serve a population of about 1 million. The information about this plant was provided by the National Water Commission (NWC) of Jamaica, whose courtesy is highly acknowledged.

The Soapberry treatment plant was inaugurated in 2008 and has currently the capacity of treating 75,000 m^3/day but does not operate yet at full capacity since not all the sewerage networks in its influence area are already connected to it. The plant consists of an interesting lagoons system composed of two modules each composed of six lagoons, all of them facultative or maturation but with no anaerobic lagoons. The lagoons effluent is subjected to the DAF treatment followed by sand filtration. The DAF is operated with the help of a cationic polyelectrolyte only (without alum or ferric chloride) at a dose of 3 mg/l. The final effluent is a transparent liquid which does not contain a green color and is of high quality. The BOD concentration in the effluent is 10 mg/l, the TSS concentration is 10 mg/l and the Phosphorous concentration is 4 mg/l. As expected, the plant achieves only partial removal of nitrogen.

The configuration of the lagoons system of the Soapberry plant is presented in Figure 1.121. This lagoon configuration was first used in the first stage of the Dan Region (Tel Aviv metropolitan area, Israel) wastewater treatment plant, where it was in operation for many years and was recently taken out of stream. The idea in this plant is to keep all the lagoons with an aerobic top layer so as to prevent generation of odours. The critical area in this respect is the first lagoon which receives the raw sewage, since in this lagoon the organic load is the highest and the perspective of propagation of anaerobic conditions is high. To avoid the generation of anaerobic conditions, the effluent of the final lagoon is recirculated to the raw sewage at a recirculation ratio of about 2:1 (effluent to raw sewage). The recirculated effluent contains high concentration of oxygen and of algae and it also dilutes the raw sewage. So the liquid that enters the first lagoon is diluted, contains a high concentration of oxygen and high concentrations of algae which immediately start producing additional amounts of oxygen. All this prevents depletion of oxygen from the first lagoon (and from the following lagoons), maintains aerobic conditions and prevents propagation of odours. The solids removed in the DAF and sand filter units are recirculated to the lagoons where they settle and undergo anaerobic digestion at the bottom. The mineral matter which remains after the digestion is removed once every several years, so the process does not generate sludge on a continuous daily basis. Sludge is only removed once every several years and this is also a great advantage of the process.

Figure 1.120 The Soapberry wastewater treatment plant in Kingston, Jamaica, based on a lagoons system followed by DAF followed by sand filtration

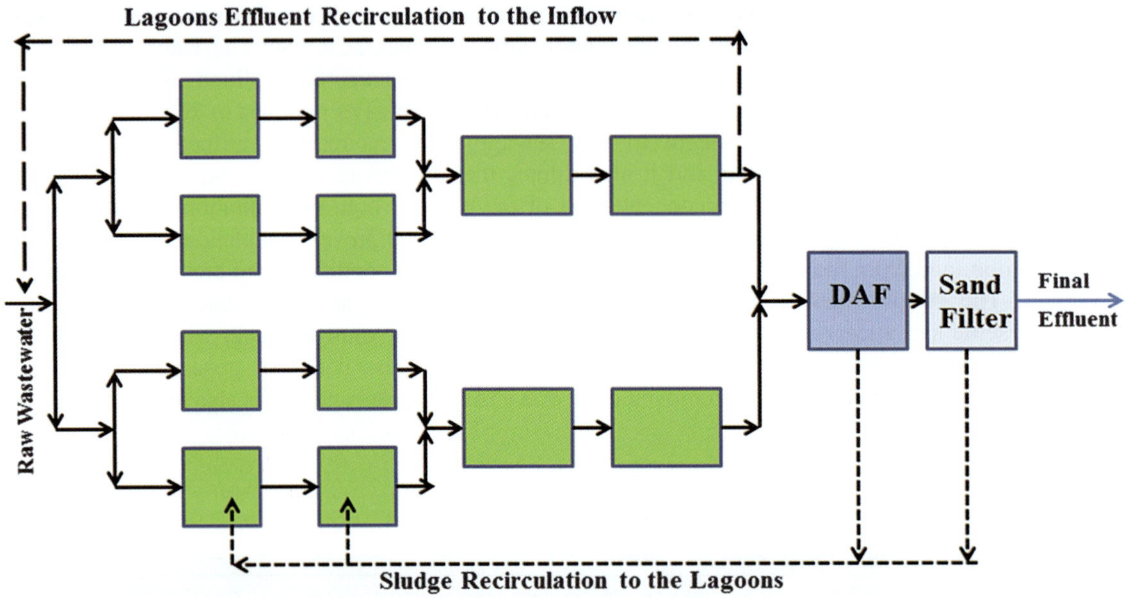

Figure 1.121 The configuration of the lagoons system in the Soapberry treatment plant

1.10 THE EFFECT OF TEMPERATURE ON WASTEWATER TREATMENT AND CLASSIFICATION OF APPROPRIATE TECHNOLOGY PROCESSES ACCORDING TO THEIR ADEQUACY FOR DIFFERENT TEMPERATURE ZONES

1.10.1 Introduction

Temperature is a very important environmental parameter that effects bacterial and physicochemical process activities and therefore also effects wastewater treatment. Most bacteria grow well in the temperature range 15–40°C, and those are termed mesophilic bacteria. Some bacteria grow best in lower temperatures, they are termed psychrophilic bacteria, and some bacteria require higher temperatures, and they are the thermofilic bacteria. In tropical and subtropical regions most bacteria present in wastewaters are mesophilic. Bacterial growth rate varies with temperature, being slower at lower temperature, and accordingly, the performance of wastewater treatment processes improves with increase in temperature. Low temperature also hinders the coagulation in physicochemical processes because almost all chemical reactions in aqueous solutions occur more slowly as the temperature drops. As a result, coagulation becomes less efficient and slower with the drop of temperature. Also, lower temperature influences floc settling. At near freezing temperatures, water becomes denser and that keeps the floc suspended in the water. To compensate for the lower coagulation efficiency and higher water density at low temperatures, higher coagulant doses must be used at lower temperatures.

The anaerobic wastewater decomposition process is more sensitive to temperature than the aerobic process. Aerobic decomposition occurs even at a temperature of close to 0°C as long as the liquid does not freeze. Anaerobic decomposition, on the other hand, becomes too slow at temperatures lower than 12°C. However, there is evidence that in anaerobic lagoons that remain long periods of time, during months, under low temperatures (air temperature below zero), anaerobic activity does not seem to be adversely affected and the lagoons continue to function well, presumably because the anaerobic bacteria have adjusted to long periods of low temperature. There is no such evidence in regards to the UASB and anaerobic filter processes but piston anaerobic reactors (PAR) are known to function well at low temperatures of about 12–13°C. The physicochemical processes also lose efficiency with the decline of temperature, however, they still continue to function even at wastewater temperatures very close to zero, and their efficiency can be recovered by adjusting the doses and combination of coagulants and flocculants and, by adjusting the pH.

As to the performance of anaerobic lagoons under low temperature, there is evidence from the city of El Alto in Bolivia, which is located at an altitude of about 4,000 meters above sea level in the Andean highlands that anaerobic lagoons function well at very low temperatures. During the low temperature season, air temperatures in the El alto zone drop at night to −7°C. El alto has a large lagoons wastewater treatment plant, the Puchukollo plant, consisting of anaerobic lagoons followed by facultative and maturation ponds. The temperature in the anaerobic lagoons never drops below +7°C even when the air temperature around them is −7°C, and even at this temperature (+7°C) the anaerobic lagoons perform well and remove 60% of the BOD contained in the raw sewage. This happens presumably because the anaerobic bacteria have adjusted to long periods of low temperature, and perhaps it has to do with the higher solar radiation at high altitudes (which in El alto is double the radiation at sea level) which keeps the lagoons warmer. This phenomenon has also been reported by Mara (2004) in reference to the highlands of Peru, but it cannot be considered a general phenomenon at all altitudes. Anaerobic lagoons are deep earth lagoons, usually of 4 meters depth or more. It seems that the heat loss from such lagoons is small, so the temperature of the incoming raw sewage, which is much higher than the air temperature, tends to remain at the same level in the liquid inside the lagoon and not drop (probably because the temperature of the soil around the lagoon is similar or even higher than the temperature of the raw wastewater). That helps the anaerobic decomposition process to continue functioning. It is estimated that covered anaerobic lagoons will maintain the heat even better than uncovered lagoons since the cover plastic sheets act as insulators so the heat loss through the surface of the lagoons will be lower.

The effect of decreased efficiency of treatment processes at lower temperature must be taken into account as part of the design of wastewater treatment plants and it is common practice to take into consideration the effect of temperature when sizing the treatment units during the design process. Treatment plants are designed to achieve the required performance even during the low temperature season, that is, the volumes of the biological reactors in treatment plants are designed for the critical conditions of the lowest temperatures in the locations of the treatment plants. Similarly, the physicochemical treatment units and the dosing systems of coagulants and flocculants are designed to cope with the low temperature conditions. In this section we expand the discussion on the effect of temperature to cover two additional aspects: (i) the effect of temperature on appropriate technology processes; and (ii) tailoring the treatment process to the specific temperature conditions in the region in which the treatment plant is located.

The general perception is that most of the developing countries are located in tropical and subtropical regions in which the air temperatures are high and do not fall below 15–20°C even during the low temperature season. Such is the case of Central and South America, the Caribbean, Africa, the Middle East and large parts of Asia. However many developing countries are located in zones in which during the cold season air temperatures drop to low or very low values, way below zero, and consequently the temperatures of the wastewater are also low and in some regions approach zero. Those include Andean Countries in South America, countries in East Europe, Central Asia and in parts of China. In Some areas of China the air temperatures drop during wintertime to very low values of about −30°C. Given that large populations reside in regions with seasonal low temperatures, we also need to present adequate wastewater treatment solution for such zones. In addition to regions with seasonal cold temperatures there are regions with perpetual cold temperatures, for example semi-polar zones or on high mountains zones like the Himalayas. However, the populations in such regions are not large so they are not of main interest. Nevertheless, some of the treatment processes proposed for the regions with seasonal low temperature can be adjusted for perpetual low temperature regions. It is important to mention that the temperature inertia of water makes it very difficult for the wastewater and the liquid in treatment plants to vary by more than 4°C, even when great variation in air temperature occurs.

For any wastewater treatment process to keep functioning, the wastewater and the content of the biological reactors and other treatment units in the treatment plant need to be liquid. If they freeze the treatment processes stops functioning and the wastewater stops flowing. Mixers in the biological reactors and other vessels in a treatment plant can prevent during periods of very low air temperatures the freezing of liquids contained in them. There is a correlation between the air and wastewater temperatures. In the high range temperatures (over 15°C), the air temperature is either higher than the wastewater temperatures or their temperatures are similar. But in the low range temperatures (below 12°C), the wastewater temperature is higher than the air temperature. The relation between the air temperature and the raw wastewater temperature (which is the temperature of the influent to the treatment plant) is presented in Figure 1.122 which shows the variation of the air temperature, of the temperature of the wastewater flowing into a constructed wetland plant and of the temperature of the effluent of this plant during a period of about two years in a constructed wetland treatment plant site in Minnesota, USA (Wallace, 2007). During the cold season the air temperature in the site drops to −25°C (but the influent wastewater temperature keeps way above zero, in the range of +5°C to +10°C, most of the time closer to +10°C) and during the hot season the air temperature climbs to +30°C. We use here the mean air temperature of the coolest month of the year as the temperature indicator since air temperature is a parameter meaningful to everybody, while wastewater temperature does not say much to most people. But we bear in mind that there is a correlation between the two parameters (for instance, Wallace, 2007 seen in Figure 1.122) and this correlation is already taken into account when discussing the adequate treatment processes for each temperature region.

For the purpose of discussing the effect of temperature on wastewater treatment we propose to consider three types of temperature regions: (i) Regions type 1: are regions of very cold temperatures in which the air temperatures drop during the cold season to below −3°C, (which would be roughly equivalent to zones with low season wastewater temperatures in the range +1°C to +7°C); (ii) Region type 2: are regions of medium cold temperatures in which the air temperatures during the cold season remain above −3°C but below +7°C (which would be roughly equivalent to zones with low season wastewater temperatures in the range +8°C to +12°C); and (iii) Region type 3: are regions of mild temperatures in which the air temperatures during the cold season remain always above +7°C (which would be roughly equivalent to zones with low season wastewater temperatures above +13°C).

For each region there is a group of adequate processes which functions well in this region. The processes adequate for each temperature region are presented in the following section. Processes which function in Region type 1 (very cold low season temperatures) will obviously function well in Region type 2 and in Region type 3 since the low temperatures in these two regions are higher than those of region type 1. And because of the same reason, processes which function in Region type 2 (medium low season temperatures) will function well in Region type 3. But processes which function well in Region type 3 (highest low season temperatures) will not necessarily function in Regions type 1 and 2 and processes which function in Region type 2 will not necessarily function in Region type 1.

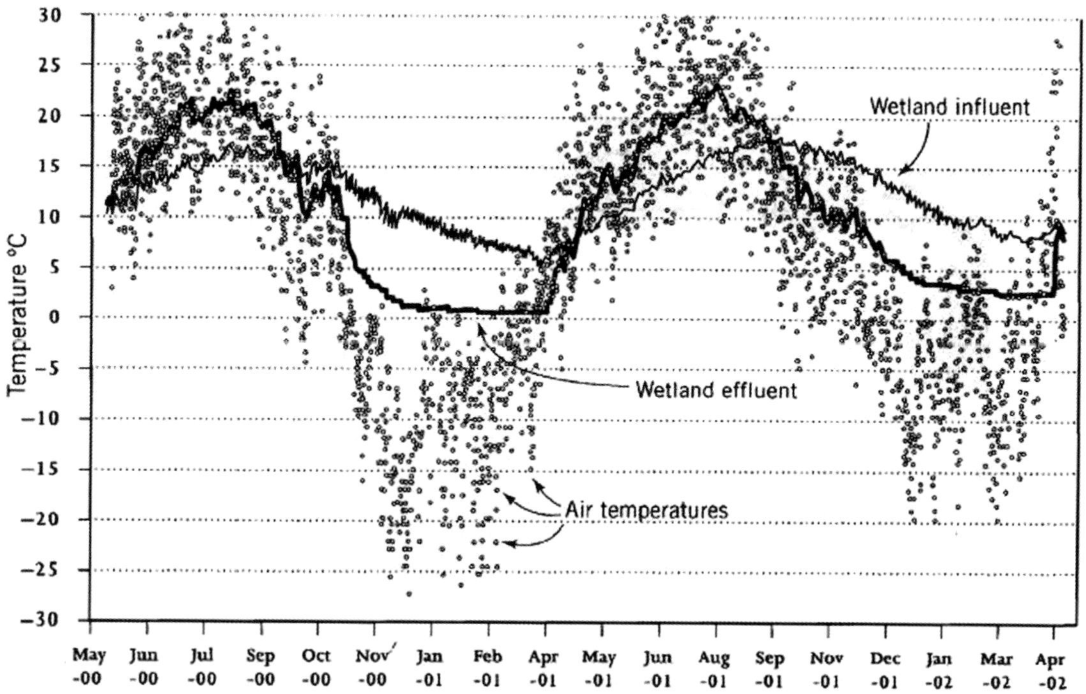

Figure 1.122 Variation of (i) air, (ii) inflow wastewater, and (iii) effluent temperatures with time in a constructed wetland treatment plant site in Minnesota, USA (*Source*: Wallace, 2007)

Regions type 1, with very low seasonal temperatures, are difficult for any treatment process, even for aerobic processes like activated sludge, and certainly for anaerobic processes. Biological processes would, in general, not be effective in these zones during the low temperature season. What can work well at such low temperatures (that is, during the low season in this type of zones) are physical and physicochemical processes which are less sensitive to low temperatures and can be made to function well under such conditions by adjusting the dose of coagulants and flocculants. The CEPT process, a physicochemical process, can be made to function well in such zones and its effluent can be polished by sand filtration or dissolved air floatation (DAF). The sand filtration and the DAF are also physicochemical processes and can be made to function at low temperatures. Constructed wetlands will also perform in regions type 1, especially as polishing units. In subsurface flow constructed wetlands (SSFCW) the water is not exposed to the air during the treatment process so the surface of such wetlands can be insulated with a layer of compost or other type of insulation to prevent loss of heat from the surface. The earth in which the wetland is excavated prevents loss of heat in other direction. In this way the constructed wetlands maintain an inside comfortable temperature at which aerobic bacteria can perform well even at very low air temperature outside the wetlands, as seen in Figure 1.122. Overland flow and Aerated Lagoon will also function well as polishing units. Infiltration can also function if the snow on the surface of recharge basins is cleared by mechanical equipment. Another process that can work in zones type 1 is stabilization reservoirs equipped with mixers. The pretreatment unit of the reservoir needs in this case to be based on a process that functions at very low air temperatures (such as

CEPT, Aerated Lagoons or Constructed Wetlands). The effluent will be stored in the reservoir in which the mixing will prevent ice formation. The aerobic biological processes will continue to function in the reservoir even at low water temperatures. When temperature will start increasing after the cold season, undigested organic matter will then be decomposed.

In regions type 2, Anaerobic Lagoons may function, especially covered lagoons, but its is not recommended to use them. Facultative lagoons will perform well, but of course they have to be designed with dimensions adjusted to the temperatures of the low season. Mixers can be added to the facultative lagoons to improve performance, and air supply can also be beneficial in low temperatures, as described in the lagoons unit process section. Other aerobic unit processes (Aerated Lagoons, Overland Flow, and Constructed Wetlands) can be used as polishing units or in other process configurations. In addition, all the unit processes mentioned as applicable in regions type 1 will function in regions type 2, as well as combinations of all these unit processes. UASB, Anaerobic Filters and even PAR will not function during the low temperature season in regions type 2.

In regions type 3 all the appropriate technology processes, including all types of anaerobic processes (Anaerobic Lagoons, UASB and Anaerobic Filters, PAR), will function well. As a precautionary measure it is recommended to use UASB and Anaerobic Filters only in cases where the wastewater temperature at the low temperature season is above 16°C. Anaerobic lagoons and PAR will perform well during all seasons, including during the low temperature season even if the wastewater temperature during this season is lower than 16°C.

The main implication of the effect of temperature on appropriate technology processes is the limitation on the use of anaerobic processes in low temperatures. In regions where the air temperature does not fall below +7°C (which correspond to regions type 3), the four types of anaerobic reactors (Anaerobic Lagoons, UASB, Anaerobic Filter and PAR) can be used, but we recommend to restrict the use of UASB and Anaerobic Filter only to regions where the wastewater temperature during the low temperature season is above +16°C. In regions where the air temperature does not fall much below zero (that is, not lower than −3°C, which correspond to regions type 2), Anaerobic Lagoons might function even in the low temperature season, especially covered anaerobic lagoons, but we do not recommend to use them. Other anaerobic processes cannot be used during the low temperature season. In regions where the temperature during the cold season is lower than 7°C (which correspond to regions type 1), anaerobic reactors of any type cannot not be used during the low temperature season.

1.10.2 Appropriate technology processes adequate for zones with seasons of very low temperatures

The processes presented in this section are processes adequate for Regions type 1 that are regions of the very cold temperatures in which the air temperatures drop during the cold season to below −7°C. That would be roughly equivalent to zones with low season wastewater temperatures in the range +1°C to +7°C. The unit processes of appropriate technology which function in such temperatures during the low temperatures season are physical and physicochemical processes which are less sensitive to low temperatures and can be made to function well under such conditions by adjusting the dose of coagulants and flocculants. The CEPT process, a physicochemical process, can be made to function well in such zones and its effluent can be polished by Sand Filtration or dissolved air floatation (DAF) which can also be made to function at low temperatures. Constructed Wetland, Overland Flow, Infiltration and Aerated Lagoons will also function in these temperature zones as polishing units. Treatment processes composed of combinations of the mentioned units will also function in these temperature zones. The processes specified as functioning in this type of regions will also function in regions of higher temperatures. In the processes presented, disinfection is not mention as a unit processes but it can of course be applied to the final effluent of any of the processes, if required.

Processes which are adequate for zones in which air temperatures fall below −7°C include:

- *Preliminary Treatment followed by CEPT*. The CEPT process produces a fair quality effluent and sometimes this quality is sufficient, for example in case that the effluent is discharged to a large flow river. The flow diagram of the CEPT process is presented in Figure 1.37.
- *Preliminary Treatment followed by CEPT followed by Sand Filtration*. The flow diagram of this process is presented in Figure 1.72.
- *Preliminary Treatment followed by CEPT followed by DAF*. The flow diagram of this process is presented in Figure 1.99.
- *Preliminary Treatment followed by CEPT followed by Overland Flow*. The flow diagram of this process is presented in Figure 1.100.
- *Preliminary Treatment followed by CEPT followed by Constructed Wetlands*. The flow diagram of this process is presented in Figure 1.101.
- *Preliminary Treatment followed by CEPT followed by Infiltration Basins*. The flow diagram of this process is presented in Figure 1.118. If the CEPT effluent presents difficulties in infiltration, it can be polished by a Sand Filter unit or a DAF unit prior to the infiltration.
- *Preliminary Treatment followed by CEPT followed by Facultative Lagoons and Maturation Lagoons*. This is similar to the treatment plant of Riviera Sao Lourenco. The flow diagram of this process is presented in Figure 1.102 but omitting the anaerobic lagoon which appears in this figure.
- *Preliminary Treatment followed by CEPT followed by Aerated Lagoons*. The flow diagram of this process is presented in Figure 1.117.

As seen, the main treatment unit process during the extreme cold season is the CEPT process. This process has relatively high O&M cost because of the consumption of chemicals for coagulation and flocculation. Such costs may be an overburden for part of the utilities in developing countries. To overcome the problem, it is possible to construct two treatment units, each based on a different process, one for the very cold season and the other for all other seasons. The principle of this approach is presented in Figure 1.123. During the very cold season the treatment process is preliminary treatment followed by CEPT followed by a polishing unit resistant to cold temperatures. During all other seasons, in which the temperatures are higher, the treatment process is preliminary treatment followed by a treatment unit adequate for higher temperatures followed by the same polishing unit used during the cold temperature season. The main treatment unit during the higher temperatures season can be one of the many processes mention above, for example any one of the anaerobic processes (Anaerobic Lagoons, UASB, Anaerobic Filter, PAR), lagoons system of various types and combinations of these unit processes. This concept is similar to the concept used in Riviera Sao Lourenco (see Figures 1.103, 1.104 and 1.105) however, in Riviera Sao Lourenco the CEPT unit kicks in during high load seasons and in this case it kicks in during the low temperature season.

As a more specific example of this concept, one of the parallel units can be a CEPT unit and the other an anaerobic unit (UASB or anaerobic lagoons). These two parallel units will be followed by a polishing unit which can be Sand Filtration or DAF. During the low temperature season the CEPT unit will be operated as the first unit process (and the anaerobic unit will be bypassed), so the treatment process will be CEPT followed by Sand Filtration (or DAF). During the medium and high temperature seasons the CEPT will be bypassed and the anaerobic unit will be put on stream (UASB or anaerobic lagoons) followed by the polishing unit (Sand Filtration or DAF). Since the anaerobic unit will be working intermittently, part of the year on and part of the year off, problems of sludge acclimatization during stat up periods might occur. This is not a serious problem in anaerobic filters or anaerobic lagoons. In the case of UASB the granular sludge might be kept active even during periods that the unit does not receive wastewater. Also, seeding from other plants can be helpful.

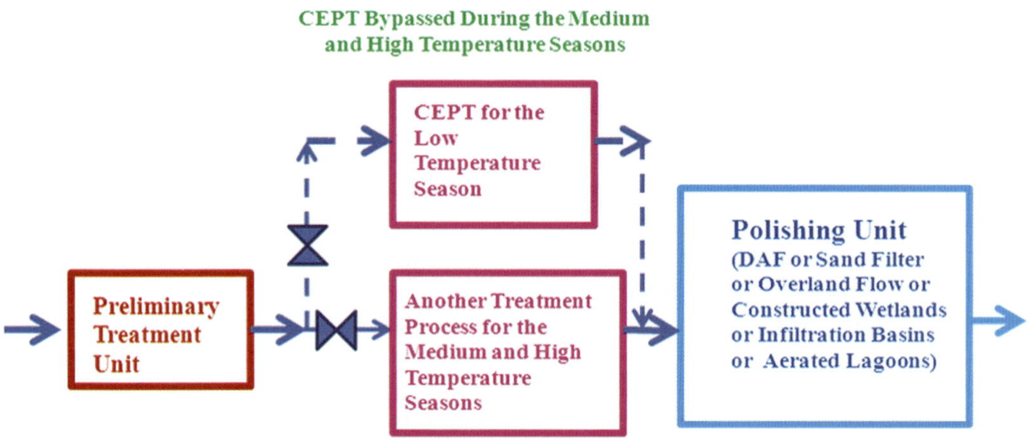

Figure 1.123 Use of two parallel treatment processes, one for the low temperature season and another for all other seasons

Under this concept the initial investment is higher since two main treatment unit need to be constructed. However, the investment is not very high and utilities can obtain grants or soft loans for investments. The higher investment is usually compensated by the reduced O&M costs. Since there are no subsidies for financing O&M costs and the utilities have to pay these costs from their own cash generation, a concept which results in reducing O&M costs is very valuable.

1.10.3 Appropriate technology processes adequate for zones with seasons of medium low temperatures

The processes presented in this section are processes adequate for Regions type 2 which are regions of medium cold temperatures in which the air temperatures during the cold season remain always above −3°C but below +7°C (which would be roughly equivalent to zones with low season wastewater temperatures in the range +8°C to +12°C). Obviously, the unit processes of appropriate technology which function in zones of very low temperatures will also function in zones of medium low temperatures. Those include physical and physicochemical processes such as CEPT, as well as the following polishing units: Sand Filtration, DAF, Constructed Wetland, Overland Flow, Infiltration and Aerated Lagoons. Additional processes which can serve as the main treatment unit are Anaerobic Lagoons, Facultative Lagoons and a complete lagoons system consisting of Anaerobic Lagoons followed by Facultative Lagoons followed by Maturation Lagoons. It is preferable to use in such low temperatures covered Anaerobic Lagoons and Facultative Lagoons in which mixers are installed. Also, for these type of zones the stabilization reservoir system can be used, with pretreatment consisting of processes adequate for these temperatures such as CEPT, or CEPT followed by one of the polishing units mentioned above, or a lagoons system. It is advisable to equip the reservoir with mixers to boost its performance during the cold season. Some of the cold temperatures regions are arid regions which can use stores effluent to irrigate field crops during the high temperature season. Alternatively, it may be beneficial to discharge the effluents stored in reservoirs to rivers during the high temperature season because of larger flows and higher dilution in the rivers during this season.

Processes which are adequate for zones in which air temperatures in the cold season are in the range −3° C to +7°C include:

- *All the processes outlines in Section 1.10.2.*
- *Preliminary Treatment followed by Facultative Lagoon with mixers (and if needed assisted aeration) followed by Maturation Lagoons.* The flow diagram of this process is presented in Figure 1.52, omitting the Anaerobic Lagoon.
- *Preliminary treatment followed by an Overland Flow System*, for case under which the effluent quality of a Overland Flow unit is sufficient. The flow diagram of this process is presented in Figure 1.41.
- *Preliminary Treatment followed by Constructed Wetlands.* A photo of this process is presented in Figure 1.36.
- *Preliminary Treatment followed by any type of adequate Pretreatment (basically CEPT or CEPT followed by polishing with DAF or CEPT followed by polishing with Sand Filter or CEPT followed by polishing with Aerated Lagoons, or a full Lagoons System) followed by Stabilization Reservoir with Mixers.* The flow diagram of this process is presented in Figure 1.23. The reservoir effluent will be either reused for irrigation during the high temperature season, or discharged to the river during the high temperature season.

The concept of using two parallel treatment processes, one for the low temperature season and another for all other seasons, as presented in Figure 1.123 can also be applied in zones with seasons of medium low temperatures. During the high temperature season in these zones, the treatment process will be based on anaerobic processes.

1.10.4 Appropriate technology processes adequate for zones with seasons of mild low temperatures

The processes presented in this section are processes adequate for Region type 3 which are regions with seasons of mild low temperatures in which the air temperatures during the cold season remain always above +7°C (which would be roughly equivalent to zones with low season wastewater temperatures above +13°C). This basically refers to regions of high temperatures all year round, like tropical and subtropical zones. Obviously, the unit processes of appropriate technology which function in zones of very low temperatures and in zones of medium low temperatures will also function well in zones of mild low temperatures, so all the processes presented in sections 1.10.2 and 1.10.3 are adequate also for the mild low temperatures zones.

Within this zone we make a distinction between regions with low season wastewater temperatures above +13°C and regions with low season wastewater temperatures above +16°C. We do not recommend using UASB and anaerobic filters in regions with low season wastewater temperatures in the range +13°C to +16° C, but rather only in regions with low season wastewater temperatures of above +16°C.

Processes which are adequate for zones in which air temperatures in the cold season are above +7°C include:

- *All the processes outlines in Section 1.10.2.*
- *All the processes outlines in Section 1.10.3.*
- *Piston Anaerobic Reactor*

- *UASB Reactor*, for cases under which the effluent quality of UASB reactor is sufficient. This process is to be preferably applied in regions with low season wastewater temperatures above +16°C.
- *Anaerobic Filter*, for cases under which the effluent quality of an Anaerobic Filter is sufficient. This process is to be preferably applied in regions with low season wastewater temperatures above +16°C.
- *Anaerobic Lagoons, preferably covered,* for cases under which the effluent quality of an anaerobic lagoon is sufficient.
- *Preliminary Treatment followed by Anaerobic Lagoon followed by Facultative Lagoon with mixers (and if needed assisted aeration) followed by Maturation Lagoons*. The flow diagram of this process is presented in Figure 1.52.
- *Preliminary Treatment followed by Anaerobic Lagoon followed by Constructed Wetlands*. The flow diagram of this process is presented in Figure 1.111, omitting the facultative lagoon.
- *Preliminary Treatment followed by Anaerobic Lagoon followed by Overland Flow*. The flow diagram of this process is presented in Figure 1.85.
- *Preliminary Treatment followed by Anaerobic Lagoon followed by Infiltration Basins*. The flow diagram of this process is presented in Figure 1.106.
- *Preliminary Treatment followed by a Lagoons System*. The flow diagram of this process is presented in Figure 1.52.
- *Preliminary Treatment followed by UASB followed by Facultative Lagoons*. This process is to be preferably applied in regions with low season wastewater temperatures above +16°C. The flow diagram of this process is presented in Figure 1.59.
- *Preliminary Treatment followed by UASB followed by Anaerobic Filter*. This process is to be preferably applied in regions with low season wastewater temperatures above +16°C. The flow diagram of this process is presented in Figure 1.64.
- *Preliminary Treatment followed by UASB followed by DAF*. This process is to be preferably applied in regions with low season wastewater temperatures above +16°C. The flow diagram of this process is presented in Figure 1.69.
- *Preliminary Treatment followed by Pretreatment followed by a Stabilization Reservoir*. The flow diagram of this process is presented in Figure 1.73 and the pretreatment processes are also specified in this figure.
- *Preliminary Treatment followed by UASB followed by Sand Filtration*. This process is to be preferably applied in regions with low season wastewater temperatures above +16°C. The flow diagram of this process is presented in Figure 1.83.
- *Preliminary Treatment followed by UASB followed by Overland Flow*. This process is to be preferably applied in regions with low season wastewater temperatures above +16°C. The flow diagram of this process is presented in Figure 1.84.
- *Preliminary Treatment followed by Anaerobic Filter followed by Overland Flow*. This process is to be preferably applied in regions with low season wastewater temperatures above +16°C. The flow diagram of this process is presented in Figure 1.86.
- *Preliminary Treatment followed by UASB followed by Constructed Wetlands*. This process is to be preferably applied in regions with low season wastewater temperatures above +16°C. The flow diagram of this process is presented in Figure 1.87.
- *Preliminary Treatment followed by UASB followed by Lagoons followed by Overland Flow*. This process is to be preferably applied in regions with low season wastewater temperatures above +16°C. The flow diagram of this process is presented in Figure 1.88.

- *Preliminary Treatment followed by UASB followed by Overland Flow followed by Lagoons*. This process is to be preferably applied in regions with low season wastewater temperatures above +16°C. The flow diagram of this process is presented in Figure 1.90.
- *Septic Tank followed by Anaerobic Filter*. This process is to be preferably applied in regions with low season wastewater temperatures above +16°C. The flow diagram of this process is presented in Figure 1.92.
- *Septic Tank followed by Constructed Wetlands*. A photo of a plant of this type is presented in Figure 1.35.
- *Preliminary Treatment followed by Anaerobic Filter followed by Lagoons*. This process is to be preferably applied in regions with low season wastewater temperatures above +16°C. The flow diagram of this process is presented in Figure 1.93.
- *Preliminary Treatment followed by Anaerobic Filter followed by Sand Filter*. This process is to be preferably applied in regions with low season wastewater temperatures above +16°C. The flow diagram of this process is presented in Figure 1.94.
- *Preliminary Treatment followed by Anaerobic Filter followed by DAF*. This process is to be preferably applied in regions with low season wastewater temperatures above +16°C. The flow diagram of this process is presented in Figure 1.95.
- *Preliminary Treatment followed by Anaerobic Filter followed by Constructed Wetlands*. This process is to be preferably applied in regions with low season wastewater temperatures above +16°C. The flow diagram of this process is presented in Figure 1.96.
- *Preliminary Treatment followed by Anaerobic Lagoons followed by Anaerobic Filter*. This process is to be preferably applied in regions with low season wastewater temperatures above +16°C. The flow diagram of this process is presented in Figure 1.97.
- *Preliminary Treatment followed by Anaerobic Lagoons followed by Anaerobic Filter followed by Sand Filter*. This process is to be preferably applied in regions with low season wastewater temperatures above +16°C. The flow diagram of this process is presented in Figure 1.98.
- *Preliminary Treatment followed by one of the Anaerobic Units (Anaerobic Lagoons or UASB or Anaerobic Filter) followed by Infiltration*. This process is to be preferably applied in regions with low season wastewater temperatures above +16°C. The flow diagrams of these processes are presented in Figures 1.106, 1.107 and 1.108.
- *Preliminary Treatment followed by Anaerobic Lagoons followed by UASB*. This process is to be preferably applied in regions with low season wastewater temperatures above +16°C. The flow diagram of this process is presented in Figure 1.109.
- *Preliminary Treatment followed by CEPT followed by UASB*. This process is to be preferably applied in regions with low season wastewater temperatures above +16°C. The flow diagram of this process is presented in Figure 1.110.
- *Preliminary Treatment followed by a Lagoons System followed by Constructed Wetlands*. The flow diagram of this process is presented in Figure 1.111.
- *Preliminary Treatment followed by one of the Anaerobic Units (Anaerobic Lagoons or UASB or Anaerobic Filter) followed by Aerated Lagoons*. This process is to be preferably applied in regions with low season wastewater temperatures above +16°C. The flow diagrams of these processes are presented in Figures 1.113, 1.114 and 1.115.
- *Preliminary Treatment followed by UASB followed by Anaerobic Filter followed by Dissolved Air Flotation followed by Membrane Filtration*. This process is to be preferably applied in regions with low season wastewater temperatures above +16°C. The flow diagram of this process is presented in Figure 1.79.

General guidance for selection of unit and combined appropriate technology treatment processes according to their resistance to temperatures during the cold season is presented in Table 1.5. For a given cold season temperature range in a planned wastewater treatment plant site, this table can help identify a series of processes which are adequate and may be considered for further, more detailed analysis.

Table 1.5 Processes selection matrix by air temperature during the cold season.

Process number	Process description	Air temperature during the cold season			Comment
		Below (−7°C)	In the range (−3°C to +7°C)	Above (+7°C)	
1	CEPT	X	X	X	
2	CEPT followed by sand filtration	X	X	X	
3	CEPT followed by DAF	X	X	X	
4	CEPT followed by overland flow	X	X	X	
5	CEPT followed by constructed wetlands	X	X	X	
6	CEPT followed by infiltration basins	X	X	X	
7	CEPT followed by facultative lagoons and maturation lagoons	X	X	X	
8	CEPT followed by aerated lagoons	X	X	X	
9	Facultative lagoon with mixers followed by maturation lagoons		X	X	
10	Overland flow		X	X	
11	Constructed wetlands		X	X	
12	Adequate pretreatment followed by a stabilization reservoir with mixers		X	X	
13	Piston anaerobic reactor			X	
14	Anaerobic lagoons, preferably covered			X	
15	Anaerobic lagoon followed by facultative lagoon with mixers followed by maturation lagoons			X	

(Continued)

Table 1.5 Processes selection matrix by air temperature during the cold season (*Continued*).

Process number	Process description	Air temperature during the cold season			Comment
		Below (−7°C)	In the range (−3°C to +7°C)	Above (+7°C)	
16	Anaerobic lagoon followed by constructed wetlands			X	
17	Anaerobic lagoon followed by overland flow			X	
18	Anaerobic lagoon followed by infiltration basins			X	
19	UASB reactor			X	Recommended only for wastewater temperatures above (+16°C)
20	Anaerobic filter			X	Recommended only for wastewater temperatures above (+16°C)
21	UASB followed by facultative lagoons's			X	Recommended only for wastewater temperatures above (+16°C)
22	UASB followed by anaerobic filter's			X	Recommended only for wastewater temperatures above (+16°C)
23	UASB followed by DAF			X	Recommended only for wastewater temperatures above (+16°C)
24	Adequate pretreatment followed by a stabilization reservoir			X	
25	UASB followed by sand filtration			X	Recommended only for wastewater temperatures above (+16°C)
26	UASB followed by overland flow			X	Recommended only for wastewater temperatures above (+16°C)
27	Anaerobic filter followed by overland flow			X	Recommended only for wastewater temperatures above (+16°C)
28	UASB followed by constructed wetlands			X	Recommended only for wastewater temperatures above (+16°C)

(*Continued*)

Table 1.5 Processes selection matrix by air temperature during the cold season (*Continued*).

Process number	Process description	Air temperature during the cold season			Comment
		Below (−7°C)	In the range (−3°C to +7°C)	Above (+7°C)	
29	UASB followed by lagoons followed by overland flow			X	Recommended only for wastewater temperatures above (+16°C)
30	UASB followed by overland flow followed by lagoons			X	Recommended only for wastewater temperatures above (+16°C)
31	Septic tank followed by anaerobic filter			X	Recommended only for wastewater temperatures above (+16°C)
32	Septic tank followed by constructed wetlands			X	
33	Anaerobic filter followed by lagoons			X	Recommended only for wastewater temperatures above (+16°C)
34	Anaerobic Filter followed by sand filter			X	Recommended only for wastewater temperatures above (+16°C)
35	CEPT followed by aerated lagoons			X	
36	Anaerobic filter followed by DAF			X	Recommended only for wastewater temperatures above (+16°C)
37	Anaerobic filter followed by constructed wetlands			X	Recommended only for wastewater temperatures above (+16°C)
38	Anaerobic lagoons followed by anaerobic filter			X	Recommended only for wastewater temperatures above (+16°C)
39	Anaerobic lagoons followed by anaerobic filter followed by sand filter			X	Recommended only for wastewater temperatures above (+16°C)
40	One of the anaerobic units followed by infiltration			X	Recommended only for wastewater temperatures above (+16°C)
41	Anaerobic lagoons followed by UASB			X	Recommended only for wastewater temperatures above (+16°C)

(*Continued*)

Table 1.5 Processes selection matrix by air temperature during the cold season (*Continued*).

Process number	Process description	Air temperature during the cold season			Comment
		Below (−7°C)	In the range (−3°C to +7°C)	Above (+7°C)	
42	CEPT followed by infiltration basins			X	
43	CEPT followed by UASB			X	Recommended only for wastewater temperatures above (+16°C)
44	Lagoons system followed by constructed wetlands			X	
45	One of the anaerobic units followed by aerated lagoons			X	Recommended only for wastewater temperatures above (+16°C)
46	UASB followed by anaerobic filter followed by dissolved air flotation followed by membrane filtration			X	Recommended only for wastewater temperatures above (+16°C)

1.11 PROCESSES ADEQUATE FOR PROJECTS IN WHICH THE LAND AREA AVAILABLE FOR WASTEWATER TREATMENT IS LIMITED

1.11.1 The size of land area occupied by various appropriate technology based wastewater treatment plants

Selecting the location of a wastewater treatment plant is an important part of project development. Availability of land for treatment plants becomes more limited as the location is closer to the city, since the cost of land is higher and the social opposition to the plant is stronger as the location gets closer to the city. The size of the area occupied by a treatment plant depends on the selected treatment process. Each unit process occupies a specific area, and the area of each combined process depends on the unit processes which compose that combined process. The selection of the location of a treatment plant and of the treatment process need to be coordinates. For example, the availability of a limited area dictates the selection of a treatment process which occupies a small area while the availability of a large area enables considering additional treatment processes which occupy larger areas. In summary, the size of land occupied by each treatment processes is an important characteristic of the process. Information on the size of land occupied by the unit processes of appropriate technology and by some of the combined appropriate technology processes is provided in this section. For purposes of comparison, the information is provided in terms of specific area in square meters occupied per served person. The information does not include all the processes discussed in the book, however, it is possible to estimate, on the basis of the information provided in this section, the area of any other combined treatment process. The information provided represents only indicative estimates based on the experience of the authors and on other sources. This information can be used for preliminary comparisons between various treatment processes and for preliminary estimates of adaptability of certain processes to an available land size; however, accurate estimates need to be undertaken for each specific treatment plant as its design is progressing.

The specific land area in terms of square meters per served person occupied by the unit appropriate technology processes and by some combined processes is presented in Table 1.6. The values of sizes of area presented in this table refer to the actual land areas occupied by the treatment installations. The total areas of the plants are somewhat larger in each case, usually by 20–50% to account for office space, maintenance shops, roads, parking area and gardens. Sometimes, the plant site includes large spare areas reserved for future expansion of the treatment installation. The table also includes data on the ratio of the area of each process to the area of the Conventional Activated Sludge process. This information is presented since the activated sludge processes is the conventional and most widely used process, so it is beneficial to compare each process to the commonly used process.

Table 1.6 Land area per served person occupied by the unit processes of appropriate technology and by some combined processes.

Process	Specific area occupied by the process (m^2/Capita)	Average ratio of process area to the area of conventional activated sludge	Comments
Appropriate Technology Processes			
Rotating micro screens	(0.0005–0.001)	0.004	Very Small Area
Vortex grit chamber	(0.0003–0.005)	0.014	Very Small Area
Anaerobic lagoons	(0.15–0.2)	0.95	Small Area
Conventional lagoons system (anaerobic + facultative + maturation)	(3.0–5.0)[1]	21.6	Large Area
Upgraded lagoons system (covered anaerobic + facultative and maturation with mixers)	(1.5–2.0)	9.5	Large area
UASB	(0.03–0.1)[2]	0.35	Small Area
Anaerobic filter	(0.03–0.1)[3]	0.35	Small Area
CEPT*	(0.002–0.005)[4]	0.02	Very Small Area
Constructed wetlands	(1.5–5.0)[1]	17.5	Large Area
Stabilization reservoir (without pretreatment area)	(0.5–4.0)	14.5	Large Area
Overland flow	(2.0–3.5)[1]	14.9	Large Area
Infiltration	(0.5–1.0)	4.0	Large Area
DAF	(0.015–0.01)	0.07	Very Small Area
Sand filter	(0.002–0.01)	0.03	Very Small Area
Septic tank	(0.03–0.05)[1]	0.21	Very Small Area
Anaerobic lagoons followed by facultative lagoons	(1.2–3.5)[4]	12.7	Large Area

(*Continued*)

Table 1.6 Land area per served person occupied by the unit processes of appropriate technology and by some combined processes (*Continued*).

Process	Specific area occupied by the process (m^2/Capita)	Average ratio of process area to the area of conventional activated sludge	Comments
UASB followed by anaerobic filter	(0.05–0.15)[2]	0.54	Small Area
UASB followed by maturation lagoons	(1.5–2.5)[1,2,4]	10.8	Large Area
UASB followed by DAF	(0.05–0.15)[2]	0.54	Small Area
UASB followed by sand filter	(0.05–0.15)	0.54	Small Area
UASB followed by overland flow	(1.5–3.0)[2]	12.1	Large Area
Anaerobic lagoons followed by Aerated lagoons	(0.3–0.5)	2.2	Large Area
CEPT followed by sand filtration	(0.004–0.015)	0.05	Very Small Area
CEPT followed by DAF	(0.012–0.02)	0.09	Very Small Area
Septic tank followed by infiltration	(1.0–5.0)[4]	16.2	Large Area
Conventional processes			
Conventional activated sludge	(0.12–0.25)[1]	1.0	Small Area
Trickling filter	(0.15–0.3)[1]	1.2	Small Area
Rotating biological contactors	(0.15–0.25)[4]	1.1	Small Area
Aerated lagoons	(0.2–0.5)	1.9	Medium Area

(1) WSP (2008) based on Von Sperling
(2) Chernicharo (2007)
(3) Based on estimates from data in Andrade Neto (1997) and Chernicharo (2007)
(4) Tsukamoto (2002)
* The CEPT area might be larger, depending on the sludge treatment method
– In processes which have no source reference, the information is based on the Authors' experience
– Comment: All the listed processes include also preliminary treatment

The last four processes in the table are conventional treatment processes and are presented as reference and for highlighting the fact that there are appropriate technology processes which occupy small areas, similar to the area of conventional processes.

As seen from the Table 1.6 the appropriate technology processes present a wide range of occupied land areas. Unit or combined processes that included units of facultative or maturation lagoons, constructed wetlands, stabilization reservoirs, overland flow and infiltration occupy large areas. On the other hand, unit or combined processes which include preliminary treatment, anaerobic lagoons, UASB, anaerobic filters, septic tanks, CEPT, DAF and Sand Filters occupy small areas, similar to the areas occupied by conventional treatment processes. Processes which include aerated lagoons as polishing units of an anaerobic process occupy medium size areas.

Comparison of appropriate technology processes to conventional treatment processes shows that although there are appropriate technology processes which occupy larger areas, there are also many appropriate technology processes which occupy areas of similar size to the area occupied by the conventional processes or even smaller areas. So limited availability of land does not disqualify the use appropriate technology processes. In other words, there are appropriate technology combined processes that occupy small size areas, comparable to those of conventional processes and, as shown, which also yield effluent qualities comparable to those of conventional processes.

In many cases where processes that occupy a large area, mainly lagoons systems, were selected and located outside the city which they serve, after years the city grew and the plants became surrounded by residential areas. As a result, social pressures mounted to move the plants to other site. At such time, the values of the lands on which the plants were located became higher than what was paid for them and the utilities could benefit from the change of site or from modifying the process at the same site to a process which occupies a smaller area.

1.11.2 Processes which occupy small land areas and are adequate for cases in which the land available for wastewater treatment is limited

Land availability for wastewater treatment in each specific project is an important consideration in selecting the adequate treatment process or the scope of potential processes. If the area available for wastewater treatment is limited, than the following appropriate technology processes may be considered since they do not occupy large land extensions. The list of processes is not extensive and represents just a part of small area processes, relatively simple ones. Additional combinations which contain more unit process, and still occupy small land areas are available.

- *Preliminary Treatment followed by one of the Anaerobic Units (Anaerobic Lagoons or UASB or Anaerobic Filter).* The effluent of an anaerobic lagoon is not of the highest quality, the effluents of a UASB plant and of an anaerobic filter plant are of better quality, but still not the highest. In certain cases the quality of the effluents of these anaerobic units is sufficiently high to be discharges to large flow rivers. The specific area of an anaerobic lagoon is (0.15–0.2) m^2/Capita (95% of the area of activated sludge), the specific area of an UASB is (0.03–0.1) m^2/Capita (35% of the area of activated sludge) and the specific area of an Anaerobic Filter is (0.03–0.1) m^2/Capita (35% of the area of activated sludge).
- *Preliminary Treatment followed by CEPT.* The effluent of a CEPT unit is not of the highest quality; however, it is quite good and in certain cases is sufficiently high to be discharged to large flow rivers. The process flow diagram of a CEPT unit is presented in Figure 1.37. The specific area of a CEPT unit is (0.002–0.005) m^2/Capita (2% of the area of activated sludge). The numbers depend also on the treatment method of the sludge produced by the CEPT unit. Certain sludge treatment installation may require larger areas, but in any event, the area occupied by a CEPT unit is much smaller than the area occupied by an activated sludge unit treating the same flow.
- *Septic Tank followed by Anaerobic Filter.* The flow diagram of this process is presented in Figure 1.92. This is a processes usually used in small communities. The specific area of this process is (0.06–0.15) m^2/Capita (57% of the area of activated sludge).
- *Preliminary Treatment followed by Anaerobic Lagoon followed by Anaerobic Filter.* The flow diagram of this process is presented in Figure 1.97. The specific area of this process is (0.18–0.3) m^2/Capita (130% of the area of activated sludge).
- *Preliminary Treatment followed by one of the Anaerobic Units (UASB or Anaerobic Filter) followed by DAF.* The process flow diagram of UASB followed by DAF is presented in Figure 1.69. The process flow diagram of Anaerobic Filter followed by DAF is presented in Figure 1.95. The specific area of the process UASB followed by DAF is (0.05–0.15) m^2/Capita (54% of the area of

activated sludge). The specific area of the process Anaerobic Filter followed by DAF is (0.04–0.12) m^2/Capita (42% of the area of activated sludge).

- *Preliminary Treatment followed by one of the Anaerobic Units (UASB or Anaerobic Filter) followed by Sand Filter*. The process flow diagram of UASB followed by Sand Filter is presented in Figure 1.83. The process flow diagram of Anaerobic Filter followed by Sand Filter is presented in Figure 1.94. The specific area of the process UASB followed by Sand Filter is (0.05–0.15) m^2/Capita (54% of the area of activated sludge). The specific area of the process Anaerobic Filter followed by Sand Filter is (0.05–0.15) m^2/Capita (54% of the area of activated sludge).
- *Preliminary Treatment followed by UASB followed by Anaerobic Filter*. The flow diagram of this process is presented in Figure 1.64. The specific area of this process is (0.05–0.15) m^2/Capita (54% of the area of activated sludge).
- *Preliminary Treatment followed by CEPT followed by Sand Filter*. The flow diagram of this process is presented in Figure 1.72. The specific area of this process is (0.004–0.015) m^2/Capita (5% of the area of activated sludge).
- *Preliminary Treatment followed by CEPT followed by DAF*. The flow diagram of this process is presented in Figure 1.99. The specific area of this process is (0.012–0.02) m^2/Capita (9% of the area of activated sludge).
- *Preliminary Treatment followed by CEPT followed by UASB*. The flow diagram of this process is presented in Figure 1.110. The specific area of this process is (0.032–0.105) m^2/Capita (37% of the area of activated sludge).
- *Preliminary Treatment followed by UASB followed by Aerated Lagoons*. The flow diagram of this process is presented in Figure 1.114. The specific area of this process is (0.23–0.6) m^2/Capita (224% of the area of activated sludge).
- *Preliminary Treatment followed by Anaerobic Filter followed by Aerated Lagoons*. The flow diagram of this process is presented in Figure 1.115. The specific area of this process is (0.23–0.6) m^2/Capita (224% of the area of activated sludge).
- *Preliminary Treatment followed by UASB followed by Anaerobic Filter followed by DAF followed by Membrane Filtration*. The flow diagram of this process is presented in Figure 1.79. The specific area of this process is (0.08–0.23) m^2/Capita (84% of the area of activated sludge).

General guidance for selection of unit and combined appropriate technology treatment processes according to the land area they occupy is presented in Table 1.7. For a given available area size for locating the treatment plant, this table can help identify a series of processes which are adequate and may by considered for further, more detailed analysis.

Table 1.7 Processes selection matrix by occupied land area.

Process number	Process description	Land area occupied by the process (m^2/Capita)		
		0.002–0.01 wery small area	0.01–0.5 small area	0.5–5.0 large area
1	Rotating micro screens	X		
2	Vortex grit chamber	X		

(*Continued*)

Table 1.7 Processes selection matrix by occupied land area (*Continued*).

Process number	Process description	Land area occupied by the process (m^2/Capita)		
		0.002–0.01 wery small area	0.01–0.5 small area	0.5–5.0 large area
3	CEPT	X		
4	Septic tank	X		
5	CEPT followed by sand filtration	X		
6	CEPT followed by DAF	X		
7	Anaerobic lagoons		X	
8	UASB		X	
9	Anaerobic filter		X	
10	UASB followed by anaerobic filter		X	
11	UASB followed by DAF		X	
12	UASB followed by sand filter		X	
13	Anaerobic filter followed by DAF		X	
14	Anaerobic filter followed by sand filter		X	
15	CEPT followed by UASB		X	
16	USAB followed by arated lagoons		X	
17	Anaerobic filter followed by aerated lagoons		X	
18	UASB followed by anaerobic filter followed by DAF followed by membrane filtration		X	
19	Conventional lagoons system (anaerobic + facultative + maturation)			X
20	Upgraded lagoons system (covered anaerobic + facultative and maturation with mixers)			X
21	Constructed wetlands			X
22	Stabilization reservoir			X
23	Overland flow			X
24	Infiltration			X
25	Anaerobic lagoons followed by facultative lagoons			X
26	UASB followed by maturation lagoons			X
27	UASB followed by overland flow			X
28	Anaerobic lagoons followed by aerated lagoons's			X

(*Continued*)

Table 1.7 Processes selection matrix by occupied land area (*Continued*).

Process number	Process description	Land area occupied by the process (m^2/Capita)		
		0.002–0.01 wery small area	0.01–0.5 small area	0.5–5.0 large area
23	Septic tank followed by infiltration's			X
24	Any combined process which includes a polishing unit of lagoons, overland flow, stabilization reservoir, constructed wetlands and infiltration basins			X

1.12 REMOVAL OF PATHOGENS, PHOSPHOROUS AND NITROGEN IN APPROPRIATE TECHNOLOGY PROCESSES

1.12.1 Removal of pathogenic organisms

In developing countries the starting point in terms of wastewater treatment is quite low and in most cities, there is no wastewater treatment at all. Usually the priority objective under such conditions is removal of organic matter and pathogenic organisms from the wastewater while nutrients removal is of secondary priority and can be undertaken at a later stage, where necessary. However, in certain cases nutrients removal is also of high priority in developing countries.

The effluent of an effective submarine outfall, when considering an outfall as a treatment process, does not need to be disinfected since pathogens reduction is achieved through dilution and the UV impact of the sun. All other appropriate technology processes which were presented do not significantly remove pathogens except for lagoon systems and stabilization reservoirs, which partially remove pathogenic organisms and when necessary can be designed to fully remove them. Infiltration of effluents into the sub-soil is also a disposal method that removes pathogens, which do not survive after the percolation process and especially after a long detention time in an aquifer. Anaerobic processes and combined appropriate technology processes, except lagoon systems, stabilization reservoirs and infiltration, remove only a half to one logarithmic unit of fecal coliforms, which from the standpoint of public health risk is not a meaningful removal. They also do not achieve a meaningful removal of helminth eggs (except for lagoons systems, stabilization reservoirs and infiltration, which do eliminate helminth eggs). There is little information on removal of other pathogenic organisms by these processes, but the removal level is most probably not significant. However, this is not different than the situation with conventional treatment processes. Activated sludge processes of various types, trickling filters, aerated lagoons etc. also do not remove pathogens and remove only a half to one logarithmic unit of fecal coliforms, which is insufficient and requires disinfection of their effluents.

Removal of pathogenic microorganism from effluents is important, especially in developing countries, but its necessity depends on the specific situation of each case. For all other mentioned appropriate processes (with the exception of submarine outfall effluents and infiltrated effluents) pathogenic microorganisms removal can be achieved using conventional disinfection methods, i.e. by chlorine and its derivative compounds (hypochlorite, chlorine dioxide and others), by UV disinfection, by ozonation, or by membrane filtration. As commonly known, it is impossible to perform effective disinfection on raw wastewater or effluents with a high content of suspended solids because the pathogenic organisms hide within the pores of the suspended solids. The effluent qualities of all the appropriate technology unit processes (except for preliminary treatment) and combined processes presented above are such that each of them can undergo an effective disinfection, including lagoons and stabilization reservoirs effluents. This is an additional advantage of the appropriate treatment processes. Disinfection by chlorine gas, by liquid hypochlorite and by UV radiation can be considered appropriate technologies since they are not expensive, relatively simple and well known to water and sanitation utilities since they are also used for disinfection of water. Disinfection by ozone, chlorine dioxide and membrane filtration are not considered appropriate technology processes. Ozone and chlorine dioxide need to be manufactured on spot since they are unstable materials. Their application as disinfectants is complicated and expensive. Membrane filtration (ultrafiltration, nanofiltration) is a good disinfection method since the membranes form a barrier to the passage of pathogenic organisms and they do not introduce chemical products into the water, but it is still and expensive method.

1.12.2 Removal of phosphorous and nitrogen

Under a wastewater reuse scheme for irrigation, removal of nutrients is not necessary since they are an asset for irrigation, serving as fertilizers. Pathogens can be removed in the stabilization reservoir when the irrigated crops require a pathogen free effluent.

The effluent of an effective submarine outfall, when considering an outfall as a treatment process, does not need to undergo nutrients removal treatment because for the level of nutrients present in typical domestic wastewater, the dilution level of an effective outfall is sufficient to reduce the nutrients concentrations around the discharge point to safe levels, except in cases of discharge near a coral reef. But discharge of effluents in the vicinity of coral reefs is a practice that should be avoided in any case.

Wetlands and lagoons may provide reasonable nutrients removal efficiencies when designed for that purpose.

All other appropriate technology processes which were presented do not remove nutrients with high efficiencies. When nutrients removal is required, additional installations need to be provided for that purpose, and this means added cost and complexity. This is not however different than in the case of conventional treatment processes, to which specific nutrients removal installations need to be added when nutrients removal is required. Such installations include providing conditions for nitrification, adding denitrification reactors with recirculation pumps and piping, adding anoxic tanks for phosphorous luxury uptake and more. However, it is easier to remove nutrients when the treatment process is based on conventional technology than when it is based on appropriate technology.

If an effluent is discharged to a river, sometimes nitrogenous BOD needs to be removed from it on top of carbonaceous BOD, so that oxygen consumption for nitrification would not occur in the river. One way to remove nitrogenous BOD is to nitrify the effluent before discharging it to the river, but removal of nitrogen is not necessary and the effluent may contain nitrogen in the form of nitrate. In several countries, the quality standard for effluents discharged to a river requires a limit on ammonia, not on nitrate, and that is a good approach. If an effluent is discharged to a closed water body (lake, bay or estuary) phosphorous and nitrogen might need to be removed to prevent eutrophication, but in many cases the limiting factor is the phosphorous and it is this nutrient that needs to be removed, not nitrogen. The removal of phosphorous is easier than the removal of nitrogen, so eutrophication can be prevented by effective removal of phosphorous even without removal of nitrogen. However, since in many cases effluent quality standards also require nitrogen removal, this matter is discussed here as well. The decision regarding the nutrient which needs to be removed has to be developed for the specific condition of each case, however, usually the effluent quality standards do not allow taking into account specific conditions and require complete nutrients removal for all cases. Such an approach should be reconsidered in developing countries since nutrients removal is costly and sometimes not necessary.

Removal of phosphorous by luxury uptake does not occur in appropriate technology processes. When removal of phosphorous is required, it can be achieved by precipitating the phosphorous with the aid of flocculants (iron salts, usually ferric chloride, or aluminium sulphate – Alum). In the CEPT processes phosphorous precipitation is an inherent part of the process. For lagoons processes or combined processes whose last unit is lagoons based, only flocculants dosing equipment will be required. Mixing can take place in line by installing static mixers in the pipelines and injecting the flocculants into the line before the mixers. Phosphorous salt precipitation is then taking place in one of the lagoons. For the rest of the processes, installations for flocculants storage, dosing into the effluent stream, rapid and slow mixing of the flocculants and phosphorous sedimentation and clarification installations need to be added.

There is no nitrogen removal in anaerobic lagoons and in other types of anaerobic reactors (UASB, Anaerobic Filter etc.) only transformation of organic nitrogen to ammonia. The mechanisms of nitrogen removal in facultative and polishing ponds are: (i) volatilization of ammonia; (ii) assimilation of ammonia by algae; (iii) assimilation of nitrate by algae; (iv) nitrification-denitrification; and (v) sedimentation of particulate organic nitrogen. Assimilation of ammonia and nitrate by algae removes only a small portion of the nitrogen from the liquid. Similarly, sedimentation removes only a small portion of the nitrogen. Natural nitrification does not seem to be common in facultative and polishing lagoons. Nitrifying populations are present in these lagoons but are small. Nitrification occurs in the Western Treatment Plant in Melbourne, Australia, which is a lagoons plant with an overall detention time of 80 days. Perhaps a long detention time is required to achieve natural nitrification in lagoons, but in most lagoons systems the detention time is shorter and nitrification does not occur. The most significant nitrogen removal mechanism in lagoons is volatilization of ammonia (NH_3).

In an aqueous solution ammonia is present as dissolved ammonia gas (NH_3) which is volatile and can leave the solution naturally or be stripped out of it, and as an ammonium ion (NH_4^+) which cannot be stripped out. The relation between the NH_3 and the NH_4^+ in the solution depends on the pH and the temperature of the solution. At low values of pH all the ammonia is present in the form of NH_4^+ and with the increase of the pH it is transformed to NH_3. At a natural pH of 7 and at 20°C all the ammonia is practically in the form of NH_4^+. At a pH of about 9.5, 50% of the ammonia is in the form of NH_4^+ and 50% is NH_3. At a pH of 11 and higher, all the ammonia is in the form of NH_3. Natural volatilization can be achieved in a lagoons system and its level depends on the pH, the temperature and the detention time in the lagoons. Von Sperling (2002) provides on page 145 a table and a graph of the theoretical efficiency of nitrogen removal in lagoons as function of detention time and pH at 20°C. At pH 8 the percentage of nitrogen removal varies from 43% at a detention time of 3 days to 55% at a detention time of 40 days. At pH 9 the percentage of nitrogen removal varies from 61% at a detention time of 3 days to 69% at a detention time of 40 days. The removal efficiencies increase with the increase of temperature. Rapid photosynthesis leads to increasing the lagoons pH, often up to values of 9–10. This induces the transformation of the ammonia to dissolved gas and tends to increase nitrogen removal by volatilization of ammonia. However, the dissolved ammonia gas, NH_3, is toxic to algae. The ammonia toxicity increase with the increase of pH since the pH increase induces the increase of the dissolved ammonia gas. So a lagoon system is actually self-regulating the pH and ammonia levels. When the pH and dissolved ammonia gas increase the toxicity to algae increases, and therefore the photosynthesis decreases. As a result, the pH falls and the toxicity decreases. This allows the photosynthesis to increase again resulting in pH increase and the cycle repeats itself. This cycle provides a limit to natural volatilization of ammonia.

When considering induced nitrogen removal from lagoons there are three options. Option 1 is actually not for nitrogen removal but rather for nitrogen transformation. It refers to induced nitrification. This process transforms ammonia and organic nitrogen to nitrate and is sufficient to protect receiving waters from nitrogenous Biochemical Oxygen Demand. As mentioned, in some countries the effluent quality standards put an upper limit on the ammonia concentration (at about 5 mg/l N-NH_3) but there is no restriction on nitrate concentration so nitrification can provide a solution to such a requirement. Nitrification is achieved in a lagoon by installing in it modules of aerated fixed film beds which provide optimal conditions for development of nitrifying bacteria. Each module is filled with thousands of pieces of bio-media where, during normal operation, a healthy biomass is formed by a supply of air produced by the fine bubble membrane diffusers located on the bottom of each module and by the contact with the wastewater. As with any fixed-film biological contactor, the units will tend to become overgrown with organisms after a period of time. When this happens the module has a separate coarse-bubble system, controlled from the shore, to slough the media bed and clear it for renewed growth. Each module can be deployed individually or in series to remove varying ammonia loads. The modules system can be sized to eliminate as much ammonia as required to meet effluent requirements. The modules diffuse a high level of dissolved oxygen into the surrounding wastewater through the media bed. Induced nitrification in lagoons is a proven technology and additional information on the method can be found at www. GurneyEnvironmental.com. The method has been successfully used to maintain nitrification in cold northern climates by creating the stable, oxygen-rich, and immobilised microenvironment favoured by the nitrifying bacteria. Successful performance of this type of systems has been achieved throughout a complete winter seasons at wastewater treatment plants in upstate New York and northern Pennsylvania in the United States. A photo of such a fixed bed nitrifying modules installed in a lagoon is presented in Figure 1.18. They are the rectangular shaped objects seen in the right hand side lagoon of the photo. The nitrifying modules add cost to the treatment plant, both in investment and in O&M cost for air supply, but as previously explained; nutrients removal adds cost in any process, even in conventional systems. However, the operation of the system is still simple and it can be considered an appropriate technology system.

The second option for nitrogen removal in a lagoons system is nitrification-denitrification. This option is proposed at a level of an idea, not known yet to be performing at a full scale plant, but theoretically it should work. It is basically the same idea of nitrification-denitrification in activated sludge systems. The system flow diagram is the following: Anaerobic lagoon followed by an anoxic reactor (retention tank or basin) in which the denitrification process takes place, followed by one or two facultative lagoons in which the nitrification process takes place, followed by polishing ponds. The nitrification process in the facultative lagoons is an induced process achieved by installing in the lagoons the modules of aerated fixed film beds as described above. A recirculation stream is installed from the last facultative lagoon to the anoxic reactor, at a rate of 1.5–2 in relation to the raw wastewater flow. This recirculation stream conveys to the anoxic reactor the nitrates and since the reactor is not aerated and is covered and contains mixers, photosynthesis will not take place in it so the nitrate will be used by the heterothrophic bacteria as the oxygen source and the denitrification process will take place in this reactor. It is necessary to provide a very short detention time in the anaerobic lagoon which precedes the anoxic reactor to restrict the BOD removal in it to no more than 50%, or perhaps even omit the anaerobic lagoon at all, so as to ensure sufficient substrate for the denitrification process. It might also be necessary to recirculate some sludge from the downstream facultative lagoons to increase the concentration of heterothrophic bacteria in the anoxic reactor. The anoxic reactor and the recirculation, which requires pumping, increase investments and O&M costs, but nitrogen removal is costly. The systems remains however simple to operate and is still an appropriate technology system.

The third option for nitrogen removal in a lagoons system is natural or induced ammonia stripping. It has two variants. Variant 1 is ammonia stripping in a polishing lagoon. This variant is proposed at a level of an idea, not known yet to be performing at a full scale plant, but theoretically it should work and it was tested on a pilot scale. As previously mentioned, it is recommended to add mixers in all the facultative and polishing lagoons. These mixers can increase the natural ammonia volatilization in the lagoons. On top of that, an additional ammonia stripping lagoon can be added as a final lagoon, or the last polishing lagoon can be used for this purpose. Variant 1 has three options: (i) natural ammonia stripping in a long detention time shallow lagoon; (ii) ammonia stripping by high lime treatment that raises pH to 11.0–11.5, and (iii) ammonia stripping by low lime treatment that raises pH to 9.0–9.5. As to natural ammonia stripping, there is evidence (Cavalcanti, 2009 and Mota and von Sperling, 2009) that Ammonia stripping can occur naturally in a long detention time shallow (0.5–1.0 meter deep) polishing lagoon, even without increasing the pH by lime. In long detention time polishing lagoons the photosynthesis rate is high, leading to high pH values of over 9.0 which stimulates ammonia stripping. Cavalcanti (2009) reports 92% $N\text{-}NH_3$ natural removal in an experimental lagoons system with a series of polishing lagoons with a total detention time of about 30 days in the polishing lagoons. Von Sperling (2002) estimates that at pH 9 and 20°C nitrogen removal can reach 69% at a detention time of 40 days. Mota and von Sperling, (2009) report that the specific area required to reduce $N\text{-}NH_3$ to less than 5 mg/l in polishing lagoons is 2–3 m^2/Capita for lagoons 0.5 m deep and 3–4 m^2/Capita for lagoons 1 m deep. This is equivalent to prolonged detention times in the polishing lagoons of 30–40 days. If land is available, ammonia stripping can then be achieved naturally in long detention time shallow polishing ponds. If land availability is restricted, ammonia stripping can be induced in a shorter detention time by raising the pH in the polishing lagoon either by high lime treatment or by low lime treatment. In the case of option 2 of Variant 1, the high lime treatment, the pH in the ammonia stripping lagoon is increased to 11.0 11.5 by using either quick lime (CaO) or hydrated lime $Ca(OH)_2$ to raise the pH. The lime cannot be added directly to the lagoon because high-lime treatment results in precipitation of large quantities of calcium carbonate which will fill the lagoon. Therefore, a reactor-clarifier must be installed prior to the lagoon. The addition of lime is done in the reactor-clarifier, in which the precipitation takes place and from which the settling sludge is removed continuously. The reactor-clarifier's effluent is then diverted to the lagoon in which the ammonia stripping takes place. Because of the large amount of lime required to raise the pH to 11.0 and above, high-lime treatment produces large quantities of sludge which need to be dewatered and disposed. The lagoon is to be equipped with mixers to obtain an intensive mixing, or if necessary it can be equipped with a combination of mixers and floating aerators. This would assist the stripping of ammonia. It is estimated that over 90% removal of ammonia can be achieved under this variant with high lime treatment and with detention times of about 10 days in the lagoon, depending on the temperature and the mixing intensity in the lagoon. Higher ammonia removal efficiencies can be achieved with larger detention times. The pH in the lagoon effluent will drop naturally to about 9.5–10.0 but discharge standards usually require a lower value so lowering the pH of the final effluent by adding to it an acid solution might be required. The high lime ammonia stripping process was tested on a pilot scale plant, the results of which are reported by Idelovitch *et al.* (1981), and it was also applied in a full scale plant, the Dan Region wastewater treatment plant in Tel Aviv, Israel and functioned successfully for many years (Idelovitch et al (1981). The detention time in the lagoon of the high lime treatment option is much lower than that of the natural ammonia stripping lagoon so the area occupied by the high lime treatment installation is much smaller. In the case of low lime treatment the pH in the ammonia stripping lagoon is to be increased to 9.0–9.5 (again by using either quick lime or hydrated lime). In this case if the alkalinity of the wastewater is not high, the lime can be added directly to the inflow stream to the lagoon. The lagoon is to be equipped with mixers to obtain an intensive mixing, or if necessary it can be equipped with a combination of mixers and floating aerators. This would assist the stripping of ammonia. It is estimated that 70–80% removal of ammonia can be achieved under this variant with low lime treatment. The required detention time in the lagoon needs to be established. It depends on the pH level, the temperature and the mixing intensity. It is estimated that it is not as long as in the case of natural ammonia stripping (Variant 1 option 1) because there is no fluctuation in the pH (which in this case does not depend on photosynthesis) and because of the mixing assistance to the stripping process. As mentioned there is a risk of calcium carbonate precipitation in the lagoon, which will have to be cleaned periodically. If the alkalinity of the wastewater is high, the precipitated calcium carbonate quantity may be too high and in this case, a reactor-clarifier must be installed prior to the lagoon. The addition of lime is done in the reactor-clarifier, in which the precipitation takes place and from which the sludge is removed continuously. The clarifier's effluent is then diverted to the lagoon in which the ammonia stripping takes place.

Variant 2 of induced ammonia stripping is based on the use of a stripping tower instead of a lagoon. This is a more conventional method which can reach up to 90–95% of ammonia removal. Removal efficiency depends on temperature and on the efficiency of the air-water contact. As the temperature decreases, the amount of air required increases significantly for the same degree of removal. Lime or caustic soda is added to the nitrogen rich effluent until the pH reaches 10.8 to 11.5 units, which converts ammonium hydroxide ions to dissolved ammonia gas. In a countercurrent stripping tower the effluent rich ammonia is pumped to the top of a packed tower and effluent drops fall through the packing. Air is drawn through openings in the bottom of the tower and flows up. Free ammonia is stripped from the falling effluent droplets into the air stream and then discharged to the atmosphere. There is another type of tower, cross flow tower, in which the air enters along the entire depth of the packing, but the principle of its performance is similar to that of the countercurrent tower. The required airflow is about 1,000 times the water flow and can be higher depending on the temperature. Problems of calcium carbonate scaling and poor performance during cold weather have been reported in regard to operation of ammonia stripping towers. Removal of ammonia in stripping towers is more complex than in lagoons.

In unit or combined appropriate technology processes which do not include lagoons, such as UASB followed by Anaerobic Filters and many others, there are two options for nitrogen removal. Option 1 is actually not for nitrogen removal but for nitrogen transformation. It refers to induced nitrification. This process transforms ammonia and organic nitrogen to nitrate and is sufficient to protect receiving waters from nitrogenous Biochemical Oxygen Demand. Nitrification can be achieved by subjecting the effluent to the same modules of aerated fixed film beds described above, which in this case can be installed not in a lagoon but rather in a retention tank or even in earth basins. Option 2 for nitrogen removal in unit or combined appropriate technology processes which do not include lagoons is induced ammonia stripping. In this case there are two variants identical to the ones presented for the of lagoons systems. Variant 1 is ammonia stripping in a lagoon or a basin, which needs to be added as the final unit of the treatment plant. Variant 2 is ammonia removal in a stripping tower, which needs to be added as the final unit of the treatment plant.

1.13 RECOVERY OF RESOURCES FROM MUNICIPAL WASTEWATER, THE POTENTIAL FOR GENERATION OF ENERGY IN WASTEWATER TREATMENT PLANTS AND ITS IMPLICATIONS REGARDING THE SUSTAINABILITY OF THEIR OPERATION

1.13.1 Introduction

Municipal wastewaters contain several resources which can be recovered and reused. First, 99.9% of the municipal wastewater is water. The rest 0.1% is partly dissolved and partly in form of suspended solids, part of both is organic matter and part mineral matter. In water scarce zones, treated effluents can be used as a source of water for irrigation. Given the global water crisis, which will become more acute with time, the reclamation of water from municipal wastewater will gain importance and become more widespread. Municipal wastewaters also contain nutrients: nitrogen, phosphorous and some potassium. When effluents are reused for irrigation, these nutrients become fertilizers. Depending on the irrigation method and type of irrigated crops, the nutrients in the effluents may partially or completely satisfy the need for fertilizers. This is very important in developing countries, where poor farmers have financial difficulties to purchase fertilizers (and this is the reason that they prefer using effluents or even raw wastewater for irrigation).

Municipal and industrial wastewaters contain organic matter. This organic matter contains some energy, which can be recovered in anaerobic processes. The fact that municipal wastewater contains 99.9% water may seem to contradict the fact that the wastewater contains energy. It is to be understood that municipal wastewater is not an oil well but it contains some energy which can be recovered and put to use and this may result to be of significant financial importance, especially in developing countries.

Wastewater contains therefore three valuable resources which can be recovered: (i) water, (ii) fertilizers; and (iii) energy. They can become sources of income to utilities through the sale of effluent and the generation of energy for internal use or for sale, however in each case it is necessary to assess the feasibility of energy recovery. Additional income flows on top of income from tariffs will have a positive financial effect on the utilities, and may result in positive institutional and sustainability impacts, as discussed below.

The successful use of these resources depends to a large extent on governments' policies and regulations. An enabling environment is required both for effluents reuse in irrigation and for generation of energy. In the case of reuse for irrigation, reasonable effluent quality standards need to be in places as well as financial support and technical assistance. In the case of energy generation adequate policies and regulation are required to stimulate the generation of energy and render it financially feasible. If the enabling environment does not exist the result is utilization of raw wastewater for irrigation, because in areas of scarcity, farmers in need of water ignore the laws and use raw wastewater. In the case of energy a lack of an enabling environment will prevent utilities from accessing this resource, losing thereby the benefits which it can provide.

Another issue related to wastewater disposal in general and to energy generation or methane flaring in particular is the reduction of emission of Green House Gases. Such reduction is beneficial by contributing to the effort of mitigating the global warming, and it can also generate an additional income to the utilities through the Carbon emission Reduction (CER) credits.

Reuse of effluents for irrigation and utilization of the nutrients as fertilizers are more straightforward resources, they are already in use and their utilization is increasing. In this chapter we expand the discussion on energy generation, which is still in limited use, especially in developing countries.

1.13.2 Effluents as a water source for irrigation

Reuse of effluents for irrigation of crops is only one of the reuse options of effluents, but as explained in Chapter 1 Section 1.7.2.7, it is an attractive option. The engineering design approach of effluent reuse for irrigation is explained also in Chapter 7. Reuse for irrigation has a significant economic value. It is part of the measures to resolve water scarcity problems in arid zones. The economic value of effluents (value and not cost) may be different in different countries. The economic value of effluent irrigation is not just the cost which can be charged for the effluent but also the economic value of generating a new water source, because it liberates fresh water which was used for irrigation, to be used for purposes that require only fresh water and that has a significant economic an environmental value. Irrigation with effluent also prevents contamination of fresh water bodies by wastewater and by effluents (even high quality effluents have a contamination effect on fresh water bodies), and that also has a significant economic and environmental value.

In countries where the law recognized that the treated effluents are the property of the water and sanitation utility and that the utilities can sell it, the sale of effluent can generate income and improve the financial situation of the utilities. It is difficult to provide specific figures regarding the financial income from sale of effluent since it is country and project specific, but the point is clear: wastewater can generate water for irrigation and in this respect it is a valuable resource in water scarce zones.

Although it might not be easily perceived, this books deals extensively with technical aspects of wastewater treatment and effluent management for reuse in irrigation of crops. Chapter 1 provides general information on reuse in irrigation, on the reuse concepts and need for seasonal storage of effluents. It concentrates on the concepts of seasonal storage, on the stabilization reservoir and on the required pretreatment prior to storing effluent in a stabilization reservoir. Chapter 7 presents the design procedure of stabilization reservoirs, and the rest of the chapters in the book provide information on appropriate technology treatment processes of wastewater which can serve as pretreatment units of a stabilization reservoir. We did not expand on other aspects of effluent irrigation, such as quality standards, social aspects and management aspect, which are also very important but for which there is a wide cover in the literature. We preferred to concentrate on the practical/technical aspects of reuse. Chapter 7 is unique in the sense that it provides an algorithmic design procedure for stabilization reservoirs, which is an essential component in reuse projects and which is not extensively dealt with in the literature.

1.13.3 Effluents as a source of fertilizers

Municipal wastewaters contain nutrients: nitrogen, phosphorous and potassium. Total Nitrogen concentration in typical raw wastewater is in the range of 20–85 mg/l, total phosphorous is in the range of 4–15 mg/l and potassium is in the range of 10–30 mg/l (see table 3.2). Potassium is not removed in wastewater treatment processes. Nitrogen and phosphorous may be removed, if the treatment process is designed for nutrients removal. When the effluent is destined for irrigation, the wastewater treatment process should not include nutrient removal, and this is an advantage since processes of nutrients removal are more costly.

When effluents are reused for irrigation, the nutrients contained in it act as fertilizers. Depending on several factors such as the concentration of the nutrients in the raw wastewater, the irrigation method and the type of irrigated crops, the nutrients in the effluents may partially or completely satisfy the crops requirement of fertilizers. It is difficult to provide general estimates as to the percentage of fertilizers requirement which is provided by effluent irrigation because the figures are project specific. The following example provides a general idea.

Libhaber (1987) presents an economic analysis of a wastewater treatment and effluent irrigation system which takes into consideration the fertilizers value in the effluent. The wastewater is treated in anaerobic lagoons whose effluent is stored in a 300,000 m^3 stabilization reservoir. The project supplies a net annual volume of 460,000 m^3 for irrigating an area of 86 ha of cotton during summer and irrigating part of this area for cultivation of wheat for silage during winter. The irrigation method used is drip irrigation. Since most of the effluent is consumed by cotton, the calculations were based on application rate for irrigation of cotton which is about 5,000 m^3/ha.year. The content of nutrients in the effluent was: 25 mg/l nitrogen (equivalent to 125 mg/l Ammonium Sulphate), 5.6 mg/l phosphorous (equivalent to 74 mg/l Superphosfate) and 24 mg/l potassium (equivalent to 45.8 mg/l Potassium Chloride). Assuming that 85% of the nutrients are available to the crops and an irrigation water use efficiency of 80% by drip irrigation, the resulting rates of nutrients application are the following:

$$\text{Ammonium Sulphate:} \quad 5{,}000 * 0.85 * 0.8 * 125 * 0.001 = 425 \text{ kg/ha} \cdot \text{year}$$
$$\text{Superphosfate:} \quad 5{,}000 * 0.85 * 0.8 * 74 * 0.001 = 252 \text{ kg/ha} \cdot \text{year}$$
$$\text{Potassium Chloride:} \quad 5{,}000 * 0.85 * 0.8 * 45.8 * 0.001 = 156 \text{ kg/ha} \cdot \text{year}$$

The required fertilizers application rates according to regular practice for cotton are 800 kg/ha.year of Ammonium Sulphate, 500 kg/ha.year of Superphosfate and 500 kg/ha.year of Potassium Chloride. This means that in this case the percentages of fertilizers applied by the effluent are:

Nitrogen: 53%
Phosphorous: 50%
Potassium: 31%

In this case we can assume that an average figure of 50% of saving on fertilizers can be considered. The cost of fertilizers for cotton is 129 US$/ha.year, thus the saving is about 65 US$/ha.year or about 5,600 US $/year for the specific project described above. The price figures refer to the time that the analysis was undertaken, which is years back, and which may be different to date.

The fertilizers value in the effluent is a very important economic factor for developing countries, where poor farmers have financial difficulties to purchase fertilizers. One of the reasons farmers prefer using effluents (or even raw wastewater) for irrigation in developing countries is the fertilizers content in the effluents, which results in better crops and higher yields.

1.13.4 Wastewater as a source of energy

Municipal and industrial wastewaters contain organic matter measured as COD. This organic matter contains some energy, which can be recovered in anaerobic processes. The end products of anaerobic decomposition of wastewater are: (i) biogas, and (ii) small amounts of excess sludge (about 5–15% of the decomposed organic matter in the raw sewage). The biogas contains about 65–70% methane, about 25–30% CO_2 and the balance is mostly N_2 and some impurities, one of which may be H_2S, (if the raw wastewater contained high concentration of sulphur). If the first treatment unit in the treatment plant is an anaerobic process unit such as anaerobic lagoons, UASB or Anaerobic Filters, the biogas can be captured, cleaned and used as an energy source. In the case of an anaerobic lagoons process, it is necessary to cover the lagoons in order to be able to capture the biogas. The appropriate technology processes described in this book have low energy consumption and in most cases it is not possible to use within the plants the entire amount of energy produced by the anaerobic process. Consumers for the energy produced need therefore be found outside the wastewater treatment plants, either within or outside the utility and the feasibility of energy production from biogas needs to be studied.

Utilization of biogas is a common practice in Sanitary landfills, in activated sludge plants treating municipal wastewater, in which the sludge undergoes anaerobic digestion and the biogas produced is captured and utilized, and in industrial wastewater treatment plants where anaerobic processes are used to treat highly concentrated organic wastes. In activated sludge plants the biogas is used to generate electricity which is consumed within the plant itself (because such plants have high electricity consumption in the aeration system of the biological rector). The fact that an activated sludge plant can be the consumer of the energy it produces from the methane it generates facilitates the decision to utilize the methane because there is no need to look for external consumers. Many industrial wastewaters are treated in various types of anaerobic reactors and the biogas is captured and utilized. Usually the industry itself is the energy consumer and the private sector which owns the industry is interested in financial efficiency and will not lose the opportunity to generate additional income (or reduce expenses). However, in municipal treatment plants which are based on anaerobic treatment processes (not of sludge treatment but of treatment of the bulk of the wastewater), mainly anaerobic lagoons and UASB plants (existing anaerobic filter plants are too small and do not generate attractive biogas quantities); the practice of utilization of the biogas is basically non-existent, although the concept is well known. To capture the biogas produced in anaerobic lagoons, those need to be covered and there are only dew covered lagoons plants (the case of the city of Santa Cruz, Bolivia, is one of them). The number of UASB plant treating municipal wastewater is meaningful and increasing, but the concept of utilization of biogas still does not form part of the UASB concept. In most cases there is no need within the treatment plant for the energy which can be produced, so there is no drive to generate the energy. The design practices of USAB reactors are not geared to capture the biogas and a great part of it is lost to the atmosphere. And finally, public utilities do not have the drive to be more efficient, reduce losses and generate income, and that is another reason for not utilizing the biogas in municipal treatment plants. Anaerobic reactors can be designed to capture the biogas (as is done in industrial treatment plants) and the technology of biogas utilization is well established. The main issue is achieving economic feasibility of the biogas utilization project. This in turn strongly depends on the energy policy, legislation and regulation and on the energy prices (electricity, gas, biofuel etc.) in the country where a project is to be executed. The same project can be feasible in one country and unfeasible in another. This point is further demonstrated in the example analysis which follows.

There are several alternatives of the utilization of the biogas: (i) generation of electricity which can be used within the treatment plant if there is consumption for it, or in other installations of the utility outside the plant site, like water or wastewater pumping station, or be connected to the electricity distribution grid; (ii) use of the biogas, after adequate treatment, as a biofuel to operate vehicles and machinery; (iii) use of the biogas as an energy source for industry if there is industrial demand in the vicinity of the treatment plants. This could include generation of vapour in boilers, use of the biogas for heating and more; (iv) use of the bio gas for home cooking, and (v) use of the biogas for other purposes. Biogas can also be used in cogeneration. Cogeneration is defined as a system for generating electricity and producing another form of energy, usually steam or hot water. The biogas can be used to power an engine-generator to generate electricity, and the jacket water from the internal-combustion engine can then be used as the other energy source, for heating buildings, or for use in industrial plants to generate steam or in heat exchangers or for other purposes. In order to determine the best alternative for each case, it is necessary to undertake a specific feasibility study under the conditions of the project. This is more applicable to industrial wastewater than to domestic wastewater because of its low COD concentration. The analysis depends to a large extent on the energy policy in the country in which the treatment plant is located and on the energy prices in the country. This point is demonstrated in the following example.

The city of Santa Cruz in Bolivia has 4 wastewater treatment plants, two located in the East site (Plant E treating municipal wastewater and plant PI treating industrial wastewater) and two located in the North site (Plants N1 and N2 both treating municipal wastewater). Another plant is under construction in the South site (Plant Sur which will treat municipal wastewater). The wastewater flows received by each of the existing plant are presented in Table 1.8. In the short term the flow will increase when the South plant comes on stream. In the longer term two additional plants will be built (totalling 7 treatment plants) and the flow will increase accordingly.

Table 1.8 Wastewater flows to the existing treatment plant of Santa Cruz.

Plant/Site	Wastewater flow (m^3/d)
Plant N1	14,000
Plant N2	42,000
Total north site	56,000
Plant E	32,000
Plant PI	8,000
Total east site	40,000
Total of the 4 existing plants	96,000

The treatment process is identical in all the plants. These are lagoon plants, each composed of a set of lagoons in series including: a covered anaerobic lagoon followed by two facultative lagoons equipped with mixers followed by a maturation lagoon. Photos of the East site lagoons are presented in Figures 1.12 and 1.13 and photos of the N2 plant are presented in Figures 1.51 and 1.53. The biogas generated in the lagoons is captured (see Figure 1.57) and is currently being flared (see Figure 1.58). The treatment process practically does not consume energy. There are two energy consumption sources in this process: (i) the rotating micro screens, which constitute the preliminary treatment in each plant; and (ii) the mixers. The energy consumed by the rotating micro screens is very small. The energy consumed by the mixers when they are operated by their motors is also small, but even this small amount of energy is not consumed because most of the time the mixers are operated by the wind power. The energy contained in the biogas produced in each treatment site is way larger than the small amounts consumed by the treatment plants. This fact is positive in terms of O&M costs of the plants but is inconvenient from the standpoint of the utilization of the biogas because consumers for the energy produced need to be found outside the wastewater treatment plants, either within or outside the utility. A study of alternatives for utilizing the biogas generated in the Santa Cruz plants was undertaken by Incisa Bioenergia (2011). Several alternatives were analyzed. Three of them, the more interesting ones, are discussed here.

It was found in field measurement in the Santa Cruz plants that the ratio of methane produced to organic matter removed was in the range 0.1–0.15 Kg Methane/Kg COD removed. The number adopted for the analysis was 0.15 Kg Methane/Kg COD removed, which is lower than the default value of 0.21 Kg Methane/Kg COD removed recommended by the International Panel on Climate Change (IPCC) and than the value of 0.25 Kg Methane/Kg COD removed considered by the USEPA as the maximum possible value, but this number depends on the composition of the organic matter (i.e., on the specific wastewater) and on the reactor's efficiency and is reasonable for an anaerobic lagoon (which is a non-mixed reactor) without temperature control. It was also found that the correlation between methane production and the contributing population is 1380 m^3 methane produced per day per 100,000 persons. Measurements showed that the biogas generated in the SAGUAPAC treatment plants contains 72–79% methane and the balance is made up by CO_2, H_2O (water vapour) and N2. The biogas also contains H_2S at a concentration lower than 100 ppm, which is a low content. In the analysis of alternatives it was considered that the biogas contains 65% methane, which is a representative value.

The methane generation forecast and energy potential of the wastewater of Santa Cruz is presented in Table 1.9. The short term stage refers to connection of additional neighbourhoods, with some 120,000 inhabitants, to the sewerage network, which will be done shortly. The medium term stage refers to inauguration of the South treatment plant, which will raise the number of plants to 5, and the long term stage refers to incorporating to the system two additional treatment plants in the future.

Table 1.9 Forecasts of methane generation and energy potential in the wastewater treatment plants of Santa Cruz.

Status of development of wastewater treatment plants	Methane production (Nm^3/d)	Electricity potential (KW)	Renewable Natural Gas (RNG) potential* (Cars fuelled/day)
Current	9,388	1,146	440
Short term, 4 treatment plants	12,500	1,526	586
Medium term, 5 treatment plants	18,000	2,198	844
Long term, 7 treatment plants	26,100	3,187	1,223

*Based on RNG consumption of 16 m^3/day/car

Two energy products were considered to be generated from the biogas: (i) generation of electricity, and (ii) generation of Renewable Natural Gas (RNG) for its use as Vehicular Natural Gas (VNG). There are other products which can be generated, such as chemical transformation of the biogas but this would not be viable for the small scale project of Santa Cruz.

Electricity can be generated in two forms: (i) by motors and turbines; and (ii) by the vapour cycle, which is simpler but less efficient. The study analyzed the motors and turbines option.

RNG is a fuel generated from biogas with a composition equivalent to that of natural gas. There is a standard which defines the composition of RNG and it can be used as a fuel for vehicles. It took time for the RNG technology to mature and become economically feasible in small scale projects such as that of Santa Cruz. However, today it is an established technology, which in many aspects is more efficient than the generation of electricity.

To manufacture each of the two products the biogas needs to be treated. For both of them the water vapour needs to be removed from the biogas by dry or wet scrubbers. H_2S in concentrations in access of 100 ppm by volume also needs to be removed. In the study undertaken it is proposed to remove the H_2S by a filter of iron fibres. The gaseous H_2S attaches to the iron. After removal of water and hydrogen sulphide the biogas (with about 65% methane content) can be used to power internal-combustion engines, which in turn can operate a turbine that generates electricity.

To manufacture the RNG, the clean dry biogas which was prepared for energy generation needs to undergo additional treatment. The main objective is to separate the CO_2 so as to reach a methane content of at least 97%. Other impurities such as nitrogen, hydrogen sulphide and humidity need also to be removed so that the final product complies with the specification of natural gas. The additional purification process is based on Pressure Swing Adsorption (PSA) equipment, which separates gaseous phases by molecular membranes that control through magnetic affinity and pore opening which atoms can cross over and which cannot. It is a cyclic process in which the cylinder of PSA is filled and emptied using the pressure of the gas generated by a compressor. This compressor takes the gas to a pressure of 10 kgf/cm^2 and at the end of the purification process the GNR is compressed to its normal pressure of $220 \times 10\ kgf/cm^2$. At this moment the gas has its dew point at $-45°C$, more than 97% of it is methane and less than 0.5% is oxygen. There are ample options for purchase of high efficiency equipment from Europe and Canada. There is Brazilian equipment of lower efficiency which produces RNG at lower cost. This equipment is manufactured locally, close to Santa Cruz and that can facilitate preventive maintenance.

The potentials of electricity and RNG generation in the Santa Cruz project are specified in Table 1.9.

To understand the potential RNG market in Santa Cruz it is important to understand its competitors. The cost of fuels in Santa Cruz is presented in Table 1.10.

Table 1.10 Fuel prices in Bolivia on March 2011.

Fuel type	Price in Bolivianos (Bs)	Price in american dollars (US$)	Unit of fuel
Special gasoline	3.74	0.53	Liter
Premium gasoline	4.28	0.61	Liter
Diesel	3.72	0.53	Liter
Butane (liquefied gas)	0.034	0.05	Kg
Natural gas	1.68	0.24	Nm^3@250 Bar

Several alternatives for biogas utilization were analyzed in the study. We discuss three of them which are considered the most interesting.

Alternative 1

This alternative consisted initially of generating electricity in each treatment plant site and connecting it to the grid of the power utility, CRE, to sell the produced energy at a fair price. This is the simplest and most straightforward solution. But it was soon found that due to the country's energy policy it is impossible to implement such alternative. First, the same entity cannot be a consumer and producer of electricity. And even if it could, CRE charges SAGUAPAC a price of 0.086 US$/KWh, but due to the existing policy can only pay to a producer 0.03 US$/KWh. The cost of generating electricity in this project under alternative 1 is 0.0294 US$/KWh so the existing policy basically eliminates the implementation of Alternative 1 as described above. This is an example of the importance of the energy policy in establishing the feasibility of a project and an example of the inexistence of an enabling environment to stimulate the utilization of biogas produced in wastewater treatment plants.

SAGUAPAC is trying to obtain a licence to generate electricity, connect it to the grid and sell it to CRE for a fair price. Another option related to this alternative is that SAGUAPAC obtains the status of a non regulated producer. In this case it needs to generate more than 1 MW and it needs to identify a consumer or consumers who would buy its electricity at a price agreed between the parties. SAGUAPAC would then connect to the grid of CRE and the identified consumers would withdraw the energy from this grid. This is an option under which the produced electricity can be sold at a fair price. The financial analysis of Alternative 1, assuming a sale price of 0.086 US$/KWh for the electricity produced, yielded the following results:

Total investment: US$ 1.73 million
Annual savings in electricity or annual income from sale of electricity: US$ 1.25 million
O&M cost: 245,600 US$/year
Net present value of savings during 10 years: US$ 5.56 million
IRR: 78%
Cost of electricity generation: 0.0294 US$/KWh

Alternative 2

This alternative consists of conveying the produced biogas by pipes to 6 large water pumping station of SAGUAPAC and generation electricity at each pumping station for operating the pumps. SAGUAPAC has some large pumping station which can be operated by the biogas. This entails construction of 2 main gas pipes with a total length of 20 Km. The benefit of this alternative is that it does not require any authorization from government or contracts with consumers. It is the total discretion of SAGUAPAC to implement it. It has an additional benefit which was not taken into consideration in the study. The pumping stations do not have currently emergency power generators. This Alternative will provide a dual power system, normal operation by the generators and emergency operation by electricity from the grid. The financial analysis of Alternative 2 yielded the following results:

Total investment: US$ 2.31 million
Annual savings in electricity: US$ 0.795 million
O&M cost: 219,500 US$/year
Net present value of savings during 10 years: US$ 1.91 million
IRR: 24%

This alternative is inferior from the financial standpoint than Alternative 1, but is easier to implement from the institutional and regulatory standpoint.

Alternative 3

This alternative consists of manufacturing RNG in the treatment plants site (East North and South) along with gas sale station for sale of RNG to the public as vehicle fuel. The amount of RNG produced is much larger than the consumption of the vehicles of SAGUAPAC, so SAGUAPAC cannot be the sole consumer of the manufactured RNG. The RNG production and number of vehicles which can be fuelled at each treatment plant site is specified in Table 1.11.

Table 1.11 RNG supply station under alternative 3.

Plant site	RNG production (Nm^3/day)	Number of cars fuelled per day
North	3,975	250
East	5,400	340
South	4,125	260

The financial analysis of Alternative 3 revealed that the manufacturing cost of the RNG is 0.24 US $/Nm3RNG@250 Bar. This basically renders Alternative 3 non-viable, because, as shown in Table 1.10 this cost is identical to the price of natural gas, so the price of the RNG cannot be higher than the manufacturing cost. This shows again the impact of the energy policy. The liquid fuel prices in Bolivia are subsidised by the government and therefore low in relation to international prices. The price of natural gas, although not subsidized is kept at levels lower than international prices. This creates an environment which is not enabling for the utilization of biogas generated in wastewater treatment plants.

A project similar to that of Alternative 3 is being undertaken in the city of Oslo, Norway, where the municipality is planning to operate at the first stage 80 public service buses (a fifth of its fleet) with RNG manufactured from the biogas generated in the anaerobic sludge digestion units of the municipal wastewater treatment plants (Johansen, 2009). At a subsequent stage the entire fleet of 400 buses will be operated with RNG. One of the drives to implement this project in Oslo is the reduced carbon emission when using RNG in comparison to the emission from liquid fuel. In addition, in Oslo an enabling environment for this type of projects probably exists.

SAGUAPAC is still considering its options and trying to change the environment to somewhat more enabling. We know that policies may change with time so there is still hope. The lesson learned from the example of SAGUAPAC is actually that the utilization of biogas can be feasible in other countries. A situation similar to that of SAGUAPAC in a country with international gas prices and with a policy of stimulation generation of energy by small producers and, in general, with an enabling environment, would render all the studied alternatives feasible.

1.13.5 Wastewater treatment for reducing green house gases emission

Flaring or use of biogas also eliminates the emission of methane which is a strong Green House Gases (GHG). The reduction of GHG emission can be sold as CER (Carbon Emission Reduction) and an additional significant income can be obtained. As mentioned, SAGUAPAC operates four lagoons wastewater treatment plants. The anaerobic lagoons in each plant are covered and the biogas is being flared. The utility is studying the options of generating energy from the biogas. SAGUAPAC has already signed a contract for sale of the biogas emission reduction to the community development carbon fund, for an expected amount US$ 2,080,000 during the period 2007–2015 (the World Bank, 2007). This is the first case in developing countries of an emission reduction contract based on capturing of anaerobic lagoons biogas. Additional information on this carbon emission reduction case is presented in Chapter 12.

1.13.6 Contribution of resources generation to sustainability and improved management of utilities

The utilization of the resources contained in the effluent (reuse for irrigation) and in the by-product of anaerobic treatment (utilization of the biogas) can generate an additional income stream for the utility, on top of the income generated by providing the water and sanitation services. The reduction of emission of methane, a strong greenhouse gas, can in certain cases generate additional income from the sale of Carbon Emission Reduction.

The appropriate technology processes described in this book have low energy consumption and in most cases it is not possible to use within the plants the entire amount of energy produced by the anaerobic process because there is no demand for it. Consumers for the energy produced need therefore be found outside the wastewater treatment plants, either within or outside the utility. If the biogas is used within the utility it will save expenses on energy purchase and if used outside the utility it will generate income from the sale of energy or biogas.

The additional income to the utilities from the resources contained in the wastewater can improve not only the financial sustainability of such utilities but also their environmental and institutional sustainability. In many cities wastewater is not treated at all but rather discharged directly to receiving bodies, and utilities are institutionally weak to even manage appropriate technology treatment plants. If the discharge of raw wastewater causes severe water contamination problems (which is the situation in many cases), an appropriate technology treatment plant can be constructed (usually with the support of government grants) to resolve the problem. Such plant can include installations for energy recovery and/or effluent reuse. The income generated from the recovery of the resources can be sufficient to cover the O&M of the treatment plant and sometimes even part of the investment. This income stream can be used to finance the hiring of a professional private operator to operate the treatment plant. This can improve the overall institutional sustainability of the utility by ensuring the sustainability of the operation of the treatment plant and by enhancing the sustainability of the environmental management through the reduction in the contamination of natural water bodies. The hiring of private operators also reduces the institutional burden and challenges of the water and sanitation utilities managements and thereby contributes to enhancement of the institutional sustainability.

As an example consider a city the size of Santa Cruz, with about 1 million inhabitants, located in a developing country and with an institutionally weak utility. The city does not have any wastewater treatment installations and discharges all its raw wastewater to receiving water bodies causing their severe contamination. Given its water supply problems, the perspective of managing the wastewater disposal problem by the utility is remote. Involving the private sector to support managing the wastewater disposal seems difficult because private operators require a guaranteed income and the financially weak utility cannot provide such reliable guarantee. Now imagine that a plan is developed regarding the construction of 5 wastewater treatment plants based on the treatment process applied in Santa Cruz and that the plan includes the utilization of the biogas to generate electricity according to Alternative 1 described in Section 1.13.4. The government will provide a grant for construction of the treatment plants and the electricity generators installed in the treatment plants can be connected to the grid. The power company will pay for the electricity produced by the treatment plants the price specified in Alternative 1. The financial analysis of alternative 1 shows that the O&M costs can be generated from the sale of the methane-produced electricity and on top of that, the energy production generates an amount of US$ 5.56 million at present value. A bidding process can be undertaken for construction and operation of the treatment plants. The Constructor-Operator will specify in his proposal which portion of the investment he is willing to provide, and he will also be responsible for the O&M cost of the constructed treatment plants, once they come on stream. The bidding variable can be the amount of investment that the Constructor-Operator requests from the public sector (national government, municipally and utility). The bidder that requests the lowest public sector investment is the winning bidder. In this case, the sale of energy is the guaranteed income of the Constructor-Operator and this is a reliable guarantee which does not depend on the financial performance of a weak utility but rather on the technical performance of the bidder, so a bidder that is self-confident will consider it as a safe guarantee.

1.13.7 Example of recovery of the resources contained in wastewater

Figure 1.77 presents the Naan wastewater treatment system and stabilization reservoir. It is a simple system, a typical system of a stabilization reservoir. The raw wastewater is pre-treated in a system of two parallel anaerobic lagoons. The effluents of the two lagoons are discharged to the Naan reservoir for seasonal storage and treatment, and the effluent of the reservoir is used to irrigate the adjacent fields during the irrigation season. This is a system that already recovers two of the wastewater resources: the water and the fertilizers. It is possible to recover in this system also the third resource: energy. The idea is presented in Figure 1.124. It is possible to cover the anaerobic lagoons, capture the biogas and then either use it for generation electricity, connecting to the power grid, or convey the biogas to an industrial park located a few kilometres from the reservoir, if a consumer can be identified in this park. In Figure 1.124 one of the anaerobic lagoons was left uncovered, as it is in reality, and the other was covered to show what needs to be done (cover the anaerobic lagoons with HDPE membranes). Also, the dimensions of the biogas storage tank and of the electricity generating system are not in scale and exaggerated in the figure. In reality they will be much smaller and will occupy a small land area.

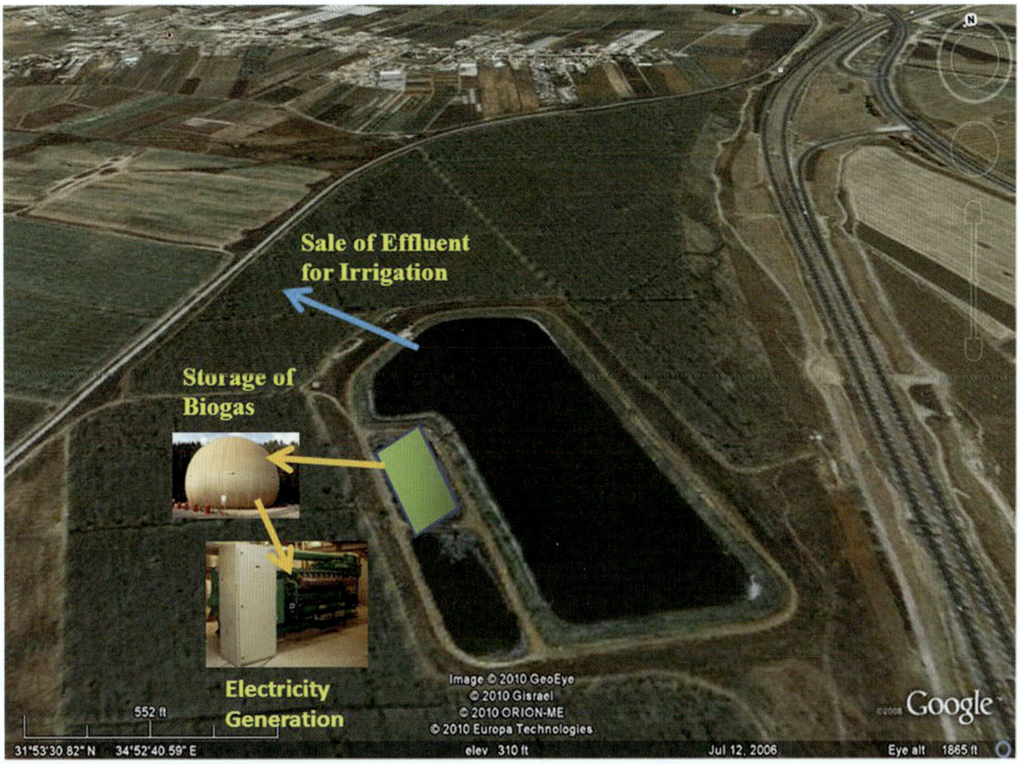

Figure 1.124 Recovery of resource from wastewater in the naan wastewater treatment and stabilization reservoir system

1.14 APPROPRIATE TECHNOLOGY TREATMENT PROCESSES CLASSIFIED ACCORDING TO THEIR ADEQUACY FOR USE IN VARIOUS CATEGORIES OF SIZE OF CITIES

The appropriate technology unit processes and some of the combined appropriate technology processes discussed in this chapter are presented in Table 1.12 classified according to their adequacy of use in cities of various sizes. The cities are classified into three categories: (i) small (with a population of up to 20,000); (ii) medium (with a population between 20,000 and 300,000); and (iii) large (with a population above 300,000). For each category a recommended list of appropriate technology processes adequate for this category is provided. This table is intended to support in taking a decision about processes which should be considered, or should be discarded for a city of a certain size. The table should be considered as a guiding instrument but not as a strict rule.

According to the table, lagoon systems should not be considered for large cities, but this recommendation refers to constructing one plant per city. If a large city needs to construct several treatment plants due to topographical considerations, each plant can be a lagoons plant if it does not serve a population larger than 300,000–400,000, and even this range is just a reference range which can be somewhat exceeded if it makes sense. Constructed wetlands are not a good solution for large cities because they consume large land areas, and even for medium cities constructed wetlands are in most cases not adequate, for the same reason. CEPT is in most cases not adequate for small cities; however, small cities which have good water utilities can use the CEPT process.

Table 1.12 Appropriate technology treatment process classified according to their adequacy for use in various categories of size of cities.

Large cities***	Medium size cities**	Small cities*
Rotating micro screens	Rotating micro screens	Rotating micro screens
	Lagoon systems of various types including mixers aided systems and covered anaerobic lagoons	Lagoon systems of various types including mixers aided systems and covered anaerobic lagoons
UASB reactors	UASB reactors	UASB reactors
Anaerobic filters	Anaerobic filters	Anaerobic filters
CEPT	CEPT	
		Constructed wetlands
Reuse for irrigation systems	Reuse for irrigation systems	Reuse for irrigation systems
Submarine outfalls	Submarine outfalls	Submarine outfalls
UASB-anaerobic filter combination	UASB-anaerobic filter combination	UASB-anaerobic filter combination
	UASB-lagoons combination	UASB-lagoons combination
CEPT-sand filtration combination	CEPT-sand filtration combination	
UASB-sand filtration combination	UASB-sand filtration combination	UASB-sand filtration combination
UASB-dissolved air flotation combination	UASB-dissolved air flotation combination	UASB-dissolved air flotation combination
Other combinations need a specific review to determine if they are adequate for large cities	Other combinations need a specific review to determine if they are adequate for medium size cities	Other combinations need a specific review to determine if they are adequate for small cities

*Small cities: Cities with population up to 20,000
**Medium size cities: Cities with populations in the range 20,000–300,000
***Large cities: Cities with populations above 300,000

1.15 PERFORMANCE AND COSTS OF APPROPRIATE TECHNOLOGY TREATMENT PROCESSES IN RELATION TO ACTIVATED SLUDGE

The treatment capacities of the appropriate technology unit processes and of some of the combined appropriate technology processes discussed in this chapter are presented in Table 1.13 in terms of ranges of removal percentages of total BOD and total suspended solids (TSS). Also presented in this table are data on ranges of investment costs and operation and maintenance costs of each of the mentioned process. Investment costs as well as O&M costs are also compared to those of secondary treatment in a conventional activated sludge process. Activated sludge was selected to represent the

Table 1.13 Treatment capacity and costs of some appropriate technology unit and combined processes.

Process	Total BOD removal capacity, %	TSS removal capacity, %	Investment cost		O&M cost	
			US $/Capita	Percentage of activated sludge cost	US$/Yr/Capita	Percentage of activated sludge cost
Conventional activated sludge (Just for reference, this is not an appropriate process)	80–90%	80–90%	100–150*	100%	4–8	100%
Rotating micro screens	0–30%	0–30%	3–10	4–10%	0.1–0.15	1.9–2.5%
Conventional lagoons systems	70–90%	70–90%	20–40	25–40%	0.2–0.4	5–8%
Mixers aided lagoons systems	70–95%	80–90%	20–40	25–40%	0.2–0.4	5%
Covered anaerobic lagoons followed by mixers aided facultative lagoons	80–95%	80–90%	20–50	25–50%	0.2–0.4	5%
UASB reactors	60–75%	60–70%	20–40	25–40%	1–1.5	19–25%
Anaerobic filters	70–80%	70–80%	10–25	10–25%	0.8–1	13–20%
CEPT	70–75%	80–90%	20–40	20–40%	1.5–2	25–38%
Constructed wetlands	80–90%	80–90%	20–30	20–30%	1–1.5	19–25%
Stabilization reservoirs systems	75–95%	75–90%	30–50	30–50%	0.2–0.4	5%
Submarine outfalls	99.9%	99.9%	3–30	3–30%	0.1–0.15	1.9–2.5%
Overland flow	70–80%	70–80%	15–30	15–30%	0.8–1.5	19–20%
UASB-anaerobic filter combination	80–90%	80–90%	20–40	20–40%	1–1.5	19–25%
UASB-lagoons combination	80–90%	70–80%	30–50	30–50%	1–1.5	19–25%
CEPT-sand filtration combination	80–90%	80–90%	40–50	40–50%	1.5–2	25–38%
UASB-sand filtration combination	80–90%	80–90%	30–50	30–50%	1–1.5	19–25%
UASB-dissolved air flotation combination	80–90%	80–90%	30–40	30–40%	1–1.5	19–25%

*The investment cost of an Activated Sludge plant used for the calculation is 100 US$/Capita

processes which are not based on appropriate technology. The investment cost in an activated sludge plant is in the range 100–150 US$/Capita. For the purpose of the comparison calculations, the figure 100 US $/Capita was used.

The cost of a combined appropriate technology process is not an arithmetic sum of the cost of the unit processes that constitute it, but is rather lower than the arithmetic sum due to economics of scale. The cost data in the table, both investment costs and O&M costs, are only indicative costs, since the costs of a wastewater treatment plant are site and country specific. However, these indicative costs allow us to draw some important conclusions.

In terms of effluent quality, most of the processes (except the Rotating Micros Screen process) achieve results that are not very different from those achieved by the activated sludge process. This is especially true for the combined treatment processes, whose effluents are of similar quality to that of an activated sludge effluent. If the dilution in the river to which the effluents are discharged is significant, than there is practically no difference between the effluents of the combined appropriate technology processes and the effluent of an activated sludge process, in terms of their impact on the oxygen content of the river.

As shown in Table 1.13, the investment costs of the presented appropriate technology treatment processes are in the range of 20–50% of investment in activated sludge. These estimates are based on the investment figure of 100 US$/Capita. If a higher figure is taken (which might well be the case) than investment cost of the appropriate technology processes will be even lower in relation to that of activated sludge. Perhaps more important is that fact that in most cases, the operation and maintenance costs of the appropriate technology processes are in the range 5–25% of the operation and maintenance cost of activated sludge.

The general conclusion that can be drawn is that by using appropriate technology treatment processes or a combination of such processes: (i) a significant amount can be saved in investment costs; (ii) a lot of O&M expenses can be saved; (iii) the operation and maintenance of the plants is much simpler than that of plants based on conventional process; and (iv) almost nothing is lost in terms of effluent quality.

All this has an outmost importance in terms of ensuring the sustainability of the utility by selecting an appropriate technology process. The much lower investment and especially operation and maintenance costs ensure financial sustainability. The simplicity of operation and maintenance ensures ease in overcoming technical difficulties and long terms sustainability in operating the plant. And all the above ensures institutional sustainability, since the wastewater management does not pose on the utility financial and technical difficulties.

1.16 SLECTION OF THE ADEQUATE TREATMENT PROCESS

A number of unit processes of appropriate technology were described in the book. Based on these processes, the idea of the use of combinations of appropriate technology unit processes was presented. Although unit processes by themselves cannot reach very high effluent qualities, specific combinations of unit processes can generate practically any required effluent quality. A plant based on a combination in series of appropriate technology unit processes is still easy to operate and is usually of lower costs than conventional processes in terms of investments and certainly in operation and maintenance. So in essence, this book shows the way to obtain high quality effluent by using treatment plants based on simple, low cost and easy to operate processes.

A large number of simple, sustainable processes (including unit processes and combinations of unit processes) were presented and the design of part of them is discussed in the following chapters. Some of the presented processes are already used in full scale plants in many parts of the world. As to the rest, it is not known if there are full scale plants based on these processes. Nevertheless, there is no reason that full scale plants using these processes will not function well, and we believe that there is room for all the processes and that there are specific conditions that justify the use of each of the them, or at least justify the consideration and analysis of each one of the processes. The processes presented constitute in fact a data base that should be considered for analysis when selecting a process for each new wastewater treatment plant.

With the relatively large number of unit processes of appropriate technology and a larger number of combined processes, the question is how to select the most adequate process for each specific project. The basis for the decision is the local conditions specific to each case. Several parameters need to be taken into account, among them: the size of the city, climatic conditions, especially variation of temperature during the day and especially during the seasons of the year, altitude above sea level, cloud cover, the flow and composition of the raw wastewater and their fluctuations (diurnal and seasonal), effluent quality required (based on local discharge standards or on the type and size of the receiving water body), the alternative sites available for location the treatment plant or treatment plants, the size of land area and topography of each site, preferences of the owner (utility or municipality) to have a very simple plant or a somewhat more complex plant, financial resources available and more.

Climate and temperature variation ranges define a series of processes that can be used. The size of land available for the treatment plant defines another series of processes which can be fit into it. If it is a large area, then most of the processes can fit in. If the area is limited, then only part of the processes can be used. The required effluent quality defines another series of processes that can yield such quality. The owner's preferred level of simplicity of the plant also defines a series of processes, for example a municipality that opts for simplicity may favour a process of covered anaerobic lagoons followed by overland flow, while a municipality that considers simplicity as a secondary objective may opt for UASB followed by DAF. Experience elsewhere with treatment plants of similar size can also be a factor in selecting the treatment process. As a result of such an analysis, a limited number of processes will be identified, and one of them will be selected. But at the end of the day, for each city a feasibility study has to be conducted by a consulting firm to select the most adequate alternative for the treatment plant. The most adequate alternative refers to location of the plant and to the treatment process. The preferences of the employer (utility or municipality) need to be defined in the Terms of Reference for the study. The large number of potential processes is reduced to a few after taking into account the local conditions of each case, so the number of alternatives which needs to be studies is not really large, and on the basis of the feasibility study the employer can select the alternative he prefers.

Tables 1.12 and 1.13 provide preliminary guidance for selection of an adequate treatment process. Table 1.12 defines process according the size of the city and Table 1.13 provides information on performance of various appropriate technology unit and combined processes in terms of BOD and TSS removal percentages and information on investment costs per Capita and operation and maintenance costs per Capita per year. General guidance for selection of several possible adequate treatment processes for each case according to its specific characteristics is provided in Tables 1.5, 1.7, 1.14 and 1.15. The possible alternatives need then to be considered in depth in a feasibility study. Table 1.14 provides guidance on selection of adequate processes according the required BOD removal efficiency, Table 1.15 provides guidance on selection according to the number of people to be served by the treatment plant, Table 1.5 provides guidance on selection according to the cold season air temperature in the proposed treatment plant site and Table 1.7 provides guidance on selection according to the available land size for location the treatment plant.

Table 1.14 Processes selection matrix by removal efficiency of DOD$_5$.

Process number	Process description	BOD_5 removal (%)				
		10–40%	40–70%	70–85%	85–95%	>95%
1	Rotating micro screens, RMS	X				
2	Anaerobic lagoons and/or facultative and/or maturation lagoons		X	X	X	
3	UASB		X	X		
4	Anaerobic filter, AF		X	X		
5	Plug flow anaerobic reactor, PAR		X			
6	Stabilization reservoirs, SR, single, continuous flow		X			
7	Stabilization reservoirs, SR, at least two, sequential batch operation				X	
8	SSFCW			X	X	
9	CEPT		X	X		
10	Combination 1: MRS + UASB + facultative lagoon				X	
11	Combination 2: RMS + UASB + AF			X		
12	Combination 3: MRS + UASB + sand filtration + UV				X	
13	Combination 4: MRS + CEPT + sand filtration + disinfection, UV				X	
14	Combination 5: RMS + UASB + FA + DAF + membrane filtration					X

Table 1.15 Processes selection matrix by population served.

Process number	Process description	Population (× 1000)				
		10–20	20–50	50–100	100–300	>300
1	Rotating micro screens, RMS	X	X	X	X	X
2	Anaerobic lagoons and/or facultative and/or maturation lagoons	X	X	X	X	
3	UASB	X	X	X	X	X
4	Anaerobic filter, AF	X	X			

(Continued)

Table 1.15 Processes selection matrix by population served (*Continued*).

Process number	Process description	Population (× 1000)				
		10–20	20–50	50–100	100–300	>300
5	Plug flow anaerobic reactor, PAR	X	X			
6	Stabilization reservoirs, SR single, continuous flow	X	X	X	X	X
7	Stabilization reservoirs, SR at least two, sequential batch operation	X	X	X	X	X
8	SSFCW	X	X			
9	CEPT		X	X	X	X
10	Combination 1: MRS + UASB + facultative lagoon	X	X	X	X	
11	Combination 2: RMS + UASB + AF	X	X	X	X	
12	Combination 3: MRS + UASB + sand filtration + UV	X	X	X	X	X
13	Combination 4: MRS + CEPT + Sand filtration + disinfection, UV	X	X	X	X	X
14	Combination 5: RMS + UASB + FA + DAF + membrane filtration	X	X	X		

Comments:
1. Anaerobic Filters are used for small towns when they are the main treatment process. When they are used as a polishing unit, they can serve larger populations.
2. There is no experience with Combination 5. It might be adequate also for larger populations than specified in the Table.

The method for a preliminary selection of the alternative unit and combined appropriate technology treatment processes which may be adequate for a specific project (Method of Process Selection – MPS) is the following:

(1) Review the matrix Tables 1.5, 1.7, 1.14 and 1.15 to identify the processes that meet the desired targets of: (i) cold season temperature range, (ii) available land size, (iii) BOD removal efficiency and (iv) design population. The processes that jointly meet the four criteria are pre-selected.
(2) Analyze whether the land area available for construction of the treatment plant is sufficient to house the pre-selected processes: for example, Lagoons, Stabilization Reservoirs and SSFCW require large land extensions which might not be available and therefore these process might be rejected even if the comply with all other requirements.
(3) Analyze other external factors as cloud cover (for facultative lagoons a low cloudiness is preferable), wind direction in the plant site (to avoid the odours transfer by the wind to population centers), altitude above the level sea, etc.
(4) Discarded from the pre-selected alternative processes the ones found inadequate and proceed to carry out a more detailed feasibility study of the remaining processes.

It is clear that nothing can replace the sound criteria and the experience of a good designer and process engineer, but the MPS is a good start for a feasibility analysis, which should be complemented by appropriate calculations.

Design models prepared for part of the unit and combined appropriate technology processes presented in the book may be used to carry out a first approximation of dimensioning the processes. A list of the respective Excel Programs is presented in Section 1.19. However, these model programs should be used with great caution. They do not replace the criteria of the design engineer who must ultimately be responsible for the design results. These programs should not be used without a full understanding of the processes, as presented in the book.

1.17 SEWERAGE NETWORKS, THE CONDOMINIAL SEWERAGE CONCEPT

Sewerage networks do not form part of wastewater treatment systems (although there are opinions that some treatment takes place in the sewerage and conveyance systems); however, sewerage and treatment systems are coupled in the sense that sewage treatment cannot take place without the existence of the sewerage network which collects the sewage and conveys it to the treatment plant. For a municipality, the investment in wastewater management includes investment in sewerage networks, wastewater conveyance and wastewater treatment. Reducing the investment in sewerage networks and conveyance is as important as reducing treatment investments.

The analysis of sewerage networks and wastewater conveyance systems is not covered in this book. Detailed information on the theory and design of these systems is widely available in the professional literature. It can for example be found in Tscobanoglous (1981).

An innovative method for reducing the investment cost of sewerage networks and conveyance was developed in Brazil (Melo, 2008) and is denominated the Condominal system. The sewerage system of the city of Brasilia, the capital of Brazil is a condominial system. It is also used in other cities in Brazil and in parts of the city of Lima, Peru and El Alto, Bolivia. The condominial sewerage technology, also called "shallow sewerage", is different from the denominated "small bore technology" or "simplified sewerage" because there is no retention of solids in the condominial systems. Consequently, there is no need to construct household retention tanks and no need for periodic sludge removal. Detailed information on the condominial technology is provided by Melo (2008) and by Vargas-Ramirez and Lampoglia (2006) and can be found also in the report of The World Bank Water and Sanitation Program (2002).

From the technical standpoint, the condominial technology simplifies the design and characteristics of pipelines, making it physically easier to connect households. Condominial sewerage considers that the network is divided into a private part - the condominial lines) and a public part - the main sewers. The condominial lines are built in areas with no road traffic such as gardens, sidewalks, etc., and are laid at a shallow depth. The diameter of these lines is usually 100 mm and it uses much smaller inspection chambers (manholes).

The condominial model also proposes the development of new relations of co-responsibility for services between the service provider and the user. From the social perspective, the model introduces a participatory component in the implementation phase that is intended to motivate users to connect to the system and to generate a commitment to keep good use practices. Consequently, one of the users' responsibilities is the maintenance of the condominial lines, especially when pipes are laid on internal areas. In those cases, the beneficiaries must receive the necessary training to perform that task. However, the utility remains responsible for the O&M of the overall system at the city level.

Condominial systems have been utilized in Brazil since the 80′s where they were first employed in the city of Natal. The conventional model of service provision has been shown to be inappropriate for the expansion of services to slums and peri-urban poor areas where chaotic occupation and complex topography constrain the installation of regular sewerage layouts. The condominial system proved to be appropriate for such areas. In the capital of Brazil, the city of Brasilia, since 1995 the condominial technology was adopted by CAESB, the local water and sanitation utility, as the only option for sewage collection, with more than 120,000 condominial connections built by the year 2000. The condominial technology had been used to expand sewerage connections to peri urban areas of cities like Salvador, Petronila, Recife and Rio de Janeiro among other. By 2000, 13.6% of the sewerage connections in Brazil were condominial. In past years the technology has been expanded from Brazil to other Latin-American countries. In Bolivia it was employed to expand sewerage services to more than 50,000 households in the city of El Alto. In Peru, SEDAPAL, the utility of Lima, implemented a program to expand condominial sewerage to 25,000 families. The City of Durban, South Africa experimented with this technology in 2000 and is introducing adaptations for scaling it up. Experiences in various countries show that an increasing number of utilities and local service providers are using this technology to reach the Millennium Development Goals, due to its significantly lower costs with identical level of service to that of conventional sewers.

Benefits of the condominial system include: (i) Financial Benefits – manifested by capital cost savings, which is one of its more appealing characteristics, since it allows the provision of services to significantly more people with the same financial resources. Cost reductions stem mainly from (a) lower excavation volumes due to more shallow location of the pipes; (b) use of simplified inspection chambers instead of costly manholes; (c) reduced pipe diameters and layout length; and (d) easiness of construction that result in reduced need for heavy machinery. The financial benefits are also manifested by reduction of O&M costs since the accessibility of the system at depths varying from 60 to 150 cm allows for easier access for manual maintenance. In case of breakage, system components are much easier and cheaper to replace; (ii) Adaptable layout - the flexible condominial layout allows working in irregular urban layout settlements, very steep slopes and rocky terrains. In fact, in many locations it may be the only feasible technical option; (iii) Better hydraulic functioning – the use of the shear stress boundary concept for design instead of the minimum velocity allows the use of lower minimum slopes. Similarly and counter intuitively, smaller diameters allow greater buoyancy and more efficient transportation of solids, especially in densely populated areas.

In addition, the condominial systems also provide social benefits such as: better and more appropriate use of services, increased rate of connections to the network, community development, and higher willingness to pay as well as higher payment rates.

Challenges to scale up the condominial technology include issues of responsibility for operation and maintenance of parts of the systems, legal constraints related to construction standards, and ownership issues. Characteristically, the condominial approaches require intensive social work with the communities. These costs –along with the steep learning curve- can be quite onerous for small pilot projects, even threatening to render system's savings negligible. The larger the project scale the easier it is for service providers to administer the condominial system. The key challenge for scaling-up is to maintain a social mobilization process on a large scale that brings about the benefits described above, while maintaining the per-capita social costs at a reasonable level.

The overall investment cost savings of implementing the condominial technology, when compared to conventional technology, are well documented, reaching 40–50%. Figure 1.125 demonstrates the reasons for cost saving. In the condominial system, most of the network consists of lower cost condominial sewers (the blue lines), while in the conventional system all the network consists of higher cost conventional sewers (the red lines).

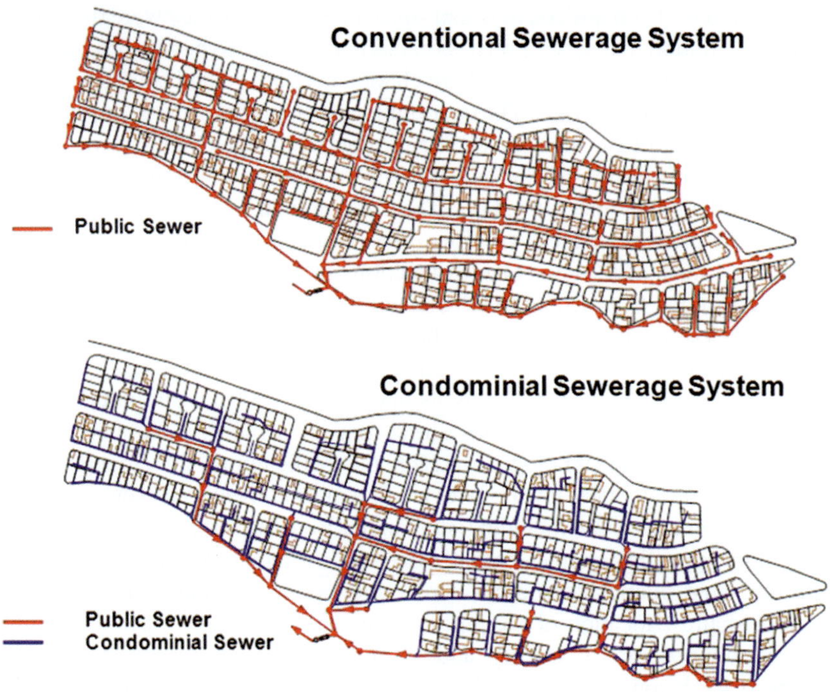

Figure 1.125 Comparison between condominial and conventional sewerage systems

1.18 WASTEWATER TREATMENT IN THE CONTEXT OF GLOBAL WATER ISSUES

1.18.1 Introduction

It is commonly accepted that adequate wastewater collection, treatment and disposal is required in order to prevent public health risks and environmental degradation. Unfortunately, in developing countries sewerage services of collection and conveyance of wastewater out of urban neighborhoods is not yet a service provided to all the population. Adequate treatment of wastewater is even less frequently practiced and is provided only to a small portion of the collected wastewater, usually covering less than 10% of the municipal wastewater generated. In slums and peri-urban areas, which constitute large portion of cities in developing countries (an average of about 38% of the total urban population in 2005 with significant differences among geographical region), it is not rare to see raw wastewater flowing in the streets. In addition to the obvious need to treat wastewater for the purposes of preventing public health risks and environmental degradation, there are also other aspects for the necessity of providing adequate treatment and disposal of wastewater, especially in developing countries. To understand these aspects we need to expand the discussion to the wider context of global water issues.

1.18.2 The global water crisis

A lot of attention is being directed in the recent years to a potential global water crisis. The UN estimates that about 1 billion people in the world do not have access to safe drinking water and 2.5 do not have access to sanitation services. These data indicate that the water crisis is not a matter of the future, it is already here.

Most of the people without access to drinking water and sanitation reside in developing countries, so for the most part; the water crisis is located in developing countries. Water is the main causes of diseases in developing countries. About 80% of the diseases in these countries result from the effects of contaminated water, lack of water and poor sanitation. About 3.3 million people die annually from water borne diseases, most of them children under the age of 5. Water Borne diseases are responsible for more deaths than any other cause. More attention is being directed to the global water crisis as a result of the ensuing global warming, which is expected to exacerbate water scarcity in some zones and flooding in others.

Water is a resource of special nature, a global common resource. Water resources transcend national boundaries. Moreover, our global interdependencies are woven through water. Water is a cross cutting natural resource upon which most of the other sectors and human activities depend, including agriculture and food production, renewable energy, industry, health, forestry, land-use, disaster risk reduction and basically most types of economic development activities. All these depend upon water for their long-term sustainability. Water as a resource is also affected by other sectors and activities, as energy costs, consumption by agriculture and industry, contamination, spread of diseases and epidemics and climate change. Good water management is one that balances water across competing demands and prioritizes water for basic human needs and ecosystem functions; as well as responding effectively to water induced hazards such as droughts and floods.

Earth is known as the Blue Planet, the Water Planet. But actually it is the Salt Water Plant. About 97.5% of its water is salt water (in the Oceans). Only 2.5% of the plant's water is fresh (sweet) and of that 69% is tied in glaciers, 30% in groundwater, part of which is unavailable, and only 0.3% (of the 2.5 %) is in rivers and lakes, so available sweet water is very limited in relation to the entire amount of water on the planet. Even so, on the average, the amount of rain water which comes down as part of the water cycle is theoretically sufficient for the needs of all the people on the plant. But this sensation of plentiful water is only a myth. The distribution of location of the population across the planet is not uniform and this is a phenomenon beyond our control. As a result the demand for water is also not uniform. Neither is the distribution of availability of fresh water uniform across the planet and along the seasons of the year. That is certainly a phenomenon beyond our control. Consequently, the relation between availability and demand of water is not uniform, resulting in areas of physical abundance of water, and also wide areas of physical scarcity. In addition to the high spatial disparities there is also a high temporal disparity.

Forecasts indicate a rapid global population growth. In 2011 we are approaching 7 billion people and within 20 years the population will reach 9 billion or more. Several additional important facts related to growth forecasts are: (i) most of the population growth is expected to take place in developing countries, while the population of developed countries will remain constant at about 1 billion, as shown in Figure 1.126; (ii) the urban population is growing and the rural population declining. As a result of this urbanization process, about 60% of the population will be urban in 2030, up from about 45% in 2010 (see Figure 1.124); and (iii) the size of cities is growing and in the future there will be much more large cities and especially mega cities (about 30 megacities with over 10 million each in 2030, up from 15 in 2010), as shown in Figure 1.127. All this implies intensification of the water crisis in the future, unless drastic actions are taken to alleviate the problem.

For the most part, developed countries have managed to avoid the direct water crisis and will most probably succeed in avoiding it also in the future, but in developing countries the crisis may intensify. However, avoidance of direct crisis will not relieve developed countries from being impacted by the crisis. The world is today a global village. Shortage of grains and rises in the prices of basic food commodities in the past few years caused social uprisings in developing countries with impacts on the entire world. Similarly will the water crisis in developing countries have an impact on developed countries through the impact of social unrest, disruption of economic, commercial and trade activities, disruption of flows of commodities, possible negative impacts on political interests of developed countries, spread of diseases, the need of emergency humanitarian support to countries in severe water crisis, stimulation of immigration out of such countries and more.

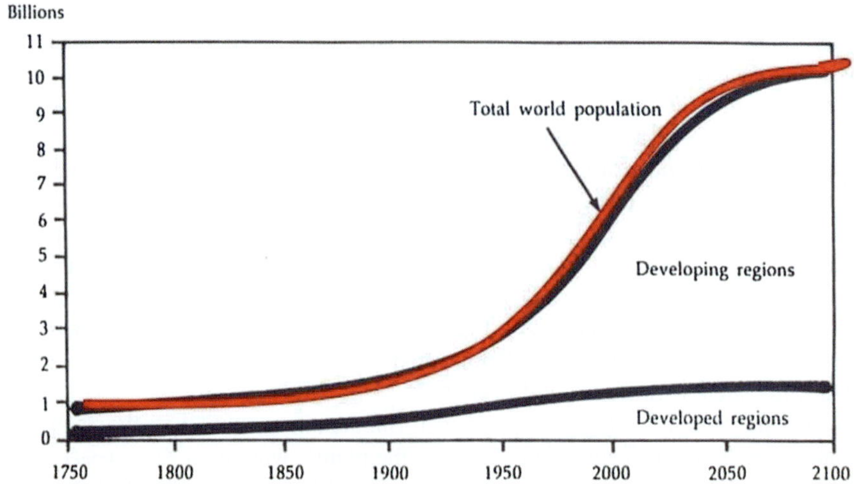

Figure 1.126 World population growth 1750–2100

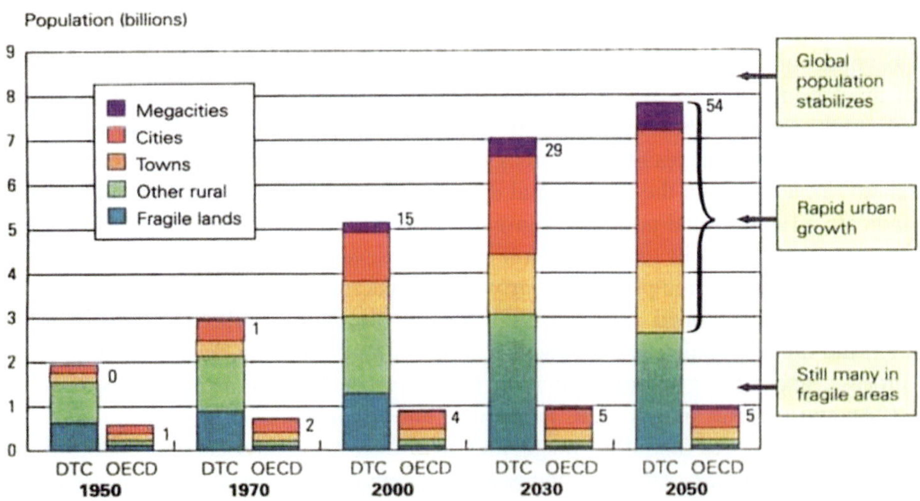

Note: DTC refers to developing and transition countries; OECD refers to high-income countries (and not all members of the Organisation for Economic Co-operation and Development). The numbers to the right of the columns show the number of megacities (cities in excess of 10 million people). Towns are classified as having a population of less than 100,000 and cities, a population of 100,000 to 10 million.
Source: Authors; global population projections are based on World Bank estimates; estimates of population shifts in urban and rural areas are based on United Nations data.

Figure 1.127 World population forecast and urbanization trends (*Source*: World Bank Development Report 2003)

According to forecasts of water availability and consumption in the Latin America Region (LAC), in the year 2030 the region will experience a shortage of 40% of its demand. The global shortage forecast is probably similar. The LAC forecast is based on certain assumptions. The scarcity trend can be mitigated by actions that can be taken and by inducing behavior changes of the consumers, rendering the assumptions incorrect and reducing or eliminating the water deficit. Technological and institutional measures are discussed below.

1.18.3 The main water consumers and the potential for water savings by consumer category

Who are the main water consumers (in terms of withdrawals of fresh water)? Forecast for 2030 generally indicate the following water withdrawers and the percentage of consumption of each one: Agriculture – about 75%, Municipal consumption – about 10%, Industrial consumption – about 10%, and Evaporation from reservoirs – about 5%. There are two additional water consumers which are sometimes omitted: (i) water serving as a receiving body of various types of pollutants, that is, water consumed for Pollution Absorption; and (ii) water required for sustaining ecosystems and life of all other non human species that share with us this planet, also known as environmental water flows. It is difficult to estimate the level of consumption of these two. Worth noting is that the largest consumer is by far the use of water for irrigation and that the current irrigation efficiency worldwide is less than 40%.

The potential for water savings, that is, for reducing water consumption by each consumer is as follows: (i) Irrigation-50%, by moving from surface irrigation to more efficient methods of micro-sprinkling and drip irrigation; (ii) Municipal consumption-(20%–30%) by conservation, using water more efficiently, managing demand and reducing losses in the piping systems; (iii) Industry-(50%–90%) by modifying production processes, moving to clean development mechanisms (cleaner production processes), using water more efficiently and recycling wastewater for use within the industries; (iv) pollution absorption consumption-100% by providing adequate treatment to all wastewaters; and (v) environmental flows-no potential for reduction, probably increase is required. As can be seen there is significant room for savings and therefore for alleviating the water crisis, all this with no sacrifice of economic output or quality of life. Achieving such savings involve application of a combination of technical measures and institutional improvements.

1.18.4 Reasons for the water crisis

Several causes contribute to the water crisis: (i) Physical Water Scarcity in certain regions, which will intensify with time due to increase of the population, the urbanization trends and probable climate change; (ii) Economic weakness which results in insufficient investments; (iii) Governance weakness at the central and local governments' levels; (iv) Institutional capacity weakness at the utilities level; and (v) various combinations of the above. The natural tendency of layman would be to consider the physical scarcity as the main reason for the water crisis. However, the main problems, especially in developing countries, are weak governance and low institutional capacity. For the most part, the current water crisis is not a scarcity crisis but rather a management (or mismanagement) crisis. With time it will transform to be also a scarcity crisis. It remains a grave moral shortcoming that about 1 billion people cannot drink water without risking disease or death and 2.5 billion do not have access to sanitation services while the reason is not so much the scarcity of water as lack of social and political commitment and capacity of governments to meet the basic needs of the poor.

The physical scarcity of water has technical solutions (which are discussed below). The economic shortcomings can also be alleviated with efforts of governments in developing countries, and with support of developed countries and the private sector. The governance and institutional problems (in other words the management crisis) are the most difficult to resolve and are the reason for having the problem concentrated in developing countries. Developed countries have managed to overcome the crisis, even in conditions of physical scarcity, through structural reforms and a combination of institutional and technical measures. Israel, Australia and Singapore can serve as examples of such countries, among others. These examples indicate that it is possible to overcome the water crisis also in many developing countries, if the institutional and management problems are resolved. However, if the management crisis is not resolved before the scarcity crisis becomes more critical, the results can be devastating.

1.18.5 Water and climate change

Climate change and its implication on the water sector is a widely discussed topic. The relation of wastewater treatment and climate changed is discussed in Chapter 12. So issues of climate change are not touched here, but it is clarified that adaptation to climate change is essentially about water management – building resilience though good water management. Global trade in food and other essential products for development depends upon water availability in producing regions – climate-related interruptions to the water cycle upon which this production depends could have potentially devastating global consequences. Climate change impacts on transboundary waters will demand a new paradigm of regional and international cooperation. Is that achievable? Given the limited international success in controlling the increase of Green House Gases, the recent events of social unrest following food increase cost crisis are just an introduction to political and economic upsets which will occur in a warming world.

1.18.6 The situation of the poor

A few words on the situation of the poor households not connected to the municipal water distribution network. The poor usually live in slums and peri-urban areas, which in most cases are not connected to water supply network. Poor households, as any other household, cannot live without water and they have to purchase small volumes of contaminated water from vendors supplying water in plastic containers or in tankers, at rates higher 2 to 10 times in relation to the tariffs charged by the water utility. The real meaning of the effect of water scarcity on the poor is therefore paying a significant portion of the household income for small volumes of contaminated water. The population which lives in slums is large, totaling about 38% of the population of developing countries (about 2.1 billion). Connecting poor households to the water supply network not only provides convenient access to adequate volumes of safe and clean water but also means a relief from an economic burden and a significant increase in the available income.

1.18.7 Water as a human right

On July 28, 2010 the UN General Assembly declared that “Safe drinking water and sanitation is a human right essential to the full enjoyment of life and all other human rights”. The declaration was not achieved in consensus. It is noted that the declaration includes sanitation services and that it does not mention economic aspects and provision of water at an affordable tariff. We profoundly agree that water is a Human Right. This is a basic right essential for sustaining human life. The meaning of this human right is that all governments have the obligation to supply potable water to all their citizens in sufficient quantities at an affordable cost (and this is not being done by most governments in developing countries). However, water is also an economic good. It is free when it falls from the sky, before it touches the ground. Once it reached the ground, there is a cost for its collection, conveyance, treatment, storage, distribution, collection of the wastewater, its treatment and safe disposal. Who should pay for all this? Potential payers are: grants, users and taxpayers. Grants alone are by far insufficient. Payment by taxpayers only is not sustainable, as shown from experience in countries that have adopted such a model. Logic provides that users pay for the water service, with taxpayer support when necessary (which means government's transfers). Those users which have the financial capacity should pay for the real cost of water. The poor that do not have the financial capacity should pay reduced, affordable tariffs and the utilities should be supported by subsidies to compensate for the reduced tariffs of the poor. There are various forms for managing such subsidies.

1.18.8 Proposed strategy options to alleviate the water crisis

During the quest for economic growth we ignore the limits of nature and over-exploit the water resources. We have to respect the limits of water resources and learn to live within these limits. Instead of continuously searching for new water sources we need to start looking inside. The efficient use of water is equivalent to creating new sources. The main additional water source is conservation, demand management (100 liter/capita/day not more), efficient water use, control of water losses, reuse and recycling. Investments in efficient water use, conservation and reuse are more beneficial than investing in conventional projects. Water cannot be used just once: it must be reused in agriculture and recycled for other purposes. Wastewater must be considered a resource containing water, fertilizers and energy. Reuse and recycling liberate more fresh water for potable use.

The strategy for alleviating the water crisis must include a combination of technical and institutional measures. It must be based on integrated water resources management, correct and flexible allocation of water between sectors (municipal, agricultural and industrial), and sometimes on introducing measures of water markets for reasonable allocation. The largest water consumer is irrigation. The rapid population growth requires increased food production, which in turn will require expansion of irrigated land. It is necessary to plan ahead and change the type and quantities of produce, moving from industrial plants to wheat, grains soy and alike, but this does not necessarily mean increase in water consumption for irrigation. A strategy approach of outmost importance for alleviating the water crisis is increasing the irrigation efficiency from the current less than 40% to the possible 90%, moving away from surface irrigation methods to micro-sprinkling, drip and subsurface drip irrigation methods.

Of great importance is incorporating in the strategy institutional strengthening measures, since institutional weakness is one of the main causes of the crisis and a prerequisite for success is to make institution work. To resolve specific water crisis issues in each country and city, specific measures need to be tailored to respond to local conditions. A general manue of activities which can be included in a crisis alleviation strategy is presented below.

For each specific case, the corresponding portion of the manue needs to be identified and adopted. The presented manue is divided to three types of activities: (i) Technical measures; (ii) Institutional measures; and (iii) Extreme measures for cases of deep crisis. Some measures appear both as institutional and technical since they have aspects of both. It is obvious that each measure presented for crisis alleviation is a wide topic in itself and it is impossible to expand the discussion on all of them in this book. Also, the measures do not appear in an order of priority, since priority is also specific to each case and local conditions. Improvements will not materialize until policies, laws and institutions begin to foster the required measures.

Technical Measures (Technical Options as part of the strategy to mitigate the water crisis)

- *Integrated Water Resources Management*
- *Application of Efficient Irrigation methods*
- *Reduction of Unaccounted For Water*
- *Water Conservation (Demand Management, flow limiting and efficient water use devices at the household level)*

- *Undertaking only Rational Investments*
- *Wastewater Treatment to prevent contamination of fresh water*
- *Reuse of municipal effluents for irrigation, the stabilization reservoir concept*
- *Effluent recycling for industrial reuse*
- *Use of simple, low cost technologies for water treatment, wastewater treatment and water and sewage networks*
- *Small scale solutions: Microdams, Shallow wells, collection of dew for providing humidity to plants, moisture conserving land techniques*
- *Use of marginal water (springs, flood waters) by seasonal storage*
- *Recharge of flood water to aquifers*
- *Use of saline water (for direct irrigation, industrial uses, diluting with sweet water, desalination of saline water)*
- *Stimulating rainfall by cloud seeding with silver iodine*
- *Increasing the irrigation efficiency by the use of efficient irrigation methods*
- *Desalination of sea water*
- *Effluent Recycling for potable municipal reuse, in certain cases*

Institutional Measures (Institutional Options as part of the strategy to mitigate the water crisis)

- *The key issue is to make institutions work*
- *Adopting an adequate and strong legal and institutional framework (institutional reforms when necessary)*
- *Wise administration and management of water resources*
- *Use of common sense, experience and innovations*
- *Integrated Water Resources Management*
- *Rational and intelligent water allocation between sectors and within sectors*
- *Modification of allocations according to needs, more for the municipal sector and less for agriculture*
- *Preparation of Master plans and conceptual plans with International Support*
- *Undertaking only Rational Investments*
- *Limiting increase of cities or even reducing population in accordance with water availability*
- *Communication campaigns and education of the public to save water (not 600 liter/capita/day, 100 liter/capita/day is enough, and even less in critical cases)*
- *Water conservation (Demand Management, flow limiting and efficient water use devices at the household level)*
- *Water Metering*
- *Efficient Management of Water Utilities*
- *Reduction of Unaccounted For Water*
- *Use of water more than just one time: reuse for irrigation, industry and non-potable municipal purposes*
- *Use of simple, low cost technologies for water treatment, wastewater treatment and water and sewage networks*
- *Efficient Management of Irrigation systems and use of efficient irrigation methods*
- *Application of adequate tariff policy (to stimulate saving and provide financing) and an effective collection system*
- *Application of adequate regulation*
- *Small Scale Solutions*
- *When applicable: Private sector participation in management of utilities and/or in financing*

- *In certain cases, creating frameworks and mechanisms for water markets*
- *Public Participation*
- *Obtaining support of the public sector through various Public Private Participation (PPP) mechanisms*
- *In developing countries, obtaining support from developed countries*

Extreme Technical Measures for cases of deep crisis

- *Limiting size of cities in accordance with availability of water*
- *Displacement within the country, when necessary, to distribute the population in accordance with availability of water*
- *Reducing Birth Rate and providing Equal Economic opportunity to Women*
- *Modification of water allocations according to needs, more for the municipal sector and less for agriculture*
- *Application of efficient irrigation methods (drip, subsurface drip, micro-sprinkles)*
- *Collection and use of dew for providing humidity to plants*
- *Use of Biogenetics to reduce plants consumption of water, resulting in allocation of less water for irrigation*
- *Desalination of seawater (50% of the world's population lives within 50 km from coastlines and can use desalinated sea water*
- *Municipal wastewater must be reused (mainly in irrigation), using seasonal storage (the stabilization reservoir concept)*
- *In-house treatment and reuse of gray water (or even wastewater) for non potable purposes using membrane techniques (Compact family size treatment plants are being developed, it is like having an additional washing machine in the basement)*
- *When necessary, wastewater reclamation for potable use (as is being done in Singapore)*

In summary, what is required is a paradigm shift from the old paradigm of: (i) Subdivided Water Management (Ag, Urban, Indus); (ii) Use of water just once; (iii) Inefficient use of water; and (iv) Polluting large quantities of water, to the emerging paradigm of: (i) Integrated Planning and Management of Water; (ii) Use Water Efficiently and Multiple Times; (iii) Matching Needs with Water Sources; and (iv) Minimizing Pollution.

Involvement of developed countries is essential to support resolution of the crisis. Developed countries need to provide support to developing countries in their attempts to resolve the water crisis because of: (i) the status of water as a global common good, as mentioned previously; (ii) the difficult situation of the poor; and (iii) the incapacity of developing countries to resolve the problems on their own. The support of developed countries should be provided especially in developing concepts and preparation of conceptual plans (most important), in assistance for governance/institutional improvements and implementation of reforms, and in financing.

The role of the private sector. The private sector can also play a role in alleviating the water crisis in two ways: (i) financing; and (ii) support in improving the performance of the water utilities. The private sector has to get involved in financing the water sector since the demand for funds is enormous and the public sector alone cannot do provide the required funding. The private sector participates in financing other sectors (transport, energy, education, health) and there is no reason to forbid it from participating in financing the water sector too, in spite of the ideological arguments used against such approach. The private sector can also participate in various models of PPP and contribute to institutional improvement of utilities in developing countries, which is the main problem of the sector and the major reason for the water crisis. This type of activity is currently quite limited. The population in developing countries served by private water operators in 2007 was only 160 million. that is, about 3% of the total population in these countries, and is not expected to ever grow beyond 10%, so the provision of water will remain mostly the responsibility of the public sector. Decision makers in developing countries need to have various options to tackle the many challenges of water utilities. Private Public Participation (PPP) can be one of them.

For many years the success in institutional improvement of water utilities in developing countries has been modest, and with time it becomes more difficult to achieve because the challenges become greater due to the increase in population. The issue of adaptation to climate change has directed attention to the water crisis and might help in identifying and implementing strategies for alleviating the problems, including addressing the institutional weakness problems, especially taking into account that the intensification of the problems caused social unrest which may push governments to take actions, as happened in the case of the recent food crises.

Will the global water crisis intensify or diminish? That is a difficult question to answer. It depends on all of us, the inhabitants of the planet. Let's hope that the human spirit will overcome the difficulties and that we will manage to alleviate water scarcity and mitigate the water crisis which may cause a retreat and decay of the human race.

1.18.9 The water crises implications on wastewater treatment

It is commonly accepted that adequate collection, treatment and disposal of municipal wastewaters is required in order to prevent public health risks and environmental degradation in general. However, wastewater management needs also to be evaluated in the context of the global water crisis. Considering the proposed strategy options to alleviate the water crisis which were discussed above, it is concluded that in addition to protection of public health and of the environment, adequate wastewater management also plays an important role in the strategy to mitigate the effect of the water crisis. Wastewater treatment is an important activity in preservation of the quality of exiting water sources by preventing contamination of water bodies by raw wastewater thus maintaining additional sources of clean water. The water crisis discussion also indicates the importance of municipal effluents reuse to irrigate farmland as both means for generating an additional source of water for irrigation (which liberates fresh water to be used for urban supply) and as a method for totally eliminating discharge of effluents to clean surface water bodies for the purpose of maintaining their quality.

In developing countries, where the coverage of wastewater treatment is very low, increasing the treatment coverage is a necessary and important goal. The rapid increase of population and the urbanization tend will result in generation of large quantities of wastewater which, if not treated, will increase the public health risks, especially in poor neighbourhoods, and will contaminate water sources. A common practice in developing countries is the uncontrolled reuse of raw wastewater to irrigate food crops, adding to the public health risks induced by untreated wastewater. Treatment of the wastewater will diminish the reuse of untreated wastewater. Untreated wastewater also contributes to the contamination of water bodies which serve as water supply sources. Consequently, increasing wastewater treatment coverage is a priority in developing countries. A key in the strategy to achieve this goal is adopting low cost and simple to operate treatment processes as means of alleviating the economic burden by lowering investments and O&M expenses, and especially as means of alleviating institutional problems by applying processes that are easily manageable.

1.19 THE PROCESSES FOR WHICH DESIGN PROCEDURES ARE PRESENTED IN THE FOLLOWING CHAPTERS

Many unit processes and combined processes of appropriate technology were presented in Chapter 1. Given the large number of processes discussed, it is impossible to provide in the book the design procedures for all the processes mentioned. In the following chapters we present the design procedure of a relatively large number of unit processes and combined processes as outlined in Table 1.16. The name of the Excel program which is used for the design of each process is also specified in this table.

Table 1.16 List of the appropriate technology processes for which the design procedures are detailed in the following chapters.

Process description	Name of excel program used for design	Comments
Rotating micros screens	CHAP 4-RM.xls	
Lagoon systems	CHAP 5-Lagoons.xls	
Piston anaerobic reactor	CHAP 6-Abaerobic-PAR.xls	
UASB followed by Anaerobic Filter	CHAP 6-Anaerobic-UASB-AF.xls	This program is the basis for calculating the UASB unit process, the Anaerobic Filter unit process and the combined process of UASB followed by Anaerobic Filter
Stabilization reservoir	CHAP 7-SR.xls	
Constructed wetlands	CHAP 8-SSFCW.xls	
Chemically Enhanced Primary Treatment (CEPT)	CHAP 9-CEPT.xls	
Sand Filter	COMB 3-UASB-SAND FILTER-UV.xls TAG:SAND FILTER	This program is the basis for calculating Combination 3. The Sand Filter unit is designed in the worksheet with Tag: Sand Filter.
Dissolved Air Flotation (DAF)	COMB 5-UASB-AF-DAF-MEMBRANES.xls TAG: DAF	This program is the basis for calculating Combination 5. The DAF unit is designed in the worksheet with TAG: DAF
UV disinfection	COMB 3-UASB-SAND FILTER-UV.xls TAG:SAND FILTER TAG: UV	This program is the basis for calculating Combination 3. The UV Disinfection unit is designed in the worksheet with TAG: UV.
Membrane Filtration	COMB 5-UASB-AF-DAF-MEMBRANES. xls TAG: MEMBRANES	This program is the basis for calculating Combination 5. The Membrane Filtration unit is designed in the worksheet with Tag: Membranes.
Combined process 1-UASB followed by Lagoons	COMB 1-UASB-MATURATION LAGOON.xls	
Combined process 2-UASB followed by Anaerobic Filter	CHAP 6-Anaerobic-UASB-AF.xls	
Combined process 3-UASB followed by Sand Filter	COMB 3-UASB-SAND FILTER-UV.xls	

(Continued)

Table 1.16 List of the appropriate technology processes for which the design procedures are detailed in the following chapters (*Continued*).

Process description	Name of excel program used for design	Comments
Combined process 4-CEPT followed by Sand Filter	COMB 4-CEPT-SAND FILTER-UV.xls	
Combined process 5-UASB followed by Anaerobic Filter followed by DAF followed by Membrane Filtration	COMB 5-UASB-AF-DAF-MEMBRANES.xls	

The model programs which are presented in Table 1.16 and are available online in: http://www.iwawaterwiki.org/xwiki/bin/view/Articles/Software+Developed+for+Sustainable+Treatment+and+Reuse+of+Municipal+Wastewater, should be used with great caution. They do not replace the criteria of the design engineer who must ultimately be responsible for the design results. These programs should not be used without a full understanding of the processes, as presented in the book.

General instructions for running de Excel sheet programs with the Orderly Design Methods (ODM)

- First, make sure that you thoroughly understand the Orderly Design Method.
- Second, get acquainted with the colour Nomenclature as follows:
- Whatever is in dark grey colour refers to titles, parameters or units, and is alphabetic. The other colours refer to numeric cells.
- The cells in light grey colour are input provided by the designer, be it (i) external parameters or (ii) primary variables given by the designer. These cells should be filled manually by the designer.
- The coloured medium grey cells are secondary variables calculated from the primary variables, which belong to the input necessary for the design. Note that the designer may choose a secondary variable as primary, and then the primary variable becomes secondary. For instance, to calculate the design flow you may select q, P and c as primary variables and from them the program calculates the process design flow (Q_{DWW}) as secondary variable; but you may know the design flow, then q becomes a secondary variable. These variations in primary variables would require changes in the program and should be avoided unless completely necessary.
- With the input defined, the program calculates the design variables in the output as determined by the ODM. Also, in this step there are variables that can be selected as input and the corresponding ones in the input become outputs. For example, in Figure 4.8, you can choose either Efficiency or screen opening (mesh size) as input. If you choose one variable as input then the other variable becomes output.
- Remember, to change primary variables to secondary variables (or input to output) in the program requires a thorough comprehension of the ODM.
- Check that the output variables, be they dimensions or specifications, are practical and are in the empirical range of design given in the Process Description section of each chapter.
- Note that the necessary steps to calculate CONCENTRATION, ORGANIC LOADS AND SPECIFC UNIT LOADS, require that the wastewater quality is given as an input.
- These general instructions apply to all programs given in the online page.

Part 2

Design

Chapter 2

Decomposition processes of organic matter

2.1 INTRODUCTION[1]

Organic Matter (OM) is the most important pollutant of natural water bodies like rivers, lakes, bays, and so on, since it depletes dissolved oxygen (DO) from such water bodies. In domestic wastewater (DWW) of typical composition, about 70% of suspended solids (SS) and 45–50% of dissolved solids is Organic matter. Organic matter is made up of Carbon, Hydrogen, Oxygen, elements common to all organic compounds, together with nitrogen in some cases. Other compounds such as phosphorus, sulphur, iron, and so on, are also often present in organic matter. For convenience organic matter is divided into different groups as follows:

- *Protein*: up from 40 to 60% of the organic matter in domestic wastewater is composed of protein. Proteins are the main constituent of animal bodies. Plants also contain protein, to a lower extent. Proteins are complex and volatile substances, and their chemistry is associated with amino acids, which are made up by the acid group, −COOH, and the base group-NH_2. Nitrogen is always present in Amino Acids at a relatively constant proportion of 16%. The molecular weight of proteins is very high, from 20,000 to 20 million. Urea, $CO(NH_2)_2$, and Protein are the main sources of nitrogen in wastewater. When Protein and Urea are present in wastewater in large quantities, the generation of odours from the wastewater is likely.
- *Carbohydrates*: constitute 25 to 50% of the domestic wastewater. Their source is mostly vegetable matter. They are widely present in nature and include sugars, starches, cellulose and wood fiber. The cellulose and wood fibers are known generically as fiber. The insoluble fibers are made up by cellulose, hemi-cellulose, lignin and certain starches. Carbohydrates are made up by Carbon, H_2 and O_2. Sugars soluble in water break down easily. Starches are more stable, but can be turned into sugars by microbial activity. Fibers are insoluble (mainly cellulose) and very resistant to decomposition in wastewater treatment plants. However in the soil they break down easily due to the action of fungi in acidic conditions.
- *Oil and Grease*: This group is the third largest component in the organic matter present in domestic wastewater. Oil and Grease (O&G) are compounds of alcohol and glycerol. The glycerides of volatile

[1]This section was prepared based on the textbook "*Bioingeniería de aguas residuales: teoría y diseño*" (2005), by *Álvaro Orozco Jaramillo*, Published by ACODAL, Bogotá, Colombia.

fatty acids (VFA) are oils, which are liquid at ordinary temperatures. The volatile fatty acids react with bases (e.g. sodium hydroxide) to form soaps, which are also very stable materials. Among the sources of the Oil and Grease in domestic wastewater are butter and vegetable oils. These compounds are very stable and difficult to break down by the bacteria present in the wastewater therefore they must be removed before treating the wastewater; otherwise they cause problems in the processes of decomposition of the organic matter.

- *Surfactants*: are large molecules slightly soluble in water, which cause foam. Known as detergents, they are used for cleaning. They can cause major problems in the aeration units of treatment plants. In the past, detergents were made up by Alkyl-benzine-sulfonates (ABS), which are non-biodegradable materials, but today they have been largely replaced by linear detergents such as Linear-alkyl-sulfonates (LAS) that are biodegradable.

The main quality parameters of wastewater are:

Dissolved Oxygen (DO)

Dissolved Oxygen is one of the important parameters in wastewater treatment since many of the organisms rely on it to maintain metabolic processes to obtain energy and carry out their reproduction. In addition, dissolved oxygen is the main indicator of contamination of a body of water, since its level has a direct effect in the dissolved oxygen uptake.

Oxygen gas is not very soluble in water, does not react with it and its solubility depends on its partial pressure in the air. The saturation concentration of oxygen in water varies from 7 mg/l at 35°C to 14.7 mg/l at 0°C, at atmospheric pressure. The measurement of DO in water is made by means of an Oxygen Meter probe or by titration using the *Winkler* method.

In waters over-saturated with oxygen to levels higher than 110% of the DO saturation concentration, fish can suffer from the disease of "gas bubble". This can occur in eutrophic waters containing an excessive algae population and at certain times of the day when oxygen production by algae is the highest. However, more frequent are circumstances of low DO concentrations resulting from the oxygen demand caused by the organic matter present in the water. In DO concentrations of above 7 mg/l a diverse population of fish, snails, insects, and so on, is present in the water. To sustain the survival of salmon fish, the DO concentration in the water must be above 5 mg/l. Most fish can however survive in water with DO concentrations of 4 mg/l, and some, like tilapia and St. Peter's fish, manage to resist DO concentrations of 3 mg/l. Lower concentrations will cause the disappearance of higher aquatic life. At an average water body DO concentration of below 1 mg/l, anaerobic zones void of oxygen are certain to exist and therefore such a water body generates some bad odours. When the DO concentration in a water body reaches zero, anaerobic conditions prevail throughout the water body causing anaerobic decomposition to takes place in the entire water body, resulting in persistent foul odours.

Biochemical Oxygen Demand (BOD)

Biochemical Oxygen Demand is caused by organic matter discharged to rivers and other water bodies. This organic matter becomes food (substrate) for bacteria that reproduce rapidly as a result of the availability of the organic matter. Under aerobic conditions these bacteria also consume oxygen, causing the decrease in the DO concentration in the water body and consequently causing the effects that are explained in the preceding paragraph. The BOD is defined as the amount of oxygen required to decompose the organic matter present in the wastewater through the action of bacteria under aerobic conditions. The BOD is caused by the respiration of the bacteria and completely ceases when the organic matter is totally consumed by the bacteria. BOD was proposed in 1912 as an indirect method to measure the content of the organic matter in water. Today, the BOD measurement is carried out during five days at a temperature of 20°C, and is denoted by the symbol DBO_5. However, measurements may be carried out during different time spans, for example, DBO_7 is the oxygen demand after 7 days, and the DBO_u (ultimate or total BOD) is a measurement which continues until the complete consumption of the organic matter, which usually lasts from 20 to 30 days. In typical domestic wastewater $BOD_5 \approx 0.70\ DBO_u$.

The BOD test is a biological test. The test simulates the conditions under which the oxygen demand occurs in nature. These conditions imply the presence of oxygen and nutrients (phosphorous and nitrogen), the absence of toxic constituents; maintaining proper pH and temperature ranges, the presence of bacteria in sufficient quantities, and so on.

The BOD test is done by measuring the DO concentration in the water sample before and after five days of incubation. Since the DO concentrations in the laboratory reached only 7 to 8 mg/l (the saturation level), and since the BOD_5 in wastewater fluctuates between 200 and 20,000 mg/l or even more, it is necessary to dilute the wastewater samples as part of the BOD analysis procedure. To make the BOD test, a sample is taken and diluted in a defined aliquot depending on the expected value of the BOD in the sample. For example, the aliquot 1:50 means that the wastewater is diluted 50 times, namely, one part of wastewater is mixed in 49 parts of distilled water. If the estimated concentration is not known, different dilutions must be prepared in the ranges in which BOD level is estimated. The dilution water used to prepare the aliquot is distilled water in which salt of potassium, sodium, calcium and magnesium are dissolved. This provides a good buffering capacity (which means that it maintains an approximately constant pH at a value close to 7.0). Each aliquot is saturated with oxygen under the laboratory conditions. Once the aliquots are prepared with the appropriate dilutions, namely, with oxygen demands not greater than 2–3 mg/l (which is a demand that can be measured without problems with a DO of 6 mg/l in the samples) the sample is poured into a Winkler bottle (wide mouth). If necessary (as in the case of some industrial wastewaters which do not contain bacteria) the dilution water is inoculated with bacteria. Since it is assumed that the dilution water which contains a bacteria inoculum may contain organic matter which, when added to the sample, will increase the content of organic matter, it is also necessary to measure the BOD of the dilution water, and for that, a bottle with exactly the same conditions of the sample is prepared, but without the wastewater. This is known as the blank. The DO concentration of the prepared sample and of the blank are then measured at zero time and they are denominated initial DO_s (DO_i is the initial DO of the sample and OD_{bi} is the initial DO of the blank) The samples are then placed in an incubator at 20°C, avoiding penetration of light (to avoid possible oxygenation by algae) and after five days the DO concentration of the samples are measured. They are denominated DO_f in the sample and DO_{bf} in the blank. The measured BOD_5 value corrected by the blank is calculated using the following formula:

$$BOD_5 = \frac{(DO_i - DO_f) - (DO_{bi} - DO_{bf})\left(\dfrac{V_m}{V_b}\right)}{D} \tag{2.1}$$

where:

BOD_5 = Biochemical Oxygen Demand after 5 days at 20°C
DO_i = initial DO in the diluted sample
OD_f = final DO in the diluted simple
OD_{bi} = initial DO in the blank
OD_{bf} = final DO in the blank
D = sample dilution, in decimals (e.g. for 2%, D = 0.02)
V_m = volume of blank minus the volume of inoculum
V_b = volume of blank

As explained, the organic matter can be (i) proteinic or nitrogenated which produces a nitrogenous BOD (NBOD); and (ii) of carbohydrates or carbonaceous which produces a carbonaceous BOD (CBOD). The oil, grease and the fibers are very stable. That means that they do not consume oxygen and therefore practically do not produce BOD in wastewater. The nitrogenous BOD can be calculated stoichiometrically from the Total Organic Nitrogen or Total Kjeldahl Nitrogen (TKN = Organic-N + Ammonia-N). While measuring the BOD content of wastewater or water, it would be convenient to inhibit the activity of nitrogenous BOD so that only Carbonaceous BOD is measured. This is done by adding to the sample bottles a nitrification inhibiting compound, such as Allylthiourea. Other compounds such as H_2S can also generate oxygen demand (in this case defined as SBOD) which can also be calculated stoichiometrically.

Carbonaceous BOD (CBOD)

Carbonaceous BOD is the main part of the oxygen demand of domestic wastewater. We define the carbonaceous BOD concentration in the wastewater as **L**. Its decay follows a first order kinetics as follows:

$$-\frac{dL}{dt} = kL \tag{2.2}$$

where k is the reaction constant, also known as bottle constant (of *Winkler*). Integrating between time 0 and time t, we obtain the equation of variation of **L**, the carbonaceous BOD concentration, with time, where L_0 is the initial carbonaceous BOD concentration or CBOD:

$$L = L_0 e^{-kt} = L_0 10^{-Kt} \tag{2.3}$$

The constant k (base e) is equal to 2.303 K (base 10). It is important to find k in laboratory tests. When measured in the laboratory the oxygen demands of several consecutive days, DBO_1, DBO_2, …, DBO_i, we find that the graph of the consumed CBOD (y axis) varies with time as presented in Figure 2.1. Note that the CBOD consumed (y) is growing every day, but eventually, after five days the NBOD also starts being consumed, causing the hump shown in the figure. This is because after five days the nitrifying bacteria appear and they consume the protein organic matter. But as noted, the nitrogenous BOD can be best measured from TKN, so inhibiting the nitrifying bacteria with Allylthiourea will stop the NBOD consumption and only CBOD will continue to be measured as shown in the bottom of the figure. If it is required to produce also the curve of NBOD, the test needs to be made twice, once with the inhibitor and once without the inhibitor. This way the two curves are produced: that of CBOD and that of (CBOD + NBOD).

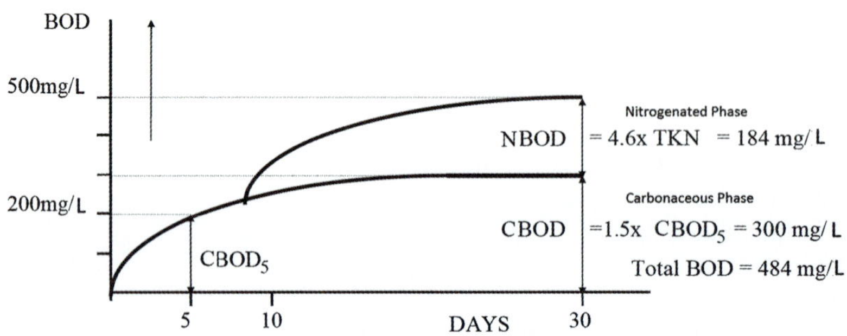

Figure 2.1 Variation of CBOD and NBOD with time in raw sewage [*Source*: http://www.petermaier.net/dl/Maier_tech.pdf (April 6/2011)]

Nitrogenous BOD (NBOD)

We have seen that for measuring Carbonaceous BOD it is required to carry out a test that lasts five days in its most simplified version. In addition, it is necessary to inhibit the Nitrogenous BOD, which is caused by organic matter of protein origin consumed by nitrifying bacteria that appear spontaneously after approximately five days. The organic nitrogen, TKN, measured in terms of NH_3, is first converted to nitrite, NO_2^-, by the bacteria *Nitrosomonas* and then to nitrate, NO_3^-, by the bacteria *Nitrobacter* as follows:

$$2NH_3 + 3O_2 \xrightarrow{Nitrosomonas} 2NO_2^- + 2H^+ + H_2O \tag{2.4}$$

$$2NO_2^- + O_2 + 2H^+ \xrightarrow{Nitrobacter} 2NO_3^- + 2H^+ \tag{2.5}$$

The entire reaction is summarized then as:

$$NH_3 + 2O_2 \longrightarrow NO_3^- + H^+ + 1/2H_2O \tag{2.6}$$

Molecular Weight: 17 2×32

This means that 64 g of O_2 are required to oxidize 17 g NH_3 (or 14 g of N-NH_3). The NBOD will then be 64 grO_2/14 grN-NH_3 = 4.57. In other words, the NBOD = 4.57 × N-TKN, and we do need not perform the NBOD test which would be very cumbersome (as explained in the section on measurement of CBOD). This way, a wastewater with a TKN concentration of 32 mg/l, will have a NBOD = 32 × 4.57 = 146.24 mg/l. This means that 146.24 mg/l of dissolved oxygen will be required to transform (to nitrify) the 32 mg/l of TKN (the organic plus ammonia nitrogen) to nitrate.

Chemical Oxygen Demand (COD)

Chemical Oxygen Demand emerged as a result of the need to measure the oxygen demand quickly and reliably. This is another way of measuring the organic matter content indirectly through the chemical oxygen demand of organic compounds. The COD is a way to measure the energy contained in the organic compounds, but initially it was conceived as a faster and more accurate substitute of the BOD. In the COD measurement, instead of decomposing the organic matter by bacterial metabolism, which uses respiration as a means of obtaining oxygen, a strong oxidizing agent in acidic environment is used for decomposing the organic matter. The most commonly used oxidizing agent is potassium dichromate at high temperature, in the presence of silver sulphate as a catalyst. The reaction of organic matter with dichromate is the following:

$$C_xH_yO_z + Cr_2O_7^{-2} + H^+ \longrightarrow Cr^{+3} + CO_2 + H_2O$$

where $C_xH_yO_z$ generically represents carbonaceous organic matter. The COD of a compound is generally greater than the BOD because many compounds that can be chemically oxidized cannot be decomposed biologically, through bacterial biodegradation. The non-biodegradable compounds are often artificial molecular substances of high molecular weight. Often for a specific wastewater, BOD can be well correlated with COD, and usually the ratio COD/BOD is constant for a specific wastewater. That means that for the wastewater of one city the ratio has a certain value, and for the wastewater of another city it has a different value, but for each city the ratio is about constant. This constitutes a great benefit because the carrying out a COD analysis requires only two to three hours while a BOD analysis requires five days. The ratio COD/BOD also determines the amount of non-biodegradable organic matter present in wastewater.

In addition, although not always necessary, it is possible to determine the COD of a specific organic matter through stoichiometric calculations. This is achieved by calculating the weight of oxygen required per unit weight of the specific organic matter or per unit liquid volume of this organic matter for the complete oxidation of this organic matter. An example of stoichiometrically calculating the COD of a known compound, in this case glucose, is the following:

$$C_6H_{12}O_6 + 6O_2 \longrightarrow 6CO_2 + 6H_2O$$

Molecular weights: (180) $6 \times (32)$ $6 \times (44)$ $6 \times (18)$

From this relationship we can conclude that $(6 \times 32)/180 = 1.066$ g O_2/g glucose are required to decompose the glucose. The COD of an aqueous solution containing 1 g/l of glucose would then be 1066 mg/l.

Solids

Solids are another very important parameter in wastewater treatment. Organic matter is often present in the form of particles in suspension, so it is necessary to distinguish between suspended solids (SS) and Dissolved Solids (DS). Also solids can be volatile (VS) indicating that they are of organic origin, or fixed which are considered inorganic solids. The general classification of solids in wastewater is presented in Figure 2.2. Total Solids (TS) are made up by suspended solids plus dissolved solids. At the same time those are sub-divided into volatile suspended solids (VSS) and fixed suspended solids (FSS), and these are sub-divided into in volatile dissolved solids (VDS) and fixed dissolved solids (FDS). The most important in wastewater treatment are the suspended solids, especially the volatile suspended solids which are conformed of organic matter present in wastewater in the form of particles. The measurement of solids is made gravimetrically, namely, by measuring weight, and consists of filtering the sample with a dry filter of known weight. After drying the filtered sample in an oven at 105°C the filter along with the dried filtered solids retained on it are weighted and by the subtracting the weight of the filter, we learn the weight of the filtered solids of a given volume of sample, and thus their concentration in the sample in units of mg/l. Volatile solids are determined by their evaporation in an oven at a temperature of at least 550°C. Inorganic or fixed solids do not evaporate at this temperature (but rather at a much higher temperature) so the solids remaining on the filter paper after heating it to 550°C are the inorganic solids and the balance are the volatile solids.

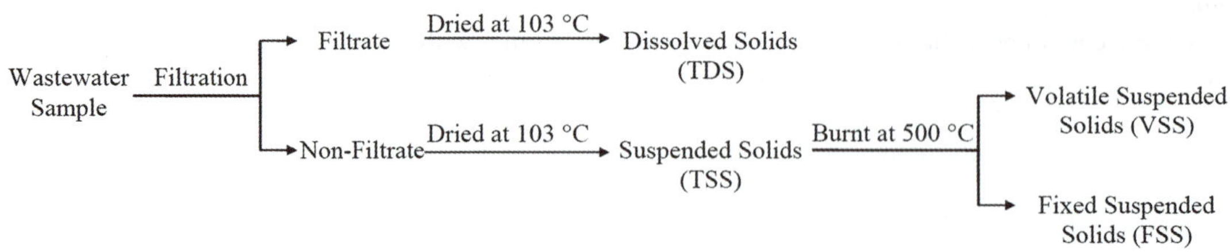

Figure 2.2 Classification of solids in wastewater [*Source:* http://www.cee.ntu.edu.sg/AboutCEE/Facilities/EnvLab/Documents/EN2703%202C-4.pdf (April 2011)]

The total solids content (soluble and particulate) in typical municipal wastewater is usually lower than 1000 mg/l. These solids basically contain all the contamination in the wastewater. This means that municipal wastewater is composed of about 99.9% of water and only about 0.1% of contaminants.

2.2 THE BIOCONVERSION EQUATION

2.2.1 Aerobic conversion

BOD and COD measure the state of "reduction" of organic matter. The BOD is used more as a criterion of water quality, while in order to interpret the kinetics of wastewater treatment, COD is a preferred indicator. Both tests estimate the amount of oxygen required for oxidizing organic matter, but the BOD only measures the biodegradable fraction, while the COD also takes into account the non-biodegradable fraction. On the other hand, the removal of substrate by bacteria is equivalent when presented in terms of BOD and of COD since the O_2 units removed are the same. In other words, if bacteria consumed 50 mg/l BOD they also consumed 50 mg/l COD. Thus, for a substrate or wastewater with initial concentrations BOD_0 and COD_0, subjected to treatment after which the final concentrations are BOD_e and COD_e, we have:

$$\Delta S = S_0 - S = BOD_0 - BOD_e = COD_0 - COD_e$$

where ΔS = Removed oxygen demand. This means that $\Delta DBO = \Delta COD$

Consequently, from the standpoint of substrate removal, BOD and COD are equivalent. COD is preferred because of the simplicity in the laboratory analysis for its measurement. As seen, COD indicates the amount of O_2 required for total oxidation of the organic matter. The higher the COD in wastewater, the greater is the contamination level in it and more O_2 will be required for its oxidation.

For purposes of biological treatment, COD tells us how much food is provided to a reactor which contains a determined biological culture. Indeed, in a wastewater treatment system (WWTS) raw wastewater or influent substrate (S_0) is the food that enters the reactor where which contains a highly concentrated culture of bacteria, measured as the Mixed Liquor Volatile Suspended Solids (MLVSS). After feeding on the organic matter flowing into the reactor, the effluent which leaves the reactor is left with a residual substrate (S). The COD removed (ΔS) during the treatment is the amount of substrate consumed by the bacteria, namely:

$$\Delta S = S_0 - S = COD_0 - COD_e \tag{2.7}$$

When consuming an organic compound with a determined COD, the bacteria use the energy available in it for their metabolic reactions, turning it into another compound with less energy, that is, with less COD. In the transformation of an organic compound with a high energy level into another with a lower energy level, a "commission" is paid in terms of entropy, in accordance with the second law of thermodynamics. In other words, there cannot be a change from one energy state to another with 100% efficiency. The energy released in the reactions is used by the microorganisms (and in general by all living beings), for the maintenance of their life. This transformation allows the employment of the liberated energy in useful work necessary for motion and all other vital functions.

COD is in fact a measure of the energy content of a compound. Simplifying, we can say that the substrate ($C_xH_yO_z$) "reacts" with O_2 to produce *microorganisms*, CO_2 and H_2O, but in fact, microorganisms consume substrate (food) respirate oxygen and transform the substrate partly to energy, used for sustaining their life, and partly for growth, producing additional microorganisms. The "chemical composition" or "formula" of aerobic bacteria in the cultures of wastewater treatment plants is $C_5H_7NO_2$ so we can express the reaction of decomposition of organic matter in the following summarized form:

$$C_xH_yO_zN + nO_2 \longrightarrow C_5H_7NO_2 + CO_2 + H_2O$$

$$\textit{Substrate} + \textit{Oxygen} \longrightarrow \textit{Microorganisms}$$

Microorganisms are measured in the laboratory gravimetrically as volatile suspended solids (mg/l), but, because they are organic matter, they have a stoichiometric oxygen demand that can be calculated as follows:

$$C_5H_7NO_2 + 5\,O_2 \longrightarrow 5\,CO_2 + 2\,H_2O + NH_3 \qquad (2.8)$$
$$(113) + 5 \times (32)$$

Which lets us calculate the theoretical COD of microorganisms as follows:

$$\text{COD (theoretical)} = \frac{5 \times 32\,\text{g}\,O_2}{113\,\text{g SSV}} = 1.42\ \text{g}\,O_2/\text{g VSS}$$

That is, the volatile organic matter formed by aerobic microorganisms with a composition of $C_5H_7NO_2$ have a COD of 1.42 g O_2/g VSS. The experimental measurement of the COD of MLVSS in activated sludge plants gives values in the range 1.40 to 1.46, which agrees with the calculation. Thus, one can express the microorganisms present in a reactor in units of oxygen or equivalent COD, simply by multiplying the volatile suspended solids concentration of aerobic sludge by 1.42, which also gives the total energy content of the microorganisms. It is worth noting that the "formula" of the anaerobic microorganisms is $C_5H_9NO_3$ so that for this type of microorganisms, the stoichiometric ratio of conversion to COD is 1.22 g O_2/g VSS.

It becomes clear from the information presented above that the substrate removed (ΔS) is converted into biomass (ΔX) and in that conversion, microorganisms obtain the necessary energy for their metabolism and pay the "commission" of entropy for the change. These two factors ultimately are counted as energy loss and can be measured, in terms of O_2, as respiration. Thus we can say that, measured as units of oxygen, the macro-level phenomenon that occurs in an *aerobic* biological reactor can be expressed in the following equation, also known as the *Equation of the Bioconversion or Biotransformation:*

$$\begin{array}{rcll} \Delta S & = 1.42\,\Delta X & + \Delta O2 & \\ \text{Substrate consumed} & = \text{biomass produced} & \mid\ \text{oxygen respired} & \qquad (2.9) \\ (\text{g}\,O_2) & (\text{g}\,O_2) & (\text{g}\,O_2) & \end{array}$$

Respiration accounts for both the energy consumed in vital functions and those of entropy losses, which close the thermodynamic balance that must be accurate and which has been well demonstrated in practice.

Equation (2.9) not only gives a thermodynamic balance of what happens in a biological reactor, but allows realizing the stoichiometric balance, in units of O_2, between the reactants and products.

Aerobic treatment can be summarized as follows: organic substrate is supplied continuously to a reactor, in which it is assimilated by a highly concentrated bacterial population. This bacterial population is provided with oxygen, which, through respiration is used by the microorganisms for obtaining energy for their own metabolisms and for the conversion of substrate into new microorganisms. Some energy is lost in the process as heat.

2.2.2 Anaerobic conversion

Bioconversion of organic matter can also occur *anaerobically*, in which case the equation for the *bioconversion* is:

$$\begin{array}{rcll} \Delta S & = 1.22\,\Delta X & + 4.00\,\Delta CH_4 & \\ \text{Substratum consumed} & = \text{Biomass produced} & + \text{Methane produced} & \qquad (2.10) \\ (\text{g}\,O_2) & (\text{g}\,O_2) & (\text{g}\,O_2) & \end{array}$$

In this case the removed substrate is converted to methane and biomass. The coefficients 1.22 and 4.00 are coefficients required to convert anaerobic biomass ($C_5H_9NO_3$) and methane into oxygen units.

In anaerobic treatment, bioconversion produces methane instead of consuming oxygen. The extraction of energy from organic matter is made by the delivery of an electron, which is received by an electron acceptor. The most common and efficient acceptor is O_2. When no molecular O_2 is present, other compounds function as electron acceptors, but are less efficient thermodynamically. This is the case of anaerobic processes, which is the reason that the energy requirements of anaerobic bioconversion are minimal and not sufficiently high to be taken into account in Equations 2.10 and 2.12.

Now we can reduce the whole phenomenon of biological treatment of wastewater to three basic points:

(1) What is the substrate removal rate and what mechanisms govern it?
(2) What is the rate of bacterial growth and what mechanisms define it?
(3) How much oxygen is needed for the process to be done? or How much methane is produced?

The answer to these questions defines:

(1) The detention time needed in the biological reactor to ensure that the substrate is removed to the desired level.
(2) The daily volume of biomass produced.
(3) The weight of oxygen which needs to be supplied each day to carry out the aerobic process or the amount of methane produced and which needs to be handled in the anaerobic process.

Considering these three points, Equation 2.9 for aerobic treatment can be transformed by dividing by Δt and taking Δt to the limit $\Delta t \to \infty$, in the following equation:

$$-\frac{dS}{X\,dt} = -\frac{dO_2}{X\,dt} + 1.42\frac{dX}{X\,dt} \tag{2.11}$$

For anaerobic treatment, the Equation 2.10 turns into:

$$-\frac{dS}{X\,dt} = 4.00\frac{dCH_4}{X\,dt} + 1.22\frac{dX}{X\,dt} \tag{2.12}$$

In Equations 2.11 and 2.12 $(-dS/dt)$ represents the removal rate of substrate or the rate at which the supplied COD is removed, and $(dS/X\,dt)$ is the removal of substrate per unit of biomass (X). The term $(dX/X\,dt)$ is the rate of growth of the biomass (VSS) per unit of biomass. Finally $(-dO_2/X\,dt)$ is the oxygen consumed per unit of bacterial mass and $(dCH_4/X\,dt)$ is the methane produced, also per unit of biomass. The constant 1.42 converts aerobic VSS into units of O_2 (COD), and for anaerobic processes the constant 1.22 does the same for anaerobic biomass and the constant 4.00 converts the grams of methane into grams of COD.

The process of substrate removal used in wastewater treatment may be carried out in reactors of suspended media or in reactors of attached media (also known as fixed media reactors). However, the microbiological process itself is identical in all types of reactors and can be explained by the same phenomena.

2.3 BACTERIAL METABOLISM

Bacterial metabolism consists of two defined processes that are represented in Equation 2.11, and carry out the process of bioconversion of assimilated food $(-dS/dt)$, namely: (i) *bioenergetics* or *catabolism* which is expressed by the term dO_2/dt and refers to the decomposition processes through which energy stored in organic matter is extracted to provide food for microorganisms. In the anaerobic process the energy extraction is minimal compared to the aerobic process, since it does not use oxygen as an electron acceptor; and (ii) *biosynthesis* or *anabolism* which is represented in Equations 2.11 and 2.12 by the term dX/dt, which corresponds to the processes by which the cellular architecture is built and the biomass is enhanced from compounds assimilated as food. The process of breakdown (catabolism) to obtain energy is done through several metabolic cycles, but it must be preceded by hydrolysis.

Hydrolysis

Particulate matter, biopolymers, and in general the complex organic compounds which are the main constituents of municipal wastewater, have to undergo initial dissolution or hydrolysis to turn them into simple organic substrates. The products of hydrolysis are sugars, amino acids, low weight volatile fatty acids and alcohols. These substrates can then be assimilated by bacteria to undergo the process of glucolysis. The hydrolysis takes place externally by the action of exoenzymes. The simple hydrolysis of some organic compounds is very fast, but that of the complex or particulate compounds can be extremely slow. A typical example is the hydrolysis of sucrose that by incorporating a water molecule is decomposed into two isomeric sugar molecules, Glucose and Fructose:

$$C_{12}H_{22}O_{11} + H_2O \longrightarrow 2C_6H_{12}O_6 \quad (2.13)$$

The exoenzyme that fosters this reaction is Glucose Hydrolase. Immediately after crossing the bacterial cell membrane the sugar molecules are incorporated into the glucolysis process.

Glucolysis

Once hydrolysis occurs, soluble compounds can be assimilated by bacteria and initiate the process of Glucolysis, which is summarized in Figure 2.3. Once the decomposition process reaches pyruvic acid, it must make a thermodynamic decision that depends on the availability of external electron acceptors: (i) if oxygen is available then the Krebs cycle, which is the main aerobic metabolic cycle, is carried out in the bacteria; and (ii) if no oxygen (or other electron acceptor) is available, then a reaction that produces Acetic Acid (Acetate) and Hydrogen, as well as other volatile fatty acids (VFA) such as propionic acid, butyric acid, and so on, occurs. This leads to anaerobic decomposition which requires the coordinated participation of five different types of microorganisms. The reaction of glucolysis is summarized as follows:

$$\underset{Glucose}{C_6H_{12}O_6} + 4H_2O \longrightarrow \underset{Acetate}{2CH_3COO^-} + \underset{Alkalinity}{2\,HCO_3^-} + 4H^+ + \underset{Hydrogen}{4H_2} \quad (2.14)$$

Bioenergetics and biosynthesis

Hydrolysis and the Glucolysis metabolic cycle need to occur prior the process of decomposition of carbohydrates by bacteria. The decomposition processes are carried out by acidogenic bacteria, which usually are *facultative*, which means that they can operate in an aerobic pathway in the presence of O_2, or in an anaerobic pathway in the absence of O_2 or nitrates. The anaerobic decomposition pathway is discussed in the next section. If molecular O_2 is available, it is then used as an electron acceptor, and the bacteria itself terminates the aerobic metabolism (i.e. the Krebs cycle). There are other metabolic cycles for different types of organic matter (lipids, proteins, and so on.).

The general plan of metabolism implies catalytic reactions and assimilation of nutrients. Small molecules which are already dissolved are directly assimilated by the cells because they can pass through the cells membrane wall without a problem. On the other hand, the polymer macromolecules once they are stopped by physical and electric barriers of the cell wall, must undergo the hydrolysis process before being assimilated.

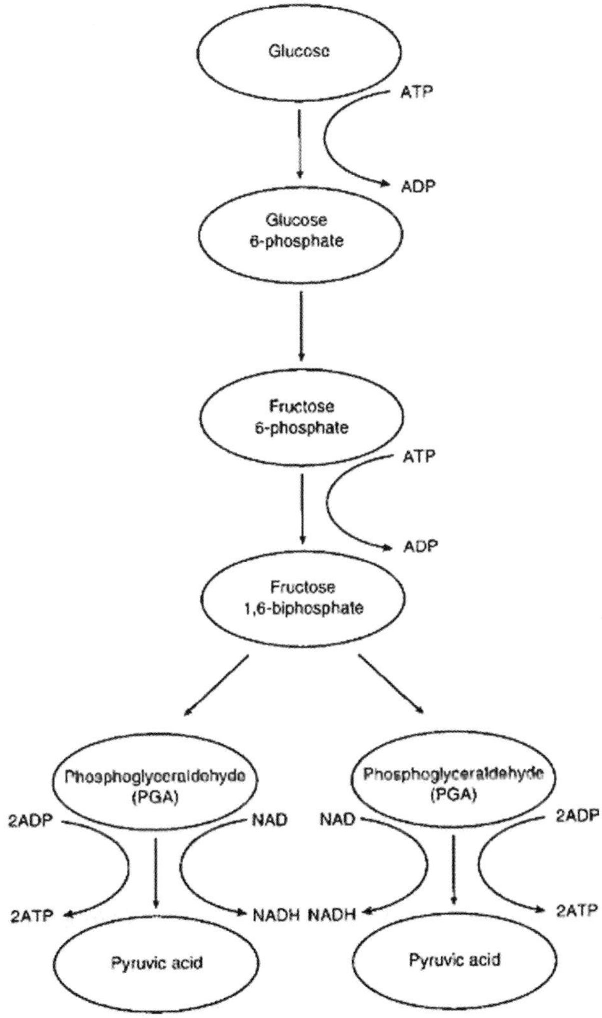

Figure 2.3 Simplified chart of glucolysis pathway [*Source:* http://dolly.biochem.arizona.edu/Bioc462b_Honors_Spring_2009/ksamsel/glycolysis%20thumb.html (April 2011)]

Catabolic (bioenergetic) and anabolic (biosynthetic) processes are shown schematically in Figure 2.4. Catabolism initially produces energy through the Glucolysis, but then the decomposition continues with the Krebs cycle which breaks down the organic matter in twelve key intermediaries, with a net production of energy of 266 KCal per molecule of glucose, energy that is stored mainly in molecules of *Adenosine tri-phosphate* (ATP). The end product is CO_2 and H_2O. The key intermediates are used in biosynthesis, through other inverse metabolic cycles that build cell matter from these fundamental building blocks. This means that organic matter decomposes with catabolism and then the cellular material is built with anabolism.

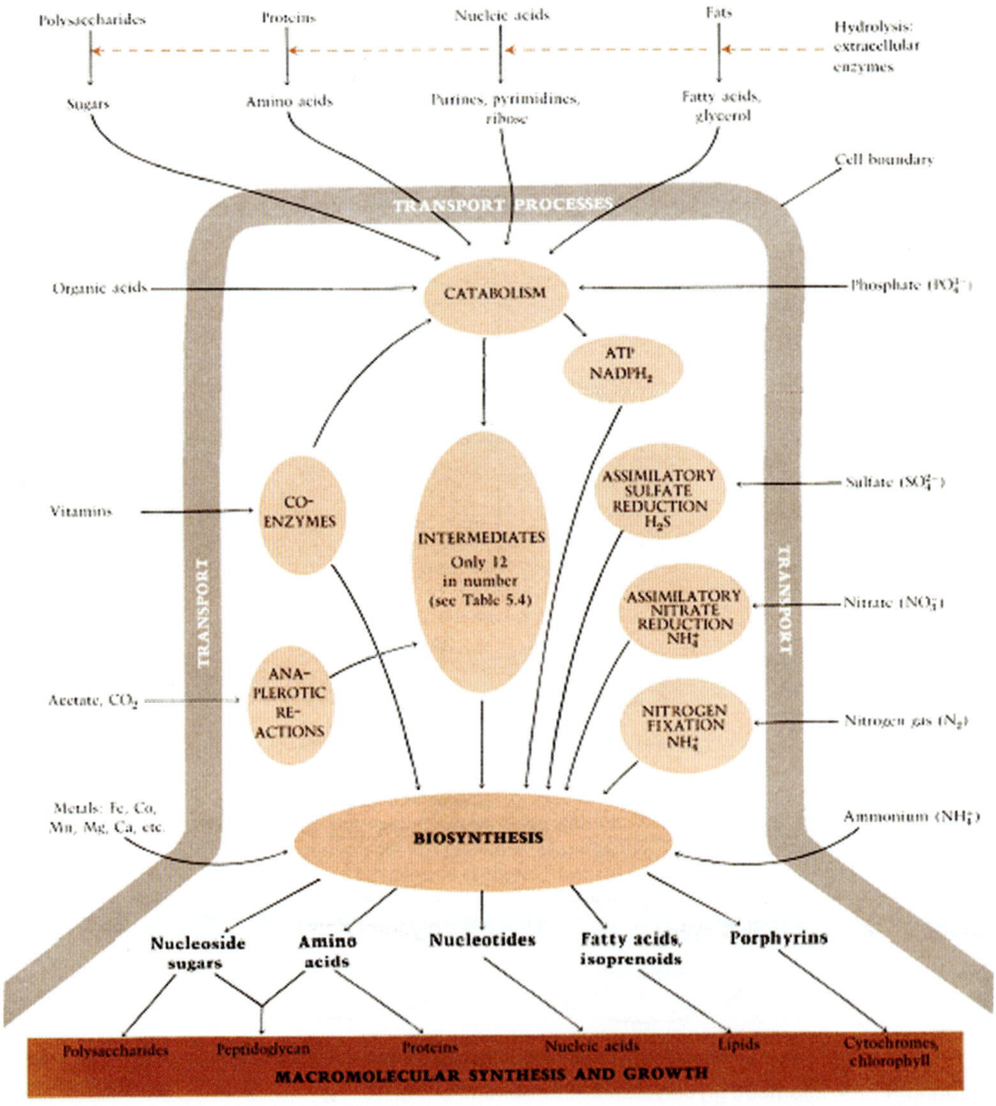

Figure 2.4 Summary of nutrition and biosynthesis (*Source:* Biology of microorganisms by brock, 1974)

The intensity of the phenomena that characterize the biological processes is reduced with the consumption of the substrate and nutrients, the cell growth and the release of products. However, bacterial growth (dX/dt) is well defined by the rate of biosynthesis, which is the rate at which cellular components are formed from precursory products. In the process of decomposition ATP, the conductor of energy for biosynthesis, and $NADPH_2$, the carrier of reducing power are produced. Also, inorganic substances such as metals, sulphur, nitrogen and phosphates are assimilated, to be used in anabolic reactions. For example, the inorganic phosphate is converted to organic phosphate by *phosphorylation* at the level of *oxidative substrate.*

In summary, it can be seen (Figure 2.4) that the polysaccharide are transformed into sugars by hydrolysis; proteins into amino acids, nucleic acids into purines, pyrimidines and ribose; and fats into fatty acids and glycerol. All these contributors to the bioenergetic decomposition produce twelve intermediate products which are listed in Table 2.1 and are the key intermediates in all biosynthetic reactions. These then are used together with organic and inorganic compounds presented in Figure 2.4 to produce the cellular constituents related to the macromolecular growth and synthesis. Thus, polysaccharides are produced from sugars; nucleic acids are produced from nucleotides and lipids are produced from fatty acids.

Table 2.1 Key intermediates and products in the biosynthesis.

Intermediary	Catabolic origin	Role in biosynthesis
Glucose-1-Phosphate	Glucose, galactose, polysaccharides	Nucleosides sugars
Glucose-6-phosphate	Glycolysis	Pentose, storage polysaccharides
Ribose-5-phosphate	Pentose phosphate pathway	Nucleotides, deoxyribonucleotides
Erythrose-4-phosphate	Pentose phosphate pathway	Aromatic amino acids
Phosphoenol pyruvate	Glucolysis	Phosphotransferase System (sugar transport), aromatic amino acids, gluconeogenesis, anaplerotic reactions (CO_2 fixation), muramic acid synthesis
Pyruvate	Glucolysis, phospho-ketolase (pentose fermentation)	Alanine, valine, leucine, anaplerotic reactions) (CO_2 fixation)
3-Phosphoglyceric Acid	Glucolysis	Serine, glycine, cysteine
α- Ketoglutarate	Tricarboxylic acid cycle	Glutamate, proline, arginine, lysine
Succinyl-CoA	Tricarboxylic acid cycle	Methionine, porphyrins
Oxaloacetate	Tricarboxylic acid cycle, anaplerotic reactions	Aspartic acid, lysine, methionine, threonine, isoleucine
Dihydroxyacetone phosphate	Glucolysis	glycerol (fats)
Acetyl-CoA	Pyruvate decarboxylation, pyrimidine breakdown	Fatty acids, isoprenoid sterols, lysine (two carbons) leucine (two carbons)

2.4 AEROBIC DECOMPOSITION

According to the metabolic processes presented, the aerobic decomposition of organic matter is carried out by facultative bacteria operating with aerobic metabolism, which are added into colonies known as *zooglea ramigera*. Carbonaceous organic matter decomposes in a different way from nitrogenous organic matter, because the degradation is produced by different bacteria fed by different substrates. The metabolic processes presented in the previous paragraph applied to the carbonaceous material.

Aerobic decomposition of carbonaceous organic matter

Carbonaceous organic matter comes from carbohydrates. The organic matter in the wastewater, which consists primarily of a mixture of carbonaceous material and protein material, has the generic "formula" $H_{10}C_{19}O_3N$. The energy reaction for the decomposition of carbonaceous organic matter of in wastewater is as follows:

$$\underset{\textit{Wastewater + Oxygen}}{C_{10}H_{19}O_3N + 12.5O_2} \xrightarrow{\text{Microorganisms}} 10CO_2 + 8H_2O + NH_3$$

Aerobic microorganisms are necessary for the reaction to be carried out and have, as seen above, the approximate "formula" $C_5H_7NO_2$. The equation that summarizes the bioenergetics and biosynthesis for the carbonaceous BOD is the following:

$$2C_{10}H_{19}O_3N + 15O_2 \xrightarrow{C_5H_7NO_2} 2C_5H_7NO_2 + 10CO_2 + 12H_2O \quad (2.15)$$

Aerobic decomposition of nitrogenous organic matter

Nitrogen is present in the wastewater in four main forms: organic nitrogen, ammonia nitrogen, nitrite nitrogen and nitrate nitrogen. The organic nitrogen is usually transformed into NH_3 and ammonium salts, by hydrolysis or bacterial action. Total Kjeldahl Nitrogen (TKN) which is the sum of organic nitrogen and ammonia nitrogen and when measured in the raw wastewater, it represents the original amount of organic nitrogen. In the presence of nitrifying bacteria, the ammonia undergoes a process called nitrification, which involves its oxidation, first to nitrite and then to nitrate. The nitrification is the cause of the Nitrogenous Oxygen Demand (NBOD). The representative reactions for energy phase are:

$$NH_4^+ + \frac{3}{2}O_2 \xrightarrow{\text{nitrosomas}} NO_2^- + 2H^+ + H_2O$$

$$NO_2^- + \frac{1}{2}O_2 \xrightarrow{\text{nitrobacter}} NO_3^-$$

The overall nitrification process can be summarized by the following reaction:

$$NH_4^+ + 2O_2 \longrightarrow NO_3^- + 2H^+ + H_2O \quad (2.16)$$

At the same time that energy is produced, the synthesis of new bacteria occurs. Nitrification produces NBOD through the oxidation to nitrites by *Nitrosomonas* and the final nitrification to nitrates by *Nitrobacter* as shown by the previous reactions. Once nitrification occurs, the denitrification process can take place, if an energy source is added (organic matter) and anoxic conditions prevail (lack of O_2 and presence of nitrates). Methanol (CH_3OH), acetic acid (CH_3COOH) or the wastewater itself can be used as a carbon source for the denitrification process. If methanol is used as the carbon source (i.e. the organic matter), the overall reaction of synthesis and energy in denitrification and is the following:

$$NO_3^- + \underset{\textbf{Methanol}}{1.08CH_3OH} + H^+ \longrightarrow \underset{\textbf{Bacteria}}{0.065C_5H_7NO_2} + 0.47N_2\uparrow + 0.76CO_2 + 2.44H_2O$$

Nitrification is required to prevent the presence of nitrogenous BOD in water bodies into which an effluent is discharged, and denitrification might be necessary to prevent eutrophication of the water body by the presence of nitrogen.

2.5 ANAEROBIC DECOMPOSITION

In anaerobic decomposition there is no free molecular O_2, which means that there are no external electron acceptors. Under these conditions an aggregate of five different kinds of bacteria is formed, which operate in a coordinated manner in a flocculant aggregate (the floc) and, very often, in a granular form (see Figure 2.5). The general biochemistry of anaerobic decomposition is explained in Figure 2.6. The processes that occur in the granules (Figure 2.5) are:

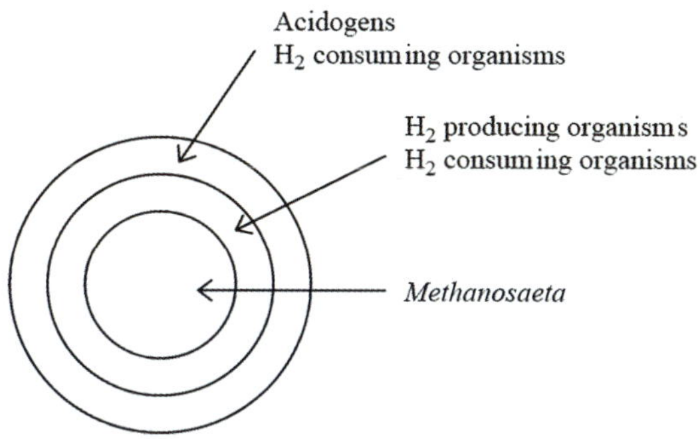

Figure 2.5 Grouping of the anaerobic bacteria in the granule (*Source:* Hulshoff *et al.* 2004)

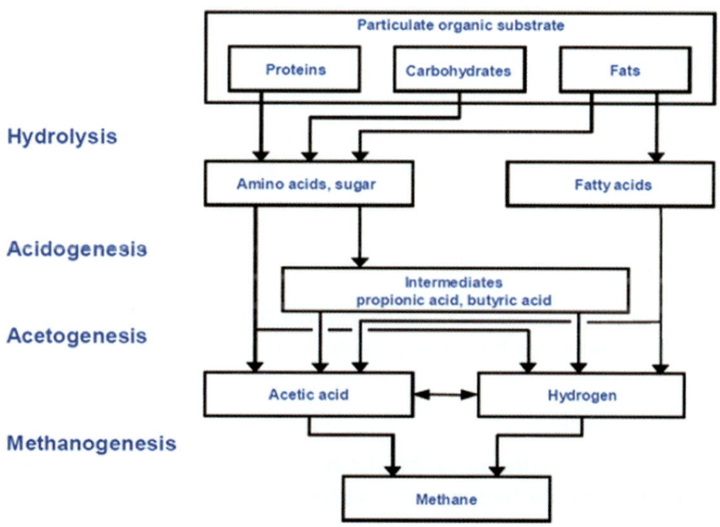

Figure 2.6 Schematic diagram of the anaerobic decomposition process [*Source:* http://www.wtert.eu/default.asp?Menue=13&ShowDok=12 (April 2011)]

Hydrolysis

As explained above, it is necessary to dissolve the particulate matter, the biopolymers, and in general the complex organic compounds, through hydrolysis. The products of hydrolysis are sugars, amino acids, volatile low weight fatty acids and alcohols. These substrates will be assimilated by the acidogenic bacteria or fermentative[2] bacteria (*eubacteria*) to undergo the process of glucolysis explained above (see Equation 2.13).

Acidogenesis or fermentation

Once the hydrolysis occurred, the glucolysis process (which has already been described) starts. When the process arrives to the pyruvate stage, and since there are no external electron acceptors, the reaction which occurs produces Acetic Acid (Acetate), other volatile fatty acids, volatile fatty acids such as propionic acid, and hydrogen. At this point the acidogenic bacteria (which are facultative) operate in anaerobic mode and deliver their metabolic products resultants of this mode (see Equation 2.14).

The accumulation of hydrogen hinders the decomposition of glucose. In fact there is a thermodynamic limit, a maximum permissible concentration of hydrogen under which the anaerobic decomposition can continue, which is given in terms of partial pressure as follows: $P_H < 10^{-4}$ atm, where P_H is the partial pressure of hydrogen. If this does not occur, that is, the partial pressure of hydrogen is above 10^{-4} atm, the reaction described in Equation 2.14 stops and only the reaction producing volatile fatty acids proceeds.

Acetoclastic acetogenesis

For the anaerobic decomposition to continue, it is necessary that the volatile fatty acids convert to acetic acid, because this is the only fatty acid that can be methanized (converted to methane). There are other compounds that can be methanized, such as methanol and formic acid, but they are not common in anaerobic decomposition.

Hydrogenoclastic acetogenesis

Although there may be *Archaebacteria* capable of methanizing Hydrogen, there are also bacteria, which under anaerobic decomposition conditions can produce acetic acid, and they are known as the *acetogenic hydrogenoclastics*. The very essential function of these bacteria is to maintain the P_H below the necessary limits.

Hydrogenoclastic methanogenesis

This is a reaction carried out by a type of *Archaebacteria* that methanize approximately 30% of the original substrate. It is a very fast reaction which competes with the previous one.

Acetoclastic methanogenesis

This is the most important reaction in the anaerobic decomposition process and is responsible for producing 70% of methane gas. There are only two types of bacteria that produce this reaction: *Methanothrix* and *Methanosarcina*.

In order for anaerobic reactions to occur in a timely manner in wastewater treatment plants, the various organisms need to grouped as a floc or granule so that the bacteria producing intermediate products remain close to the bacteria consuming the products and that all the reactions function in a sequence without accumulation of intermediate products, all in accordance with Figure 2.5. Granules formed in an anaerobic reactor treating municipal wastewater are shown in Figure 2.7 (Orozco, 1988).

[2]They are actually facultative bacteria which in presence of O_2 are aerobic and produce CO_2 and H_2O, and in absence of O_2 are acidogenic or fermentative and produce volatile fatty acids and H_2.

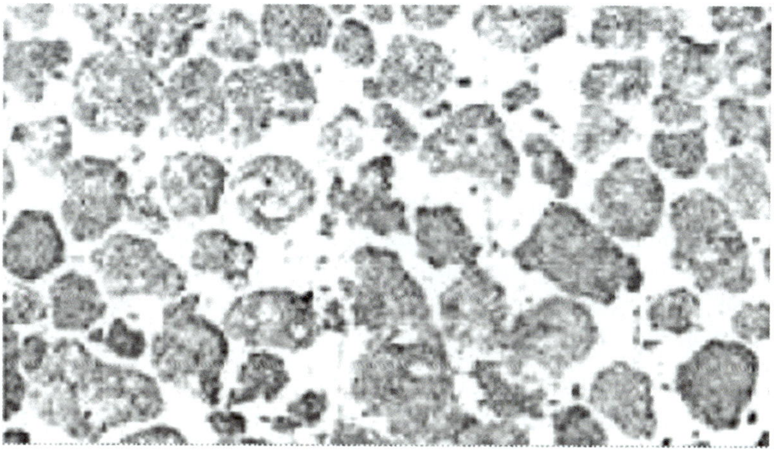

Figure 2.7 Granulated sludge of domestic wastewater (*Source:* Orozco, 1988)

2.6 DIFFERENCES BETWEEN AEROBIC AND ANAEROBIC TREATMENT

The anaerobic process is sensitive to low temperatures and its rate is greatly reduced at temperatures below 12°C (in the wastewater not in the air). However, it works well at higher temperatures and is therefore appropriate for developing countries, many of which are located in warm climate zones. Anaerobic processes were traditionally used for treatment of excess sludge and not for treatment of wastewater. Only in recent years the anaerobic process being considered in some countries as a process also suitable for treating municipal waste water, not only for treatment of sludge. The capacity and the rate of decomposition of organic matter in raw sewage under anaerobic conditions at temperatures above 12°C are similar to those of the aerobic process (as evidenced by the fact that the hydraulic detention times in anaerobic reactors for wastewater treatment are similar to the detention time in aerobic reactors of activated sludge), so an anaerobic process can perform well as the main treatment unit of wastewater in warm climates.

The fact that the overall rate of anaerobic decomposition of organic matter in raw wastewater is similar to the rate of decomposition under aerobic conditions seems to contradict the fact that anaerobic decomposition of sludge is a slow rate process. The slow rate of anaerobic stabilization of sludge made many professionals to believe that because anaerobic decomposition of excess sludge is a slow rate process, then all anaerobic processes are slow rate processes for any type of substrate. The fact is that excess sludge, which is basically excess bacterial cells, is a very difficult matter to decompose due to the need to break the cell membranes, and even aerobic digestion of excess sludge is an equally slow process. So in anaerobic digestion of sludge the problem is the type of substrate (which in this case is the excess bacterial cells) and not the anaerobic process itself (Libhaber, 2007; Van Handel & Lettinga, 1994).

An anaerobic treatment process has some advantages over an aerobic process: (i) it does not require oxygen thus it does not consume energy; (ii) excess sludge quantities are much smaller than those generated in aerobic process, thus expenses for sludge treatment and disposal are much smaller and so are negative environmental effects generated by the sludge; (iii) the gas it produces, mainly methane, can be collected and used to generate energy or be burned, reducing the emission of greenhouse gases, and carbon fund credit can be obtained for the emission reduction, increasing the financial benefits of the process; (iv) smaller amounts of nutrients are required in anaerobic processes because less biomass is produced; (v) the occupied area of an anaerobic reactor is small, except in the case of anaerobic lagoons, which require a somewhat larger area; (vi) electromechanical equipment is not required for these processes; (vii) construction is simple and uses mostly local materials; (viii) operation is quite simple; and (ix) both investment cost and operation and maintenance costs are way smaller than those of aerobic processes. Anaerobic treatment also has some disadvantages: (i) it needs longer start-up time to develop the necessary amount of anaerobic biomass in the reactor; (ii) it is more sensitive to lower temperatures; (iii) it is more sensitive to toxic substances; (iv) biological nitrogen and phosphorus removal cannot be achieved in an anaerobic process; (v) it may require alkalinity addition; (vi) it may produce odours and corrosive gases, however those can be controlled; and (vii) the effluent quality of a one stage anaerobic treatment process may not meet discharge standards and, depending on the specific project conditions, a subsequent treatment stage might be necessary.

2.7 KINETICS AND STOICHIOMETRY OF CARBONACEOUS BOD DECOMPOSITION[3]

To develop procedures for calculating the different wastewater treatment processes, it is necessary to provide answers to the three questions raised in section 2.2, namely:

(1) What is the substrate removal rate ($-dS/dt$)?
(2) What is the bacterial growth rate (dX/dt)?
(3) How much oxygen ($-dO_2/dt$) is necessary to supply for maintaining the process? Or How much methane (dCH_4/dt) is produced by the process?

To this end the kinetic and stoichiometric equations associated with the decomposition of carbonaceous organic matter need to be consulted. Following are the differential equations that represent each of the processes:

Substrate removal

When the concentration of biomass in a reactor is small compared to the substrate concentration and therefore is the limiting factor, the familiar *Monod* equation is applied for estimating the bacterial growth (see Shuler & Kargi, 1992, pp 170). For such conditions, the removal of soluble substrate can be derived from the following equation:

$$\frac{dS}{X\,dt} = \frac{kS}{K_m + S} \tag{2.17}$$

where:

X = biomass concentration in the reactor (mg VSS/l)
S = substrate concentration in the reactor (mg COD/l)
k = kinetic constant, maximum rate of unit removal (d^{-1}).
K_m = Monod's kinetics saturation constant (mg COD/l).

On the other hand, if the concentration of biomass in a reactor is high compared to the substrate concentration, the *Contois* equation is applied (see Shuler & Kargi, 1992, pp. 171). This equation can be

[3]Based on "*Bioingeniería de aguas residuales: teoría y diseño*", Orozco (2005).

transformed into the following equation for the removal of substrate (see Orozco, 1976):

$$\frac{dS}{X\,dt} = \frac{kS/X}{K_c + S/X} \tag{2.18}$$

where:

k = constant kinetic, maximum rate of unit removal (d^{-1})
K_c = *Contois'* kinetics saturation constant (mg COD/mg SSV)

This situation (high biomass concentration per unit substrate concentration) is most common in the wastewater treatment plants, specifically in aerobic treatment with high sludge ages, and in anaerobic treatment. Therefore equation 2.18 should be applied for the design of wastewater treatment plants. For further discussion on this equation see Orozco (1977). Table 2.2 presents typical values of k and K_c.

Table 2.2 Kinetic and stoichiometric constants.

Constant	Units	Aerobic	Anaerobic
k	mg COD/mg VSS · d	5.0	2.5
K_c	mg COD/mg VSS	0.02	0.04
Y	mg VSS/mg/COD	0.5	0.08
k_e	d^{-1}	0.05	0.03
T	°C	20	25
θ		1.04	1.04

Source: Orozco (2005), Metcalf & Eddy (2003), Pavlostathis & Giraldo-Gómez (1991)

When the relation S/X is small compared with K_c ($K_c \gg S/X$) Equation 2.18 becomes the following equation proposed by *McKinney* (1962):

$$\frac{dS}{dt} = k_L S \tag{2.19}$$

where:

k_L = Synthesis Factor (d^{-1})
$k_L \approx k/K_c$

Biomass production

The production of biomass or bacterial growth has been universally interpreted as follows:

$$\frac{dX}{X\,dt} = Y\frac{dS}{X\,dt} - k_e \tag{2.20}$$

where:

$dX/X\,dt$ = biomass net growth rate (d^{-1})
Y = stoichiometric coefficient of production (mg VSS/mg COD), typically 0.5 for aerobic decomposition.
k_e = endogenous respiration coefficient (d^{-1}), typically 0.05

For anaerobic treatment the same equation is applied, but the coefficient Y is replaced by the anaerobic production ratio, Y_{an}, whose value is in the range 0.5–0.1, with a typical value presented in Table 2.2.

The value of the coefficient Y represents the portion of the energy contained in the incoming substrate which is transformed to new bacterial cells, and therefore defined the excess sludge quantity of the process. In aerobic processes, about 50% of the energy of the incoming substrate is consumed by the aerobic bacteria for their life maintenance and 50% is transformed to new bacterial cells. That means that about 50% of the substrate comes out of the aerobic process in the form of excess sludge, and this is a lot of sludge. In anaerobic processes, 90–95% of the energy of the incoming substrate is consumed by the anaerobic bacteria for their life maintenance only 5–10% is transformed to new bacterial cells. This is the reason that anaerobic processes produce much less excess sludge.

The cell retention time of bacteria in the reactor (i.e. the average time in which bacterial biomass remains in the reactor before being discharged from it as excess sludge, in days), also known as sludge age (θ_c) is defined as:

$$\theta_c = \frac{X}{dX/dt} \tag{2.21}$$

So Equation 2.20 becomes,

$$\frac{1}{\theta_c} = Y\frac{dS}{X\,dt} - k_e \tag{2.22}$$

Oxygen uptake and methane production

The oxygen uptake required for maintaining the aerobic biological treatment process is calculated from the biotransformation Equation 2.11 as follows:

$$-\frac{dO_2}{X\,dt} = \frac{dS}{X\,dt} - 1.42\frac{dX}{X\,dt} \tag{2.23}$$

where:

$-dO_2/dt =$ Oxygen uptake (mg O_2/l)
1.42 is the stoichiometric conversion factor of mg VSS/l to mg O_2/l

In anaerobic treatment, methane gas (CH_4) is produced instead of oxygen being consumed. The equation for the production of methane derived from Equation 2.12 is as follows:

$$4.00\frac{dCH_4}{X\,dt} = \frac{dS}{X\,dt} - 1.22\frac{dX}{X\,dt} \tag{2.24}$$

where 4.00 is the stoichiometric conversion factor of CH_4 to O_2 and 1.22 is the stoichiometric of conversion of anaerobic biomass to oxygen units. For a broader discussion of these equations see Orozco (2005).

The kinetic constants k and k_e must be corrected for variation of temperature according to the equation:

$$k(T) = k(T_0)\theta^{(T_0-T)} \tag{2.25a}$$

T_0 being 20°C for aerobic treatment and 25°C for anaerobic treatment. However, in the anaerobic treatment the following equation is used more often:

$$k(T) = k(25^\circ C)e^{0.1(T-25)} \tag{2.25b}$$

Completely mixed and plug flow reactors

Equations 2.18, 2.22, 2.23 and 2.24 are always applied in environment of the bacterial reaction processes, but the end result of these processes depends on the hydraulic regime of the reactor in which they occur. Reactors with completely mixed hydraulic regime are the most common in wastewater treatment, especially in anaerobic treatment and in natural treatment processes (lagoons, wetlands and so on). Although the hydraulic regime of the reactor is important in any treatment system, its importance is most obvious in the suspended media type treatment, where bacterial biomass is intimately mixed with the wastewater, resulting in a mixed liquid known as Mixed Liquor (ML). The mixed liquor has the same hydraulic properties as water (except for a slightly higher density because of the biomass concentration ranging between 1,000 and 10,000 mg/l Mixed Liquor Suspended Solids (MLSS), which ranges from 0.1 to 1% concentration by weight).

The reactor in which the biological reactions take place is a basin where the mixed liquor biomass carries out the biotransformation of organic substrate and has two extremes of fully differentiated hydraulic regimes: (i) the completely mixed regime; and (ii) the plug flow regime. In the completely mixed regime there is a total and complete mixing of mixed liquor, so that there is no variation in the concentration of any parameter at any point in the reactor. Mathematically this is defined by the concentration gradient, which is the spatial variation of the concentration of the parameters being measured (e.g. the substrate, S). If the spatial variable is ordinate z, then the variation of the substrate with the length (or depth, it does not matter which ordinate is chosen) must be zero, or:

$$\frac{dS}{dz} = 0$$

This means that S is constant throughout the reactor volume. Conversely, in a reactor under a plug flow regime there is a defined spatial gradient of all its constituents. As suggested by its name, the plug flow reactor resembles a piston that travels through a long tank undergoing in its way the biotransformation in the z direction. As the wastewater travels in the tank at speed u, which depends on the wastewater flow which enters the reactor (according to the continuity equation $Q = A_T \cdot u$, where Q is the flow rate and A_T the cross section area of the tank) it is clear that each position of the water "piston" z_i is associated with a transit time t_i comparable to the detention time that the mixed liquor stayed in the reactor up to that point, and the biotransformation took place during this time. We can assume then that $dS/dt = dS/dz$, and if we also adopt, for example, the kinetics of McKinney (Equation 2.19), the resulting gradient in the plug flow reactor can then be represented by the following equation:

$$-\frac{dS}{dz} = k_L S$$

This equation means that at each point z there will be a definite concentration S different from the others in other positions the z-axis, as shown in Figure 2.9.

Completely mixed reactors

For a completely mixed reactor the following equation holds:

$$-\frac{dS}{dt} = \frac{S_0 - S}{t_d} \tag{2.26}$$

Which when combined with Equation 2.22, allows presenting the equation of the concentration of substrate (X) in the reactor as follows:

$$X = \frac{\theta_c Y(S_0 - S)}{t_d(1 + k_e\theta_c)} = \frac{Y(S_0 - S)}{t_d\left(k_e + \frac{1}{\theta_c}\right)} \quad (2.27)$$

Likewise if we combine in Equation 2.22 with Equation 2.18 we receive:

$$\frac{S}{X} = \frac{K_c(1 + k_e\theta_c)}{\theta_c(Yk - k_e) - 1} = \frac{K_c\left(k_e + \frac{1}{\theta_c}\right)}{Yk - \left(k_e + \frac{1}{\theta_c}\right)} \quad (2.28)$$

Now, if we combine Equation 2.27 with Equation 2.28, we obtain the equation for the substrate in the effluent, as follows:

$$S = \frac{S_0}{1 + \left[\frac{kY - (\frac{1}{\theta_c} + k_e)}{K_c Y}\right] t_d} \quad (2.29)$$

And if we define:

$$K_O = \frac{kY - \left(\frac{1}{\theta_c} + k_e\right)}{K_c Y} \quad (2.30)$$

Then we obtain:

$$S = \frac{S_0}{1 + K_O t_d} \quad (2.31)$$

where K_O depends only on the sludge age θ_c, since Y, k_e, k and K_c are constants. When the sludge age is large enough ($1/\theta_c \rightarrow 0$), then K_O is reduced to $\left[\frac{kY - k_e}{K_c Y}\right] \approx k/K_c$, which is the same k_L of Equation 2.19

Extremely High Sludge Age (EHSA)

In conditions when the sludge age, θ_c, is very high, with a tendency to reach infinity, that is, $(1/\theta_c) \rightarrow 0$, the sludge age is denominated **Extremely High Sludge Age (EHSA)** which is the case in a UASB reactor or in a facultative lagoon. Then we get:

$$S = \frac{S_0}{1 + \left[\frac{kY - k_e}{K_c Y}\right] t_d} \quad (2.32)$$

and

$$X = \frac{Y(S_0 - S)}{t_d k_e} = \frac{Y K_O S_0}{k_e(1 + K_O t_d)} \quad (2.33)$$

This implies that for a given S_0, S depends only on t_d. Under these conditions X also depends on t_d, which means that the lower t_d, the greater X. Thus, we find that t_d can be as small as wanted provided that X is large enough. It is noteworthy that in this case, the lesser the detention time, the higher the effluent substrate, as shown in Equation (2.32). Another finding of great importance under EHSA conditions is that, theoretically, the production of biomass can be zero. As $1/\theta_c \rightarrow 0$, then we get that:

$$\frac{dX}{X\,dt} = \frac{1}{\theta_c} = 0 \qquad (2.34)$$

This happens because the biomass produced, $Y \cdot (dS/dt)$, is equal to that consumed in the endogenous respiration phase, k_eX.

Reactors in series

Often the possibility of designing several completely mixed reactors in series occurs in practice. This case is interesting from an analytical point of view, because it allows differentiating between the plug flow and the completely mixed reactors. Referring to Figure 2.8, let us study, for simplicity, only two completely mixed reactors in series without recycle from the second one to the first. We assume that both reactors have the same volume, V/2, and that $t_d = (V/2)/Q_0$. The total volume of the system is V and the total detention time is V/Q_0.

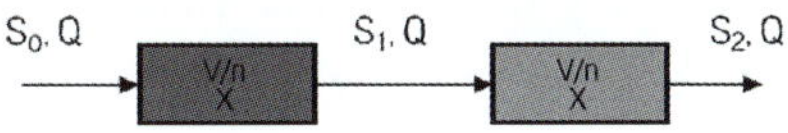

Figure 2.8 Two completely mixed reactors in series

Using Equation 2.31 we have for the reactor (1):

$$S_1 = \frac{S_0}{1 + K_O \frac{t_d}{2}}$$

For the reactor (2), the influent is the reactor's (1) effluent, namely, S1. Then:

$$S = \frac{S_1}{1 + K_O \frac{t_d}{2}} = \frac{S_0}{\left(1 + K_O \frac{t_d}{2}\right)^2} \qquad (2.35)$$

From this it follows that for n reactors in series of equal volume (V/n), where V is the total volume of system and $t_d = V/Q_0$ is the total detention time in the system, the substrate concentration in the final effluent is:

$$S = \frac{S_0}{\left(1 + K_O \frac{t_d}{n}\right)^n} \qquad (2.36)$$

It is clear that n reactors in series produce a much lower substrate concentration in the effluent than a single completely mixed reactor, with all other circumstances maintained equal.

Plug flow reactor

Let us consider the plug flow reactor presented in Figure (2.9) and assume that the recirculation from the secondary sedimentation tank is zero. This reactor can be considered as an infinite series of completely mixed reactors. Indeed, if we take Equation (2.35) and make in it n equal to infinity, we have:

$$S = \lim_{n \to \infty} \frac{S_0}{(1 + K_O \frac{t_d}{n})^n}$$

But, as we know: $\lim_{n\to\infty}(1 + K_O \frac{t_d}{n})^n = e^{K_O t_d}$, so we obtain:

$$S = S_0 e^{-K_O t_d} \tag{2.37}$$

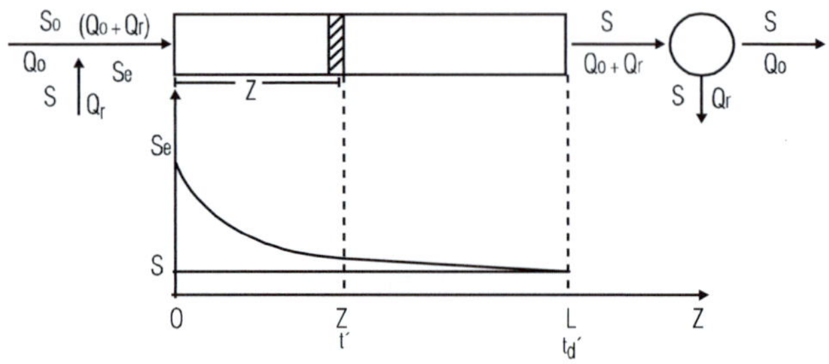

Figure 2.9 Representation of a plug flow reactor

In conclusion, from the analytical standpoint, there are two extremes in the hydraulics of treatment systems: completely mixed reactors and plug flow reactors. Between the two, there are all the systems in series, improving its efficiency as the flow stops being completely mixed and becomes plug flow.

Efficiency in anaerobic treatment

It is clear from the previous analysis, summarized in Equation (2.37), that the plug flow gives the maximum possible efficiency. In practice, the plug flow does not behave as efficiently because of microbiological and operational reasons, since the environment is continuously changing during the passing of a fluid element through the reactor. So the microbial population is forced to continuously readjust, losing efficiency in its metabolic function. The efficiency (E) can be calculated by the equation:

$$E = (S_0 - S)/S_0 \tag{2.38}$$

For anaerobic treatment, which is in EHSA conditions ($1/\theta_c \to 0$), using the coefficients from Table 2.2 we can calculated K_O:

$$K_O(25°C) = \left[\frac{kY - k_e}{K_c Y}\right]$$

$$K_O(25°C) = \left[\frac{2.5 \times 0.08 - 0.03}{0.04 \times 0.08}\right] = 53\,d^{-1} \tag{2.39}$$

This can be corrected for temperatures other than 25°C by the Equation 2.25b. Now, in anaerobic treatment the most widely used empirical design parameter is the volumetric loading (L_v):

$$L_v = S_0/t_d \tag{2.40}$$

It follows that $t_d = S_0/L_v = 0.5/L_v$, because for municipal wastewater the COD is usually around 500 mg/l (see Table 3.2), which is equivalent to 0.5 kg COD/m^3.

For *completely mixed reactors* Equation 2.32 applies, in which case Equation 2.38 combined with Equations 2.32, 2.39 and 2.40 becomes:

$$E = E_{max}\{(26.5/[26.5 + L_v])\} \tag{2.41}$$

where E_{max} is the maximum possible efficiency in anaerobic treatment, namely 90% (0.9). See Figure 2.10 to observe its behaviour. For *plug flow reactors* Equation 2.37 applies and Equations 2.38, 2.39 and 2.40 become:

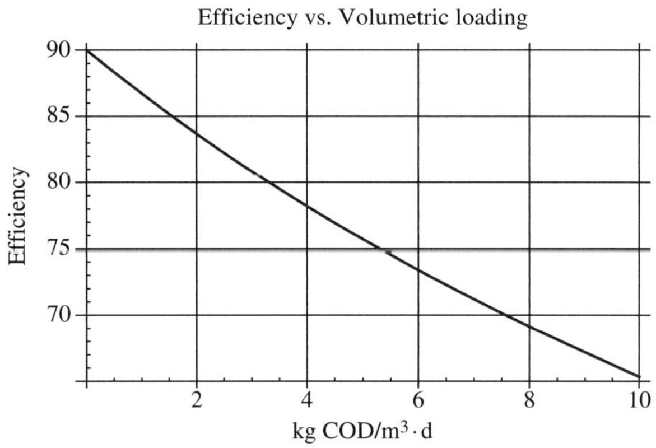

Figure 2.10 Variation of efficiency with volumetric loading for a completely mixed reactor

$$E = E_{max}(1 - e^{-26.5/Lv}) \tag{2.42}$$

See Figure 2.11 to observe the behaviour, which is very common in UASB reactors. This suggests that the UASB reactor may behave sometimes as a plug flow reactor.

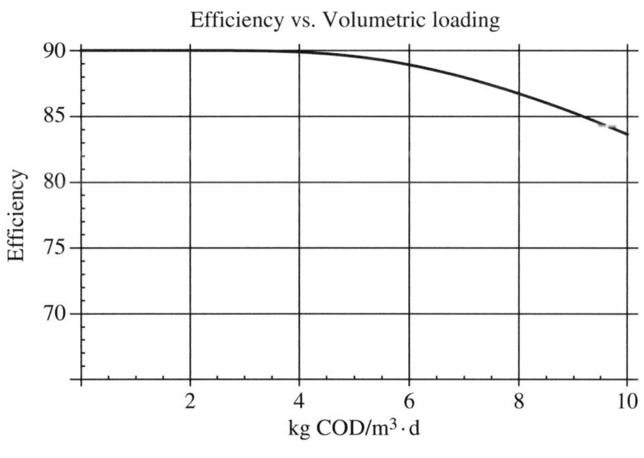

Figure 2.11 Variation of efficiency with volumetric loading for a plug flow reactor

Chapter 3

Calculation of the wastewater flow and BOD load

3.1 DESIGN FLOW[1]

The design flow is one of the essential basic data required for the preparation of the process design of a wastewater treatment system (WWTS). There are two main flow values for design of a wastewater treatment plant: (i) the average flow, and (ii) the peak flow or maximum flow. The average flow is used for the design of the treatment units such as the biological reactors and others. The peak flow is used for the design of all the hydraulic structures of the treatment system.

The calculation of wastewater flow depends on two variables: the design population (P) and the per capita water consumption (q) of this population in $l/\text{Capita} \cdot d$. The design population is estimated by one of the methods of population projections during a given design period and per capita water consumption is measured directly, when possible.

There are two types of water consumption: (i) domestic consumption (q_{dom}) that defines the consumption per household (which is 20–30 m^3/month, or about 150 to 250 $l/\text{Capita} \cdot d$) and total consumption (q_T) which is calculated by dividing the treated flow of the water purification plant by the number of inhabitants supplied with potable water. This latter provision covers all uses, including commercial and industrial, but it also includes *non-revenue water*, which comprises the physical losses of water in the pipe system and commercial losses resulting from fraud and erroneous connections. In general, q_{dom} is preferably measured separately from the other types of water consumption such as commercial, industrial and institutional, setting goals for the non-revenue water with an attempt to keep it low. The non-revenue water is usually very high in cities of developing countries (40 to 50% and even more) while the desirable goal should be between 15 and 25%.

The q_{dom} value refers to the consumption of municipal drinking water and it includes the consumption of commercial and industrial activities. We are interested in determining the flow of wastewater (WW). To calculate it we use a constant c, the *coefficient of return*, which defines the fraction of the domestic water consumed which is discharged into the sewerage system. The value of c is usually between 0.8 and 0.9. So the wastewater flow generated by a population P is:

$$Q_{ww} = \frac{c q_{\text{dom}} P}{86400} \tag{3.1}$$

[1]This paragraph was prepared based on the textbook "*Bioingeniaría de aguas residuales: teoría y diseño*" (2005), Alvaro Orozco Jaramillo, published by ACODAL, Bogotá, Colombia.

where the coefficient 86,400 is the number of seconds in the day. Dividing by this coefficient is required in order to present the result in liters per second. In addition to the domestic sewage that is discharged to the sewerage system, this system also continuously receives infiltration water which is always present in the soil matrix, either from groundwater or from the physical loss of water supply piping network. The amount of infiltration water that penetrates the sewerage system depends not only on the water present in the underground but also on the soil characteristic, mainly its transmissibility. The following formula applied for estimating the infiltration flow:

$$Q_I = q_I A_a \tag{3.2}$$

where:

Q_I = infiltration flow
q_I = infiltration unit flow, usually in the range 0.1 to 0.2 lps/ha
A_a = effective area of infiltration into the sewerage system, ha.

Consequently, the wastewater process design flow or the average flow of a domestic wastewater treatment plant (Q_D) for the processes design is:

$$Q_D = \frac{cq_{dom}P}{86400} + q_I A_a = Q_{ww} + Q_I \tag{3.3}$$

For design of the pre-treatments units and of the hydraulic of the plant the peak flow (Q_{DH}) is used. The peak flow is calculated as follows:

$$Q_{DH} = k_1 k_2 \frac{cq_{dom}P}{86400} + q_I A_a \tag{3.4}$$

where k_1 is the factor which defines the ratio of the flow of maximum day to the average flow (Q_{maxd}/Q_D), and its value is usually between 1.5 and 1.8, and k_2 is the peak hourly flow during the day (Q_{maxh}/Q_{maxd}), usually between 1.3 and 1.7. A typical value of total peak facto is k_1k_2 is 2.7. The value of this peak factor depends on the population, the smaller the population the higher the peak factor. There are several equations to calculate the total peak factor ($PF = k_1k_2$) depending on the population (P), in thousands, namely:

Babbit Equation:

$$PF = \frac{5}{P^{0.2}} \tag{3.5}$$

Harmon Equation:

$$PF = 1 + \frac{14}{4 + \sqrt{P}} \tag{3.6}$$

"Ten States" Equation:

$$PF = \frac{18 + \sqrt{P}}{4 + \sqrt{P}} \tag{3.7}$$

From these equations we can obtain the maximum peak flow if we know the population served. Figure 3.1 presents a comparison among the three: the equations, from which it is noted that Equations

3.6 and 3.7 overlap. *If possible, the PF should be calculated from real data measured during two or three years* (See Freni *et al.* 2003).

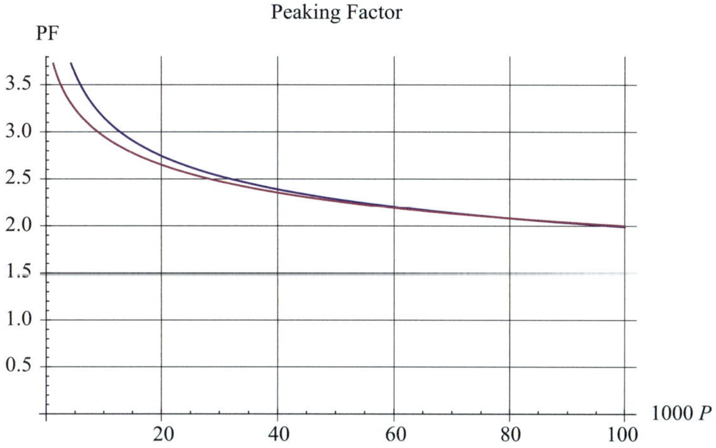

Figure 3.1 Comparison of the maximum peak factor according to Babbit, Harmon and "*Ten States*", the last two equations are equivalent and overlap PF is maximum peak factor and P is the population in thousands

Table 3.1 Flows of municipal wastewater as recommended by *RAS (2000)*, Colombia.

Range		Popul. hab.	Complexity level	Specifc Comsumption (l/person · d)	Peak factors			Flow (lps)		
					k_1	k_2	$k_1 k_2$	Average	Daily maximum	Hourly maximum
Range 1	*Minimum*	500	Low	100	1.3	1.6	2.08	0.58	0.75	1.21
	Average	**1500**	Low	100	1.3	1.6	2.08	1.74	2.26	3.62
	Maximum	2500	Medium	120	1.3	1.5	1.95	3.47	4.51	3.77
Range 2	*Minimum*	2500	Medium	120	1.3	1.5	1.95	3.47	4.51	6.77
	Average	**3750**	Medium	120	1.3	1.5	1.95	5.21	6.77	10.16
	Maximum	5000	Medium	120	1.3	1.5	1.95	6.94	9.02	13.53
Range 3	*Minimum*	5000	Medium	120	1.3	1.5	1.95	6.94	9.02	13.53
	Average	**8750**	Medium	120	1.3	1.5	1.95	12.15	15.80	23.69
	Maximum	12500	Medium	120	1.3	1.5	1.95	17.36	22.57	33.85
Range 4	*Minimum*	12500	Medium	120	1.3	1.5	1.95	17.36	22.57	33.85
	Average	**36250**	Medium high	130	1.2	1.45	1.74	54.54	65.45	94.90
	Maximum	60000	Medium high	130	1.2	1.45	1.74	90.28	108.34	157.09
Range 5	*Minimum*	60000	Medium high	130	1.2	1.45	1.74	90.28	108.34	157.09
	Average	**180000**	High	150	1.2	1.45	1.74	312.50	375.00	543.75
	Maximum	300000	High	150	1.2	1.45	1.74	520.83	625.00	906.24

Coefficient of return: 80%.

The maximum flow of wastewater in a town is a somewhat probabilistic value. At certain instances it can be larger than the maximum flow value used for the hydraulic design of the treatment plant, for instances in periods of increase water consumption or in cases severe floods which result in overflow of flood water to the sewerage system. However, a treatment plant cannot handle of flow larger than the maximum flow used for its hydraulic design. It is therefore important to include in each plant overflow weirs which prevent excess flows higher that the flow calculated by Equation 3.4 from entering the plant. The excess flow which overflows the weir may be stored in a reservoir for processing at a later time, when the incoming flow decreases, or directly discharged to a nearby stream, depending on quality considerations.

As an additional reference, we present in Table 3.1 the values recommended by the RAS 2000 (Technical Norms for the Water and Sanitation Sector of Colombia), which are a bit low compared to norms. As stated, direct measurements are preferred for the design of a wastewater treatment plant. Infiltration rates into the sewerage system also need to be taken into account when establishing the average and hydraulic design flows.

3.2 BOD DESIGN LOAD

To calculate the design load of organic matter and nutrients in the raw wastewater, which are the basic date for the treatment process design, it is necessary to know the quality of the wastewater (or its composition). The parameters which need to be measured in wastewater are presented in Table 3.2, along with typical values for each one. For design purposes, the quality parameters need to be measured in composite samples collected at different parts of the city. A further comparison of typical values of the wastewater quality parameters are presented in Table 3.3.

Table 3.2 Concentrations of contaminants in typical municipal wastewater.

Contaminant	Unit	Weak	Medium	Strong
Total solids (TS)	mg/l	350	720	1200
Dissolved solids (DS)	mg/l	250	500	850
Suspended solids (SS)	mg/l	100	220	350
Volatile suspended solids (VSS)	mg/l	80	165	275
Fixed suspended solids (FSS)	mg/l	20	55	75
Settleable suspended solids (SSed)	ml/l	5	10	20
DBO_5	mg/l	110	220	400
COD	mg/l	250	500	1000
N-Total	mg/l	20	40	85
N-Org	mg/l	8	15	35
N-NH_3	mg/l	12	25	50
P-Total	mg/l	4	8	15
P-Org	mg/l	1	3	5
Cl_-	mg/l	30	50	100
SO_4^{-2}	mg/l	20	30	50
G&A	mg/l	50	100	150
Coli-Total	NMP/100 ml	10^6–10^7	10^7–10^8	10^7–10^9

Source: Orozco (2005), "*Bioingeniería de aguas residuales: teoría y diseño.*"

Table 3.3 Comparison of typical values of wastewater quality parameters proposed by different authors.

Parameter	Unit	Reference			
		a	b	c	d
Total solids	mg/L	720	403	600	
Dissolved solids	mg/L	500	243	308	
Volatile dissolved solids	mg/L	200	278		
Suspended solids	mg/L	220	160	291	416
Volatile suspended solids	mg/L	165	129		
Sedimentable solids	mg/L	10		154	
DBO	mg/L	220	190	257	416
COT	mg/L	160			
COD	mg/L	500	382	520	
Coliform	NMP/100 mL	10^7–10^8	8.23×10^6		1.6×10^8
Total Nitrogen	mg/L-N	40		27	
Organic Nitrogen	mg/L-N	15			
Ammoniacal Nitrogen	mg/L-N	25			
Nitrites	mg/L-N	0			
Nitrates	mg/L-N	0			
Total Phosphor	mg/L-P	8			
Organic Phosphor	mg/L-P	3			
Inorganic Phosphor	mg/L-P	5			
Chlorides	mg/L-Cl	50			
Alkalinity	mg/L-CaCO3	100			
Greases	mg/L	100			

Note: Table prepared on the basis of Sources Metcalf & Eddy (2003), Freni *et al.* (2003), Botiva (1997), Uribe (1981), RAS (200)

In the design procedure, we need to calculate the *Specific unit load* of each contaminant, for example q_{DBO_5} in kg DBO_5/Capita · d since the quality of the effluent of the treatment plant depends on the total flow, which in turn depends on the infiltration rate into the sewerage system, it is clear that the BOD_5 measured in the different sectors is not the same as the BOD_5 which arrives to the plant. On the basis of the specific unit load, the calculation of the concentration of a parameter (C) in the raw wastewater is as follows:

$$C = P \cdot q_C/Q_D[\text{Capita} \times (\text{kg/Capita} \cdot \text{d})/(\text{m}^3/\text{d})] \quad (3.8)$$

The result is the concentration in kg/m^3, which is the same in g/l.

This calculation is made for BOD_5, COD, Nitrogen, Phosphorus, and so on. To design a wastewater treatment plant it is necessary to have: (i) the flow rate of the raw wastewater, calculated as explained in Paragraph 3.1, (ii) the quality of the raw wastewater, which is calculated by Equation 3.8; and (iii) the total load of each parameter (e.g. L_{DBO_5}: total load of BOD_5, kg/d) estimated from the multiplication of the specific unit load by the population:

$$L_{DBO_5} = P \times q_{DBO_5} \quad (3.9)$$

The discussion presented above is best clarified by an example.

3.3 SAMPLE CALCULATION

Next, we will develop the calculation procedure for a town with a population of 20,000, which will be the basis for all the examples presented in the following chapters of this book.

3.3.1 Solution

Flow Calculation

The flow for the design of a wastewater treatment plant for a population of 20,000 is obtained by adding flows from municipal inputs and infiltration. In these calculations, the flows of infiltration include contributions by erroneous connections of drainage water. All the excess water contributions from storm water drainage need to be prevented from reaching the treatment plant through the use of overflow weirs. The average design flow is calculated as:

$$Q_D = Q_{ww} + Q_I$$

The average flow is used for the design of biological processes, and the maximum flows used for the hydraulic design of the plant. If we have a known industrial wastewater flow Q_{IWW}, it should be included in the total flow. Since this is not the case of a small typical population, we will omit the separate industrial flow and assume that in the example, the industrial and commercial wastewater flows are includes in the domestic wastewater flow.

To calculate the maximum design flow, the domestic and industrial contributions need to be multiplied by their respective peak factors. When possible, it is recommended to undertake a consumption study in residential sectors (not necessarily for the entire city) so that the per capita consumption, the peak factor and the net flow of infiltration can be calculated. For our example, we use the data shown in Table 3.4, which are date of a *real-life case*, obtained with direct measurements for a period of six months in the city of Tunja, Colombia (Orozco, 1999).

Table 3.4 Calculation of the per capita contribution (q) the infiltration flow (qi) and the peaking factor $k_1 \times k_2$.

Calculation of the per capita consumption								
Neighborhood	**Population (persons)**	**Area (ha)**	**Population density (persons/ha)**	**Medium flow (m^3/d)**	**Maximum flow (m^3/d)**	**Minimum flow (m^3/d)**	**Peak factor**	**Infiltration flow (l · s · ha)**
1	2699	6,89	391,79	243,50	393,49	144,23		
2	185	1,69	109,17	100,68	148,20	59,96		
3	734	3,40	215,74	260,01	483,06	110,28		
4	230	0,74	310,14	34,86	90,46	8,73	2.59	0.136
5	257	0,80	320,63	21,38	62,90	9,46		
6	1553	5,04	308,04	256,18	438,08	69,49	1.71	0.160
7	938	5,74	163,46	203,93	316,86	53,25	1.55	0.107
Total	6596	24,30	259,85	1120,55		**Average, Q_D**	1.95	0.13
						Standard deviation	0.56	0.03
Unit flow, l/capita · d				127,08		**Infiltration, Q_I**		3.27

It can be seen that, with the composite sampling in seven districts with a total of only 6,596 inhabitants, we can conclude that: (i) the wastewater per capita flow: $q_{WW} = c\ q_{dom} = 1000\ (Q_D\text{-}86 \cdot 4Q_I)/P = 127 \approx$ 130 l/Capita · d. This unit flow (per capita flow) excludes infiltration (which was subtracted), but it takes into account the return factor (c); (ii) the net infiltration flow is $q_I = 0.13$ l/s · ha; and (iii) the average peak factor for that population is $PF = k_1 k_2 = Q_{maxh}/Q_D = 1.95$, as an average over the districts 4, 6 and 7, which had no leaks or special losses.

If we have had no measurements, we would use typical values with safety factors: $q_{dom} = 150$–200 l/capita · d; $q_I = 0.2$ l/s · ha; and the PF in accordance with Figure 3.1. Now, considering P = 20,000 inhabitants, we receive from Equation 3.6:

$$PF = 1 + \frac{14}{14 + \sqrt{20}} = 2.65$$

A value that is close to the maximum standard deviation: 1.95 + 0.56 = 2.51. The value of *Harmon* is preferred so as to be on the side of safety. We adopted the parameters which are presented in Table 3.5.

Table 3.5 Parameters and design flows.

- Population: P = 20,000 inhabitants (Example definition. This must calculate in each case)
- Population density: d = 259.85 ≈ 260 hab/ha (measured)
- WW per capita flow: $q_{WW} = cq_{dom} = 130$ L/hab · d. (measured)
- Peak factor: PF = 2.65 (*Harmon* Equation)
- Infiltration flow: $q_I = 0.13$ l/s · ha (measured)
- Area of influence (of infiltration): $A_a = P/d = 20{,}000/260 = 77$ ha
- Process design flow: $Q_D = \frac{cq_{dom}P}{86400} + q_I A_a = 40$ l/s $= 3456$ m^3/d
- Hydraulic design flow: $Q_{DH} = 2{,}65\frac{cq_{dom}P}{86400} + q_I A_a = 90$ l/s $= 7776$ m^3/d

Wastewater quality and contaminants loads calculations

Data from the same real-life case are also be used for wastewater quality and contaminants loads calculations. These are typical data. It is emphasized that *nothing can replace direct measurement.*

The specific unit loads are now used to calculate the load of each contaminant. For example, the unit load of BOD (q_{DBO_5}, $kgDBO_5$/Capita · d) is used to calculate the total load of BOD. Table 3.6 presents the results of wastewater sampling and quality analyses in seven sectors. In the table, the measured concentration of each parameter is divided by the population and multiplied by the flow, to obtain the specific unit load in kg/person · d.

To calculate the quality of the wastewater at the treatment plant site, we calculate the concentration of each parameter using Equation 3.8, and the total load using Equation 3.9. A summary of the results for the example used is presented in Table 3.6. The flow data, population, wastewater quality and contaminants loads presented in Table 3.7 are used for the examples of design calculations of some selected appropriate technology treatment processes presented in the following chapters of the book.

Table 3.6 Quality parameters of the wastewater and calculation of the specific unit load for each quality parameter.

Calculation of organic load				Calculation of physico-chemical parameters											
	Population (hab)	Average Flow (m^3/d)	BOD_5 Concentration (kg/m^3)	BOD_5 Unit load (kg/capita · d)	COD Concentration (kg/m^3)	COD Unit load (kg/capita · d)	SS Concentration (kg/m^3)	SS Unit load (kg/capita · d)	VSS Concentration (kg/m^3)	VSS Unit load (kg/capita · d)	TKN Concentration (kg/m^3)	TKN Unit load (kg/capita · d)	PT Concentration (kg/m^3)	PT Unit load (kg/capita · d)
1	2699	243.50	0.284	0.026	0.49	0.045	0.25	0.022	0.21	0.018			0.007	0.001
2	185	100.68	0.075	0.041	0.19	0.104	0.04	0.019	0.03	0.015			0.002	0.001
3	734	260.01	0.190	0.067	0.54	0.190	0.27	0.096	0.15	0.051			0.004	0.001
4	230	34.86	0.217	0.033	0.51	0.078	0.11	0.017	0.36	0.054	0.05	0.007	0.003	0.000
5	257	21.38	0.293	0.024	0.53	0.044	0.15	0.012	0.11	0.009			0.006	0.001
6	1553	256.18	0.449	0.074	0.67	0.111	0.21	0.035	0.16	0.026	0.10	0.016	0.010	0.002
7	938	203.93	0.316	0.069	0.69	0.150	0.19	0.042	0.15	0.032	0.10	0.021	0.009	0.002
Average	6596.00	1120.55	0.26	0.048	0.518	0.103	0.174	0.035	0.165	0.030	0.080	0.015	0.006	0.001
Standard	deviation		0.12	0.022	0.163	0.054	0.082	0.029	0.101	0.018	0.029	0.007	0.003	0.001

Note: $kg/m^3 = 10^{-3}$ mg/l

Table 3.7 Calculation of wastewater total load and rounded concentration at the treatment plant.

Flow: $Q_D = 40$ lps		**Population: P = 20,000**	
Parameter	**Unit load kg/Capita · d**	**Concentration mg/l**	**Total Load kg/d**
BOD_5	0.048	277.78	860
COD	0.103	596.06	2060
SS	0.035	202.55	700
VSS	0.030	173.61	600
TKN	0.015	86.814	300
Total Phosphorus	0.001	5.79	20

Chapter 4

Rotating Micro Screens – RMS

4.1 PROCESS DESCRIPTION

4.1.1 Introduction

When financial resources are limited it is advisable to use simple treatment processes that are highly cost-effective in terms of kgBOD removed per US$ invested. In many cases, for a given investment, a higher environmental benefit is achieved at lower cost by removing organic matter by means of mechanical microscreening processes which have a proven track record and are reliable. For example, rotating micro screen (RMS) with a screen opening of 0.2 mm can reduced the COD of municipal wastewater by almost 35%, and with the addition of coagulants before the micro screens treatment, a total COD removal of 60% can be reached (Koehler & Rainer, 2003).

Fine screening is achieved in screen with openings in the range 0.2–2 mm while coarse screening is done in screens with openings in the range 2–10 mm and even larger. Coarse screens remove floating materials, paper and plastics. Rotating Micro Screens remove fine material, sand, and some suspended organic matter. A rotating micro screen with a screen opening of 0.2 and 1.5 mm can be compared with a primary treatment by sedimentation with BOD removal of around 15% (or 35% of COD) and SS removal of about 50%. The efficiency can be significantly improved by the use of coagulants prior to the screening operation (Orozco, 2005).

Treatment by a rotating micro screen is a mechanical process for removal of suspended matter from wastewater. It is used to also to remove some organic load. Microscreening operates similarly to the rough screening, but at a finer level. Screens can be static or rotational. The static screens are self-cleaning due to the slope of water flow, while the rotating screens are cleaned by a scraper in the inner side of the screen which peals of the solids retained on it, or by backwashing, which is the more modern method.

A self explanatory diagram of a typical rotating micro screen is shown in Figure 4.1: raw wastewater flows into the inner part of a drum which is covered by the screen and has to flow through the screen, coming out filtered downstream the channel. The system is backwashed by the jet stream located at the upper side of the drum. The removed material falls into a screw pump located inside the drum, which carried it up into a compactor which compresses the solid material and reduces its volume. The compacted material is then conveyed to a container located on top of the channel, from which it is removed for its final disposal, usually to a sanitary landfill. The screen opening varies with its application. If the rotating micro screen is the sole treatment unit prior to discharge of the effluent to a water body, for example to the sea, a fine screen is used so as to remove a significant portion of the suspended matter from the raw sewage. If the rotating micro screen is part of a preliminary treatment unit in an activated sludge plant, there is no need to remove all the fine material because downstream units can handle the fine suspended solids, so the screen does not need to be very fine.

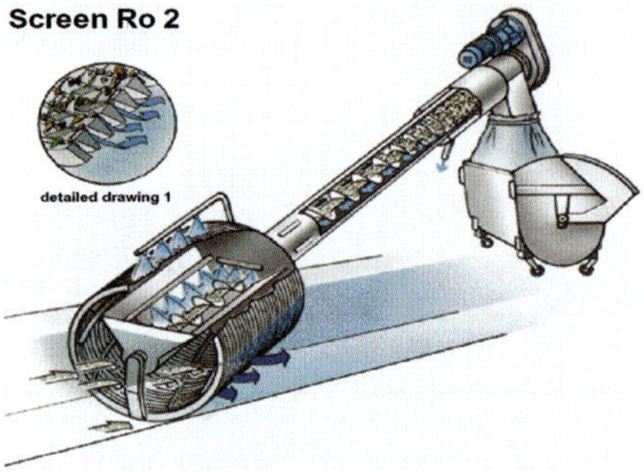

Figure 4.1 Schematic presentation of the operation principle of a typical rotating micro screen (Courtesy Huber Technology)

A rotating micros screen with fine screen opening (0.2 to 1.5mm) can be used as part of a preliminary wastewater treatment unit often in conjunction with a preceding rough screening unit (which can also be a rotating screen with a wider opening of 6 to 10 mm) and a preceding grit chamber (sometimes the grit chamber is located following the micro screen). A typical scheme of such a preliminary treatment unit is shown in Figure 4.2. The rough screening unit is not shown in this figure because it is located in the pumping station which conveys the wastewater to the plant. The unit that follows the rotating fine screens is, in the Cartagena treatment plant, a Vortex grit chamber.

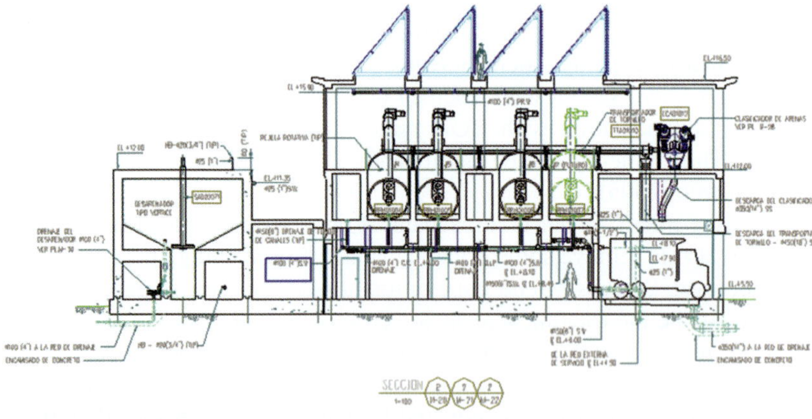

Figure 4.2 Typical rotating micro screen system, the wastewater treatment plant of Cartagena, Colombia (Courtesy Aguas de Cartagena)

In Latin America, rotating fine screens have been installed in the four treatment plants of the city of Santa Cruz, Bolivia, in the treatment plants of Cartagena, Colombia, La Plata and Bahia Blanca, Argentina and in several plants in Brazil, as well as in other cities. Information about the plants of Santa Cruz is given by Libhaber (2007), Annex 15 and by Blanco *et al.* (2007).

4.1.2 Process basics

The Coanda effect

The functioning of the screens is based on the so-called *Coanda effect*[1]: The best way to explain it is with an example: suppose we have a curved surface, such as a cylinder, as presented in Figure 4.3. If something solid is pour onto it (rice, for instance) it will bounce to the right. The cylinder, by the principle of action-reaction, will tend to go left. This can be seen in the first part of the illustration. If we repeat this experiment with a liquid, because of the liquid's viscosity, it will tend to adhere to the curved surface and the fluid will go in the opposite direction, to the left. In this case, the cylinder will be attracted to the fluid. If we imagined the liquid falling like water layers, the layers touching the cylinder will stick to it. The adjacent layers, by friction, will stick to it and will deviate a bit and so on (Wikipedia, 2007).

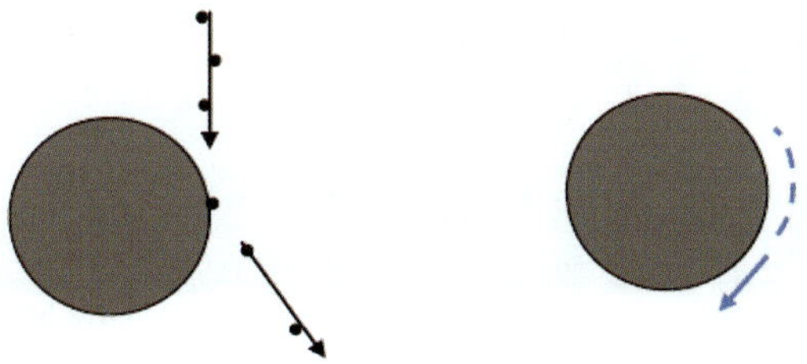

Figure 4.3 The *Coanda* effect scheme

This effect has been used in the manufacturing of screens that maximize the effect attracting the water flowing on the screen thereby facilitating the filtration process easier. The floating materials larger than the size of the screen opening will remain caught on the screen and can thus be separated from the water. This principle was first developed in the "self-cleaning" static screens, but eventually, it was found that the material that adhered to the screen required an external method of cleaning such as a scraper or, what is in most used today, a backwash liquid jet (see Figure 4.1). The hydraulic theory of the *Coanda* effect in self cleaning screens is presented in the studies Wahl (1995) and Wahl (2000). Figure 4.4 provides a schematic description of the operation of the screens showing the importance of the *warp* and *weft* (fibers that make up the fabric of the screen).

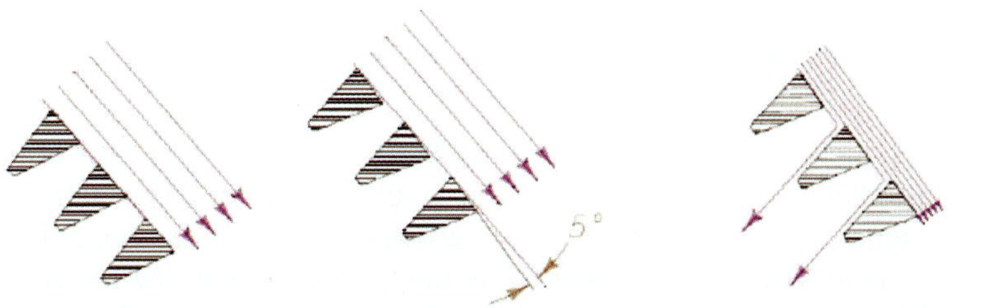

Figure 4.4 *Coanda* effect in static screens (*Source:* Wahl, 1995)

[1]Henri Coandă (1886–1972): Romanian inventor, aerodynamics pioneer and builder of the first jet aircraft

Hydraulic and mechanical considerations

Manufacturing of micro screens has been refined and today they can be supplied by various reliable manufacturers. Various types of screens are presented in Figure 4.5. The modern design of rotating micros screens has focused primarily on the operation of the hydraulic and mechanical systems, with emphasis on: (i) obtaining a low head loss; (ii) improving the screens cleaning capacity; and (iii) facilitating the evacuation of the removed solids. Within this quest numerous proposals have arisen, of which the most frequently used today are the rotating micro screens. In the past, the most common used were the tray screen, the disk screen and a certain type of rotating screen, all of them generating a significant hydraulic head loss (about 1.5 m), difficult to clean and easily clogged. These types of screens are presented in Figure 4.6 and are still in use today, although less frequently.

Figure 4.5 Various types of screen (*Source:* Huber Technology, 2006)

In the recent years the modern designs which resolved the operational problems of micro screens have emerged, providing installations with low head loss coupled with efficient self-cleaning and waste disposal. The basic design of a rotating Micros Screen consists of placing the rotating drum with the axis inclined in the direction of the flow in the channel, forming an angle with the horizontal. A typical case is shown in Figure 4.1. This provision has small head loss, efficient cleaning of the screen and an effective removal of the separated solid materials. Also, depending on the screen opening, the efficiencies of removal of suspended solids vary and can reach levels similar to those of primary treatment. The primary treatment level is sometimes sufficient for discharge to certain water bodies (such as discharges to large rivers or to the sea). In other cases it is convenient to implement the treatment in stages so as to optimize the return on investments in accordance with the availability of funding.

Figure 4.6 Other types of micro screens (*Source:* García Agudo & Rusell, 1992)

Design considerations

One of the main design parameters of a rotating micro screen is the hydraulic head loss which depends on three factors: (i) the velocity of the wastewater; (ii) the screen opening; and (iii) the mechanical design of the screen. The most commonly used equation for estimating the head loss (h_f) is the following (Orozco, 2005):

$$h_f = \frac{1}{C(2g)}\left(\frac{Q}{A}\right)^2 \tag{4.1}$$

where:

h_f = head loss, m
C = discharge coefficient of a clean screen. With large screen openings (>6 mm) the value of C is approximately 0.6. But C must be obtained for each screen opening. Also, the load loss that really matters is that which occurs during operation, with the filtered material attached to the inner side of the screen. Such information is obtained from experiments during operation.
A = effective area of the screen, m^2
Q = flow rate, m^3/s
g = gravity acceleration coefficient, 9.81 m/s^2.

In Figure 4.7, which is used for the design of rotating micro screens, the variation of the drum diameter with the wastewater flow for different screen openings can be observed.

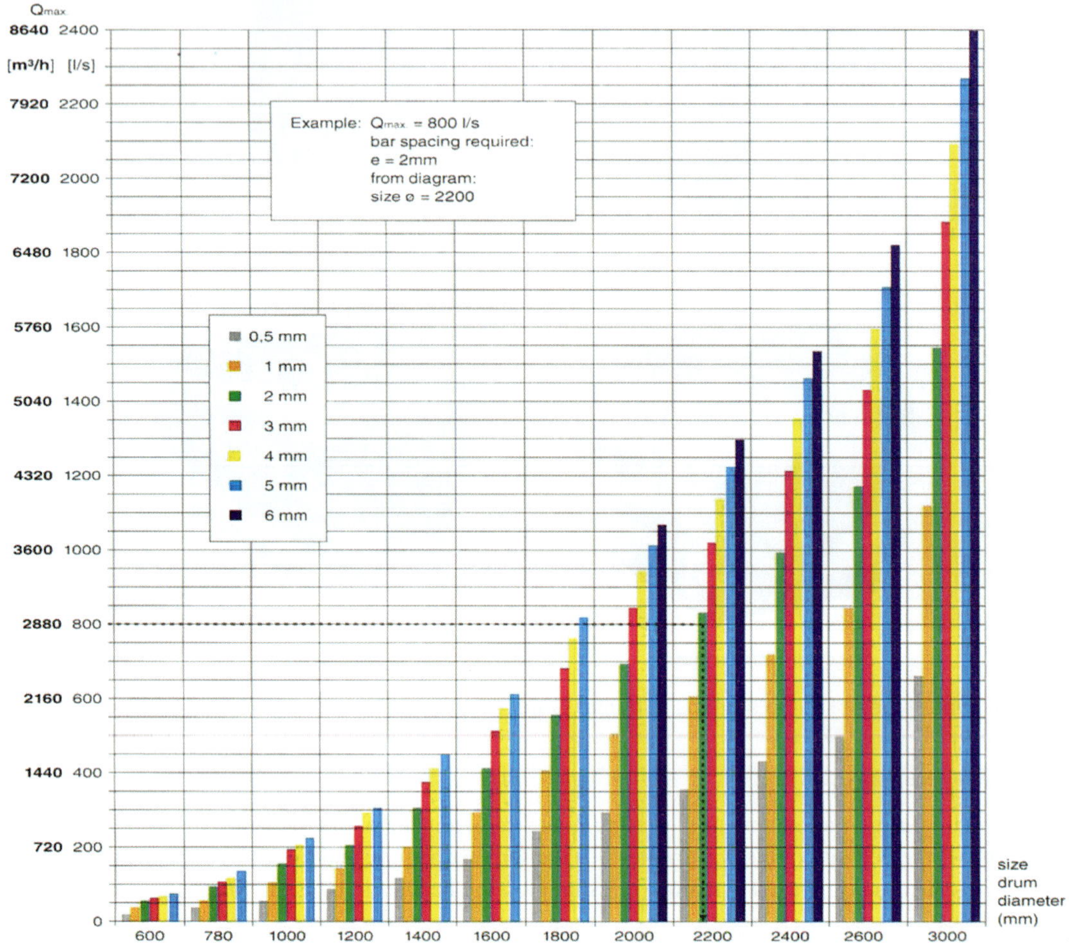

Figure 4.7 Dimensions of the screen drum diameter VS wastewater flow rate for different screen openings of a Rotamat type rotating micro screen (*Source:* Catalog of Rotamat Micro Screens, Courtesy Huber Technology)

The head loss of a clean screen is in fact very low so the true operating hydraulic head loss depends on the material which is collected in the screen, on the cleaning efficiency of the screen (which is usually achieved by backwash) and on the disposal of the material removed by the screen. For design purposes, the maximum operating flow and the minimum cleaning cycle should be considered.

The design must take into account: (i) the nature of the suspended matter; (ii) the corrosiveness of the wastewater; (iii) the effect of chlorination when it is required as a pre-treatment; (iv) the duplication of units for continuous operation during equipment maintenance periods; and (v) the backwash operation. In addition, the following provisions should be installed: (i) a durable corrosion resistant screen; (ii) a by-pass; (iii) hydraulic seals when using potable water for backwashing; and (iv) proper disposal of solids retained by the screen (Indian & Northern Affairs Canada, 2006). In addition, if increase of the process efficiency is required, application of coagulant agents should be considered to promote flocculation of the suspended solids before passing through the micro screens. This issue is further discussed in the chapter on chemically enhanced primary treatment (CEPT). Table 4.1 presents typical design parameters for rotating micro screens.

Table 4.1 Typical design parameters.

Item	Unit	Typical value
Screen opening	mm	0.2–2 mm
Hydraulic load	$m^3/m^2 \cdot min$	3–6
Head loss	mm	75–150
Submergence	% area	60–70
Drum diameter	m	0.6–5.0
Drum speed	m/min	4.5
Backwash	% of flow @ 350 and 500 kPa	2% and 5%
Waste removed	$l/1000\ m^3$	100–250

Source: Prepared from Metcalf & Eddy (2003) and Catalog of HUBER Rotamat (2007)

Management and disposal of the solid wastes generated in preliminary treatment

The appropriate technology treatment processes generate two types of solid wastes. Rotating screens with coarse openings larger than 0.7 mm and grit chambers produce residual solids which are mainly mineral, consisting of sand, grit, coarse particles and large materials like plastics, bottles, cans and other floatable material. This material is defined as solid wastes from preliminary treatment. Rotating fine screens with openings less than 0.7 mm and most other appropriate technology treatment processes (such as CEPT, UASB, Anaerobic Filter etc.) produce an additional stream of residual solids which consist of a combination of mineral and organic matter, defined as sludge. These two types of solids, solid wastes and sludge, are handled and disposed of differently. The management and disposal of the solid wastes generated in the preliminary treatment is discussed below. The management of the sludge is discussed in Chapter 9.

Disposal of screenings (solid wastes collected in preliminary treatment plants) may include: (1) removal by hauling to disposal areas (sanitary landfill); (2) disposal by burial on the plant site (small installations only); (3) incineration alone or in combination with sludge and grit (large installations only); (4) disposal with solid wastes; or (5) discharge to grinders or macerators where they are ground and returned to the wastewater.

Removal to a sanitary landfill is the most common disposal method. In a rotating fine screen unit the solids are compacted prior to disposal. A photograph of typical solid wastes of a preliminary treatment unit is presented in Figure 4.8. The solid wastes container fills automatically without manual intervention. When the container is full it is put on a removal truck by a crane built onto the truck. The same truck leaves an empty container at the plant for filling and then travels to the landfill site to dispose of the solid wastes.

Figure 4.8 Solid residues from a preliminary treatment unit

The quantities of solid wastes generated in preliminary treatment units are generally smaller than in primary and secondary treatment units. Estimated volumes of screened material under hot climate conditions, based on data from preliminary treatment plants in Florida, USA, are presented in Table 4.2.

Table 4.2 Volume of screened material for various screen openings.

Type of screen	Screen opening (mm)	Volume of screened material	
		(m^3 per m^3 wastewater)	(ft^3 per MG wastewater)
Mechanical bar	19	0.90×10^{-5}	1.2
Mechanical bar	13	1.4×10^{-5}	1.8
Mechanical bar	8	1.7×10^{-5}	2.3
Step	6	2.2×10^{-5}	3.0
Step	2	3.4×10^{-5}	4.5
Rotating micro	1.5	4.7×10^{-5}	6.3
Rotating micro	1.0	8.2×10^{-5}	11.0
Rotating micro	0.5	13.2×10^{-5}	17.7

MG: Million U.S. gallons

4.1.3 Performance

Efficiency

The performance efficiency of rotating micro screens with a screen opening in the range 1–1.5 mm is 5–10% removal of BOD. The efficiency of rotating micro screens with a screen opening of 0.2–0.3 mm is 20–30% removal of BOD_5 without chemical aid (i.e., without adding coagulants to the incoming wastewater) and up 50–60% removal with the application of coagulants (Koppl & Frommann, 2004). This makes microscreening competitive with primary treatment. In addition, the cost-effectiveness indicator (CEI, which defined as US Dollars invested for each kg of COD removed – $USD_i/kgCOD_r$) of microscreening is high in comparison to other treatment processes. When using coagulants and/or other type of chemicals, the screen opening must be 0.2 or 0.3 mm. In this case the quantities of retained solids are large; the solids are similar to primary sludge and need to be treated prior to their disposal.

For Total Suspended Solids (TSS) the removal efficiency without chemicals can reach a maximum of 60% depending on the screen opening size and the composition of the wastewater. According to Koppl and Frommann (2004) a screen with an opening of 0.2 mm removes, without chemicals, 30–60% of TSS (40% on average), and a screen with an opening of 1 mm removes 20–40% of TSS. Figure 4.9 shows the curve of TSS removal efficiency versus screen opening size calculated from data of Koppl and Frommann (2004) and Libhaber (2004). It also shows the exponential equation fitted to the curve.

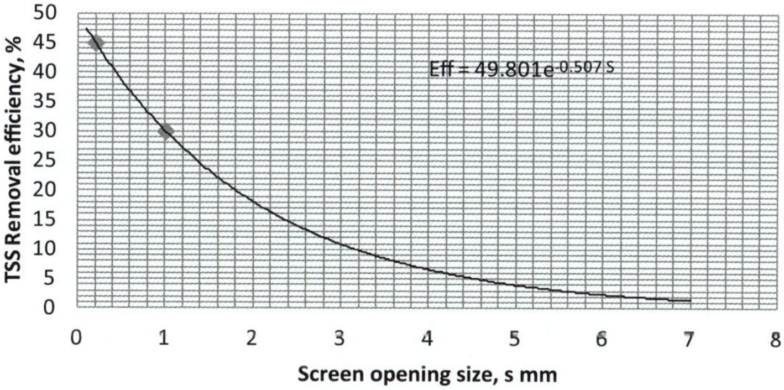

Figure 4.9 TSS removal efficiency as function of the micro screen opening size [*Source:* Graph built on the basis of data from Koppl & Frommann (2004); Libhaber (2004)]

Table 4.3 presents a comparison of the CEI (cost-effectiveness indicator) and efficiencies of COD removal for different arrangements of rotating micro screens and for a conventional biological wastewater treatment plant. As can be seen, the capacity of rotating micro screens to remove COD at a low cost is high, or in other words, its cost-effectiveness level is high and its cost-effectiveness indicator is low. It is within this context that the applicability of the screening process has to be evaluated.

Preliminary treatment will often suffice as the only treatment required for discharges of effluent to the sea or to large rivers. Even for small rivers, if the budget is limited it is better to provide preliminary treatment to several communities rather than to provide secondary treatment to one community. The rationale for this is explained in Table 4.4.

A seen in Table 4.4, for a fixed budget of €1 million (about US$1.5 million), preliminary treatment could serve a population of 140,000, eliminating about 3,400 kg/day of COD. But if secondary treatment were mandated, the same budget could serve only a population of about 8,000, eliminating about 1,270 kg/day of COD. Therefore, to protect the river and the environment with a limited budget (which is usually the situation in developing countries), it is better to start with a lower treatment level serving a wider population, rather than advanced treatment serving a smaller population.

Table 4.3 Comparison of COD removal efficiencies and CEI for rotating micro screens and conventional wastewater treatment.

Parameter	Unit	RMS	RM + Flocculation	RM + Settling	WWTP
CEI	USD_i/kgCOD	270	480	545	2420
Applied load	kgCOD/d	25680	5760*	5280	600
Efficiency	%	20	50	65	95
COD removed	kg/d	5136	2880	3432	570

*Does not include costs of sludge disposal generated by flocculation

Source: Prepared from "COD/BOD_5 Reduction with ROTAMAT® Fine and Micro Screens": Catalog Huber Technology, 2005

Table 4.4 Comparisons of the performance of rotating fine screens and secondary biological treatment for a fixed budget of €1 million.

Parameter	Unit	Rotamat high removal screen	Secondary biological treatment
Per capita investment	€/Capita	7	125
Connected population	Population	142,000	8,000
Influent COD load	kg O_2/d	17,000	960
Influent N-NH_4 load	kg O_2/d	6,700	380
Removal efficiency of COD	%	20	95
Removal efficiency of N	%	0	95
Eliminated COD load	kg O_2/d	3,400	910
Eliminated N load	kg O_2/d	0	360
Eliminated oxygen consuming substances load	kg O_2/d	3,400	1,270
Investment per daily unit of oxygen removed	€/(kg O_2/d)	294	787

(Data based on information from Huber Technology)

Source: Roberts *et al.* 2010

Advantages and disadvantages

The main advantages of the rotating micro screens are: (i) a relatively low head loss (between 30 and 50 cm); (ii) a high level of cost-effectiveness; (iii) low power consumption; and (iv) the system functions automatically and is very easy to operate and maintain. The disadvantages are: (i) usually low efficiency in removal of TSS (the efficiency of a rotating Micros Screen is determined by the hydraulic and solids loads as well as by the characteristics of the influent stream); (ii) micro screens do not remove colloidal material (or small size algae); (iii) micro screens require an efficient self-cleaning system; and (iv) micro screens are affected by fluctuations in the composition and quality of the influent.

The application of coagulants improves the operating efficiency of micro screens but also increases the production of solid residues (in this case sludge) which require disposal. The addition of chemicals modifies the type of solids retained on the screen, which in this case is practically sludge that is a material more difficult to handle. The investment costs of rotating micro screens are in the range of and 3–10 US$ per Capita served and the operating costs are in the range 0.10 and 0.15 US$ per Capita per year (Libhaber, 2007).

4.2 BASIC DESIGN PROCEDURE

4.2.1 General design considerations

Pretreatment is the preparatory process necessary prior to subjecting municipal wastewater to secondary treatment. Coarse screening is the first of the pretreatment units. Screens are classified as coarse when the screen opening is in the range of 2.0 to 6.0 mm (while in fine screens the opening is between 0.2 and 2 mm). Screens are used to remove materials present in the wastewater, such as paper, plastics and other floating materials. The objective of screens is to protect the operation of valves, pumps, aerators and other equipment included in a wastewater treatment plant. The fine screens are used to remove fine suspended material, among which is the particulate organic matter. One type of very fine screen used is the rotating micro screen (RMS) that can remove, under certain circumstances, up to 90% of TSS and about 35% of the COD. The RMS are considered a sustainable Appropriate Technology process because of its ease of operation and cost-effectiveness (Koehler, 2003; Orozo, 2005).

To specify a RMS equipment it is most important to select the screen opening size. This is a *designer variable* (variable to be proposed by the designer, based on his experience and knowledge of the process). Some guidelines for making this selection are: (i) if the RMS is a pretreatment preceding biological treatment, aerobic or anaerobic, the screen opening does not have to be very small and can be in the range of 2–10 mm since the organic matter will be removed in the biological process; (ii) if the RMS is used as the main treatment unit prior to direct discharge of the effluent to a water body through a subaquatic outfall, then, since it is the sole treatment unit, a smaller screen opening in the range 0.2–2.0 mm should be used, however, the fine RMS should be protected by a preceding coarse screening, which can be a coarse rotating screen unit; (iii) when making the selection it is necessary to analyze the cost, taking into account that when the *screen opening is smaller the cost is higher*, so it is necessary to assess the cost-effectiveness of screening unit. In practice, there is no procedure to determine the type of RMS to be used, except the designer's experience (Indian & Northern Affairs Canada, 2006).

4.2.2 Orderly design method (ODM)

In general, the design methodology of a RMS is the following:

(1) *Determine the external variables* (or independent variables), which, among others, are the following:

- *Flow rate*: since this process is hydraulic, we must use as design flow the peak flow or the hydraulic design flow (Q_{HD}) calculated by Equation 4.4 (below) using the appropriate units (lps, m^3/h, etc):

$$Q_{HD} = k_1 k_2 \frac{c q_{dom} P}{86400} + q_I A_a \tag{4.4}$$

- Geographical and altitude conditions: such as maximum and minimum temperatures of air and water, solar radiation and cloudiness, wind rose, altitude above sea level, proximity to population centers, and so on.
- Concentration of Contaminants in the wastewater: at the influent to the treatment plant, taking into account the infiltration water according to the methodology presented in Chapter 3. This information is required for determining the loads (kg/d) and the specific unit loads (kg/Capita · d) of the contaminants.
- Contaminants Loads: of the important constituents as BOD_5, COD, TSS, pH, N, P, and so on. The contaminants loads can be calculated from direct measurements or estimated according to the methods described in Chapter 3.

(2) *Determine the screen opening* of RMS using Figure 4.9. Determine the required TSS removal percentage and from the curve obtain the screen opening size. In performing this task, it is necessary to consider the treatment processes following the pretreatment, in accordance with the criteria presented in Paragraph 4.2.1. If a fine screen is selected, a preliminary coarse screen and probably a grit chamber must be placed a before the fine screen (Hans Huber, 2007). The equation to relate TSS removal efficiency (Eff) to screen opening size (s) according to Figure 4.8 is:

$$\text{Eff} = 49.8\,e^{-0.51s} \tag{4.1b}$$

(3) *Determine the type of RMS to be used*: Each manufacturer provided a design procedure specific to his equipment. For instance, if the selected equipment is a Rotamat type RMS manufactured by Huber Technology, then with Figure 4.7, the diameter of the drum (D_T) is determined for a given screen opening size and a given hydraulic design flow rate.

(4) *Determine the approximate width of the inflow channel* (W) based on the drum diameter (D_T) and the expression: $W = 1.1\ D_T$. If the flow is large, several drums in parallel might be needed, and then the width of the channel is based on the sum of the diameters of the individual drums.

(5) *Determine the channel depth* (h) is calculated using the equation of continuity, based on a stream velocity of between 0.6 and 1.0 m/s (larger than the particles drag velocity): $Q_{HD} = v \cdot (W \cdot h)$, from which we get:

$$h = Q_{HD}/W \cdot v \tag{4.2}$$

(6) *Determine the channel slope* using the Manning's Equation in SI (International System) units:

$$s = (v \cdot n)^2/R^{4/3} \tag{4.3}$$

where:

S = slope, in decimal
V = cross-sectional velocity
R = hydraulic radius, equal to wet area/wet perimeter. For a rectangular channel
R = Wh/(W + 2 h).
N = roughness coefficient: 0.012 (finished concrete), 0.013 (rough concrete), 0.014 (unfinished concrete), 0.015 (iron sheets).

(7) *Determine the head loss* with Equation 4.1:

$$h_f = \frac{1}{C(2g)}\left(\frac{Q_{DH}}{A}\right)^2$$

C ≈ 0.6. The manufacturer curve can be used if C is not known

(8) *Determine the backwash volume and pressure* from Table 4.1.

(9) *Determine the amount of solids to remove.*

- The amount of excess solids from RMS is estimated in Table 4.1 (between 30 and 60 l/1000 m^3).
- The study of Aguas de Cartagena (2007) proposes an excess solid generation rate of 4.7 m^3 per 100,000 m^3 of treated wastewater (equivalent to 47.5 l/1000 m^3).
- If the screen opening is 0.7 mm and higher, the solids retained on it are mostly made of inert mineral matter and can be directly transported to disposal in a sanitary landfill. If the screen opening is smaller than 0.7 mm, the retained solids contain a higher portion of organic matter. In this case it is preferable to provide a stabilizing treatment prior to their final disposal. The stabilization is carried out in the containers which temporarily store the material, applying in the container a layer of saturated lime every time a layer 20 cm thick of retained solids is deposited in the container. The high level of pH of the caused by the lime ensures the destruction of bacteria and pathogenic microorganisms. The generation of hydrogen sulphide (H_2S), which causes odour problems, is also prevented this way, favouring the drying process of the solids and reducing the possibility of attracting vectors.
- Should this measure be insufficient and offensive odours persist, the solids containers needs to be stored in s closed structure equipped with odour control installations (such as air aspirators and treatment of the air for odour removal before its discharge to the atmosphere. The treatment can be based on passing the air through a biological filter or through an absorption tower, using a chemical solution as an odour absorbent).

(10) *Design a by-pass* to divert the influent to the discharge of the RMS unit. The by-pass is used during maintenance operation and in cases of emergency. This can be avoided by using two parallel RMS units.

4.3 BASIC DESIGN EXAMPLE

(The model program for this Example is available online at http://www.iwawaterwiki.org/xwiki/bin/view/Articles/Software+Developed+for+Sustainable+Treatment+and+Reuse+of+Municipal+Wastewater)

In continuation we present an example that follows the Orderly Design Method (ODM) of Paragraph 4.2.2. The example is developed for a population of 20,000, as presented in Chapter 3. The ODM is developed step by step and is calculated by a program developed in Excel, available online as explained elsewhere. One way to solve the design example is presented by the ODM, but the reader should be aware that there might be several different methods to achieve a good design. If properly understood, the model could be changed when the *a priori* (external) conditions demand a different approach, leaving a proper back up of the original program, called CHAP 4-RM.xls. Please note that the following Tables identified by alphabetic order contain computer calculated results.

General instructions for running de Excel sheet program with the ODM

- First, make sure that you thoroughly understand the Orderly Design Method.
- Second, get acquainted with the colour Nomenclature as as shown in Table 4A.

Table 4A

Nomenclature	
Dark grey color	Titles, parameters, units
Light grey color	Input
Grey color	Output
Medium grey color	Calculated variables

- Whatever is in dark grey colour refers to titles, parameters or units, and is alphabetic. The other colours refer to numeric cells.
- The cells in light grey colour are input provided by the designer, be it (i) external parameters or (ii) primary variables given by the designer. These cells should be filled manually by the designer.
- The coloured medium grey cells are secondary variables calculated from the primary variables, which belong to the input necessary for the design. Note that the designer may choose a secondary variable as primary, and then the primary variable becomes secondary. For instance, to calculate the design flow you may select q, P and c as primary variables and from them the program calculates the process design flow (Q_{DWW}) as secondary variable; but you may know the design flow, then q becomes a secondary variable. These variations in primary variables would require changes in the program and should be avoided unless completely necessary.
- With the input defined, the program calculates the design variables in the output as determined by the ODM. Also, in this step there are variables that can be selected as input and the corresponding ones in the input become outputs. For example, in Figure 4.8, you can choose either Efficiency or screen opening (mesh size) as input. If you choose one variable as input then the other variable becomes output.
- Remember, *to change primary variables to secondary variables (or input to output) in the program requires a thoroughly comprehension of the ODM.*
- Check that the output variables, be they dimensions or specifications, are practical and are in the empirical range of design given in the Process Description section of the chapter.
- Note that the necessary steps to calculate CONCENTRATION, ORGANIC LOADS AND SPECIFC UNIT LOADS, require that the wastewater quality is given as an input.
- These general instructions apply to all programs given in the attached CD.

(1) *Calculation of external variables*

1.1 *Design flows*: The calculation related to the design flows are presented in Table 4B.

The final design flows are obtained from Equation 3.4, and are presented in different units as an *output*. The average flow Q_D is used for the design of the biological process and the maximum flow Q_{HD} for the hydraulic designs. The resulting design flows are presented in Table 4C.

In the case of RMs the design flow is Q_{HD} because the process is basically a physical process depending on the hydraulics.

1.2 *Determination of the external geographical and environmental variables*: These variables depend solely on the geographic location. They values used in the example are presented in Table 4D.

Table 4B

Design flow				
Variable	**Value**	**Unit**	**Value**	**Unit**
Per capita consumption, q	162.5	L/capita · d		
$PF = 1 + 14/(4 + \sqrt{P})$	2.65			Hamon equation
Population, P	20000	inhab		
Population density, d	260	inhab/ha		
Return coefficient	0.8			
DWW mean flow, Q_{DWW}	30.09	L/s	108.33	m^3/h
DWW maximum flow, Q_{maxDWW}	79.82	L/s	287.35	m^3/h
Infiltration unit flow, q_I	0.13	L/s · ha (o km)		
Afferent area, A_a	76.92	ha		
Infiltration flow, Q_I	10.00	L/s		

Notes: (i) Per capita consumption is a measured value, (ii) Peak Factor is calculated by Equation 3.6 (*Harmon* Equation), (iii) the population density should be the maximum acceptable density according to city planning standards, in order to be able to serve the people at saturation, (iv) flows are calculated based on the population and the return coefficient, c, (v) infiltration must be measured or alternatively a value chosen between 0.10 and 0.15 l/s · ha or l/s · km.

Table 4C

Design flows						
Variable	**Value**	**Unit**	**Value**	**Unit**	**Value**	**Unit**
Process design flow, $Q_D = Q_{DWW} + Q_I$	40.09	L/s	144.3333333	m^3/h	3464.00	m^3/d
Hydraulics design flow, $Q_{HD} = Q_{DWWmax} + Q_I$	89.82	L/s	323.3515496	m^3/h	7760.44	m^3/d

Table 4D

Variable	Value	Unit	Value	Unit
Minimum temperature of water, T_w	25	°C	77.0	°F
Minimum temperature of air, T_a	18	°C	64.4	°F
Altitude above sea level, hasl	350	masl		
Mean wind velocity	6	kph		
Predominant wind direction	NW			
Minimum solar radiation, **S**	230	$Cal/cm^2 \cdot d$		
Daily fraction of sunlight hours	0.8	decimal		

Notes: (i) For the design of RMS these variables do not have a significant importance, (ii) The minimum water temperature, as well as the minimum air temperature are the averages of the coldest month, (iii) the speed and direction of the prevailing or predominant wind is important for locating the wastewater treatment plant in relation to the location of the town or the urban center, (iv) the fraction of hours of sunlight is the ratio of actual hours of sunlight to the possible hours of sunlight.

1.3 *Concentration and pollutant loads*: The quality of the wastewater may be different in the urban areas and at the entrance to the treatment plant if the conveyance system is long and receives an important infiltration inflow. In the example, the wastewater treatment plant is located close to the town and there is no significant variation of the wastewater composition during its conveyance to the plant.

The following table presents as outputs the calculated *loads* (in kg/d) and the *Specific unit loads* (in kg/Capita · d) of each contaminant. These outputs are calculated from the *concentrations* given as *inputs*. This is calculated for BOD_5, COD, TSS, and so on. In cases of significant infiltration due to long conveyance to the wastewater treatment plant, it is necessary to calculate the additional infiltration flow from the conveyance influence area (A_a) as follows: $Q_{Iemi} = q_I \cdot L$, where L is the length of the conveyance pipeline in km. The concentration of a contaminant at the influent to the wastewater treatment plant is calculated as its specific unit load (q_C) multiplied by the population (P) and divided by the total flow: $Q_T = (Q_D + Q_{Iemi})$. For example: the concentration of COD is $COD = q_{COD} \cdot P/Q_T$. The contaminants concentrations in the raw wastewater, their loads and the per capita specific unit loads are presented in Table 4E. Note that in Table 4E the load and the per capita load (specific unit load) of each contaminant are *outputs* while only the concentrations are an *inputs*, presented in the separately Excel sheet.

(2) *Variables selection by the designer*: The main variable which the designer needs to provide in this process is the screen opening size. For this, directions given in Section 4.2.1 should be followed, but they should not replace the experience of the designer. Figure 4.9 gives the relation between the percentage of TSS removal and the screen opening size. Take into account that 30% removal of TSS provides a 10% removal of BOD_5.

If we propose a 30% removal of TSS then, from Figure 4.9, we find that the screen must have an opening of 1.0 mm. The results that the program calculates automatically are presented in the Table 4F. The results include also the calculation of other variables that are required later for the design of other treatment processes.

(3) *Specification of the rotating micro screens*: At this point we proceed to carry out Steps 3 to 8 of ODM which is summarized in the Table 4G. Note that the table gives the equations (Parameter column) and the number of equation or table in the chapter (Observation column). Following, the Orderly Design Method is presented step by step.

THE ORDERLY DESIGN METHOD APPLIED TO ROTATING MICROSCREEN

Step 1: Determine External Variables as explained in Paragraph 1 of Section 4.3, above.

Step 2: Propose Designer Variables as explained in Paragraph 2 of Section 4.3, above.

Step 3: Select a brand of Rotating Micro Screens and use design instructions of the manufacturer. For example, for a *Huber Rotamat* use Figure 4.7 to determine the required diameter of the drum based on the screen opening size and the hydraulic design flow. In our example $Q_{HD} = 90$ lps and the screen opening size is 1 mm, from which we conclude, using the figure, that the drum diameter should be 600 mm, that is, $D_T = 0.6$ m.

Step 4: Determine the width of the approach channel, W, using the diameter of the drum, D_T, (step 3) by the equation: $W = 1.1, DT = 1.1 \times 0.6 = 0.66$ m

Table 4E

	Concentration, load and per capita consumption					
Variable	**Concentration**	**Unit**	**Load**	**Unit**	**Per capita**	**Unit**
BOD_5	277.800	mg/L	962.3	kg/d	0.048	kg/capita · d
COD	596.100	mg/L	2064.9	kg/d	0.103	kg/capita · d
COD/BOD_5	2.146					
TKN	40.000	mg/L	138.6	kg/d	0.007	kg/capita · d
N-Nitrate	2.000	mg/L	6.9	kg/d	0.000	kg/capita · d
Total Phosphorus	5.800	mg/L	20.1	kg/d	0.001	kg/capita · d
pH	7.100	UN				
Alkalinity	100.000	mg/L	346.4	kg/d	0.017	kg/capita · d
TSS	202.600	mg/L	701.8	kg/d	0.035	kg/capita · d
VSS	173.600	mg/L	601.4	kg/d	0.030	kg/capita · d
O&G	100.000	mg/L	346.4	kg/d	0.017	kg/capita · d
Fecal Coli	10000000.000	NMP/100 mL	346400000000000.0	NMP/d	17320000000.000	NMP/capita · d

Table 4F

Designer variables				
Variable	**Value**	**Unit**	**Value**	**Unit**
Eficiency TSS	30	%		Figure 4.8
Screen opening	1.01	mm		Equation 4.1b
Screen opening (rounded)	1.00	mm	0.039	inch
Flow velocity in the channel	0.6	m/s		0.6 to 1.0 m/s
Backwashing pressure	350	kPa		Table 4.1
Solids removed	60	$L/1000\ m^3$		Table 4.1

Table 4G

Design			
Parameter	**Value**	**Unit**	**Observation**
Flow, Q_{HD}	90.000	L/s	
Screen opening, a	1.000	mm	
***Drum diamater,* D_T**	0.600	m	Figure 4.7
Channel width, $W = 1{,}1\ D_T$	0.660	m	
Channel depth, $h = Q_{DH}/W \cdot v$	0.230	m	Equation 4.2
Hydralics radius, $R = Wh/(W+2h)$	0.136	m	Equation 4.3
Manning's n	0.012		Equation 4.3
Hydrulics slope, $s = (v \cdot n)^2/R^{4/3}$	0.074	%	Equation 4.3
Head loss, $h_f = (1/C \cdot 2g)(Q_{DH}/A)^2$	0.030	m	Equation 4.1
Total head loss, $h_f + 0.05$	0.080		Plus 5 cm
Backwashing flow	1.800	L/s	Table 4.1
Solids removed	466.560	L/d	Table 4.1

Step 5: Determine the depth of the channel (h) with the continuity equation for a cross-sectional velocity of between 0.6 and 1.0 m/s (greater than the drag velocity of the particles): $Q_{DH} = v\ .\ (W \cdot h)$. Choosing $v = 0.6$ m/s we obtain:

$$h = Q_{DH}/W \cdot v = 0.23$$

Step 6: Determine the hydraulic slope with the Manning's Equation.

First calculate R, $R = Wh/(W + 2\ h) = 0.136$. Then select Manning's coefficient n. We propose Manning's $n = 0.012$. Inserting in the Manning equation we obtain:

$$s = (v \cdot n)^2/R^{4/3} = 0.00074 = 0.074\%$$

Step 7: Determine the hydraulic head loss with Equation 4.1:

$$h_f = \frac{1}{C(2g)}\left(\frac{Q_{DH}}{A}\right)^2$$

Using $C = 0.6$ we get $h_f = 0.03$ m. We add 5 cm for safety, then $h_T = 0.08$ m.

Step 8: Determine the backwashing flow and pressure and the amount of solids collected using Table 4.1. We chose a pressure of 350 kPa which requires for backwashing 2% of Q_{DH} ($Q_{DH} = 90$ lps). The volume of solids removed is 60 L/1000 m^3, which for a flow of 90 lps results in 467 liter per day of retained solids.

(4) *Comments to the design*: The presented ODM can be done in its entirety using the Excel program CHAP 4-RM, with only two designer decisions to make: (i) the required removal efficiency of TSS; and (ii) the wastewater flow velocity in the approach channel. It is also necessary to make a small calculation with Figure 4.7 to determine the drum diameter. The rest of the calculations are made by the program.

In the design example, since the rotating micro screen opening is 1 mm it performs as a fine screen. The dimensions of the selected rotating micro screen, Rotamat 600 mm, manufactured by Huber Technology can be found in the manufacturer's catalogue, which provides the dimensions of the unit and the manufacturer's instructions for its installation.

Figure 4.10 provides the manufacturer's method for estimating the head loss in the screening unit. This is another way to assess the head loss. The curves in the figure provide the depths of the water in the channel upstream (h_u) and downstream (h_v) with the Rotamat 600 mm micro screen unit for various values of wastewater flow. For example, in Figure 4.10 for a Ro2 a drum diameter of 600 mm and an aperture size of 2 mm, in the curve of 50 Lps: if the control point downstream gives an $h_v = 100$ mm, the h_u will be 130 mm for a head loss of 30 mm. In some cases, to obtain the h_v it may be necessary to have a checkpoint or weir downstream to ensure the h_v, what ensures depth, but should not alter the calculated head loss.

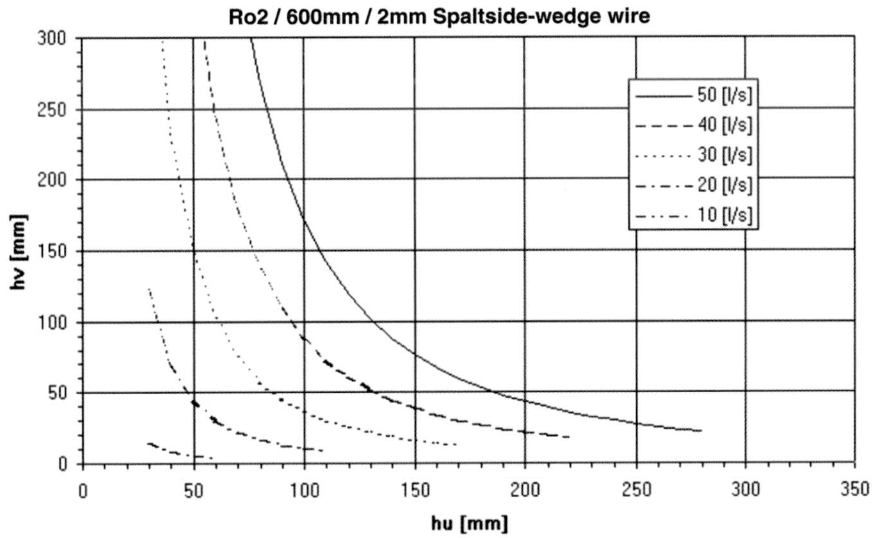

Figure 4.10 Difference of heads for different flow rates in a screen of 600 mm

This type of information should be requested from the manufacturer as a method to check the head loss calculations and to ensure the proper installation of the equipment.

For large flow treatment plant, a series of RMS drums in parallel is usually required. Figure 4.11 presents a drawing of a rotating micro screens installation consisting of four drums in parallel.

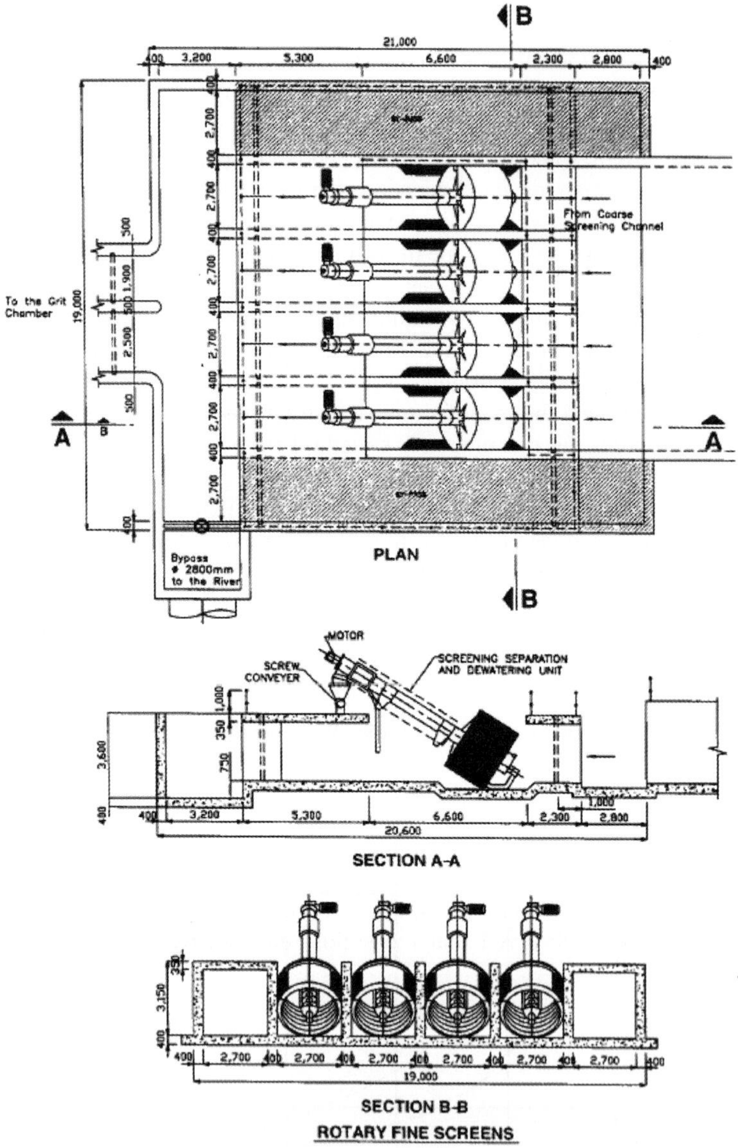

Figure 4.11 Plan and sections of a typical rotating micro screen unit consisting of four drums in parallel

Figure 4.12 shows a plan and a section of the preliminary treatment unit as that proposed in the design example. It consists of first coarse rotating screen with an opening of 6 mm, followed by a unit of conventional grit chambers to remove heavy particles such as sand, which in turn is followed by the RMs unit with a screen opening of 1 mm to remove organic particles (VSS and some BOD_5).

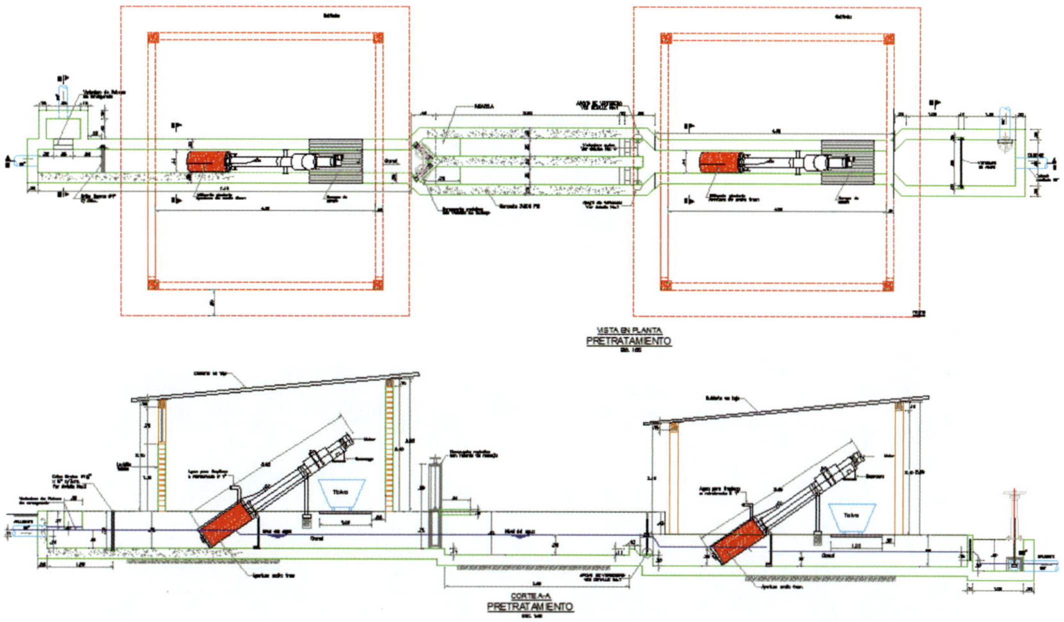

Figure 4.12 A typical preliminary treatment unit consisting of a coarse and a fine rotating screens (*Source:* MAVDT, 2008)

Figure 4.13 shows a photo of a series of rotating Micros Screen drums ready for installation in the treatment plants of Santa Cruz, Bolivia. Figure 4.14 shows a photo of a small diameter drum (1 meter diameter) of a rotating micro screen which functions as a pretreatment unit in the industrial park wastewater treatment plant in Santa Cruz. The wastewater flow of this plant is small but the BOD load is large, since industrial wastes contain highly concentration of organic matter. The unit is installed in the canal which conveys the wastewater to the plant. It can be observed in this photo that the RMS unit is simple to operate. It can also be observed that the solids removed by the RMS are preliminary solid residues which are different than sludge and can be disposed in a landfill without further treatment.

Figure 4.15 shows photos of the RMS treatment unit in La Plata, Argentina, which treats large flow and consists of three parallel large diameter drums of RMS.

Most of the rotating micro screen units are installed in the open (not inside a building), as is the case in the plants of Santa Cruz and La Plata, and do not cause environmental problems. However, when the treatment installations are located within a city, it is sometimes necessary to install the RMS units inside a building. Such is the case of the treatment plant of the city of Cartagena, Colombia. Figure 4.16 shows the preliminary treatment plant of this city, which has a fixed population of over 1 million and an additional floating population of tourists, since it is a vacation resort city. The peak wastewater flow in Cartagena is close to 4 m^3/s. The treatment plant includes 8 rotating micros screens each consisting of a 2.6 meters diameter drum, all of them located inside the modern architecture building shown in Figure 4.16. Inside the building the drums are located in two rows, each containing four drums. A photo of the drums inside the building is shown in Figure 4.17.

Figure 4.13 Rotating micro screen drums ready for installation

Figure 4.14 A small diameter drum of a rotating micro screen unit functioning as a pretreatment unit in the industrial park wastewater treatment plant in Santa Cruz, Bolivia

Figure 4.15 A preliminary treatment plant in La Plata, Argentina consisting of 3 parallel large diameter drums of rotation micro screens

Figure 4.16 The wastewater treatment plant of the city of Cartagena, Colombia, consisting of rotating micro screens and vortex grit chamber, located inside a modern architecture building

Figure 4.17 The rotating micro screen drums located inside the building of the treatment plant of Cartagena

The largest microscreening plant is now (the year 2011) being constructed in the Taboada wastewater treatment plant in Lima, Peru. This plant will treat an average flow of 14 m^3/s and consists of 24 large diameter drums of rotating micro screens.

Chapter 5

Treatment in stabilization lagoons

5.1 PROCESS DESCRIPTION

5.1.1 Introduction

Stabilization Lagoons (SL) consist of artificially constructed ponds, in which a microbial population develops naturally, mainly composed of bacteria and algae that live in symbiotic union under which the algae consume the CO_2 generated by the bacteria and the bacteria consume the oxygen generated by the algae. In addition, also by protozoa and fungi develop in the lagoons. Figure 5.1 presents the main relationships between algae and bacteria.

The microbial population develops gradually until it reaches the balance point of the various species according to the organic loads applied to the lagoon. The role of the algae in active photosynthetic lagoons is to produce the oxygen required by bacteria for respiration. The bacteria in turn are responsible for removing the organic substrate contained in the wastewater. Oxygen production is variable during the day, depending on solar illumination, and totally ceases in the absence of light. Aeration caused by wind also contributes somewhat to the oxygen supply to the lagoon, and occasionally additional oxygen is provide by mechanical aeration equipment when photosynthesis is greatly reduced due to bad weather (Orozco, 2005). Stabilization lagoons have become very attractive with the appearance of low-cost mixer systems in the market (Gurney type and similar), and the application of ultrasound systems for the removal of algae from the effluent, which is the main problem presented by the stabilization lagoons (LG Sonic, 2003). In addition, CO_2 production by bacteria and even by algal respiration causes variation in the acidity of the lagoon when it is not consumed by algae in the photosynthesis. Thus, the pH tends to be low during the afternoon and evening, giving the lagoon a green appearance. A declining pH is recognized by the yellow colour of the lagoon.

Stabilization lagoons remove from the wastewater organic matter (BOD), Fecal Coliforms (FC) and other organisms such as helminth eggs (worms), protozoa of the strain *Leishmania*, and so on. Several environmental factors are involved in the performance of a lagoon, among them dissolved oxygen (DO), pH, temperature, wind velocity, solar luminosity, and so on. Since the performance of the lagoon is defined by the environmental conditions, these factors must be taken into account during stage of design of the lagoon. During the operation stage only the flow of wastewater from one lagoon to the next can be controlled by the operator. In other words, once a lagoons is constructed, there are almost no means to control its performance, so it needs to be designed in such a manner that it would perform well under all the expected environmental conditions. One operating tool which helps to control a lagoons system is recirculation of the effluent, through which it is possible to decrease or increase the detention time in the lagoons and to distribute organic shocks loads.

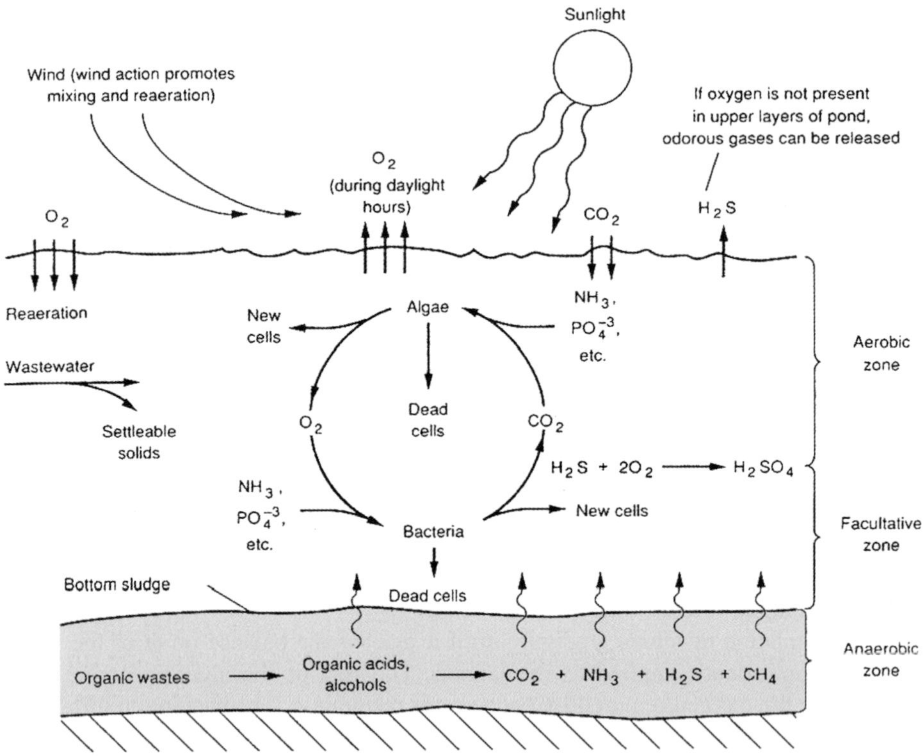

Figure 5.1 Symbiotic union between algae and bacteria in stabilization lagoons (*Source:* Aktug, 2002)

Lagoons are classified to several types as follows:

- *Anaerobic Lagoons:* These are lagoons subjected with a high volumetric organic load (L_V) (g $DBO_5/m^3 \cdot d$) of approximately 100–500, in which anaerobic fermentation dominates. Anaerobic lagoons can be covered with plastic geomembrans to control odours and recover CH_4. A photo of a covered anaerobic lagoon is presented in Figure 5.2.
- *Facultative Lagoons:* This type of lagoon is stratified into three zones: aerobic at the upper part, facultative beneath the aerobic zone and anaerobic at the bottom. The design parameter is the surface organic load (L_S) (kg DBO_5/ha · d), because the functioning of the lagoon depends on the sunlight and so the area is the defining parameter. To maintain an aerobic upper layer this type of lagoon needs to be operated with a surface organic load in the range 100–350 kg DBO_5/ha · d.
- *Aerobic Lagoons:* This type of lagoons is constructed as shallow lagoons to maintain dissolved oxygen throughout the entire depth of the lagoon and the entire body of water of the lagoon. The performance of this lagoon also depends on the sunlight and consequently on the surface organic load (L_S) which in this type of lagoon is usually in the range 20–200 kg DBO_5/ha · d. Aerobic

lagoons are also used as maturation lagoons (see below) and the design of such lagoons is not based on surface organic loading but rather on removal of pathogenic microorganisms, which require longer detention time and therefore a larger volume and a larger surface area. As a result, the effective loading in aerobic lagoons is usually in the range 20–100 kg DBO_5/ha · d.

- *Maturation or Polishing Lagoons:* The objective of this type of lagoons is to improve the quality of treatment plant's effluent. In a lagoons treatment plant maturation lagoons are aerobic lagoons which receive the effluent of the final facultative lagoon, and their main objective is to reduce the number of excreted pathogenic microorganisms, mainly faecal bacteria and viruses, present in the effluent of facultative lagoons. This is done by using shallow lagoons which function with low surface organic loading, in which the pathogens are removed mainly due to greater light penetration. Sometimes maturation lagoons only act as a settling basin.
- *Mixed Lagoons (Lagoons assisted by mixers):* These are lagoons in which mechanical mixers are installed to mix the upper aerobic layer of the lagoon without disarranging the stratification zones. The mixing eliminates short circuits in the lagoon, increases the volume of the aerobic zone and distributes efficiently the oxygen throughout the entire aerobic zone, transforming thereby the lagoon to a more efficient biological reactor which can sustain higher surface organic loads of up to 450 kg DBO_5/ha · d, or even more at very high temperatures. In this type of lagoons, the Gurney mixers (see Figure 5.7) or equivalent mixers can function very well. It is emphasized that mixers are not aerators and they consume very little energy (a motor of only 0.75HP is required to operate a mixer). Mixers can be operated by the power of wind, when the wind velocity is 6 km/hr or higher, and some mixer models include the dual operation mode, electric operation at low wind velocity and wind operation at high wind velocity. Other models of mixers use energy from solar panels installed on the them as the power source for their operation.

Figure 5.2 A covered anaerobic lagoon

Stabilization lagoons with assisted mixing (through the use of Gurney or equivalent mixers) allows lagoons which operate photosynthetically to maintain sufficient oxygen in periods of low oxygen production, such as during cloudy days and to an extent during the night time. Gurney and similar type mixers consume very little energy and their low power level promotes an "optimal mixing", a mechanism by which the oxygen in the lagoon is kept uniformly mixed but the critical biological layers (aerobic, facultative, anaerobic) do not mix, nor is the benthic sludge suspended. An optimized mixed facultative lagoon maintains the three bacterial zones differentiated and that allowed taking advantage of the many benefits of the facultative process (low operating costs, sludge auto digestion and more), as detailed by Gurney Environmental (www.gurneyenvironmental.com).

The lagoons are always constructed in series (so that the effluent of one lagoon is the influent to the next) to improve the removal of organic matter and faecal coliforms. A typical diagram of a configuration of lagoons in series consisting of anaerobic-facultative-maturation lagoons is shown in Figure 5.3.

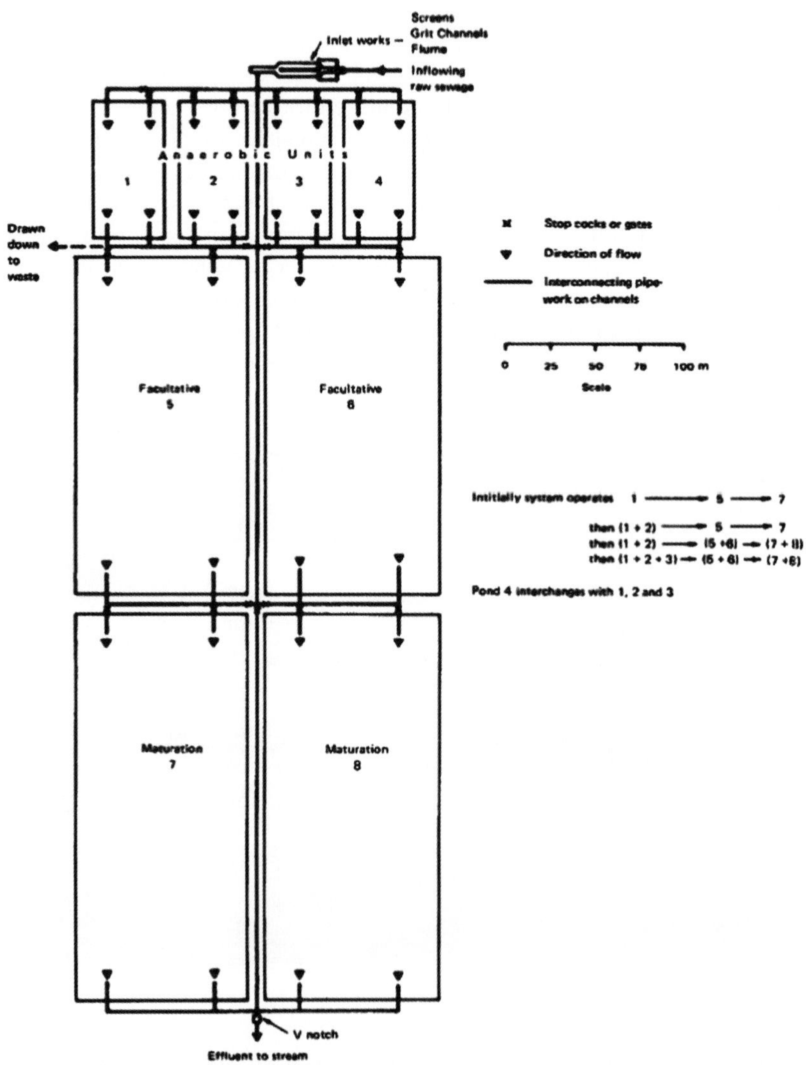

Figure 5.3 A typical layout diagram of stabilization lagoons in series (*Source:* Arthur, 1983)

5.1.2 Basics of the process

According to Arthur (1983) the mechanisms of stabilization in lagoons are: (i) treatment of the organic wastes by aerobic bacterial oxidation in the presence of oxygen, or by anaerobic bacterial decomposition, in the absence of oxygen; (ii) the sedimentation of suspended matter, allowing settleable solids to sink to the benthic sludge layer where it undergoes anaerobic stabilization; and (iii) the reservoir effect that enables lagoons to absorb both organic shocks loads (L_{DBO_5}) (kg DBO_5/d) and hydraulic shock loads, q_H ($m^3/m^2 \cdot d$).

The hydraulic flow in the lagoons depends, in addition to wastewater inflow rate, also on evaporation from the surface of the lagoon and on the percolation from the bottom and side walls of the lagoon, so to date attention is directed to minimize percolation and evaporation. Percolation is reduced by placing impermeable material at the bottom of lagoons, either clay or geomembrans resistant to microbial action. Evaporation can be controlled in anaerobic lagoons by covering such lagoons. Covering anaerobic lagoons has additional benefits such as odour control and providing means for capture of the biogas produced in the anaerobic process.

There are some empirical design parameters of lagoons which prescribe, based on experience, for each type of lagoon the range of values of these parameters which are to be used in order to obtain proper performance of the lagoons. Table 5.1 provides values for photosynthetic lagoons (aerobic, facultative and maturation).

Table 5.1 Design criteria of photosynthetic stabilization lagoons.

Type of lagoon	Surface organic load kg BOD_5/ha · d	Detention time days	Depth meter	BOD removal efficiency* %
Facultative	100–250	5–50	1.5–3.0	90
Facultative with mixers	100–450	2–10	2.0–5.0	90
Maturation	20–200	2–5	1.0–1.8	80

*Efficiency does not include algae removal

Following is a discussion on the most important design parameters:

Surface organic load

The surface organic load is the most important parameter for the design of photosynthetic stabilization lagoons. This parameter measures the daily organic loading per unit area of the lagoon. Stabilization lagoons operate on the basis of active photosynthesis of algae, which depend on sunlight and consequently on the surface area of photosynthesis. In technical language, the surface organic load is the daily amount of food per hectare received by te microorganisms in the lagoon, in terms of kg DBO_5/ha · d. It is calculated as follows:

$$L_s = \frac{DBO_5 \cdot Q}{ha} = \frac{S_0 \cdot Q}{A} = \frac{S_0\, h}{t_d} \tag{5.1}$$

where:

L_s = Surface organic load, kg DBO_5/ha · d
S_0 = influent substrate concentration, kg DBO_5/m^3

Q = flow rate, m^3/d
A = surface area of the lagoon, ha
h = lagoon depth, m
t_d = detention time, (V/Q), in *d*

The Surface organic load of facultative lagoons needs to be maintained between 100 and 350 kg DBO_5/ha · d and between 20 and 200 kg DBO_5/ha · d for aerobic shallow lagoons, like maturation lagoons. According to Orozco (2005), the applied surface organic loading affects the depth of the aerobic layer of the lagoon (and consequently the depth of the facultative or anaerobic layer). It is clear that the higher the load, the smaller and shallower is the aerobic zone, which totally disappears in the case of anaerobic lagoons. This is well illustrated in the performance curve of various lagoon plants operating under different surface organic loading conditions, as presented in Figure 5.4.

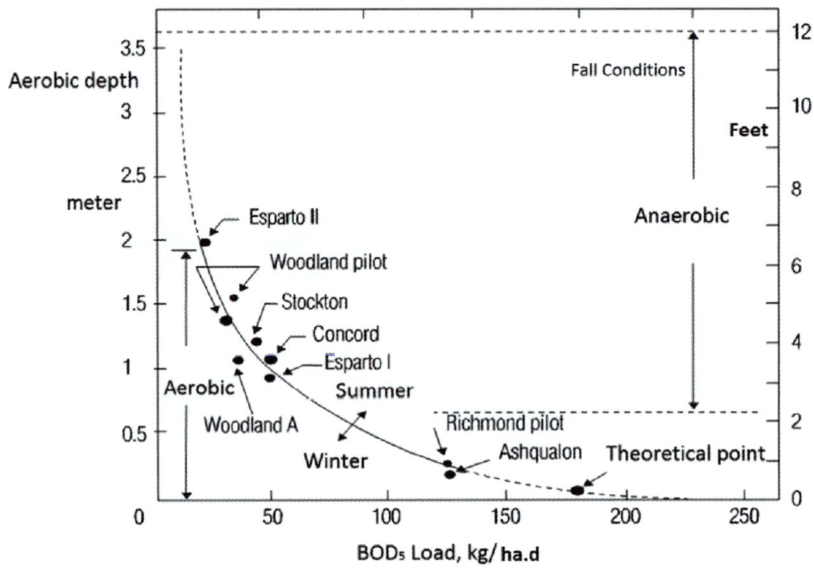

Figure 5.4 Relationship between the surface organic load and the depth of the aerobic zone in various lagoon plants (*Source:* Orozco, 2005)

Volumetric organic load

The main design parameter for anaerobic lagoons is the volumetric organic load, L_V (g $DBO_5/m^3 \cdot d$), which is calculated as follows:

$$L_v = \frac{DBO_5 \cdot Q}{V} = \frac{S_0 \cdot Q}{V} = \frac{S_0}{t_d} \tag{5.2}$$

The anaerobic lagoons are designed with an L_V in the range 100–500 g $DBO_5/m^3 \cdot d$ or even higher, depending on the temperature and desired efficiency (Libhaber, 2007; Mara, 2006).

Mixing

When the lagoons are mechanically mixed the Power Level (PL) becomes an important parameter. The mixed lagoons have to continue operating as facultative lagoons. Facultative lagoons have three layers zones: an aerobic layer on the top, a facultative layer in the center and an anaerobic layer at the bottom. The settleable solids should be left to sink to the bottom of the lagoon where they undergo anaerobic decomposition. Therefore the mixer must sufficiently powerful to provide effective mixing which is able to distribute the oxygen throughout the aerobic layer but should not be powerful enough to suspend the bottom sediments but rather keep them at the bottom. The Gurney and equivalent mixers provide a low enough power level to maintain the lagoons in facultative conditions. The required power level is 0.75 HP per 0.65–3 hectares, depending on the surface organic loading on the lagoon. Higher loaded lagoons require more mixing power. That translates to power per unit area in the range of 0.25–1.2 HP/ha.

The hydraulic power applied for mixing becomes then the design element. In addition, the temperature and the pH in the lagoon must be taken into account during the design, as discussed below. The power level units are W/m^3 (or $kW/1000\ m^3$) and for maintaining facultative conditions, the required power level is about 0.5. At power levels above 1.0 W/m^3 the lagoon becomes an aerated lagoon and particles will become suspended and discharged with the effluent.

Removal of algae

It is important to note that often the amount of algae produced in a lagoon and discharged with the effluent is so large that if included in the effluent BOD measurement, the effluent might contain more organic load than the influent. However, if the body receiving the discharge is an open water body (flowing river, open sea) algae produce more oxygen than they need for respiration and are incorporated into the food chain, are consumed by higher life form and improve the productivity of ecosystems. Algal BOD and Suspended solids are therefore different from non-algal BOD and SS. This fact should be taken into account by regulators when setting effluent quality standards. The EU standard requires that lagoon systems effluents achieve a Filtered BOD quality of 25 mg/l or less and a total suspended solids concentration of 150 mg/l or less. This standard allows the algal SS to be discharged to receiving bodies. Such an approach should also be adopted in developing countries.

If the effluent is discharged to closed water bodies such as lakes and bays, algae can sometimes contribute to eutrophication of the receiving body, in which case it might be necessary to remove the algae with the help of a chemical coagulant followed by flotation followed by a filtration system (either sand filters or rotating micro screen with a very small screen opening). Filtration of a lagoons effluent by a rock filter can partially remove algae. Recently a different method of algae removal by ultrasonic sound waves was developed. This method allows solving the algae problem effectively and economically (see LG Sonic, 2003). However, the front part of the transducer (ultrasound emitting instrument) needs to be cleaned at least once a week, which for many operators seems inconvenient. An automatic cleaning system has been developed to solve this problem. If mixers are used to improve the performance of a lagoon plant, the lagoons, including maturation lagoons, can be deeper and then the effluent can be withdrawn from a deeper layer in which the algae content is low.

Design considerations

The design of stabilization lagoons is one of the most studied subjects in the field domestic wastewater treatment. It is therefore natural that different methodologies and conceptual approaches have emerged, all useful for the design of lagoon systems. The most representative and successful of these methodologies are presented in this chapter, as follows:

Kinetics approach

(i) *Equations:* For the design of *completely mixed* lagoons first-order kinetic equations can in general be used. The constant K_L for the removal of BOD depends on the lagoon type. The following equation (similar to Equation 2.31) can be used for a single lagoon (aerobic, anaerobic or facultative):

$$S = \frac{S_0}{1 + K_L t_d} \tag{5.3}$$

where:

S = effluent substrate, mg/l (g/m^3) of DBO_5
S_0 = influent substrate, mg/l (g/m^3) of BOD_5
K_L = kinetic constant of BOD_5 removal, d^{-1}
t_d = V/Q, hydraulic detention time, d

Equation 5.4 can be used for n lagoons in series (similar to Equation 2.36), where n is the number of lagoons:

$$S = \frac{S_0}{\left(1 + K_L \frac{t_d}{n}\right)^n} \tag{5.4}$$

For a lagoon operating under a plug flow regime the equation that applies is the following (similar to Equation 2.37):

$$S = S_0 e^{-K_L t_d} \tag{5.5}$$

Note that the flow regime in a system of many lagoons in series approaches the plug flow regime (Orozco, 2005). In general, the fluid element (ΔV) in an elongated reactor tank is not transported in a *perfect plug flow*. A perfect *completely mixed* reactor neither exists in practice. This means that there is always dispersion in a reactor, which can be measured by the dispersion factor (***d***). Wehner and Wilhelm (1958) found that the expression for the substrate in the effluent of a tank with dispersion ***d*** is:

$$\frac{S}{S_0} = \frac{4a\,e^{1/2d}}{(1+a)^2 e^{a/2d} - (1-a)^2 e^{-a/2d}} \tag{5.6}$$

where:

$a = \sqrt{1 + 4K_L t_d}$
$d = D/u \cdot L$ = dispersion factor
D = axial dispersion coefficient, m^2/h
u = fluid velocity, m/h
L = characteristic length, m
K_L = reaction constant of 1st order, h^{-1}

From this equation it is possible to draw a curve of the value $K_L t_d$ versus S/S_0. Figure 5.5 presents such a set of curves, in which variation parameter is the dispersion factor ***d***. When $\boldsymbol{d} = 0$ we have completely mixed conditions, and for $\boldsymbol{d} = \infty$ we have perfect plug flow conditions. As seen in the figure these are the tow extreme cases.

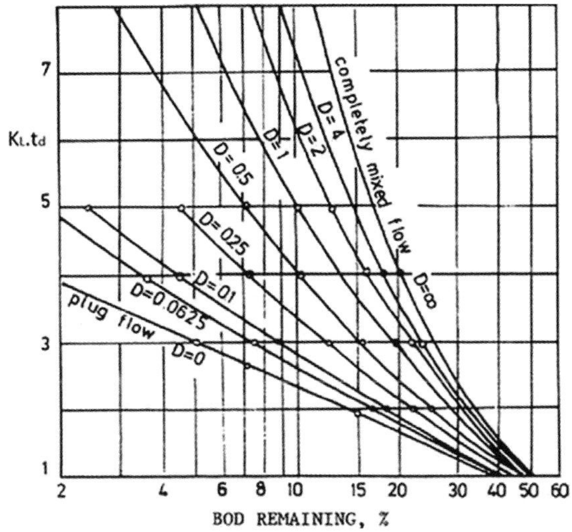

Figure 5.5 K_Lt_d *vs* BOD remaining values according to the Wehner and Wilhelm equation (*Source:* Metcalf & Eddy, 2003)

The value of K_L corresponds to the first-order substrate removal rate and t_d is the detention time in the reactor. K_L has reported values of between 0.01 and 1.50 d^{-1}, and typically 0.35 d^{-1} at 20°C. K_L fluctuates significantly with temperature, according to Equation (5.7)

$$K_L = K_L(20^\circ C) \cdot \theta^{(T-20)} \tag{5.7}$$

where the value of θ (temperature correction coefficient) varies between 1.05 and 1.09 for the first order BOD_5 reaction constant.

The kinetics of removal of faecal coliforms (B) has a different kinetics constant (K_B), but still follows the first order kinetics as presented above. The typical value of K_B (20°C) is 2.6 d^{-1}, and of θ is 1.19.

(ii) *Temperature:* Due to the dependence of the design equation on the temperature of the liquid in the lagoon, it is of utmost importance to establish the heat balance of the lagoon, which defines the operating temperature of the lagoon. According to Mancini and Barnhart, the heat balance is given by (see Yanez, 2003):

$$T_L = \frac{A_s f T_a + QT_i}{A_s f + Q} \tag{5.8}$$

where:

T_L = final temperature of the lagoon (°C)
T_a = environmental air temperature (°C)
T_i = temperature of influent wastewater (°C)
A_s = surface area of the lagoon (m^2)
Q = wastewater flow rate (m^3/d)
f = proportionality factor, 0.5 m/d

As shown in Table 5.1 and Figures 5.4 and 5.5, the depth is important for the mode of operation of the lagoon since the penetration of light defines the depth of the layer in which the photosynthetic action occurs. Other important parameters are the detention time of the wastewater in the lagoon and surface organic load. For the design of a lagoon Equations 5.1, 5.3 and 5.7 can be used, taking into account that the first order constant K_L for this type of treatment varies between 0.01 and 1.50 d^{-1}. The impact of temperature variation needs to be taken into account according to Equations 5.8 and 5.7.

Photosynthetic approach

For the design of photosynthetic lagoons we can use Oswald's approach which found that the production of oxygen by photosynthetic action is:

$$Y_{O_2} = 0.25 \cdot F_{O_2} \cdot \mathbf{S} \tag{5.9}$$

where:

Y_{O_2} = production of O_2 (kg O_2/ha · d)
F_{O_2} = oxygenation factor
S = solar radiation (cal/cm^2 · d)

The oxygenation factor varies between 0 and 4 and is the ratio of the weight of O_2 produced to the DBO_u that must be removed. McGauhey (1968) recommended a value of 1.6 for approximately 90% BOD_u removal. The variation of F_{O_2} with the BOD removal percentage in relation to the oxygenation level is shown in Figure 5.6 (Alley *et al.* 2007, Shammas *et al.* 2009).

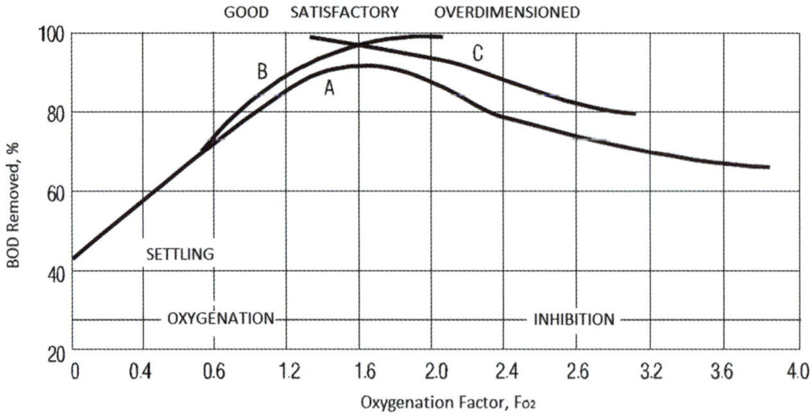

Figure 5.6 Relation F_{O_2} *vs* the percentage of BOD_u removed (*Source:* Orozco, 2005)

The solar radiation, **S**, varies with months and latitude. Table 5.2 presents the maximum and minimum values of **S** as function of latitude and day of the year. The average value is calculated as:

$$\mathbf{S} = S_{min} + P(S_{max} - S_{min}) \tag{5.10}$$

where:

S: average value of solar radiation in cal/cm^2/d
P: actual hours of sunshine divided by the potential sunshine hours at the site

Table 5.2 Probable values of visible solar energy based on month and latitude.

Latitude		Month[b]											
		Jan.	Feb.	Mar.	Apr.	May	Jun.	Jul.	Aug.	Sept.	Oct.	Nov.	Dec.
0	max	255	266	271	266	249	236	238	252	269	265	256	253
	min	210	219	206	188	182	103	137	167	207	203	202	195
10	max	223	244	264	271	270	262	265	266	266	248	228	225
	min	179	184	193	183	192	129	158	176	196	181	176	162
20	max	183	213	246	271	284	284	282	272	252	224	190	182
	min	134	140	168	170	194	148	172	177	176	150	138	120
30	max	136	176	218	261	290	296	289	271	231	192	148	126
	min	76	96	134	151	184	163	178	166	147	113	90	70
40	max	80	130	181	181	286	298	288	258	203	152	95	66
	min	30	53	95	125	162	173	172	147	112	72	42	24
50	max	28	70	141	210	271	297	280	236	166	100	40	26
	min	10	19	58	97	144	176	155	125	73	40	15	7
60	max	7	32	107	176	249	294	268	205	126	43	10	5
	min	2	4	33	79	132	174	144	100	38	26	3	1

[a]After Table II, W. J. Oswald and H. B. Gotaas, "Photosynthesis in Sewage Treatment" *ASCE Proceedings* 81, Separate No. 686 (May, 1955).
[b]Values of S in Langleys, cal/(cm^2)(d)
Correction for cloudiness: $\mathbf{S}_e = S_{min} + r(S_{max} - S_{min})$

Equation 5.9 is used to determine the oxygen production (Y_{O_2}, kg/ha · d). The surface organic load (L_S, in kg BOD_5/ha · d) is obtained from Equation 5.1, using the appropriate units for design:

$$L_s = 10(h/t_d)DBO_5 = 10(h/t_d)DBO_u/1.5 \quad (5.11)$$

where:

h = aerobic zone depth in the lagoon, m
t_d = detention time, *d*
$DBO_u = 1.5\ DBO_5$

Equalizing the O_2 production to the total oxygen demand, that is, $Y_{O_2} = L_S$ (in terms BOD_u), we obtain:

$$10(h/t_d)(1.5DBO_5) = 0.25\, F_{O_2}\mathbf{S} \quad (5.12)$$

Which by rearranging becomes:

$$h/t_d = 0.017\, F_{O_2}\mathbf{S}/DBO_5 \quad (5.13)$$

The above relations are essentially valid for aerobic stabilization lagoons. To design a facultative lagoon more depth needs to be added, since h is calculated in the above equations only for the aerobic zone.

Duncan Mara's approximation method

A summary of the methodology proposed by Duncan Mara based on Mara (2006) and Pena-Varon and Mara (2005) is presented below.

(i) *Anaerobic lagoons:* The volumetric load (L_V) given by Equation 5.2 is used for design. The recommended range of volumetric loads is $100\ g/m^3 \cdot d$ for temperatures below 10°C, increasing linearly to $300\ g/m^3 \cdot d$ at 20°C and then more slowly up to $350\ g/m^3 \cdot d$ at 25°C. In this case, the lagoon is designed with the *average environment temperature* in the coldest month. The removal of BOD_5 is 50% (<10°C) to 80% (at 25°C and above). The recommended depth of the anaerobic lagoon is between 3 and 5 m.

(ii) *Facultative lagoons:* The surface organic load (L_S, on terms of BOD_5) is used as the design parameter. It is given by Equations 5.1 and 5.11 (using the appropriate units). The following equation can be employed using as the sole design parameter the average temperature of the coldest month, T:

$$L_s = 350(1{,}107 - 0{,}002T)^{(T-25)} \tag{5.14}$$

This equation assumes that $L_S = 350$ $kgDBO_5/ha \cdot d$ is the optimal surface organic loading at 25°C.

(iii) *Minimum detention time:* The proposed minimum detention time, t_{dmin} is 1 d in anaerobic lagoons, 4 d in facultative lagoons and 3 d in maturation lagoons.

(iv) *Removal of helminth eggs:* For the removal of helminth eggs (see Mara, 2006) the following equation is used:

$$E(\%) = 100(1 - 0.41 \exp[-0.49\, t_d + 0.0085\, t_d^2]) \tag{5.15}$$

The concentration of helminth eggs in domestic wastewater of low income developing countries is between 300 and 800 Ova/l. For high and middle income developing countries the concentration is between 1 and 300 Ova/l [see http://www.iwawaterwiki.org/xwiki/bin/view/Articles/Helmintheggs (May 12, 2011)].

(v) *Removal of fecal coliforms:* The removal of fecal coliforms (FC) in a series of maturation lagoons of equal dimensions can be estimated using Equation 5.4, applied to FC:

$$B = \frac{B_0}{\left(1 + K_B \frac{t_d}{n}\right)^n} \tag{5.25}$$

With B is the FC concentration. The constant K_B is temperature dependent. According to Arthur (1983) the variation of K_B with temperature is given by:

$$K_B = 2.7 \times 1.19^{(T-20)} \tag{5.26}$$

Recent innovations in design of lagoons systems

(i) *Use of mixers in facultative and maturation lagoons:* Mechanical mixers which employ the concept of "optimal mix" can be installed in facultative lagoons to improve their performance. The mixer can also be installed in maturation lagoons. Several manufacturers supply this type of mixers, for example, Gurney Environmental. According to the "optimal mix" concept a low power level is applied by the mixers in such a way that in a properly optimized mixed facultative lagoon the three distinctive layers (upper aerobic, intermediary facultative and lower anaerobic) are maintained. However, the depth of the upper aerobic layer increases due to the mixing and the upper aerobic zone is well mixed through the function of the mixers. The gentle mixers pump liquid from the depth of the lagoon, without breaking the facultative boundary layer or disturbing the anaerobic layer at the bottom of the lagoon, and discharge the liquid near the surface. The mixers prevent short-circuits, especially in the upper layer, bring to the upper layer organic matter from deeper layers for processing, distribute the oxygen generated by the algae in the upper layer to the entire volume of the aerobic layer thus creating uniform oxygen profiles from the top of the lagoon to the facultative layer, prevent stratification, break oil films, break snow and ice layers, and actually resolves all the problems of facultative lagoons and thereby improve the performance of such lagoons. The mixers used for this purpose are not aggressive aerators which consume large amounts of energy, but rather gentle mixers, each operated by a small motor of 0.75 HP that consumes little energy and can be operated by wind when the wind velocity in the treatment plant site is sufficiently high (over 5–6 km per hour). The mixer switches automatically to wind operation when the wind velocity passes a selected threshold. Each mixer has an influent range of 0.65–3 hectare, depending on the type of lagoon. In higher loaded lagoons the influence area of each mixer is smaller, between 0.65 to 1 hectare and in maturation ponds the influence area of each mixer is between 0.75 to 3 hectares, so higher loaded lagoons require more mixers. When designing a new lagoons plant with mixers, the depth of the lagoons should be 2–3 meters and even up to 4 meters, that is, deeper than lagoons without mixers. The installation of the mixers in a facultative lagoon can significantly increase its treatment capacity. Organic loads on mixers equipped lagoons can reach values of 450 Kg BOD/d/ha in the high temperatures season of 25–30°C and even higher, with lower loading values at lower temperatures. A mixer for facultative lagoons manufactured by Gurney is presented in Figure 5.7.

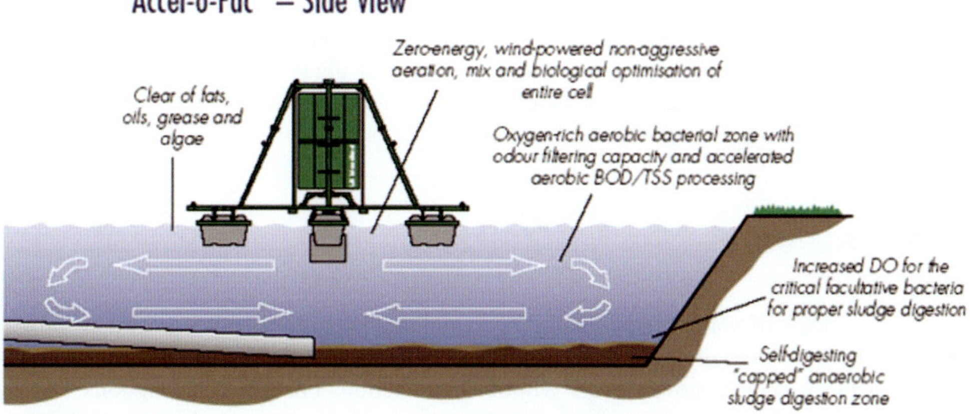

Figure 5.7 Gurney type mixers (Courtesy Gurney Environmental)

Another innovation in treatment by lagoons is injecting air into a facultative lagoon through blowers (not compressed air) to augment the oxygen quantity produced by the algae without changing the flow regime, that is without transforming the lagoon to an aerated lagoon. For that purpose, a combination of blowers with air distributors and gentle mixers can be installed in the lagoon. The addition of air to the lagoons increases its treatment capacity to a required level. Injection of air requires more energy so this process is not strictly an appropriate technology process. However, it is still a simple process and is not very costly. It can be used to for increasing the capacity of an existing lagoons plant if additional area is not available and the required capacity increase cannot be reached by just adding mixers.

(ii) *Control of odours in anaerobic lagoons:* Using an anaerobic lagoon as the first of a series of lagoons in a lagoons treatment system is a cost-effective and practical approach to treat municipal wastewater. An anaerobic lagoon with a detention time of 1–2 days can remove between 50 and 70% of the BOD contained in the raw wastewater. The main problem of anaerobic lagoons, which is a problem difficult to control, is that under certain conditions they generate odours.

When the concentration of sulphate in the raw wastewater is low it is possible to design anaerobic lagoons which do not generate odours. However, when sulphate concentrations are high, emissions of odour producing gases (sulphide gases) and of greenhouse gases (methane and carbon dioxide) can be significant. The emission of odours intensifies when high organic loads are applied on anaerobic lagoons. An effective way to prevent odours is to cover the anaerobic lagoons and collect the gases produced in the anaerobic process. The collected gases can then be either flared or used to generate energy. Covering of the anaerobic lagoons totally eliminates the odour problem. The lagoons can be covered by geomembrans as depicted in Figure 5.2. The cost of geomembrans has dropped in Latin America to levels which make covering of anaerobic lagoons affordable. The geomembrane presented in Figure 5.3 is an HDPE geomembrane resistant to sun light.

According to Libhaber (2007), covering anaerobic lagoon has many benefits, namely: (i) capturing odorous gases to remove, flare or treat them, eliminating thereby odour problems; (ii) preventing emission of greenhouse gases into the atmosphere, contributing thereby to the effort of mitigating the global warming; (iii) generating additional funds for the water and sanitation utility from the sale of Carbon Emission Reductions (CER) to industrialized countries; (iv) reducing the loss of heat from the anaerobic reactor and improving thereby its performance; (v) reducing the evaporation and loss of water from the anaerobic lagoon; (vi) blocking sunlight from reaching the lagoon thus inhibiting the growth of algae in it; and (vii) generating energy from the collected methane gas, in case it is economically feasible under the energy policy of the country in which the lagoons are located.

5.1.3 PERFORMANCE

Efficiency

The removal efficiencies of some contaminants by lagoons compared to other treatment processes are presented in Table 5.3. The lagoons can achieve a very high removal of pathogens while the other process cannot achieve such removal levels without disinfection. For example, lagoons can achieve a reduction of six orders of magnitude in fecal coliforms (equivalent to removal of 99.9999%, since one cycle of $\log_{10} = 90\%$). Similarly, lagoons can achieve a reduction of four orders of magnitude of viruses, and a 100% removal of helminth eggs, as well as about 90 percent of cysts and oocysts of protozoa. Table 5.4 presents removal levels of BOD and fecal coloiforms in various types of single lagoons.

Table 5.3 Removal efficiencies of some contaminants by several wastewater treatment process.

Process	% Removal BODs	SS	Cycles log_{10} removal			
			Virus	Bacteria	Helmints ova	Protozoa cysts
Primary settling	20–40	40–70	0–1	0–1	0–1	0–2
Activated sludge	55–95	55–95	1–2	0–2	0–1	1–2
Trickling filters	50–95	50–90	1–2	0–2	0–1	1–2
Chlorine disinfection	NA	NA	0–4	2–6	0–1	0–3
Lagoons in serie	70–95	55–95	2–4	2–6	2–4 (100%)	2–4 (100%)

Source: Rodríguez (2007)

Table 5.4 Expected efficiencies of BOD and fecal coliforms removal in various types of single* lagoons.

Lagoon type	% BOD_5 removal			% Fecal coliform removal		
	12°C	20°C	25°C	12°C	20°C	25°C
Anaerobic	55	62	70	60	86	93
Facultative	75	80	85	91	97	98
Aerobic	70	80	82	72	93	96

*Data refer to a single lagoon, not to a system of lagoons in series
Source: Arthur (1983)

Several lagoons in series (the series may consist of different types of lagoons) improve the performance efficiency of the system, in accordance with Equation 5.4, which can be combined for different types of lagoons according to the generalization of Equation 5.5, as follows:

$$S = \frac{S_0}{\left(1 + K_{L1}\frac{t_{d_1}}{n_1}\right)^{n_1}\left(1 + K_{L2}\frac{t_{d_2}}{n_2}\right)^{n_2}\cdots\left(1 + K_{Lj}\frac{t_{d_j}}{n_j}\right)^{n_j}} \qquad (5.16)$$

for *j* types of lagoons (i.e. anaerobic, facultative, etc.).

The BOD removal efficiencies of completely mixed lagoons with first-order kinetic approximation, for different temperatures and detention times are presented in Figure 5.8.

Figure 5.9 shows generalized removal curves of BOD, Helminth Eggs, excreted bacteria and viruses in Waste Stabilization Lagoons at Temperature above 20°C, in terms of removal levels as function of the detention time in the lagoons system. This figure gives a general idea of the efficiency of treatment in lagoons as function of the detention time in the lagoon system. To obtain performance data of a specific lagoons system, calculations need to be carried out according to the methodology presented below.

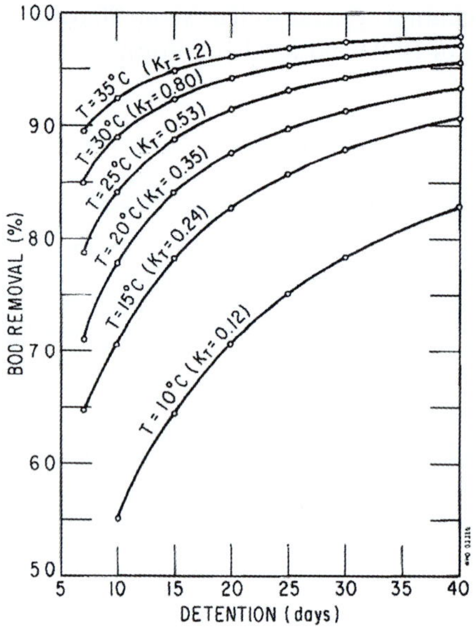

Figure 5.8 Percentage removal of BOD as function of detention time in the lagoons system at different temperatures (*Source:* Gloyna, 1971)

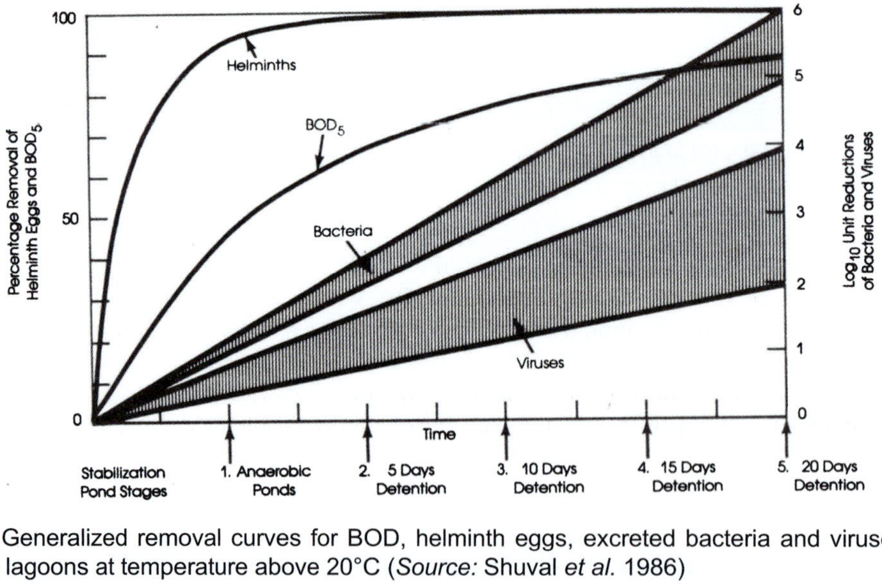

Figure 5.9 Generalized removal curves for BOD, helminth eggs, excreted bacteria and viruses in waste stabilization lagoons at temperature above 20°C (*Source:* Shuval *et al.* 1986)

Advantages and disadvantages

As explained above, the anaerobic and facultative lagoons are used primarily for removing BOD, while the removal of pathogens occurs mainly in facultative and maturation lagoons. Hydraulic detention time in a lagoon system is between 5 and 50 days. This is much longer than the detention time in other process in which detention times are usually less than one day. In practical terms it means that the area occupied by lagoons plants is much larger than that occupied by plants based on other processes. Large areas are not available inside cities, so lagoon plants are usually located outside the limits of the city, so an additional investment in conveyance of the wastewater is required. On the other hand, the operation and maintenance costs are much smaller than those of conventional processes, so in the case of lagoons a significant portion the money is invested in tangible assets and can be recovered after years, while in the case of conventional processes the money is invested in paying electricity bills and maintenance of equipment, and can never be recovered.

The advantages of stabilization lagoons are the followings (OPS, 2005):

- Because of the large volume of the lagoons they have the capacity to handle well organic and hydraulic shock lads.
- No additional mechanical facilities are required for the production of oxygen (except in the case of aerated lagoons, which we do not consider as an appropriate technology process when they serve as the main treatment unit). The oxygen is produced naturally in the system by photosynthetic action.
- Utilization of mixers in facultative lagoons ensures the oxygen supply will be secured even during unfavourable climate conditions.
- Due to the mechanisms of the processes occurring in the lagoons and the long detention times in lagoons systems they are (when properly designed) highly effective in removing pathogenic organisms such as bacteria, viruses and parasites, compared to other treatment processes. Consequently there is no need to disinfect lagoons effluents.
- Lagoons require a low level of maintenance.
- Lagoons do not require high level qualified personnel for operation.
- Lagoons do not consume energy during their operation, and in certain cases can produce energy.
- The O&M costs of lagoons are very low compared to other processes.

The disadvantages of the lagoons are:

- They occupy large land areas.
- A lagoons system is sensitive to climatic conditions.
- Once constructed, lagoons systems are difficult to control since only little can be done to change the process operation. This problem is mitigated in a system with covered anaerobic lagoons and mixers in the facultative lagoons. Covering the anaerobic lagoons eliminated the odour problems, and the operation of the mixers in facultative lagoons provides a measure of control on their performance.
- Lagoons may support the spread of disease vectors.
- Lagoons may promote eutrophication of closed water bodies (lakes, bays) because of the continuous supply of algae with the discharged effluent. However, this problem can be mitigated by the removal of algae from the effluent, for instance with ultrasonic waves.

Anaerobic lagoons may generate offensive odours. This problem can be resolved by covering the lagoons.

5.2 BASIC DESIGN PROCEDURES

5.2.1 General design considerations

For the design of stabilization lagoons, it is important to bear in mind that the main mechanisms of stabilization, according to Section 5.1.2, are: (i) Removal of BOD by aerobic bacterial oxidation in the presence of oxygen, or by anaerobic bacterial decomposition, in the absence of oxygen; (ii) the sedimentation of suspended matter, allowing settleable solids to sink to the benthic sludge layer where it undergoes anaerobic stabilization; and (iii) the reservoir effect that enables lagoons to absorb both organic and hydraulic shock loads.

In fact a single lagoon is rarely designed. Lagoons are rather built in series to maximize efficiency (as seen from Equation 5.4). The series most commonly used is shown in Figure 5.3 (Anaerobic Lagoon → Facultative Lagoon (preferably mixed) → Maturation Lagoon) because such a configuration allows expanding the range of controlled parameters and improving the efficiency of removal of BOD, fecal coliforms and helminth eggs. Mixers can be installed in the facultative lagoons to improve their performance.

A problem with photosynthetic lagoons is that the production of algae converts CO_2 released from bacterial respiration (and also CO_2 from the atmosphere) into biomass, so that sometimes the removal of organic matter from municipal wastewater is achieved at the cost of producing living organic matter in the form of algae. This transformation does not cause negative impacts when the effluent is discharged into open water bodies such as rivers, sea, and so on, so that the algae are incorporated into the food chain to improve productivity of aquatic ecosystems. Different is the case of effluent discharge into closed systems such as lakes and bays, since the nutrient (nitrogen and phosphorus) contained in the algae enter the receiving body promoting eutrophication, thus damaging more than protecting the water body. This is the situation in the Federal District of Brasilia, the capital of Brazil, where it is necessary to remove algae from the lagoons effluent using Chemically Enhanced Primary Treatment (CEPT), in which the separation of algae is made by dissolved air flotation (DAF). It also requires the handling of the removed sludge, so this algae removal treatment becomes a highly mechanized tertiary treatment.

A simpler algae removal system is the rock filter lagoons polishing system, which does not completely remove the algae but can achieve 50–60% removal, as presented in Figure 5.10 (Moreira *et al.* 2006). In addition, algae can be removed by ultrasonic waves equipment designed for algae control (LG Sonic, 2003). Figure 5.11 demonstrates the impact of the ultrasonic waves instrument on algae removal.

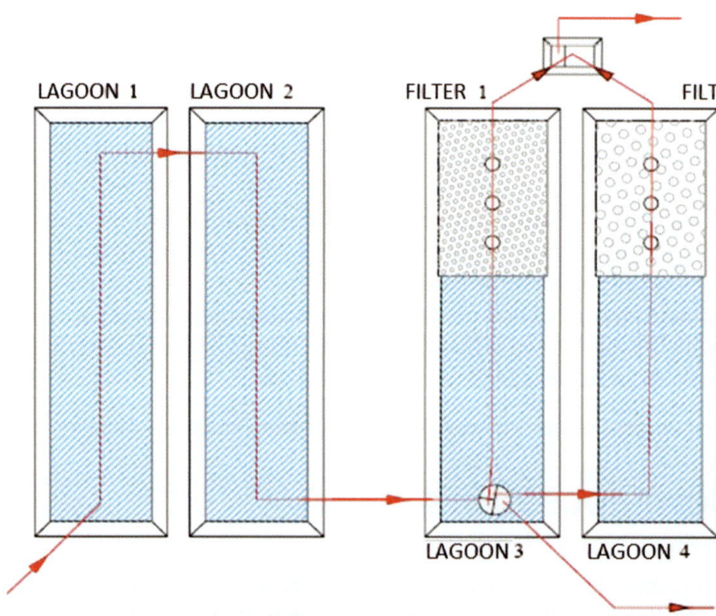

Figure 5.10 Maturation lagoons with rock filters for algae removal (*Source:* Moreira *et al.* 2006)

Figure 5.11 The effect of ultrasonic waves on algae: (a) Day 1 after installing the ultrasonic waves instrument; and (b) Day 21 after installing the ultrasonic waves instrument (*Source:* LG Sonic, 2003)

5.2.2 Orderly design method – ODM

In general the design methodology of a lagoon system consists of the following:

(1) *Determine the external variables* (or independent variable), which, among others, are the following:
- *Flow:* average design flow (or process design flow, Q_D) and peak discharge or hydraulic design flow (Q_{HD}) with Equations 3.4 and 3.5 (below) in appropriate units (lps, m^3/h, etc):

$$Q_D = \frac{cq_{dom}P}{86400} + q_I\,A_a \tag{3.3}$$

$$Q_{HD} = k_1k_2\frac{cq_{dom}P}{86400} + q_I\,A_a \tag{3.4}$$

- *Geographical and altitude conditions:* such as maximum and minimum temperatures of air and water, solar radiation and cloudiness, wind rose, altitude above sea level (masl in meters), proximity to population centers, and so on.
- *Concentration of contaminants in the raw wastewater:* at the influent to the wastewater treatment plant, taking into account water infiltration into the pipelines according to the methodology explained in Chapter 3. From the concentrations data coupled with the flow data it is possible to determine the load of each contaminant (in kg/d) and the specific unit load per Capita (kg/Capita · d) as output.
- Pollutant loads: of the parameters as BOD_5, COD, TSS, pH, N, P, and so on. This can be calculated or directly measurement according to the method described in Chapter 3.

(2) *Determine the flow diagram of the lagoons system:* the components and layout of a typical lagoons system are presented in Figure 5.3. However, other layouts can also be used. The maturation ponds can be equipped with a rock filter to remove algae, as shown in Figure 5.11 or with ultrasonic waves generation instruments an another method for algae removal.

(3) *Design the preliminary treatment system*

3.1 *Select a screening system:* it may be a rough rotating screen with a mesh size of 6 mm. From Figure 4.8 we obtain that a micro screen with an opening of 6 mm achieves a suspended solids removal efficiency of 3%. The design flow of the screening system is Q_{HD} since the screening installation is a hydraulic unit. The method of specification of the screening unit is identical to the one developed in Chapter 4.

3.2 *Design the grit chamber:* use a ratio length/width (L/W) of at least 4, with a surface overflow rate ($SOR = Q_{HD}/A_{des}$) in the range 600–1200 m/d and a water cross-sectional velocity in the range 0.15–0.60 m/s. Large plants can contain several grit chamber units in parallel. Allow for one stand-by grit chamber unit to enable cleaning of one unit while the others are in operation. The design flow of the grit chamber is also Q_{DH} since the grit chamber is also an hydraulic unit.

To size the great chamber it is necessary to:

- Calculate the surface area of the grit chamber: $A_{des} = Q_{DH}/SOR$.
- Compute L and W (the length and width of the grit chamber): if $L/W = 4$; then $A_{des} = L.W = 4W^2$. Then find $W = \sqrt{(A_{des}/4)}$ and $L = 4w$.
- Calculate the depth (H_{des}) of the grit chamber: If the cross-sectional velocity $v = Q_{HD}/W \cdot H_{des}$, then the grit chamber depth is: $H_{des} = Q_{HD}/W \cdot v$. Add 0.20 m for freeboard and 0.30 for grit storage between cleaning cycles. Then the total depth is: $H = H_{des} + 0.50$.

3.3 *Select a Parshall flume:* The Parshall flume, used for measurement of the wastewater inflow, is thoroughly discussed in Chapter 9. The following equation is used to select the adequate parshall flume:

$$h_0 = K\,Q^m \tag{9.6}$$

- This equation allows obtaining the flow of the wastewater by measuring the depth of the flowing wastewater in the flume. The adequate flume throat width (w, in inches) is selected on the basis of with Q_{HD} using Table 5.5 shown below (which is identical to Table 9.2).
- With w known select the values of D, N, K and m defined in Figure 9.5 and whose values are given in Tables 9.1 and 9.3. These values define the partial flume. On the basis of these figures it is possible to obtain from manufacturers or suppliers the required parshall flume. It is also possible to obtain an adequate parshall flume just by the selecting w, the flume throat.

Table 5.5 Flow rate measured by a parshall flume as function of the throat diameter.

Diameter throat inch	Minimum flow lps	Maximum flow lps
1	0.13	5.36
2	0.19	12.24
3	0.25	32.04
6	1.45	87.18
9	2.59	176.69
12	9.21	462.75
18	14.45	696.40
24	18.67	936.93
30	23.09	1180.92

(Continued)

Table 5.5 Flow rate measured by a parshall flume as function of the throat diameter (*Continued*).

Diameter throat inch	Minimum flow lps	Maximum flow lps
36	27.38	1426.74
48	35.77	1922.93
60	62.76	2424.04
72	74.50	2928.87
84	115.31	3437.17
96	130.95	3948.24
120	162.43	10245.08
144	192.84	17381.13
180	238.44	28899.47

(4) *Anaerobic Lagoon:* For the design of an anaerobic lagoon it is best to use the method of Duncan Mara (Mara 2004, Mara 2006, Arthur 1983) which specifies the formulae for estimating the BOD volumetric loading on the lagoon and BOD removal efficiencies it can achieve as function of the mean temperature (T, °C) of the coldest month of the year in the site in which the lagoon is located, as presented in Table 5.6.

After establishing the volumetric BOD loading and the BOD removal efficiency proceed as follows:

4.1 *Calculate the lagoon volume:* the volume of the anaerobic lagoon (V_A) is calculated from the BOD_5 volumetric load ($L_{V,}$ $g/m^3 \cdot d$) as follows:

$$V_A = Q_s/1000L_v = (Q_D \cdot S_0)/1000 \cdot L_v \tag{5.17}$$

where:

$Q_s = BOD_5$ load, kg/d, (while 1000 the conversion factor from g to kg)
$Q_D =$ Process design flow rate, m^3/d
$S_0 = BOD_5$ concentration in the inflow to the anaerobic reactor, g/m^3.

Table 5.6 Design values of volumetric BOD loading and percentage BOD removal in anaerobic lagoons at various temperatures.

Temperature °C	Volumetric BOD_5 loading $g/m^3 \cdot d$	BOD_5 removal %
≤ 10	100	40
$10 < T \leq 20$	$20T - 100$	$2T + 20$
$20 < T \leq 25$	$10T + 100$	$2T + 20$
> 25	350	70

Source: Mara (2004)

The minimum detention time (t_{dmin}) in an anaerobic lagoon is usually taken as one (1) day and the final volume of the lagoon (V_F) is established as the higher of the two values: (i) volume V_A calculated by applying Equation 5.17; and (ii) that calculated with detention time of one day. In other words, if the calculation using Equation 5.17 yields a value lower than 1 day, than the volume is taken as 1 day (although a detention of somewhat lower than 1 day might also be sufficient).

4.2 *Calculate the filling time of the lagoon by excess sludge:* This is the time required to fill half of the lagoon volume with the excess sludge generated by the anaerobic process. This time, denominated n years, is calculated as follows:

$$n = \frac{1}{2} V_F / v_L \cdot P \tag{5.18}$$

where:

N = number of years to fill half of the lagoon with excess sludge
V_F = final volume of the lagoon, m^3
v_L = sludge accumulation rate, $m^3/\text{Capita} \cdot yr$, between 0.05 in the tropics and 0.08 in temperate climate.
P = population

4.3 *Calculate the biogas production:* The biogas produced in the anaerobic process is mainly composed of methane and carbon dioxide (Usually about 70% CH_4 and the balance is CO_2, water and other impurities). According to Mara (2006) the volume of the biogas (V_g, m^3/d) is approximately 30% of the influent BOD_5, that is:

$$V_g = 0.30\, S_0 \cdot Q_D / 1000 \tag{5.19}$$

Odours may become a problem if the BOD volumetric loading is too high, $L_V > 500\ g/m^3 \cdot d$ or if there is excess of sulphate in the influent. To prevent odours and to control greenhouse gases emission it is recommended to cover the anaerobic lagoon and provide a gas flaring system with a capacity to handle a flow two times larger than the flow calculated by Equation 5.19, so as to include a safety factor.

4.4 *Calculate the inflow and outflow structures:* for the dimensioning of the inflow and outflow structures it is recommended to use an average cross-sectional velocity of 1 m/s, and, and under all circumstances remain in the range 0.6 and 2.2 m/s.

(5) *Design the Cover of the anaerobic Lagoon for odour control*[1]: to control odours produced by anaerobic lagoons and to be able to implement the Clean Development Mechanism (CDM) of the Kyoto Protocol, it is important to cover anaerobic lagoons, and to collect and flare the biogas.

5.1 *Cover:* The cover of anaerobic lagoons must be a geomembrane of plastic material, smooth texture and resistant to sunlight exposure, with a thickness of 1.5 mm (60 mil). High density polyethylene (HPDE) made of non-recycled polyethylene resins are an example of a material used for manufacturing geomembrane for anaerobic lagoons covering. The polyethylene sheets are welded together by heat fusion or extrusion. The HDPE geomembrane must have a density greater than or equal to 0.95 g/cm^3, which is waterproof and applicable to storage of liquids and solids. It must have high resistance to chemicals, ultraviolet

[1]Based on a report by SAGUAPAC (2007), "Project of capture and burning of biogas in the wastewater treatment plants – Technical Studies," The Water and Sanitation Utility of Santa Cruz, Bolivia.

radiation and excellent mechanical properties. The geomembrane cover is secured by anchoring trenches on the sidelines of the lagoon to get a firm grip. The material for filling the anchoring trenches can be the same material excavated from the trenches. The filling needs to be done in three layers and should be compacted.

5.2 *Floats:* Floats are fixed to the upper side of the geomembrane by extrusion welding prior to deployment of the membrane over the lagoon according to the installation plans. The purpose of the float is to support the floating of the geomembrane over the water surface. They are set out in the two directions of the covered lagoon. Each float is 2 meters long, 0.50 meters wide and 0.1 meter high and they are spaced every two meters on the same line. They are made of expanded polystyrene (Plastoform) and covered by a smooth waterproofed geomembrane envelope 0.75 mm thick. Details of the goemembrane cover are shown in Figure 5.12.

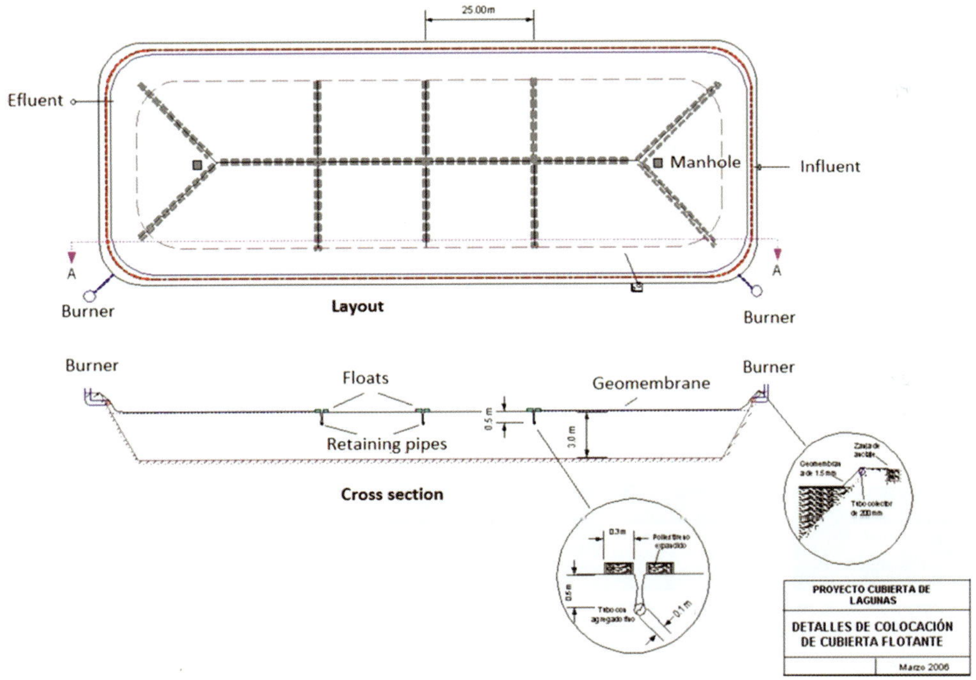

Figure 5.12 Details of the floating geomembrane cover of an anaerobic lagoon including the floats (*Source:* SAGUAPAC, 2007)

5.3 *Counterweight to stabilize the geomembrane (fastening pipe):* The counterweights control the inflation of the cover due to the emission of gases and allow the channeling of rain water which accumulate on cover surface. There are two types of counterweights. The main ones weigh 125 kg each and the secondary, 95 kg each. The counterweights or ballast consist of a smooth sleeve of HDPE geomembrane 0.75 mm thick within which inert aggregates are filled.

5.4 *Gas collection:* The biogas is collected by plastic pipes which are located around of the lagoon at the top of the lagoon dykes. The pipes can be HDPE of double thickness wall, 6″ in diameter. The pipes are perforated along the circumference but not in the section that conveys the biogas from the lagoons to the flaring system. The perforations are done by Drilling holes ½″ in diameter 20 cm apart.

5.5 *The biogas transport and fleuring systems:* A review of technological alternatives of biogas burning identified two types of equipment:

- Burning in elevated torches
- Burning in incinerators (thermal or catalytic)

Torch systems provide a means for the disposal of the gaseous waste streams with an almost unlimited range of flows and pressure drops. There are at least three fundamental aspects which are taken into account when deciding how to burn a residual gas. These are: (1) the variability of flow, (2) the maximum expected volume and (3) the heat content of the gas. High flow variability is the most important factor. A torch is designed to operate under virtually an infinite range of flows ranging from low to high flows. Alternative systems of disposal of waste gases, such as incinerators or other combustion devices need proper monitoring of the flow and can be used only in cases of reasonably continuous gas flow. The components of a fleuring system which must be considered are:

- Suction pipes from the covered lagoons to the burning site
- Liquid or hydraulic seal
- Suction spool
- Vacuum suction pump
- Lung tube
- Gas washer (when necessary)
- Flow meter
- Flame trap
- Gas burner

Figure 5.13 shows the components of the biogas burning system of one of the lagoons treatment plants of Santa Cruz, Bolivia.

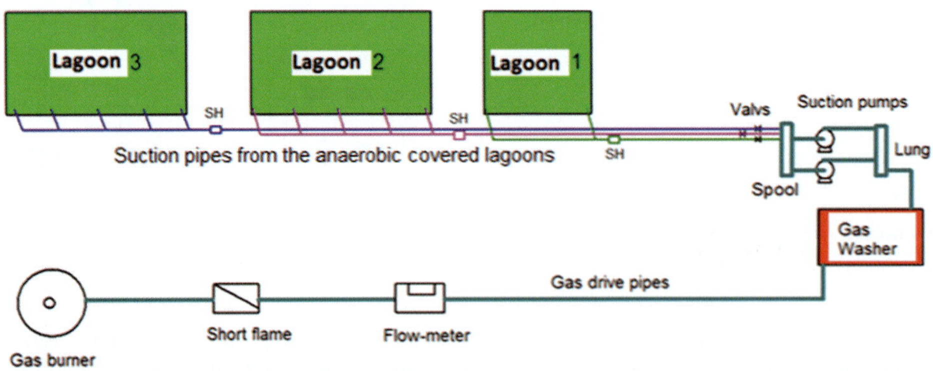

Figure 5.13 The biogas burnings system of one of the lagoons treatment plant in Santa Cruz, Bolivia (*Source:* SAGUAPAC, 2007)

(6) *Facultative Lagoon:* There are a number of methods that can be used for designing facultative lagoons. In this chapter, a combination of the methods is presented, for better application of the phenomena involved. The method includes the use of mixers in the facultative lagoons.

6.1 *Calculation of the detention time,* t_d*:* The surface organic load can be calculated from Equation 5.11:

$$L_s = 10(h/t_d)DBO_5 = 10(h/t_d)DBO_u/1.5$$

Another convenient way to calculate the surface organic load is by using Equation 5.14 (the Mara's approximation method):

$$L_s = 350(1.107 - 0.002T)^{(T-25)}$$

T is the average temperature of the coldest month of the year. Combining Equations 5.11 and 5.14 yields a new form of obtaining the ratio h/t_d:

$$h/t_d = (35/DBO_5)(1{,}107 - 0.002T)^{(T-25)} \quad (5.20)$$

Note that Equation 5.20 allows calculating the detention time (t_d) from the value of L_s, for a given temperature (based on Mara's method).

The value h is proposed as a variable of the designer. Using Table 5.1 and Equations 5.11 or 5.20 we calculate the t_d. *If we install mixers in the facultative lagoon (for example Gurney type mixers), we can adopt an organic surface load* (L_s) *of 400 kgBOD*$_5$*/ha · d at 20°C. The proposed minimum detention time* (t_{dmin}) *for a facultative lagoon is 4 days.* The larger of the two detention time values is selected as the design value. In other words, if the value calculated by Equation 5.20 is higher than 5 days, it becomes the selected value. If the value calculated by Equation 5.20 is lower than 5 days, than 5 days is the selected value.

6.2 *Calculation of the volume* (V_F) *and the efficiency of the lagoon:* With the t_d calculated as in the previous section, we can calculate the volume of the lagoon (V_F) and the surface area (A_s) as follows:

$$V_F = Q_D \cdot t_d \quad (5.21)$$

$$A_s = V_F/h \quad (5.22)$$

To calculate the efficiency we use Equation 5.4 from the kinetic method for a series of lagoons:

$$S = \frac{S_{0_A}}{\left(1 + K_L \frac{t_d}{n}\right)^n}$$

where S_{0_A} is the BOD concentration in the effluent of the Anaerobic Lagoon and n is the number of facultative lagoons in series. The final water temperature in the lagoon is obtained from Equation 5.8. If the value of K_L is not known from pilot plant testing, we can use the typical value of K_L (20°C) = 0.35 d^{-1}, corrected for temperature using Equation 5.7. These values are also given in Figure 5.3.

To improve the performance of a facultative lagoon, mixers can be installed in it. The required power level is 0.75 HP per 0.65–3 hectares, depending on the surface organic loading on the lagoon. Higher loaded lagoons require more mixing power. It is proposed to use a value of 0.75 HP for every 1 ha in facultative lagoons and 0.75 HP for every 3 ha in maturation lagoons. The installation of mixers in facultative lagoons allows using greater lagoons depths (of up to 5 m). Figure 5.14 shows a Gurney mixer before its installation in a lagoon.

Figure 5.14 Photo of a mixer before its installation in a lagoon

6.3 *Calculation of accumulation of excess sludge in facultative lagoons*[2]: The sludge produced by bioconversion is calculated by the following equation (Orozco, 2005):

$$X = \frac{\theta_c Y(S_0 - S)}{t_d(1 + k_e\theta_c)} = \frac{Y(S_0 - S)}{(1 + k_e t_d)} \tag{5.23}$$

where:

X = biomass concentration in the lagoon, g/m^3
θ_c = sludge age, d. In this case $\theta_c = t_d$
Y = production constant, 0.5
k_e = endogenous respiration constant, 0.05 d^{-1}

[2]Based on Orozco (2005), pages 253–258.

Taking into account the sludge accumulating in the lagoon is build-up of endogenous sludge (dX_e/dt), it can be calculated as follows (Orozco, 2005):

$$dX_e/dt = 0.1\ k_e\,X$$

And the rate of sludge accumulation (v_F, kg/year) is:

$$v_F = V_F(dX_e/dt) = V_F(0{,}1\,K_e)\frac{Y(S_0 - S)}{(1 + k_e t_d)}\frac{365d}{year}\frac{kg}{1000g}, \text{ or:}$$

$$v_F = 0.00091 V_F \frac{(S_0 - S)}{(1 + 0.05\,t_d)} \qquad (5.24)$$

V_F being the volume of the lagoon, $Y = 0.5$ and $k_e = 0.05$.

A facultative lagoon also removes Fecal Coliform and helminths ova. The removal levels of these organisms can be calculated using Equations 5.25 (see below) and 5.15.

(7) *Maturation lagoons*: The purpose of maturation lagoons is to remove pathogenic microorganism, therefore they are constructed as a series of lagoons which is a configuration that achieves higher removal efficiencies. The removal of fecal coliforms (FC) in a series of maturation lagoons can be estimated using Equation 5.4, applied to FC:

$$B = \frac{B_0}{\left(1 + K_B \frac{t_d}{n}\right)^n} \qquad (5.25)$$

With B is the FC concentration. The constant K_B is temperature dependent. According to Arthur (1983) the variation of K_B with temperature is given by:

$$K_B = 2.7 \times 1.19^{(T-20)} \qquad (5.26)$$

Also, the removal of helminth eggs is estimated by Equation 5.15:

$$E(\%) = 100(1 - 0.41 \exp[-0.49\,t_d + 0.0085\,t_d^2]).$$

The minimum detention time for maturation or polishing lagoons is three (3) days and the depth of such lagoons is between 1 and 1.5 m (See Peña-Baron, 2005). However, if mixers are installed in maturation lagoons, they can be deeper.

The design method used is to target to a percentage removal of helminths and find the t_d required to achieve such removal using Equation 5.15. Then, this detention time is used to calculate the removal of FC by Equations 5.25 and 5.26, and if resulting FC removal is satisfactory, the design is adopted. If the resulting FC removal is too low, the helminths removal efficiency is increased and the calculations repeated successively until the two required removal criteria (of helminthes and FC) are met.

(8) *Inlet, outlet and interconnection structures, and construction issues:* Typical inlet structures to lagoons, interconnection between lagoons and typical outflow structures from lagoons proposed by Gloyna (1971) and others are presented in Figures 5.15, 5.16, 5.17, 5.18 and 5.19.

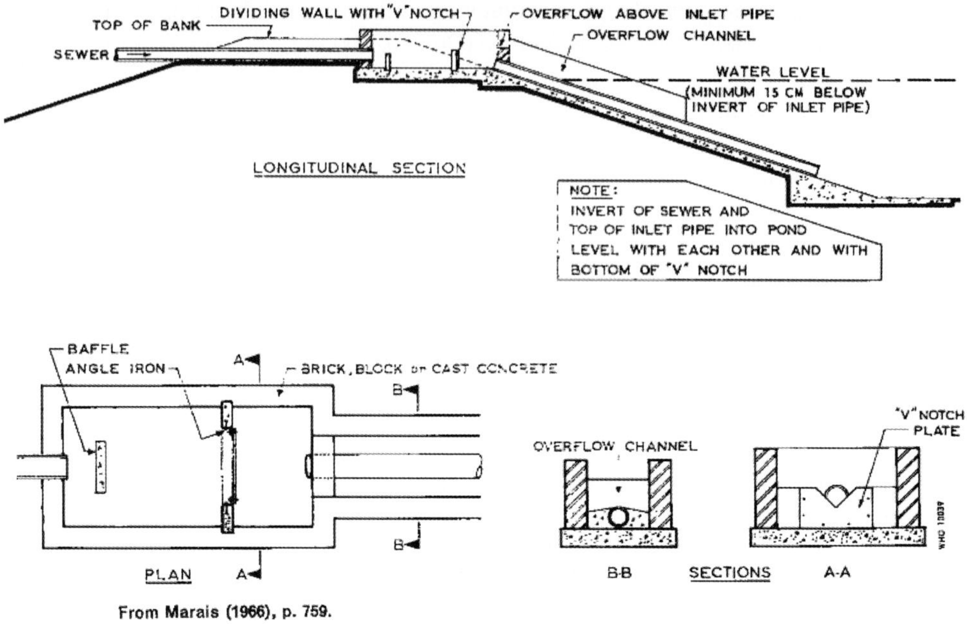

Figure 5.15 Typical inlets structure to lagoons

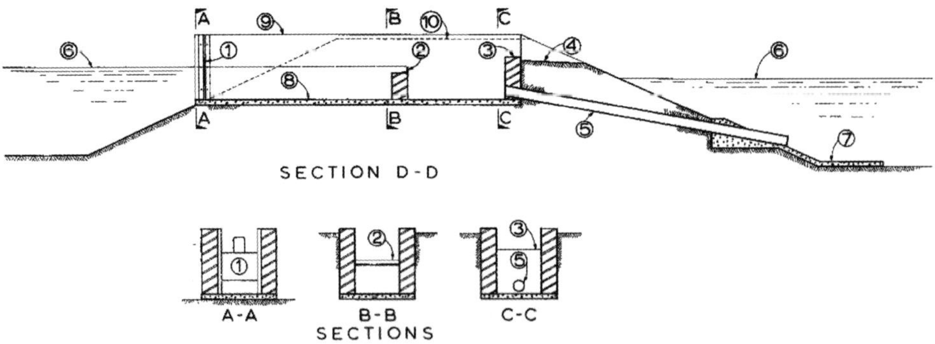

Figure 5.16 Typical interconnection structure between lagoons

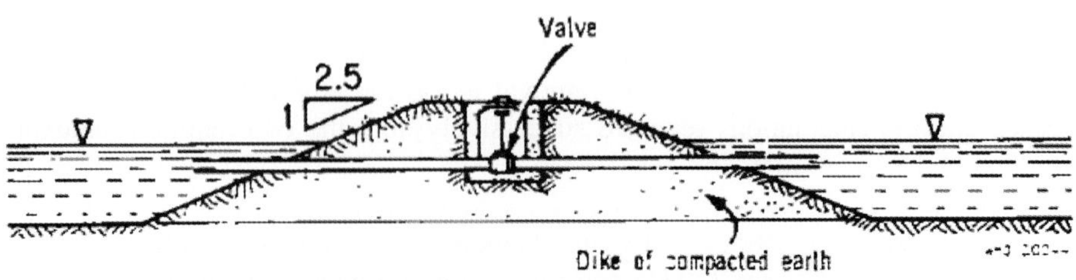

Figure 5.17 Another typical interconnection structure between lagoons

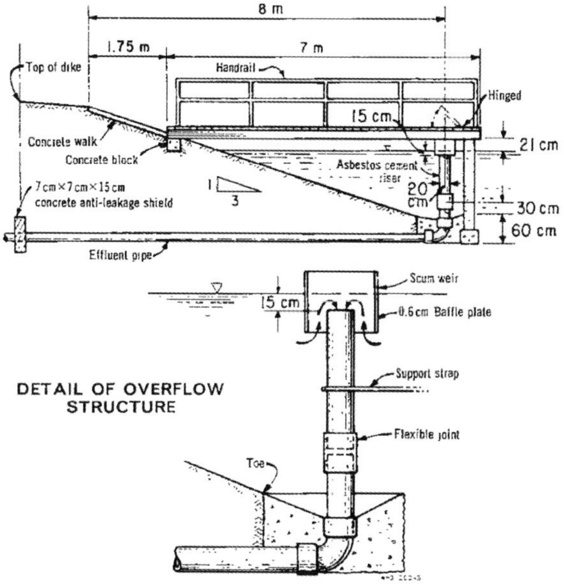

Figure 5.18 Typical outlet structure from lagoons

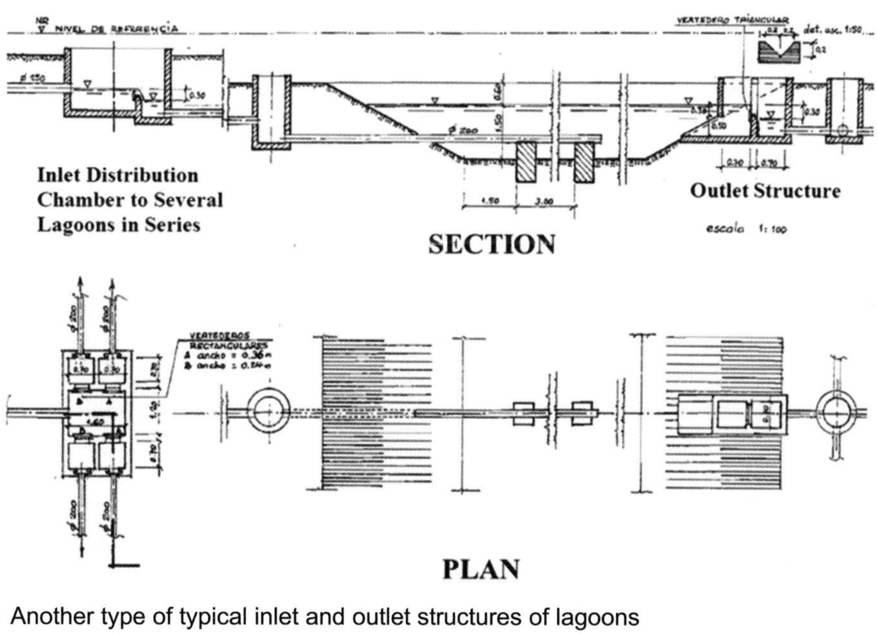

Figure 5.19 Another type of typical inlet and outlet structures of lagoons

A photo of a lagoon outlet structure of the type presented in Figure 5.18 is shown in Figure 5.20. This figure also portrays the effluent of a lagoons system consisting of a series of anaerobic-facultative-maturation lagoons.

Figure 5.20 A photo of a lagoon outlet structure of the type presented in Figure 5.18

The lagoons treatment plant process design is the basis for the physical design of the treatment plant. The physical design is critical for the success of lagoons treatment plant project. Many aspects need to be taken into account when preparing the physical design and most of them are not discussed here. Special attention should be directed to geotechnical aspects of a lagoons plant design. Detailed geotechnical studies need to be carried out prior to the physical design so as to ensure the correct design of embankments and to determine whether the soil in the proposed plant site is sufficiently impermeable or needs to be lined to prevent infiltration from the bottom of the lagoons. A variety of lining materials can be used for lining the lagoons when lining is required, including plastic geomembranes or a 300 mm thick liner. Selection of the impermeabilization method depends on local availability of materials and their cost.

Embankment slopes are usually 1 to 3 on the internal side of the lagoon and 1.5 to 2 on the external side. Slope stability should be ensured in accordance with standard soil mechanics methods for small earth dams. External embankments should be protected from stormwater erosion by providing adequate drainage. Internal embankments need to be protected against erosion by wave action. This can be achieved by lining with geomembranes, covering the embankments with lean concrete, with precast concrete slabs, or with stone rip-rap at the top water level. In most cases embankments are constructed from the soil excavated at the lagoons site and there should be a balance between the cut and fill, so as to reduce construction costs. The actual ratio between cut and fill depends on local civil work costs, that is on the cost of excavation and of embankments construction. Whenever possible, embankment design should allow for vehicular access to facilitate maintenance.

The most common shape of lagoons is rectangular with varying ratios of length to width, however, lagoons do not need to be strictly rectangular, and they can have other shapes. One of the main problems of lagoons is hydraulic short-circuiting which occurs because they are large water bodies. Short-circuiting reduces the effective volume of the biological reactor and results in reducing its treatment efficiency. The optimal geometry, which should also take into account the relative position of the inlet and outlet structure, is the geometry which minimized hydraulic short-circuiting. Installing mixers in the lagoons resolves the problem of short-circuiting (in addition to also providing other benefits as discussed in previous sections). A single inlet and outlet are usually sufficient but they have to be located in diagonally opposite corners of the lagoon. Baffles are used sometimes to reduce short-circuiting but they should be used with caution. Care must be taken to avoid a too high BOD loading in inlet zones, which can lead to generation of odours. Baffles are sometimes used in maturation ponds in an attempt to induce a plug flow in such ponds. That approach can cause the problem of a high BOD loading in inlet zone. A better way to overcome all the short-circuiting problem is the use of mixers.

The areas of the various types of lagoons calculated by the process design procedures outlined in this chapter refer to mid-depth areas so the dimensions calculated from them are mid-depth dimensions. These need to be corrected taking into account the slope of the embankment, to calculate the dimensions of the lagoon at the bottom and at the top water level (TWL), as shown in Figure 5.21. The dimensions and levels that the contractor (hired for constructing the lagoons plant) needs to know are those of the base and the top of the embankment, the latter including the effect of the freeboard. On the basis of these data he can calculate the volume of the excavation. The volume of a lagoon can be calculated using the following equation (see Crites *et al.* 2006)

$$V = [(LW) + (L - 2nD)(W - 2nD) + 4(L - nD)(W - nD)](D/6)$$

where:

V = the lagoon volume, m^3
L = Lagoon length at TWL, m
W = Lagoon width at TWL, m
n = horizontal slope factor (a slope of 1 in n)
D = the lagoon liquid depth, m

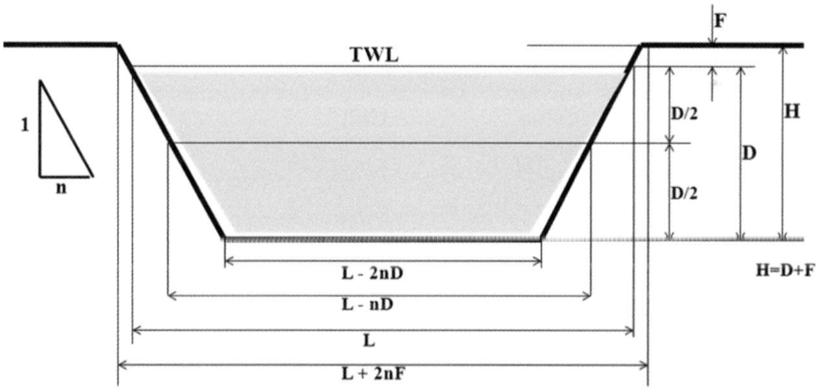

Figure 5.21 Relations between top, bottom and mid-depth dimension of a lagoon

The minimum freeboard is decided on the basis of preventing wind-induced waves from overtopping the embankments. For small ponds with areas of less than 1 ha, a 0.5 m freeboard should be provided. For ponds of up to 3 ha, the freeboard should be 0.5–1 m. For larger ponds the freeboard can be calculated from the following equation (Mara, 2004):

$$F = (\log_{10} A)^{1/2} - 1$$

where: F is the freeboard in meters and A is the lagoon area at TWL, m^2.

Lagoons liquid depths are usually in the following ranges:

- Anaerobic lagoons 2–5 m

- Facultative ponds 1–2 m
- Maturation ponds 1–1.5 m

However, if mixers are installed in facultative and maturation lagoons their depth can be larger, 2.5–4 meters.

5.3 BASIC DESIGN EXAMPLE

(The model program for this Example is available online at http://www.iwawaterwiki.org/xwiki/bin/view/Articles/Software+Developed+for+Sustainable+Treatment+and+Reuse+of+Municipal+Wastewater)

In continuation we present an example that follows the Orderly Design Method (ODM) of paragraph 5.2.2. The example is developed for a population of 20,000, as presented in Chapter 3. The ODM is developed step by step and is calculated by a program developed in Excel, which can be downloaded from the page shown elsewhere. One way to solve the sample design is presented by the ODM, but the reader should be aware that there might be several different methods to achieve a good design. If properly understood, the model could be changed when the *a priori* (external) conditions demand a different approach, leaving a proper back up of the original program, called CHAP 5-Lagoons.xls. Please also read the *General instructions for running de Excel sheet program with the ODM* are presented in Section 4.3 of Chapter 4. The following Tables identified by alphabetic order contain computer calculated results.

(1) *Calculation of external variables*

1.1 *Design flow:* The inputs required for flow calculations are presented in Table 5A:

Table 5A

Design flow				
Variable	**Value**	**Unit**	**Value**	**Unit**
Per capita consumption, q	162.5	L/capita · d		
$PF = 1 + 14/(4 + \sqrt{P})$	2.65			Hamon Equation
Population, P	20000	inhab		
Population density, d	260	inhab/ha		
Return coefficient	0.8			
DWW mean flow, Q_{DWW}	30.09	L/s	108.33	m^3/h
DWW maximum flow, Q_{maxDWW}	79.82	L/s	287.35	m^3/h
Infiltration unit flow, q_I	0.13	L/s · ha (o km)		
Afferent area, A_a	76.92	ha		
Infiltration flow, Q_I	10.00	L/s		

Notes: (i) Per capita consumption is calculated for drinking water consumption. With the return factor of 0.8, the wastewater per capita flow of $c \cdot q_{dom} = 0.8 \times 162.5 = 130$ l/Capita · d, as measured in the example in Chapter 3, (ii) the Peak Factor is calculated by Equation 3.6 (*Harmon* Equation), (iii) the population density should be the maximum acceptable density according to planning standards, in order to be able to serve the population at saturation conditions, (iv) the flows are calculated based on the population and the and the return coefficient – c, and (v) infiltration must be measured or alternatively a value of between 0.10 and 0.15 l/s · ha or l/s · km is to be chosen.

The final design flows are obtained from Equations 3.3 and 3.4, and are presented in Table 5B in different units as an *output*. The average flow Q_D is used for the biological process design and the maximum flow Q_{HD} for hydraulic design.

Table 5B

Design flow						
Variable	**Value**	**Unit**	**Value**	**Unit**	**Value**	**Unit**
Process design flow, $Q_D = Q_{DWW} + Q_I$	40.09	L/s	144.33	m^3/h	3464.00	m^3/d
Hydraulics design flow, $Q_{HD} = Q_{DWWmax} + Q_I$	89.82	L/s	323.35	m^3/h	7760.44	m^3/d

In the case of preliminary treatment the design flows for the coarse screen, the grit chamber and the *Parshall* flume is the maximum flow Q_{HD} since they are hydraulic units. On the other hand, the lagoons are designed on the basis of the average flow Q_D since they are biological process with detention times of days, during which all variations are averaged.

1.2 *Determination of the geographical and external environmental variables:* These variables depend solely on the geographic location. The values used in the example are presented in Table 5C.

Table 5C

Variable	Value	Unit	Value	Unit
Minimum Temperature of water, T_w	25	°C	77.0	°F
Minimum Temperature of air, T_a	18	°C	64.4	°F
Altitude above sea level, hasl	350	masl		
Mean wind velocity	6	kph		
Predominant wind direction	NW			
Minimum solar radiation, **S**	230	$Cal/cm^2 \cdot d$		
Daily fraction of sunlight hours	0.8	decimal		

Notes: (i) for the design of the Preliminary Treatment process these variables do not have a significant importance, for the design of the biological processes they are important, (ii) the minimum water temperature as well as the minimum air temperature are the average of the coldest month, (iii) the speed and direction of the predominant wind is important for locating the wastewater treatment plant in relation to the location of the town or the urban center, (iv) the fraction of hours of sunlight is the ratio of actual hours of sunlight to possible hours of sunlight.

1.3 *Concentration and pollution loads:* The quality of the wastewater may be different in the urban areas and at the entrance to the treatment plant if the conveyance system is long and receives an important infiltration inflow. In the example, the wastewater treatment plant is located close to the town and there is no significant variation of the wastewater composition during its conveyance to the plant.

The following table presents as *outputs* the calculated *loads* (in kg/d) and the *Specific unit loads* (in kg/Capita · d) of each contaminant, calculated from the *concentrations* given as *input*. This is calculated for BOD_5, COD, TSS, and so on. In cases of significant infiltration due to long conveyance to the wastewater treatment plant, it is necessary to calculate the additional infiltration flow from the conveyance influence area (A_a) as follows: $Q_{Iemi} = q_I \cdot L$, where L is the length of the conveyance pipeline in km. The concentration of a contaminant at the influent to the wastewater treatment plant is calculated as its specific unit load (q_C) multiplied by the population (P) and divided by the total flow: $Q_T = (Q_D + Q_{Iemi})$. For example: the concentration of COD is $COD = q_{COD} \cdot P/Q_T$. Note that in Table 5D the load the load and the per capita load (specific unit load) of each contaminant are *outputs* while only the concentrations are an *inputs*, presented in the Excel sheet separately.

Table 5D

DWW Quality				
Variable	**Value**	**Unit**	**Value**	**Unit**
BOD_5	277.8	mg/L	0.28	kg/m^3
COD	596.1	mg/L	0.60	kg/m^3
COD/BOD_5	2.1			
TKN	40.0	mg/L	0.04	kg/m^3
N-Nitrate	2.0	mg/L	0.00	kg/m^3
Total Phosphorus	5.8	mg/L	0.01	kg/m^3
pH	7.1	UN		
Alkalinity	100.0	mg/L	0.10	kg/m^3
TSS	202.6	mg/L	0.20	kg/m^3
VSS	173.6	mg/L	0.17	kg/m^3
O&G	100.0	mg/L	0.10	kg/m^3
Coli-Fecal	10000000	NMP/100 mL	100000000000.00	NMP/m^3

(2) *Determining the flow diagram for the lagoon system:* A typical flow diagram of a lagoons system is presented in Figure 5.3 (see Arthur, 1983). As shown, it consists of preliminary treatment (coarse screening → grit chamber → Parshall flume) followed by three units of lagoons in series (Anaerobic Lagoon → Facultative Lagoon with Gurney type mixers → Maturation Lagoon).

This is the most logical configuration. Other configurations can also be used, for example several facultative lagoons in series (instead of one) and several maturation lagoons in series (instead of one). Large plants are usually constructed in a modular fashion, that is, several parallel modules are constructed in the first phase and additional ones in the second phase. When several modules are constructed in one phase the flow diagram may be modified, for example, constructing

4 anaerobic lagoons which distribute their effluent to 2 trains of facultative followed by maturation lagoons. The only governing guideline is to set up a flow diagram which makes sense.

(3) *Sizing the pretreatment system*

3.1 *Coarse screening*

Variables selection by the designer: The coarse screening system can be a rotating screen. If a target of 3% solids removal efficiency (coarse screening has very low efficiency to remove TSS) is selected, than according to Figure 4.8 a rotating screen with an opening of 6 mm will be adequate. The design flow is the maximum flow Q_{HD} since screening is a hydraulic unit. The results that the program calculates automatically are presented in Table 5E. They include in addition to the solids removal efficiency also the water velocity in the channel and the backwashing pressure. The module CHAP 4-RM.xls of the excel program is to be used for these calculations.

Table 5E

Designer variables				
Variable	**Value**	**Unit**	**Value**	**Unit**
Eficiency TSS	3	%		Figure 4.8
Screen opening	5.54	mm		Equation 4.1b
Screen opening (rounded)	6.00	mm	0.236	inch
Flow velocity in the channel	0.6	m/s		0.6 to 1.0 m/s
Backwashing pressure	350	kPa		Table 4.1
Solids removed	60	L/1000 m^3		Table 4.1

Design: A coarse rotating screen of 600 mm in diameter is obtained from the calculations. The final output data are presented in the table below. The design of the coarse rotating screen is similar to that of the rotating micro screen so the reader is referred to Chapter 4 in which the design procedure is explained in detail. Only course screening is used in the preliminary treatment unit because further processing of the fine material that will be contained in the effluent of this unit will be processed in the series of lagoons into which the preliminary effluent will be discharged. For the purpose of clarification, the preliminary treatment unit calculations which are produced by the Excel program are presented in Table 5F. The design support is presented in the column entitled OBSERVATION.

Table 5F

Design			
Parameter	**Value**	**Unit**	**Observation**
Flow, Q_{HD}	90.000	L/s	
Screen opening, a	6.000	mm	

(*Continued*)

Table 5F (*Continued*).

Design			
Parameter	**Value**	**Unit**	**Observation**
Drum diamater, D_T	0.600	m	Figure 4.7
Channel width, $W = 1{,}1\ D_T$	0.660	m	
Channel depth, $h = Q_{DH}/W \cdot v$	0.230	m	Equation 4.2
Hydralics radius, $R = Wh/(W + 2h)$	0.136	m	Equation 4.3
Manning's n	0.012		Equation 4.3
Hydrulics slope, $s = (v \cdot n)^2/R^{4/3}$	0.074	%	Equation 4.3
Head loss, $h_f = (1/C \cdot 2g)(Q_{DH}/A)^2$	0.030	m	Equation 4.1
Total head loss, $h_f + 0.05$	0.080		Plus 5 cm
Backwashing flow	1.800	L/s	Table 4.1
Solids removed	466.560	L/d	Table 4.1

3.2 *Grit chamber*

Variables selection by the designer: According to the ODM in Section 5.2.2, part 3 on design of preliminary treatment systems, we propose the design variables presented in Table 5G.

Table 5G

Grit channel					
Variable	**Valuc**	**Unit**	**Value**	**Unit**	**Observation**
$SOR = Q_{DH}/A_{des}$	800.0	m/d			600–1200 m/s
$v = Q_{DH}/W \cdot H_{des}$	0.2	m/s			0.15–0.60 m/s
# Grit channels	2.00				
L/w	4				

Design: Applying these variables to the equations presented in Section 5.2.2 Paragraph 3.2 which are shown in the left column of Table 5H, we obtain the results presented in Table 5I. Note that there are two grit chambers.

3.3 *Parshall flume channel:* For a $Q_{HD} = 90$ lps follows the throat width for measuring the flow is $W = 9''$. Rapid mixing is not required because there is no injection of chemicals.

(4) *Anaerobic lagoon*

Variables selection by the designer: Using the methodology proposed in the ODM, Section 5.2.2, Paragraph 4, and with the guidance of Table 5.6 we propose the input variables presented in Table 5I.

We proposed a n_A of four parallel lagoons and a sludge accumulation rate of 0.05 m^3/Capita · yr. The L_V is automatically calculated by the Excel program CHAP 5-Lagoons-01, according to Table 5.6.

Table 5H

Grit channel			
Parameter	**Value**	**Unit**	**Observation**
Area, $A_{des} = Q_{HD}/SOR$	9.7	m^2	Section 5.2.2 Paragraph 3.1
Width, $w = \sqrt{(A_{des}/4)}$	1.6	m	Section 5.2.2 Paragraph 3.1
Length, $L = 4w$	6.2	m	Section 5.2.2 Paragraph 3.1
Depth, $H_{des} = Q_{DH/}v \cdot W$	0.3	m	Section 5.2.2 Paragraph 3.1
Total Depth, $H = H_{des} + 0.50$	0.8	m	Section 5.2.2 Paragraph 3.1
Parshall flume			
Throat, W	9.0	inch	See "PARSHALL" sheet

Table 5I

Designer variables					
Variable	**Value**	**Unit**	**Value**	**Unit**	**Observation**
Anaerobic lagoon (A)					
Depth, H_A	4.0	m			3–5 m
T_{min} air	18.0	°C			Coldest month
Sludge accumulation, v_L	0.04	m^3/capita · yr			Range 0.04–0.08
# Lagoons A, n_A	4				Parallel
L/w (ratio length to width)	1				Between 1 and 3
Velocity in inlet pipe, v	1	m/s			Range 0.6–2.0 m/s
Volumetric load, L_V	260	$g/m^3 \cdot d$			Table 5.5

Design: From the input date we obtain the results presented in Table 5J.
Following are clarifications to the calculations of the figures in Table 5J:

- $V_A = Q_D \cdot S_0/1000\,L_V = Q_s/L_V = 3701.2\ m^3$: preliminary volume of the lagoon calculated using Equation 5.17.
- $t_d = V_A/Q_D = 1.1$ d, which is larger than the minimum recommended detention time $t_{dmin} = 1$ d, so we select the larger of the two numbers as final detention time, which in this case is $t_d = 1.1$ d.
- $V_F = t_d \cdot Q_D = 3701.2\ m^3$: the final selected volume of the lagoon.
- $A_s = V_F/H_A = 3701.2/4 = 925.3\ m^2$: total surface area of the lagoons at the surface level of the lagoon.
- $a_s = A_s/n_A = 231{,}3\ m^2$: area of each of the 4 (n_A) anaerobic lagoons.
- The total width and length of the squared lagoons ate $w = \sqrt{(a_s)} = \sqrt{(231.3)} = 15.2$ m. There are other possible geometric forms (for instance with a $L/w = 2$). Then the calculation of the dimensions would change.

- The time required to fill half of the lagoon with accumulated sludge is n = 2.3 years, calculated according to Equation 5.18.
- The volume of biogas produced is calculated by Equation 5.19 and is 288.7 m^3/d.
- The BOD_5 removal efficiency and effluent BOD_5 concentration are automatically calculated by the program CHAP 5-Lagoons-01, according to Table 5.6.
- For input and output structures use an average liquid velocity of 1 m/s, (this value can be different but preferably is in the range 0.6 and 2.0 m/s) which results in a diameter of $\varphi = 13''$ and $4''$ for inlet pipe conveying the total wastewater flow and the inlet pipe to each lagoon respectively. Note that the design of the inflow pipe is based on the maximum flow Q_{HD} since there are no flow equalization structures prior to entering the plant but the outlet flow is Q_D because the flow is equalized in the lagoon. Odours become a problem if $L_V > 400\ g/m^3 \cdot d$ or if there is excess sulphate in the influent. To prevent odours and to control emission of greenhouse gases it is recommended a cover the anaerobic lagoons, and burn the collected biogas in a gas burning system of a capacity that includes a safety factor of two (2) times the biogas flow calculated by Equation 5.19: $V_{gD} = 2\ V_g = 577.5\ m^3$. See Paragraph 5 of Section 5.2.2 for details of cover design.

Table 5J

	Design				
Parameter	**Value**	**Unit**	**Value**	**Unit**	**Observation**
Anaerobic lagoon					
Volume, $V_A = (Q_D \cdot S_0)/1000 \cdot L_V$	3701.2	m^3			Equation 5.17
Detention time, $t_d = V_A/Q_D$	1.1	d			
Final Volume, $V_F = Q_D \cdot t_d$	3701.2	m^3			Larger of t_d and $t_{dmin} = 1\,d$
Total Area, $A_S = V_F/H_A$	925.3	m^2			
Area per lagoon $a_s = (A_s/n_A)$	231.3	m^2			
Lagoon Width, $w = [a_s/(L/w)]^{0,5}$	15.2	m			
Lagoon Length, $L = (L/w)w$	15.2	m			
Sludge filling time, $n = ½\ V_F/V_L \cdot P$	2.3	years			Equation 5.18 (half filled)
Gas production					
Volume, $V_g = 0.30\ S_0 \cdot Q_D/1000$	288.7	m^3/d			Equation 5.19
Efficiency of anaerobic lagoon					
Efficiency, % BOD_5	56	%			Table 5.6
Effluent BOD_e, S_A	122	mg/L			
Inlet and outlet structures					
Diameter of inlet pipe $\varphi = (4 \cdot PI \cdot v \cdot Q_{DH}/1000)^{0.5}$	0.338	m	13	inch	Calculated with Q_D
Diameter of pipe in Lagoon $\varphi_L = \varphi = [4 \cdot PI \cdot v \cdot (Q_D/4)/100]$	0.113	m	4	inch	Calculated with $Q_D/4$

(5) *Facultative lagoon*

Variables selection by the designer: Use the methodology proposed in the ODM, Section 5.2.2, Paragraph 6, and with the guidance in Table 5.1, as shown below, propose the designer variables.

The issues that require consideration are: (i) are we choosing to install mixers in the facultative lagoons? If we do, the design needs to take that fact into account, (ii) selecting the number of facultative lagoons (n_F) in series, (iii) adopting the value of K_L (20°C) = 0.35 d^{-1}, (iv) selecting the ration of Length to width of the lagoon, and (v) taking into account that the influent of the facultative lagoon is the effluent of the anaerobic lagoon. Other input data are based on external conditions such as air and water temperatures, altitude, sun radiation, and so on.

In the design example we chose facultative lagoons with mixers, bearing in mind that under such conditions we can choose a larger than usual lagoon depth. We propose to use the input variables presented in Table 5K.

Table 5K

Designer variables					
Variable	**Value**	**Unit**	**Value**	**Unit**	**Observation**
Facultative lagoon					
¿ Mixer installed?	1				YES:1; NOT:0
Depth, H_F	3	m			2–5 m
T_{min} air	18.0	°C			Coldest month
T_i water	25.0				
# Lagoons F, n_F	2				Parallel
L/w (ratio length to width)	2				Between 1 and 3
Velocity in inlet pipe, v	1	m/s			Range 0.6–2.0 m/s
$L_s = 350\ (1.107–0.002T)^{(T-25)}$ or 400	400.0	$kgBOD_5/ha \cdot d$			Equation 5.14 or 400
K_L (20°C)	0.35	d^{-1}			Typical
θ (BOD_5)	1.06				Range 1.04–1.09
Influent BOD_{5i}, $S_{OF} = S_A$	122.2	mg/L			
K_B (20°C)	2.7	d^{-1}			Typical for FC
θ (CF)	1.19				Typical for FC
Influent FC: B_0	10000000	NMP/100 mL			
Influent Helminthes	500	Ova/L			Between 300–800

Design: From input data, the design values are obtained by following the ODM, the results obtained are presented in Table 5L.

- The equation number used for each calculation is given in the right column, in accordance with the procedures of Section 5.2.2, Paragraph 6. The equations are named in the right column.

Table 5L

Design					
Parameter	**Value**	**Unit**	**Value**	**Unit**	**Observation**
Facultative lagoon					
L_s	400.0	$kgBOD_5/ha \cdot d$			Selected in input
$t_d = 10 \cdot H_F BOD_5/L_S$	9.2	d			Equation 5.11
Volume, $V'_F = Q \cdot T_D$	31755.9	m^3			Equation 5.21
Final Volume V_F	31755.9	m^3			Greater of t_d and $t_{dmin} = 4$ d
Area, $A_s = V_F/h$	10585.3	m^2			Equation 5.20
Lagoon Area $a_s = (A_s/n_F)$	5292.6	m^2			
Lagoon Width, $w = [a_s/(L/w)]^{0.5}$	51.4	m			
Lagoon Length, $L = (L/w)w$	102.9				
# of mixers LAS $= A_s/(3 \times 10000)$	2.0				0.75 hp/ha
$T_L = (A_s f T_a + Q T_i)/(A_s f + Q)$	18.0	°C			Equation 5.8
$K_L = K_L(20°C)\ \theta^{(T\text{-}20)}$	0.31	d^{-1}			Equation 5.7
Effluent BOD_5 $S_F = S_0/(1 + K_L \cdot t_d)$: lagoons in parallel	31.70	mg/L			Equation 5.4, parallel
Effluent BOD_5 $S_F = S_0/(1 + K_L \cdot t_d/n)^n$: lagoons in series	20.74	mg/L			Equation 5.5, series
$v_F = 0.00091 V_F(S_0 - S)/(1 + 0.05 t_d)$	1793.86	kg/yr	1793.86	m^3/yr	Equation 5.24
Filling years $= 0.5\ V_F/v_F$	8.9	years			½ Volume
Facultative lagoon: Fecal coliform and helminthes removal and efficiencies					
$K_B = K_B(20°C)\ \theta^{(T\text{-}20)}$	6.4	d^{-1}			Equation 5.26
Effluent FC: $B = B_0/(1 + K_B \cdot t_d)$	166480.8	NMP/100 mL			Equation 5.25
Effluent helminthes	9.8	Ova/L			
Efficiency FC $= 100 \cdot (B_0 – B)/B_0$	98.3	%			From Equation 5.25
Efficiency Helminthes $= 100\ (1 - 0.41\ \text{Exp}\ [-0.49\ t_d + 0.0085\ t_d^2])$	98.0	%			Equation 5.15
Efficiency BOD_5 removal facultative	74.1	%			
Efficiency BOD_5 removal (Anaerobic + Facultative)	88.6	%			
Inlet and outlet structures					
Diameter in inlet pipe, $\varphi = (4 \cdot PI \cdot v \cdot Q_D/1000)^{0.5}$	0.226	m	9.00	in	With Q_D
Diameter of pipes to lagoons, $\varphi_L = \varphi = [4 \cdot PI \cdot v \cdot (Q_D/4)/1000]^{0.5}$	0.160	m	6.00	in	With $Q_D/4$

- Note that the substrate concentration to the influent of the first facultative lagoon is that of the effluent of the anaerobic lagoon and the substrate concentration in the influent to the second facultative lagoons is that of the effluent of the first facultative lagoon.
- The organic surface load (L_S, kg BOD_5/ha · d) is calculated from Equation 5.14 and the detention time t_d is calculated by solving Equation 5.11. Then t_d is chosen as the higher of the calculated detention time and the minimum acceptable of 4 days.
- Then we calculate the final volume (V_F), total surface area (A_S), the area of each facultative lagoon (a_s) and the dimensions of each facultative lagoon.
- Facultative lagoons can be operated in series or in parallel so the effluent BOD_5 is calculated for both cases: S (parallel) and S′ (in series).
- Then the removal efficiencies of the BOD_5 (of F and A + F), Fecal Coliforms and Helminthes eggs are calculated and also the concentrations in the effluent.
- The dimensions of the lagoon are calculated using in this case a ratio of $L/w = 2$.
- The diameters of the inflow pipelines to the facultative lagoons in series (calculated using Q_D) or in parallel (calculated using Q_D/number of lagoons) are presented in the last rows of the table.

(6) *Maturation lagoons*
Variables selection by the designer: First, define the desired targets of removal of Fecal Coliforms (FC) AND helminthes eggs. According to procedure outlined in the ODM of Section 5.2.2, Paragraph 6, it is also necessary to choose de lagoon depth (H_M) and the L/w ratio. The influent of the Maturation lagoon is the effluent of the Facultative lagoon. The design procedure is similar to that of the facultative lagoon but the objectives are different and in this case are: (i) Remove Fecal Coliforms and additional BOD_5, and (ii) remove helminth eggs, and in general, obtain the desired quality of the final effluent as expected from the complete lagoons system (and usually, as required by the prevailing effluent discharge standards). The variables selected by the designer are presented in Table 5M.

Table 5M

Designer variables					
Variable	**Value**	**Unit**	**Value**	**Unit**	**Observation**
Maturation lagoon					
Depth, H_M	1.5	m			2–5 m
T_{min} air	18.0	°C			Coldest month
T_i water	25.0				
# Lagoons M, n_M	2				Parallel or series
L/w (ratio length to width)	2				Between 1 and 3
Water velocity in pipe, v	1	m/s			0.6–2.0 m/s
$L_s = 350\,(1.107–0.002T)^{(T-25)}$ or 400	216.5	kgBOD_5/ha · d			Equation 5.14 or 400
K_L (20°C)	0.35	d^{-1}			Typical
θ (BOD_5)	1.06				1.04–1.09

(Continued)

Table 5M (*Continued*).

Designer variables					
Variable	**Value**	**Unit**	**Value**	**Unit**	**Observation**
Influent BOD_{5i}, $S_{OM} = S_F$	31.7	mg/L			
K_B (20°C)	2.7	d^{-1}			Typical FC
θ (CF)	1.19				Typical FC
Influent FC: B_{OM}	166481	MPN/100 mL			
Influent helminthes	10	Ova/L			Between 300–800
Target FC	10000	MPN/100mL			Between 1000 and 10000
Target helminthes eggs	1	Ova/L			<1 Ova/L

Design: The design outputs are calculated by the Excel program CHAP 5-Lagoons.xls and are presented in the following Table 5N.

Table 5N

Design					
Parameter	**Value**	**Unit**	**Value**	**Unit**	**Observation**
Maturation lagoon					
L_s	216.5	$kgBOD_5/ha \cdot d$			Selected in input
$t_d = 10 \cdot H_F BOD_5/L_s$	2.2	d			Equation 5.11; $t_{dmin} = 3$ d
Volume, $V'_F = Q \cdot t_d$	7607.1	m^3			Equation 5.21
Final Volume V_F	10392.0	m^3			Greater of t_d and $t_{dmin} = 3$ d
Area, $A_s = V_F/h$	6928.0	m^2			Equation 5.20
Lagoon area $a_s = (A_s/n_F)$	3464.0	m^2			
Lagoon width, $w = [a_s/(L/w)]^{0.5}$	41.6	m			
Lagoon length, $L = (L/w)$ w	83.2				
$T_L = (A_s f T_a + Q T_i)/(A_s f + Q)$	21.5	°C			Equation 5,8
$K_L = K_L(20°C)\ \theta^{(T-20)}$	0.38	d^{-1}			Equation 5.7
Effluent BOD_5 $S_M = S_0/(1 + K_L \cdot t_d)$	17.24	mg/L			Equation 5.4, Parallel
Effluent BOD_5 $S_M = S_0/(1 + K_L \cdot t_d/n)^n$	15.74	mg/L			Equation 5.5, series
$V_F = 0.00091 V_F(S_0 - S)/(1 + 0.05 t_d)$	123.23	kg/yr	123.23	m^3/yr	Equation 5.24

(*Continued*)

Table 5N (*Continued*).

Design					
Parameter	**Value**	**Unit**	**Value**	**Unit**	**Observation**
Filling years = 0.5 V_F/v_F	42.2	years			1/2 Volume
Maturation lagoon: Fecal coliform and helminthes removal and efficiencies					
$K_B = K_B(20°C)\ \theta^{(T-20)}$	6.4	d^{-1}			Equation 5.26
Effluent FC: $B = B_0/(1 + K_B \cdot t_d)$	10989.3	NMP/100 mL			Equation 5.25
Effluent Helminthes	1.7	Ova/L			
Efficiency FC = 100. $(B_0 - B)/B_0$	93.4	%			From Equation 5.25
Efficiency Helminthes = 100 (1 – 0.41 exp $[-0.49\ t_d + 0.0085\ t_d^2]$).	82.6	%			Equation 5.15
Efficiency BOD_5 removal Maturation	45.6	%			
Total efficiency BOD_5	**93.8**	**%**			**Lagoon system**
Total efficiency fecal coliform	**99.9**	**%**			**Lagoon system**
Total efficiency helminthes	**99.7**	**%**			**Lagoon system**
Inlet and outlet structures					
Diameter of inlet pipe, $\varphi = (4 \cdot PI \cdot v \cdot Q_D 1000)^{0,5}$	0.226	m	9.00	in	With Q_D
Diameter of pipes to lagoons, $\varphi_L = \varphi = [4 \cdot PI \cdot v \cdot (Q_D/4)/1000]^{0.5}$	0.160	m	6.00	in	With $Q_D/4$

- The removal efficiency of helminthes is calculated using Equation 5.15:

$$E(\%) = 100(1 - 0.41 \exp[-0.49\, t_d + 0.0085\, t_d^2]).$$

- Removal of Fecal Coliforms (FC) is calculated using Equations 5.4 and 5.5:

$$S = \frac{S_0}{\left(1 + K_L \frac{t_d}{n}\right)^n}$$

- If the target for helminthes removal is not reached, the design needs to be changed in order to increase detention time. The same goes for FC if the target effluent concentration is not reached. In this example the target were very close to be reached, so it was not necessary to do any modification. However, the calculations were made for lagoons operating in parallel. Changing the design to operate the maturation lagoons in series, would result in achieving both targets.
- The final detention time t_d, used to calculate the volumes, areas and dimensions of the lagoons, is the larger of the calculated detention time and the minimum proposed for maturation lagoons which is 3 days. Furthermore, maturation lagoons are more effective when functioning, but can also be parallel so that the effluent BOD_5 is calculated for both cases with the program: B (parallel) and B′ (in series).

- Ultrasound waves generating instruments for removal of algae can be installed at the exit section of the maturation lagoons. *LGSonic* instruments are an example of such devices. According to the manufacturer's recommendation one instrument should be installed per acre of lagoon area. A photo of the instrument is presented in Figure 5.22.
- The diameters of the inflow pipe for lagoons in series mode (calculated using Q_D) or in parallel mode (calculated using with Q_D/number of lagoons) are presented in the last rows of the table. The input and output structures are constructed in accordance with Paragraph 7 below.

Figure 5.22 Ultrasonic waves generating instrument (Manufactured by LGSONIC)

(7) *Inlet, outlet and interconnection structures:* In the Orderly Design Method, Section 5.2.2 Paragraph 8, different designs of inlet, outlet and interconnection structures are presented. Each tab of the CHAP 5-Lagoons-01program, "Anaerobic", "Facultative" and "Maturation", brings the diameters of the required inlet pipes. The slopes of the lagoons levee depend on soil and geological conditions which needs to be established through corresponding studies, but as a first approximation a slope with a gradient of 2:1 can be proposed (see Figures 5.14–5.19 and 5.23).

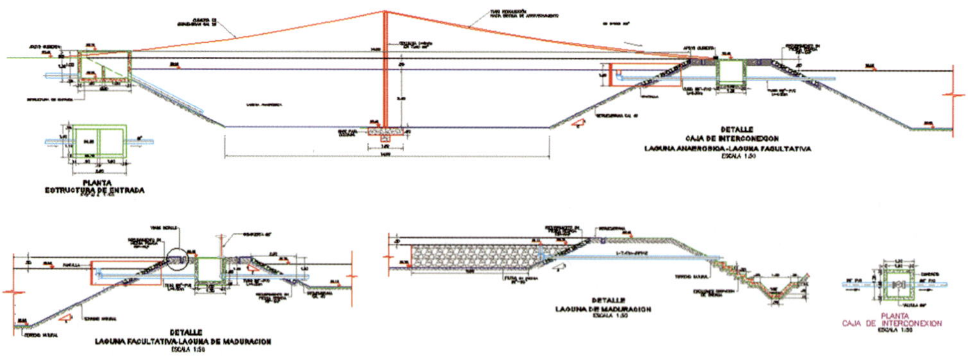

Figure 5.23 Cross section of a typical lagoons system

Chapter 6

Anaerobic treatment

6.1 PROCESS DESCRIPTION

6.1.1 Introduction[1]

Anaerobic treatment emerged as an option for wastewater treatment in the 1960s. In the recent years anaerobic processes have proved to be cost-effective for the treatment of BOD removal in the range of 40 to 85%. Anaerobic treatment involves the bioconversion of substrate (in this case the municipal wastewater), while passing through a biomass medium (i.e. the anaerobic bacterial population) mainly to methane (CH_4), CO_2 and H_2O and an additional small amount of more biomass. In this case the bioconversion does not require a high energy loss because most of the energy remains stored in methane. So the energy balance between the substrate and product only varies by less than 5% in terms of calories, and is equivalent in terms of COD. The energy balance in anaerobic bioconversion is presented by Equation 2.10 as follows:

$$\Delta S = 4.00\Delta CH_4 + 1.22\Delta X \tag{2.10}$$

where:

ΔS = substrate removed, mg/l COD
ΔCH_4 = methane produced, mg/l CH_4
ΔX = anaerobic biomass produced, mg/l VSS.

The factors are 4.00 and 1.22 are converting the mass of CH_4 and anaerobic biomass (X) to mass in terms of COD. The main treatment occurs by the conversion of COD into methane, which is then separated from the water, and into the small amount of anaerobic biomass, which are the excess sludge removed from the process.

There is a large variety of types of anaerobic reactors for treatment of wastewater including: anaerobic digesters of excess sludge (of low loading, high loading and two stages), septic tanks, anaerobic lagoons, high loading reactors including fixed bed reactors also known as anaerobic filters (of ascending or descending flows, Andrade Neto, 1997), baffled reactor (BR), piston or plug-flow anaerobic reactor (PAR) described by Orozco (1997) and Metcalf and Eddy (2003), rotating bed reactor, expanded bed

[1]This paragraph is based on the textbook "*Bioingeniería de aguas residuales: teoría y diseño*" (2005), Alvaro by Orozco Jaramillo, published by ACODAL, Bogotá, Colombia and the article "*Appropriate Technology for Wastewater Treatment and Reuse in Developing Countries*" Menahem Libhaber (2007).

reactor, fluidized bed reactor, two stage anaerobic reactor, upflow Anaerobic sludge blanket reactor (UASB), expanded bed granular reactor, internal recirculation anaerobic reactor and various combinations of these reactors. Detailed information on all these types of reactor is presented by Chernicharo (2007) and Campos (2009). Some of these reactors are used to treat industrial wastewater. The three anaerobic reactors (or anaerobic unit processes) which are used most frequently for treatment of municipal wastewater are: Anaerobic Lagoons, UASB and Anaerobic Filters. The piston anaerobic reactor also has a potential for being used in municipal wastewater treatment plants. Wastewater processing in the four anaerobic reactors used for treatment of municipal wastewater fall into the category of Appropriate Technology (AT) according to the classification of Libhaber (2007), which was discussed in Chapter 1.

According to Libhaber (2007): "Anaerobic processes achieve organic matter removal in the range 40%–80%, depending on the type of process, and as such they are most valuable for developing country since they have the potential to achieve such a meaningful removal at low cost and simple to operate installation. If a higher level of organic matter removal is required, anaerobic processes can be the first stage of treatment followed by a variety of polishing processes, such as lagoons of various types, activated sludge, constructed wetlands, filtration, dissolved air floatation and others. It is also possible to use one anaerobic process as the first stage treatment and another as a following polishing process, for instance, a UASB followed by an anaerobic filter, or even anaerobic lagoons followed by UASB followed by an anaerobic filter."

Treatment in Anaerobic Lagoons is discussed in Chapter 5. The current chapter deals with treatment in a UASB reactor, Anaerobic Filter and Piston Anaerobic Reactor.

6.1.2 Basics of the processes

Anaerobic digestion fundaments

The biochemistry of anaerobic decomposition is presented in paragraph 2.4, Chapter 2, and is summarized in Figures 2.6 and 6.1 (which is a simplified diagram). Libhaber (2007) provides a summary of the microbiology of the above processes, as follows: "Anaerobic decomposition is a two-stage process, each stage being carried out by a different group of bacteria, with the second stage bacteria being more sensitive than the first to environmental conditions such as pH. In the first stage acid-forming or nonmethanogenic bacteria convert the organic matter present in the sewage to organic acids, whereas in the second stage, methane forming-bacteria or methanogens convert the organic acids to methane gas and carbon dioxide. For efficient performance, the methanogens require a pH in the range 6.5–7.5, and they cannot develop at all below a pH of 6.2. However, if too many acids are produced by the acid-forming bacteria, which develop and multiply easily, the result is a low pH which may impede the production of methanogens. In the presence of high concentrations of sulphates, the methanogens also have to compete with sulphur-reducing bacteria. The main consequences of such a situation are the appearance of unpleasant odours and the reduction in the efficiency of the anaerobic process."

The reactions which take place in each stage of the anaerobic process are presented in a more detailed form in Table 6.1, along with the energy required for each reaction.

For these sequential reactions to be carried out under such defined and complex environmental conditions, it is necessary that a consortium of different types of bacteria form into a floc or granule, so as to optimize the metabolic processes which dependent on one of another. In the granules, which form spontaneously in the UASB and the PAR reactors (see Figure 2.7), three concentric layers are formed: the acidogenic bacteria in the outside layer, the acetogenic in the middle layer and in the middle of methanogenic in the center of th granules (see Figure 2.5).

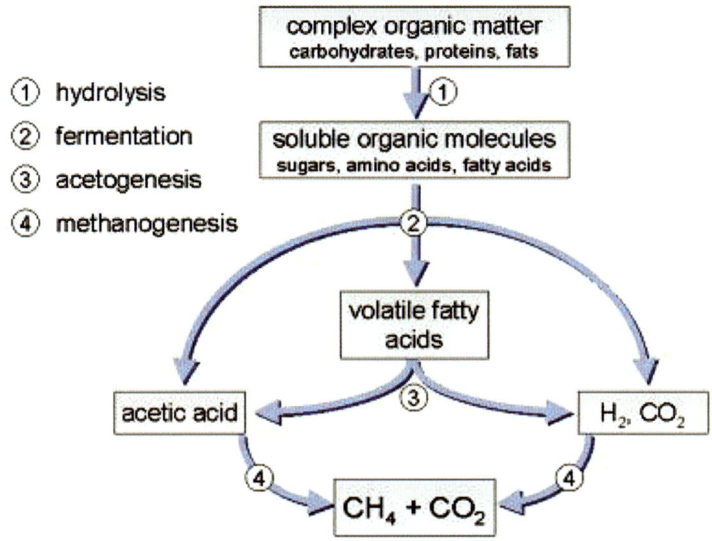

Figure 6.1 Schematic diagram of the anaerobic decomposition process [*Source*: http://water.me.vccs.edu/courses/ENV149/lesson4.htm (May 22, 2011)]

Table 6.1 Representative reactions of the anaerobic decomposition of organic matter.

Reaction	Reactives		Products	Kj/Reaction	
				$\Delta G^{\circ\prime}$	$\Delta G'$
Complete conversion from glucose to methane and CO_2	$C_6H_{12}O_6$	$\longleftrightarrow$	$3CH_4 + 3HCO_3^- + 3H^+$	−403.6	−399.1
Acidogenesis of carbohydrates to acetic acid	$C_6H_{12}O_6 + 4H_2O$	$\longleftrightarrow$	$2CH_3COO^- +$ $2HCO_3^- + 4H^+ + 4H_2$	−206.3	−318.5
Acetogenesis of propionate	$CH_3CH_2COO^- + 3H_2O$	$\longleftrightarrow$	$CH_3COO^- + H^+ + 3H_2$	76.1	−5.4
Hydrogenoclastic Acetogenesis	$4H_2 + 2HCO_3^- + H^+$	$\longleftrightarrow$	$CH_3COO^- + 2H_2O$	104.6	−7.0
Acetoclastic Methanogenesis	$CH_3COO^- + H_2O$	$\longleftrightarrow$	$CH_4 + HCO_3^- + H^+$	−31.0	−24.5
Hydrogenoclastic Methanogenesis	$4H_2 + HCO_3^- + H^+$	$\longleftrightarrow$	$CH_4 + 3H_2O$	−135.6	−31.6

Notes: 1) Calculation of $\Delta G^{\circ\prime}$ (standard) is based in the following conditions: pH = 7; solutes conc. 1 molar; gases at 1 atm., 25°C.
2) Calculation of $\Delta G'$(real) for an anaerobic decompostion; pH = 7; glucose = 10 micro-mole; acetate, propionate = 1 mili-mole; HCO_3^- = 20 mili-mole; CH_4 = 0.6 atm.; H_2 = 10^{-4} atm., 37°C.
Source: Orozco (2005)

Anaerobic metabolism is measured by the *methanogenic activity*, MetA, which in the anaerobic flocs and granules is in the order of 0.4 to 1.5 g COD/gVSS · d. The MetA and the other anaerobic metabolic reactions such as the substrate removal rate (r_a), depend, in general, on the temperature according to the following equation:

$$r_a(T) = r_a(T_0)e^{0.1(T-T_0)} \quad (6.1)$$

where T_0 is the reference temperature, usually 20°C, and T the temperature in °C. According to this reaction, the metabolic rates are doubling every 7°C (Orozco, 2005).

UASB

In order to take advantage of the capacity of bioconversion of organic matter to methane gas with the bacterial consortia formed in the anaerobic granules, a treatment system known as the Upflow anaerobic Sludge Blanket – UASB – reactor was developed (known in Portuguese as RAFA – Reactor Anaerobio de Fluxo Ascedente). A diagram of an UASB reactor is presented in Figure 6.2. It is a reactor which contains a sludge blanket held up by the upflow wastewater and gas bubbles which are produced in the reaction. The gas bubbles also provide a fairly good mixing of the liquid and biomass contained in the reactor. A device known as Gas-Solid-Liquid-Separator (GSLS) or Three Phases Separator is located at the top of the reactor. This device serves to separate the gas bubbles, which carry up the biomass granules, from the liquid flowing up, and also minimizes the loss of biomass trapped in the liquid outflow. In order for the treatment to be able to proceed properly, it is necessary that the bacteria be grouped in a compact form, either in a floc or granule form as mentioned.

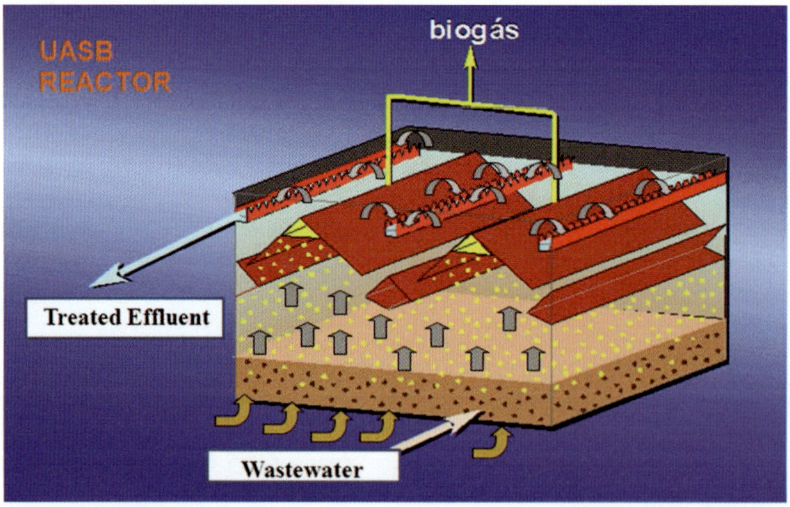

Figure 6.2 Schematic diagram of a UASB reactor

The main *hydraulic parameter* for design of the UASB reactors for municipal wastewater treatment is the ascending velocity or the hydraulic load in the reactor, $v_r = Q_D/A_{UASB}$ ($m^3/m^2 \cdot h \equiv m/h$), where A_{UASB} is the reactor surface area. Additional important parameters are: (i) the velocity in the settling zone within the GSLS, $v_s = Q_D/A_s$, where A_s is the settler surface area; (ii) the velocity of the liquid passing through the entrance zone to the Gas-Solid-Liquid-Separator, $v_p = Q_D/a_p$, where a_p is the entrance area perpendicular to the flow; and (iii) the biogas load $v_g = Q_g/A_g$, where Q_g is the biogas flow and A_g is the outlet area of the gas, (v_g in m^3 biogas/$m^2 \cdot h$) all applied at the locations shown in the diagram in Figure 6.3 (which is only a schematic diagram, not to scale). The suggested values of these parameters for municipal wastewater shown in Table 6.2 (using the above nomenclature) are the maximum values required to avoid granule or floc overflow with the effluent, and in the case of the gas, the minimum value to break the scum layer or cake that forms on the surface, at the exit of the biogas to the gas collectors (see Orozco, 2005). Sometimes is necessary to add a secondary clarifier in order to decrease the TSS in the effluent.

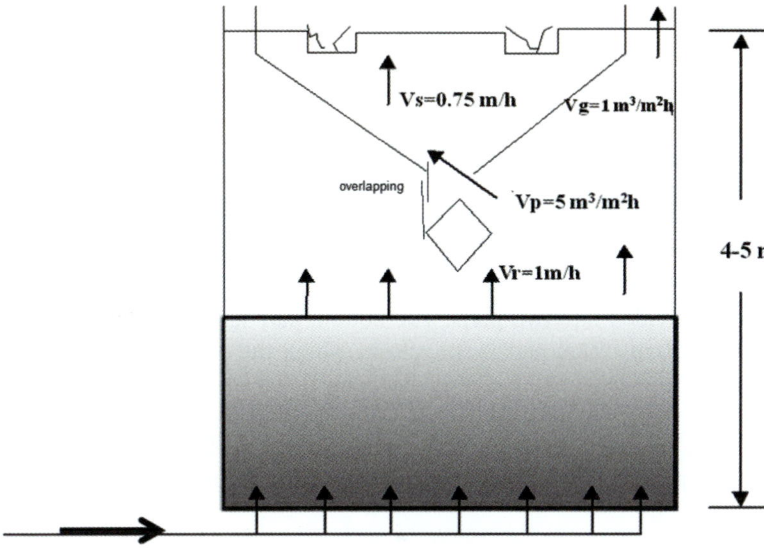

Figure 6.3 UASB schematic diagram with typical design parameters for municipal wastewater (not a scale) (*Source*: Orozco, 2005)

Table 6.2 Design parameters of a UASB reactor for municipal wastewater at 25°C.

Parameter	Formula	Unit	Value
Maximum TSS	–	mg/L	1000
TSS/COD	–	–	<0.5
v_r	Q/A_{UASB}	m/h	$<1.00 < v_s$
v_s	Q/A_s	m/h	0.75–1.00
v_p	Q/a_p	m/h	<5
v_g	Q_g/a_g	m3/m2 · h	>1.00
SOR secondary clarifier (If built)	Q/A_{sed}	m3/m2 · h	0.5
Baffles inclination		Degrees	45 a 60
Overlapping		m	>0.20

Note: Nomenclature used as explained in the paragraph above

The main *process parameter* for design of the UASB reactors is the organic volumetric load, $Lv = S_0/t_d = S_0 \cdot Q/V$ (kg $COD/m^3 \cdot d$). This volumetric load should be maintained at values in the range 2–30, depending on the temperature, on the design of the reactor and on the concentration of volatile fatty acids in the wastewater. A value of $L_v = 2$ kg $COD/m^3 \cdot d$ can be used for municipal wastewater at 15°C, adjusted for the temperature change using Equation 6.1.

The hydraulic load or Surface Overflow Rate (SOR), equivalent to the rising water velocity in the reactor (v_r) ($m^3/m^2 \cdot h = m/h$) should be less than 1 m/h, so as to prevent trapping particles into the effluent and their washout with the effluent stream leaving the reactor. Since the typical height of a UASB reactor is 4–5 m, for a vr of 1 m/h the detention time in the UASB reactor is $t_d = 4$–5 hours (0.17 to 0.21 d). So for municipal wastewater which received wastewater with a COD concentration of 400 mg/l (0.4 kg/m^3) or less, the L_v will be around (0.4 kg COD/m^3)/(0.2 d) = 2 kg $COD/m^3 \cdot d$. It is clear that, for municipal wastewater, the hydraulic criterion $v_r \leq 1$ m/h automatically implies that it meets the criterion for volumetric load (which is the fundamental criterion), so the design of a UASB reactor for municipal wastewater becomes a purely hydraulic exercise. As a result the hydraulic criteria for design of a UASB reactor for municipal wastewater which are summarized in Table 6.2 are the ones to be applied. For instance, the settler in the GSLS could be wider than the reactor, if needed, in order to meet the criteria $v_r < 1.00 < v_s$. The overlapping is defined as the horizontal length (in the settler entrance area) used to avoid the direct vertical flow of water from the reactor into the settler of the GSLS.

The UASB reactor is divided into two parts. The bottom part serves as the liquid volume storage and has a height H_L (1.5 to 2.0 m). This part is the site in which the sludge blanket material is stored. An extra height of 0.5 to 1.0 m is left to take care of the peak flows so the total liquid height is $H_L \approx 2.5$ m. The upper part, with a height H_G of also about 2.5 m, comprises the GSLS, with an angle of plate inclination of 45 degrees. Thus, the typical total height of a UASB reactor is $H_T = H_L + H_G = 5.0$ m (Metcalf & Eddy, 2003, pp. 1009). The UASB has proven to be a simple, cost-effective and reliable reactor.

Anaerobic filter (AF)[2]

The anaerobic filters (AF) are fixed medium (attached growth) reactors in which anaerobic bacteria grow attached to a filter medium of rock or plastic material with a net area (a) of between 50 and 200 m^2/m^3 of media, and mostly in the porous zone of the media. Treatment occurs as the wastewater passes through the filter medium, where the bacteria come into close contact with the wastewater and consume the BOD contained in it. Fixed bed reactors have a variety of geometrical forms such as circular and rectangular, and are of two kinds in terms of the direction of flow: (i) up-flow or ascending flow reactor; and (ii) down-flow or descending flow reactor. See Figures 6.4 and 6.5.

Anaerobic microorganisms are formed in flocs or granules and are trapped in the pores of the media. With the increasing accumulation of biomass, part of it is separated from the filter media and is carried out with the effluent. Despite this process, the average residence time of the biomass in the reactor is over than 20 days. Up-flow filters are more common than down-flow filters. Down-flow filters can work with a floating medium. In Down-flow filters, recirculation is commonly used. The need for a uniform distribution along the bottom or surface, limit their size to a manageable hydraulic flow (Libhaber Menahem, 2007).

[2]This paragraph is mainly based on the Book "Sistemas Simples para Tratamento de Esgotos Sanitarios, Experiencia Brasileira" by Cicero Onofre de Andrade Neto (1997) and on the article "Appropriate Technology for Wastewater Treatment and Reuse in Developing Countries" by Menahem Libhaber (2007).

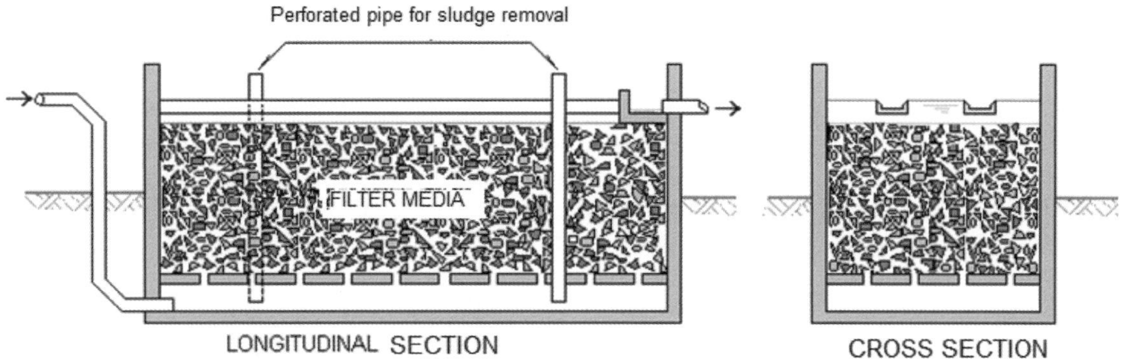

Figure 6.4 Up-flow anaerobic filter (*Source*: Andrade Neto *et al.* 2009)

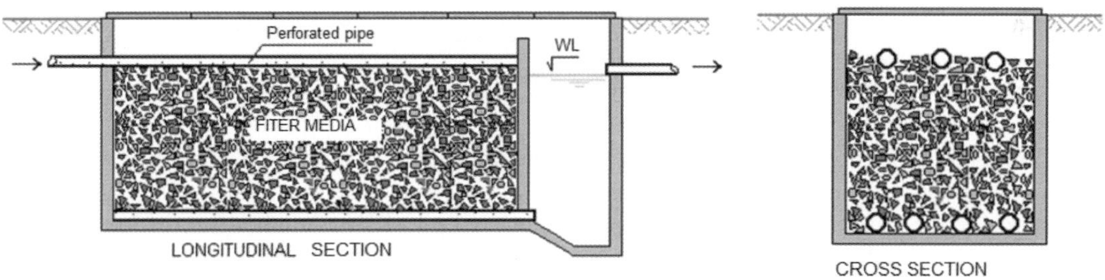

Figure 6.5 Down-flow anaerobic filter, with inflow and outflow by perforated tubes (*Source*: Andrade Neto *et al.* 2009)

To prevent clogging of the filter, effective removal of suspended solids from the raw sewage is required prior to its inflow to the filter. Such removal of suspended solids can be achieved by rotating micro screens. For small treatment plants, decanter-digesters (otherwise known as large septic tanks or Imhoff tanks) are used to remove suspended solids prior to an anaerobic filter. In addition to the use of anaerobic filters as the main treatment unit, they can be used as polishing unites for improvement of the quality of the effluent of a preceding unit. They are in fact more suitable to perform as polishing units since when fed with treated effluent of a preceding unit they are less bound to clog by suspended solids and have to treat a lower content dissolved organic matter. Anaerobic filters are used as the main treatment unit mainly for small and medium size municipalities, while for polishing application anaerobic filters are used in the entire range of cities size, including in treatment plants of large cities.

With time, the anaerobic filter becomes clogged because of the growth of the biomass. When the pressure drop on the filter becomes too high, it needs to be washed. Similarly to sand filters for water treatment, the anaerobic filter is washed by a stream of water or treated effluent which is passed through it in a counter wise direction to its operation flow. The wash water, which contains the excess sludge, is directed to the sludge handling facilities, which are usually just drying beds. Since the process is anaerobic, the excess sludge quantities are small and the washing, which is specific to each systems, is required infrequently, once every several days to once every few months. Backwashing becomes more complicated as the anaerobic filter system is larger. Therefore when they are the main treatment unit, anaerobic filters are more frequently used for plants serving small populations (in which the backwashing is simpler). In larger plants, anaerobic filters serve as a polishing unit, where they produce less sludge and need less frequent back washing. All anaerobic filters must have efficient backwashing systems and convenient arrangements for replacing the filtering media.

Design parameters are the detention time t_d (V/Q, h), volumetric organic load L_v (S_0/t_d, kg COD/$m^3 \cdot d$) and the hydraulic load, q_a ($m^3/m^2 \cdot d$). The up-flow filter should have a height between 1.2 and 3.0 m, and the filter medium may be of rock or plastic with a net area (a) preferably larger than 100 m^2/m^3 of filter material. Table 6.3 presents typical design criteria for a temperature of 25°C (Andrade Neto, 1997; Wikilibros 2007; Libhaber 2007).

Table 6.3 Design criteria of the up-flow anaerobic filter at 25°C*.

	Detention time, t_d	Volumetric load, L_V	Hydraulic load, q_a	Height, h
Units	hours	Kg COD/$m^3 \cdot d$	$m^3/m^2 \cdot d$	m
Range	4–12	1.0–1.6	6–15	1.2–3.0
Typical value	6	1.2	10	2.5

*The temperature correction of L_V is done using Equation 6.1
Source: Prepared based on Andrade Neto, Cicero Onofre (1997), pp. 55–60

Anaerobic Filters are often used as a Polishing treatment system for Septic Tanks (ST) with removal efficiencies of about 50% in the septic tank (and even higher at high temperatures), for a net efficiency of the system (Septic Tank followed by Anaerobic Filter) of up to 85%. Another way of looking at such a system is considering the septic tank as the pretreatment unit of the anaerobic filter. At any rate, such a system is basically a system of two anaerobic reactors in series. This type of system is used in small towns of up to 10,000 inhabitants, because of the limitation of the septic tank in handling larger flows. The detention times in an anaerobic filter for a given population and temperature, in a system of a septic tank followed by an anaerobic filter, are shown in Table 6.4 (based on Andrade Neto, 1997). Figure 6.6 shows an innovative design of a septic tank combined with an up-flow anaerobic filter followed by a down-flow anaerobic filter (Andrade Neto, 1997).

Table 6.4 Variation of detention time t_d in days, in an anaerobic filter with population size and with temperature in a system of a septic tank followed an anaerobic filter.

Population inhabitants	Temperature °C		
	< 15°C	15 to 25°C	>25°C
1500	1.17	1.00	0.92
1501–3000	1.08	0.92	0.83
3001–4500	1.09	0.83	0.75
4501–6000	0.92	0.75	0.67
6001–7500	0.83	0.67	0.58
7501–9000	0.75	0.58	0.50
>9000	0.75	0.50	0.50

Source: Andrade Neto (1997) p. 56

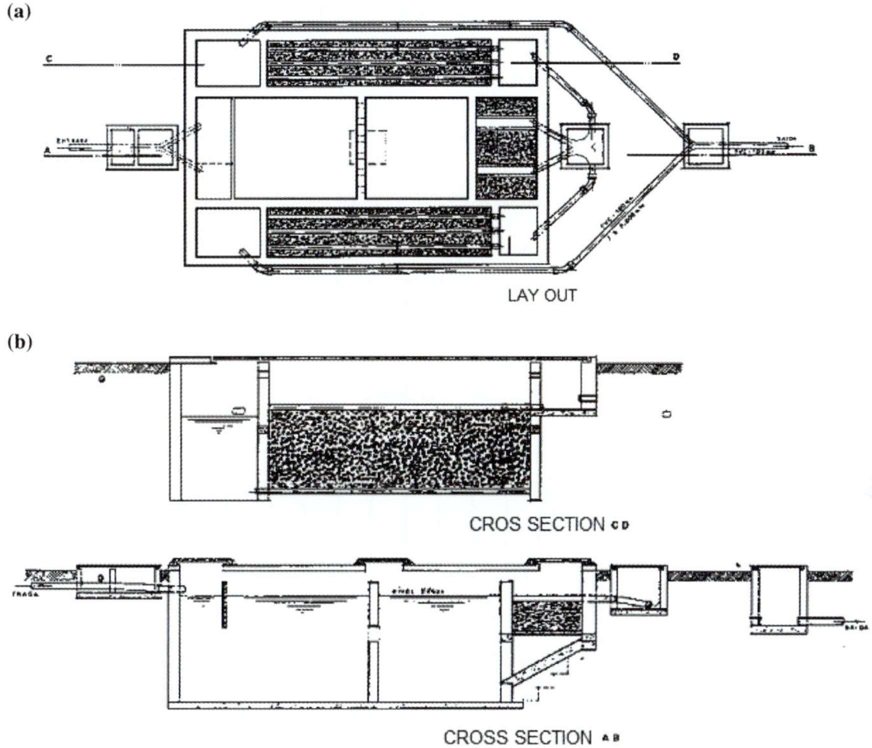

Figure 6.6 Septic tank with an up-flow anaerobic filter followed by a down-flow anaerobic filter (a) lay out and (b) cross sections (*Source*: Andrade Neto, 1997 pp. 69–71)

Piston Anaerobic Reactor (PAR)

The piston (or plug-flow) anaerobic reactor (PAR) is a variant of baffled reactors which uses a Gas-Solid-Liquid-Separator (GSLS) in the final chamber to improve the entrapment of solid particles so as to prevent their washout with the effluent and maintain thereby the sludge age at a level as high as possible. The PAR was developed at the Los Andes university in Bogota, Colombia, (Orozco, 1988, 1997) for temperatures below 20°C, taking advantage of the higher efficiency of substrate removal obtained with the more perfect plug flow regime with baffles (see Chapter 2, Section 2.7). Full scale plants based on RAP reactors were constructed and operated achieving COD removal efficiencies of 70% at temperatures between 13 and 17°C (See Metcalf & Eddy, 2003, pp. 1005, 1018). These efficiencies can be improved to the extent that the retention of solids becomes more effective, because the effluent contains a high level of suspended solids which represent most of the COD in it.

Figure 6.7 shows the design of a PAR reactor, including plan and sections. The baffles along the reactor and the GSLS at the end of it can be seen in the figure. The design parameters of the three phase separator SGSL is of the same type explained in Table 6.2 for the UASB reactor. In a PAR reactor it is not necessary to maintain a volume for the sludge beneath the GSLA because the sludge accumulates in the chambers preceding the GSLS. The flow velocity between baffles (v_b) up-flow and down-flow, should be maintained at less than or equal to 3 m/h because at such velocities the product $[S \cdot v_b]$ in kg $COD/d \cdot m^2$ (of surface area) reaches high values h, favouring the formation of granules with high sedimentation velocities (Orozco, 1988). A media of plastic material can be added at the top of each chamber to assist in the retention of solids in the chambers, because the ascending granules or flocs attached to gas bubbles find an obstacle (the plastic media) that forces the gas bubbles to release the flocs, which settle in the same chambers. The recommended detention time in the RAP reactor is 10 h at 15°C. This value should be corrected for the temperature, in the range 15–20°C, using Equation (6.1).

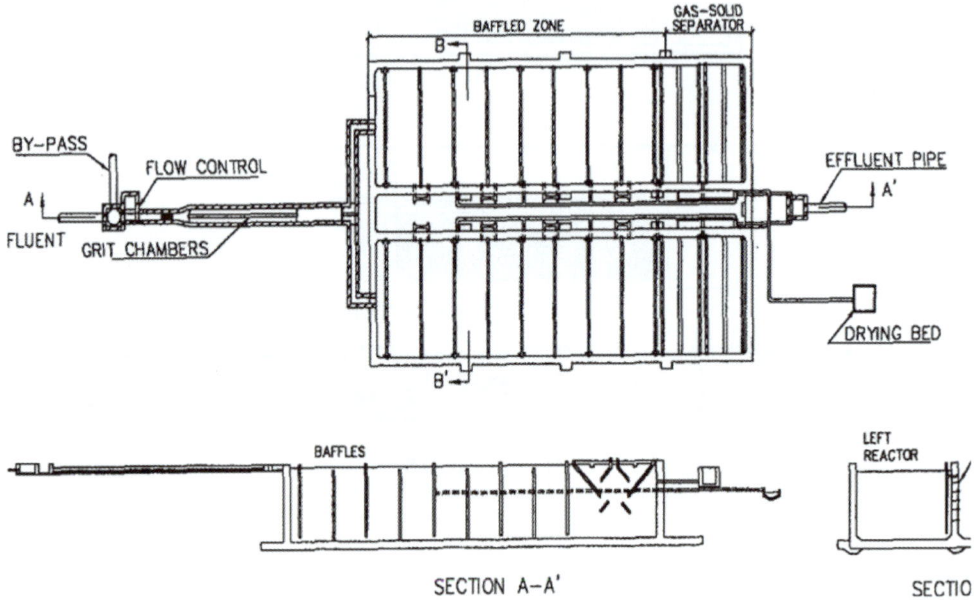

Figure 6.7 Plan and sections of a PAR type baffled reactor (*Source*: Orozco, 2005)

The PAR reactor was the first baffled reactor constructed in the world in a full scale plant (Metcalf & Eddy, 2003, pp. 1005). Moreover, their use for operation at sub-optimal temperatures (lower than 15°C) has influenced the development of other baffled reactors that have been built at an accelerated pace in several countries, especially in China (Zonglian She *et al.* 2006), New Zealand (Foxon *et al.* 2003) and Bolivia (Medina *et al.* 1993).

The quantity of the excess biomass is minimal and is slowly washed out with the effluent, but it could be trapped with a secondary clarifier. As the PAR is an open to air reactor, a possible improvement is to cover it with a geomembrane to capture the biogas for burning o use.

6.1.3 Performance

Efficiency

UASB

UASB reactors treating municipal wastewater produce an effluent that usually contains 35–100 mg/l of total DBO_5, 30–40 mg/l of soluble DBO_5 and 50–130 mg mg/l TSS. The removal efficiencies are 60–75% of the total BOD_5 and 65–80% for TSS. Construction costs of a UASB reactor are in the range 20–40 US$/Capita. Even after including additional treatment for effluent polishing, the investment cost still compares favourably with the investment of 100–150 US$/Capita considered typical for a conventional secondary treatment plant. The operation and maintenance costs (O&M) are in the range 1.0–1.5 US$/Capita · yr, which is much lower than those of conventional treatment plants. Several UASB plants under operation in Brazil are successfully serving populations of about 1 million each, making the UASB process adequate for cities of all sizes (Libhaber, 2004; Libhaber, 2007).

Anaerobic filter

Chernicharo (2000) reports Total BOD removal efficiencies of anaerobic filters in the range of 70–80% when the anaerobic filter is the main treatment unit and 75–95% when it used as a polishing units. Andrade Neto (1997, 2000) reports results of a research on an anaerobic filter system treating municipal wastewater which was in operation for six years. This system, consisting of a septic tank followed by an up-flow anaerobic filter followed by a down-flow anaerobic filter, achieved a COD removal of up to 90%. The effluent of this anaerobic filter system contained less than 30 mg/l of BOD and less than 20 mg/l total suspended solids. Sludge withdrawal from the filter was undertaken once every six months. The investment cost in anaerobic filter systems is 10–25 US$/Capita and the O&M are in the range 0.8–1.0 US$/Capita · yr. (Libhaber, 2007).

Plug flow anaerobic reactor

The typical COD removal efficiency of a PAR reactor at 15°C is 60–70%, if an additional solids separation system after the GSLS system in last chamber is not used. An additional solids separator can improve the removal efficiency (Orozco, 2005).

Reasons for the difference in performance of the various anaerobic reactors types

The three anaerobic reactors (or anaerobic unit processes) which are used most frequently for treatment of municipal wastewater are: Anaerobic Lagoons, UASB reactors and Anaerobic Filters. Anaerobic filters achieve the highest organic matter removal, in the range 70–80% of BOD_5 removal, followed by UASB with BOD_5 removal levels in the range 60–75%. Anaerobic lagoons achieve a lower BOD_5 removal level, in the range 40–70%. The reason for the different removal level of organic matter in each of these reactors is that the mixing mechanisms and the intensity of the contact of the wastewater with the bacterial biomass are different in each type of reactor. In anaerobic lagoons the mixing in the reactor is not effective and the contact of the organic matter dissolved in the raw sewage flowing into the lagoon with the biomass which develops on the bottom of the lagoon is not intense, so only partial removal of dissolved organic matter is achieved. The removal of the suspended solids contained in the raw sewage, which settle in the lagoon, is high, but as a whole, the organic matter removal in an anaerobic lagoon is low in comparison to other anaerobic processes. In a UASB reactor the granular biomass particles float in the liquid due to the flow pattern and the action of the biogas released in the reactor. Consequently, the contact of the liquid with biomass is more intense than in anaerobic lagoons and as a result, the organic matter removal is higher. However, there is a limit to the capacity of the UASB reactor to remove organic matter since a higher concentration of biomass is required for achieving higher removal efficiencies, but it is impossible to achieve higher sludge concentrations since when the concentration increases, more sludge is carried away with the effluent, so the organic matter removal in UASB is still not very high. In an anaerobic filter, where the biomass is attached to the filter media and the reactor is flooded with the wastewater, the biomass concentration levels in this reactor are high and the contact of the liquid that flows into the reactor with the biomass is more effective, and as a result, the level of organic matter removal is higher than in UASB and certainly higher than in anaerobic lagoons.

Advantages and disadvantages

BOD removal efficiencies in anaerobic systems are in the range 40–90% COD. The investment costs of these systems are in the range 20–40 US$/Capita and they employ a variety of flow patterns which include completely mixed, plug flow, filters, and so on. These systems represent innovative technologies with a high perspective to be used in developing countries. The various anaerobic systems have basically the same advantages and disadvantages, which are summarized below (Libhaber, 2007).

The main advantages are:

- They do not require oxygen, and thus has low power consumption.
- The gas produced, primarily methane, can be collected and used to generate energy, or can be burned, reducing greenhouse gas emissions. Carbon emission reduction credits can be obtained, increasing the financial benefits of the project. The carbon credits may expire in 2012 with the end of the Kyoto Protocol.
- The excess sludge quantities produced are much smaller than those generated by aerobic processes, thus the treatment and disposal costs of sludge are small and the negative environmental effects of the sludge are minor.
- The area occupied by an anaerobic reactor is small, except for anaerobic lagoons which require a somewhat larger area.
- The electromechanical equipment required for anaerobic processes is minimal.
- Construction is simple and uses mostly local materials.
- The operation is simpler than mechanized systems.
- The investment and O&M costs of anaerobic processes, are much lower than those of aerobic processes.

The main disadvantages are:

- Anaerobic processes need a longer start-up time to develop the necessary amount of anaerobic biomass in the reactor.
- They are more sensitive to lower temperatures.
- They are more sensitive to toxic substances.
- Biological nitrogen and phosphorus removal cannot be achieved in an anaerobic process.
- They have a potential of generating odours and corrosive gases, however those can be controlled.
- They have a low removal efficiency of BOD compared to aerobic systems so the effluent quality of a one stage anaerobic treatment process may not meet discharge standards and, depending on the specific project conditions, a subsequent treatment stage might be necessary.

6.2 BASIC DESIGN PROCEDURE

6.2.1 General design considerations

As explained, the basic principle of anaerobic treatment is the concentration of a large amount of biomass, whether in a sludge blanket or attached to a fixed medium, which is a consortium of several types of microorganisms cooperating metabolically to produce the bioconversion of BOD to biogas.

The main type of anaerobic treatment is the *UASB*, along with the *baffled reactor* (BR), one of whose forms is the *PAR*. These types of treatment are known as "Suspended Growth Anaerobic Treatment". Such treatment consists basically of a tank where biomass accumulates and reacts with the wastewater, and a three phase separation system, the gas-solid-liquid separator (GSLS). The other type of anaerobic treatment is the "Attached Growth Anaerobic Treatment" of which the main type is the Anaerobic Filter, often used as a polishing treatment, consisting of tanks filled with crushed rocks or a plastic filter medium, to which biomass is attached and grows, on the surface and within the pores of the medium. Attached growth anaerobic systems are mainly used in treatment plants of small towns and small communities, because they require filters backwashing and cleaning, which something are more difficult to perform in larger scale plants. Also, the homogeneous hydraulic distribution system required for proper operation, limits the size of anaerobic treatment plants. However, large plants can be built of smaller modules which enable overcoming these problems. The anaerobic treatment system can be classified as Sustainable Innovative Appropriate Technology processes as defined by Libhaber (2007) since they are low-priced treatments processes, simple to operate, and can be constructed using local materials, without major or equipment requirements.

The basic process design parameter of the UASB reactors is the volumetric load, $L_v = S_0/t_d = S_0 \cdot Q/V$ (kg COD/$m^3 \cdot d$) which must be maintained in the range 1–30, depending on the design of the reactor, temperature and concentration of volatile fatty acids. For municipal wastewater at 15°C it is recommended to use $L_v = 1$ kg COD/$m^3 \cdot d$, corrected for the temperature using Equation 6.1.

The other parameter of importance for design of a UASB reactor is the hydraulic load or Surface overflow rate (SOR) which is the up-flow velocity in the reactor, v_r ($m^3/m^2 \cdot h = m/h$). This parameter must be kept at a value lower than 1 m/h, to avoid dragging of suspended solid particles (the biomass granules) out of the reactor with the effluent. For municipal wastewater, the hydraulic criterion $v_r < 1$ m/h automatically implies that it meets the criteria for volumetric load so the design of a UASB for municipal wastewater becomes a hydraulic exercise. However, often, for certain geometric design of the UASB reactor and typical GSLS, the detention time, t_d, can be used as the main design parameter[3]. In fact it has been shown in practice that a reactor 5 m high with a GSLS with baffles at a 45° angle and 2.5 m high, works well with municipal wastewater. The geometry of the reactor can vary. Here we analyze the standard module (denominated GSLS-SM) presented in Figure 6.9, but there are other typed of modules, with circular reactor tank, triangular cross section, and so on.

For the anaerobic filter, the main design parameters are the t_d (V/Q), L_v (S_0/t_d, kgCOD/$m^3 \cdot d$) and the hydraulic load, q_a ($m^3/m^2 \cdot d$). The up-flow anaerobic filter should have a height of between 1.2 and 3.0 m, and the filter medium may be rock or a plastic medium with a specific net area (**a**) greater than 100 m^2/m^3 of filter material.

The PAR treatment presented above is usually designed for temperatures below 20°C, with chambers (between baffles) water velocity of less than 3 m/h and a GSLS-SM located in the last chamber.

In general, a main drawback of the anaerobic process is that the solids separation units are of low efficiency, and so it is recommended whenever possible to design a final secondary clarifier to improve the TSS (and BOD) removal efficiency in the effluent.

6.2.2 Orderly design method, ODM

The overall design methodology for anaerobic treatment consists of the following:

(1) *Determine the external or independent variables*: Same as in Chapter 5, Section 5.2.2, Paragraph 1.
(2) *Determine the Flow Diagram of the Anaerobic Treatment*: The components of a typical UASB system are shown in Figure 6.8.
(3) *Design the Preliminary Treatment System:* This design is the same to the design of the preliminary treatment in Chapter 5, Section 5.2.2, Paragraph 3.

[3]In effect the component that carries out the bioconversion is the sludge blanket, and hence its volume determines the t_d, what makes this measure a design parameter. The GSLS must meet the hydraulic relationships shown in Table 6.2.

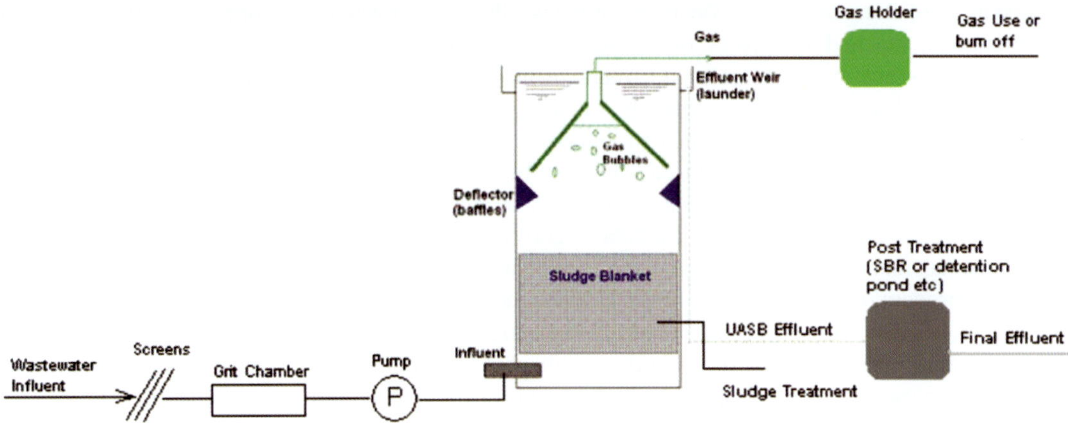

Figure 6.8 Flow diagram of a typical UASB plant [*Source*: http://wastewaterengineering.com/uasb_upflow_anaerobic_sludge_blanket.htm (May 25, 2011)]

(4) *UASB*

4.1 *Reactor design:* To design the UASB with Standard Model GSLS, or GSLS-SM, we start with the parameters presented in Figure 6.3 and Table 6.2. A typical GSLS-SM has 45° inclination of the baffles, and a settling basin 4 meters wide and 2.0 m high, plus 0.5 m needed for overlapping structures to assure gas separation (0.5 m is recommended and at least 0.20 m is needed), giving a 2.5 m SLGS-SM total height, as shown in Figure. 6.9. For this typical design, an additional 2.5 m is required to allow for a volume to contain the liquid sludge blanket at the bottom, resulting in a total UASB reactor height of 5 m.

This GSLS proved to be reliable in its operation, which is the reason that it is called the "Standard Module" (GSLS-SM) and is proposed as typical design geometry. However, different geometric models for the UASB reactor have also been developed. CAESB, the water and sanitation utility of Brasilia, developed a UASB module with sloped walls which is structurally very economical, because the inclination of the outer walls allows the reactor to be an excavated tank supported by the ground. In addition, the design facilitates the even distribution of wastewater at the bottom. A photo of this type of reactor, constructed in Uberaba, Minas Gerais, Brazil, is presented in Figure 1.8.

Continuing with the standard design geometry, the UASB module is divided vertically into two parts. The bottom section which serves as the nominal liquid volume and is H_L meters high. This is the site in which the sludge blanket is stored and it is calculated from the volumetric load (L_V) at the design temperature. A height of 0.5 m needs to be added to allow for additional storage at peak flows so the effective height of the sludge blanket becomes $H_L \approx 2.5$ m. The upper section of the UASB reactor is composed by the GSLS-SM, whose height H_G is also about 2.5 m. Thus, the total height of a UASB reactor is:

$$H_T = H_L + H_G \qquad (6.2)$$

Figure 6.9 shows for clarity two adjacent modules, separated by the gas outlet channel, which has a width (w_g) that must enforces the criteria of $v_g > 1\ m^3/m^2 \cdot h$ to be met. Figure 6.10 is shows a diagram of a pipe distribution system for a UASB reactor.

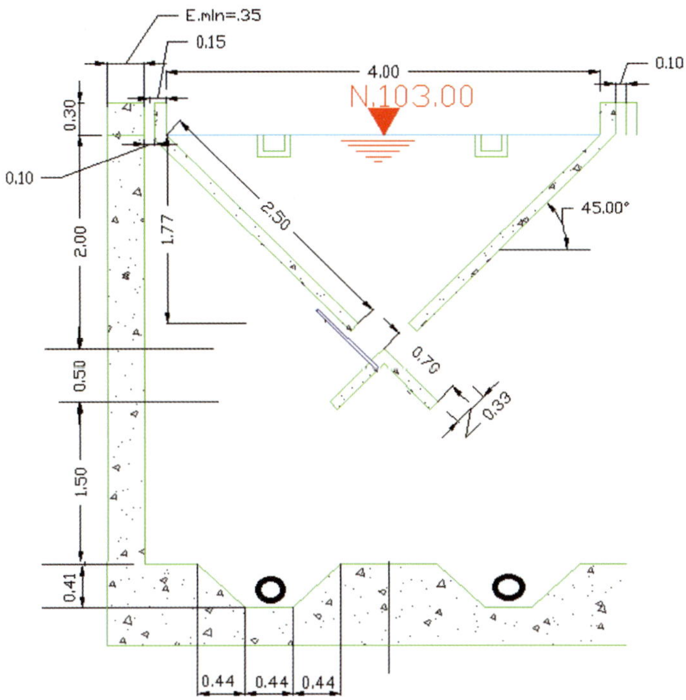

Figure 6.9 Standard module of a gas solid liquid separator (GSLS-SM) of a UASB reactor (*Source*: Orozco, 2005)

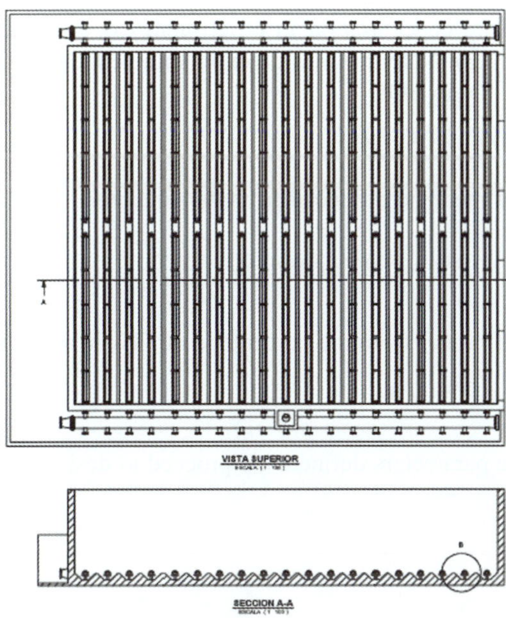

Figure 6.10 Plan and section of a manifold for inflow distribution at the bottom of the UASB reactor

The design is made with the help of the Excel program CHAP 6-UASB-AF-01 which calculates the UASB reactor and the Anaerobic Filter as a second stage (polishing) treatment. To prepare the design with the help of Table 6.2 and Figure 6.9, we select the following parameters:

Designer variables

- Width of the settler at the top level of the SGSL-SM: $w_s = 4.00$ m (Figure 6.9)
- Height of GSLS-SM, $H_G = 2.5$ m (Figure 6.9)
- Thickness or width of the baffles walls that compose the bell of gas separation of the GSLS-SM, w_b: which can be between 0.10 and 0.20 m.
- The values of v_r and v_g are selected from Table 6.2. Note that with the GSLS-SM v_s should be larger than v_r, so this value should also be checked in the Table 6.2.
- We adopt the anaerobic yield coefficient $Y = 0.08$ (Table 2.2) and a volumetric loading applied to the liquid volume of the sludge blanket of between 2.0 and 4.0 kg $BOD_5/m^3 \cdot d$ or between 2.0 and 4.0 kg $sCOD/m^3 \cdot d$ (where sCOD is soluble COD).
- For the UASB (see UASB tag in program), other inputs are calculated by the program as follows:

(1) The efficiency is obtained from Equation 2.42, assuming that the UASB works as a plug flow with a possible maximum efficiency of $E_{max} = 90\%$ (0.9), as follows:

$$E = E_{max}(1 - e^{-26.5/Lv}) \tag{6.3}$$

(2) Volumetric load at actual water temperature corrected by Equation 6.1:

$$L_V(T) = L_V(15^\circ)e^{0,1(15-T)} \tag{6.4}$$

(3) The substrate concentration in eh effluent (S) calculated from efficiency equation (Equation 6.3). Note that S is given in soluble COD which in municipal wastewater is about equal to BOD_5.

$$S = (100 - E)S_0/100 \tag{6.5}$$

(4) Volume of methane produced from the equation:

$$V_{CH_4} = 0.35(1-1.22Y)(dS/dt)V_L\left[\frac{273}{273 + T^\circ C}\right] \tag{6.6}$$

(5) Volume of biogas from the percentage of CH_4 in the biogas, $\eta = 0.65$:

$$V_{biogas} = V_{CH_4}/\eta \tag{6.7}$$

Design: With these parameters defined, we proceed to design the reactor:

- Area of reactor:

$$A_r = Q_D/v_r \tag{6.8}$$

- Net height:

$$H_N = Q_D \cdot S_0/A_r \cdot L_v(T) \tag{6.9}$$

- Effective liquid height (at least 2.5 m):

$$H_L = H_N + 0.5 \tag{6.10}$$

- UASB height:

$$H_T = H_L + H_G \tag{6.11}$$

- Detention time:

$$t_d = H_T/v_r \tag{6.12}$$

- UASB volume:

$$V_{UASB} = Q_D \cdot t_d = H_T \cdot A_r \tag{6.13}$$

- Net volume of sludge blanket

$$V_L = A_r \cdot H_L \tag{6.14}$$

- Total gas extraction area:

$$A_g = V_{biogas}/v_g \tag{6.15}$$

- UASB length in the direction perpendicular to Figure 6.9, so that the length to width ratio is 3/2:

$$L_{UASB}/W_{UASB} = 3/2 = 1.5\text{: } L_{UASB} = \sqrt{(1.5A_r)} \tag{6.16}$$

- UASB width:

$$W_{UASB} = A_r/L_{UASB} \tag{6.17}$$

- Total width of the gas outlet throats ($W_g = n_{MP} \cdot w_g$), w_g being the throat width of each of the n modules:

$$W_g = A_g/L_{UASB} \tag{6.18}$$

- Preliminary number of modules (n_{MP}):

$$n_{MP} = W_{UASB}/(w_g + w_s + 2w_b)$$

From which:

$$n_{MP} = (W_{UASB} - W_g)/(w_s + 2w_b) \tag{6.19}$$

It is noted from Figure 6.9 that the module width is ($w_g + w_s + 2w_b$). At this point, all the figures necessary to compute the dimensions of the UASB reactor are calculated, but it should be remembered that the number of modules (n_M) must be an integer and therefore n_{MP} must be rounded up. In this way all the hydraulic parameters are on the safe side.

- Gas throat width:

$$w_g = W_g/n_M \tag{6.20}$$

- Total UASB width:

$$W_{UASBD} = n_M(w_g + w_s + 2w_b) \quad (6.21)$$

- Total Area of settlers:

$$A_s = n_M \times 4{,}00 \times L_{UASB} \quad (6.22)$$

- Then, the settling velocity (or SOR): $v_s = Q_D/A_s$, should be around 0.75 (or less). *Note that for the geometric design proposed* $v_r < v_s$. The subscript D is for *design* or *definite* values. The reactor final area is calculated as:

$$A_{rD} = W_{UASBD} \times L_{UASBD} \quad (6.22a)$$

- Final reactor velocity:

$$v_{rD} = Q_D/A_{rD} \quad (6.23)$$

- Net volume of sludge blanket:

$$V_{LD} = A_{rD} \cdot H_L \quad (6.24)$$

- Design volume of UASB:

$$V_{UASBD} = A_{rD} \cdot H_T \quad (6.24a)$$

- A system to remove the excess sludge should be provided. Sludge production (Q_x):

$$Q_x = V_{UASBD}\, Y(S_0 - S)/t_{dD} \quad (6.25)$$

4.2 *Inlet and outlet structures: Inlet structures* should ensure the proper effective flow the distribution in the sludge blanket volume and this is achieved by placing outlets on the floor, one at every 1 to 2 m^2 of tank bottom. One way to do it is by using an external manifold that feeds a series of perforated pipes laid throughout the floor area, ensuring an approximately uniform distribution as shown in Figure 6.10. To meet proper distribution objectives, the design of the feed pipelines is the following:

- Place two parallel branch pipes each driving a flow of $Q_m = Q_D/2$ with a velocity that meets following restriction: the area (a_o) of each outlet orifice should have a diameter of at least 2″ for the wastewater to flow freely. Then calculate the pipe diameter (φ_m) of each branch manifold of area a_m for a given pipe velocity v:

$$a_m = Q_D/2v \quad (6.26)$$

- Then calculate the preliminary number of outlet orifices (n_o) selecting either 1 or 2 for each m^2 of area. n_o is the orifices' density (ε_o) or number of outlet orifices per square meter of floor area:

$$n_o = ArD/\varepsilon_o \quad (6.27)$$

- Every manifold branch (there are two manifolds, one at each side of the reactor) feeds $n_o/2$ orifices, each orifice of an area a_o (see Figure 6.10). The orifices are distributed in a number of distribution pipes n_d, leaving the principal manifold branch and going inside the reactor, with an approximate number of orifices per each distribution pipe (n_{do}) of :

$$n_d = \sqrt{n_o}. \quad (6.27a)$$

$$n_{do} = n_o/(2n_d) \quad (6.28)$$

- The area a_d of each branch distributor with diameter (φ_d), must meet the condition:

$$\Sigma a_d \leq 0.4a_m \quad (6.29)$$

- The area a_o of each hole with diameter (φ_o) must meet the condition:

$$\Sigma a_o \leq 0.4a_d \quad (6.30)$$

However, there is an additional constraint. The diameter of area a_o should be minimum $\varphi_o = 2''$ in order for the wastewater to flow freely. This is achieved by varying the velocity (v) in the inlet manifold. In the Excel program it can be done with function "Goal Seek".

Outlet structure as shown in Figure 6.9 is a channel per module, which is designed with the following equation:

$$Q_m = 1.38\,b \cdot h^{1.5} \quad (6.31)$$

with:

Q_m = flow per module, Q_D/n_M, m^3/s
b = channel width, m
h = maximum depth of water in the channel m.

The depth of the channel is then set as, $h_c = h + 0.10$, taking in account an 0.10 m channel freeboard.

4.3 *Gases and sludge disposal:* The volume of gases produced must be calculated in order to specify the gas burner. This calculation is performed using Equations 6.6 and 6.7. For the calculation of sludge produced a variant of Equation 6.25 is used:

$$Q_x = Y(dS/dt) \cdot V_L \quad (6.32)$$

where $dS/dt = (S_0 - S)/t_d$. With the production of sludge (Q_x, kg/d) and the application rate of excess sludge to the drying beds of (q_x, $kg/m^2 \cdot yr$) the required area of the drying beds (A_{bed}, m^2) is calculated as follows:

$$A_{bed} = Q_x/q_x \quad (6.33)$$

(5) *Anaerobic filter:* The anaerobic filters are mainly used as post-treatment, most commonly of a septic tank (ST) in small populations, but lately they have been used in Brazil as a UASB post-treatment to obtain an overall efficiency of BOD_5 removal between 80 and 95%. Anaerobic filters can also serve as the main treatment unit, preceded by preliminary treatment which included fine screening. Since the most frequent application of anaerobic filters is a UASB post-treatment, this is the application presented below. The anaerobic filter referred to is a up-flow type filter (Figure 6.4). The design parameters are presented in Table 6.3. The main

design parameter is the detention time, corrected for temperature by Equation 6.1, modified as follows:

$$t_d(T) = t_d(25°C)e^{-0.1(T-25)} \tag{6.34}$$

The design procedure, performed with the Excel program CHAP 6-Anaerobic-UASB-AF.xls (tag AF) is as follows:

Designer variables

- Select a detention time at 25°C based on Table 6.3.
- Choose the filter height (H_F) from Table 6.3, depending on the selected filter medium material. A freeboard of 0.50 m is added for to a final construction height (H_T) (see Figures 6.4 and 6.5).
- Select a yield coefficient (Y) from Table 2.2.
- Propose the fraction of CH_4 in the biogas (η), usually 0.65.
- Select the relation length/width (L/w) of the anaerobic filter, usually 2:1.
- Select the surface overflow rate (SOR) or v_s for the design of the secondary settler, if there is one
- Propose porosity (n) for the support media. If the media is gravel the porosity should be between 35 and 40%. (See Table 8.3).
- Determine the water temperature.
- Calculate detention time for the determined temperature using Equation 6.34.
- Calculate the BOD concentration in the effluent (S) assuming a 50% BOD_5 removal in the Anaerobic Filter and taking into account that the effluent of the UASB is the influent to the Anaerobic Filter.
- Calculate the production of CH_4 using Equation 6.6.
- Calculate the biogas production using Equation 6.7.
- Calculate the input and output structures for a selected number of distribution manifolds (n_M) (See Figure 6.5). The flow per manifold is:

$$Q_D/n_M \tag{6.35}$$

- Select a preliminary water velocity (1 m/s) for the inlet manifolds. This velocity will be corrected later with the "Goal Seek" function of Excel or by trial and error.
- Propose an application rate (q_x) for drying beds.

Design

- Calculate the nominal volume of the anaerobic filter:

$$V_{AF} = Q_D \cdot t_d \tag{6.36}$$

- Calculate the area of the reactor:

$$A_r = V_{AF}/H_F \tag{6.37}$$

- Calculate the total height of the reactor: $H_T = H_F + 0.50$
- Calculate the actual L_V using Equation 5.2 and then correct for the temperature using Equation 6.4. Compare with the limits defined in Table 6.3.
- Calculate the hydraulic load:

$$q = Q_D/A_r \tag{6.38}$$

- Calculate filter width:

$$w = \sqrt{[(A_r/(L/w)]} \quad (6.39)$$

- Calculate filter length:

$$L = A_r/w \quad (6.40)$$

- Calculate the area of the secondary clarifier :

$$A_s = Q_D/v_s \quad (6.41)$$

- Calculate the daily volume of sludge (Q_x), taking into account the porosity (n) using the equation:

$$Q_r = Y(dS/dt)(n \cdot V_{AF}) \quad (6.42)$$

A system to remove the excess sludge should be provided.
- Calculate the daily volume of gas (V_g) with Equations 6.6 and 6.7, taking into account that the void volume of the anaerobic filter is (nV_{AF}).
- Calculate the number of inflow orifices to the anaerobic filter at a density of one every meter, and calculate their diameters taking into account the Equation 6.30. The velocity in each manifold (v) is calculated so that the minimum diameter of the outlet orifice is 1″. The Excel program can use the function "Goal Seek" to perform these calculations.
- Calculate the number and dimensions of drying beds using Equation 6.33.

(6) *Piston anaerobic reactor (PAR):* The piston anaerobic reactor (PAR) is a type of *baffled reactor* which uses a GSLS at the final chamber to prevent suspended solids washout with the effluent and to maintain the sludge age at an as high as possible value. The PAR is open to the air, so no gas collecting system is required and the GSLS is used only for separation the suspended solids from the liquid. Figure 6.7 shows the drawing of a PAR in plan and sections, including the baffles and the SGSL at the end chamber. The design parameters for this GSLS are the same of those for the UASB reactor described in Section 6.2.2, Paragraph 4.1. The liquid volume (and the sludge blanket) is maintained in the baffle chambers, so that the final SGSL only operates with a total depth, $H_T = H_G + 0.50 = 2.5 + 0.5 = 3.0$ m, which is the same depth of the PAR (the 0.50 m added to H_G is the inlet structure below GSLS). A system to remove the excess sludge needs to be provided. In this reactor the GSLS-SM works only as a particle remover, not as a gas collection system, so no specific gas throats are needed in this case. A $w_g = 0.10$ m will works satisfactorily y (See Figure 6.9).

The water velocity (v_b) up and down the baffles must be less than or equal to 3 m/h. The main difference between the UASB reactor and the PAR is that the plug flow in the PAR makes this system kinetically more efficient (see Chapter 2). For this reason its best use is at low temperatures of below 20°C. The detention time is the design parameter and for municipal wastewater the recommended value is 10 hours at 15°C (the detention time calculation includes the volume of the GSLS structure). The design Equation 6.35 is modified to be based on a t_d of 10 h at 15°C:

$$t_d(T) = 10e^{-0.1(T-15)} \quad (6.43)$$

The sludge blanket height in the baffle chambers is 2.00 m (except in the GSLS), whereas the physical height of the PAR chambers is 3.00 m, leaving 1.00 m for flow fluctuations. A free board of 0.20 m needs also to be added. The distance between baffles is given by the design velocity, $v_b = 3.00$ m/h. The detention time takes into account the volume of the calculated

GSLS. The design procedure is performed with the Excel program CHAP 6-Anaerobic-PAR-01. The procedure is the following:

- Define the designer variables for the GSLS, which may be the standard module with $w_s = 4.00$ m and thickness of the 45° inclined baffles (w_b) of 0.15 m, with a depth (H_G) of 2.50 m, adding 0.50 m as the inlet structure at the bottom and 0.20 m of freeboard at the top. The length of the GSLS (transversal to the flow) will be same than the PAR width. In other words, one GSLS standard module similar to the one used in the UASB is also used for the RAP but with a different liquid height (0.50 m instead of 2.5 m in the UASB, but can 1 m). A GSLS-SM with a gas throat of $w_g = 0.20$ m, a $v_r = 1$ m/s and a $v_s = 0.84$ m/h will do. The length of the GSLS (which is equal to the width of the PAR) would be the only variable to be calculated. It is noted that that there is only one (1) GSLS per PAR.
- Define the number of PAR modules, which in this case becomes the number of PAR in parallel (n_{par}) as each PAR can have only one SGSL.
- Calculate the area of GSLS:

$$A_r = (Q_D/n_{RAP})/v_r \quad (6.44)$$

- Calculate the total PAR height:

$$H_{PAR} = H_T = H_{SGSL} + 0.5 \quad (6.45)$$

- Calculate the detention time for the design temperature using Equation 6.43.
- Calculate the GSLS volume:

$$V_{SGSL} = A_r \cdot H_T \quad (6.46)$$

- Including the GSLS volume, calculate the total volume of each PAR:

$$V_{PAR} = Q_D \cdot t_d/n_{RAP} \quad (6.47)$$

- Calculate the width ("length") of the GSLS:

$$W_{GSLS} = w_s + w_g + 2w_b \quad (6.48)$$

- Calculate the length ("width") of GSLS which is the same width for each PAR:

$$L_{GSLS} = A_r/W_{GSLS} \quad (6.49)$$

- Calculate the distance between baffles (W_b) for the selected v_b:

$$W_b = Q_D \cdot v_b/n_{PAR} \cdot L_{GSLS} \quad (6.50)$$

- Calculate the net total volume of the baffles by subtracting the volume of GSLS (without considering the baffle thickness):

$$V_b = V_{PAR} - V_{GSLS} \quad (6.51)$$

- Calculate the net total length of baffles:

$$L_b = V_b/(L_{GSLS} \cdot H_T) \quad (6.52)$$

- Calculate the number of baffles:

$$n_b = L_b/W_b \quad (6.53)$$

- Calculate total baffle length:

$$L_{bD} = L_b + (n_b + 1) \cdot W_b \quad (6.54)$$

- Calculate total PAR length:

$$L_{PARD} = L_{bD} + W_{GSLS} \quad (6.55)$$

- Calculate sludge production using Equations 6.25 or 6.32.
- Calculate the outlet channel using Equation 6.31.
- Calculate production of sludge using Equation 6.32 and the drying beds dimensions using Equation 6.33, as is done in the UASB example.

6.3 BASIC DESIGN EXAMPLE

(The model program for this Example is available online at http://www.iwawaterwiki.org/xwiki/bin/view/Articles/Software+Developed+for+Sustainable+Treatment+and+Reuse+of+Municipal+Wastewater).

A Design example that follows the Orderly Design Method (ODM) of Paragraph 6.2.2 is presented in continuation. The example is developed for a population of 20,000, with the calculation of flow as presented in Chapter 3. The example refers to a UASB reactor followed by an Anaerobic Filter. The ODM is developed step by step in a program developed in Excel, CHAP 6-Anaerobic-UASB-AF.xls, which is posted online as presented elsewhere. The preliminary treatment preceding the UASB reactor is also calculated by this and others programs. A PAR system is also calculated by a different program: CHAP 6-Anaerobic-PAR.xls, for temperatures below 20°C. Please note that the following Tables identified by alphabetic order contain computer calculated results.

(1) *Calculation of external variables*

1.1 *Design flow:* The basis for calculation the design flow is presented in Table 6A.

Table 6A

Design flow				
Variable	**Value**	**Unit**	**Value**	**Unit**
Per capita consumption, q	162.5	L/capita · d		
$PF = 1 + 14/(4+\sqrt{P})$	2.65			Eq. Hamon
Population, P	20000	inhab		
Population density, d	260	inhab/ha		
Return coefficient	0.8			
DWW mean flow, Q_{DWW}	30.09	L/s	108.33	m^3/h
DWW maximum flow, Q_{maxDWW}	79.82	L/s	287.35	m^3/h
Infiltration unit flow, ql	0.13	L/s · ha (o km)		
Afferent area, A_a	76.92	ha		
Infiltration flow, QI	10.00	L/s		

Notes: (i) Per capita consumption is calculated for domestic water consumption. With the return factor of 0.8, it has a wastewater per capita flow of $c \cdot q_{dom} = 0.8 \times 162.5 = 130$ L/hab · d, as measured in the example in Chapter 3; (ii) the peak factor is calculated by Equation 3.6 (*Harmon*); (iii) the population density is selected to be the maximum acceptable according to planning regulations in order to serve the people at saturation; (iv) the flows are calculated based on the population and the return factor (c); and (v) it is preferable that infiltration value be obtained from field measurements. Alternatively a value of between 0.10 and 0.15 l/s · ha or l/s · km can be used.

The final design flows are obtained from Equations 3.3 and 3.4, and they are presented in different units in Table 6B. The Q_D is used for the biological process design and the Q_{DH} for hydraulic design.

Table 6B

Design flows						
Variable	**Value**	**Unit**	**Value**	**Unit**	**Value**	**Unit**
Process design flow, $Q_D = Q_{DWW} + QI$	40.09	L/s	144.33	m^3/h	3464.00	m^3/d
Hydraulics design flow, $Q_{DH} = Q_{DWWmax}$	89.82	L/s	323.35	m^3/h	7760.44	m^3/d

In the case of preliminary treatment the design flow for the rough screen, the grit channel and the Parshall flume is the Q_{DH} since they are hydraulic treatment units. On the other hand, the anaerobic treatment units are designed based on Q_D since they are biological processes units with detention times of several hours during which all variations are averaged.

1.2 *Determination of the geographical and external environmental variables:* These variables depend solely on geographic location. They are presented in Table 6C for the conditions of the example under consideration.

Table 6C

Variable	Value	Unit	Value	Unit
Minimum temperature of water, T_W	25	°C	77.0	°F
Minimum temperature of air, T_a	18	°C	64.4	°F
Altitude above sea level, hasl	350	masl		
Mean wind velocity	6	kph		
Predominant wind direction	NW			
Minimum solar radiation, S	230	$Cal/cm^2 \cdot d$		
Daily fraction of sunlight hours	0.8	decimal		

Notes: (i) These variables are not very important for the preliminary treatment unit but are important for the biological processes; (ii) The minimum water temperature as well as the minimum air temperature is the average of the coldest month; (iii) the speed and direction the prevailing wind is important for locating the wastewater treatment plant in relation to the urban population centers; (iv) the fraction of hours of sunlight is the ratio of actual hours of sunlight to potential sunlight hours.

1.3 *Concentration and pollution loads:* The quality of the municipal wastewater may be different within the city and at the entrance to the wastewater treatment plant if the conveyance system is long and significant inflow infiltration takes place. In the case of the example, the wastewater treatment plant is located nearby the city and there is no significant variation in the wastewater quality along the conveyance pipeline.

Table that calculates as output the load (kg/d) and the specific unit load per Capita (kg/Capita · d) of each pollutant from the concentration of each pollutant is Table 6D. In this table, the concentration of each pollutant is given as an input. The calculation is performed on each pollutant such as BOD_5, COD, TSS, and so on. If there is a significant infiltration into the conveyance pipeline to the treatment plant it is necessary to calculate the additional infiltration flow of the sewerage afferent area (A_a) as follows: $Q_{Iemi} = q_I \cdot L$, where L is the length of the pipeline in km. The concentrations at the entrance to the treatment plant are calculated as the specific per Capita load (q_C) multiplied by the population (P) and divided by the total flow: $Q_T = (Q_D + Q_{Iemi})$. For example: the Concentration of COD $= q_{COD} \cdot P/Q_T$. Note that in Table 6D the load and the per Capita load (specific unit load) is an *output* while the concentrations are an *input*, presented in the Excel sheet separately.

Table 6D

Concentration, load and per capita consumption						
Variable	**Concentration**	**Unit**	**Load**	**Unit**	**Per capita**	**Unit**
BOD_5	277.8	mg/L	962.3	kg/d	0.048	kg/capita · d
COD	596.1	mg/L	2064.9	kg/d	0.103	kg/capita · d
COD/BOD_5	2.1					
TKN	40.0	mg/L	138.6	kg/d	0.007	kg/capita · d
N-Nitrate	2.0	mg/L	6.9	kg/d	0.000	kg/capita · d
Total Phosphorus	5.8	mg/L	20.1	kg/d	0.001	kg/capita · d
pH	7.1	UN				
Alkalinity	100.0	mg/L	346.4	kg/d	0.017	kg/capita · d
TSS	202.6	mg/L	701.8	kg/d	0.035	kg/capita · d
VSS	173.6	mg/L	601.4	kg/d	0.030	kg/capita · d
O&G	100.0	mg/L	346.4	kg/d	0.017	kg/capita · d
Fecal Coli	10000000.0	NMP/100 mL	346400000000000.0	NMP/d	17320000000.000	NMP/capita · d

(2) *Determining the flow diagram of the anaerobic system*

- For the UASB system: The flow diagram of the UASB system is shown in Figure 6.8. It consists of preliminary treatment (course screening → grit Chamber → Parshall flume) followed by a UASB reactor.
- For the Anaerobic Filter system: The flow diagram is identical to that of the UASB system but is also followed by an Anaerobic Filter.
- For the RAP system: the flow diagram consists of a preliminary treatment as shown in Figure 6.8, followed by a RAP reactor (instead of the UASB reactor).

(3) *Sizing the preliminary treatment system*

3.1 *Course screening*

Selection of designer variables: The course screening system can be a Rotating Screen. If the target efficiency for solids removal is 3%, we obtain from Figure 4.8, that a Rotating Screen with 6 mm screen opening will be required. The design flow is the Q_{DH} since the screening is a hydraulic unit. The designer variables are the water velocity in the inflow channel and the screen backwashing pressure, in addition to the removal efficiency of TSS. The Excel program CHAP 4-RM.xls is used to perform these calculations. The values of the selected designer variables are presented in the Table 6E.

Table 6E

Designer variables				
Variable	**Value**	**Unit**	**Value**	**Unit**
Eficiency TSS	3	%		Figure 4.8
Screen opening	5.54	mm		Equation 4.1b
Screen opening (rounded)	6.00	mm	0.236	Inch
Flow velocity in the channel	0.6	m/s		0.6 to 1.0 m/s
Backwashing pressure	350	kPa		Table 4.1
Solids removed	60	L/1000 m^3		Table 4.1

Design: The diameter obtained for the course rotating screen is 600 mm. The final output data are presented in Table 6F. The course rotating screen design is similar to that of the fine rotating screen so the reader is referred to Chapter 4, in which the design process is explained in detail. A coarse screening is sufficient in this case because further treatment is be performed by the UASB-Anaerobic Filter system which follows the preliminary treatment. For better clarity we show in Table 6F the calculations of the preliminary treatment unit which is produced by the Excel program. The design support is presented in the column entitled OBSERVATION.

Table 6F

Design			
Parameter	**Value**	**Unit**	**Observation**
Flow, Q_{HD}	90.000	L/s	
Screen opening, a	6.000	mm	
Drum diamater, D_T	0.600	m	Figure 4.7
Channel width, $W = 1,1\ D_T$	0.660	m	
Channel depth, $h = Q_{DH}/W \cdot v$	0.230	m	Equation 4.2

(*Continued*)

Table 6F (*Continued*).

Design			
Parameter	**Value**	**Unit**	**Observation**
Hydralics radius, $R = Wh/(W + 2h)$	0.136	m	Equation 4.3
Manning's n	0.012		Equation 4.3
Hydrulics slope, $s = (v \cdot n)^2/R^{4/3}$	0.074	%	Equation 4.3
Head loss, $h_f = (1/C \cdot 2g)(Q_{DH}/A)^2$	0.030	m	Equation 4.1
Total head loss, $h_f + 0{,}05$	0.080		Plus 5 cm
Backwashing flow	1.800	L/s	Table 4.1
Solids removed	466.560	L/d.	Table 4.1

3.2 *Grit chamber*

Selection of designer variables: According to the ODM in Chapter 5, Section 5.2.2, Paragraph 3 related to preliminary treatment design, the proposed design variables are presented in Table 6G.

Table 6G

Grit channel					
Variable	**Value**	**Unit**	**Value**	**Unit**	**Observation**
$SOR = Q_{DH}/A_{des}$	800.0	m/d			600–1200 m/s
$V = Q_{DH}/w \cdot H_{des}$	0.2	m/s			0.15–0.60 m/s
# Grit channels	2.00				
L/w	4				

Design: Applying these variables to the equations presented in Chapter 5, Section 5.2.2 Paragraph 3.2 and shown in the left column of Table 6H, we obtain the results for the two grit chambers as shown in Table 6H.

Table 6H

Grit channel			
Parameter	**Value**	**Unit**	**Observation**
Area, $A_{des} = Q_{DH}/SOR$	9.7	m^2	Section 5.2.2 Numeral 3.1
Width, $w == \sqrt{(A_{des}/4)}$	1.6	m	Section 5.2.2 Numeral 3.1
Length, $L = 4w$	6.2	m	Section 5.2.2 Numeral 3.1

(*Continued*)

Table 6H (*Continued*).

Grit channel			
Parameter	**Value**	**Unit**	**Observation**
Depth, $H_{des} = Q_{DH}/v \cdot w$	0.3	m	Section 5.2.2 Numeral 3.1
Total Depth, $H = H_{des} + 0.50$	0.8	m	Section 5.2.2 Numeral 3.1
Parshall flume			
Throat, W	9.0	inch	See "PARSHALL" tag

3.3 *Parshall flume:* For a $Q_{DH} = 90$ Lps the required throat width for flow measurement is $W = 9''$. Rapid mixing is not required.

(4) *UASB*

4.1 *Reactor design*

Designer variables: Using the methodology proposed in the ODM, Section 6.2.2, Paragraph 4.1 and with the guidance of Table 6.2, Figure 6.9 and Equations 6.2 to 6.25 we propose the input variables presented in Table 6I:

Table 6I

Designer variables					
Variable	**Value**	**Unit**	**Value**	**Unit**	**Observation**
UASB standard					
Settling basin width, W_s	4.0	m			Figure 6.10: UASB-SM
GSLS heigth, H_G	2.5	m			Figure 6.10: UASB-SM
Baffle width, w_b	0.15	m			Figure 6.10: Baffle width
Reactor velocity, v_r	0.75	m/h			<1,00 m/h; $V_r < v_s$; Table 6.2
Gas velocity, V_g	1.0	$m^3/h \cdot m^2$			>1,00 m/h; Tabla 6.3
Yield coefficient, Y	0.08				Table 3.2: 0.08
Fraction of CH_4, η	0.65				Typical
Maximum Efficiency, E_{max}	90.00	%			80 to 90%
Volumetric load, Lv at 15°C	2.0	$kg/m^3 \cdot d$			2.0 to 4.0 $kg/m^3 \cdot d$ at 15°C,applied to liquid volume
Efficiency, E	89.31	% DBO_5 or sCOD			Equation 6.3
Actual volumetric load, L_v (T)	5.44	$kg/m^3 \cdot d$			Equation 6.4

(*Continued*)

Table 6I (*Continued*).

Designer variables					
Variable	**Value**	**Unit**	**Value**	**Unit**	**Observation**
Water Temperature	25.00	°C			Coldest month of year
$sCOD_e$ or BOD_5, S	29.69	mg/L	0.03	kg/m^3	Equation 6.5
Methane production, V_{CH4}	124.34	m^3/d			Equation 6.6
Biogas production, V_{biogas}	191.29	m^3/d	7.97	m^3/h	Equation 6.7
Inlet structures					
Branch manifold flow	0.020	m^3 /S			$Q_D/2$
Manifold velocity, v	0.03	m/s			Goal seek: see note in cell N38
orifices per m^2, ε_o	2.00				1 a 2 per m^2
# orfices, preliminary, n_o	114.00				Equation 6.27
# distribuitors, n_d	11.00				Equation 6.27a
# orifices per distributor, n_{od}	5.00				Equation 6.28
Outlet structures					
Module flow, $Q_M = Q_D/n_M$	0.01	m^3/s			
Channel widthl, b	0.30	m			Select
Draying bed					
Applying rate, q_x	150	$kg/m^2 \cdot yr$			Between 120–150

The origin of the data is shown in the right column, OBSERVATION. With these data we have the information necessary to determine the dimensions of the UASB reactor. It is important to remember that the number of modules (n_M) must be an integer. The proposed volumetric load (L_V) is between 2 and 4 kg $BOD_5/m^3 \cdot d$ (or kg $sCOD/m^3 \cdot d$, where sCOD is soluble COD) for the liquid volume, but should be corrected for temperature using Equation 6.4. The COD removal efficiency is calculated using Equation 6.3.

Design: With the designer variables defined, we proceed to design the UASB reactor, applying the methodology and equations presented in the ODM. The results are presented in Table 6J. The right column "OBSERVATION" shows how to calculate each parameter. When calculating the settling velocity (v_s), it must meet the following conditions: (i) be within the limits defined in Table 6.2; and (ii) be higher than v_r. If the second condition is not met (the program produces a zero: 0) it is an indication that adjustment in the design need to be made until the tow conditions are complied and the results are appropriate. Table 6J also shows the gas production (calculated using Equations 6.6 and 6.7), which is the information required for specifying the gas burner. It also shows the excess sludge amounts (calculated using Equation 6.25) required for the design of the drying beds.

Table 6J

	Design				
Parameter	**Value**	**Unit**	**Value**	**Unit**	**Observation**
UASB Standard					
Reactor Area, Ar	192.4	m^2			Equation 6.8
Net height, H_N	0.9	m			Equation 6.9
Effective liquid height, H_L	2.5	m			Equation 6.10 (minimum 2.5 m)
UASB heigth, H_T	5.0	m			Equation 6.11
Detention time, t_d	6.7	h			Equation 6.12
UASB total volume, V_{UASB}	962.2	m^3			Equation 6.13
Sludge blanket volume V_L	481.1	m			Equation 6.14
Total gas extraction area, A_g	8.0	m^2			Equation 6.15
Design UASB length, L_{UASB}	17.0	m			Equation 6.16
Net UASB width, W_{WASB}	11.3	m			Equation 6.17
Total gas width, $W_g = n_{MP} \cdot W_g$	0.47	m			Equation 6.18
# preliminar modules, n_{MP}	2.52				Equation 6.19
# modules, n_M	3.00				Integer rounded up
Gas throat width, W_g	0.16	m			Equation 6.20
Design UASB total width, W_{UASBD}	13.4	m			Equation 6.21
Settling basisn total area, A_s	204.0	m^2			Equation 6.22
Design SOR or $v_{sD} = Q_D/A_s$	0.71	m/h			$<$1,00 and $> v_{rD}$. If otherwise change v_r or/and v_g
Design UASB area, $A_{rD} = W_{UASBD} \times L_{UASB}$	227.27	m^2			For n_M modules: Equation 6.22a
Design v_r, $v_{rD} = Q_D\ A_{rD}$	0.64	m/h			Equation 6.23
Design UASB volume, $V_{UASBD} = A_{rD} \cdot H_T$	1136.35	m^3			6.24a
Design detention time, $t_{dD} = V_{UASBD}/Q_D$	7.87	H			
Sludge production					
$Q_x = V_{UASBD}\ Y\ (S_0 - S)/t_{dD}$	68.76	kg/d			Equation 6.25

(*Continued*)

Table 6J (*Continued*).

	Design				
Parameter	**Value**	**Unit**	**Value**	**Unit**	**Observation**
Biogas production					
Volume per day, V_g	191.3	m^3/d			Equations 6.6 and 6.7
Inlet structure					
Manifold area a_m	0.696	m^2			Equation 6.26, with $Q_D/2$
Manifold diameter, $\varphi_m = (4\ a_m/\pi)\wedge(1/2)$	0.942	m	37.0	Pulg	
Distribuitor area, $a_d = 0.4\ a_m/n_d$	0.025	m^2			From Equations 6.29 and 6.27a
Distributor diameter, $\varphi_d = (4\ ac/\pi)\wedge(1/2)$	0.180	m	7.0	Pulg	
Orifice area, $a_o = 0.4\ a_d/n_o$	0.002	m^2			From Equations 6.30 and 6.28
Orifice diameter, $\varphi_o = (4\ a_o/\pi)\wedge(l/2)$	0.051	m	2.00	Pulg	Goal seek: set cell L38 to 2 by changing C14
Outlet structures					
Channel water depth, h	0.102	m			Equation 6.31
Total channel depth, h_c	0.202	m			$h + 0.10$
Drying bed					
$A_{bed} = Q_x/q_x$	167.3	m^2			Equation 6.33

A summary of the dimensions of the designed UASB reactor are presented in the Table 6K.

Table 6K

Summary of the designed UASB reactor dimensions		
Parameter	**Value**	**Unit**
Designed UASB area, A_{rD}	227.27	m^2
Designed UASB total width, W_{UASBD}	13.4	m
Designed UASB length, L_{UASB}	17.0	m
Designed UASB heigth, H_T	5.0	m

(*Continued*)

Table 6K (*Continued*).

Summary of the designed UASB reactor dimensions		
Parameter	**Value**	**Unit**
# modules, n_M, with GSLS-SM	5.00	
Designed detention time, t_{dD}	7.87	h
Designed Gas throat width, W_g	0.16	m
Designed v_n	0.64	m/h
Designed SOR or v_{sD}	0.71	m/h

4.2 *Inlet and outlet structures and drying beds:* The inlet and outlet structures need to be calculated with care, according to the ODM procedure. Specifically, the design must meet the following conditions: (i) The area (a_d) of each distributor with diameter (φ_d) must satisfy the condition: $\Sigma\, a_d \leq 0.4\ a_m$, where a_m is the area of the manifold with diameter (φ_m); (ii) The area (a_o) of each orifice with diameter (φ_o) must fulfil the condition: $\Sigma\, a_o \leq 0.4\ a_d$; and (iii) The diameter (φ_o) of each orifice should be minimum 2″ to allow the wastewater to flow freely. The "Goal Seek" function in Excel helps to computes the velocity of each manifold pipe so that the diameters of the outlets are 2″. The design diagram of the inlet distribution system is presented in Figure 6.10.

The outlet structures are designed as a water receiving channel using Equation 6.31. As shown in the tables above: b = 0.30 m and h = 0.20 m. See outlet channel cross section in Figure 6.9. The area needed for the drying beds is calculated using Equation 6.33. A drawing of a drying bed in shown in Figure 6.11. A drawing of a UASB reactor with inlet and outlet structures is shown in Figure 6.12.

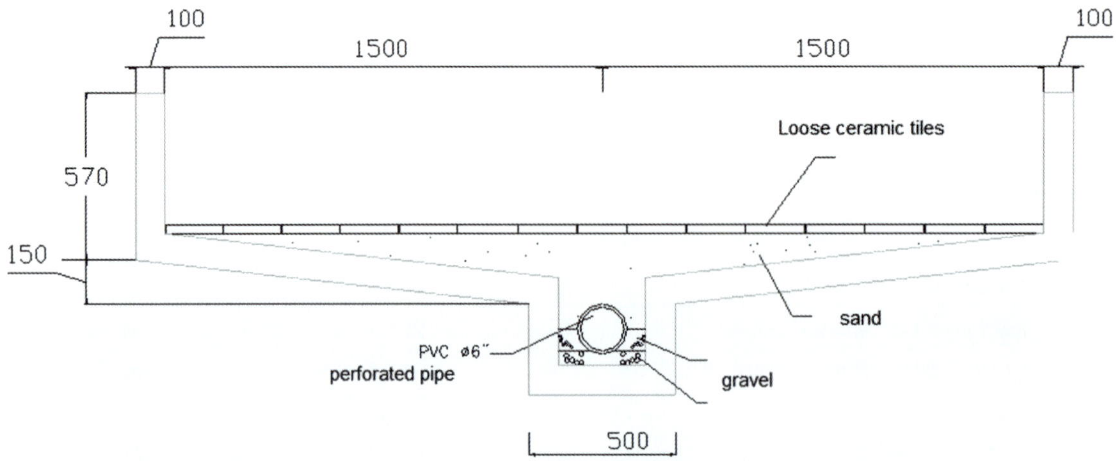

Figure 6.11 Cross section of a typical drying bed (*Source*: Orozco, 2005)

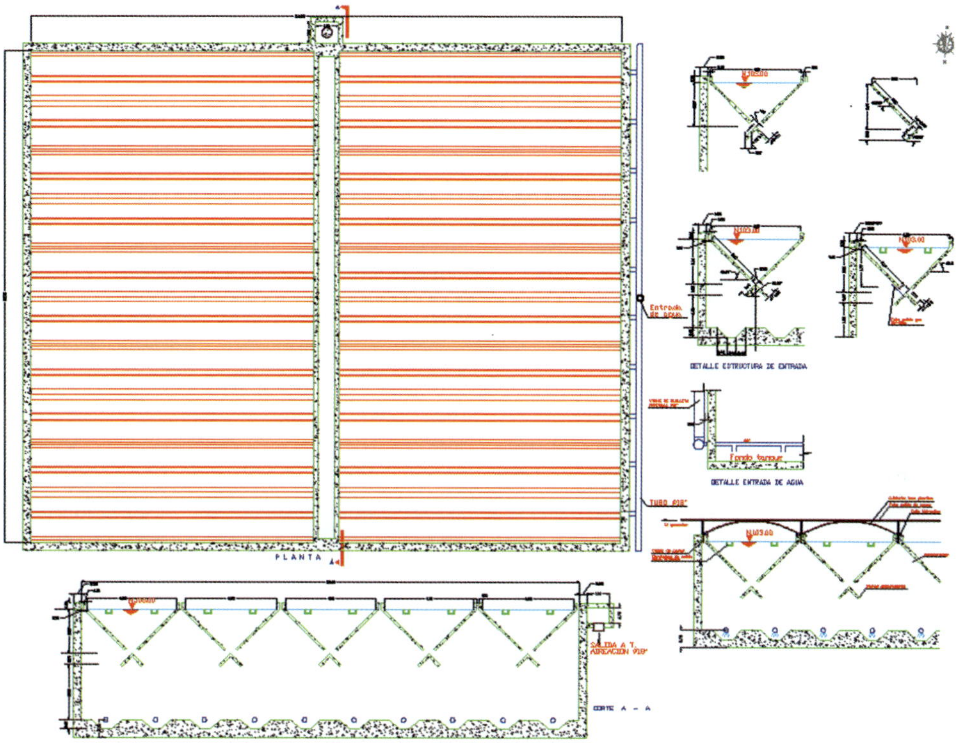

Figure 6.12 Plan and sections of a typical UASB reactor with inlet and outlet structures

(5) *Anaerobic filter*

Designer variables: According to the methodology outlined in the ODM, we propose the designer variables, as shown in Table 6L. The origin of the data is explained in the column "OBSERVATION."

Table 6L

Designer variables					
Variable	**Value**	**Unit**	**Value**	**Unit**	**Observation**
Anaerobic filter					
Detention time at 25°C, t_d	6.0	h			Table 6.3
Filter height, H_F	2.00	m			Table 6.3
Yield coefficient, Y	0.08				Table 2.2
Fración de CH_4, η	0.65				0.65
Length/Width, L/w	2.0				Select 2:1
SOR or v_s	1.0	m/h			Between 0.5 to 1.0

(*Continued*)

Table 6L (*Continued*).

	Designer variables				
Variable	**Value**	**Unit**	**Value**	**Unit**	**Observation**
Medium porosity, n	0.4				0.35 to 0.40. Table 8.4
Water Temperature °C	25.00	°C			Coldest month of year
Detention time at T°C	6.00	h			Equation 6.34
BOD_5 or sCOD, S_e	14.85	mg/L	0.01484503	kg/m^3	0.5 S_{UASBe}
V_{CH4}	5.95	m^3/d			Equations 6.6 and 6.7. Volume $= n \cdot V_{AF}$
V_{biogas}	9.16	m^3/d	0.38	m^3/h	Equation 6.7
Inflow structures					
# manifolds, n_M	3				Select. See Figure 6.5
Flow per manifold, Q_D/n_M	0.01	m^3 s			Equation 6.35
Velocity, V	1.000	m/s			Goal seek
Drying bed					
Application rate, q_x	150	$kg/m^2 \cdot yr$			Between 120–150

The most important parameters are based on Tables 6.3 and 2.2, and the selected values are corrected from 25°C to actual process temperature. It is assumed that the COD removal efficiency in the anaerobic filter acting as a post-treatment of the UASB reactor is 50%. The free volume of the anaerobic filter is calculated by multiplying the filter volume by the porosity (n) of the media.

Design: With the designer variables defined, we proceed to design the Anaerobic Filter of the type presented in Figure 6.5, applying the methodology and equations presented in the ODM. The results are presented in Table 6M. The right column "OBSERVATION" shows how to calculate each parameter.

Table 6M

	Design				
Parameter	**Value**	**Unit**	**Value**	**Unit**	**Observation**
Anaerobic filter					
Nominal volume AF, $V_{AF} = Q_D \cdot t_d$	866.0	m^3			Equation 6.36
Reactor area, $A_r = V_{FA}/H_F$	433.00	m^2			Equation 6.37
Total heigth, H_T	2.5	m			$H_T = H_F + 0,5$
Actual volumetric load, L_V	0.06	$kg/m^3 \cdot d$			Equation 5.3
Volumetric load at T °C, $L_V(T)$	0.06	$kg/m^3 \cdot d$			Equation 6.4
Carga Hidráulica, $q_a = Q_D/A_r$	8.0	m/d			Equation 6.38: between 6 and 15

(*Continued*)

Table 6M (*Continued*).

	Design				
Parameter	**Value**	**Unit**	**Value**	**Unit**	**Observation**
Filter width, w	14.7	m			Equation 6.39
Filter length, $L = A_r/w$	29.4	m			Equation 6.40
Settling area, $A_s = Q_D/V_s$	144.3	m^2			Equation 6.41
Sludge production					
Dayly volume, Q_x	1.6	kg/d			Equation 6.42
Biogas production					
Volume per day, V_g	9.2	m^3/d			Equations 6.6 and 6.7
Inlet structures					
Orifices per manifold	29.00				1 cada m
Manifold area a_m	0.013	m^2			Equation 6.26, with Q_D/n_M
Manifold diameter, φ_m	0.130	m	5.0	inches	
Orifice area, a_o	0.000	m^2	1.8	cm^2	Equation 6.30
Orifice diameter, φ_d	0.015	m	1.00	inches	Goal seek: set L25 to value 1 changing C22
Drying bed					
A_{bed}	4.0	m^2			Equation 6.33

For the inlet structure that requires uniform distribution, it is necessary to apply the criterion of Equation 6.30. The "Goal Seek" modality in Excel helps to computes the water velocity of each manifold pipe so that the diameters of the orifice outlets are 2″ (See ODM). The calculations of the drying bed (based on Equation 6.33) and of the biogas quantity directed to the burner (Equation 6.7) are explained in the section describing the UASB reactor. The settler is designed according to Figure 6.5 with the area calculated with Equation 6.41.

A summary of the dimensions of the designed Anaerobic Filter are presented in Table 6N:

Table 6N

Summary of designed anaerobic filter dimensions		
Parameter	**Value**	**Unit**
Designed AF area, A_{rD}	433.00	m^2
Designed AF total width, w_D	14.7	M
Designed AF length, L	29.4	m
Designed AF heigth, H_T	2.5	m

(*Continued*)

Table 6N (*Continued*).

Summary of designed anaerobic filter dimensions		
Parameter	**Value**	**Unit**
# manifolds n_M	3.00	
Designed detention time, t_{dD}	6.00	h
Designed SOR or v_{sD}	1.00	m/h

(6) *Piston anaerobic reactor, PAR*
Designer variables: Since the PAR is used in areas of low temperatures of between 15 and 20°C, in this example we use for the RAP a wastewater temperature of 16°C, which is different than the temperature used from the UASB example. For this to be consistent other external variables can change but they are not of significant importance for the design. The geographical and external environmental variables are presented in Table 6O. A PAR system is also calculated by a different program: CHAP 6-Anaerobic-PAR.xls, for temperatures below 20°C.

Table 6O

Variable	Value	Unit	Value	Unit
Minimum Temperature of water, T_w	16	°C	60.8	°F
Minimum Temperature of air, T_a	8	°C	46.4	°F
Altitude above sea level, masl	1850	masl		
Mean wind velocity	6	kph		
Predominant wind direction	NW			
Minimum solar radiation, S	230	$Cal/cm^2 \cdot d$		
Daily fraction of sunlight hours	0.8	decimal		

With this new temperature, and using the methodology proposed in the ODM, Section 6.2.2 Paragraph 6, we obtain the input data presented in Table 6P (the selected designer variables), whose origin can be seen in the column denominated "OBSERVATION".

Table 6P

Designer variables					
Variable	**Value**	**Unit**	**Value**	**Unit**	**Observation**
Piston anaerobic reactor, PAR					
Settling basin width, W_s	4.0	m			Figure 6.10: UASB-SM
GSLS heigth, H_G	2.5	m			Figure 6.10: UASB-SM
Baffle thickness, w_b	0.15	m			Figure 6.10: Baffle width
Flow velocity in reactor v_r	0.75	m/h			$<$1.00 m/h; $V_r < V_s$; Table 6.3

(*Continued*)

Table 6P (*Continued*).

Designer variables					
Variable	**Value**	**Unit**	**Value**	**Unit**	**Observation**
Settling velocity, SOR, v_s	0.84	m/h			Calculated for a gas throat of $W_g = 0.10$ m
Gas throat, w_g	0.10	m			Given
Detention time at 15°C	10.0	h			Typical
Yield coefficient, Y	0.08				Table 3.2: 0.08
Maximum Efficiency, E_{max}	90.00	%			80 to 90%
Efficiency, E	85.19	% DBO_5 or sCOD			Equation 6.3
Water Temperature	16.00	°C			Coldest month of year. Should be less than 20°C
$sCOD_e$ or BOD_5, S	41.15	mg/L	0.04	kg/m^3	Equation 6.5
Flow velocity in baffles, v_b	3.0	m/h			≤ 3.0 m/h
PAR units, n_{PAR}	3				Select
T agua	16.00	°C			Below 20°C
Outlet structures					
Flow per PAR	48.111	m^3/h	0.0133642	m^3/S	Q_D/n_{RAP}
Channel width, b	0.30	m			Select
Drying bed					
Application rate	150	$kg/m^2 \cdot yr$			120–150

Design: Note that 3 PAR reactors were selected, so the design flow of each reactor is $Q_D/n_{PAR} = Q_D/3$. Also note that there is one (1) GSLS-SM for each PAR reactor, and the "GSLS length" is the same as the PAR reactor width. From the input data, we obtain the results shown in Table 6Q following the ODM. The calculations were made with the equations presented in the ODM and identified in the right column denominated "OBSERVATION".

Table 6Q

Design					
Parameter	**Value**	**Unit**	**Value**	**Unit**	**Observation**
Plug-flow anaerobic reactor, PAR					
Area GSLS A_r	64.1				Equation 6.44
Total PAR height	3.0				Equation 6.44
Baffles height, $H_L = H_T$	3.0				Equation 6.45

(*Continued*)

Table 6Q (*Continued*).

	Design				
Parameter	**Value**	**Unit**	**Value**	**Unit**	**Observation**
Actual detention time, $t_d(T)$	9.0	h			Equation 6.43
Total Volume GSLS, V_{GSLS}	192.4	m^3			Equation 6.46
Volume PAR, V_{PAR}	435.3	m^3			Equation 6.47
GSLS width, W_{GSLS}	4.4	m			Equation 6.48
GSLS length, L_{GSLS}	14.6	m			Equation 6.49
Baffles' distance, W_{BAFLE}	1.1	m			Equation 6.50
Net Baffle total Volume, V_b	242.9	m^2			Equation 6.51
Net Baffle total lengths, L_b	5.6	m^2			Equation 6.52
# baffles, n_b	5.00				Equation 6.53
Baffle total legth, L_{bD}	6.40	m			Equation 6.54
PAR total length, L_{PARD}	10.80	m			Equation 6.55
Sludge production					
Volume per day, Q_x	21.9	kg/d			Equation 6.32
Outlet structure					
Water depth, h	0.102	m			Equation 6.31
Total channel depth, h_T	0.202	m			$h_T + 0.10$
Drying bed					
A_{BED}	53.2	m^2			Equation 6.33

A summary of the designed PAR reactor is presented in Table 6R.

Table 6R

Summary of designed par dimensions		
Parameter	**Value**	**Unit**
Designed GSLS area, A_{rD}	3.00	m^2
Designed GSLS total width, W_{UASBD}	4.4	m
Designed GSLS length or PAR width, $L_{UASB} = W_{PA}$	14.6	m
Designed PAR heigth, H_T	3.0	m
Baffle units, n_b	5.00	
Baffle total legth, L_{bD}	6.40	m
PAR total length, L_{PARD}	10.80	m
Designed detention time, t_{dD}	9.05	h

Chapter 7

Stabilization reservoirs, concepts and application for effluent reuse in irrigation

7.1 PROCESS DESCRIPTION

7.1.1 Introduction

All municipal wastewater treatment processes depend on: (i) natural phenomena such as gravity which induces sedimentation, and (ii) of natural biological compounds such as organisms. However, in the case of conventional wastewater treatment plants, these natural components are supported by a series (often complex) of mechanical equipment intensive in energy consumption. In contrast, natural treatment processes, many of which are discussed in this book, are processes that depend mostly on natural components to perform the treatment. A natural system may include pumps and pipes to transport the wastewater, but it does not depend solely on external energy sources to carry out the treatment process (Crites, Middlebrooks & Sherwood, 2006).

The land application of the municipal wastewater (also known as overland flow) was the first "natural" technology (recently re-discovered). In the nineteenth century it was the only method used for treatment of sewage, but it was gradually replaced by other, more modern methods. Studies and research established that land application treatment can obtain many desirable treatment goals, while at the same time, significant benefits are obtained by the reuse of minerals and organic matter applied on land. Other "natural" concepts such as lagoon systems and the application of sewage sludge on land have never fallen into disuse. The stabilization lagoons, including stabilization reservoirs, simulate physical and biochemical interactions that occur in natural lakes, while the application of sewage sludge on land mimics conventional agricultural practices of utilization of manures. The concepts of surface and subsurface artificial wetlands are new developments related to reuse of wastewater and sludge.

Of these natural treatment processes, the Stabilization Reservoirs (SR) for wastewater treatment and reuse for irrigation, presented in this chapter, represents an interesting cost-effective option for municipal wastewater post-treatment and utilization. Other natural systems, such as stabilization lagoons and constructed wetlands are discussed in other chapters of the book and other publications (such as Crites, Middlebrooks & Sherwood, 2006).

According to Libhaber (2007), reuse of wastewater for irrigation is a combined concept of treatment and disposal. Stabilization Reservoirs offer an alternative disposal of municipal wastewater which is better than conventional ones because it generates economic and agricultural benefits and eliminates the negative impacts of discharge of effluents to receiving water bodies. A stabilization reservoir system using simple pretreatment processes can produce effluents with the required quality for both restricted irrigation (i.e. adequate for irrigation of industrial crops) and for unrestricted irrigation (i.e. for irrigation of food crops consumed uncooked).

The stabilization reservoirs concept was developed in Israel in the early 1970s and is in wide spread use in this country. Over 250 stabilization reservoirs are operated today in Israel and their number is increasing with time. Therefore, the design methodology and most of the information on the process and related examples come from Israel. The use of reservoirs to store and treat wastewater is an old practice, but the development of deep reservoirs to reduce evaporation losses, the widespread use of reservoirs for long term-storage of wastewater and the massive research undertaken to better understand the behaviour of this type of reservoirs and improve their design and operational criteria are relatively new. As explained by Libhaber (1996), stabilization reservoirs are deep lagoons (with depth of 4–12 meters, but usually in the range 8–12 meters) with a variable water level, that are used for a combination of two objectives: (i) storing partially treated sewage over a long period in order to use it for irrigation or discharge it to a receiving water body during a specific period of the year, under strictly controlled optimal conditions; and (ii) improve the quality of effluents stored during the long residence time within the reservoir. The controlled release is possible thanks to climates with "rains-dry" seasons which occurs in several regions of the world and generates a demand for large volumes of water for irrigation during the summer.

In some cases, wastewater can be stored in reservoirs during the summer in order to be discharge into the sea during the rainy season, which is not a bathing season. Also, in cases of effluent discharge into rivers that affect their use, reservoirs can store the effluent during the season of minimum flow in the river (dry season), in order to discharge it to the river during the high flow season when the dilution capacity is high (usually during the rainy season).

The stabilization reservoir process is presented and discusses as one of the unit processes of appropriate technology in Paragraph 1.7.2.7 of Chapter 1. As part of this discussion, it was clarified that before discharging the wastewater to the reservoir for storage, it needs to be treated to remove a part of the organic matter contained in it; so as to prevent development of anaerobic conditions in the upper layer of the reservoir. Prevention of such conditions is important since anaerobic condition may generate odours. So in fact, the reservoir stores effluent and not raw wastewater. The reservoir system is therefore a combined system consisting of a pretreatment unit followed by the reservoir. The required pre-treatment of the wastewater before it is stored in the reservoir is not a specific treatment processes. It can be achieved using a variety of treatment process, among them; all the appropriate technology processes or combination of processes described in this book, as long as they remove the required load of organic matter and ensure that the surface organic loading on the reservoir does not exceed the permitted limit. The reservoir itself, in addition to its function seasonal storage of effluent, is also a treatment unit. It functions as a facultative lagoon with a varying water depth. The upper layer of the reservoir (whose depth depends on the specific design of the reservoir and on the effectiveness of the pretreatment unit) is an aerobic layer in which algae and bacteria develop. The algae generate oxygen which is consumed by the bacteria that decompose the organic matter present in the reservoir. An anaerobic layer is formed beneath the upper layer and anaerobic bacteria develop in this layer and decompose the organic matter which arrives in it. Pathogenic organisms decay in the reservoir by similar mechanisms of decay in conventional lagoon. However, the detention time in a stabilization reservoir is much larger than in regular lagoons so the removal of pathogens can be much more effective. The actual removal of pathogenic organisms depends on the mode of operation of the reservoir, as discussed below. The combination of the pretreatment unit and the reservoir form as a whole an appropriate technology process, if the pretreatment unit is based on appropriate technology.

The schematic process flow diagram of pretreatment followed by a stabilization reservoir is presented in Figure 7.1. This figure is not the flow diagram of just one combined process but rather of a large series of combined processes, since the pretreatment unit is not just one specific unit, but can be one of a variety of unit processes or combination of unit processes.

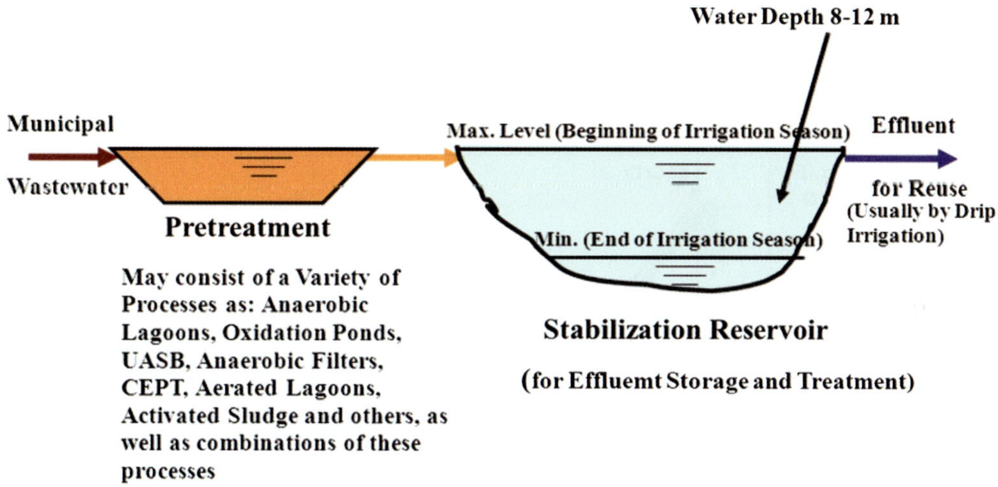

Figure 7.1 The stabilization reservoir reuse concept and schematic process flow diagram (*Source:* Libhaber Menahem, 1996)

The aspects of wastewater reuse for irrigation, the characteristics of stabilization reservoirs and reservoirs as a combined appropriate technology systems are discussed in Chapter 1 (Paragraphs 1.7.2.7 and 1.8.8). The current chapter deals with the design procedure of the stabilization reservoir. We believe that every wastewater reuse for irrigation project needs to include a stabilization reservoir in order to be successful. Inspection of the inflow pattern, which is the constant raw wastewater flow along the year, and the outflow pattern which is the variable pattern of consumption for irrigation, as shown in Figure 1.20, explains the need for the seasonal storage, that is, the need for a storage reservoir in a reuse project.

7.1.2 Basics of the process

Fundamentals

The design concepts of Stabilization Reservoirs (SR) are similar to those of stabilization lagoons, presented in Chapter 5, namely on the following parameters: (i) the organic surface loading (L_S) in (kg BOD_5/ha · d); (ii) the detention time (t_d) days; and (iii) the temperature (T) °C. The hydraulic regime is very important and due to the impact of the wind and the long fetch resulting from the large dimensions of reservoirs, they usually act as completely mixed reactors. During periods of stratification the upper layer acts as completely mixed. The equations which apply for the design of a reservoir, as discussed below, are similar to those used in lagoons, but are modified for the hydraulic regime. In addition, there are limnological components to be taken into account in the operation of reservoirs. For basic kinetic equations refer to Chapter 2. The modified kinetic equations are presented below.

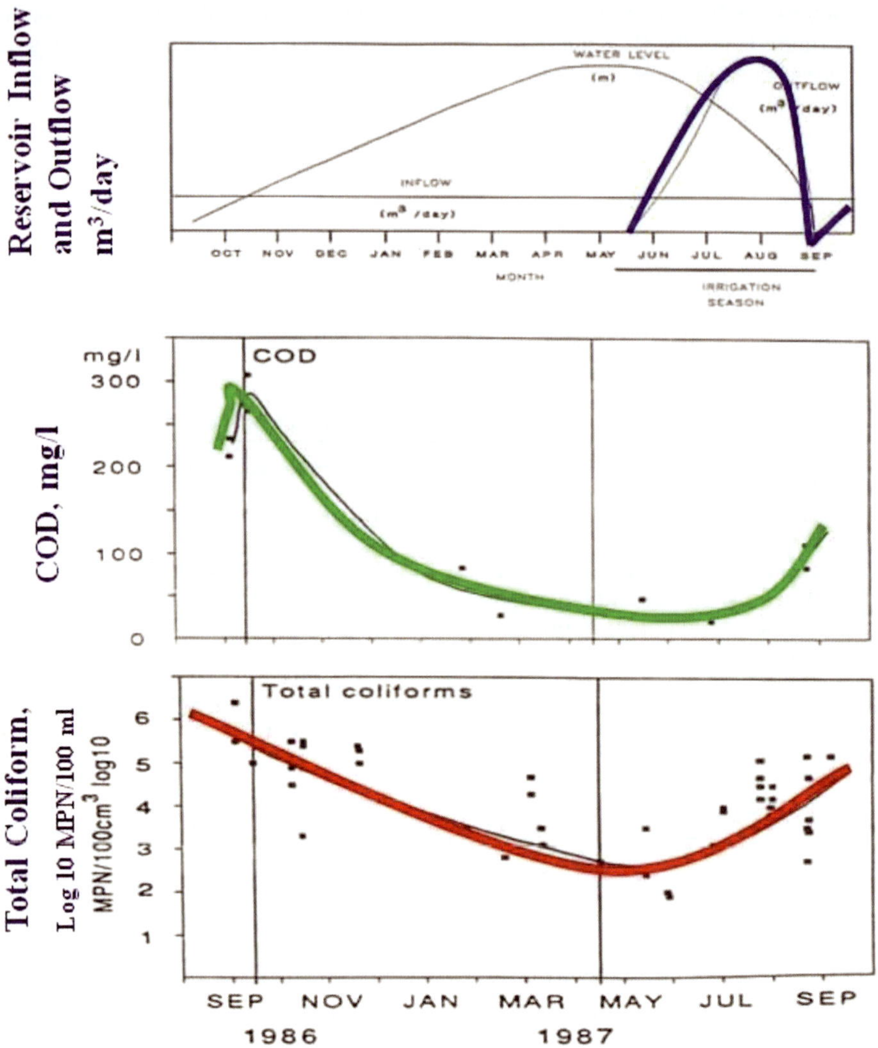

Figure 7.2 Typical annual variations of inflow, outflow, effluent COD and effluent total coliforms in a typical single continuous flow stabilization reservoir (the Getaot Reservoir) (*Source:* Adapted from Juanico & Shelef, 1994)

In terms of public health aspects of wastewater reuse, the guidelines of the World Health Organization[1]. (WHO) need to be addressed. It has been shown that the if these recommendations are applied, reuse of treated municipal wastewater presents a public health risk of between 10 and 1000 times lower than the than the risk posed by consumption of drinking water. WHO recognizes that the treatment of wastewater

[1]Namely: (i) intestinal nematode eggs: $\leq$1 per liter; and (ii) for unrestricted irrigation: fecal coliforms $\leq$1000 per 100 mL. In 2006 the WHO issued new guidelines, but they are less stringent.

in stabilization lagoons is an effective and economical way of obtaining the microbiological quality limits for the reuse of wastewater. However, the effluents of the stabilization lagoons can be used to irrigate crops only during the dry season. In other seasons of the effluent must be discharged into surface streams or other receiving water bodies, unless it is stored in a Stabilization Reservoir to be used during the irrigation season. The use of stabilization reservoirs is also beneficial due to the fact that significant additional treatment occurs in them (Mara & Pearson, 1999, pp. 591).

According to Libhaber (1996), the priority criteria for treatment of wastewater for the specific objective of reuse of effluent for irrigation are: (i) maximum removal of helminths and protozoa, (ii) effective removal of bacterial and viral pathogens, and (iii) effective reduction of BOD to prevent unpleasant doors and unpleasant appearance of the effluent in the reservoir. This priority order differs from that used for the control of BOD and COD pollution in surface water. The optimal treatment strategies for reuse are therefore different from strategies for control of contamination of surface waters.

The operating regime used in a single continuous flow stabilization reservoirs is such that during the irrigation season when the withdrawal of effluent from the reservoir starts, the treated effluent from the lagoons or any other pretreatment system which feeds the reservoir keeps flowing, so that an increasing proportion of pretreated effluent mixes with the initial content of the well treated liquid contained in the stabilization reservoir. This causes that towards the end of the irrigation season the effluent withdrawn from the reservoir for irrigation has an increasingly deteriorated quality, especially in terms of pathogenic organisms, as shown in Figure 7.2 (referring to the single continuous flow stabilization reservoir at Getaot, Israel). The lower quality of effluent has no negative impact when the effluent is used for restricted irrigation. However, when it is used for restricted irrigation (i.e. irrigation of crops that require a high quality effluent all the time) the quality deterioration towards the end of the irrigation season presents a problem. A strategy to overcome the problem by the use of two reservoirs is being practices in Israel for many years and is explained by Mara & Pearson (1999) as discussed below.

Operation regimes

There are many ways of operating a reservoir system which consists of a pretreatment unit followed by the stabilization the reservoir (a lagoon system as the pretreatment unit is a preferred solution, but there can be others, practically all the appropriate technology unit processes and combinations of these processes) but they can be reduced to three basic operation modes, as shown in Figure 7.3 (Mara & Pearson, 1999; Mara, 2004), which is based on lagoons as pretreatment:

(i) *Operation mode of Figure 7.3(a)*: This is the conventional mode. In this case the pre-treatment is an anaerobic lagoon but can be different. Under this mode of operation there is one single stabilization reservoir which is continuously filled up with the pretreated effluent all the year round and discharges effluent only during the irrigation season. The effluent quality deteriorates at the end of season, as seen in Figure 7.2, because the good water quality is used up at the beginning of the irrigation season and then. Finally, the lower quality of water from the bottom is mixed with the anaerobic pond effluent. In consequence the effluent of this system can be use only for restricted irrigation, mainly of industrial crops. This system is efficient and is in widespread use, but is limited only to restricted irrigation. The main problem of this system is that the inflow stream to the reservoir continues to flow even during the irrigation season. We saw that the main requirement for unrestricted irrigation is to achieve a low content of coliforms in the effluent. The inflow stream to the reservoir contains high concentrations of coliforms. Due to the wind action the content of the reservoir is well mixed and coliforms which enter with the inflow easily reach the outflow and render the effluent to be of restricted quality.

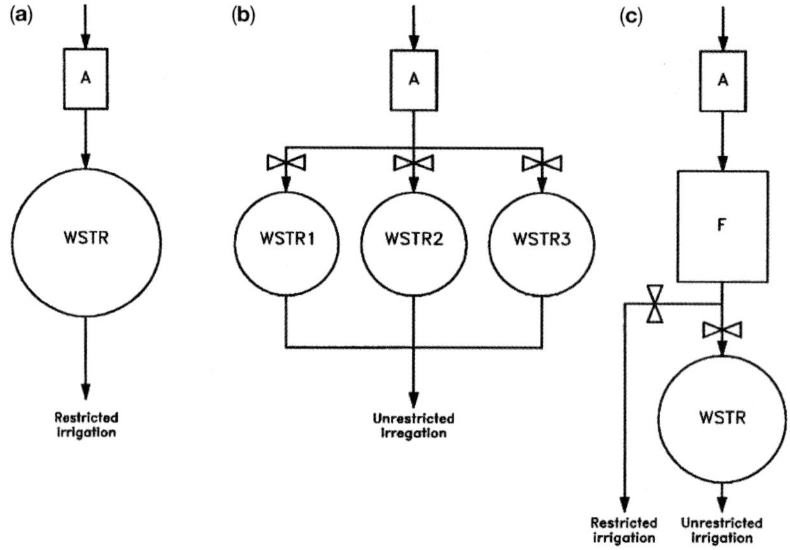

Figure 7.3 Operation options of stabilization reservoirs (a) single SR for restricted irrigation; (b) sequential batch-fed for unrestricted irrigation; (c) hybrid system for both restricted and unrestricted irrigation A: anaerobic lagoon; F: facultative lagoon, WSTR: Waste Storage and Treatment Reservoir or Stabilization Reservoir (*Source:* Mara & Pearson, 1999)

(ii) *Operation mode of Figure 7.3(b)*: Under this mode the pretreatment unit is followed by several stabilization reservoirs, usually two, which are located in parallel. The reservoirs are filled one after the other. When the first reservoir becomes full, the second one starts to be filled. This mode of operation is a sequential batch feeding mode. The principle of operation is that effluent is withdrawn for irrigation only from a reservoir which does not receive an inflow stream. That means that coliforms are not coming into that reservoirs, coliforms which have previously entered have decayed with time and its effluent contains a low level of coliforms and is therefore fit for unrestricted irrigation. Even of food crops that are eaten raw. During the irrigation season, the pretreated effluent fills the second reservoir and will be used in the following year's irrigation season, so this system provides not only seasonal storage but actually multiannual storage. The storage volume in this system is larger than in the previous one so it is more expensive, but it also generates a higher quality of effluent which, allows producing higher value crops.

(iii) *Operation mode of in 7.3(c)*: This is the hybrid system, in which the pretreatment is of higher level (in the figure it consists of an anaerobic lagoon followed by a facultative lagoon, but it can also be of a different type, albeit of a level higher than just an anaerobic lagoon). Before the irrigation season, the pretreated effluent fills the reservoir. When the irrigation season starts, the inflow to the reservoir is stopped. The effluent of the reservoir is then used for unrestricted irrigation, while the effluent of the pretreatment unit is used for restricted irrigation. The difference between this operation mode and the previous mode is that in this case the effluent of the pretreatment unit is not stored in a second reservoir for use of unrestricted irrigation during the next season, but is rather used during the same season albeit only for restricted irrigation. This system is of intermediate cost and its use is broad and serves for all types of crop.

Additional explanations on the various modes of operation and information on the effluent quality yielded by each mode are presented in Chapter 1 and in Figures 1.26, 1.27 and 1.28. A photo of a system type (a) of Figure 7.3 is presented in Figure 1.77, which is the system of the Naan reservoir in

Israel and includes the anaerobic pretreatment unit and the reservoir. This system can also be operated as system type (c) in Figure 7.3. In fact, systems (a) and (c) are similar in terms of infrastructure. The difference between them is in the mode of operation. A photo of a system type (b) of Figure 7.3 is presented in Figure 1.29, which is the Maale Hakishon reservoir system consisting of two reservoirs. The pretreatment system is not shown in Figure 1.29 because it is located in a different site, close to the city which is the source of the wastewater.

Design parameters of a stabilization reservoir

The design parameters of a stabilization reservoir are the same as those of a lagoon system, namely: volume (V), surface organic loading (L_s), detention time (t_d) and depth (H.) However, as there are differences in the flow regimes of a lagoon and a reservoir, the calculation of the parameters is done differently. In the case of a stabilization reservoir it is also necessary to include other design parameters.

Percentage of Fresh Effluents, PFE_i: This is a most important parameter, even more than the detention time. The ***PFE_i*** (percentage of "fresh" pretreated effluent residing in the reservoir between 1 to *i* days), is expressed as the percentage of THE volume of pretreated effluent residing in the reservoir 1 to *i* days, in relation to the total volume of water in the reservoir (Juanicó & Dor, 1999). For example, PFE_{30} is the percentage of pretreated effluent residing in the reservoir less and up to 30 days (expressed as a percentage of the total Reservoir content). The volume of effluent residing in the reservoir up to 30 days is the volume of effluent which resides from 1 to 30 days. This means the volume of pretreated effluent residing 1 day plus the volume residing 2 days plus the volume residing 3 days and so successively until plus the volume residing 30 days. The total fresh pretreated effluent is the portion that is still not well treated in the reservoir and since, because of the mixing characteristics in the reservoir (due to the wind effect), part of it is discharged with the reservoir effluent; it is the parameter that affects negatively the quality of the reservoir's effluent. So the smaller the value of the fresh pretreated effluent in the reservoir, the better is the quality of the content of the reservoir and of its effluent. According to Juanico and Dor, 1999, it is the *PFE* that governs the effluent quality of any reactor. In lagoon, which perform under steady state flow conditions (the inflow and outflow are continuous and equal), the ratio PFE/t_d (*PFE* to Detention Time) is constant, so t_d can substitute for *PFE* in the analysis of reactor performance and hence the close relationship between the performance of perfectly mixed steady state reactor (the lagoon) and the detention time. So for a lagoon, it can be said that it is the detention time that determines the quality of the lagoon effluent. In a reservoir, which is not a steady state reactor (the flow is not continuous and the inflow and outflow are not equal), the performance is related to *PFE* and it is the *PFE* (the percentage of fresh pretreated effluent in the reservoir) which determines the quality of the reservoir's effluent. The content of each specific quality parameter in the effluent of the reservoir depends on the kinetics of the removal of this parameter in the reservoir. It was found that the value of PFE_{30} has the highest correlation to the removal of BOD and that the values PFE_1 and PFE_5 have the highest correlation to the removal of faecal coliforms in a reservoir. Hence the importance of PFE_{30} and of $PFE_1 - PFE_5$.

Mean Detention Time (or Mean Residence Time): the *mean time of detention* (t_{dmj}) is used in the design of a stabilization reservoir, but unlike the detention time in lagoons which function under stable conditions of continuous flow ($t_d = V/Q$), the detention time of a stabilization reservoir that operate under *non-continuous conditions* (the reservoir is filled throughout the year and discharged during only a few months, more like a Sequential Batch Reactor, SBR) is calculated on a daily basis, day by day, with the following equation:

$$t_{dmj} = \frac{[(t_{dm(j-1)} + 1)\,V_{(j-1)}] + 0.5\,v_j}{V_{(j-1)} + v_j} \tag{7.1}$$

where:

j = day from the start of filling
t_{dmj} = mean detention time up to day j
v_j = volume of inflow into the reservoir during day j
V_{j-1} = volume of the reservoir on day $j-1$

Equation 7.1 defines the average detention time on days j and (j-1). Storage in stabilization reservoirs is a seasonal storage and that implies that the mean detention time in a stabilization reservoir is long, reaching values of a few months. Usually it is in the range 100–200 days. Figure 7.4 shows an example of typical annual variation of the parameters of interest in a stabilization reservoir. It is noteworthy that the variation of *PFE* is inversely proportional to the average detention time (increasing average detention time decreases the fraction of "fresh" pretreated effluent in the reservoir), but is proportional to the organic load, since increasing the "fresh" pretreated effluent portion increases BOD. Other quality parameters behave the same way. This shows that the *Percentage of Fresh Effluent* is the parameter that governs the behaviour of a stabilization reservoir. Consequently it is clear that one way to control the effluent quality of the reservoir is by making the *PFE* minimal, as in the design shown in Figure 7.3(b).

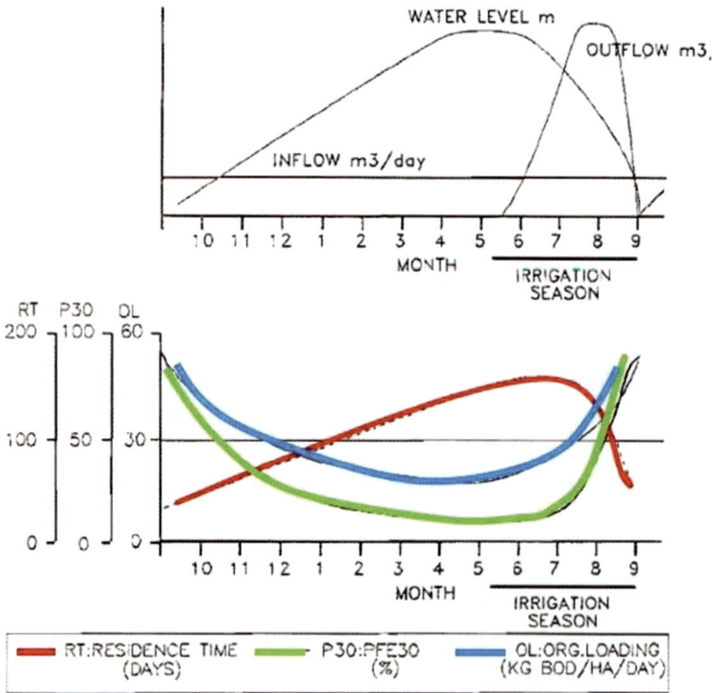

Figure 7.4 Operational parameters (Flow, t_{dmj}, PFE_{30} and organic loading) of a typical stabilization reservoir in Israel, when operated as a continuous flow reactor, that is, influent continues to enter the reservoir during the irrigation season (*Source:* Adapted from Juanico, 1993)

This is achieved by constructing several stabilization reservoirs so that when one is filled and reaches a low *PFE* the filling of this reservoir is stopped and the filling of another one starts. In this way the *PFE* of a full reservoir does not deteriorate. When the irrigation season arrives, the withdrawal of effluent is done from the reservoir with the oldest water in which the *PFE* is the smallest, the t_{dm} is the highest and the water quality is the best. A dual or multiple reservoir system, such as system (b) in Figure 7.3 is more expensive than a single reservoir system, as systems (a) and (c), but the cost effectiveness analysis of each system needs to take into account the income from agriculture. System (b) which is of higher investment might turn out to be more cost-effective because of its higher effluent quality (adequate for unrestricted irrigation) which may generate more income. In addition, the issue of regulation needs to be taken into account. In certain countries reuse is permitted only when the effluent quality is adequate for unrestricted irrigation. In all cases, an economic analysis based on local conditions and regulations needs to be undertaken to decide which reservoir system is the optimal for each specific case,

Volume: The volume of a stabilization reservoir is specific to each case. It depends on the flow of the wastewater, on climate (the length of the irrigation season) and on the type of the irrigated crops (which define the variation with time of consumption of effluent for irrigation). Stabilization reservoirs are used for seasonal storage so they are water bodies of large volume. In Israel, reservoirs volume range from 20,000 m^3 (the smallest) to 12 million m^3 (the largest which are two reservoir in the Maale Hakishon project receiving the treated effluent of the city of Haifa, each with a volume of 12 million m^3 and are presented in Figure 1.29). Most of the reservoirs have a volume in the range 200,000–6,000,000 m^3. The calculation of the volume of the stabilization reservoir is a very important stage in the design. To design the reservoir it is necessary to take into account the topography of the site, the geometry of the reservoir, the geology at the location of the reservoir and the depth of the reservoir, which is usually between 4 and 12 m. Deeper reservoirs with an active depth of up to 20 meters are being considered. Dissolved oxygen in reservoirs comes from photosynthetic activity of algae and from diffusion of atmospheric oxygen. Thus deeper reservoirs with a small area/volume ratio have inferior oxygen supply conditions than shallower reservoirs. Consequently, deeper reservoir need to be operated with lower surface organic loading (which means a higher level of pretreatment) or with mixers and even aerators, to maintain in them aerobic conditions. Once the depth and embankments slopes are defined, the volume is calculated based on the incoming flow and the pattern of the outflow for irrigation, taking into account a dead volume of 1 meter deep at the bottom for sediments storage, and a freeboard, which is around 1 m, but should be calculated for each case on the basis of the intensity of local winds and the size of the reservoir (from which the fetch is calculated).

The volume of wastewater stored in the stabilization reservoir system is calculated on the basis of the flow of the incoming wastewater and the pattern of consumption of effluent for irrigation, which in turn depends on the type of crop to be irrigated. The monthly precipitation-evaporation balance must be taken into account in the volume calculation procedure. Similarly, water loss by infiltration from the bottom of the reservoir must be taken into account in permeable sites, if impermeabilization is not done. Due to the slope of the embankments the surface area of the water in the reservoir varies significantly with the water level, as can be appreciated from Figure 1.25.

***Surface Organic Loading* (L_s):** The dimensions of the reservoir, including its surface area, are established on the basis of the raw wastewater flow and the consumption of water for irrigation, that is, on the basis of hydraulic data without taking into account the quality of the wastewater. However, if the organic matter concentration in the inflow stream entering the reservoir will be too high, the reservoir will become anaerobic up to the surface layer and may consequently generate offensive odours. To avoid such conditions, the raw wastewater must be pretreated prior to its inflow to the reservoir so as to decrease the organic loading on the reservoir to levels which will ensure the preservation of an aerobic top layer in the reservoir. Such aerobic layer prevents generation of odours. Oxygen balance models of reservoirs, as well as experience, established that the maximum permissible organic loads which still prevent generation of anaerobic conditions in the upper layer of stabilization reservoirs are 30–50 kg BOD/d · ha during the winter low temperature season and 60–100 kg BOD/d · ha during autumn and high temperature summer season. In Figure 7.4 the variation of the organic loading on a continuous flow single reservoir is in the range 15–60 kg BOD/d · ha. The control of the organic loading on the reservoir is achieved by providing a pretreatment unit which reduces the organic matter content in the wastewater before it is discharged to the reservoir. The pretreatment unit is designed in such a manner that it removes sufficient organic matter from the raw wastewater so that the organic loading on the reservoir will be maintained within the permissible limits. Another way to prevent odour generation in the reservoir is to install in it mixers (not aerators) identical to the mixers for facultative lagoons discussed in Chapters 1 and 5. These mixers perform in the reservoir in the same form they perform in lagoons, increasing the capacity of the reservoir as a biological reactor and allow for increasing the maximum permissible organic load on a reservoir. Preliminary results indicate that an increase in the organic load of up to 250 kg BOD/d · ha is possible for a reservoir equipped with mixers. Most of the reservoirs in Israel are not equipped with mixers and their performance is controlled by the pretreatment unit. The practice is that if a reservoir starts to generate odours after years of odourless performance (usually due to increase of contributing population and therefore wastewater flow) mixers are used to resolve the problem. Figure 1.32 shows installation of mixers in a reservoir and Figure 1.33 shows the reservoir operating with the mixers in it and Figure 1.78 shows a satellite photo of this reservoir with the mixers operating in it.

In the design process, the area considered for calculating the average surface organic loading is taken as the area at the mid-depth of the reservoir without taking into account the dead volume of 1 m depth. To facilitate the calculations curves of: (i) Area vs reservoir water level, and (ii) Volume *vs* reservoir water level are prepared based on the topography and designed slopes.

In terms of pretreatment units, many systems in Israel are based on scheme (a) of Figure 7.3 under which the pretreatment is only anaerobic lagoons, as can be seen in Figure 1.77. Many others are based in scheme (b) of Figure 7.3, in which the pretreatment unit is a combination of anaerobic lagoons followed by facultative lagoons. Another pretreatment systems used frequently in Israel is the system presented in Figure 7.5 (Juanico & Milstein, 2004), which consists of anaerobic ponds followed by aerated lagoons. Photos of such a system are shown in Figure 1.74 and Figure 1.82 (in both cases it is the system of the city of Nazareth). As discussed in Chapter 1, aerated lagoons treatment is not an appropriate technology process when it is the first and main treatment stage, however when aerated lagoons are used as a polishing process, it can be considered an appropriate technology process.

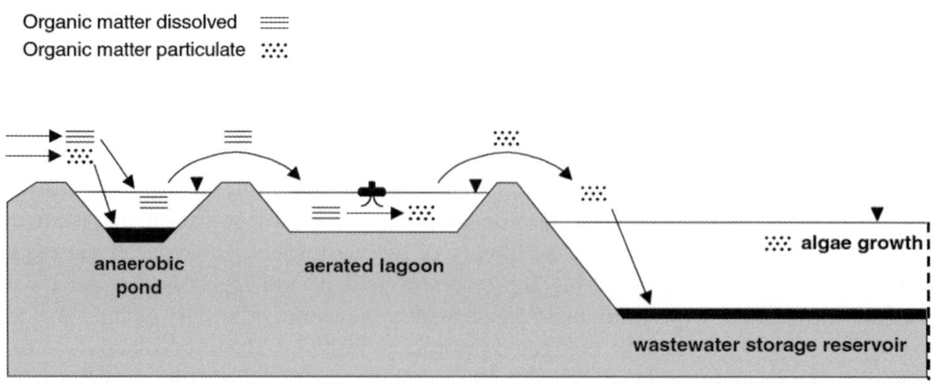

Figure 7.5 A pretreatment unit of a reservoir composed of an anaerobic lagoon followed by an aerated lagoon (*Source:* Juanico & Milstein, 2004)

During the past few years, many pretreatment units in Israel were replaced by activated sludge units due to regulatory requirement. The activated sludge units achieve the required levels of organic matter removal but are not appropriate technology units so their use turns the entire reservoir system to a complex process which is not based on appropriate technology. The use of activated sludge plants as pretreatment for a stabilization reservoir is not recommended for developing countries.

We reiterate that the prtreatment units do not need to based only on lagoons treatment and can be based on a variety of appropriate technology unit processes such as UASB, Anaerobic Filters, CEPT and of combined processes (and also on processes which are not appropriate technology processes). UASB, Anaerobic Filters and CEPT are not in use in Israel as pretreatment to stabilization reservoirs, but there is no reason that such systems would not successfully fulfil the pretreatment function. The combination of CEPT as a pretreatment unit followed by a stabilization reservoir can be advantageous from the standpoint of resistance to low temperatures and can be used in regions with very low temperatures during the cold season. The CEPT unit performs well in low temperatures, and the CEPT effluent can be stored in a reservoir without causing nuisance until temperatures rise to the level that enables resumption of bacterial activity.

An important consideration in the design of a reservoir system is the location of the pretreatment unit. The reservoir is always located in the vicinity of the irrigated farmland, since the area it occupied is large and its use for treatment and supply of irrigation water can only be done in locations outside city limits, where the value of land is lower. The pretreatment unit can be located either inside the city limits or next to the reservoir. The location of the pretreatment unit is related to institutional issues and cooperation between the city and the farmer's organization. When located inside the city, the pretreatment unit is usually operated by the municipal utility and when located next to the reservoir it may be operated by the utility or, in many cases, by the farmers' organization. When the pretreatment unit is located inside the city, it is usually necessary to base it on a process that occupies a small size of land. In this respect unit processes such as UASB, Anaerobic Filters and CEPT are advantageous since they occupy small land areas. In the reservoir systems of Nazareth, Naan and Og (Figures 1.75, 1.77 and 1.78 respectively), the pretreatment units are located next to the reservoirs, where availability of land is not restrictive and therefore all of them are lagoons based systems. But there are many reservoir systems in Israel in which the pretreatment units are located inside the cities, at a large distance from the reservoirs. The systems of the city of Haifa and of the western basin of Jerusalem can be mentioned as such examples. In each of these cities the pretreatment is an activated sludge plant located inside the respective city limits and operated by the municipal water and sanitation utility. The effluents of these pretreatment units are conveyed to the reservoirs located outside the cities, next to the farmlands. In the case of Haifa the reservoir is the Maale Hakishon reservoir, shown in Figure 1.29, located 36 km away from the city. The west Jerusalem Treatment plant feeds a series of over 10 reservoirs located at distances of 20–40 km from the city. One of them is the Kfar Menahem Resrevoir shown in Figure 1.31.

It would be beneficial to have on the same figure the variation of the operational parameters and the quality parameters of a single continuous flow stabilization reservoir along the yea. For that purpose, Figures 7.2 and 7.4 were merged and are presented jointly in Figure 7.6. This figure shows that as PFE_{30} and the organic surface loading decrease from September on, when the reservoir is just filled with pretreated effluent, without withdrawal of effluent for irrigation, the water quality in the reservoir significantly improves, the COD drops and the total coliforms count falls. With the start of the irrigation season on May, as effluent is withdrawn from the reservoir for irrigation, the *PFE* and the surface loading change direction and start to increase at a rate which accelerates with time. This causes deterioration of the water quality in the reservoir, which also accelerates with the progress of the irrigation season. The quality of the effluent withdrawn from a single continuous flow stabilization reservoir is not constant throughout the irrigation season. It is of good quality at the beginning of the season and deteriorates along the season. To obtain a good and stable effluent quality during the entire irrigation season, the reservoir needs to be operated under a sequential batch mode which means that the inflow to the reservoir needs to be stopped a short time before initiating the withdrawal of effluent for irrigation. Under such an operation mode the effluent quality remains constant, at the level it had before the initiation of effluent withdrawal. This has been measured in full scale systems operating under a sequential batch mode, for example, a reservoir named Reservoir III which produced an effluent containing a BOD concentration of less than 10 mg/l and practically void of Fecal coliforms (Juanico & Dor, 1999, pp. 75–76). A photo of a Stabilization reservoirs system consisting of a pretreatment unit of anaerobic lagoons followed by aerated lagoons (as presented in Figure 7.5) and of two stabilization reservoirs, is presented in Figure 7.7. This system can be operated as a sequential batch system under an operation mode presented in Figure 7.3 (b) or in Figure 1.26 (B) or 1,26 (C).

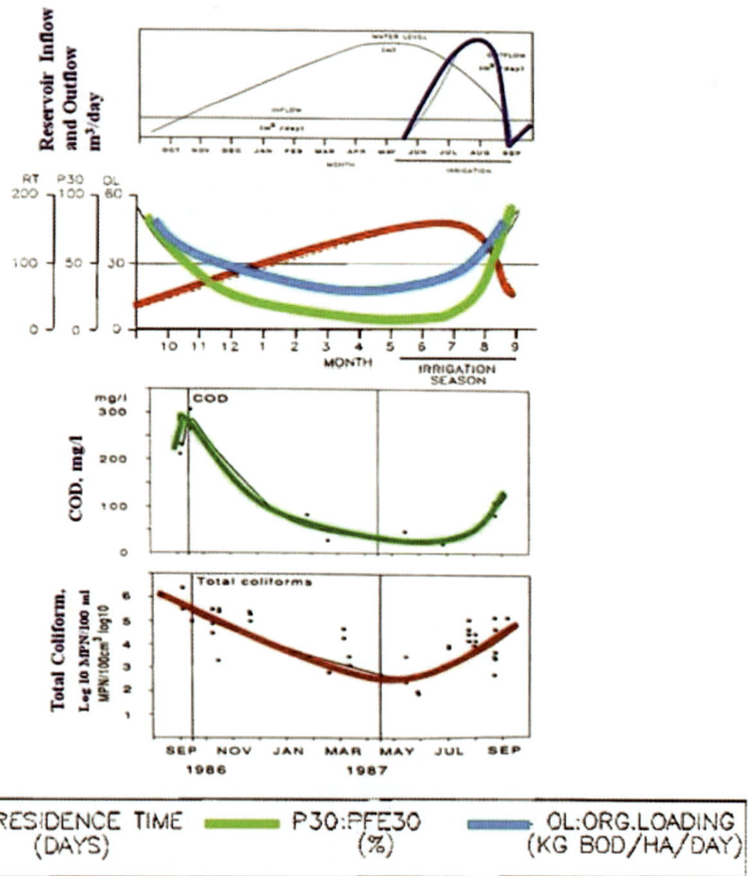

Figure 7.6 Typical annual variations of some operational parameters and main effluent quality parameters in a single continuous flow stabilization reservoir in Israel (*Source:* Adapted from Juanico, 1993; Juanico & Shelef, 1994)

Figure 7.7 A stabilization reservoirs system consisting of pretreatment by anaerobic lagoons followed by aerated lagoons and two reservoirs which can be operated as a sequential batch system

Irrigation methods

The topic of methods of irrigation with effluent is a wide topic that cannot be considered here in detail. Any irrigation method can be used to spread the effluent in the fields, however the irrigation methods has to be selected as part of the public health protection strategy. There are three measures of health protection when using effluents for irrigation: (i) selection of Crops; (ii) selection of level of treatment (i.e. selection of the treatment process); and (iii) selection of the irrigation method. Low quality effluent with surface irrigation may be adequate for irrigation of industrial crops like cotton, but inadequate for irrigating food crops like tomatoes. Drip irrigation of orchard threes may be acceptable even when the effluent quality is not the highest, because there is no contact between the effluent and the fruit.

In Israel, the effluent irrigation method used is, for the most part, drip irrigation, with some use of micro sprinkling. These are pressure irrigation methods, so the effluents are withdrawn from the reservoirs by pumps, which in addition to the conveyance to the fields provide the pressure for the drip and micro-sprinkling irrigation. Reservoir's effluents contain small levels of algae which need to be removed prior to the drip irrigation to prevent clogging of the drippers. The algae are removed by a combination of chlorination and filtration (sand filtration, disk filtration and similar methods). Usually the effluent withdrawn from the reservoir undergoes filtration in a central unit, as shown in Figure 1.31. This is a pressure filtration, which is the preferred method because the irrigation systems needs to be under pressure because of the irrigation methods used. In addition, individual filters are installed on the distribution pipes at the entrance to each plot. Addition of hypochlorite is done either at a central level or individually by each farmer. The chlorination is done mainly to facilitate the filtration and removal of algae for preventing clogging of the drippers; however it also helps to complement the disinfection of the effluents.

7.1.3 Performance

Efficiency

The performance of a reservoir system and the quality of effluent it produces is highly dependent on the mode of operation of the system and on the operational parameters. As explained above, a single continuous flow reservoir does not produce a high quality effluent. The effluent quality is not constant and it deteriorates along the irrigation season. In regions with an irrigation season of 4–5 months like Israel, the annual mean detention time in a reservoir is 80–130 days. The annual mean total BOD removal percentage is 70–85%, total COD removal 50–85% and TSS removal is 40–80%. The system also achieves 25 to 50% removal of N and P. Higher nutrient removal percentages are achieved in system with a longer detention time (the Maale Hakishon reservoirs, with a joint mean detention time of 180 days achieve 70% removal of N and 60% removal of P). The total coliforms removal in continuous flow reservoirs is only one order of magnitude, a low value for reactors with such a long mean detention time. However, as explained, the effluent quality is not constant during the irrigation season and the maximum effluent quality, which is the quality at the beginning of the irrigation season, is much higher, with removal values of BOD, COD in the range 85–95% and TSS in the range 90–98%. Effluents of single continuous flow reservoir systems are usually used for restricted irrigation.

A very high effluent quality is achieved in reservoir systems operated under a sequential batch mode. Such systems can achieve BOD, COD and TSS removals of 85–95% each, continuously during the irrigation season. They produce an effluent with a BOD_5 concentration of 5–10 mg/l and a TSS concentration of 5–20 mg/l. Moreover, the faecal coliforms removals in such systems are very high, producing effluents with faecal coliforms levels of 0–100 MPN/100 ml. This means that sequential batch reservoir systems perform disinfection without the use of chemicals. In addition, the stabilization reservoirs completely remove helminth eggs and nematodes without difficulty. The effluents of sequential batch reactors are adequate for unrestricted irrigation.

An important characteristic of the stabilization reservoir is the fact that they do not produce excess sludge. The only unit that may generate excess sludge in a reservoir system is the pretreatment unit. If this unit is a stabilization lagoons unit, it produces very small quantities of excess sludge so the entire system is practically a sludge free system. If the pretreatment unit is based on other anaerobic processes, it generates small quantities of excess sludge. However, if the pretreatment is based on aerobic treatment processes, it generates larger amounts of excess sludge.

A comparison of some basic performance parameters of a Stabilization Reservoir system and a facultative lagoons system is presented in Table 7.1.

Advantages and disadvantages of a reservoir system

The main purpose of using stabilization reservoir is to recover wastewater for its efficient reuse for irrigation. According to Libhaber, 1996, this brings the following benefits

- Irrigation is usually needed in areas where water tends to be scarce and wastewater reuse is necessary for supplementing available fresh water sources. In terms of availability it is also a safe water source, independent on climatic conditions.

Table 7.1 Basic parameters of facultative lagoons and reservoir systems.

Process	Treatment goal	Required climate	Mean detention time (days)	Depth (m)	Surface organic loading (kg/ha · d)	Effluent quality (mg/l)
Facultative lagoons	Secondary to advanced	any	10–50	1.5 a 2.5	50–250	DBO: 30–40 SST: 40–100
Stabilization reservoirs	Secondary to advanced	any	80–200	4 a 12	<50	Continuous flow: Varying quality BOD: 15–30 TSS: 20–60 Sequential batch: BOD: 5–10 TSS: 5–20

Sources: Crites, Middlebrooks & Sherwood (2006); Juanicó & Dor (1999); Orozco (2005)

- Agriculture requires large amounts of water, which are used only once, since irrigation is a consumptive use. Thus, there is no danger of gradual accumulation in the water of undesirable substances by continuous recycling.
- Agriculture can use not only the water, but also additional resources found in wastewater, such as organic matter or nutrients (nitrogen and phosphorous), which are thus converted from an environmental nuisance to an asset.
- Irrigation is relatively flexible with respect to water quality requirements. Some crops can be irrigated with low quality water, and some water quality problems can be overcome by suitable agricultural practices.
- Use of wastewater for irrigation prevents its discharge to receiving water bodies and pollution of water resources.

The disadvantages associated with the reuse of treated wastewater for irrigation can be divided into two categories:

- Public health is the issue of greatest concern, and it is related to wastewater reuse that can endanger consumers of crops irrigated with effluents, to farmers, and to residents in the neighbourhood of the irrigated fields.
- The agro-technical problems that may be caused by inappropriate reuse of effluent, relate primarily to damage to crops and soils.

These two problems can be managed through the proper design of reuse systems (conveyance, treatment and irrigation) and through the adoption of adequate agro-technical practices during the cultivation of crops.

7.2 BASIC DESIGN PROCEDURES

7.2.1 General design considerations

Two basic factors need to be considered prior to the design of a stabilization reservoir: (i) the type of crop which will be irrigated with the reservoir's effluent and the quality of effluent they require, that is, is it a quality for restricted or for unrestricted irrigation; and (ii) the required volume of the reservoir, which depends on the irrigation needs of crops. Stabilization reservoirs are deep lagoons that are used for two purposes: (i) storing partially treated wastewater over a long period of time in order to use it during a specific period of the year, under optimal and controlled conditions; and (ii) improving the quality of stored effluents during their long residence time in the reservoir. The stabilization reservoir receives effluents of primary or secondary treatment units, stores them and improves their quality exercising the following priorities of treatment: (i) maximum removal of helminths and protozoa, (ii) effective removal of bacterial and viral pathogens and (iii) Effective reduction of BOD to avoid unpleasant smell unpleasant appearance of the effluent (Libhaber, 2007).

There are basically three modes of operating stabilization reservoirs system, as shown in Figure 7.3. The selection of the mode of operation is determined on the basis of the planned objective of use of the effluent: either for restricted irrigation, unrestricted irrigation, or a combined of the two. Each use objective prescribes a different mode of operation.

7.2.2 Orderly Design Method, ODM

In general the design methodology of a Stabilization Reservoir consists of the following:

(1) *Determine the external or independent variables*, which among others, include the following:

- Wastewater flows, geographical conditions pollutants concentrations and loads are handles using the methodology described in Chapter 5, Section 5.2.2, Paragraph 1. In the case of reservoirs is necessary to calculate the municipal wastewater flows available for irrigation, as well as the amounts and patterns of water required by the crops. Two specific issues are important for these calculations: (i) the estimates of availability of municipal wastewater for irrigation, both in terms of quantity and quality, using the methodology presented in Chapter 2 and the efficiencies of the treatment process employed in pretreatment of the wastewater prior to its discharge into the reservoir; and (ii) the estimated water requirements for irrigation and for, and its distribution along the year, for which it is necessary to have the curves of monthly precipitation and evapo transpiration for the type of crop in the specific project area (which needs to be determined by a specialist in agronomy). The monthly distributions of precipitation and evapo-transpiration and consumption of water for irrigation depends on the climate on the type of soil, on the type of crops to be cultivated, and so on. These data properly processed provide the irrigation requirements and the net balance needed to meet the demand. The net demand is calculated in Paragraph 4 below. Figure 7.8 shows an example a summary of the results of such calculations.
- Geographical conditions, such as maximum and minimum temperatures of air and water, solar radiation and cloud cover, wind rose, altitude above sea level, proximity to population centers, and so on. These data are very important from the agricultural standpoint, as well as from the wastewater treatment standpoint.
- In addition, the data regarding potential evapo-transpiration (ET_0) and monthly precipitation must be obtained. Usually they are provided by the regional hydrological authority. An example of this type of data is presented in the Table 7.2 below and in Figure 7.9.

(2) *Determine the flow diagram of the stabilization reservoir system:* The components of a typical stabilization reservoir system are shown in Figure 7.5, but obviously, the pretreatment unit can be different than the one presented in Figure 7.5.

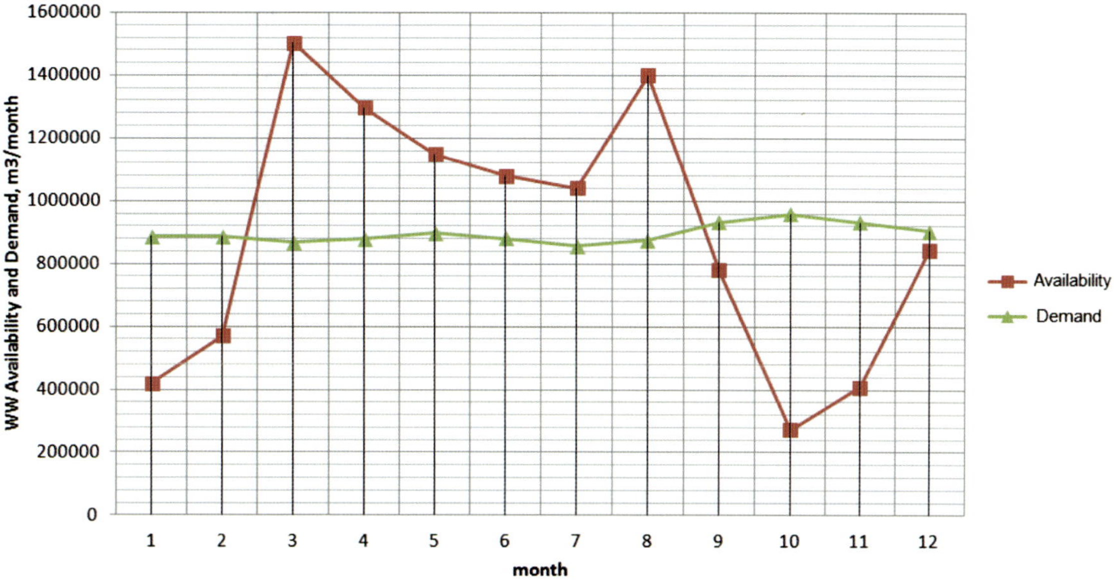

Figure 7.8 Example of the distribution of availability and demand of treated wastewater for irrigation along the months of the year (Availability-▲, Demand-■)

Table 7.2 An example of data of monthly distribution of evapo-transpiration and precipitation.

Month	Jan	Feb	Mar	April	May	June	July	August	Sept	Oct	Nov	Dec
ET_0, mm/month	145.08	156.52	190.03	177.00	163.68	173.70	193.50	197.78	155.10	139.81	125.40	139.19
Precipitation, mm/month	5.80	8.80	4.10	24.00	89.60	27.00	4.60	29.00	102.00	167.00	115.00	26.00

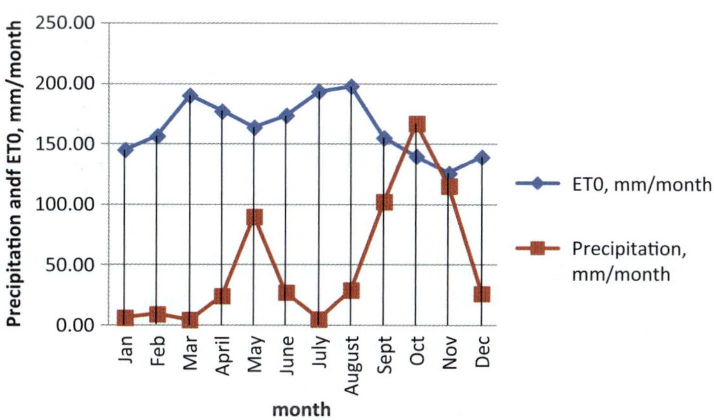

Figure 7.9 Example of the variation of potential evapo-transpiration (ET_0) and precipitation along the months of the year

(3) *Identify the pretreatment system.* The stabilization reservoir design depends on the pretreatment unit. Usually both systems (the pretreatment unit and the reservoir) are designed and constructed jointly. The design of the pretreatment unit is performed in accordance with the methodologies of design of the appropriate technology processes presented in other chapters of the book. The level of pretreatment and the design of the pretreatment unit are adjusted to the design of the reservoir, taking into account its dimensions of the reservoir and the surface organic loading which will be applied to it. The objective of the pretreatment is to reduce the organic matter content in the wastewater so that the effluent discharged to the reservoir would not cause the entire reservoir to become anaerobic, but will rather leave an aerobic upper layer in the reservoir. It is again emphasized that the pretreatment process is not just one specific process, but can be one of a variety of unit processes or combination of unit processes, which can be based appropriate technology processes and also on processes which are not appropriate technology processes. The purpose of the pretreatment unit is to reduce the organic matter content in the wastewater to a desired level and this objective can be achieved through the use of a variety of processes. We obviously recommend employing appropriate technology processes as pretreatment units, especially in developing countries.

Hydraulic design

(4) *Determine the net demand of water for irrigation.* The participation of an expert agronomist is essential for the estimation of the demand of water for irrigation. This expert has to determine the type of crops that can be cultivated, depending on the availability and quality of the wastewater. In particular, once the crops have been defined, the monthly demand of water for irrigation needs to be calculated. The formula for estimation the monthly water demand is the following (Gurovich & Ramos, 1990).

$$D_w = 10\,ET_0 \Sigma k_c A_c / E_{Ta} \tag{7.2}$$

where:

D_w = total monthly demand of water for agriculture, $m^3/ha \cdot month$. The value 10 is a conversion factor from mm/month to $m^3/ha \cdot month$.

ET_0 = potential evapo-transpiration, mm/month. In general, the hydrological control authority provides these data[2].

A_c = area of each crop, ha.

k_c = coefficient of each crop (ratio of ET_0 to actual ET), that is, actual evapo-transpiration (ET) is a fraction of the potential (ET_0). For example, for corn it is 0.55. If there are multiple crops there are different k_c (per crop), and they can vary month to month, depending on the harvest season. Therefore, the expression $\Sigma\ k_c A_c$ is used.

E_{Ta} = overall efficiency of water use in the conveyance system, in the application method of irrigation (the irrigation method), and so on. It does not necessarily include infiltration losses (which can be treated as a sensitivity parameter, for example, for 1 mm infiltration, 5 mm, and so on).

[2]This is a climate parameter and it can also be calculated by the *Penman* method, by the *Radiation* method or by *Tank Class A* Method. The results are similar, and for their calculation the following data are required: (i) temperature, (ii) wind velocity, (iii) relative humidity, (iv) cloudiness, (v) radiation, and (vi) effective number of hours of sunshine.

With these data, for the crops that are to be developed using the reservoir's effluent, the monthly water demand for agriculture is determined by the use of Equation 7.2.

To calculate the net demand (V_w) it is necessary to take into account the monthly precipitation (P) that sometimes may be sufficient to meet the irrigation demand, and the infiltration (I) which is a water loss.

$$V_w = D_w - 10 \cdot (P - I) \cdot A_c \tag{7.3}$$

The factor 10 is again a conversion factor used to convert "mm/month" to "m^3/ha · month". During the rainy months, the rainfall may be sufficient to meet the water demand of the crops and even excessive (negative demand), but since the rainwater is not stored, for calculation purposes V_w is considered zero in the rainy season months.

When the demand is positive, the domestic wastewater flow (Q_{WW}) has to provide this demand. A properly designed stabilization reservoir provides the necessary irrigation water in the most efficient manner, and it is the available storage volume which determines the area A_c. that can be irrigated.

(5) *Determine the volume of the stabilization reservoir.* Taking into consideration the available supply of wastewater and the distribution of net monthly demand of effluent for irrigation (i.e., the required water volume every month), and considering a single continuous flow reservoir (operation mode of Figure 7.3a), which means restricted irrigation, the volume of the reservoir required to cover the irrigation water deficit is calculated using the mass curve method.

The mass curve is calculated using the following methodology: (i) build the consumption line for irrigation or aggregate demand (ΣV_w) adding demands every month, taking care to sum zero when the demand generated by the formula is negative (i.e. there is sufficient rainwater and no additional demand from the reservoir, and the excess rainwater is not stored), (ii) connect the string joining the first and last point of the series, (iii) draw the tangents parallel to the string of the curve through the points of maximum (maximum deficit) and minimum (minimum deficit), and (iv) calculate the vertical distance between the two parallel tangent lines and this distance gives the net operative volume of the reservoir (V_{SR}) required to provide effluent for irrigation when required.

If it is decide to provide effluent for unrestricted irrigation, it is necessary to construct one, two or more reservoirs which will function in parallel. This is done using the calculated volume V_{SR} and dividing it by the number of reservoirs to be built. After one reservoir is filled, the pretreated effluent is diverted to another reservoir. If irrigation water is needed before the first reservoir (SR_1) has been filled up then its filling is stopped and only then is his effluent conveyed to irrigation, while the pretreated effluent is diverted to the second reservoir (SR_2). This ensures that the *PFE* of the first reservoir is not increased during the irrigation period, because irrigation is applied on crops which require unrestricted effluent quality.

(6) *Calculate the irrigated area* (A_c). Since the crop area (which is the same irrigate area) depends on the available water for irrigation, it is calculated by trial and error from the following equation:

$$\Sigma\left[Q_{WW} - 10(E - P) \cdot A_{SR}\right] = \Sigma V_w \tag{7.4}$$

with:

ΣQ_{ww} = the amount of wastewater produced during the year, $\Sigma m^3/month = m^3/year$.
A_{SR} = Area of the stabilization reservoir, ha. Initially this number needs to be estimated.
E = evaporation from the reservoir, mm/month
P = the precipitation over the SR, mm/month

ΣV_w = The annual net water demand for irrigation, which depends on crop area, A_c (m^3), as shown in Equation 7.3.

Again 10 is the conversion factor from mm/month to $m^3/ha \cdot month$. Note that the evaporation (E) is from the reservoir surface and is different from the ET for the crop area. If E is not known, the Visentini's Equation can be applied:

$$E = CT + M \quad (7.5)$$

where T is the mean annual temperature, in degree centigrade, and C and M are constants that depend on the altitude (meters above sea level, masl) as follows:

Altitude, masl	C	M
<200	75	0
200 to 500	90	0
>500	90	300

Kinetics design

(7) *Calculate of the quality of the reservoir's effluent:* according to Juanico (2005), the efficiencies obtained by stabilization reservoirs (additional to pretreatment) with continuous flow-single reservoir and with a sequential batch flow reservoir system (two or more stabilization reservoirs) are presented in Table 7.3. This table can be use as a first approximation; however, a direct kinetics design will provide more accurate results.

Table 7.3 Efficiencies in continuous flow and sequential batch fed stabilization reservoirs.

Parameter	Continuous flow reservoir	Sequential batch flow reservoirs with batch from 30 to 50 Days
BOD_5	70%	90%
COD	50%	80–90%
Detergents	50%	90%
Nitrogen	–	70–80%
Phosphorus	<30%	10–60%
Fecal Coliforms	90–99%	99.99%
Streptococcus	–	90%
Giardia	–	80–90%

Source: Juanico (2005)

Instead of using Equation 7.1 to calculate the mean residence time, which would be impractical for design, it is better to use, as an approximation, the average detention time (t_d) of the reservoir calculated with the half (or approximate "average") reservoir volume ($V_{1/2}$). The same goes for the "average" surface area (A_s). Note that the net volume $V_N = 2 \cdot V_{1/2}$:

$$V_{1/2} = Q \cdot t_d \quad (7.6)$$

$$A_s = V_{1/2}/h \quad (7.7)$$

To calculate the efficiency by the kinetics method, Equation 7.8 is used:

$$S = \frac{S_{0_A}}{\left(1 + K_L \frac{t_d}{n}\right)^n} \tag{7.8}$$

where S_{0A} is the pretreatment effluent BOD, and "n" the number of reservoirs in series. Note that the detention time is the cycle's "average" detention time.

The final water temperature in the reservoir is calculated as follows:

$$T_L = \frac{A_{EE} f\, T_a + Q_{ww} T_i}{A_{EE} f + Q_{ww}} \tag{7.9}$$

where

T_L = Final reservoir temperature (°C)
T_a = Air temperature (°C)
T_i = Influent wastewater temperature (°C)
A_{EE} = Surface area (m^2)
Q_{ww} = Mean wastewater flow (m^3/d)
f = Proportionality factor, 0.5 m/d

If K_L is unknown the typical K_L (20°C) = 0.35 d^{-1}, can be used corrected for temperature with Equation 7.10:

$$K_L = K_L(20^\circ C)\, \theta^{(T-20)} \tag{7.10}$$

where θ is a constant and its value is between 1.05 and 1.09.

The above approximation of the average detention time (t_d) was used as the kinetic approach in the Excel program CHAP 7-SR for calculating the reservoir. However the following additional information is presented in regard to the dynamic solution analysis.

If it is desired to calculate the Stabilization Reservoir (SR) effluent quality on a dynamic basis (i.e. day by day) instead of the approximated (or "averaged") calculation proposed in Equation 7.6 using the "average" detention time ($t_d = Q_D/V_{1/2}$), it is necessary to apply a dynamic mass balance in order to obtain the exact solution to be used in the kinetics equations governing the water quality.

For instance if we assume for simplicity a Sequential Batch (SB) regime (i.e. the Stabilization Reservoir is filling during the rainy season without outflow, and the wastewater inflow is cut off just prior to starting the use of treated water outflow for irrigation during the dry season) the dynamic mass balance is ($V_i.dS/dt) = Q_D.S_0 - V_i.r(S)$, where S is the BOD_5 (or any other parameter) concentration in the reservoir, S_0 is the influent concentration, $r(S)$ is the reaction rate of the BOD_5 inside the reservoir, and V_i is the instantaneous volume at time t. For simplicity let's assume that Q_D is constant, and $V_i = Q_D.t$, where t is the elapsed time. Note that $r(S)$ acts as a sink, while the inflow BOD_5 ($Q_D.S_0$) acts as a source in the mass balance. Now, the reaction rate for BOD in its most common form can be expressed as $r(S) = K_L S$. Then the mass balance becomes:

$$\frac{dS}{dt} V_i = Q_D S_0 - K_L S V$$

or

$$\frac{dS}{dt} t = S_0 - K_L S t$$

The solution of this differential equation according to Wolfram's *Mathematica* 8.0 is:

$$S[t] = \theta^{-K_L t} C[1] + \theta^{-K_L t} S_0 \mathbf{ExpIntegralEi}[K_L t]$$

Where:

$$\mathbf{ExpIntegralEi}[K_L.t] = \int_{-K_L t}^{\infty} \frac{\theta^{-K_L t}}{K_L t} \mathrm{dt}$$

and C is the integration constant,

which at $t = 0$ becomes $(-\infty)$. Then, the **mathematical exact solution** even for the simplest case becomes intractable.

On the other hand, it is possible to reach another "exact" approach with the mean detention time (t_{dmj}) as calculated on a daily base from Equation 7.1:

$$t_{dmj} = \frac{[(t_{dm(j-1)} + 1)V_{(j-1)}] + 0.5 v_j}{V_{(j-1)} + v_j}$$

For simplicity let's assume that Q_D is constant and is given in m^3/d, then replacing $V_{(j-1)} = Q_D \cdot (j-1)$, $v_{(j-1)} = Q_D$ and "j" by "t" (remember: t is the elapsed time and t_{dmj} is the mean detention time at $t = j$), Equation 7.1 becomes:

$$t_{dmt} = \frac{[(t_{dm(t-1)} + 1)(t-1)] + 0.5}{t}$$

Using this approach we can simulate day by the detention time as presented in the table below:

***t*, days**	***t*_{dmt}, days**
1	0.5
2	1.0
3	1.5
4	2.0
5	2.5
6	3.0
7	3.5
8	4.0
9	4.5
10	5.0
11	5.5
12	6.0
13	6.5
14	7.0
15	7.5

(*Continued*)

t, days	t_{dmt}, days
16	8.0
17	8.5
18	9.0
19	9.5
20	10.0
21	10.5
22	11.0
23	11.5
24	12.0
25	12.5
26	13.0
27	13.5
28	14.0
29	14.5
30	15.0

From this simulation, it is clear that the proposed "average" detention time $t_d = (V/2)/Q_D = V_{1/2}/Q_D$ is close enough to an "exact" simulated solution.

Civil engineering design

(8) *Design the stabilization reservoir.* The engineering design of a reservoir is a very wide topic which cannot be discussed here in detail. A detailed manual for civil engineering design of stabilization reservoirs was prepared by Romem (1991). Some engineering design aspects of reservoirs are discussed in this section.

The design of a reservoir is similar to the design of a lagoon, however reservoirs are much larger water bodies, with higher embankments and their design is more complex. The physical design is critical for the success of a reservoir project. Many aspects need to be taken into account when preparing the physical design of a reservoir. The Reservoir's site topography is an important basis for the physical design of the reservoir. The shape of reservoirs depends to a large extent on the topography of the site allocated for their construction. The most common shape is rectangular with varying ratios of length to width, however, reservoirs do not need to be strictly rectangular, and can have other shapes. There are several types of reservoirs, as presented in Figure 7.10 and the feasibility of their use depends on the topography of the site allocated for the construction of the reservoir. If the site is plane, the reservoir is a peripheral embankment reservoir, which is a reservoir with dikes constructed on its four sides (Figure 7.10a). In a hilly area, one side of a hill can serve as an embankment and in this case the reservoir is a long embankment reservoir (Figure 7.10b). If a valley can be fitted to house a reservoir, than the embankment can be short, as shown in Figure 7.10c, in which case the reservoir is a short embankment reservoir.

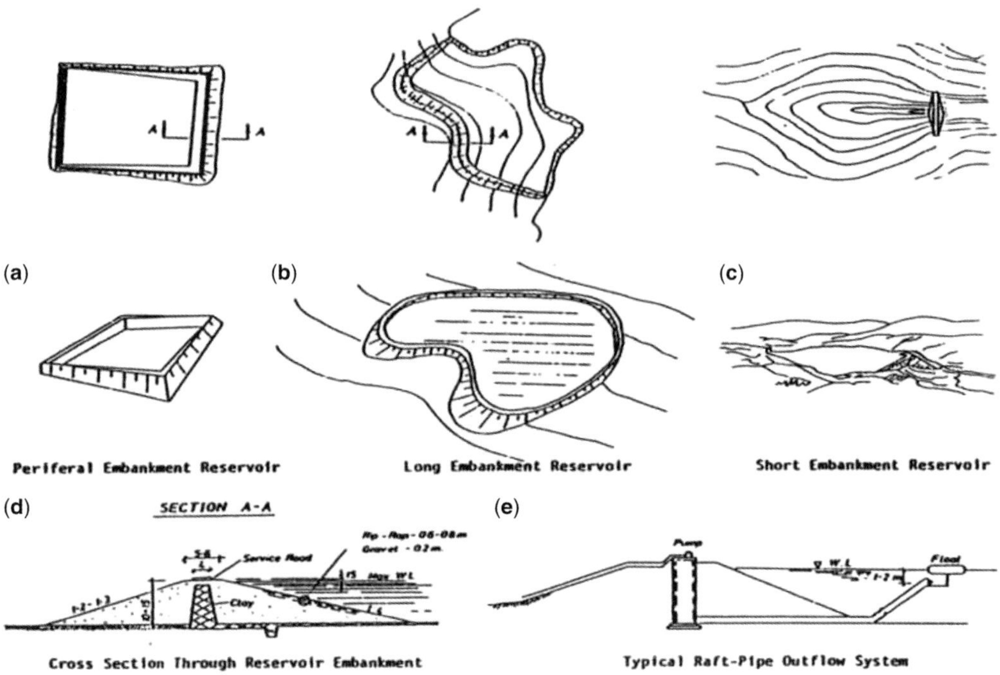

Figure 7.10 Types and construction details of stabilization reservoirs

Special attention should be directed to geotechnical aspects of the design. Detailed geotechnical studies need to be carried out prior to the physical design so as to ensure the correct design of embankments and to determine whether the soil in the proposed reservoir site is sufficiently impermeable or needs to be lined to prevent infiltration from the bottom of the reservoir. In most cases embankments are constructed from the soil excavated at the lagoons site and there should be a balance between the cut and fill, so as to reduce construction costs. The actual ratio between cut and fill depends on local civil work costs, that is, on the cost of excavation and of embankments construction. Embankment design should allow for vehicular access to facilitate maintenance. Reservoir's embankments are high, in many cases reaching a height of up to 10 m. They still classified as small dams, but are higher than embankments of stabilization lagoons. It is important to ensure that they are impermeable because infiltrated water can cause their collapse. If the soil used for constructing the embankments is not permeable, it is needed to take corrective measures such as using clay or geomembrane lining to prevent water transport through the embankments. Figure 7.10d shows a cross section through a reservoir embankment. In this case the embankment impermeabilization was achieved through the use of a clay core inside it. Embankment slopes depend on the type of soil. They are usually 1:3 to 1:4 on the internal side of the reservoir (and 1:5 to 1:6 when impermeabilization is done by geomembrane lining) and 1:2 to 1:2.5 on the external side. Slope stability should be ensured in accordance with standard soil mechanics methods for small earth dams. External embankments should be protected from stormwater erosion by providing adequate drainage. Internal embankments need to be protected against erosion by wave action. This can be achieved by lining with geomembranes, covering the embankments with lean concrete, with precast concrete slabs, or with stone rip rap as is the case in Figure 7.10d. A variety of lining materials can be used for lining the reservoir when lining is required to prevent infiltration of water from the bottom of the reservoir including a clay layer, plastic geomembranes. Selection of the impermeabilization method depends on local availability of materials and their cost. Figure 1.25 shows an empty reservoir (at the end of the irrigation season) lined with an HDPE sheets. Figure 7.11 shows a clay lined empty reservoir. Safety and security issues need to be taken into consideration. An emergency overflow weir needs to be included in the design, so as to prevent the possibility of water overflowing from the reservoir through the top of the embankments and reaching the external side of the embankments since that can cause a collapse of the embankments. The reservoir site should be fenced all around to prevent entrance of unauthorized guests, since the reservoir is a deep water body and people can drown in it. Inflow and outflow structures are discussed in the following section.

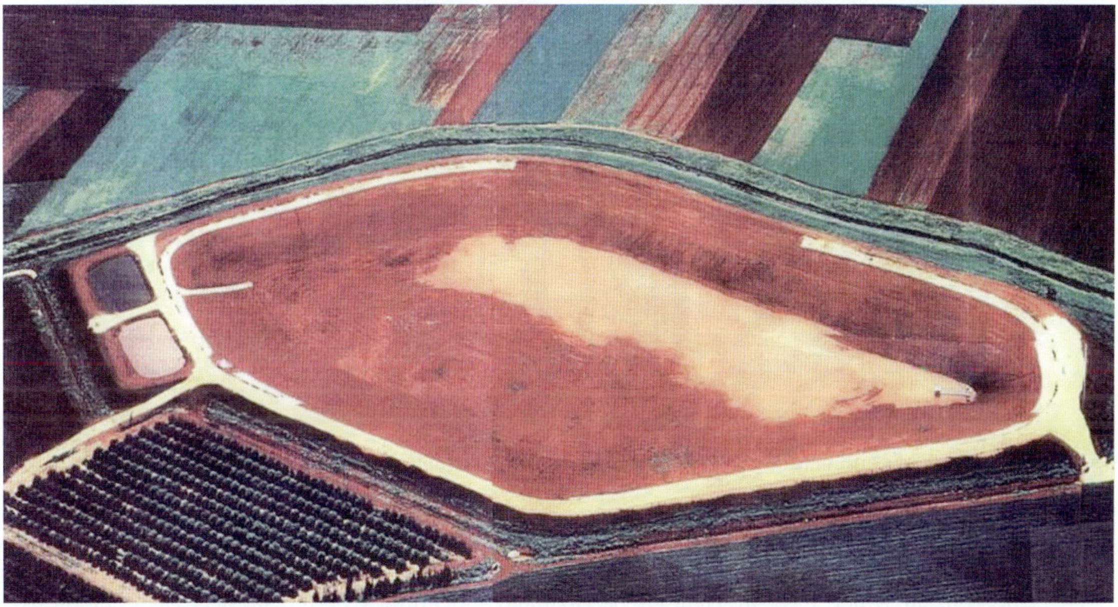

Figure 7.11 A clay lined empty reservoir at the end of the irrigation season

When the irrigation method used is pressure irrigation (sprinkling, micro-sprinkling or drip) the withdrawal of effluent from the reservoir is done by the use of pumping. Even with gravity irrigation, topography and hydraulics may require the use of pumping. Pumping stations that withdraw water from a reservoir can be quite large if the irrigation season concentrates in a few months only, since in such a case the withdrawn flows can be large. If the irrigation method used is drip irrigation, the reservoir effluent needs to be filtered before reaching the drippers, to prevent their clogging bay algae residues that are contained in the reservoir. A sample of a reservoir effluent is presented in Figure 1.30 and the green colour can be noted. This effluent is void of coliforms but still has algae and therefore needs to be filtered. In Israel, the main irrigation method is drip irrigation so pressure the filtration method mostly used is pressure filtration since the effluent system is a pressure system. A typical reservoir effluent filtration system is shown in Figure 7.12 (Lehavim Reservoir) and also in Figure 1.31 (Kfar Menahem Reservoir).

Figure 7.12 A typical sand filtration system for reservoir effluent filtration at the lehavim reservoir

Stabilization reservoirs are always located outsides cities, close to the irrigated fields. The pretreatment units are located wither within the cities and operated by the utilities, or near the reservoir, in which case they might be operated the utilities or the farmers' association. The reservoirs are mostly operated by the farmers' association. Operation and maintenance of a reservoir system includes mainly the operation and maintenance of the pumping station and the filtration system and periodical preventive maintenance of the embankments. Operation and maintenance is similar to that of a lagoons system, with an additional aspect of operation and maintenance of the system of effluent supply for irrigation.

(9) *Inflow and outflow structures.* A single inlet and outlet in a reservoir are usually sufficient but they have to be located in diagonally opposite corners of the reservoir to prevent short circuiting (as can be seen in Figure 1.25 and in Figure 7.11). Hydraulic short-circuiting is a series problem and should be avoided. An effective way to avoid this problem is to properly locate the inflow and outflow

structures. Installing floating mixers in the reservoir can also resolve the problem of short-circuiting, but this method is not in frequent use.

The inflow structure to a reservoir is similar an inflow to a lagoon. Figure 7.13 shows a such typical inflow structure. The pipe diameter is calculated for a flow velocity of between 0.6 and 5.0 m/s.

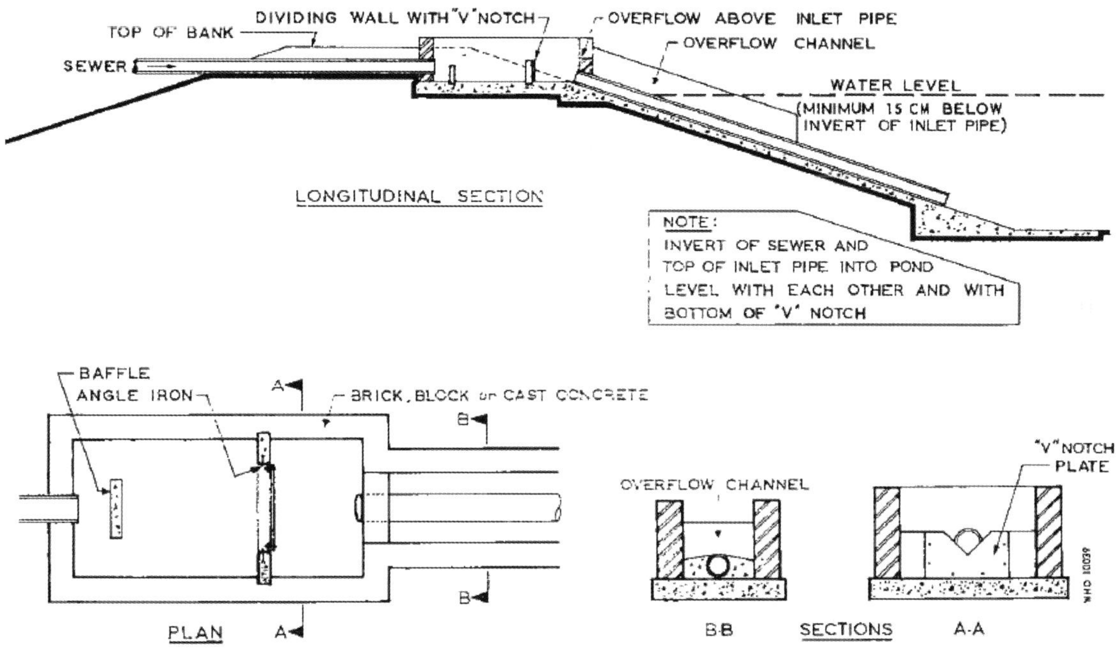

From Marais (1966), p. 759.

Figure 7.13 A typical inflow structure to a stabilization reservoir with an outlet measuring "V" notch plate

There are various types of outflow structures. On type extracts effluent from a fixed point above the bottom of the reservoir. Another type extracts effluent from a vertical perforated pipe which actually withdraws a mixture of effluent from various depths. The best point for withdrawal of effluent is at a constant depth of about 1 meter beneath the surface water level of the reservoir. The reason is that at such a point the algae concentration is low and the effluent quality is the highest. The extraction of effluent from such a fixed point beneath the water surface, which is the most frequently used extraction method in the reservoirs of Israel, is done using a raft-pipe outflow system which is a floating extraction device functioning under the principle described in Figure 7.10e. Details of such a floating extraction device which withdraws water from a constant depth beneath the water surface is presented in Figure 7.14. This device can be seen in Figure 1.25 and in Figure 7.11. A Plan and Section of a typical stabilization reservoir is presented in Figure 7.15.

The mean irrigation flow, Q_R, is calculated as the aggregate demand of effluent for irrigation during the year divided by the number of days in the months of the dry season:

$$Q_R = \Sigma V_{ww}/(n_d \times 30.3) \tag{7.11}$$

where n_d is the number of dry months and 30.3 is the average days per month and ΣV_{ww} is the total annual wastewater supply for irrigation. The maximum irrigation flow needs to be determined from pattern of monthly distribution of irrigation of water consumption. The maximum month's consumption divided by 30.3 yields the maximum daily flow for irrigation.

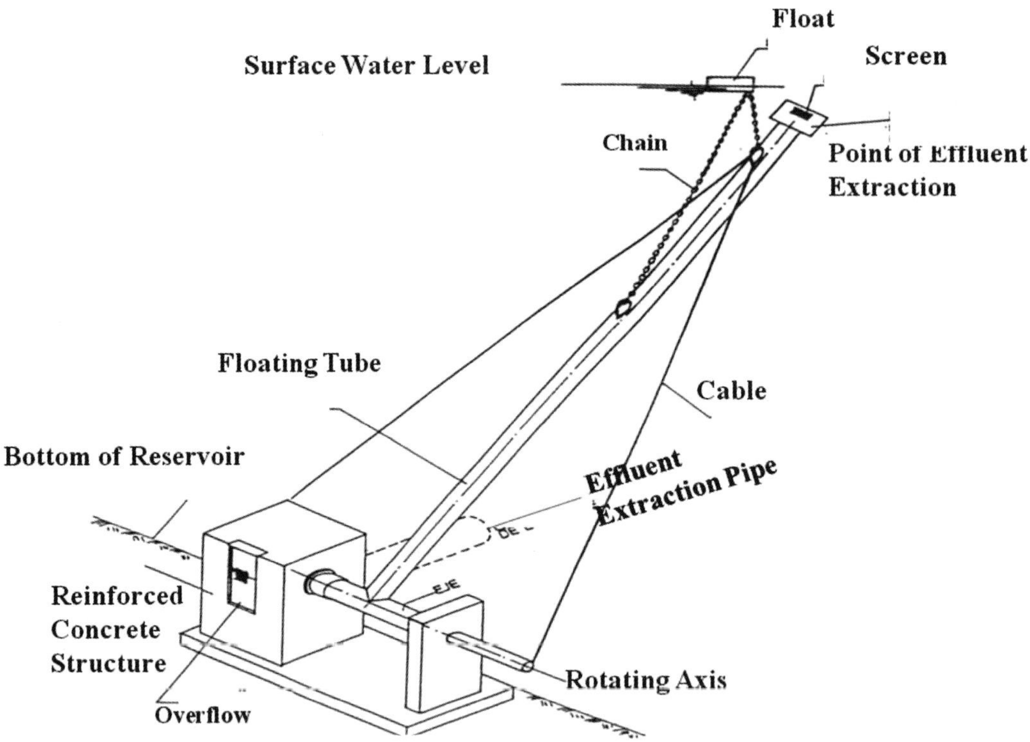

Figure 7.14 Details of a reservoir outflow structure withdrawing effluent from a constant depth beneath the water surface

(10) *Economic and Management Aspects.* There are some conditions for achieving success in projects of wastewater reuse for irrigation. First, a long term commitment of consumption of the effluent is required. This means that there are well established farmers who have the required agricultural infrastructure and are willing to use their effluent as the source of water for irrigation. Attempts to mobilize farmers because there is a new water source are usually not successful, especially if the mobilized farmers are not experienced agricultural workers. Second, the proposed reuse project must be located in water scarce regions. If a region has water in abundance, farmers will always prefer using fresh water over reclaimed wastewater. And finally, for a reuse project to be economically feasible, the effluent must be used efficiently. This means that the project must include seasonal storage (a stabilization reservoir system) and that an efficient irrigation method is used, preferably drip irrigation.

Since the pretreatment and stabilization reservoir system are part of an agricultural production unit, the economic analysis should be done on the entire agricultural production system, for which the reservoir system is just one of the inputs – the water production and

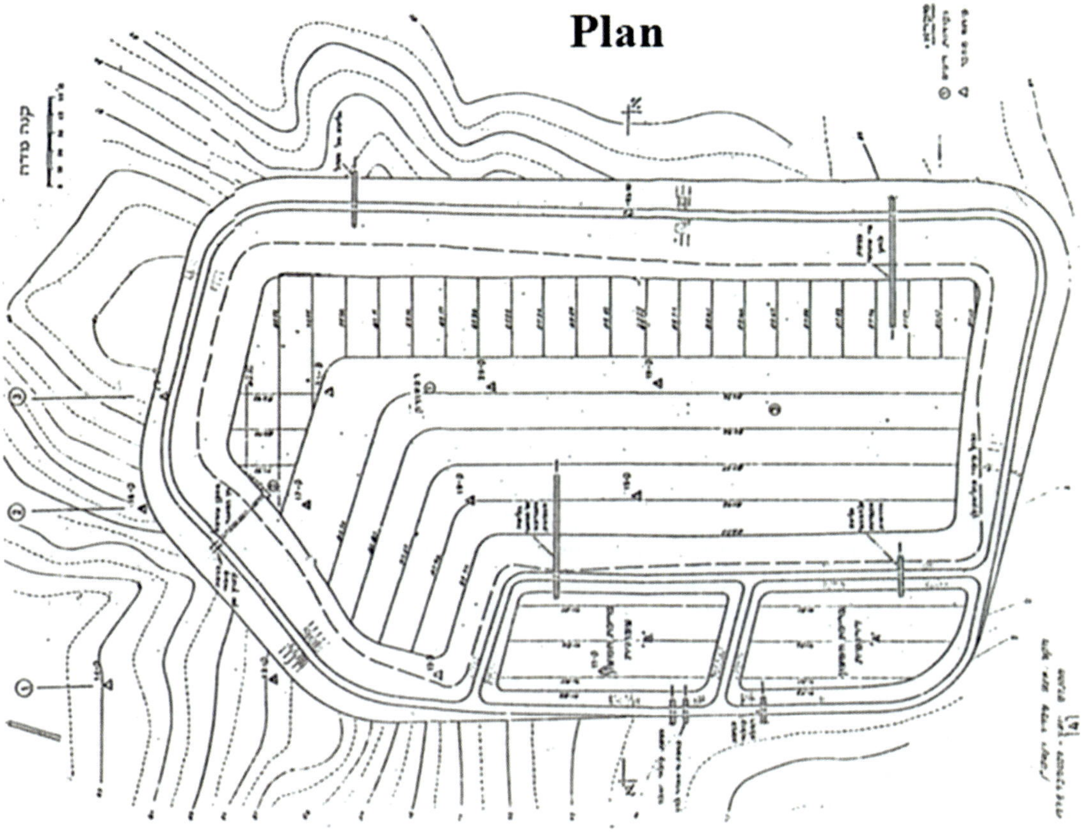

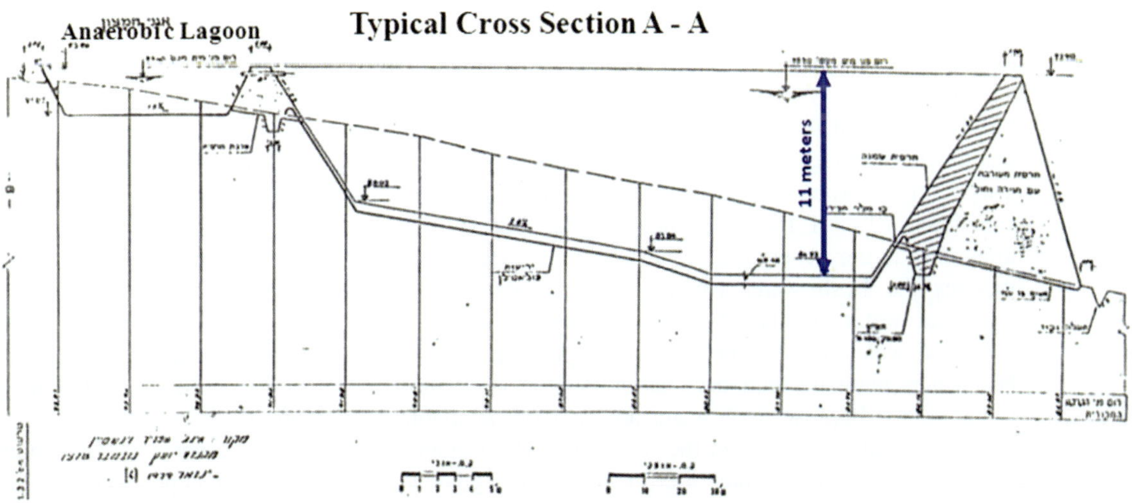

Figure 7.15 Plan and section of a typical stabilization reservoir system with pretreatment by anaerobic lagoons (*Source:* Tahal 1981)

supply input. The investment and O&M costs of the water production system should not be the entire costs of the pretreatment and stabilization reservoir system. The cost of the alternative wastewater treatment which the wastewater contribution city has to provide under the prevailing regulations should be deducted from the cost of the reservoir system since it is the obligation of the city to provide such investments and O&M costs. Sometimes the alternative cost which the city should have invested is larger than the entire costs of the pretreatment and reservoir system because discharge of effluents to water streams (which would have had to be done if the wastewater would not have been used for irrigation) requires a higher quality effluent and a higher and more costly level of treatment. Such type of analysis was undertaken on a pretreatment and single continuous flow reservoir producing effluent for restricted irrigation under the conditions of Israel at the period which de analysis was undertaken (Libhaber, 1987). The effluent was used to irrigate cotton and wheat and the outcome of the analysis depends on the market price of cotton and wheat. The results (valid for the agricultural inputs and produce prices used in the calculations) show that if the city does not contribute anything to the investment costs (which means that the farmers provide the entire funds for the pretreatment and for the reservoir), the IRR (Internal Rate of Return of the project) would be 16%. If the city provides 50% of the investment costs of the water production system (pretreatment and reservoir, including O&M costs) the IRR would be 28%, and if the city provides 100% of the investment costs of the water production system (pretreatment and reservoir, including O&M costs), the IRR would reach a value of 57%.

This type of economic analysis is usually not undertaken prior to taking decisions on reuse projects. Wastewater reuse is usually practiced in scarce water zones, where the demand for irrigation water is high and the driving force for investments is the generation of an additional water source. The city provides the investment and O&M costs of the pretreatment installations and the farmers' organization provides the investment and O&M costs of the reservoir. Governments usually provide grants to render reuse projects feasible. Such grants may be directed to the farmers, to the city or to both.

In Israel, the cities finance the wastewater treatment installations and the farmers associations (the group of users in each case) finance the stabilization reservoir investment and O&M. The government provides a grant to the farmers association to partially finance the investment of the reservoir system (including pumping stations and filtration systems) so that the resulting cost of water for the farmers is reasonable and somewhat lower than the cost of fresh water supplied by the national water company. Government grants reach sometimes 70% of the investment in the reservoir system. The government's interest in doing that is the generation of an additional water source. The use of effluent for irrigation liberates the fresh water which was previously used by the farmers to be directed for municipal water supply. It is a win-win situation for all. The government generates an additional water source; the national water company increases its fresh water reserves and the farmers maintain their occupation as farmers (without effluent they would have lost their waster source because fresh water are no more available for irrigation).

The way all this works id the following. The farmers to whom the wastewater of a certain city was allocated constitute an effluent use management unit. This is an independent entity in charge of constructing and managing the reservoir system. This unit retains the consulting engineer to design the reservoir and the construction of the reservoir. It uses the government fund and bank loans to finance the investment. Then it sells the water to the farmers, each farmer pays monthly for the measured water he consumed, at a tariff set by the management unit. The monthly income of the unit is used to repay the bank loans and to cover O&M costs (mainly electricity costs). The level of the government grant is set on such a level that the tariff for the farmers would be about 0.2–0.3 US$/m^3, which is somewhat lower than the tariff of fresh water for irrigation.

The tasks of the management unit after the reservoirs construction has been completed is to operate the reservoir system (mainly the pumping station, filtration unit and embankments periodical maintenance) and sometimes also to operate the pretreatment installations, and to conduct the financial management (collection of payments from the farmers, repayment of loans and payment of O&M costs). A management unit can be in charge of one reservoir system and of several reservoirs, depending on the size of the wastewater contributing city. If the number of reservoirs handled by a management unit is small (one or two) the management is apart time job. For larger systems, management can be a full time job. As an example, in one of the largest systems in Israel, which consists of about 10 reservoirs (the number increases with time because of the increase of wastewater flow), the management unit consists of two full time employees: the manager, who is mainly in charge of the financial management, and the technician, who is in charge of operation and maintenance.

When appropriate technology processes are used for pretreatment, the investment cost in the system of pretreatment followed by a Stabilization Reservoir is in the range of 30–50 US$/Capita and the O&M cost is in the range 0.2–0.4 US$/Year/Capita.

The cost of a complete stabilization reservoir system, including pretreatment, depends to a large extent on the pretreatment system. The additional investment cost of the reservoir system is about 20–30 US$/Capita. When the pretreatment is based on appropriate technology, the investment cost of a complete reservoir system, including pretreatment, is about 30–50 US$/Capita. When the reservoir is in the same site of the pretreatment installations and both are operated by the same operator (usually the farmers' organization) the additional operation and maintenance cost of the reservoir is small. In many cases, the pretreatment and the reservoir are not in the same site and are operated by different entities, the pretreatment by the municipality or the water utility of the city and the reservoir by a farmers association. In this case, the operation and maintenance costs of the reservoir are similar to those of operating a stabilization lagoons system, that is, in the range of 0.2–0.4 US$/Year/Capita. The cost of construction of reservoirs in Israel is 1–2 US$/m^3 of reservoir volume.

7.3 BASIC DESIGN EXAMPLE

(The model program for this Example is available online at http://www.iwawaterwiki.org/xwiki/bin/view/Articles/Software+Developed+for+Sustainable+Treatment+and+Reuse+of+Municipal+Wastewater)

Following is a design example based on the Orderly Design Method (ODM) presented in Section 7.2.2. The example is developed for a population of 20,000, with a raw wastewater flow and quality as presented in Chapter 3. The ODM is applied step by step and is specified in the Excel program CHAP 7-SR. The calculations presented below are the outputs of this program. An anaerobic lagoon (A) is selected as the treatment prior to the reservoir (for details of design, see Chapter Five). The crop area has the Evapo-transpiration (ET_0) and Precipitation given, and E is the Evaporation in the Stabilization Reservoir area (without vegetation). The data are given in Table 7A.

The area is of flat topography and the expected infiltration rate is 1 mm/month. The agricultural plan is to cultivate corn all year around. The crop coefficient is $k_c = 0.55$ every month of the year (the k_c may vary monthly depending on local conditions; the values of k_c should be provided by the agronomy/agricultural expert). The efficiency of transport of water from the reservoir to the irrigation fields is estimated at 65% (which means that 35% of the water is lost). It is required to design a continuous flow stabilization reservoir which will store the pretreated effluent and will produce effluent for restricted irrigation, and to calculate the area of the corn fields which can be irrigated with the stored effluent. Please note that the following Tables identified by alphabetic order contain computer calculated results.

Table 7A

Month	Jan	Feb	Mar	April	May	June	July	August	Sept	Oct	Nov	Dec	Annual
ET_0, mm/month	100,0	44,9	67,2	78,1	131,6	152,0	112,0	146,2	214,2	139,3	98,9	45,6	1330,0
Precipitation, mm/month	90,0	70,0	127,3	144,7	106,0	120,2	166,4	174,4	161,1	101,6	106,9	80,9	1449,5
E, Evaporation	180,0	180,0	180,0	180,0	180,0	180,0	180,0	180,0	180,0	180,0	180,0	180,0	2160,0

(1) *Calculation of external variables*

1.1 *Design flows:* The flows are calculated according to the methodology presented in Chapter 3, what yield the results presented in Table 7B.

Table 7B

Design Flow				
Variable	**Value**	**Unit**	**Value**	**Unit**
Per capita consumption, q	162.5	L/capita · d		
$PF = 1 + 14/(4 + \sqrt{P})$	2.65			Eq. Harmon
Population, P	20000	inhab		
Population density, d	260	inhab/ha		
Return coefficient	0.8			
DWW mean flow, Q_{DWW}	30.09	L/s	108.33	m^3/h
DWW maximum flow, Q_{maxDWW}	79.82	L/s	287.35	m^3/h
Infiltration unit flow, q_I	0.13	L/s · ha (o km)		
Afferent area, A_a	76.92	ha		
Infiltration flow, Q_I	10.00	L/s		

Notes: (i) Per capita consumption is calculated for drinking water consumption. With the return factor of 0.8, the wastewater per capita flow of $c \cdot q_{dom} = 0.8 \times 162.5 = 130$ L/capita · d, as measured in the example in Chapter 3, (ii) the Peak Factor is calculated by Equation 3.6 (*Harmon* Equation), (iii) the population density should be the maximum acceptable density according to planning standards, in order to be able to serve the population at saturation conditions, (iv) the flows are calculated based on the population and the return coefficient - c, and (v) infiltration must be measured or alternatively a value of between 0.10 and 0.15 L/s · ha or l/s · km is to be chosen. *In the case of a stabilization reservoir, the design should be based on the average flow* (Q_D).

The design flow data are presented in Table 7C.

Table 7C

Design Flows						
Variable	**Value**	**Unit**	**Value**	**Unit**	**Value**	**Unit**
Process design flow, $Q_D = Q_{DWW} + Q_I$	40,09	L/s	144,33	m^3/h	3464,00	m^3/d
Hydraulics design flow, $Q_{DH} = Q_{DWWmax} + Q_I$	89,82	L/s	323,35	m^3/h	7760,44	m^3/d

1.2 *Determination of the geographical and external environmental variables:* These variables depend solely on geographic location. They are presented in Table 7D for the example.

Table 7D

Variable	Value	Unit	Value	Unit
Minimum temperature of water, T_w	25	°C	77.0	°F
Minimum temperature of air, T_a	18	°C	64.4	°F
Mean annual air temperature, T	24	°C	75.2	°F
Altitude above sea level, masl	350	masl		
Mean wind velocity	6	kph		
Predominant wind direction	NW			
Minimum solar radiation, **S**	230	$Cal/cm^2 \cdot d$		
Daily fraction of sunlight hours	0.8	decimal		

Notes: (i) For the design of the anaerobic lagoon process these variables are important; (ii) the minimum water temperature as well as the minimum air temperature are the averages of the coldest month; (iii) the mean annual air temperature is needed in order to calculate the Evaporation over the water surface of the reservoir; (iv) the velocity and direction of the predominant wind is important for correctly locating the reservoir system in relation to the location of the town or the urban center; and (v) the fraction of hours of sunlight is the ratio of actual hours of sunlight to possible hours of sunlight.

1.3 *Pollutants concentrations and loads:* The calculation of the contaminants loads (in kg/d) and of the specific unit loads (kg/per capita · d) as *outputs* base on the *concentrations* given as *inputs* is presented in Table 7E. This is calculated for each contaminant, such as BOD_5, COD, TSS, and so on. If there is significant infiltration of groundwater into the sewerage and conveyance system it is required to calculate the additional infiltration flow from the influence area of the conveyance system (A_a) as follows: $Q_{Iemi} = q_I \cdot L$, where L is the length of the sewerage pipes in km. The concentrations at the entrance to the wastewater treatment plant are calculated as the net load (q_C) multiplied by the population (P) and divided by the total flow: $Q_T = (Q_D + Q_{Iemi})$. For example: $COD = q_{COD} \cdot P / Q_T$. Note that in Table 7E the load and the specific per capita unit load is an outputs while only the concentrations are an inputs, presented in the Excel sheet separately.

Table 7E

Concentration, Load and per Capita Consumption To WWTP						
Variable	**Concentration**	**Unit**	**Load**	**Unit**	**Per Capita**	**Unit**
BOD_5	277.8	mg/L	962.3	kg/d	0.048	kg/hab · d
COD	596.1	mg/L	2064.9	kg/d	0.103	kg/hab · d
COD/BOD_5	2.1					
TKN	40.0	mg/L	138.6	kg/d	0.007	kg/hab · d
N-Nitrate	2.0	mg/L	6.9	kg/d	0.000	kg/hab · d
Total Phosphorus	5.8	mg/L	20.1	kg/d	0.001	kg/hab · d
pH	7.1	UN				
Alkalinity	100.0	mg/L	346.4	kg/d	0.017	kg/hab · d
TSS	202.6	mg/L	701.8	kg/d	0.035	kg/hab · d
VSS	0.0	mg/L	0.0	kg/d	0.000	kg/hab · d
G&O	100.0	mg/L	346.4	kg/d	0.017	kg/hab · d
Coli-Fecal	10000000.0	MPN/100 mL	346400000000000.0	MPN/d	17320000000.000	MPN/hab · d

(2) *Determine the flow diagram of the stabilization reservoir system:* The components of a typical stabilization reservoir system are shown in Figure 7.5. In this example the pretreatment was selected to be an anaerobic lagoon.

(3) *Identify or design the pretreatment system:* This is an external condition, that is, the Stabilization Reservoir design depends on the pretreatment which in many cases is already built and operating before the initiation of the reuse project. This identification allows calculating the quantity and quality of the treated wastewater which will be received by the stabilization reservoir system. In other case, a pretreatment system does not yet exist has to be planned along with the reservoir system. The design of the pretreatment system is carried out in accordance with the methodologies presented in the other chapters of the book. For this example an anaerobic lagoon is selected as the pretreatment unit and its designed is presented in Chapter 5. The expected efficiency and effluent quality of the anaerobic lagoon unit is used as an input for calculating the quality of the pretreated effluent flowing into the Stabilization Reservoir.

Variables selection by the designer: Give the expected efficiency of the pretreatment process, in this case the anaerobic lagoon. The expected removal efficiencies of the important contaminants of this process are presented in Table 7F. These values may vary depending on the type of the pretreatment process, so they are variables of the designer.

These values may vary depending on the type of pretreatment process, so they are variables of the designer. Also included in this table is the number of dry months (n_d) based on the water balance as shown in the Figure 7A "Water Requirements" (there are four month with demand) in which the blue columns are the demand for irrigation, positive or negative (in the latter case no need for irrigation water). The red columns are the net wastewater influent (Equation 7.4).

Table 7F

Variable	Pretreatment efficiency			Observation
	Value	Unit	Value	
BOD	70.0	%		Based on pretreatment
TSS	90.0	%		Based on pretreatment
Fecal-Coliforms	0.0	%		Based on pretreatment
Helminths	80.0	%		Based on pretreatment
O&G	60.0	%		Based on pretreatment
nd, number of dry months	4.0			Based on water balance

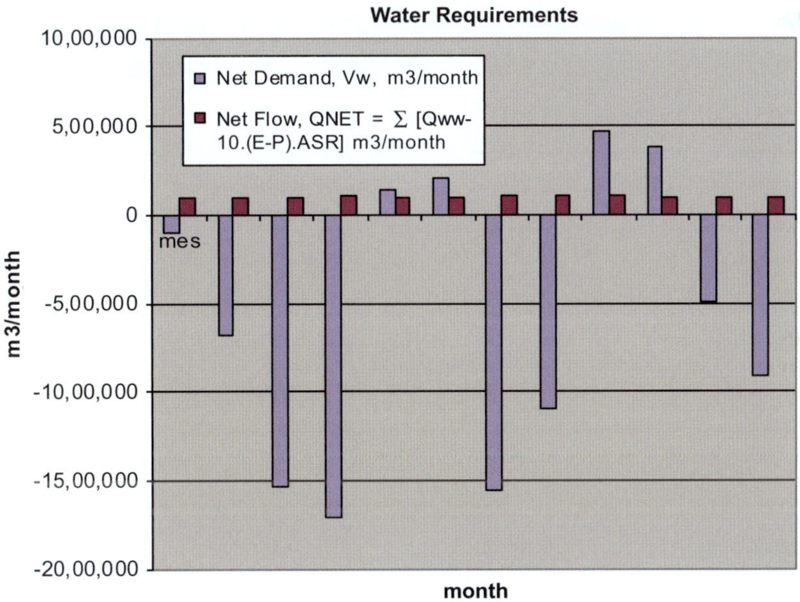

Figure 7A

Design: Applying these efficiencies to the raw wastewater entering the CEPT pretreatment unit, we obtain the quality of the effluent of the pretreatment system which flows into the stabilization reservoir. The program calculates the quality which is presented in Table 7G.

Hydraulic design

(4) *Determine the net demand for irrigation:*
Variables selection by the designer: The designer variables are: (i) the type of crop that will be irrigated with the effluent, in this case is *corn*, (ii) the crop constant for each month: in this

case $k_c = 0.55$ throughout the year (these are data supplied by the agronomy/agricultural expert, this constant may vary monthly), (iii) the total water use efficiency: $E_{Ta} = 0.65$ (65%); this means that the water losses in the conveyance system from the stabilization reservoir to irrigation area is $(1-E_{Ta})$ or 35%, (iv) precipitation (P) and potential evapotranspiration (ET_0). The data for this example are given at the beginning of section 7.3 above, in the statement defining the example. This type of data is obtained from the regional hydrological institution. For design purposes, these date are by the agricultural expert of the design team; (v) infiltration rate at the reservoir site. This information is an outcome of the soil and geotechnical study at the reservoir site. It can also be handles as a sensitivity parameter, that is, a range of probable infiltration rate values is used and the impact of their variation on the design is analyzed.

Table 7G

Concentrations and Loads Flowing into the Stabilization Reservoir				
Variable	**Concentration**	**Unit**	**Load**	**Unit**
BOD_5	83	mg/L	289	kg/d
COD	179	mg/L	–	kg/d
COD/BOD_5	2		–	
TKN	–	mg/L	–	kg/d
N–Nitrate	–	mg/L	–	kg/d
Total phosphorus	–	mg/L	–	kg/d
pH	7	UN	–	
Alkalinity	–	mg/L		kg/d
TSS	20	mg/L	70	kg/d
VSS	–	mg/L	–	kg/d
O&G	40	mg/L	0	kg/d
Helminths	100	Ova/L		
Fecal coliforms	10000000	MPN/100 mL	0	MPN/d

Design: With the data mentioned in the previous paragraph and applying Equation 7.2 for the data of each month, the total monthly demand of water for irrigation, D_w is calculated Then the net monthly demand (V_w) is calculated using Equation 7.3 with the data for each month. The meaning of a negative value of a calculated monthly value of V_w is that the precipitation on that month is sufficient to provide the water needed by the plants, so the requirements of effluent during such a month is zero. The results of the net monthly water demand after determining the possible cultivation areas for irrigation with the effluent under the example case, as explained below, are presented in Table 7H.

The Table also shows the availability of effluent for irrigation (corrected by evaporation, E, and precipitation, P, in the reservoir) on a monthly and total basis $[\Sigma(Q_{WW} - 10\ (E - P) \cdot A_{SR})]$, and the aggregate irrigation water demand (ΣV_w) that results from adding the demand of each

Table 7H

Irrigation Area (A_c) ha	2209													
Crop	**Corn**													
Month	**Jan**	**Feb**	**Mar**	**April**	**May**	**June**	**July**	**August**	**Sept**	**Oct**	**Nov**	**Dec**	**Average**	**Observation**
ET_0, mm/month	100	44.9	67.2	78.1	131.6	152	112	146.2	214.2	139.3	98.9	45.6	111	Water Authority
Precipitation, P, mm/month	90.0	70.0	127.3	144.7	106.0	120.2	166.4	174.4	161.1	101.6	106.9	80.9	121	Water Authority
Infiltration, I, mm/month	1	1	1	1	1	1	1	1	1	1	1	1	1	Soil study
Total water Efficciency, E_{ta}	0.65	0.65	0.65	0.65	0.65	0.65	0.65	0.65	0.65	0.65	0.65	0.65	1	Transport and irrigation
Crop Coefficient, k_c	0.55	0.55	0.55	0.55	0.55	0.55	0.55	0.55	0.55	0.55	0.55	0.55	1	Corn, given by agronomist
$ET = k_c ET_0$, mm/month	55	25	37	43	72	84	62	80	118	77	54	25	61	Actual ET
Agricultural Demand D_w, m^3/month	1,868,956	839,161	1,255,938	1,459,655	2,459,546	2,840,813	2,093,231	2,732,414	4,003,304	2,603,456	1,848,397	852,244	2071426	Equation 7.2
(P - I), m^3/month	1,965,802	1,524,049	2,789,672	3,173,997	2,319,204	2,632,849	3,653,299	3,830,000	3,536,234	2,222,019	2,339,083	1,764,804	2645918	
Net Demand, V_w, m^3/month	−96,846	−684,887	−1,533,733	−1,714,342	140,342	207,964	−1,560,068	−1,097,587	467,069	381,437	−490,686	−912,560	−574492	Equation 7.3
Net Flow, $Q_{NET} = \sum [Q_{WW} - 10 \cdot (E\text{-}P) \cdot A_{SR}]$ m^3/month	95,756	93,172	100,575	102,823	97,823	99,658	105,627	106,660	104,942	97,255	97,939	94,580	1196811	$\sum [Q_{WW} - 10 \cdot (E\text{-}P) \cdot A_{SR}]$
Cumulative Demand, $\sum V_w$, m^3	0	0	0	0	140,342	348,305	348,305	348,305	815,374	1,196,811	1,196,811	1,196,811	0.0	Equation 7.4
Volume in Reservoir, m^3	287,967	381,139	481,714	584,537	542,019	433,713	539,340	646,000	283,873	−309	97,630	192,211		
Total Volume in Reservoir m^3	646,000													

Note: Apply "Goal Seek" in cel 016

month to that of the previous month, taking the precaution to sum zero in a month of negative demand (since a negative demand means that rainfall satisfies the demand on such month and there is no need for supplying effluent). The curves of monthly water demand for irrigation and monthly supply of effluent are shown in Figure 7A (not include the correction of evaporation from and precipitation over the reservoir).

It is noted that there is demand for effluent only during the dry months of May, June, September and October. During the other months there is excess of water due to heavy rainfall. The number of months in which irrigation is required is $n_d = 4$ months. The supply of pretreated effluent into the reservoir is continuous and approximately constant throughout the year. The cumulative demand is presented in Figure 7B.

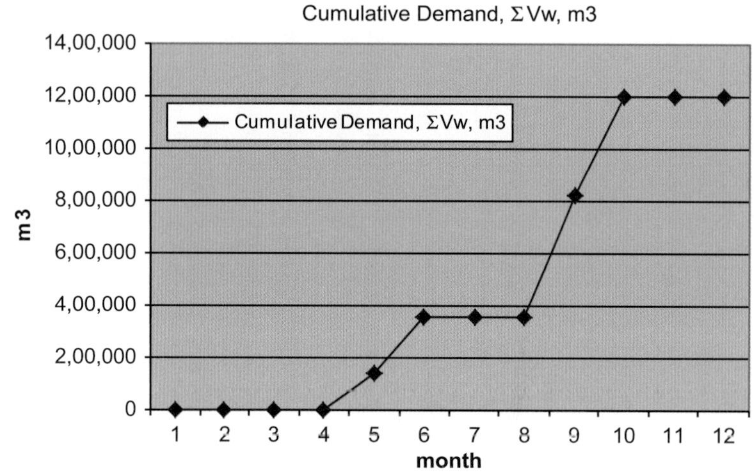

Figure 7B

(5) *Determine the volume of stabilization reservoir:* The calculation of the stabilization reservoir's volume is performed taking into account the monthly distribution of the net effluent the net and the monthly distribution of the water demand for irrigation. In the case of this example a single reservoir is to be constructed (according to the scheme of Figure 7.3a), to provide effluent for restricted irrigation. The volume of the reservoir is calculated using the method of mass curve described in the ODM. Figure 7B shows the demand curve, which, after being incorporated into the mass curve procedure.

The SR volume calculation is made taking the net monthly supply of waste water and the demand for irrigation water every month. In this case it is wanted to build a single SR (case of Figure 7.3a), that is, to make irrigation with restrictions. The reservoir volume necessary to cover the deficit is calculated using the method of mass curve described in the ODM. Figure 7B shows the demand curve. After applying on it the mass curve procedure, Figure 7C obtained.

Based on Figure 7C it is concluded that the volume of the required reservoir is: 846,000–200,000 = 646,000 m^3 (approximately).

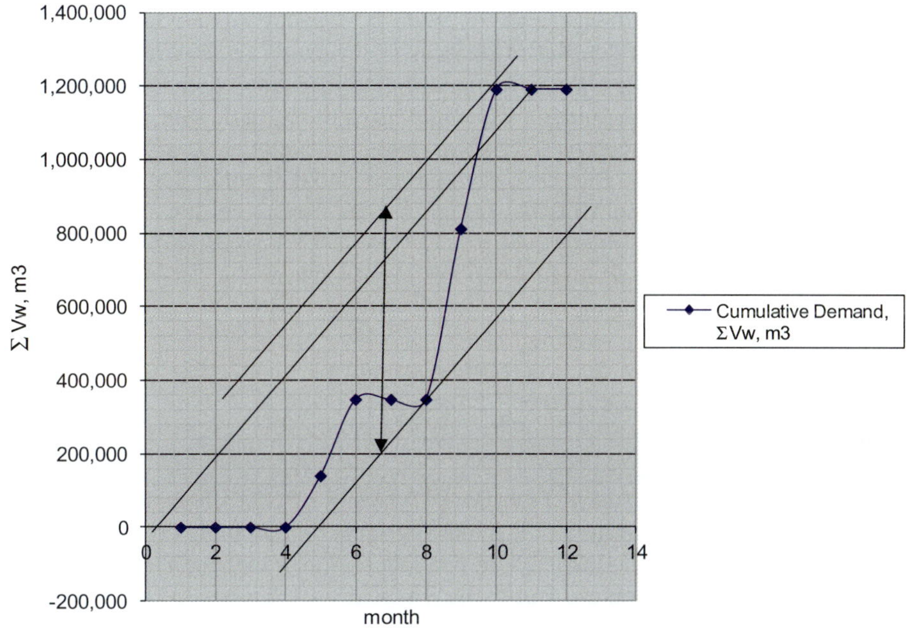

Figure 7C

This value of 646,000 m^3 for the volume of the reservoir is compared with the numeric calculation presented in the two last lines of the design table (Table 7H) and both are identical. The variation in the reservoir's volume each month (one line before last in Table 7H) is presented in Figure 7D. According to this figure, the month in which the required stored volume is the largest is August and the volume needs to be stored in this month, 646,000 m^3, is the total volume of the reservoir.

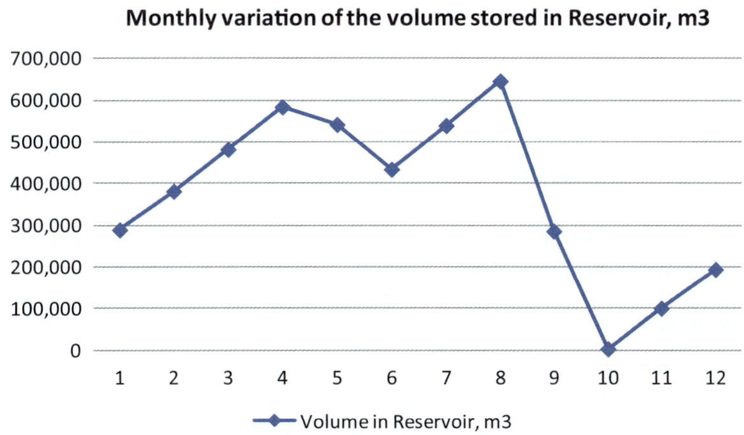

Figure 7D

Note that the month in which the reservoir is completely empty (October) coincides with the start of the heavy rainy season (November, December, January, February, March and April). In May, with the start of the dry season, the reservoir is almost filled, and in August it is completely filled ready for the heaviest demand of effluent for irrigation in September and October.

If it is required to produce effluent for unrestricted irrigation, three (3) stabilization reservoirs can be constructed to function under a sequential batch mode, each with a volume of 323,000 m^3. This ensures that a full reservoir of 323,000 m^3 filled and well treated effluent (void of coliforms) will be ready when needed. The storage volume of 646,000 cubic meters ensures the required supply of effluent during the irrigation season but the quality of this effluent is only adequate for restricted irrigation. A storage volume of 969,000 m^3 guarantees the continuous supply during the irrigation season of high quality effluent, adequate for unrestricted irrigation.

The procedure of calculating the reservoir volume using the method of mass curve is performed manually on the drawing board, or using a computer as shown in Figure 7.D, based on the explanation in Section 7.22 regarding the ODM. The reservoir volume is also calculated by the Excel program CHAP 7-SR.

(6) *Calculate the irrigated area, A_c*

Variables selection by the designer: From the ODM Section 7.22 Paragraph 5 it is clear that the total annual water demand for irrigation (ΣV_w) must be equal to the supply of wastewater corrected for the losses for evaporation and the gains of precipitation over the reservoir [$\Sigma(Q_{WW} - 10\ (E - P) \cdot A_{SR})$]. Since ΣV_w depends on the area to be irrigated (A_c), Equation 7.4 is applied to calculate A_c. In the Table 7H this area is already determined ($A_c = 2209$ ha). However, to start we must use an approximate value, and then proceed as indicated in the design section below. Also, to initiate the calculations a starting value of the area of the reservoir (A_{SR}) is required, and an approximate value is used, to be corrected once the design is performed. Some trial and error iterations for A_c and A_{SR} may be necessary to obtain the right design. These calculations are facilitated through the use of the Excel program CHAP 7-SR.xls.

Design: After calculating all the designer variables we proceed to calculate the exact area of irrigated crops applying a varying A_c values in Equation 7.3 until the Equation 7.4 is met (which is the trial and error procedure). In the example it is possible to irrigate 2,199 ha with the wastewater of the population of the contributing town and with the given values of evaporation and precipitation over the reservoir. In the Excel spreadsheet the exact value is calculated using the "*goal seek*" function in *Data's "What if analisys"* (go to *Data*, then open *What if Analisys* and select the function "*goal seek*", at the cell "*Evap-T' O16*". All graphs can be generated automatically by the Excel program except the mass curve to calculate V_{SR}, which must be prepared manually. Every assumption on A_c produces a different A_{SR}, and some trial and error is required to close the differences (note that A_c define the reservoir volume, which affects A_{SR}, which in turn affects A_c: the circle in the operations is cut because the total volume *is defined*, but in any case it is necessary to observe how changes in A_{SR} causes changes in A_c. It is considered that the trial and error ends when the minimum volume of water in reservoir is close to zero).

Civil engineering design

(7) *Design the stabilization reservoir*

Variables selection by the designer: Topographical maps of the site allocated for constructing the reservoir are required for the design. In the example it is assumed that the reservoir site is flat. We select an effective depth of 5 m (the range is 4 to 12 m. It is necessary to leave a dead reservoir volume of about 1 m high and a freeboard of about 1 m so the total depth of the reservoir is 7 m. Assuming an internal embankment slope of 1:1, the angle of the embankment is 45 degrees. Also, a geometric shape of an inverted truncated cone is proposed for the reservoir in the example. When the topography is complex, it is necessary to do volume calculations based on the topographic surveys.

Design: The net volume of the truncated cone is $V_{SR} = \frac{1}{3} h (A_1 + A_2 + \sqrt{[A_1 A_2]}) = 646{,}000\ m^3$, with A_i the areas of the bases, r_i their radii and h the net depth (selected as 5 m). Calculating for a 1 : 1 slope and solving for the height of 5m as the useful volume of the reservoir we find $r_1 = 200.3$ m and $r_2 = 205.3$ m. Taking into account a dead reservoir volume of 1 m depth and a freeboard of 1 m (which gives a total depth of $h_T = 7$ m): then R_1 (at the bottom of dead volume) is 199.3 m ($A_1 = 124{,}786\ m^2$) and R_2 (at the top of freeboard) is 206.3 m^2 ($A_2 = 133{,}705\ m^2$) for a total volume of 904,539 m^3. The variation of reservoir's water volume with depth, taking into account the first dead meter (water depth of 6 m), is as presented in Figure 7E and the variation of the reservoir's surface area with the depth is presented in Figure 7F.

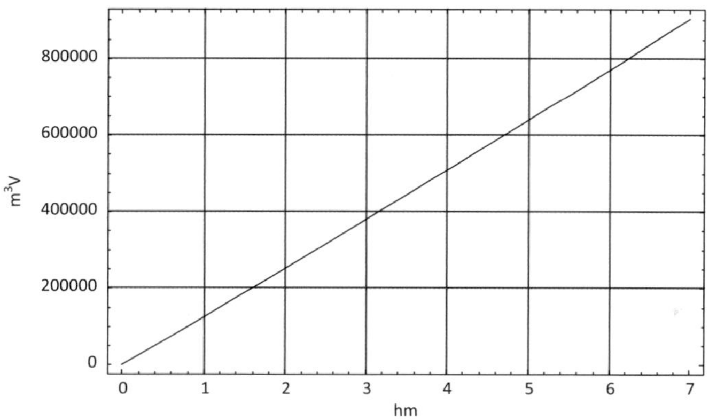

Figure 7E

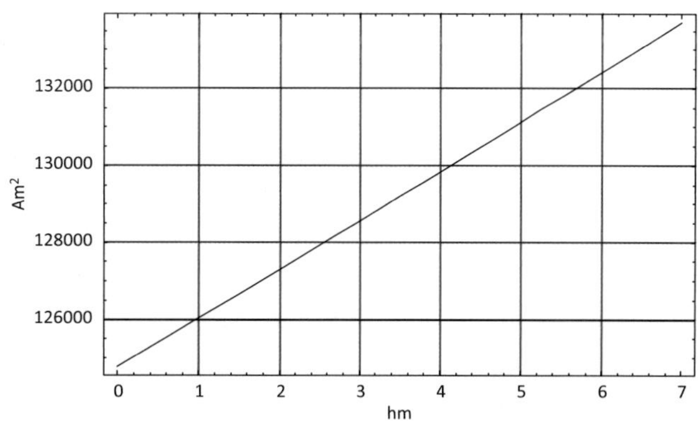

Figure 7F

The depth of the average volume of the reservoir (taking into account the first meter that is a dead volume) is 4 m. The corresponding surface area at that depth is 129845 m^2 or 13 hectares.

The BOD_5 of the pretreated effluent that reaches the reservoir is 83.3 mg/l and the organic load is 288.7 kg DBO_5/d. This implies that the average surface organic loading on the stabilization reservoir is $L_s = 288.7$ (kg DBO_5/d)/12.9 ha $= 22.4$ kg BOD_5/ha · d, which is within the acceptable range of loading. The pretreated effluent that enters the reservoir is of a very good quality and therefore the surface organic loading is low and this reservoir will not generate odour problems.

Kinetics design

(8) *Calculation of the quality of the effluent used for irrigation*

Variables selection by the designer: For the calculation of quality of the effluent used for irrigation, we use the geometric and kinetics values presented in Table 7I. It is noted that the reservoir volumes and areas for different depths have already been calculated in the previous paragraph on "Civil Engineering Design" (see Figures 7E and 7F) and the kinetic coefficients proposed are similar to those of facultative lagoons.

Table 7I

Stabilization reservoir net dimensions and coefficients				
V_{SR}, net volume	646000	m^3		
SR net depth, H	5.0	m		
A_{SR} at total depht h = 6 m	129200.0	m^2	12.92	ha
$V_{1/2}$, net	323000	m^3		
$A_{1/2}$ at h = 4 m (1 m death reservoir)	129845.0	m^2	12.98	ha
$K_L(20°C)$	0.35	d^{-1}		Typical
θ (BOD_5)	1.06			1,04–1,09
K_B (20°C)	2.7	d^{-1}		FC typical
θ (FC)	1.19			Typical

Design: The kinetics constants need to be corrected according to the actual temperature, and once applied to the stabilization reservoir conditions the results presented in Table 7J are obtained:

Table 7J

Stabilization reservoir kinetics				
Variable	**Value**	**Unit**	**Value**	**Observation**
$t_d = V_{1/2}/Q_D$	93	d		Equation 7.6
$T_L = (A_s f T_a + Q_D T_i)/(A_s f + Q_D)$	18	°C		Equation 7.9
$K_L = K_L(20°C)\ \theta(T-20)$	0.32	d^{-1}		Equation 7.10
$S = S0/(1+KL^*t_d/n)^n$	2.72	mg/L		Equation 7.8
$K_B = K_B(20°C)\ \theta^{(T-20)}$	2.03	d^{-1}		Equation 7.10
$B = B_0/(1+K_B{}^*t_d)$	52590	MPN/100 mL		Equation 7.8
$L_S = Q_D{}^*DBO_5/A_{1/2}$	22.23	kg DBO_5/d · ha		<50

The equations applied to calculate the principal quality parameters in the effluent for irrigation result in 3 mg/l BOD_5, 5275 NMP/100 ml of Fecal Coliforms and zero Helminthes. Other quality parameters are shown in Table 7K.

Table 7K

		Irrigation water quality			
Variable	**Value**	**Unit**	**Value**	**Unit**	**Observation**
BOD_5	3	mg/L	–	kg/d	
COD	54	mg/L	–	kg/d	
COD/BOD_5	–		–		
TKN	–	mg/L	–	kg/d	
N–Nitrate	–	mg/L	–	kg/d	
Total Phosphorus	–	mg/L	–	kg/d	
pH	7	UN	–		
Alkalinity	–	mg/L	–	kg/d	
TSS	10	mg/L	–	kg/d	
VSS	–	mg/L	–	kg/d	
O&G	16	mg/L	–	kg/d	
Helminths	0	Ova/L			
Fecal coliforms	52590	MPN/100 mL	–	MPN/d	

The removal efficiencies for the main contaminants are summarized in Table 7L.

Table 7L

	Stabilization reservoir efficiency		
BOD	99.47	%	
COD	70	%	Typical
TSS	50	%	Typical
Fecal–Coliforms	99	%	
Helminths	100	%	Equation 3.10
G&O	60.00	%	

(9) *Inflow and outflow structures:* The inflow structures (see Figure 7.13) are designed for the wastewater flow ($Q_D = 40$ Lps) and the outflow structures for irrigation are calculated for the maximum flow. Considering the reservoir's filling and withdrawal cycle it can be observed that from August to September the complete reservoir is emptied (646,000 m^3/2 months =

343,000 m^3/month = 11,320 m^3/d = 131 Lps, which is the maximum flow for irrigation). The calculated diameters of the inflow and outflow pipes are presented in Table 7M.

Table 7M

	Maximum irrigation flow					
Variable	**Value**	**Unit**	**Value**	**Unit**	**Value**	**Unit**
Irrigation flow, Q_R	131	L/s	472	m^3/h	11320.00	m^3/d
Inlet and oulet pipes						
v_e, water velocity	1	m/s				0,6–5,0 m/s
A_e, pipe area	0.040	m^2				
D_e, inlet pipe diameter	0.226	m	9.000	inch		
v_s, water velocity	1	m/s				0,6–5,0 m/s
A_s, irrigation pipe are	0.131	m^2				
D_s, irrigation pipe diameter	0.408	m	16.000	inch		

The design inlet flow for is $Q_D = 40$ lps, which for a flow velocity of 1 m/s results in a 9″ pipe diameter. The maximum design outflow for irrigation is $Q_R = 131$ Lps (as calculated above) and the resulting outlet pipe diameter for a flow velocity of 1 m/s is 16″.

Chapter 8

Sub-Surface Flow Constructed Wetlands (SSFCW)

8.1 PROCESS DESCRIPTION

8.1.1 Introduction

The concepts of artificial wetlands, aquatic and subsurface, are essentially new advances in wastewater reuse. Subsurface Flow Constructed Wetland (SSFCW) represents a cost-effective option for post-treatment of municipal wastewater of small populations (Crites, Middlebrooks & Sherwood, 2006). The SSFCW is discussed in this chapter.

A wetland is defined as an area where the water table is at or above the soil surface for a long enough period to maintain saturation conditions and allow the growth of related vegetation. The capacity of the improvement of the quality of wastewater in wetlands has been verified in studies in a variety of geographical locations. Wetlands used for wastewater treatment include swamps, mangroves, and so on. Ultimately wetlands systems are built especially for the treatment of wastewater (artificial wetlands).

A major constraint in the use of many natural wetlands for treatment of wastewater is the fact that they are considered by most regulatory authorities as part of the receiving water bodies. Consequently, the wastewater discharged to them must comply with the standards of use of the wetland. In these cases, the potential capacity of the wetland for improving the quality of wastewater is not utilized completely. Constructed wetlands avoid these special requirements of influent quality and can also ensure a better control on the hydraulic regime in the system; therefore, they are more reliable than natural wetlands.

The mostly used two types of constructed wetlands are: (i) surface flow constructed wetland, which is a system similar to a natural swamp because the water surface is exposed to the atmosphere; and (ii) subsurface flow constructed wetland, which uses a permeable medium in which the wastewater flows and the water level remains below the level of the bed. Table 8.1 presents the basic parameters of the main types of wetlands, natural and constructed (Reed *et al.* 1995; Crites *et al.* 2006).

This chapter focuses on the Subsurface Flow Constructed Wetlands (SSFCW) which are those that can better handle higher organic loads and generate fewer environmental problems common in surface wetlands, such as the proliferation of mosquitoes and the generation of unpleasant odours. A schematic cross section of a SSFCW unit which shows its internal structure is presented in Figure 8.1. During its transport through the matrix of saturated soil or rock filter, the inflow stream undergoes treatment processes and in addition, the plant roots are the media of adherence to multiple microorganisms that support the bioconversion of the organic matter contained in the inflow. The effluent out flowing from a constructed wetland has the water quality shown in Table 8.1.

Table 8.1 Basic parameters of the main types of wetlands.

Concept	Treatment goal	Climate	Detention time d	Depth m	Surface organic load kg BOD/ha · d	Effluent quality Mg/l
Natural swamps	Polishing	Warm	10	0.2–1	100	BOD: 5–10 TSS: 5–15 TN: 5–10
Horizontal flow constructed wetland	Secondary to advanced	Any	7 to 15	0.1–0.6	200	BOD: 5–10 TSS: 5–15 TN: 5–10
Subsurface flow constructed wetland	Secondary to advanced	Any	3 to 14	0.3–0.6	600	BOD: 5–40 TSS: 5–20 TN: 5–10

Source: Crites, Middlebrooks & Sherwood (2006, Table 1.2, pp. 5)

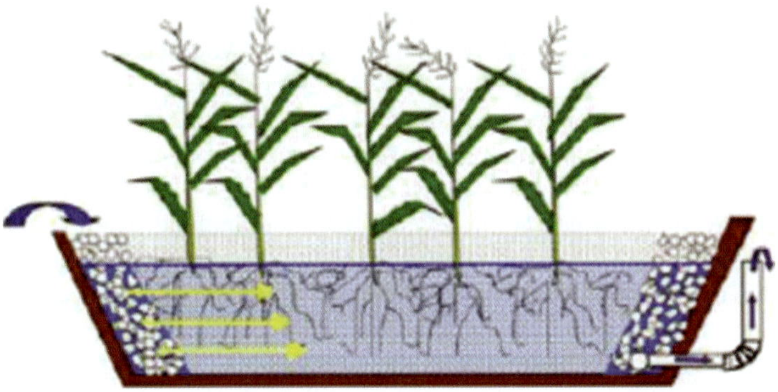

Figure 8.1 A Schematic Cross Section of a SSFCW Unit [*Source*: WSP (2008)]

Constructed wetlands plants which are used as the main treatment unit of municipal wastewater are usually limited in size to serve populations of up to 20,000. The reason for this limit is that for larger populations the area occupied by constructed wetlands becomes very large and sometimes prohibitive. Also, the cost becomes prohibitive due to the required large quantities of gravel or rock media. However, constructed wetlands can serve as polishing units of other secondary treatment processes and in that capacity they can be used in plants serving larger populations because in this case they function under lower surface organic loadings.

According to Mara (2004) a constructed wetland occupies more area than a secondary facultative lagoon and it is of higher investment cost because of the required gravel media, so detailed comparison calculations need to be undertaken to demonstrate the preference of constructed wetlands over stabilization lagoons. However, the cost issue is site specific and for small populations constructed wetlands may provide more benefits.

8.1.2 Basics of the process

Fundamentals

There are three natural systems for effective treatment of municipal wastewater: aquatic treatment (lagoons), wetlands treatment and land treatment (land application). One of the interesting systems is the subsurface flow constructed wetland (SSFCW). The mechanisms of processing organic matter which occur in wetlands are similar to those occurring in lagoons systems, but in wetlands attached growth of also play an important role. This means that microorganisms that grow attached to the roots of the aquatic plants and rocks and gravels that serve as a matrix for support of the plants also take part in the organic matter decomposition processes. Therefore it is necessary to take into account in the case of SSFCW the porosity of the support material. However, the main design parameters continue being the hydraulic load (q_a) and the detention time t_d. The basic kinetic equations are those presented in Chapter 2.

According to Menahem Libhaber (2007): "A constructed wetland bio-filter is a gravel or volcanic rock aerobic biological filter sown with marsh plants, through which preliminary treated wastewater is flowing horizontally or vertically. A film of aerobic bacteria is formed on the filter bed and consumes organic matter dissolved in the wastewater. To avoid clogging of the filter bed, large suspended solids need to be removed from the inflowing wastewater before it reaches the wetland. This can be achieved by preliminary treatment, preferably by rotating fine screens, or by the use of settling units such as Imhoff tanks or settling tanks". The SSFCW has a good capability for removing BOD, but is less effective in removing pathogens and nutrients.

The components of a subsurface flow constructed wetland are:

- Waterproof lagoon
- Filtering materials
- Swamp plants: marsh cane, ginger, Typha, reed (Phragmites), papyrus, heliconia, and so on.
- Structures of inlet flow distribution and outlet flow collection.

The removal mechanisms of various contaminants in a Subsurface Flow Constructed Wetland and the efficiency of their removal are presented in Table 8.2. Note the high organic matter (BOD) removal efficiency and low removal efficiency for pathogens and nutrients such as nitrogen and phosphorous (WSP, 2008).

Table 8.2 Mechanisms and efficiencies of contaminants removal in subsurface flow constructed wetlands.

Type of contaminant	Main removal mechanism	Removal efficiency
Organic matter	Bacterial degradation	80–90%
Suspended solids	Sedimentation, filtration	80–90%
Nitrogen	Ammonification, nitrification-denitrification	20–40%
Phosphorus	Absorbing processes in the filtrating material	20–30%
Pathogenic	Predation, natural death (for fecal coliforms)	1–3 log units
Microorganisms	Sedimentation, filtration (for helminths)	

Source: WSP (2008)

Figure 8.2 shows a flow diagram of a complete SSFCW system, including preliminary and primary treatment. Figure 1.35 shows a photo of a constructed wetland plant in Copacabana, Bolivia, consisting of pretreatment in a septic tank and three constructed wetlands lagoon in series.

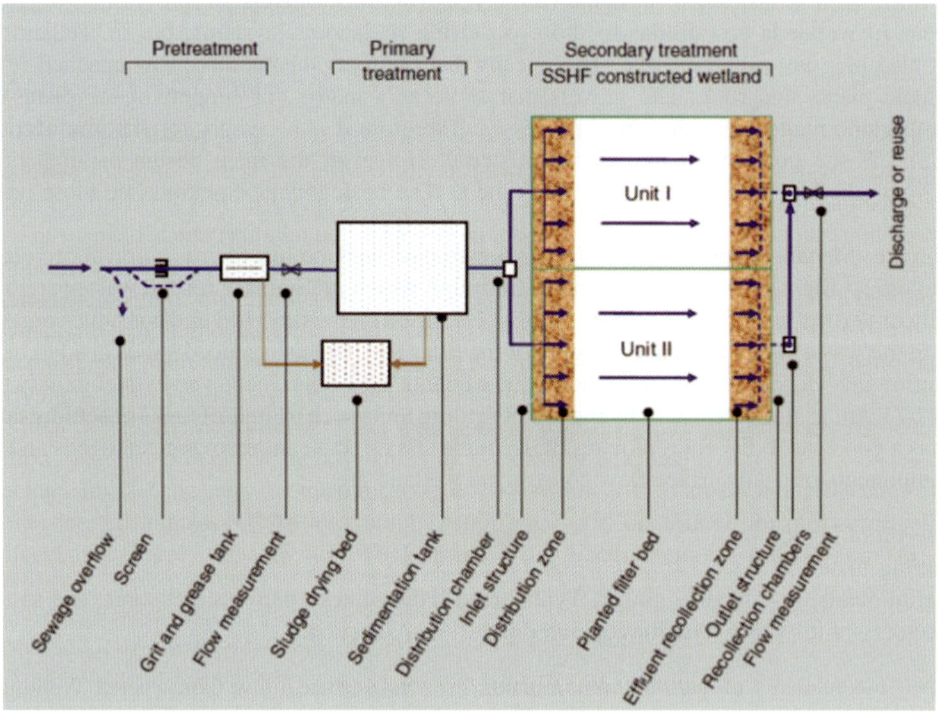

Figure 8.2 Flow Diagram of a Complete Subsurface Flow Constructed Wetlands System [*Source*: WSP (2008)]

The specific unit area of SSFCW is 1.5 to 5.0 m^2 per Capita. Construction of SSFCW requires access to good building material (fine gravel), and stable land with adequate drainage so that rainwater is be deviated away from the wetland. The level of the water table at the plant's site should be below the wetland and, if the soil is not impermeable, impermeabilization is required.

It is important to take into consideration possible negative impacts of the pretreatment unit, especially the generation of excess solid and sometimes unpleasant odours. The generation of odours from the wetland unit itself, as well as proliferation of insects in this unit are moderate, but precautions must taken.

Design considerations

Hydraulics

The main hydraulic design parameters are Darcy's law for calculating the flow through the filter bed and porosity (n) for correctly calculating the of hydraulic detention time in the wetland. The recommended hydraulic slope of the wetland is between 0.5 and 1%. To avoid going deepening too much at the end part of the wetland because the bottom slope, it is recommended to limit its length (50 m is an appropriate value) and divide the system into parallel treatment units. The hydraulic design is based on Darcy's law:

$$v = k_s s \equiv Q_D/B \cdot H \tag{8.1}$$

where:

v = Darcy's velocity, m/d
s = terrain slope, m/m
k_s = hydraulic conductivity, $m^3/m^2 \cdot d$
Q_D = average flow, m^3/d
B = width of the wetland, m
H = average depth of the wetland, 0.3 to 0.6 m

The hydraulic load is defined by the following equation:

$$q_a = Q_D/A_s \tag{8.2}$$

where:

q_a = Hydraulic load, between 0.04 and 0.75 m/d
Q_D = Design flow, m^3/d
A_s = Surface area of the wetland, m^2 (width × length)

The SSFCW's length is obtained from the hydraulic load. The length (L) obtained from Equation 8.2 is therefore:

$$L = Q_D/B \cdot q_a \tag{8.3}$$

The ratio length/width (L/B) is usually between 3/1 and 0.5/1.

The *hydraulic load* (q) correlates with total suspended solids (TSS) according to the following equation (Crites *et al.* 2006):

$$TSS_e = TSS_0(0.1058 + 0.000011q_a) \tag{8.4}$$

where:

TSS_e = Total Suspended Solids in the Effluent, mg/l
TSS_0 = Inffluent TSS, mg/l
q_a = hydraulic load, m/d

The calculation of the detention time in the wetland is done using the equation:

$$t_d = L \cdot B \cdot H \cdot n/Q_D \equiv V_N/Q_D \tag{8.5}$$

where:

L = filter length, m
B = filter width, m
H = effective depth, m
N = porosity of the filter bed as a decimal fraction, see Table 8.3
V_N = net volume, m^3
Q_D = flow rate, m^3/d

Characteristics of the filter medium of a SSFCW is given in Table 8.3.

Table 8.3 Characteristics of the filter media of the SFCW.

Filtrating bed	Effective size D_{10}, mm	Porosity n, %	Hydraulic conductivity k_s, m/d
Thick Sand	2	28–32	100–1000
Gravel Sand	8	30–35	500–5000
Fine Gravel	16	35–38	1000–10000
Gravel	32	36–40	10000–50000
Thick Gravel	128	40–45	50000–250000

Kinetic equations

The kinetic equations of a SSFCW give another approach to the design of the wetland. The BOD removal kinetics is similar to that of a stabilization lagoon under a plug flow regime (see Equation 2.37), but the kinetic of removal of other contaminants are different. It is also necessary to consider the hydraulic equations that have been explained in the previous section. The kinetic equations proposed for other contaminants are those used by Reed *et al.* (1995) as quoted by Merz (2000) in his article "Guidelines for Using Free Water Surface Constructed Wetlands to Treat Municipal Sewage", complemented by the book of Crites, Middlesbrooks and Sherwood (2006. The equations are written using the following not previously defined nomenclature:

Nomenclature:

Sub-indexes "0" and "e" (S_0, S_e): influent, effluent
K_{Hi}: first order wetland constant for parameter i = DBO_5, NTK, NO_3 and CF, d^{-1}
K_p: first order area constant for P: $K_p = 0.0273$ m/d
Q_D: wetland average flow, m^3/d
t_{med}: average detention time, VN/Q_{med}, d
N: Fecal coliforms, MPN/100 ml, for the influent (0) and the effluent (e)
N_c: number of wetland units in series
P_j: Phosphorus, mg/l, for the influent (0) and effluent (e).

Equations

For removal of DBO_5, NTK and NO_3:

$$S_e = S_0 \exp - \left[K_{Hi} \cdot n \cdot (A_s \cdot H)/Q_D\right] \tag{8.6}$$

From where we conclude:

$$A_s = Q_D L \cdot n\left(S_0/S_e\right)/K_{Hi} \cdot n \cdot H \tag{8.7}$$

Also, the average hydraulic load is:

$$q_a = Q_D/A_s \tag{8.8}$$

For removal of SST: use Equation 8.4.

For removal of Fecal Coliforms (FC): we use the following equation:

$$N_e = N_0/(1 + t_{med} \cdot K_{HCF})^{Nc} \quad (8.9)$$

For Removal of Total Phosphorus: the flow regime is plug flow, but the reaction area constant correlates inversely with the hydraulic load, as follows:

$$P_e = P_0 \exp(-K_p/q_a) \quad (8.10)$$

Note that the constant of removal of the first order varies with temperature, for any sub index (i) as follows:

$$K_i(T°C) = K_i(20°C)\theta^{(T-20°C)} \quad (8.11)$$

Table 8.4 shows the common values of θ (temperature correction coefficient) and Kj.

Table 8.4 Constants of first order reactions for the different parameters.

Contaminant removal process	K_H d^{-1}	θ
DBO_5	0.678	1.06
TKN	0.2187	1.048
NO_3	1.00	1.15
Faecal Coliforms	2.6	1.19

Source: Reed *et al.* (1995)

8.1.3 Performance

The SSFCW are "natural" systems which behave in manners that are not simple to describe, and adequate knowledge is required to perform its functional and effective design. However, once a good design is made, its operation is simple and the results are cost-effective and satisfactory.

Because the water is not exposed during the treatment process, subsurface flow constructed wetlands can be insulated with a layer of peat or compost. This allows SSFCW to be utilized in cold-climate applications, and perform well during very low air temperature conditions of up to −25°C (Wallace, 2007). This makes the SSFCW an attractive process in regions with very low temperature seasons.

Efficiency

The typical contaminants removal efficiencies of constructed wetlands are presented in Table 8.2. The SSFCW with a nominal detention time greater than four days can achieve a very good BOD removal with effluent quality of less than 20 mg/l based on the 50 percentile (P_{50}) and often at 80 percentile (P_{80}). However, if the influent BOD concentration is lower than 5–10 mg/l, the BOD removal efficiency becomes low or negative (Reed *et al.* 1995).

Advantages and disadvantages

The advantages of SSFCW are the following (Adapted from Merz, 2000):

- Ability to deal with high organic loads.
- Tolerance to cold weather.

- More treatment per unit land area compared the Surface Flow Constructed Wetlands.
- Proliferation of mosquitoes and generation of odours are not usually a problem.
- SSFCW do not pose a public health problem because they are not open surface water.
- There is less need for weed control compared to Surface Flow Constructed Wetlands.
- Insulated SSFCW can be utilized in cold-climate applications and perform well during very low air temperature conditions. This makes the SSFCW an attractive process in regions with very low temperature seasons.

The main disadvantages are:

- High capital cost due to the value of the filter media.
- It has a tendency to clogging, especially in the entrance area.
- It is a treatment method limited to small communities. Most of the existing constructed wetland plants are designed for a flow of up to 2,000 m^3/d, serving populations of up to 20,000.

8.2 BASIC DESIGN PROCEDURE

8.2.1 General design considerations

A flow diagram of a SSFCW is shown in Figure 8.2 and a section of a plant in Figure 8.1. The inflow to the wetland unit undergoes treatment during its transport through the matrix of saturated soil or rock filter artificially constructed, where in addition the plant roots are a basis for adherence of microorganisms that aid in the bioconversion of organic matter. The removal mechanisms of various contaminants and their removal efficiencies in a SSFCW are presented in Table 8.2. The SSFCW occupies a specific unit area of 1.5 to 5.0 m^2 per Capita and requires the availability of good building material (fine gravel), as well as a stable soil with adequate drainage.

The design of a SSFCW needs to take into account the combination of hydraulic and kinetic aspects, because the goals of treatment may be incorporated in the hydraulic variables of the system. The proposed design procedure is based on both aspects.

The SSFCW unit must be preceded by a pre-treatment unit, preferably primary treatment, to avoid rapid clogging of the filter.

8.2.2 Orderly Design Method, ODM

In general the design methodology of a SSFCW plant consists of the following:

(1) *Determine the external or independent variables:* which is the same as is presented in Chapter 5, Section 5.2.2, Paragraph 1.

(2) *Determine the SSFCW flow diagram:* A typical Flow Diagram is presented in Figure 8.2. The components of a subsurface flow constructed wetland according to WSP (2008) are:
   - *Pretreatment:* Usually Primary Treatment but may also be a septic tank or a Imhoff tank or anaerobic lagoon.
   - *Impermeable Lagoon:* One or more parallel impermeable lagoon are required. Impermeability is required in order to prevent groundwater contamination by infiltration from the wetland lagoon. The impermeabilization can be achieved by lining with layers of compacted clay (less than 0.30 m in 2 layers with a hydraulic permeability equal to or less than 10^{-7} m/d). Alternatively, lining with plastic membranes of HDPE (high density polyethylene) can be used for impermeabilization of the lagoon.

- *Rock filtering material:* The filter material plays an important role in the treatment process in a wetland. It retains solids of the pretreated influent, whose organic fraction is biologically degraded in the pores of the bed. The media provides surface area for the development of a bacterial film which has an essential role in the degradation of organic pollutants and in the transformation of the various forms of nitrogen compounds. The wastewater level is always kept a few millimeters below the surface of the filter bed. Coarse sand and fine gravel are preferred materials for the filling of wetland. Selection criteria include: (i) size and material granulometry, which determine its hydraulic permeability and porosity; and (ii) resistance to chemical and physical wear caused by wastewater.
- *Marsh plants:* Plants frequently used are, among others, canes, ginger, Typha, reed (Phragmites), papyrus, heliconia, and so on.
- *Inflow distribution and outflow collection structures:* In addition to the principal filtering material, a coarser filling material (diameter 5–10 cm) is required to facilitate the distribution of the inflow in the entry area and the collection of the effluent in the outflow zone. Simple structures for even distribution of the inflow across the width of the wetland include buried perforated pipes or channels with weirs.

Additionally, the coarse filter material placed in the entrance area facilitates even distribution of the inflow, as presented in Figure 8.3.

Figure 8.3 Inflow Distribution Structures [*Source*: WSP (2008)]

To collect the effluent outflow, a drainage pipe is usually installed, placed at the bottom of the exit area of the system, filled with coarse material similar to that at the entrance area. The perforated pipe is connected to a central collection box of treated effluent installed outside the wetland. A flexible hose is installed in the collection box for the adjustment of water level in the system.

(3) *Design the pretreatment system:* A pretreatment init need to be included in the SSFCW system in order to control the TSS in the influent and avoid clogging. Usually the pretreatment unit is a Conventional Primary Treatment (CPT) that is designed exactly like the Chemically Enhanced Primary Treatment (CEPT described in Chapter 9), but without chemical dosing. In small systems, a septic tank is sometimes used as the pretreatment treatment unit. An anaerobic lagoon can also be used as pretreatment.

3.1 *Design the preliminary treatment:* The preliminary treatment system which precedes the primary sedimentation is designed in accordance with the methodology presented in Chapter 5, Section 5.2.2, Paragraph 3.

3.2 *Design primary clarifier:* The main design parameter is the Surface Overflow Rate (SOR) measured $m^3/d \cdot m^2$ of m/d. The average SOR (i.e. SOR based on the average flow) is in the range 24–48 m/d and the peak SOR is in the range 48–120 m/d. The weir load should be in the range 125–500 $m^3/d \cdot m$(linear), and the detention time in the range 1.5 to 4.0 h, as shown in Table 9.4. Consequently:

- $A_{sed} = Q_D/SOR$, $A_{sedmax} = Q_{DH}/SOR_{max}$. Select the largest area of the two, and define it as A_s.
- Calculate the net diameter (D_N) of the clarifier (between the ends of weirs) by using the following equation:

$$D_N = \sqrt{\frac{4A_s}{\pi}} \tag{8.12}$$

- Calculate the clarifier's perimeter, $C = \pi D_N$, and estimate the weir load, $q_v = Q_D/C$. The number must be between 125–250 $m^3/d \cdot m$ (linear weir length), or less.
- Calculate the sludge produced, Q_x (kg/d), using Figure 8.4 and Table 9.4. Defining E as the TSS removal efficiency then $Q_x = E \cdot TSS \cdot Q_D$. The removal efficiency of BOD_5 and TSS is calculated, respectively with:

$$E(DBO) = -0.29\,SOR + 42.48 \tag{8.13}$$

$$E(SST) = -0.4\,SOR + 73.78 \tag{8.14}$$

- Define the sludge concentration at the bottom in the settler, for example, 5%.
- Calculate the volume of sludge produced at 5% (0.05): $V_{5\%} = Q_X/0.05$ (m^3/d).
- Determine the mean depth (H_{Sed}) of the settler, usually between 3.5 and 4.5 m. Then calculate the settler volume, $V = H_{Sed} \cdot A_s$. Calculate the settling detention time, $t_d = V/Q_D$.

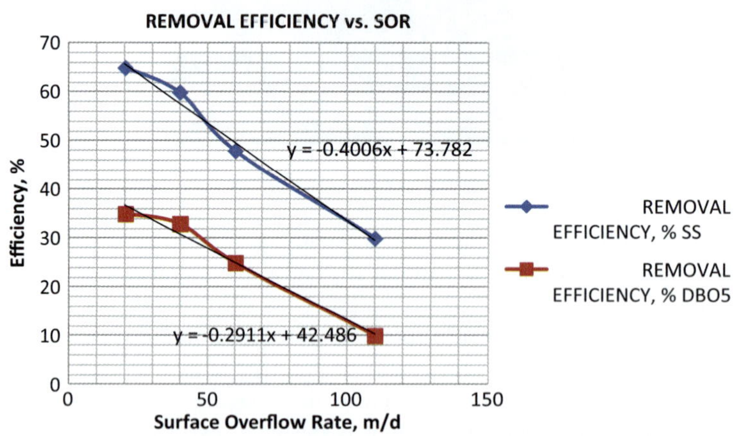

Figure 8.4 Efficiencies of BOD and TSS Removal in a Primary Settling Basin [*Source*: based on Harleman (2004)]

(4) *Define the type of filter bed to be used:* Table 8.3 is used for this purpose, and the selection depends on the designer's experience, the availability of materials in the project zone and type of marsh plant to be planted. The parameters specifying the filter bed are the effective size (D_{10}, mm), which in turn

determines the porosity (n,%), and the hydraulic conductivity (k_s, m/d). The values of these parameters are very important to design.

(5) *Calculate the size of the wetland*

5.1 *First define the designer variables*: There are several possibilities for selecting the variables since some variables depend on others. We present a methodology to calculate the surface area of the wetland, A_s (m^2), from the desired efficiency of BOD_5 removal. The other variables to be selected are geometric variables and their selection is based on the experience of the designer. Following are some guidelines:

- *Hydraulic gradient:* s, m/m, should be between 0.3 and 1.0%.
- *Required effluent BOD_5 concentration:* Table 8.1 can be used as a guide to select the BOD_5 concentration.
- *Surface Area*: An initial surface area is first proposed, and then using Equations 8.6 and 8.7 it is possible to perform the exact calculation of A_s.
- *The Kinetics Constant of BOD_5*: $K_{HDBO} = 0.678$ at 20° C, according to Table 8.4.
- *Depth of the wetland:* H, should be between 0.3 and 0.6 m.
- *Length/width ratio of each wetland lagoon unit:* (L/b), where L is the length and b the width of each wetland unit. The ratio should be between 0.5 and 4.0. L should not exceed 100 m.

5.2 *Calculate the dimensions of the wetland*: With the proposed designer variables proceed to calculate all the dimensions of the wetland lagoon as follows:

- *Darcy Velocity:* is calculated using Equation 8.1:

$$v = k_s s \tag{8.1}$$

where:

v = Darcy velocity, m/d
s = bottom wetland slope, m/m
k_s = hydraulic conductivity, $m^3/m^2 \cdot d$

- *Water temperature in the wetland:* there are several methods of calculation. In tropical regions Equation 5.8 can be used (the same equation used in the design of stabilization lagoons):

$$T_L = \frac{A_s f T_a + Q_D T_i}{A_s f + Q_D} \tag{5.8}$$

where:

T_L = Final temperature of the wetland water (°C)
T_a = Ambient air temperature (°C)
T_i = Influent wastewater temperature (°C)
A_s = Surface area of the wetland (m^2)
Q_D = Wastewater Flow (m^3/d)
f = Proportionality factor, 0.5 m/d

- *Correction of Kinetics the Constant of BOD_5 and the temperature:* calculate K_{HDBO} at the wetland temperature (T_L), using Equation 8.11:

$$K_{Hi}(T°C) = K_{Hi}(20°C)\theta^{(T-20°C)} \tag{8.11}$$

The value of θ is obtained from Table 8.4.

- *Transversal Area:* A_T (m^2), using the equation:

$$A_T = Q_D/v \tag{8.15}$$

- *Hydraulic load:* q_a (m/d o $m^3/d \cdot m^2$) using equation:

$$q_a = Q_D/A_s \tag{8.2}$$

The hydraulic load should be between 0.04 and 0.75 m/d (based on the definitive surface area is once it has been calculated).

- *Minimum width of the wetland unit:* B_{min} (m) using the equation:

$$B_{min} = A_T/H \tag{8.16}$$

For the purpose of providing a safety factor to the infiltration capacity to absorb the peaks in inflow, the total final width of the wetland unit (B) must be at least twice $B_{min,}$ (WSP, 2008). In addition, since the system is designed with at least two parallel SSFCW units to facilitate their maintenance, a single wetland unit will receive all the flow during the duration of the maintenance, and even the maintenance period is short, the unit must have the capacity to handle the entire flow. Therefore:

$$B = 2B_{min} \tag{8.17}$$

One strategy is to design each wetland unit for width B_{min} and build two identical parallel units of SFFCWs

- *Wetland length:* The length L (m) of the wetland unit is calculated using the equation:

$$L = A_s/B_{min} \tag{8.18}$$

- *Number of wetland units:* The number of units N, is calculated using the equation:

$$N = B_{min} \cdot (L/b)/L \tag{8.19}$$

It is clear that the length (L) calculated using Equation 8.18 does not have to adjust to the selected length to width ratio (L/b), so to achieve required (L/b), the SSFCW is divided into several parallel channels, each fulfilling the required (L/b). It is important to maintain the length (L) not larger than 100 m in order to avoid excessive depth at the end of the wetland, which might be caused by the bottom slope.

- *Width of the wetland unit:* each parallel channel will then have a width (b) which is calculated using the equation:

$$b = B_{min}/N \tag{8.20}$$

- *Detention time:* The detention time is calculated using Equation 8.5.
- *TSS concentration in the effluents:* TSS_e is calculated using equation 8.4:

$$TSS_e = TSS_0(0.1058 + 0.000011\ q_a)$$

where:

TSS_e = total suspended solids concentration in the effluent, mg/l
TSS_0 = total suspended solids concentration in the inflow, mg/l
q_a = hydraulic load, m/d

- *Calculate the Surface Area of the wetland for the required efficiency:* the efficiency obtained is calculated using the equation:

$$S_e = S_0 \exp(-n \cdot A_s \cdot H \cdot K_{Hi}/Q_D) \tag{8.6}$$

It is then necessary to vary A_s in Equation 8.6 to obtain the result: $S_e = DBO_{5e}$ required as designer variable. Because A_s depends on the K_{HDBO}, and K_{HDBO} depends on the temperature which in turn depends on A_s, we face a circular calculation which is resolved by trial and error. In Excel this is solved very easy, first proposing an initial value for A_s and then using the function "goal seek", which performs the trial and error procedure, so that the S_e calculated with Equation 8.6 is exactly equal to the required value selected as a designer variable.
- *Calculation of removal efficiency of other parameters:* These calculations are performed using Equations 8.9, 8.10 and 8.11. The removal efficiencies of other parameters are calculated as follows:

For removal of fecal coliforms, FC: Use the following equation,

$$N_e = N_0/(1 + t_{med} \cdot K_{HCF})^{Nc} \tag{8.9}$$

For Total Phosphorus removal: similar to the above equations, but the area rate constant correlates inversely with the hydraulic load as follows:

$$P_e = P_0 \exp(-K_p/q_a) \tag{8.10}$$

Note that variation with temperature of any first order removal constant(i) is given by (following the Nomenclature presented above):

$$K_{Hi}(T°C) = K_{Hi}(20°C)\theta^{(T-20°C)} \tag{8.11}$$

The values of K_{Hi} are given in Table 8.4.

(6) *Inflow and Outflow structures:* the inflow and outflow structures are usually perforated pipes embedded in a more coarse rock media, as shown in Figures 8.5 and 8.6.

Large wetland systems have usually concrete inlet and outlet structures. Outflow structures usually have a device that enables controlling the water level in the wetland cell, as shown in Figure 8.6.

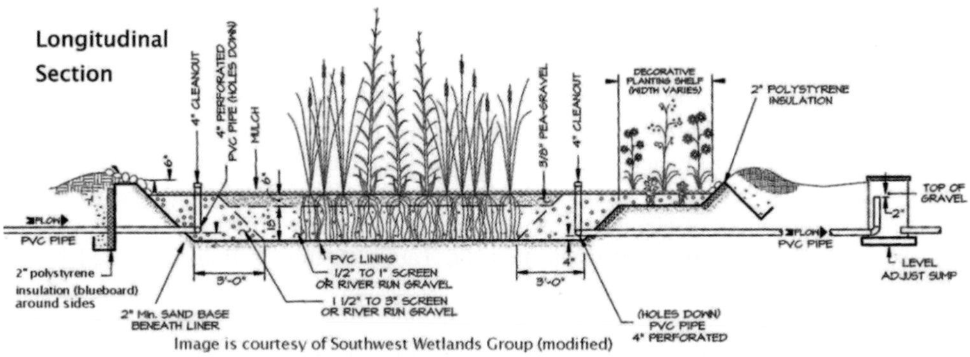

Figure 8.5 Inflow and Outflow Structures of a SSFCW (*Source*: https://engineering.purdue.edu/~frankenb/NU-prowd/images/image09_05b.jpg)

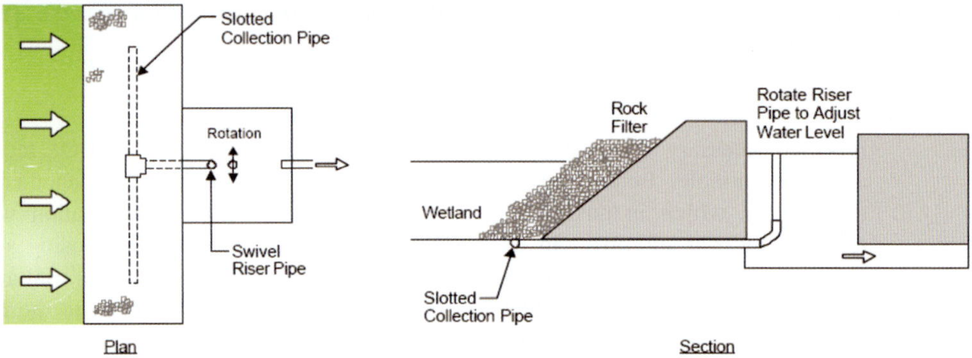

Figure 8.6 Concrete Outflow Structure with a variable level withdrawal device (*Source*: Orozco, 2005)

To achieve a proper distribution of the inflow and outflow it is recommended to use inlet and outlet pipes with a diameter of about 8″ and that the distance between each two outlet orifices (d_p) be set at 1–5 meters. The diameter of the outlet orifices has to be such that:

$$\Sigma a_i = 0.40A_p \tag{8.21}$$

where:

A_p = main pipeline area
a_o = area of each outlet orifice = 0.40 A_p/n_o.

The diameter of each orifice, φ_o is then calculated by:

$$\varphi_o = \sqrt{[4 \times (0.4A_p/n_o)/\pi]} \tag{8.22}$$

The length (L_p) of the distribution pipes should be more than 10 m. Consequently, a wetland unit of width b will have several distribution pipes (n_p), whose number is given by:

$$n_p \approx b/L_p \tag{8.23}$$

8.3 DESIGN EXAMPLE

(The model program for this Example is available online at http://www.iwawaterwiki.org/xwiki/bin/view/Articles/Software+Developed+for+ Sustainable+Treatment+and+Reuse+of+Municipal+Wastewater)

The design example presented follows the Orderly Design Method (ODM) of Section 8.2.2. The example is developed for a population of 20,000, using the procedure for the flow calculation as presented in Chapter 3. The ODM is developed step by step in the example and the calculations are performed by the Excel program, CHAP 8-SSFCW, available on line. The Primary Treatment (PT) which precedes the wetland is sized similarly the procedure of sizing the sedimentation unit in the CEPT process 9 (see Chapter 9), but without the use of coagulants. Please note that the Tables in this section, identified by alphabetic order contain computer calculated results.

The design is made for the minimum wetland width (Bmin) but two (2) SFFCWs units are built in parallel, in order to obtain the total width B = 2Bmin

(1) *Calculation of external variables*

1.1 *Design flows*: The input required for flow calculations are presented in Table 8A and are identical to those presented in the examples in previous chapters.

Table 8A

Design flow				
Variable	**Value**	**Unit**	**Value**	**Unit**
Per capita consumption, q	162.5	L/capita · d		
$PF = 1+14/(4+\sqrt{P})$	2.65			Eq·Harmon
Population, P	20000	inhab		
Population density, d	260	inhab/ha		
Return coefficient	0.8			
DWW mean flow, Q_{DWW}	30.09	L/s	108.33	m^3/h
DWW maximum flow, Q_{maxDWW}	79.82	L/s	287.35	m^3/h
Infiltration unit flow, q_I	0.13	L/s · ha (o km)		
Afferent area, A_a	6.92	ha		
Infiltration flow, Q_I	10.00	L/s		

Notes: (i) Per capita consumption is calculated for drinking water consumption. With the return factor of 0.8, the wastewater per capita flow of $c \cdot q_{dom} = 0.8 \times 162.5 = 130$ L/capita · d, as measured in the example in Chapter 3, (ii) the Peak Factor is calculated by Equation 3.6 (*Harmon* Equation), (iii) the population density should be the maximum acceptable density according to planning standards, in order to be able to serve the population at saturation conditions, (iv) the flows are calculated based on the population and the return coefficient – c, and (v) infiltration must be measured or alternatively a value of between 0.10 and 0.15 L/s · ha or l/s · km is to be chosen.

The final design flows are obtained from Equations 3.3 and 3.4, and are presented in Table 8B in various units. The Q_D is used for the biological process design and the Q_{DH} for hydraulic design.

Table 8B

Design flows						
Variable	**Value**	**Unit**	**Value**	**Unit**	**Value**	**Unit**
Process design flow, $Q_D = Q_{DWW} + Q_I$	40.09	L/s	144.33	m^3/h	3464.00	m^3/d
Hydraulics design flow, $Q_{DH} = Q_{DWWmax} + Q_I$	89.82	L/s	323.35	m^3/h	7760.44	m^3/d

In the case of a Primary Treatment the design flow for screening, grit channel, the Parshall flume is the Q_{DH} because they are hydraulic units. The primary settler, the thickener and the drying beds are designed with Q_D.

1.2 *Determination of the external geographical and environmental variables*: These variables depend solely on geographic location. The values used in the example are presented in Table 8C.

Table 8C

Variable	Value	Unit	Value	Unit
Minimum temperature of water,	25	°C	77.0	°F
Minimum temperature of air, T_a	18	°C	64.4	°F
Altitude above sea level, masl	350	masl		
Mean wind velocity	6	kph		
Predominant wind direction	NW			
Minimum solar radiation, S	230	$Cal/cm^2 \cdot d$		
Daily fraction of sunlight hours	0.8	decimal		

Notes: (i) for the design of the Preliminary Treatment process these variables do not have a significant importance, for the design of the biological processes (in this case the SSFCW) they are important; (ii) the minimum water temperature as well as the minimum air temperature are the averages of the coldest month; (iii) the speed and direction of the predominant wind is important for correctly locating the wastewater treatment plant in relation to the location of the town or the urban center; (iv) the fraction of hours of sunlight is the ratio of actual hours of sunlight to possible hours of sunlight.

1.3 *Pollutant concentration and loads*: The quality of the wastewater may be different in the urban areas and at the entrance to the treatment plant if the conveyance system is long and receives an important infiltration inflow. In the example, the wastewater treatment plant is located close to the town and there is no significant variation of the wastewater composition during its conveyance to the plant.

Table 8D presents as *outputs* the calculated *loads* (in kg/d) and the *Specific unit loads* (in kg/Capita · d) of each contaminant, calculated from the *concentrations* given as *input*. This is calculated for BOD_5, COD, TSS, and so on. In cases of significant infiltration due to long conveyance to the wastewater treatment plant, it is necessary to calculate the additional infiltration flow from the conveyance influence area (A_a) as follows: $Q_{Iemi} = q_I \cdot L$, where L is the length of the conveyance pipeline in km. The concentration of a contaminant at the influent to the wastewater treatment plant is calculated as its specific unit load (q_C) multiplied by the population (P) and divided by the total flow: $Q_T = (Q_D + Q_{Iemi})$. For example: the concentration of COD is $COD = q_{COD} \cdot P/Q_T$. Note that in Table 8D the load and the per capita load (specific unit load) of each contaminant are *outputs* while only the concentrations are an *inputs*, presented in the Excel sheet separately.

(2) *Determine the flow diagram of the SSFCW:* The flow diagram is presented in Figure 8.2. As shown, it consists of a conventional primary treatment as the pretreatment unit and two parallel SSFCW units. In this example the pretreatment unit is primary treatment. In many cases, however, the pretreatment consists of a septic tank, and an anaerobic lagoon can also be used as pretreatment.

(3) *Sizing the pretreatment system*

3.1 *Coarse screening*

Variables selection by the designer: For rough screening a coarse rotating screen with 6 mm screen opening size is proposed. Apply the same methodology explained in the design example in Chapter 4 for a 3 % removal of TSS.

Table 8D

Concentration, load and per capita contribution						
Variable	**Concentration**	**Unit**	**Load**	**Unit**	**Per capita**	**Unit**
BOD_5	130.0	mg/L	962.3	kg/d	0.048	kg/capita · d
COD	279.0	mg/L	2064.9	kg/d	0.103	kg/capita · d
COD/BOD_5	2.1		0.0	kg/d	0.000	
TKN	18.7	mg/L	138.6	kg/d	0.007	kg/capita · d
N-Nitrate	0.9	mg/L	6.9	kg/d	0.000	kg/capita · d
Total Phosphorus	2.7	mg/L	20.1	kg/d	0.001	kg/capita · d
pH	7.1	UN				
Alkalinity	46.8	mg/L	346.4	kg/d	0.017	kg/capita · d
TSS	94.8	mg/L	701.8	kg/d	0.035	kg/capita · d
VSS	81.2	mg/L	601.4	kg/d	0.030	kg/capita · d
O&G	46.8	mg/L	346.4	kg/d	0.017	kg/capita · d
Fecal Coli	10000000.0	MPN/100 mL	346400000000000.0	MPN/d	17320000000.000	MPN/capita · d

Design: A coarse rotating screen of 600 mm in diameter is obtained from the calculations. The final output data are presented in Table 8E. The design of the coarse rotating screen is similar to that of the rotating micro screen so the reader is referred to Chapter 4 in which the design procedure is explained in detail. Only coarse screening is used in the preliminary treatment unit because further processing of the fine material that will be contained in the effluent of the preliminary treatment unit will be processed in the following primary sedimentation tank into which this effluent is discharged. For the purpose of clarification, the preliminary treatment unit calculations which are produced by the Excel program are presented in Table 8E. The design support is presented in the column entitled OBSERVATION.

Table 8E

Design			
Parameter	**Value**	**Unit**	**Observation**
Flow, Q_{HD}	90.000	L/s	
Mesh size, a	6.000	mm	
Drum diamater, D_T	0.600	m	Figure 4.7
Channel width, $W = 1{,}1\ D_T$	0.660	m	
Channel depth, $h = Q_{DH}/W \cdot v$	0.230	m	Equation 4.2
Hydralics radius, $R = Wh/(W + 2h)$	0.136	m	Equation 4.3

(Continued)

Table 8E (*Continued*).

Design			
Parameter	**Value**	**Unit**	**Observation**
Manning's n	0.012		Equation 4.3
Hydrulics slope, $s = (v \cdot n)^2/R^{4/3}$	0.074	%	Equation 4.3
Head loss, $h_f = (1/C \cdot 2g)(Q_{DH}/A)^2$	0.030	m	Equation 4.1
Total head loss, $h_f + 0,05$	0.080		Plus 5 cm
Backwashing flow	1.800	L/s	Table 4.1
Waste solids excess	466.560	L/d	Table 4.1

3.2 *Grit chamber*
Variables selection by the designer: According to the ODM numeral 3, propose the following design variables:
According to the ODM in Section 8.2.2, Part 3 on design of preliminary treatment systems, we propose the design variables presented in Table 8F.

Table 8F

Designer variables			
Variable	**Value**	**Unit**	**Observation**
Grit channel			
Sor	800	m/d	600–1200
v	0.20	m/s	0,15–0,60
# Grit Channels	2.00		
L/w	4		3 a 6

Design: Applying these variables to the equations presented in the ODM Section 5.2.2 Paragraph 3.2 which are shown in the left column of Table 8G, we obtain the results presented in Table 8G. Note that there are two grit chambers.

3.3 *Parshall flume channel*: For a $Q_{HD} = 90$ lps the throat width for measuring the flow is $W = 9''$ (see Table 9.2). Rapid mixing is not required because there is no injection of chemicals.

3.4 *Primary clarification*
Variables selection by the designer: According to the ODM, it is necessary to propose for detailed design the average and peak Surface Overflow Rate (SOR) and a weir loading rate, $q_{v.}$ It is also necessary to select the sludge concentration required at the bottom of the clarifier and to select the depth of the clarifier. The selected variables are shown in Table 8H:

Table 8G

Grit channel			
Parámetro	**Valor**	**Unit**	**Observation**
Area, $A_{des} = Q_{DH}/SOR$	9.7	m^2	Section 5.2.2 Numeral 3.1
Width, $w == \sqrt{(A_{des}/4)}$	1.6	m	Section 5.2.2 Numeral 3.1
Length, $L = 4w$	6.2	m	Section 5.2.2 Numeral 3.1
Depth, $H_{des} = Q_{DH}/V \cdot W$	0.3	m	Section 5.2.2 Numeral 3.1
Total depth, $H = H_{des} + 0{,}50$	0.8	m	Section 5.2.2 Numeral 3.1
Parshall flume			
Throat, W	9.0	inch	See “Parshall” tag

Table 8H

Primary clarifier			
SOR	32.00	m/d	Table 9.4
SOR_{max}	60.00	m/d	Table 9.4
X_R	5	%	typical
H_{sed}	4	m	≥3.5 m

The SOR values for conventional clarification are selected from Figure 8.4 and Table 9.4. *Design*: By applying the methodology proposed in the ODM we obtain the dimensions of the primary clarifier, as well as the sludge production and the sludge volume. For a 4 m clarifier depth (H_{sed}) the average detention time, a t_d, is 3.0 hours. The sludge production is 8.6 m^3/d at a 5% TSS concentration ($X_R = 50.000 mg/l = 0.05\%$). The net clarifier diameter (D_N) is 12.8 m. The weir hydraulic load (q_v) is well in the required range (less than 125 m3/d · m). The BOD and TSS removal efficiencies are calculated using Equation 8.13 and 8.14, obtained from Figure 8.4. The resulting design figures for the primary clarifier are presented in Table 8I. The right column indicates how the calculations are performed. A more detailed explanation of primary treatment is given in Chapter 9.

Table 8I

Primary clarifier			
A_{sed}	108.3	m^2	Q_D/SOR
A_{sedmax}	129.3	m^2	Q_{DH}/SOR_{max}
A_s	129.3	m^2	Larger A_{sed} and A_{sedmax}
D_N	12.8	m	Equation 8.12
C	40.3	m	$C = \Pi D_N$
$q_v = Q_D/C$	85.9	$m^3/d \cdot m$	125–250 or less

(Continued)

Table 8I (*Continued*).

Primary clarifier			
E (BOD)	29.7	$\%BOD_5$	Equation 8.13
E (TSS)	61.0	% TSS	Equation 8.14
$Q_x = E \times SST \times Q_D$	428.0	kg/d	
$V_x = Q_x/X_r$	8.6	m^3/d	at 5%
$V = H_{sed} \times A_s$	517.4	m^3	
$t_d = V/Q_D$	3.6	h	Between 1.5–4.0 h

3.5 *Excess sludge thickener and drying beds*

Variables selection by the designer: The most important parameter for calculating the dimensions of the excess sludge thickener is the solids load on the thickener, Q_s (kg TSS/m^2·d) from which the area and the diameter of the thickener are obtained. The designer variable for Drying Beds is the rate of the sludge application, which should be in the range 120–150 kg/yr·m^2. The proposed designer variables and their sources are presented in Table 8J:

Table 8J

Thickener			
Q_s	100	$kg/m^2 \cdot d$	Table 9.5
H_e	3.5	m	Table 9.5
X_e	10	%	5–10%
Drying beds			
Application rate	150	$kg/m^2 \cdot yr$	120–150

Design: Application of The ODM methodology presented in Chapter 9, Section 9.2.2, Paragraph 6 and Paragraph 7, yields the design values presented in Table 8K.

Table 8K

Thickener			
A_e	4.3	m^2	$Q_s = Q_{XT} \cdot Q_D / A_e$
D_e	2.3	m	$D == (4A_s/\pi)^{(1/2)}$
V_{xe}	4.3	m^3/d	$V_{xe} = Q_{xT}/0.10$
Drying beds			
A_{lecho}	1041.4	m^2	Equation 6.33

To design the thickener we use the surface load (Q_s) recommended in Table 9.5. We use the equation $Q_s = Q_{XT} \cdot Q_D / A_e$. The total volume of sludge produced at 10% concentration is 4.3 m^3/d. The resulting diameter of the thickener is 2.3 m. The sludge needs to be chemically stabilized (for instance by adding to it lime). Using Equation 6.33 and the recommended application rate of 150 $kg/m^2 \cdot$ year yields the sludge drying beds area of 1041.4 m^2.
The calculation up to this point referred to the design of the Primary Treatment, which is the pretreatment unit of the constructed wetland. From this point on, we handle the design of the constructed wetland.

(4) *Define the gravel bed or matrix to be used*

Variables selection by the designer: From Table 8.3 we select for the bed gravel with an effective size, $D_{10} = 8$ mm. This selection defines to the designer the variables, which are presented in the first three rows in the Table 8L.

Table 8L

Designer variables			
Variable	**Value**	**Unit**	**Observation**
SSFCW			
Bed gravel effective size, D_{10}	8	mm	Table 8.3
k_s	3000	m/d	Table 8.3
n	35	%	Table 8.3
s	0.5	%	0.5–1.0%
BOD_{5e}(Req)	40	mg/L	Table 8.1
A_s	42687	m^2	Start with approximation
K_{HDBO}(20°C)	0.68	d^{-1}	Table 8.4
K_{HCF}(20°)	2.6000	d^{-1}	Table 8.4
K_p(20°C)	0.0273	m/d	See Nomenclature
H	0.30	m	0.3–0.6 m
L/b	0.50		0.5–4.0
N_c	1.00	UN	Number of SSFCW in series?

Design: These variables allow us to select from Table 8.3 the porosity, n and the hydraulic conductivity k_s. They also defines the slope s = 0.5%.

(5) *Calculation of the wetland size*

Variables selection by the designer: According to the ODM, Section 8.2.2, Paragraph 5.1, we propose the design variables presented in Table 8L.

Note that the initial A_s is an approximate value and that all the constants values are for the temperature of 20°C so it is necessary to correct them to the calculated wetland temperature.

Design: By strictly following methodology presented in Section 8.2.2, Paragraph 5.2, the variables selected by the designer, which are presented in Table 8L allow us to calculate the dimensions of the SSFCW and its performance in treating the wastewater. The calculation results are presented in Table 8M. The calculations of the variables presented in the left column are performed using the formulae and/or methodologies presented in the right column, all following the ODM methodology presented in Section 8.2.2. The coliforms and phosphorus removal efficiencies are obtained from Equations 8.9 and 8.10.

The calculation should start with an approximate value of A_s, the surface area of the wetland unit (which can be any value). A_s depends on K_{HDBO} (Equation 8.6), and this constant depends on the temperature in the wetland (Equation 5.5) which in turn depends on A_s. It results in a circular calculation which is solved by trial and error. The BOD_5 calculated by Equation 8.6 must coincide with the required BOD_5 set as a target to be achieved by the wetland unit (which means that BOD_5 required − BOD_5 calculated = 0) and for this condition the calculated A_s is the right one. The surface area shown in the Table 8M is the definitive area obtained after performing the trial and error calculations. In Excel, these calculations are easily made using the "goal seek" function.

Table 8M

Output					
Variable	**Value**	**Unit**	**Value**	**Unit**	**Observation**
HAFSS					
v	15	m/d			Equation 8.1
Temperature	18.98	°C			Equation 5.8
A_T	231	m^2			Equation 8.15
q_a	0.08	m/d			Equation 8.2
B_{min}	770	m			Equation 8.16
L	55	m			Equation 8.18
# parallel channels in wetland	7				Equation 8.19
b (channel width)	110	m			Equation 8.20
t_d	1.29	d			Equation 8.5
K_{HDBO}	0.64	d^{-1}			Equation 8.11
K_{HCF}	2.18	d^{-1}			Equation 8.11
K_p	0.0273	m/d			Typical
TSS_e	3.91	mg/L			Equation 8.4
BOD_{5e}(Calc)	40.00	mg/L			Equation 8.6
BOD_{5e}(Req) = BOD_{5e}(Calc)	0.00				Goal seek
P_e	1.94	mg/L			Equation 8.10
Fecal coliforms	2620552	MPN/100 mL			Equation 8.9

> Bear in mind that program SSFCW-01 designs the wetland unit with Bmin, so to meet the condition B = 2Bmin, two parallel identical SSFCW units need to be constricted.

(6) *Inflow structures*: For the design of the inflow and outflow structures follow the recommendations of the ODM, Section 8.2.2, Paragraph 6. The following variables are defined in the first three rows in Table 8N, according to the recommendation of the right column of this table: (i) Inflow pipe diameter (φ) (ii) Length of the inflow pipe (L_P) and (iii) distance between the orifice holes on the inflow pipe (dp). Using these input parameters it was found that $n_p = 14$, that is, 14 inflow distribution pipes are required in each parallel unit of wetland, each inflow pipe 10 m long with $n_o = 4$ orifices in each pipe. The diameter of each orifice is 2.53″ (that is 2 ½ inches).

Table 8N

Inflow pipes					
Diameter, ϕ	8	pulg	0.2032	**m**	>8"
L_p	8.00	m			<10 m
d_p	2.00	m			1–5 m
n_p	14.00				Equation 8.23
A_p	0.03	m^2			pi $\Phi^2/4$
n_o	4				# orifices = L_p/d_p
ϕ_o	0.06	m	2.53	pulg	Equation 8.22
a_o	0.003	m^2			

A cross section of a typical SSFCW unit is presented in Figure 8.7.

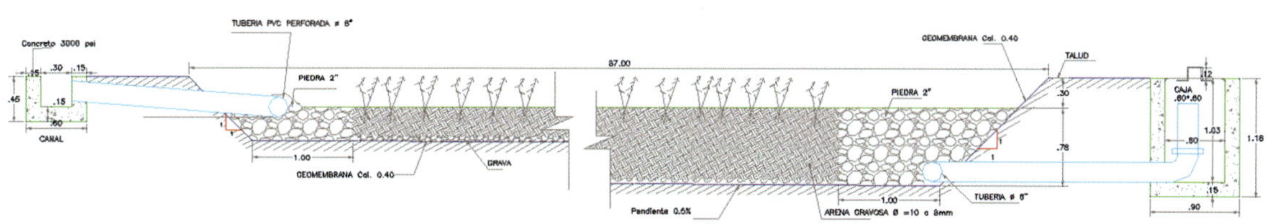

Figure 8.7 Longitudinal cross section of a typical SSFCW unit

A photo of a SSFCW plant is presented in Figure 8.8. This plant is located near Oak Grove, Minnesota (Wallace, 2007). Its surface is insulated with a layer of compost to minimize heat loss from the system, enabling it to function well even under the extremely low air temperatures in the plant site, which reach −25°C during December–March. The variation of the air temperature, of the temperature of the wastewater flowing into the this SSFCW plant and of the temperature of the effluent of this plant during a period of about two years is shown in Figure 1.122.

Figure 8.8 Insulated subsurface flow constructed wetland near Oak Grove, Minnesota (*Source*: Wallace, 2007)

Chapter 9

Chemically Enhanced Primary Treatment (CEPT)

9.1 PROCESS DESCRIPTION

9.1.1 Introduction[1]

Chemically Enhanced Primary Treatment (CEPT) is a process known over one hundred years. However, its application to municipal wastewater was consolidated only in the past two decades, based on technological development of coagulation, flocculation and sedimentation stages for wastewater. It's most common applications are: (i) treatment prior to discharge into oceans, seas or large rivers, (ii) removal of phosphorus, and (iii) as a component in compact treatment units. The CEPT is a process by which wastewater is treated with chemical coagulants as ferrous chloride or aluminium sulphate (alum), to which a flocculating agent (polymer) is added after coagulation, before the primary sedimentation tanks. The coagulation and flocculation cause particles to cling together to larger flocs which are heavier and settle more rapidly. In this way higher removal rates than in conventional primary treatments are reached for both TSS (about 80–85% in CEPT compared to about 40–50% in conventional primary sedimentation) and for BOD (BOD 50–70% in CEPT compared to 20–35% in conventional primary sedimentation). CEPT is operated at twice the Surface Overflow Rate (SOR) used in conventional primary tanks, reducing thereby the area of the sedimentation tanks by half. Figure 9.1 shows a typical flow diagram of the CEPT process (Libhaber, 2007).

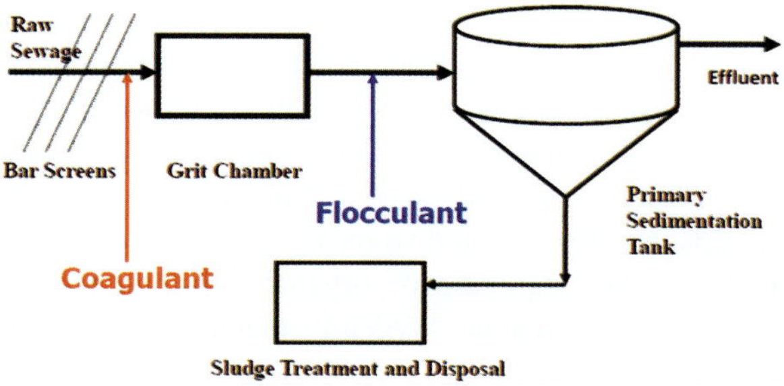

Figure 9.1 A typical flow diagram of the CEPT process [*Source:* Libhaber (2007)]

---

[1]Section adapted mainly from Menahem Libhaber (2007, 2004).

To convert a conventional primary treatment plant into a CEPT plant, all that is needed is the addition of a chemical coagulant and optionally addition of flocculants. The chemical dosing systems are inexpensive and simple to install. CEPT effluent quality approaches that of an effluent of a secondary biological treatment plant in terms of removal efficiency of TSS and BOD. In addition, the CEPT process removes phosphorus. Because of the high TSS removal in the CEPT process, its effluent is clear can be effectively disinfected, and this is an important feature for developing countries, in which high levels of morbidity result from waterborne diseases. A primary treatment facility upgraded to CEPT has double the capacity of the original conventional settling system so it is able to serve future flows without an additional upgrade. If a CEPT unit is designed from scratch, the sedimentation tanks will be smaller and cheaper than primary sedimentation tanks for the same wastewater flow without chemicals addition. In this case, the additional investment required for treating the increased sludge generated is offset by the lower cost of the sedimentation tanks. When a primary treatment unit is upgraded to CEPT, however, this offset does not exist and additional investment for sludge treatment is required, but an upgraded system is able to serve future flows without an additional expansion.

The chemistry of coagulation and flocculation is widely known, so that incorporation of the CEPT to engineering practice is relatively easy. However, CEPT produces more sludge than conventional primary treatment because it is through the production of sludge that the organic matter is removed in this process. The greatest portion of the increased sludge production in CEPT is the result of the increased solids (and therefore also organic matter) removal in the settling tank through the physicochemical (not biological) process, which is precisely the CEPT's goal. The chemicals themselves add only a slight contribution to the sludge quantities. An adequate treatment and disposal of sludge needs to be part of the CEPT system. The main options for sludge treatment, which must be part of the installation, are lime stabilization and use of drying beds for drying the sludge, anaerobic digestion, and composting, as discussed below.

CEPT treatment does not preclude additional polishing treatment such as secondary or tertiary. It reduces the size and cost of these subsequent treatment units due to increased removal of suspended solids and organic matter. CEPT is a relatively simple technology, which provides low cost, easy to implement and effective treatment.

9.1.2 Basics of the process

Fundamentals[2]

Tsukamoto (2002) classifies into six classes contaminants which need to be removed from municipal wastewater before their discharge to receiving waters:

- *Pathogenic Organisms:* These are the most important pollutants from the standpoint of risk to human health since they generate water-borne diseases (which are transmitted through water). Among the causes of these diseases are the **worms** (earthworm, tapeworm, schistosoma and others), **protozoa** (amoebas, Giardia, Cryptosporidium, and so on.), **Bacteria** (cholera, Salmonella, E. coli, and so on.), and **viruses** (hepatitis, enterovirus, and so on.).
- *Organic Matter (BOD):* is the aggregate of organic substances present in wastewater whose biodegradable part produces Biochemical Oxygen Demand. The BOD in the municipal wastewater

[2]This paragraph is based on the article by Tsukamoto (2002), "*Tratamiento físico-químico de aguas residuales municipales: estrategia eficiente, rápida y de bajo costo para Latinoamérica*" (Physicochemical Treatment of Municipal Wastewater: Efficient, Rapid and Low Cost Strategy for Latin America).

is only about half of the organic matter contained in this type of wastewater. The rest is non-biodegradable organic matter. The sum of biodegradable and non-biodegradable organic matter is measured by the COD test.

- *Suspended Solids (SS):* is the aggregate of all particles in the wastewater. It is one of the parameters of major environmental and health impacts since it encompasses all organic and inorganic matter in suspension. The SS include 100% of pathogenic organisms present in the wastewater, 50% to 70% of the organic matter present in the wastewater and 100% of the colloidal matter in the wastewater.
- *Nutrients (P and N):* Organic and inorganic compounds of phosphorus (P) and nitrogen (N) contained in wastewater stimulate the uncontrolled reproduction of micro algae in aquatic environments. Phosphorus is the major contaminant responsible for the effects of eutrophication and water poisoning. Its elimination is an important goal of wastewater treatment. In fresh water, blue-green algae (cyanophyte) and some diatoms produce carcinogenic toxins and generate in the water unpleasant taste and odour.
- *Fetid Odour (H_2S):* The smell of "rotten egg" that characterizes wastewater is caused by hydrogen sulphide gas produced by anaerobic bacteria from the sulphur present in the wastewater. This gas is toxic to plants, animals and humans. It is also corrosive to metal and concrete pipes and to construction materials. The human smell sense perceives H_2S in concentrations in the order of a few ppt (parts per trillion $= 10^{-12}$).
- *Heavy metals:* These toxic chemicals released into municipal wastewater by industries with inadequate effluents treatment (mercury, chromium, nickel, lead, arsenic, silver and others) have, in general, a cumulative and often carcinogenic effect, being toxic to all living beings. Contamination of sediment in the bottom of receiving water bodies is prolonged, since heavy metals do not degrade in the environment, and therefore have a cumulative effect.

According to Tsukamoto (2002) the conventional classification of municipal wastewater treatment is: (i) **primary treatment**: which is the separation of the SS by settling or flotation, which can be chemically enhanced, (ii) **secondary treatment**: which is the removal of soluble and insoluble organic matter by biological treatment, and (iii) **tertiary treatment**: which is the removal of N and P, (P is present mainly in colloidal form). The suspended solids contain 100% of the pathogens, 50% to 70% of the organic matter and 100% of the colloidal matter present in wastewater. Accordingly, the CEPT removes very effectively, SS, phosphorus, heavy metals and H_2S. It also removes a large amount of BOD (it does not reach the efficiency of BOD removal in secondary treatment, but it is close) and part of the pathogens. To obtain the same degree of removal of these parameters by conventional treatment primary treatment, secondary treatment and advanced treatment (to remove nutrients heavy metals) is required. This is much more complex and costly than CEPT. However, if a high efficiency of BOD removal (over 80%) is required, biological treatment is inevitable. In this case, a good alternative is biological treatment as post-treatment (polishing) of a CEPT effluent. In terms of cost-effectiveness, CEPT emerges as a new paradigm because it achieves a lot more treatment goals for less cost, as shown in Figure 9.2. It is important to emphasize that the CEPT system must include installation for management of the sludge generated in this process.

Coagulation by addition of metallic electrolytes

The processes by which the coagulation of colloidal suspensions and particulate matter is facilitated for their subsequent removal by sedimentation are known as coagulation and flocculation.

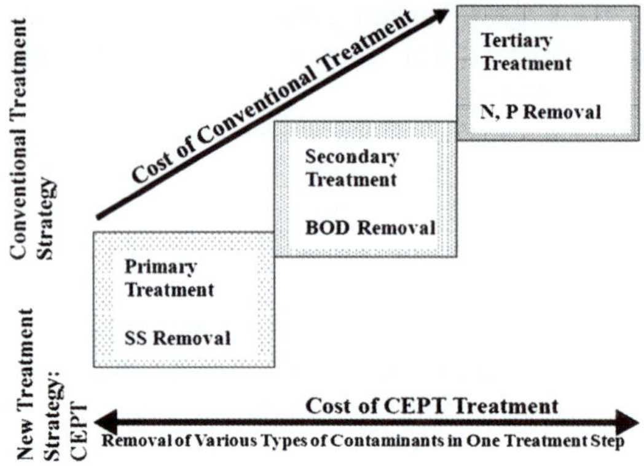

Figure 9.2 Cost-effectiveness of CEPT compared to conventional treatment [*Source:* Tsukamoto (2002)]

CEPT relies on small doses of metal salts. The most commonly used coagulants in water treatment are aluminium sulphate (alum) and ferric chloride. The dose of coagulant is in the range of 10–50 mg/l, combined with addition of 0.3–1.0 mg/l of organic polyelectrolyte as flocculation aid (Mrriesey & Harleman, 1992), but in most cases, less than 20 mg/l ferric chloride and less than 0.2 mg/l polyelectrolyte is used. The dispersion of coagulants is done in three completely different stages: (i) hydrolysis of the ions Al^{+3} or Fe^{+3} carried out in a very short time, (ii) diffusion of the formed compounds and their adsorption in the colloidal particles, and (iii) polymerization or reaction of the hydrated ions to form dimeric or polymeric compounds. The time during which these processes occur is less than one second, and it is advisable to do it with the help of a polyelectrolyte (Arboleda, 1992).

Colloidal suspensions are stable since the uniform electric charge of colloids keeps them in stable suspension due to the repulsion between charges. Coagulation-flocculation involves three stages: the addition of coagulant, destabilization of the colloidal particle and floc formation. The addition of coagulants salts such as aluminium sulphate, ferric sulphate or ferric chloride produces polymeric cations such as $[Al_{13}O_4(OH)_{24}]^{+7}$ and $[Fe_3\ (OH)_4]^{+5}$, whose positive charges neutralize the negative charges of the colloids, allowing the particles to aggregate and form flocs (see Figure 9.3 and "Potabilización de Aguas Naturales").

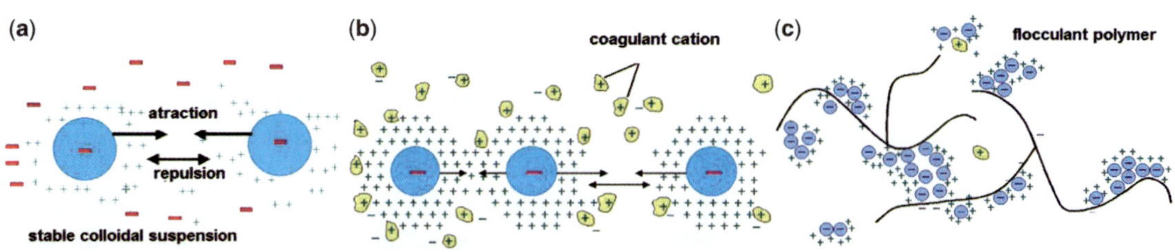

Figure 9.3 Coagulation-flocculation: (a) stable colloidal suspension, (b) destabilization of the suspension by coagulating agent upon neutralizing the charge, (c) floc formation with the addition of polymers that trap several clots (*Source:* http://www.puc.cl/quimica/agua/potabiliz.htm)

There are two basic mechanisms of floc formation: (i) neutralization of the charge of the colloids. Upon solubilization the coagulant releases into the water positive ions with sufficient charge density to attract colloidal particles and neutralize their charge. It has been observed that the effect increases sharply with the number of charges of the coagulant ions[3], and (ii) coalescence of the neutralized colloids into a floc. The coagulants form in water certain low solubility products which precipitate. The colloidal particles serve as the nucleus of precipitation and remain immersed within the precipitate.

The pH is a critical factor in the coagulation process. There is always a pH range in which a specific coagulant works best, which coincides with the range of the minimum solubility of the metal ions of the coagulant used. Whenever possible, the coagulation should be made within this optimum pH range since otherwise chemicals are wasted and a decrease in plant performance can be experienced. If the pH of the water is not adequate, it can be modified through the use of coagulant aids, among which are lime, quick lime, sodium carbonate, caustic soda and mineral acids. Other factors which also influent coagulation are the alkalinity of the water, the quantity of coagulant and the mixing intensity.

Coagulants

Alum

The alum, a hydrated aluminium sulphate with n molecules of H_2O (n is between 13 and 18), reacts with alkalinity in the water according to the following equation:

$$\underset{\textbf{Alum}}{Al_2(SO_4)_3\,nH_2O} + \underset{\textbf{Alkalinity}}{3Ca(HCO_3)_2} \longrightarrow \underset{\textbf{Al Hydroxide}}{2Al(OH)_3} \downarrow + 6CO_2 + 3CaSO_4 + xH_2O$$

The formation of aluminium hydroxide leads to floc precipitation. If the water does not have enough alkalinity, then it should be increased by adding calcium carbonate. The range of pH for optimum coagulation is between 5 and 7.5. The Alum dose for wastewater treatment is from 100 to 300 g/m^3, depending on the type of wastewater and the effluent quality requirement. The pH regulation can be done with:

- Lime: it takes about 30% of the dose of commercial aluminium sulphate, according to the following reaction:

$$Al_2(SO_4)_3 + Ca(OH)_2 = 2Al(OH)_3 + 3CaSO_4$$

- Sodium carbonate: it takes between 50 and 100% of the dose of commercial aluminium sulphate:

$$Al_2(SO_4)_3 + 3H_2O + 3Na_2CO_3 = 2Al(OH)_3 + 3Na_2SO_4 + 3CO_2$$

Ferric chloride

Ferric chloride, $FeCl_3$, is the most widely used coagulant in wastewater treatment. It reacts with natural alkalinity as follows:

$$\underset{\textbf{Fe Chloride}}{2FeCl_3} + \underset{\textbf{Alkalinity}}{3Ca(HCO_3)_2} = 3CaCl_2 + \underset{\textbf{Fe Hydroxide}}{2Fe(OH)_3} + 6CO_2$$

The range of pH for optimum coagulation is between 4 and 6, and above 8. The dose of Ferric Chloride when using lime (to provide alkalinity when needed) is from 5 to 160 g/m^3 of commercial reagent $FeCl_3 \cdot 6H_2O$, and it reacts with the lime as follows:

$$2FeCl_3 + 3Ca(OH)_2 = 2Fe(OH) + 3CaCl_2$$

[3]For negatively charged colloids, divalent ions (Ba and Mg) are 30 times more effective than monovalent Na, and, in turn, trivalent Fe and Al, are about 30 times more effective than divalent ions. For positively charged colloids, the same approximate ratios exist between the monovalent chloride ion, Cl^-, the divalent sulfate $(SO_4)^{-2}$, and the trivalent phosphate $(PO_4)^{-3}$ (Wikibooks, 2007).

The selection of coagulants and the determination of their doses for each specific case should be done in practice with Jar test.

Design considerations

Hydraulic jump mixing

In the CEPT process, the agitation required to facilitate coagulation can achieved using hydraulic mixing, as shown in Figure 9.4. The crucial design parameter in this type of mixing is the hydraulic gradient, which in its most general form is calculated as:

$$G = \sqrt{\frac{P}{V \cdot \mu}} \tag{9.1}$$

where:

G = velocity gradient, s^{-1}
V = volume, m^3
μ = dynamic viscosity, $N \cdot s/m^2$
P = power, W

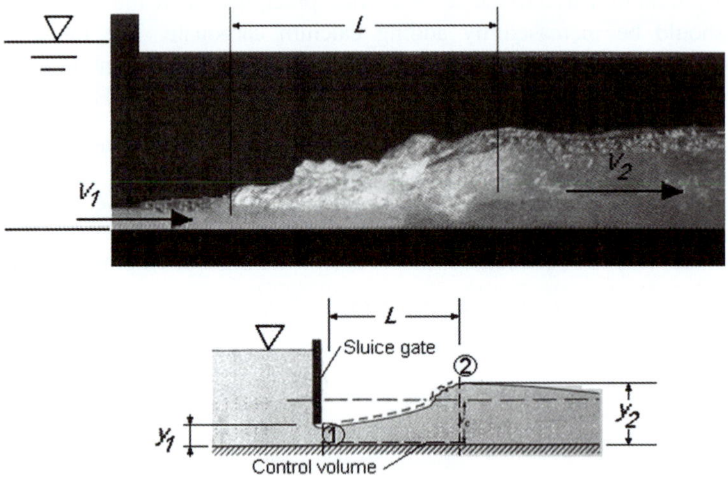

Figure 9.4 Hydraulic jump [*Source:* Sincero & Sincero (2003)]

In hydraulic mixing, the power (P) is calculated using the following equation:

$$P = \gamma Q h \tag{9.2}$$

where:

γ = water specific weight, N/m^3, (9789 at 20°C)
Q = flow, m^3/s
μ = hydraulic head, m

The applicable gradients are:

- $G = 10–30\ s^{-1}$ for fragile floc
- $G = 20–50\ s^{-1}$ for turbidity floc
- $G = 40–100\ s^{-1}$ for floc precipitation

Figure 9.4 shows the mixing produced by a hydraulic jump. Taking the figure as a reference, we can calculate the volume of the jump (V) and the detention time in it (t_d), as follows:

$$V = 1/2\,(y_1 + y_2)\,L \cdot W \tag{9.3}$$

where L is the length of the jump and w the width of the channel.

The height y_1 is controlled by the gate and the height y_2 of the hydraulic jump is calculated as follows:

$$y_2 = \frac{y_1}{2}\left(\sqrt{1 + 8\frac{v_1^2}{g \cdot y_1}} - 1\right) \tag{9.4}$$

where:

g = gravity acceleration, 9.8 m/s^2
v_1 = velocity at point y_1: $v_1 = Q/w \cdot y_1$
$L = 6y_2$: length of the jump

t_d is then calculated from Equations 9.3 and 9.4 ($t_d = V/Q$), as follows:

$$t_d = 3y_2 w(y_1+y_2)/Q \tag{9.5}$$

Parshall flume[4]

Usually the Parshall flume is used with the dual purpose of flow measuring and rapid mixing. For this purpose, it works with free flow discharge. The liquid stream passes from a supercritical to a subcritical condition, causing the hydraulic jump. It was invented by R.L. Parshall. It is presented in Figure 9.5 and has the dimensions shown in Table 9.1. The maximum and minimum flows according to the throat width are presented in Table 9.2. The height of water in the measuring section, h_0, is calculated from Equation 9.6 using the parameters in Table 9.3, for Q in m^3/s.

$$h_0 = KQ^m \tag{9.6}$$

The water velocity in the measured sector is then:

$$v = Q/h_0 \cdot [2/3(D - W) + W]$$

Which is:

$$v = (1/K\,Q^{m-1})[2/3(D - W) + W] \tag{9.7}$$

The specific energy will be:

$$E = v^2/2g + h_0 + N \tag{9.8}$$

[4]Paragraph based on Chapter 2 of CEPIS (2004).

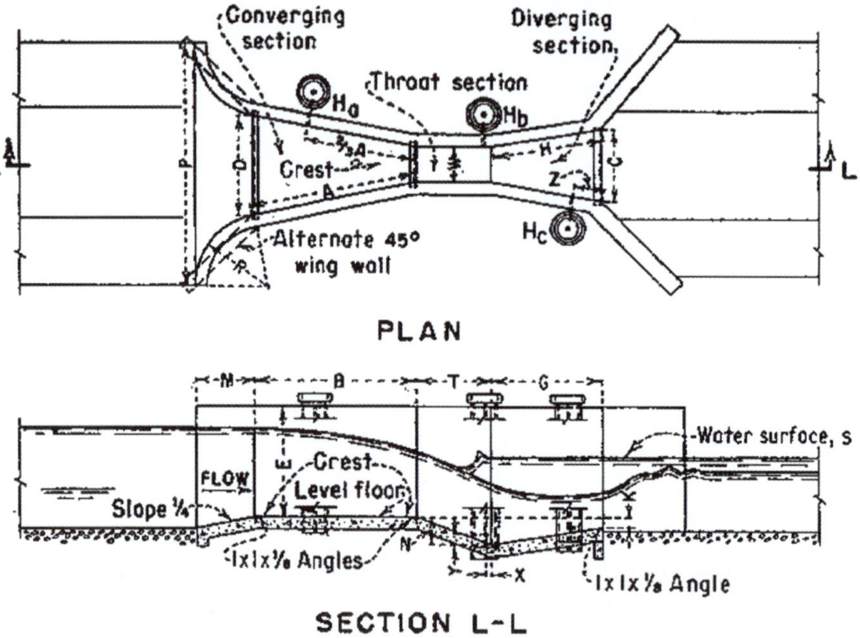

Figure 9.5 Basic dimensions of Parshall flumes (*Source:* http://www.usbr.gov/pmts/hydraulics_lab/pubs/wmm/chap08_10.html)

Table 9.1 Dimensions of the parshall flumes (see Figure 9.5).

W		A	B	C	D	E	F	G	K	N
inches	(cm)	(cm)	(cm)	(cm)	(cm)	(cm)	(cm)	(cm)	(cm)	(cm)
1″	2,5	36,3	35,6	9,3	16,8	22,9	7,6	20,3	1,9	2,9
3″	7,6	46,6	45,7	17,8	25,9	45,7	15,2	30,5	2,5	5,7
6″	15,2	61,0	61,0	39,4	40,3	61,0	30,5	61,0	7,6	11,4
9″	22,9	88,0	86,4	38,0	57,5	76,3	30,5	45,7	7,6	11,4
1′	30,5	137,2	134,4	61,0	84,5	91,5	61,0	91,5	7,6	22,9
1 1/2′	45,7	144,9	142,0	76,2	102,6	91,5	61,0	91,5	7,6	22,9
2′	61,0	152,5	149,6	91,5	120,7	91,5	61,0	91,5	7,6	22,9
3′	91,5	167,7	164,5	122,0	157,2	91,5	61,0	91,5	7,6	22,9
4′	122,0	183,0	179,5	152,5	193,8	91,5	61,0	91,5	7,6	22,9
5′	152,5	198,3	194,1	183,0	230,3	91,5	61,0	91,5	7,6	22,9
6′	183,0	213,5	209,0	213,5	266,7	91,5	61,0	91,5	7,6	22,9
7′	213,5	228,8	224,0	244,0	303,0	91,5	61,0	91,5	7,6	22,9
8′	244,0	244,0	239,2	274,5	274,5	91,5	61,0	91,5	7,6	22,9
10′	305,0	274,5	427,0	366,0	475,9	122,0	91,5	183,0	15,3	34,3

Table 9.2 Flows as function of the throat width.

Throat aiameter inches	Minimum flow lps	Maximum flow lps
1	0.13	5.36
2	0.19	12.24
3	0.25	32.04
6	1.45	87.18
9	2.59	176.69
12	9.21	462.75
18	1.45	696.40
24	18.67	936.93
30	23.09	1180.92
36	27.38	1426.74
48	35.77	1922.93
60	62.76	2424.04
72	74.50	2928.87
84	115.31	3437.17
96	130.95	3948.24
120	162.43	10245.08
144	192.84	17381.13
180	238.44	28899.47

Source: Orozco (2005)

Table 9.3 Values of K and m in Parshall flumes for different throat widths.

Throat width	K	M
3″	3.704	0.646
6″	1.842	0.636
9″	1.486	0.633
1′	1.276	0.657
1 1/2′	0.966	0.650
2′	0.795	0.645
3′	0.608	0.639
4′	0.505	0.634
5′	0.436	0.630
6′	0.389	0.627
8′	0.324	0.623

For N see Table 9.1. The velocity before the jump, according to Arboleda (1992), is obtained by solving Equation 9.9:

$$v_1^3 - 2gv_1E + 2Qg/W = 0 \tag{9.9}$$

In order for the Parshall flume to produce good mixing, it is required that the velocity at the throat be maintained between 2 and 7 m/s (Orozco, 2005).

Sedimentation[5]

Sedimentation is a fundamental process in CEPT. According to Orozco (2005) Sedimentation is generally classified into four types, namely:

(1) *Discrete Sedimentation:* or Sedimentation Type I. This refers to sedimentation of discrete particles in low concentrations, where sedimentation occurs individually by each particle, without interference of one particle to others. The theory of discrete sedimentation is explained by the Stokes equation; however when it refers to wastewater treatment, the approximation through the Surface Overflow Rate (SOR), which is defined as the flow divided by the sedimentation area (Q/A_s), is more frequently used.

(2) *Flocculent Sedimentation:* or sedimentation Type II. This refers to the sedimentation of particles which are not highly concentrated but which tend to flocculate, so that the settling velocity of the particle increases with the sedimentation process. This is what happens in primary clarifiers and in the CEPT process. The design parameter of this type of sedimentation is SOR.

(3) *Hindered or Zone Sedimentation:* also known as Sedimentation Type III, occurs with in cases where the particles concentrations is of intermediate levels. In this case at the end of the sedimentation the particles form a defined solid-liquid interface. The forces between particles are sufficient to hinder each other, remaining in fixed positions one in relation to the others. This occurs in secondary clarifiers in biological wastewater treatment plant, and the parameters governing the design are SOR and the solids loading, Q_s ($Q \cdot X/A_s$), where X is the TSS concentration in the mixed liquor (also known as MLSS – Mixed Liquor Suspended Solids), and A_s is the surface area of the clarifier.

(4) *Thickening:* also known as Sedimentation Type IV, is what happens when the particles already underwent a sedimentation process and have reached a structure of formed particles. This happens when the particles concentration is very large, and additional "sedimentation" can occur only by compression. This is what happens in thickeners and at the bottom of secondary clarifiers. The design parameter of this type of sedimentation is the solids loading, Q_s.

Sedimentation tanks can be designed as tanks of horizontal flow, of vertical up flow, with inclined parallel plates, or as circular sedimentation tanks with variable horizontal flow. The latter is preferred in the wastewater treatment industry because it allows continuous removal of solids by mechanical rotating scrapers for sludge removal, as shown in Figure 9.6.

The CEPT process is used in the flocculent or type II sedimentation and the parameter of design is the Surface Overflow Rate (SOR). Table 9.4 shows the SOR design values used for conventional primary sedimentation without chemical enhancement. When coagulants are used, the sedimentation efficiency substantially improves, as shown in Figure 9.7 prepared based on Harleman (2004). Because of its much

[5]Paragraph based on Chapter 7 of Orozco (2005).

higher sedimentation efficiency, SOR values used in CEPT are double of those used in conventional primary sedimentation, that is, in the range 60–100 $m^3/d \cdot m^2$. This implies that for treatment of the same flow of wastewater a much smaller CEPT settler can be used in relation to a settler required for conventional sedimentation, while obtaining a much higher TSS and BOD removal efficiencies than those which can be obtained in conventional sedimentation without chemical enhancement (Libhaber, 2007).

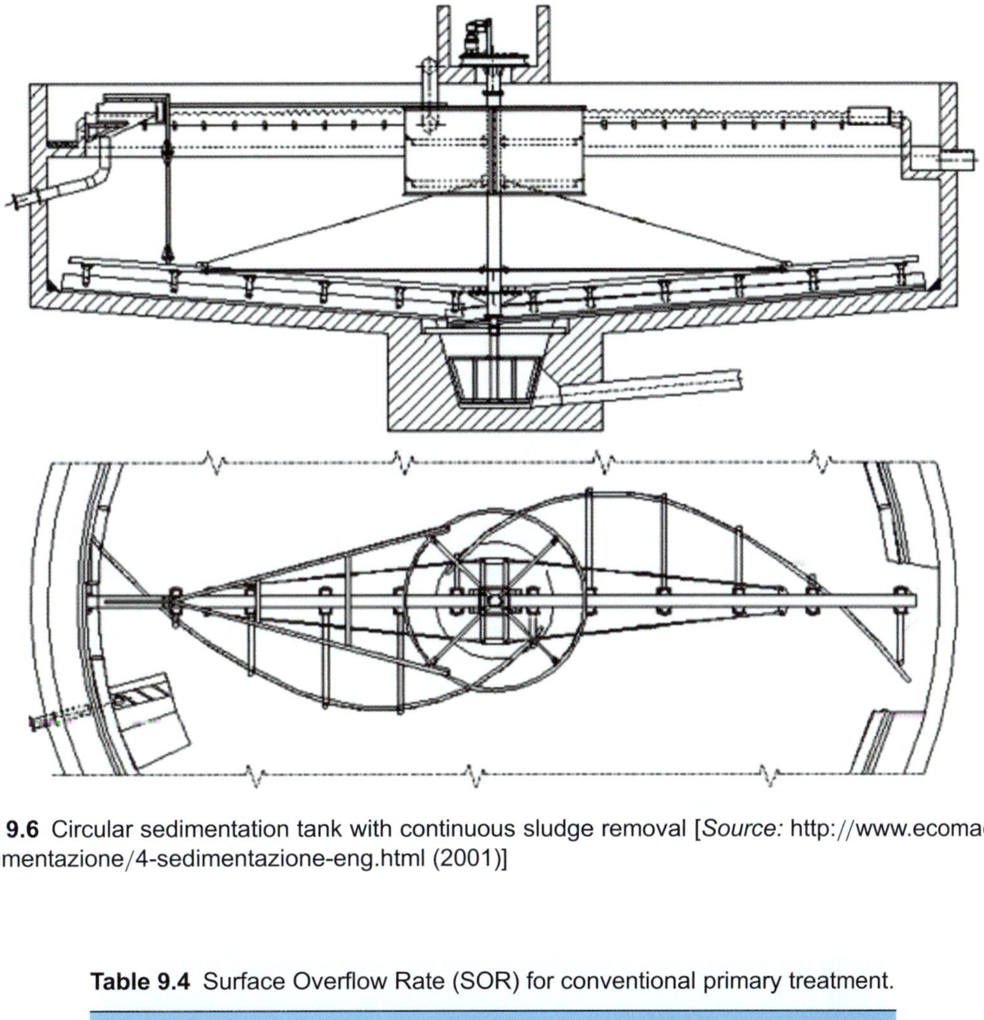

Figure 9.6 Circular sedimentation tank with continuous sludge removal [*Source:* http://www.ecomacchine.it/documentazione/4-sedimentazione-eng.html (2001)]

Table 9.4 Surface Overflow Rate (SOR) for conventional primary treatment.

Parameter	Units	Range	Typical
t_d	hour	1.5–2.5	2
SOR mean	$m^3/d \cdot m^2$	24–48	32
SOR peak	$m^3/d \cdot m^2$	48–120	60
Weir Hydraulic Load	$m^3/d \cdot m$	125–500	250

Source: Orozco (2005)

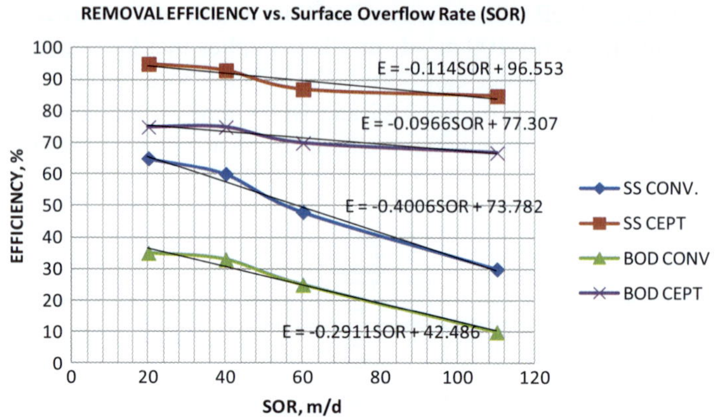

Figure 9.7 Comparison of BOD and TSS removal efficiencies of conventional primary treatment and CEPT [*Source:* Harleman (2004)]

Chemically Enhanced Rotating Microscreens (CERM)

Multiple-stage screening and the addition to flocculants can significantly increase contaminant removal levels. The rotating fine screen devices discussed in Chapter 4 can remove up to 50% of particulate COD and up to 30% of total COD with screen openings of around 0.2 mm. Investigations by Huber Technology (2003) indicate that addition of organic coagulant to the raw sewage entering a fine rotating screen of 0.2 to 0.3 mm opening achieves up to 60% reduction of total COD and BOD, and TSS reduction from 50% to 95%. This process is still under development and seems to be based on two-stage screening where the first screen has an opening of about 6 mm and the second is a fine screen with an opening of 0.2 mm, as shown in Figure 9.8. This process is named Chemically Enhanced Rotating Microscreens (CERM). Thc coagulant is added to the wastewater before the second unit Libhaber (2007).

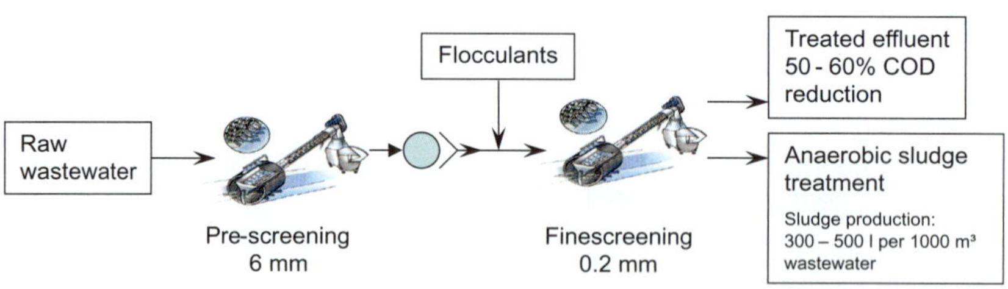

Figure 9.8 Chemically enhanced solids separation in rotating micro screens (CERM)

The CERM process is achieved by a simple machine that removes a significant amount of solids and organic matter. It might not reach the level of solids and BOD removed by CEPT, however, it is less costly and simpler to operate. It is certainly more affordable for developing countries. The level of organic matter removal achieved in CERM might be sufficient to prevent environmental problems if the effluent is discharged to a high flow river or to other receiving bodies with high assimilation capacity. The chemically enhanced rotating fine screens might be a better first stage goal than CEPT, especially in low income developing countries.

The high removal of solids in CERM generates larger quantities of sludge those generated in preliminary treatment without chemicals addition, so sludge treatment facilities need to be included as part of the CERM concept. Sludge treatment options are similar to that of CEPT sludge, as described below.

Since the CERM is still a new concept, more research is required to establish its performance and its cost. A comparison among the different options of screening, with and without addition of coagulants, according to Köppl & Frommann (2004), is shown in Figure 9.9.

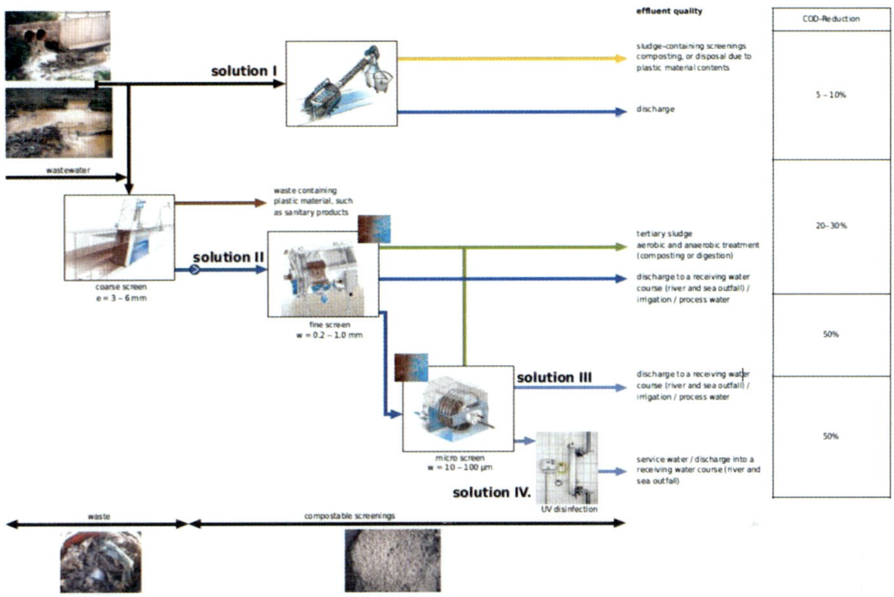

Figure 9.9 Various innovative chemically enhanced screening options [*Source:* Köppl & Frommann (2004)]

Sludge management

As mentioned, CEPT produces more sludge than conventional primary treatment because it is through the production of sludge that the organic matter is removed in this process. Therefore, adequate treatment and disposal of sludge needs to be part of the CEPT system. In most cases the doses of chemicals in the CEPT process are less than 20 mg/l of metal salt and less than 0.2 mg/l of polyelectrolyte. Most of the sludge consists of the solids settled from the raw wastewater. The chemicals used for coagulation add only a slight contribution to the sludge quantities.

The CEPT process, as well as most of the appropriate technology processes described in the book, generates two types of solid wastes. Rotating screens with coarse openings larger than 0.7 mm, or preliminary treatment in general, produce residual solids which are mainly mineral, consisting of sand, grit, coarse particles and large materials like plastics, bottles, cans and other floatable material. This material is defined as solid wastes from preliminary treatment. Rotating micro screens with openings less than 0.7 mm, CEPT, CERM, UASB, Anaerobic Filters and some of the other appropriate technology processes produce residual solids which consist of a combination of mineral and organic matter, defined as sludge. These two types of solids, that is, solid wastes and sludge, are handled and disposed of differently. The management and disposal of the solid wastes generated in the preliminary treatment is discussed in Chapter 4. The treatment and disposal of sludge generated by rotating fine screens with openings of less than 0.7 mm, by CEPT, CERM and some of the other appropriate technology processes is basically identical to the treatment of sludge generated by conventional secondary treatment. Management option of this sludge is discussed below.

Estimation of the sludge quantity produced in the CEPT process: In the case of use of Ferric Chloride ($FeCl_3$) as a coagulant, Morrisey & Harleman (1992) propose the following simplified relation for estimating the quantity of sludge produced:

$$S = TSSrem + 1.42Prem + 0.66FeCl_{3in}$$

where:

S = Total Sludge quantity produced kg/day
$TSSrem$ = Total suspended solids removed from the wastewater, kg/day
$Prem$ = Total phosphorous removed from the wastewater, kg/day
$FeCl_{3in}$ = Total $FeCl_3$ added, kg/day

The estimation of the sludge quantity presented above refers only to the use of Ferric Chloride ($FeCl_3$) as a coagulant. In the ODM presented below we use an approximation, assuming that the quantity of sludge produced is the sum of the total suspended solids removed from the wastewater and the total coagulant added.

The main options for treating CEPT sludge are shown in Figure 9.10. They are anaerobic digestion, lime stabilization drying beds, and composting. Details on the unit treatment process within each option, such as thickening, stabilization, dewatering and drying were briefly discussed in Chapter 1 Section 1.2.3.2. Composted and lime stabilized sludge are used as organic fertilizers in agriculture. Anaerobic digested sludge has also been used for many years as an organic fertilizer, in liquid or dewatered form. It is still used in this manner in many parts of the world, although its use has recently been prohibited in some countries due to public health risks from pathogenic organisms. Anaerobic digested sludge should therefore be disinfected prior to agricultural use, disposed of in a landfill, or reused for other purposes such as a supplement to construction materials. Biogas which contains a significant portion (65–85%) of methane is a by-product of the anaerobic digestion process. In large CEPT treatment plants it might be economically feasible to generate energy from the methane produced in the digestion process. The CEPT process does not consume a lot of energy so it does not require the energy which can be produced from the biogas, but other uses can be found for the energy in the water and sanitation utility, if feasible.

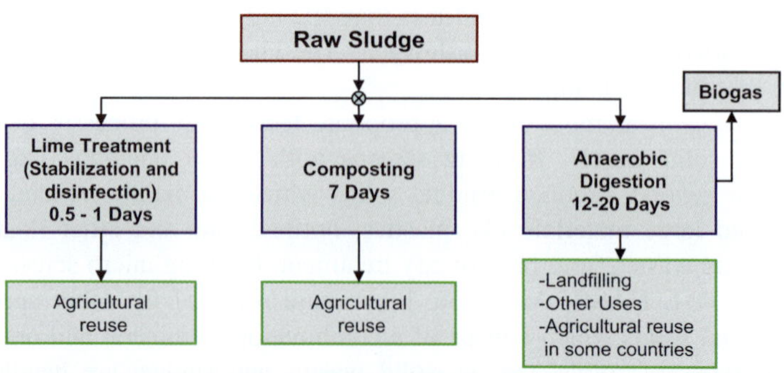

Figure 9.10 Sludge handling option in a CEPT plant [*Source:* Libhaber (2007)]

9.1.3 Performance

Efficiency

The BOD removal efficiency in the CEPT process is 70–75% and the TSS removal efficiency is 80–95%, depending on the SOR employed (see Figure 9.7). The removal CERM efficiency is in the range of 50–60% for COD and 70–80% for TSS.

Construction costs of the CEPT system are in the range of 20–40 USD/Capita. Operating and maintenance costs are in the range of 1.5–2 USD/Capita · yr. The construction costs of CERM are estimated to be in the range of 15–25 USD/capita and O & M costs in the range of 1.5–2 USD/Capita · yr.

Advantages and disadvantages

The advantages of CEPT are:

- Achieves removal of all important contaminants present in wastewater (SS, BOD, phosphorus, H_2S, heavy metals and partial removal of pathogens) in a single treatment stage.
- Because it uses installations of primary treatment, it is of low investment cost. Existing primary treatment installations can easily be converted to CEPT.
- The wastewater treatment is quick (detention time of about an hour) and yields a transparent and deodorized treated effluent.
- Since it is not a biological treatment process, the CEPT process is less sensitive to temperature variations. The flocculation process loses some efficiency at low temperatures but the efficiency can be recovered by adjusting the doses of coagulant and flocculant. Consequently, the CEPT process can perform well in all the temperature range of wastewater and therefore in all types of conditions and zones, from tropical to sub-polar.
- Because of its low SS content, the CEPT effluent can be effectively disinfected, and that is an important feature for developing countries, where due to the high level of morbidity originating in water borne diseases, disinfection is a priority.
- It achieves high phosphorous removal.
- It is a very compact treatment process which occupies a very small land area, so it is advantageous for projects with restricted land availability.
- It reduces the size of subsequent treatment units if polishing of the CEPT effluent is required.

The main disadvantages of the CEPT are:

- It produces more sludge than conventional primary treatment and even more sludge than that produced by the conventional activated sludge process.
- The sludge produced is more difficult to handle than biological sludge.

9.2 BASIC DESIGN PROCEDURES

9.2.1 General design considerations

The CEPT can be presented as the original design choice or as a conversion of a conventional primary treatment plant. The latter modification is easy, since it only requires incorporation of installations for adding a coagulant and a polymeric flocculant. The injection of the chemicals must be made at appropriate points to meet the design variables corresponding to the coagulation/flocculation in a way that promotes a good mixing for these processes, namely: (i) the adequate velocity gradient, and (ii) the required detention time. The CERM process is presented as an additional design option because although there is no experience yet regarding its operation, some commercial firms already have it among their products.

The CEPT process flow diagram is presented in Figure 9.1, which for municipal wastewater can consist of Screening (coarse Rotating Screens) → Grit channel → Parshall flume → Primary Sedimentation. The excess sludge flow diagram in turn consists of: Sludge Thickening → Sludge Stabilization → Drying Beds.

If the design is originally a CEPT process design (not a converted conventional Primary treatment dsesign), a Parshall flume should be used as mixing system, but if the design refers to a conversion, it is preferable to use an hydraulic jump generated by a sluice gate as presented in Figure 9.4. For design purposes it is possible to use with a high level of certainty a surface overflow rate, SOR, twice as large as the SOR for conventional primary treatment, which is presented in Table 9.4. However, the hydraulic load of the weir recommended in Table 9.4 should be maintained. After thickening, the sludge can be stabilized with lime and dried in a drying bed or a similar system; or it can be digested and dehydrated.

9.2.2 Orderly design method, ODM

In general, the design methodology of a CEPT consists of the following:

(1) *Determine the external or independent variables*, which among others, are the following:
- Average flow (process design flow, Q_D) and peak flow (hydraulic design flow, Q_{DH}), calculated using Equations 3.3 and 3.4 (below) in appropriate units (Lps, m^3/h, and so on):

$$Q_{DH} = \frac{cq_{dom}P}{86400} + q_1 A_a \quad (3.3)$$

$$Q_{DH} = k_1 k_2 \frac{cq_{dom}P}{86400} + q_1 A_a \quad (3.4)$$

- Geographic conditions such as maximum and minimum air and water temperatures, solar radiation and cloud cover, wind rose, altitude above sea level, proximity to population centers, and so on. (see Chapter 3).
- Contaminants loads to be treated, primarily BOD_5, COD, TSS, N, P.
- Concentrations of contaminants at th entrance to the treatment plant taking into account the infiltration effect in accordance with the methodology presented in Chapter 3.

(2) *Design the screening system* which may be a coarse RM of 6 mm screen opening. From Figure 4.8 we obtain that a Rotating Screen with 6 mm screen opening will removal 3% of TSS. The design flow for the screen is Q_{DH} since it is a hydraulic unit (see explanations in Chapters 4 and 5).

(3) *Design the grit chamber:* the main parameters for the grit chamber design are: (i) the ratio length/width (L/w) of at least 4, (ii) the surface overflow rate ($SOR = Q_{DH}/A_{des}$) in the range 600–1200 m/d, and (iii) the horizontal velocity of 0.15–0.60 m/s. Always design a stand by grit chamber. The design flow is Q_{DH} since it is a hydraulic unit. To size the grit chamber it is necessary to:
- Calculate the surface area of the grit chamber $A_{des} = Q_{DH}/SOR$.
- Calculate L and w: if $(L/w) = 4$, $A_{des} = L \cdot w = 4w^2$, then $w = \sqrt{(A_{des}/4)}$ and $L = 4w$.
- Calculate the depth (H_{des}): the transversal velocity should be in the range 0.20–0.40 m/s, and $v = Q_{DH}/w \cdot H_{des}$; then the depth of the grit chamber is: $H_{des} = Q_{DH}/w \cdot v$. We add 0.20 m of freeboard and 0.30 for sand storage between cleanings. The total depth is then $H = H_{des} + 0.50$.

(4) *Design the Parshall flume for flow measure and rapid mixing:* the following procedure is used:

- Q_{DH} is estimated from Equation 3.4.
- Throat (W) is chosen from in Table 9.2 using Q_{DH}.
- With W known, D and N are selected from Table 9.1. K and m are selected from Table 9.3.
- h_0 is calculated with:

$$h_0 = KQ^m \text{ with Q en } m^3/s \tag{9.6}$$

- v is calculated with

$$v = 1/KQ^{m-1}[2/3(D - W) + W] \tag{9.7}$$

- E is calculated with

$$E = v^2/2g + h_0 + N \tag{9.8}$$

- The equation

$$v_1^3 - 2gv_1E + 2Qg/W = 0 \tag{9.9}$$

is solved for a velocity v_1 between 2 and 7 m/s at the channel throat; this way the channel serves as a mixer and as a flow meter. In an Excel program is possible to use the "goal seek" function to solve Equation 9.9. If the calculation does not provide the desired value, change W.

(5) *Design the primary clarifier.* As design parameters, we use double the average SOR and double the peak SOR from Table 9.4 (between 48–96 and 96–240 m/d, respectively) but we maintain the same weir hydraulic load (125–500 $m^3/d \cdot m$) and detention time (1.5–2.5 h) as in the table. Then:

- $A_{sed} = Q_D/SOR$ y $A_{sedmax} = Q_{DH}/SOR_{peak}$. Select the largest area of the two, and name it as $\mathbf{A_s}$.
- Calculate the clarifier's net diameter, D_N, (between weir ends) with:

$$D_N = \sqrt{\frac{4A_s}{\pi}} \tag{9.10}$$

- Calculate the clarifier circumference $C = \pi D_N$, and calculate the weir hydraulic load, $q_v = Q_D/C$ (the weir has a circumference C, see Figure 9.6). The hydraulic load should be between 125–250 $m^3/d \cdot m$ (cubic meters per day per lineal meter of weir) or less.
- Calculate the sludge produced, Q_x (kg/d) using Figure 9.7: If E is the TSS removal efficiency for CEPT then $Q_x = E \cdot TSS \cdot Q_D$. The removal efficiencies of BOD_5 and TSS are calculated, respectively using:

$$E(DBO_5) = -0.096\ SOR + 77.30 \tag{9.11}$$

$$E(TSS) = -0.114\ SOR + 96.553 \tag{9.12}$$

- Calculate the coagulant dosing: $Q_D \times$ Dose. The sum of TSS and coagulants gives the total amount of sludge produced, Q_{XT}.
- Define the concentration of the sludge at the bottom of the clarifier, for example, 5%.
- Calculate, from the dry sludge weight produced the sludge volume at 5%, as followes: $V_x = Q_{XT}/0.05$ (m^3/d).
- Determine an average depth of the clarifier (H_{SED}) usually in the range 3.5 to 4.5 m (but can be more) and calculate the clarifier's volume, $V = H_{SED} \cdot A_s$
- Calculate detention time, $t_d = V/Q_D$.

(6) *Design the sludge thickener:* to design a sludge thickener use the surface load (Q_s) from Table 9.5. Select $Q_s = q_{XT}/A_e$. The total solids load (Q_{XT}) in the thickener should take into consideration the applied coagulant: $Q_{XT} = Q_x + \text{Dose} \times Q_D$. ($\text{Dose} \times Q_D = Q_{FeCl_3}$)

Table 9.5 Design parameters of a sludge thickener.

Parameter	Sludge type	Recommended values
Surface load, Qs	Primary sludge	90–120 $kg/m^2 \cdot d$
	Secondary sludge	20–30 $kg/m^2 \cdot d$
Average depth, h	–	3–4 m

Source: Orozco (2005)

Then, calculate the thickener surface area (A_e) as follows:

$$A_e = Q_{XT}/Q_s \quad (9.13)$$

- Calculate the thickener diameter: $D_e = \sqrt{4A_s/\pi}$
- Then choose the thickener depth, and propose an outlet sludge concentration that should be between 5–10%, and calculate the volume of the thickened sludge, say 10% sludge concentration by:

$$V_{xe} = Q_{XT}/0.10 \quad (9.14)$$

- The total volume of chemically stabilized sludge is the sum of total sludge (Q_{XT}) and lime added to stabilize the sludge (50% of Q_{XT}): $V_{stab} = 1.5\ V_{xe}$.

(7) *Design the sludge drying beds:* Once the sludge is stabilized with a lime application rate of 50% of the total excess sludge (dry weight), it needs to be dried in drying beds. The drying beds are designed with an applied rate (q_x) of between 120–150 $kg/m^2 \cdot yr$. The area is the ratio of sludge (plus the added lime) produced and the selected application rate:

$$A_{bed} = 1.5Q_{XT}/q_x \quad (9.15)$$

9.3 BASIC DESIGN EXAMPLE

(The model program for this Example is available online at http://www.iwawaterwiki.org/xwiki/bin/view/Articles/Software+Developed+for+Sustainable+Treatment+and+Reuse+of+Municipal+Wastewater)

In continuation we present an example that follows the Orderly Design Method (ODM) of Paragraph 9.2.2. The example is developed for a population of 20,000 using the recommendations for the calculation of flow, as presented in Chapter 3. The ODM is developed step by step and is calculated a program developed in Excel, CHAP 9-CEPT, available on line as explained elsewhere. Please note that the Tables in this section identified by alphabetic order contain computer calculated results.

(1) *Calculation of external variables*

1.1 *Design flows:* The inputs required for flow calculations are as presented in Table 9A.

The final design flows are obtained from Equations 3.3 and 3.4, then and they are presented in different units below. The Q_D is used for the biological process design and the Q_{DH} for hydraulic design.

Table 9A

Design flow				
Variable	**Value**	**Unit**	**Value**	**Unit**
Per capita consumption, q	162.5	L/capita·d		
$PF = 1 + 14/(4 + \sqrt{P})$	2.65			Eq. Harmon
Population, P	20000	inhab		
Population density, d	260	inhab/ha		
Return coefficient	0.8			
Wastewater mean flow, Q_{DWW}	30.09	L/s	108.33	m^3/h
Wastewater maximum flow, Q_{maxDWW}	79.82	L/s	287.35	m^3/h
Infiltration unit flow, q_I	0.13	L/s · ha (o km)		
Afferent area, A_a	76.92	ha		
Infiltration flow, Q_I	10.00	L/s		

Notes: (i) Per capita consumption is calculated for drinking water consumption. With the return factor of 0.8, the wastewater per capita flow of $c \cdot q_{dom} = 0.8 \times 162.5 = 130$ l/Capita · d, as measured in the example in Chapter 3, (ii) the Peak Factor is calculated by Equation 3.6 (*Harmon* Equation), (iii) the population density should be the maximum acceptable density according to planning standards, in order to be able to serve the population at saturation conditions, (iv) the flows are calculated based on the population and the and the return coefficient - c, and (v) infiltration must be measured or alternatively a value of between 0.10 and 0.15 l/s · ha or l/s · km is to be chosen.

Table 9B

Design flows						
Variable	**Value**	**Unit**	**Value**	**Unit**	**Value**	**Unit**
Process design flow, $Q_D = Q_{DWW} + Q_I$	40.09	L/s	144.33	m^3/h	3464.00	m^3/d
Hydraulics design flow, $Q_{HD} = Q_{DWWmax} + Q_I$	89.82	L/s	323.35	m^3/h	7760.44	m^3/d

The final design flows are obtained from Equations 3.3 and 3.4, and are presented in Table 9B in different units as an output. The average flow Q_D is used for the biological process design and the maximum flow Q_{HD} for hydraulic design.

In the case of a CEPT the design flow for screening, grit chamber and Parshall flume is Q_{HD} because they are hydraulic units. The primary clarifier, thickener and drying beds are designed with Q_D.

1.2 *Identification of geographic and environmental external variables:* These variables depend solely on geographic location. They are presented in Table 9C for the conditions of the example under consideration.

1.3 *Concentration and pollution loads:* The quality of the municipal wastewater may be different within the city and at the entrance to the wastewater treatment plant if the conveyance system is long and significant inflow infiltration takes place. In the case of the example, the wastewater treatment plant is located nearby the city and there is no significant variation in the wastewater quality along the conveyance pipeline.

Table 9C

Variable	Value	Unit	Value	Unit
Minimum temperature of water, T_w	25	°C	77.0	°F
Minimum temperature of air, T_a	18	°C	64.4	°F
Altitude above sea level, hasl	350	masl		
Mean wind velocity	6	kph		
Predominant wind direction	NW			
Minimum solar radiation, **S**	230	$Cal/cm^2 \cdot d$		
Daily fraction of sunlight hours	0.8	decimal		

Notes: (i) These variables are not very important for the CEPT process because it is not a biological processes and therefore it is less sensitive to external environmental conditions; (ii) The minimum water temperature as well as the minimum air temperature is the average of the coldest month; (iii) the velocity and direction the prevailing wind is important for locating the wastewater treatment plant in relation to the urban population centers; (iv) the fraction of hours of sunlight is the ratio of actual hours of sunlight to potential sunlight hours.

The table that calculates as output the load (kg/d) and the specific unit load per Capita (kg/Capita · d) of each pollutant from the concentration of each pollutant is Table 9D. In this table, the concentration of each pollutant is given as an input. The calculation is performed on each pollutant such as BOD_5, COD, TSS, and so on. If there is a significant infiltration into the conveyance pipeline to the treatment plant it is necessary to calculate the additional infiltration flow of the sewerage afferent area (A_a) as follows: $Q_{Iemi} = q_I \cdot L$, where L is the length of the pipeline in km. The concentrations at the entrance to the treatment plant are calculated as the specific per Capita load (q_C) multiplied by the population (P) and divided by the total flow: $Q_T = (Q_D + Q_{Iemi})$. For example: the Concentration of $COD = q_{COD} \cdot P/Q_T$. Note that in Table 9D the load and the per Capita load (specific unit load) is an *output* while the concentrations are an *input*, presented in the Excel sheet separately.

(2) *Screening*

Variables selection by the designer: For coarse screening a coarse rotating screen is proposed with an opening of 6 mm. We apply the same methodology explained in the design example in Chapter 4 for a 3% removal of TSS.

Design: The diameter obtained for the course rotating screen is 600 mm. The final output data are presented in Table 9E. The course rotating screen design is similar to that of the fine rotating screen so the reader is referred to Chapter 4, in which the design process is explained in detail. A coarse screening is sufficient in this case because further treatment is be performed by the CEPT system which follows the preliminary treatment. For better clarity we show in Table 9E the calculations of the preliminary treatment unit which is produced by the Excel program. The design support is presented in the column entitled OBSERVATION.

(3) *Grit chamber*

Variables selection by the designer: According to the ODM Section 9.2.2, Paragraph 3 related to preliminary treatment design, the proposed design variables are presented in Table 9F.

Table 9D

Concentration, load and per capita contribltion						
Variable	**Concentration**	**Unit**	**Load**	**Unit**	**Per capita**	**Unit**
BOD_5	277.8	mg/L	962.3	kg/d	0.048	kg/capita·d
COD	596.1	mg/L	2064.9	kg/d	0.103	kg/capita·d
COD/BOD_5	2.1					
TKN	40.0	mg/L	138.6	kg/d	0.007	kg/capita·d
N-Nitrate	2.0	mg/L	6.9	kg/d	0.000	kg/capita·d
Total Phosphorus	5.8	mg/L	20.1	kg/d	0.001	kg/capita·d
PH	100.0	UN	0.0			
Alkalinity	100.0	mg/L	346.4	kg/d	0.017	kg/capita·d
TSS	202.6	mg/L	701.8	kg/d	0.035	kg/capita·d
VSS	173.6	mg/L	601.4	kg/d	0.030	kg/capita·d
O & G	100.0	mg/L	346.4	kg/d	0.017	kg/capita·d
Fecal Coli	0.0	mpn/100mL	346400000000000.0	mpn/d	17320000000.000	mpn/capita·d

Table 9E

Design			
Parameter	**Value**	**Unit**	**Observation**
Flow, Q_{HD}	90.000	L/s	
Screen opening, a	6.000	mm	
Drum diamater, D_T	0.600	m	Figure 4.7
Channel width, $W = 1{,}1\ D_T$	0.660	m	
Channel depth, $h = Q_{DH}/W \cdot V$	0.230	m	Equation 4.2
Hydralics radius, $R = Wh/(W + 2h)$	0.136	m	Equation 4.3
Manning's n	0.012		Equation 4.3
Hydrulics slope, $s = (v \cdot n)^2/R^{4/3}$	0.074	%	Equation 4.3
Head loss, $h_f = (1/C{\cdot}2g)(Q_{DH}/A)^2$	0.030	m	Equation 4.1
Total head loss, $h_f + 0{,}05$	0.080		Plus 5 cm
Backwashing flow	1.800	L/s	Table 4.1
Solids removed	466.560	L/d	Table 4.1

Table 9F

Designer variables			
Variable	**Value**	**Unit**	**Observation**
Grit channel			
SOR	800	m/d	600–1200
v	0.20	m/s	0.15–0.60
# Grit channels	2.00		
L/w	4		3 a 6

Design: Applying these variables to the equations presented in the ODM in Chapter 5, Section 5.2.2 Paragraph 3.2 and shown in the left column of Table 9G, we obtain the results for the two grit chambers as shown in Table 9G.

(4) *Parshall flume*

Variables selection by the designer: According to the methodology explained in the ODM, the following variables are selected from Tables 9.1, 9.2 and 9.3 and are presented in Table 9H. Note that the $Q_{DH} = 89.82$ Lps define the throat width (W) for flow measurement, which is $W = 9''$. In the Excel program CHAP 9-CEPT, the Pharsall flume is calculated in Tag "Parshall", for its tow functions: (i) a flow meter, and (ii) a rapid mixing structure.

Design: It is necessary to verify that the flume provides sufficiently intense mixing for coagulation. This requires the use of Equations 9.6 to 9.9, as explained in the ODM. The application of the equations yields the results presented in Table 9I.

Table 9G

Parameter	Valor	Unit	Observation
Grit channel			
Area, $A_{des} = Q_{DH}/SOR$	9.7	M^2	Section 5.2.2 Numeral 3.1
Width, $w == \sqrt{(A_{des}/4)}$	1.6	m	Section 5.2.2 Numeral 3.1
Length, $L = 4w$	6.2	m	Section 5.2.2 Numeral 3.1
Depth, $H_{des} = Q_{DH}/v \cdot w$	0.3	m	Section 5.2.2 Numeral 3.1
Total Depth, $H = H_{des} + 0{,}50$	0.8	m	Section 5.2.2 Numeral 3.1
Parshall flume			
Throat, W	9.0	inch	See "Parshall" tag

Table 9H Parshall flume deimensions.

Variable	Value	Unit	Value	Unit	Source	Hint
Caudal	89.82	Lps	7760.44	Lps	Equation 2.4	
W	9.00	inch	0.2286	m	Table 9.2	
K	1.84				Table 9.3	
m	0.64				Table 9.3	
D	57.50	cm	0.575	m	Table 9.1	
N	11.40	cm	0.114	m	Table 9.1	

Table 9I Parshall flume design as rapid mixer.

h_0	0.40	m		Equation 9.6	
v	0.49	m/s		Equation 9.7	
E	0.52	m		Equation 9.8	
V_i	3.20	m/s	−0.001	Equation 9.9	Goal seek: "0"

Note that the throat velocity (v_1) of 3.2 m/s is within the required range for good rapid mixing (which is between 2 and 7 m/s) and this ensures a sufficiently high mixing intensity. To calculate the throat velocity through Equation 9.9 it is necessary to use the Excel function "goal seek" to obtain the value "0" for the sum in Equation 9.9. Note that the function approached to −0.001 at end of calculation (close enough to zero). If the calculation does not yield a value of v_1 within the required range which ensures good mixing, W needs to be changed and the calculation repeated.

The coagulant is applied first before the flume and the flocculant polymer is then applied after a short coagulation time (see Figure 9.1).

(5) *Primary clarifier*
Variables selection by the designer: According to the ODM, it is necessary to propose a SOR for average and peak flows, and a weir hydraulic load. The selected values are shown in Table 9J.

Table 9J

Primary clarifier			
SOR	64.00	m/d	Table 9.4 (double)
SOR_{peak}	120.00	m/d	Table 9.4 (double)
X_R	5	%	typical
H_{sed}	4	m	$\geq$3.5 m
$FeCl_3$dose	50	g/m^3	

The selected SOR values are double the values proposed in Table 9.4 for conventional sedimentation. A value for the concentration of excess sludge (X_R) leaving the bottom of the clarifier is also selected. In this case we assume that the coagulant is $FeCl_3$ and that its dose is 50 mg/l. This is actually a high dose. In real cases, jar test experiments need to be undertaken to select the best coagulant and its optimal dose. The selected dose value is applied to calculate the production of sludge. Explanations for the selected values of the other parameters are given in the right column.

Table 9K

Primary clarifier			
A_{sed}	54.1	m^2	Q_D/SOR
A_{sedmax}	64.7	m^2	Q_{DH}/SOR_{peak}
A_s	64.7	m^2	Larger A_{sed} and A_{sedmax}
D_N	9.1	m	Equation 9.10
C	28.5	m	$C = \pi D_N$
$q_v = Q_D/C$	121.5	$m^3/d \cdot m$	125–250 or less
E (BOD)	71.1	$\%BOD_5$	Equation 9.11
E (TSS)	89.3	% TSS	Equation 9.12
$Q_x = E \cdot SST \cdot Q_D$	626.4	kg/d	
Q_{FeCl_3}	173.2	kg/d	Dose $\times$ Q_D
Q_{XT}	799.6	kg/d	$Q_x + Q_{FeCl_3}$
$V_{XT} = Q_X/X_R$	16.0	m^3/d	
$V = H_{sed}A_s$	258.7	m^3	
$t_d = V/Q_D$	1.8	h	Between 1.5–4.0 h

Design: Applying the methodology of the ODM, we obtain the dimensions of the primary clarifier and calculate the sludge production and its volume. The results are presented in Table 9K. The calculation of the clarifier area, A_{sed}, is made using the equations at the right hand column. For a clarifier average depth of 4 m we obtain a detention time t_d of 1.8 hours. The production of excess sludge is 16 m^3/d at 5% solids concentration. The clarifier net diameter (D_N) is 9.1 m and this diameter meets the design requirements for average and peak flows. Explanations for the calculated values of the other parameters are given in the right column.

(6) *Thickener*

Variables selected by the designer: The most important parameter for calculating the thickener is the surface solids load Q_s (kg $SST/m^2 \cdot d$). From this parameter the required thickener area and its diameter are calculated. The selected designer variables are presented in Table 9L.

Table 9L

Thickener			
Q_s	100	$kg/m^2 \cdot d$	Table 9.5
H_e	3.5	m	Table 9.5
X_e	10	%	5–10%
Drying beds			
Application rate	150	$kg/m^2 \cdot yr$	120–150

Design: Applying the ODM for thickener design presented in Section 9.22 Paragraph 6, we obtain the thickener design values presented in Table 9M.

Table 9M

Thickener			
A_e	8.0	m^2	$Q_s = Q_{XT} \cdot Q_o/A_e$
D_e	3.2	m	$D_e = (4A_s/n)^{(1/2)}$
V_{xe}	8.0	m^3/d	Equation 9.14
V_{stab}	12.0	m^3/d	1.5 V_{xe}
Drying beds			
A_{bed}	1945.7	m^2	Equation 9.15

The resulting diameter of the thickener is 3.2 m. The total volume of sludge produced at 10% concentration is 8 m^3/d. Before conveying the sludge to the drying bed it needs to be chemically stabilized. In this case the stabilization is obtained by adding lime to the sludge. The lime quantities are quite large. In this case we choose to add to the sludge 50% lime by volume. This increases the excess sludge volume significantly, from 8 m^3/d to 12 m^3/d (1.5 $Q_{XT} = 1.5 \times 799.6$ kg/d = 1199.4 kg/d that is, about 1.2 ton/day sludge on a dray basis, which at a 10% concentration is 12 ton/day liquid sludge, which is equivalent to 12 m^3/d liquid sludge).

(7) *Drying beds*

Variables selected by the designer: The variable selected by the designer is the sludge application rate. The select value 150 kg/year · m^2, as presented in Table 9L.

Design: This rate applied to the stabilized sludge volume (799.6 kg/d) results in a required drying beds area of 1945.7 m^2 as presented Table 9M.

A cross section of a typical drying bed is presented in Figure 9.11.

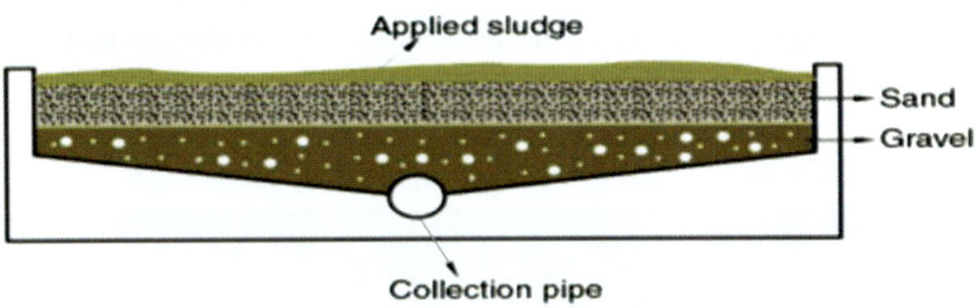

Figure 9.11 Cross section of a typical drying bed [*Source:* http://www.unep.or.jp/ietc/publications/techpublications/techpub-15/2-5/5-5.asp (2011)]

A typical drawings of a CEPT treatment plant is presented in Figure 9.12.

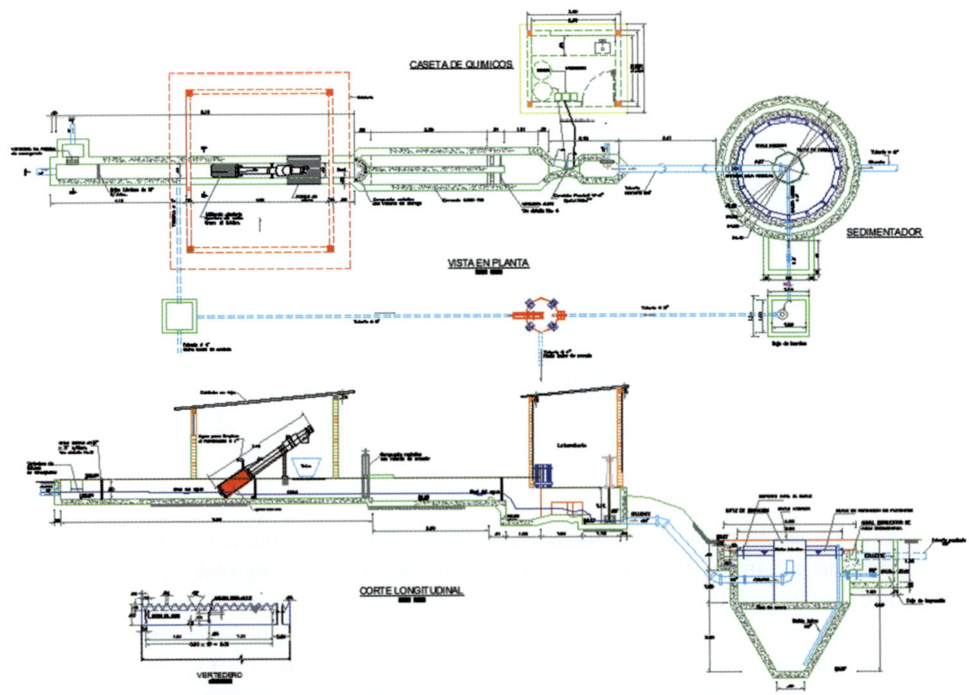

Figure 9.12 Plan and section of a typical CEPT plant

Chapter 10

Complementary processes to combine with appropriate technology processes

10.1 INTRODUCTION

In cases that require higher quality of effluent than those possible with unit processes of appropriate technology it is possible to subject the effluent of a unit process to one or more polishing stages. In this way, combination of processes can yield higher quality effluents. The polishing units can be other appropriate technology unit processes such as those which have already been discussed in the second part of the book, and also additional processes which have not yet been discussed and which are nominated as complementary processes. However, the use of a complex process as a polishing step of effluent of a unit appropriate technology process should be avoided, since it transforms the entire treatment plant into a complex plant (Libhaber, 2007). There are certain treatment technologies which perhaps are not appropriate technologies when they are used as the main treatment unit, but they meet the requirements of being simple to operate and easy to maintain, and when they are used as polishing units of a preceding main treatment step they can be considered appropriate technology processes.

Complementary treatment unit which can be mentioned include Sand Filtration (SF), Dissolved Air Flotation (DAF), Disinfection with ultraviolet rays (UV disinfection), Overland Flow, Infiltration-Percolation, Septic Tanks, Submarine and Large River Outfalls, and under certain circumstances, Aerated Lagoons. Among those Sand Filtration, Dissolved Air Flotation, UV disinfection, and Infiltration-Percolation can only be used as polishing units since they require a pretreatment stage and are not able to perform as a main treatment unit. Septic tanks usually perform as a pretreatment unit and sometimes as the main treatment unit, but are not used as a polishing unit. Aerated lagoons can perform both as a main treatment unit and as a polishing unit. As discussed in Chapter 1, they can be considered appropriate technology units when they are employed as polishing units.

It was not possible to present in this book the design procedures of all the mentioned complementary processes. We chose to present the following: Sand Filtration, Dissolved Air Flotation and UV disinfection. This does not mean that the others are of a lower level of importance. All of the complementary processes are worthy processes and are discussed in Chapter 1.

The Membrane Filtration Technology [Micro Filtration (MF), Ultra Filtration (UF), Nanofiltration (NF) and Reverse Osmosis (RO)] is not an appropriate treatment technology. However, in certain cases, a very high effluent quality is required even in developing countries. In such cases the use of Membrane Filtration is advantageous because it is not too complex to operate and it yields effluent qualities which are higher than those that can be achieved by any other known technology. Membrane Filtration can be combined with appropriate technology processes which can serve as its pretreatment units. For these reasons it was decided to present in this chapter also the Membrane Filtration Technology.

These complementary processes are introduced only briefly in this chapter, but there is an extensive bibliography which can be used by readers interested to deepen their knowledge, for example Metcalf & Eddy (2003).

10.2 SAND FILTRATION

10.2.1 Introduction

Sand filtration is one of the oldest and most wildly used water treatments processes. It is used primarily for potable water treatment but is recently enployed as a polishing process for wastewater treatment. Filtration is the removal of suspended and colloidal particles present in an aqueous suspension by passing the suspension through a porous medium (Arboleda, 1992; Metcalf & Eddy, 2003).

Filtration is the result of two distinct but complementary mechanisms: transport and adhesion. Initially, particles which are to be removed are carried from the aqueous suspension to the surface of filter medium grains. The particles remain attached to the grains, provided that they resist the action of shear forces due to the hydrodynamic flow conditions. The particle transport is a hydraulic phenomenon, governed by the parameters that define the mass transfer. The adhesion between particles and grains is a phenomenon of surface action, which is influenced by physical and chemical parameters.

The mechanisms that can carry out the transport are: (i) straining, (ii) sedimentation, (iii) interception, (iv) diffusion, (v) inertial impact, (vi) hydrodynamic action, and (vii) combined transport mechanisms. The adhesion mechanisms are: (i) Van der Waals forces, (ii) electrochemical forces, and (iii) chemical bridges (Maldonado-Yactayo, 2007).

Filtration through granular media can be simple filtration (with sand as the sole filtration medium) or mixed filtration (with two or more filter media usually sand and anthracite). Design elements of a sand filter are:

- Hydraulic loading rate or filtration rate (q_F), $L/min \cdot m^2$, (the filtration rate can be slow, rapid and high)
- Type of flow (continuous, driven, intermittent)
- Flow direction (up flow, down flow).

Mixed filters consist of two or more filter media. A multi-media filter works with finer and denser media in the top and middle part and thicker and less dense media at the lower part. This is achieved when gravel is used at the bottom of the bed as support, sand in the middle and anthracite coal on top as filter media. The most common flow of multi-media filters is gravity down flow. However, other filter types with up flow, horizontal flow and double flow directions are also in use. The granular media filters require a periodic backwash cleaning to maintain their efficiency. In rapid filters this is done by reversing the direction of flow, injecting water and air through a false bottom to expand the filter medium, with upper collection of water washout.

When hydraulic losses caused by trapped particles reduce the wastewater rate of filtration through the filter bed, it reaches the end of the "filter cycle" and it requires backwashing to remove the suspended solids trapped in the filter bed. Also, the bed may be shaken mechanically or with air to help remove solids. The wash water is recycled back to the inflow wastewater stream. Filtration through granular media is most often used for tertiary treatment of sewage and for additional suspended solids removal from effluents of chemical treatment processes such as CEPT. Figure 10.1 shows a typical diagram of a gravity sand filter, used in large towns. The use of pressure filters it is becoming more widespread in small communities (Maldonado-Yactayo, 2007).

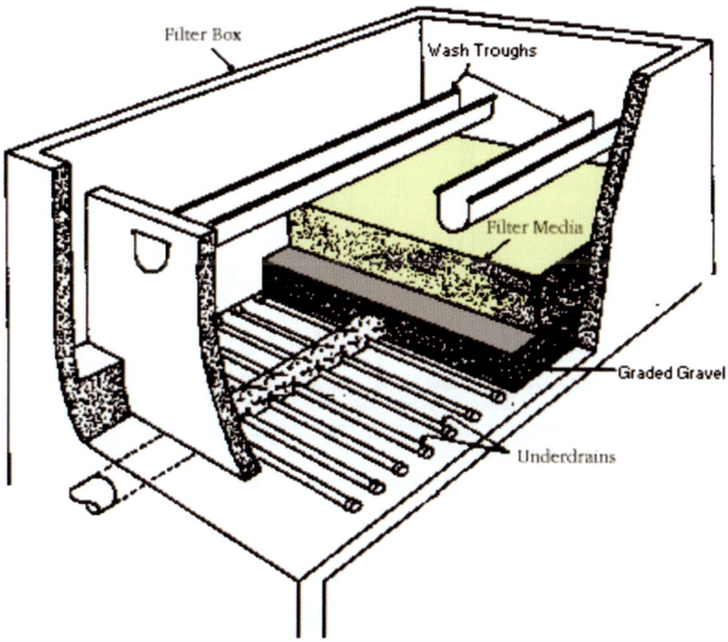

Figure 10.1 A Diagram of a rapid sand filter [*Source:* http://water.me.vccs.edu/concepts/filters.html (2011)]

Table 10.1 Classification of filters.

	Slow sand filter	Fast sand filter	Pressure filter
Filtration rate, L/min · m^2	0.006–0.024	80–200	80–160
Filtrating media	Sand	Sand Sand + Anthracite	Sand Sand + Anthracite
¿Gravity or pressure?	Gravity	Gravity	Pressure
Washing	Top layer removal (5 cm)	Backwash	Backwash
Advantages	Reliable, little operation	Reliable, compact	Cheap for small plants
Disadvantages	Requires large area, Manual cleaning	Requires chemical treatment	Mechanized
Application	Small populations	Surface water of all kinds	Industry and small communities

Source: Prepared from http://water.me.vccs.edu/concepts/filters.html

The filteration parameters for the three most common types of filters (slow, rapid and pressure filters) are presented in Table 10.1. Rapid gravity filters can be of constant or declining rate. Declining rate filter use small amounts of equipment such as valves and automatic controls, which are replaced by hydraulic-type controls, either on the inflow or outflow. In this type of filtration at least one battery of four filters is required so that three of the filters produce the required backwash flow for the fourth filter. It is used for large populations (more than 50,000 inhabitants) and the hydraulic is complex. Figure 10.2 presents a cross section of a filter battery of variable filtration rate with hydraulic control at the outflow.

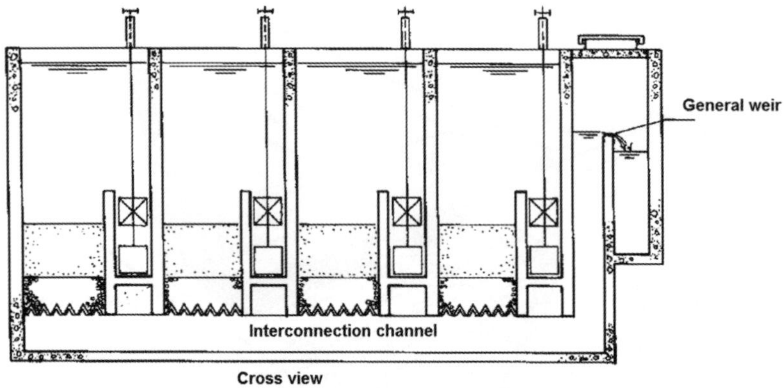

Figure 10.2 Cross section of a filter battery of variable rate filtration [*Source:* CEPIS-MAUAL II (2005)]

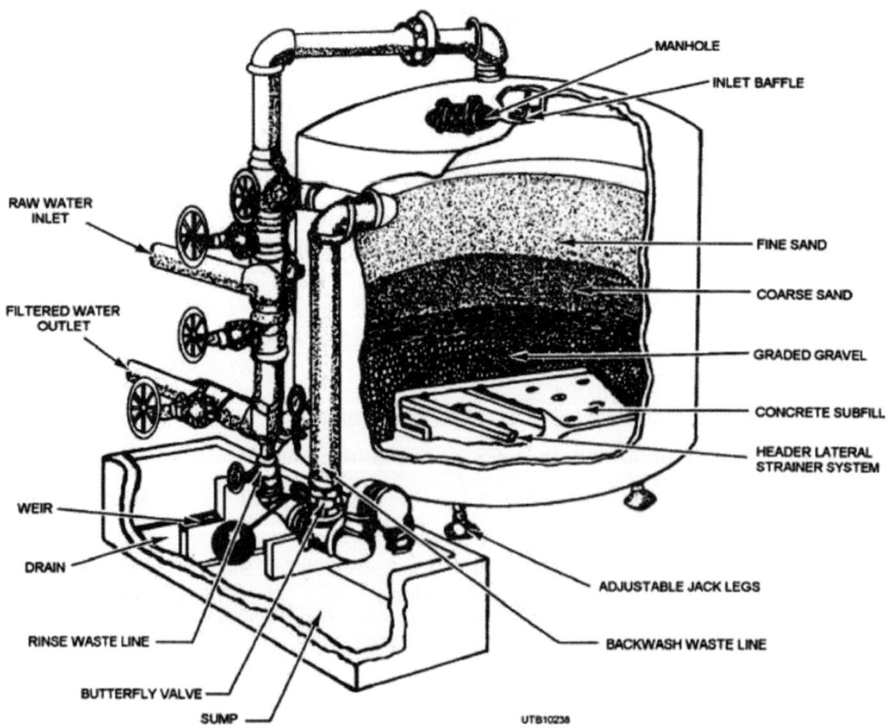

Figure 10.3 Details of a pressure filter [*Source:* http://www.tpub.com/content/construction/14265/css/14265_293.htm (2011)]

For smaller populations and small flows it is more convenient to use pressure filtration because although it consumes energy for pumping (which creates the pressure) it is a simpler system in design and operation. A typical diagram of this type of filters is presented in Figure 10.3. Figure 7.12 shows a photo of a pressure sand filtration system in which the valves are automatically operated (i.e the back washing operation is automatic). The system is suitable for medium-size applications (up to 100 Lps). In general, it is desirable to undertake pilot filtration test before applying it to full scale filtration of municipal wastewater.

10.2.2 Basics of the process

Although filtration theory is essentially the same for different types of granular media filters, in this section declining rate filters for large populations and pressure filters for small populations are discussed separately.

Declining rate filtration

Declining rate filtration compared with constant rate filtration has the following advantages (CEPIS-MANUAL II, 2005): (i) The plants do not require a very large hydraulic head to operate. Constant rate filters operate with a hydraulic loading of 1.80 to 2 meters to complete an operation cycle of 40–50 hours, on average. Under these circumstances, a battery of declining rate filters requires a hydraulic head four (or the number of filters that make up the battery) times less than operating with constant rate, (ii) It requires no pipes gallery (or filters gallery). The transport of decanted water, filtered water, Backwash water and the drain of filters backwash water is carried out through channels, (iii) The filtered water is also transported through channels, with an insulation channel and another channel that links together the output of all units. These channels are located immediately after the filters boxes, (iv) It does not require an elevated tank or pumping equipment to perform the filter backwashing. Through the interconnection channel and due to a special hydraulic design, the water produced by at least three filters are used to backwashes another filter (that is the reason for the required minimum filter battery size of 4 filters), (v) Due to hydraulic design, it is only necessary to close the inflow of clarified water and open the outflow of backwash water for the washing to occur automatically and with the correct sand backwash expansion (25–30%), (vi) Sophisticated instruments are not required for the operation, (vii) The filters battery operates under the principle of communicating vessels. The units are interconnected at the water input through the input channel and through the interconnection channel at the water outlet. Figure 10.4 shows a diagram of a declining rate filter.

In this operation system the available hydraulic head is fully applied from the beginning to the end of the filter run (or cycle), which implies a gradual decrease in the filtered flow rate over time. The average effluent quality of a declining rate filter is higher and with longer filter runs, than those obtained in filters operated with constant rate.

Nevertheless, the declining rate has several disadvantages compared to the constant rate filtration: (i) the variable rate subjects the filter bed to different filtration rates, and in general to different conditions that can affect the filter bed, (ii) if the design flow is not applied because the operation flow is different from the design flow or because one filter is out of operation, then the hydraulic design conditions are not applied and that can lead to inadequate backwash and damages to the filter bed due to formation of sand balls in the filter bed.

Filtration rate

The filtration rate (Q_F) depends on the type of water and fluctuates between 120 and 360 $m^3/d \cdot m^2$. Table 10.2 presents the rate ranges applied depending on the water quality and filter media. To properly define this parameter it is often advisable to perform pilot testing. Note that the minimum number of filters in a declining rate battery is four, so that three of the filters produce the required backwash water to wash the fourth filter.

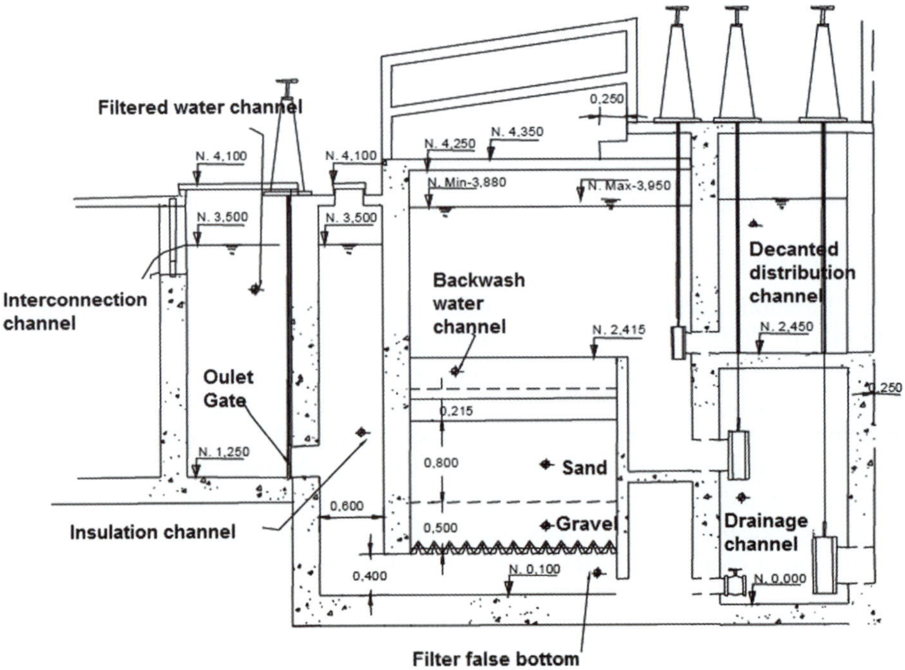

Figure 10.4 Typical cross section of a declining rate filter [*Source:* CEPIS-MANUAL II (2005)]

Table 10.2 Filtration rates in declining rate filters.

Type of filter		**Filtration rate ($m^3/m^2 \cdot d$)**
Rapid declining rate filtration		
a)	Of decanted water, with a single sand filtration media with effective size (T.E.) of 0.50 to 0.60 mm and filter depth around 0.80 meters.	120–150
b)	Of flocculated or pre-flocculated water, in a single coarse filtration media, with a depth of more than one meter and use of polymer as flocculant (direct filtration)	240–250
c)	Of decanted water in a double filtration media, with total depth of less than 0.80 meters and a good level of operation and maintenance.	240–360

Source: CEPIS-MANUAL II (2005)
Note: $m^3/m^2 \cdot d \equiv 1.44\ L/min \cdot m^2$

Filtration media

The selection of the filter bed material is of great importance for the operation and durability of the filter. The filter material determines the characteristics of the bed. Usually the filter bed materials are: (i) Sand bed, and (ii) Combined beds of sand and anthracite. To select the filter material it is necessary to apply the following parameters (Maldonado-Yactayo, 2007):

Effective size (D_{10}) is defined as the size of the sieve in mm through which ten percent of the sample of sand by weight will pass.

Uniformity coefficient (UC): is defined as the ratio of the sieve size in mm through which 60 percent of the sample of sand will pass, to the effective size of the sand.

Form: the shape of the grains is usually evaluated based on sphericity coefficient (Ce). The *sphericity coefficient* of a particle is defined as the result of dividing the surface area of the sphere of equal volume to the grain by the surface area of the particle considered.

Minimum size: size below which no grain is to be found in the filter medium.

Maximum size: size above which no grain is to be found in the filter medium.

Table 10.3 presents the desirable characteristics for the two types of beds used in declining rate filtration.

Table 10.3 Recommended beds for (a) sand bed and (b) combined sand and anthracite beds.

Characteristics	Symbol	Criteria	
(a)			
Depth (cm)	L_1	60–80	
Effective size (mm)	D_{30}	0.50–0.80	
Uniformity coefficient	CU	≤1.5	
Minimum size (mm)		0.42	
Maximum size (mm)	D_{90}	2.0	
Characteristics	**Symbol**	**Sand**	**Anthracite**
(b)			
Depth (cm)	L_1	15–30	45–60
Effective size (mm)	D_{30}	0.50–0.60	0.80–1.10
Uniformity coefficient	CU	≤1.5	≤1.5
Minimum size (mm)		0.42	0.59
Maximum size (mm)	D_{90}	1.41	2.0

Source: CEPIS-MANUAL II (2005)

Filtration hydraulics

The hydraulic head available in the system is calculated so that the relationship between the highest filtration rate (q_{Fmax}) occurring at the start of the filtration cycle (filter recently washed) and the average filtration rate (q_F), does not exceed 1.5. A careful calculation of the head loss in the filter needs to be undertaken to define this head (H) because: (i) if it is insufficient, the filter runs between washings will be very short, and (ii) if it is exaggerated, initial filtration rates (in a freshly washed filter) will be very high and this will cause deterioration of the effluent quality. The following equation is used for calculating the hydraulic head:

$$H = \Sigma k_i(q_F)^2 + \Sigma c_i(q_F) + Z_v \tag{10.1}$$

where:

H = hydraulic head with a clean bed, m
q_F = average filtration rate ($m^3/m^2 \cdot d \equiv m/d$)
Σk_i = sum of the constants corresponding to the calculated head losses in the entrance gate and drainage orifices
Σc_i = sum of the constants corresponding to the calculated head losses in the sand and/or anthracite
Z_v = height of the filter battery outlet weir

It should be kept in mind that all losses can be given in terms of the filtration rate, since the total flow is $Q = q_F \cdot A_F$, where A_F is the filtration area. Now, if we assume that a_j is the section area being analyzed, (for example. the area of the inlet gate), then:

$$h_f = kv^2/2g = k(q_F \cdot A_F/a_j)^2/2g = k_i q_F^2$$

where:

v = velocity through the section (for example the velocity through inlet valve)
k = the constant of loss in the inlet valve
$k_i = k\,(A_F/a_j)^2/2g$

The calculation of the head loss in the clean filter bed (h_F) can be done with the following equation (the Carmen-Kozeny equation, Metcalf & Eddy, 2003):

$$h_F = \frac{f}{\varphi}\frac{(1-\varepsilon_0)}{\varepsilon_0^3}\frac{L \cdot q_F^2}{g}\frac{\rho_p}{d} \tag{10.2}$$

where:

Φ = granule form factor, if it is spherical the factor is 1
ε_0 = bed porosity
ρ_p = density of the particles
L = bed depth, m
d = effective size of the particle
g = gravity acceleration coefficient, 9.8 m/s^2

If there are multiple layers in the filter bed, Equation (10.2) is applied to each layer and the head losses are added. In this case the c_i in Equation 10.1 is Equation 10.2 without the filtration rate square, $q_F{}^2$. Table 10.4 provides some common parameters for the mostly used beds.

When filtered material begins to accumulate on the filter bed, a layer is formed on top of the bed and consequently the head loss begins to increase according to the equation:

$$\Delta h = a q_F{}^b \tag{10.3}$$

Where a and b are constants that need to be calculated empirically. Typical values of a and b are presented in Table 10.5 (Sincero & Sincero, 2003).

Table 10.4 Typical properties of commonly used filtering granular beds.

Characteristic	Silica sand	Anthracite coal	Granular activated coal	Granate idaho
Material density (ρ_s) (g/cm^3)	2.65	1.45–1.73	1.3–1.5	4.0–4.2
Bed porosity (ε_0)	0.42–0.47	0.56–0.60	0.50	0.45–0.65
Sphericity (Ce)	0.7–0.8	0.46–0.60	0.75	0.60

Source: CEPIS-MANUAL II (2005)

Table 10.5 Typical values of constants in the formula of head loss in the filter bed.

Δh	Bed material	Diameter
$0.051\ q_F^{2.22}$	uniform sand	0.5 mm
$0.022\ q_F^{2.22}$	uniform sand	0.7 mm
$0.0133\ q_F^{2.22}$	uniform sand	1.0 mm
$0.00114\ q_F^{2.10}$	uniform anthracite	1.0 mm
$0.00069\ q_F^{2.10}$	uniform anthracite	1.6 mm
$0.00057\ q_F^{2.10}$	uniform anthracite	2.0 mm

Thus, the total hydraulic head (H_T) is calculated using the following equation:

$$H_T = H + \Delta h \tag{10.4}$$

Backwashing hydraulics

For a filter battery to provide sufficient water for backwashing a filter, it is necessary that two conditions are met: (i) The flow should be sufficient to produce a washing velocity necessary to expand the filter bed to between 25 and 30%, and (ii) The level of the outflow weir should provide the required hydraulic head to overcome the head losses that occur during the operation.

The length (L_e) of the bed expanded by the backwash is calculated by the following equation using the expanded porosity (ε_e):

$$\frac{L(1-\varepsilon_0)}{(1-\varepsilon_e)} \tag{10.5}$$

The expanded porosity is obtained by the empirical formula:

$$\varepsilon_0 = (q_{RL}/v_s)^{0.22} \tag{10.6}$$

where:

q_{RL} = backwash rate, equal at least to $3q_F$.
v_s = settling velocity of sand particle (Stokes' law)

Stokes's law in laminar flow is as follows:

$$v_s = \frac{d_p^2 g(\gamma - 1)}{18\upsilon} \tag{10.7}$$

where:

v_s = sedimentation velocity
d_p = particle diameter
γ = specific gravity of the particle
υ = kinematics viscosity of water

Finally, the head loss in the backwash, H_{RL} can be calculated as follows (Sincero & Sincero, 2003):

$$h_{RL} = L_e(1 - \varepsilon_e)(\gamma_p - \gamma_w)/\gamma_w \tag{10.8}$$

γ_p, γ_w are the specific weights of particle and water respectively.

Pressure filters

Rapid pressure filters are generally used for small and medium flows. They are easy to install and operate and usually present little maintenance problems. They are the preferred solution for industrial applications and for small and medium towns. Pressure filters are contained in steel or fiberglass tanks. There are two basic structure types: of pressure filters: vertical and horizontal.

They have the following characteristics:

- The water flows into and out of the filter under pressure. The filter bed rests on a bed support composed of several layers of decreasing particle size (coarse material at the bottom and becoming finer towards the top).
- Normally the filter bed is one single material, sand or anthracite. The filtered water collector is a grid of perforated pipes, embedded within the gravel support bed, or alternatively, a false bottom of perforated metal with plastic or metal nozzles. Details of the internal structure of a pressure filter are shown in Figure 10.5.

The same principles of rapid sand filtration apply to pressure filtration but the operation of pressure filters is simpler. Some types of internal structure of pressure filters are manufacturers' property, but a simple system as shown in Figure 10.5 is available in the market supplied by numerous workshops at competitive prices. The construction of this filter is very similar to the gravity filter. The sand rests on the gravel that serves as a bed support. Raw water enters the unit through the inlet of raw water, using a pump or other means to increase its pressure. It passes through the filter bed and then reaches the outlet pipe. The unit operates under pressure, so the filter medium is encapsulated. When the filter bed is clogged by particles entering with the water, it is necessary to clean it by backwashing. The operation of the filter needs to be guaranteed by the manufacturer and can be manual or automatic.

10.2.3 Basic design

The following procedure is used to calculate the dimensions of a filter, applied to the case of a population of 20,000.

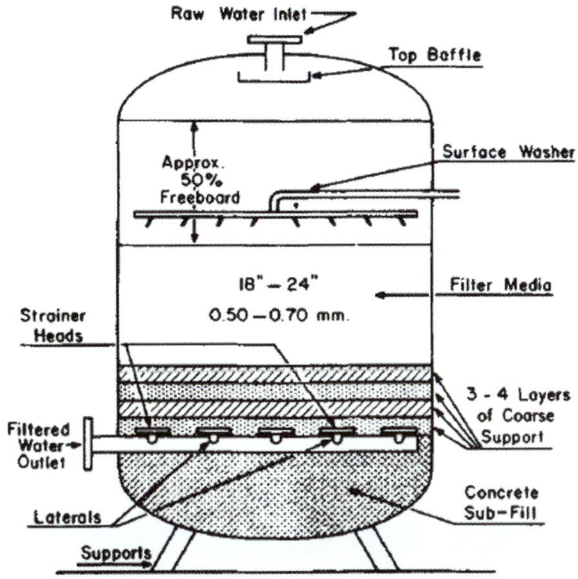

Figure 10.5 Internal structure of a pressure filter [*Source:* Nalco (1988)]

Variables selected by the designer

The main variables selected by the designer are: (i) the flow, which is calculated using Equation 3.4, (ii) The filtration rate (q_F) which is determined from Tables 10.1 and 10.2, (iii) the backwash flow rate (Q_{BW}), which is selected as at least $3q_F$, (iv) the number of filters (n_F) which should be at least two. The design variables applied to a population of 20,000 are presented in Table 10A. (Please note that Tables identified by alphabetic order contain computer calculated results).

Table 10A

Designer variable					
Variable	**Value**	**Unit**	**Value**	**Unit**	**Observation**
Filtro de arena					
Q_D	89.8	L/s			Equation 3.4
Pressure filter?	Yes				Yes for P < 100,000 inhab
Q_F	120.00	$L/min \cdot m^2$	2.94	gpm/ft^2	Tables 10.1 and 10.2
q_{BW} (select)	480.00	$L/min \cdot m^2$	11.76	gpm/ft^2	Minimum $3q_F$
n_F	3.00				>2

Basic design

Having defined the design variables the sizing of the filter units is as follows:

- Calculate the flow per filter,

$$Q_F = Q_D/n_F \tag{10.9}$$

- Calculate the filter area (F_A),

$$A_F = Q_F / q_F \tag{10.10}$$

- Calculate the filter diameter (D_F),

$$D_F = \sqrt{(4A_F/\pi)} \tag{10.11}$$

- Calculate backwash flow of (Q_{BW}) with,

$$Q_{BW} = 3Q_F \tag{10.12}$$

Select the filter bed from Table 10.3

The dimensions of each of the three pressure filters for 20,000 inhabitants are presented in Table 10B.

Table 10B

Design			
Parameter	**Value**	**Unit**	**Observation**
Sand filter			
Q_F	29.9	L/s	Equation 10.10
A_F	15.0	m^2	Equation 10.11
D_F	4.4	m	Equation 10.12
Q_{BW}	119.8	L/s	Equation 10.13

An outline of a typical installation of a pressure filters system is shown in Figure 10.6.

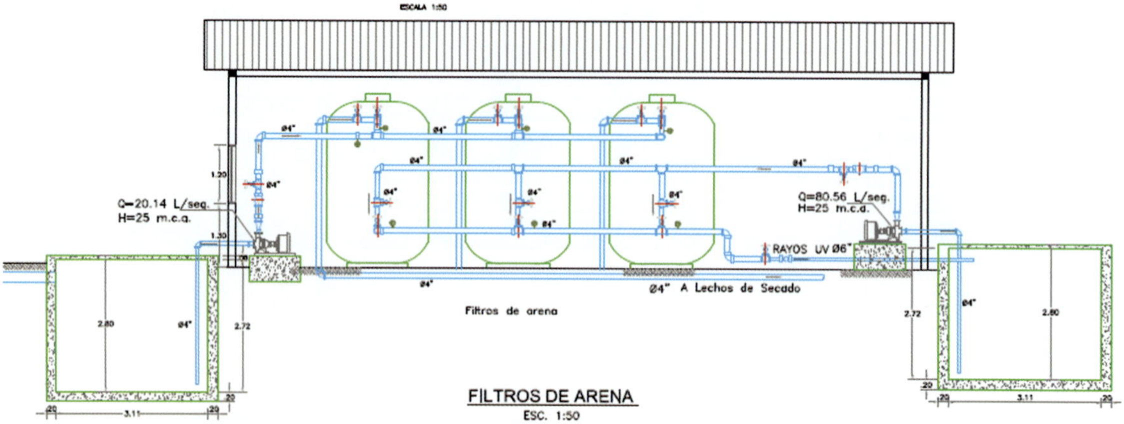

Figure 10.6 A Typical installation of pressure filters

10.3 DISSOLVED AIR FLOTATION (DAF)[1]

10.3.1 Introduction

DAF (Dissolved Air Flotation) is used to remove particles and oil and grease (O&G) by flotation. Flotation is a primary clarification process, which is also used as post treatment in wastewater treatment plants. To separate suspended solids from wastewater the wastewater needs to first be coagulated with aluminium or iron salts, and flocculated with a polyelectrolyte. When air bubbles are applied to a coagulated-flocculated solution they clog to the flocs and create a net flotation action which carries the floc-particles upwards. The same procedure applies to oil and grease, but in this case less time is required for flotation since oils are lighter than water and therefore float easily. The following activities are necessary to produce the flotation process: (i) produce bubbles of suitable size for the particles which need to be removed, (ii) produce an effective adhesion between the air bubbles and suspended particles, which is achieved by coagulation-flocculation, and (iii) ensure an adequate separation of floating material.

There are basically two types of DAF processes: (i) *dispersed* air flotation which usually employs air nozzles that produce dispersion, aeration and agitation of the suspension, producing air bubbles that are about 1 mm in diameter, and (ii) *dissolved* air flotation which has been one of the most studied processes in wastewater treatment and this is the process discussed below. There are three basic types of dissolved air flotation systems: with partial influent pressurization, with total influent pressurization and with recirculation.

The DAF process is used in several applications: (i) DAF is applied as pretreatment when the raw wastewater contains high concentrations of oil and greases that needs to be removed prior to the biological treatment of the wastewater since they can interfere with the performance of the biological reactor, (ii) DAF is also used as pretreatment when the raw wastewater contains large amounts of SS and there is not enough space to remove them by conventional primary treatment, and (ii) DAF is used as post treatment when it is necessary to polish an effluent of a main treatment process by removing suspended solids from its effluent.

A diagram of a DAF process with partial air pressurization of the effluent is shown in Figure 10.7.

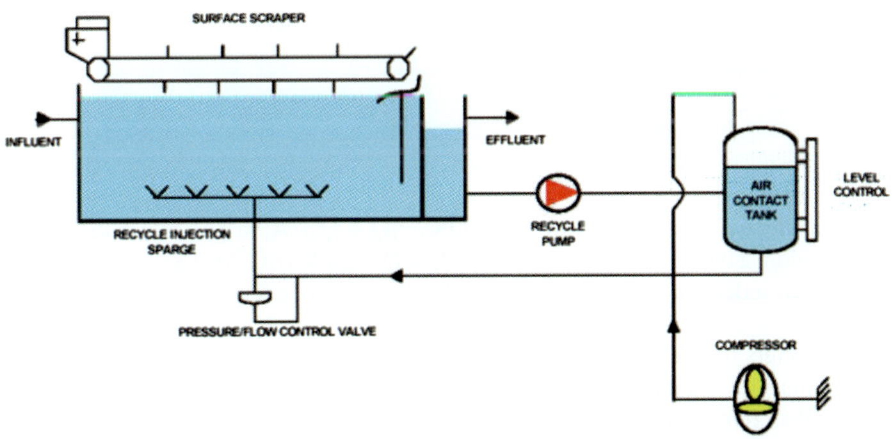

Figure 10.7 A Diagram of the dissolved air floating system with partial pressurization of the effluent (*Source:* http://www.esi.info/detail.cfm//stetfield-separators-ltd/stetfloat-dissolved-air-flotation-DAF-unit/_/R-26789_TR109EP#)

[1]Based on Orozco (2005), Chapter 7

10.3.2 Basics of the process

The design parameters of a DAF process are the surface overflow rate, the solids loading and the air to solid ratio (A/S). The ratio (A/S) is the most influential parameter in the process and is calculated by following equation:

$$\frac{A}{S} = \frac{1.3\, s_a(f\,P - 1)Q_r}{S_a\, Q} \tag{10.13}$$

where:

A/S = ratio of air to solids (mL air/mg SS)
s_a = air solubility, mL/L
f = fraction of dissolved air pressurized (P) normally 0.5 a 0.6
$P = \frac{p + 101.35}{101.35}$ = pressure, atm
P = manometric pressure, kPa
S_a = influent suspended solids, mg/L
Q_r = pressurized recycled flow, m^3/d
Q = Municipal wastewater flow

The value 1.3 in the numerator of Equation 10.13 is the specific weight of air in g/L. Thus, the numerator is the weight of air and the denominator the weight of solids. The recommended values of A/S are between 0.006 to 0.02 kg of air per kg of TSS. If all the water is pressurized, then in Equation 10.13: $Q = Q_r$. Table 10.6 presents the solubility of air in water at different temperatures.

Table 10.6 Solubility of air in water at different temperatures.

Temperature, °C	0	10	20	30
s_a, mL/L	29.2	22.8	18.7	15.7

The DAF must achieve a surface overflow rate (SOR = $(Q + Q_r)/A_s\ m^3/h \cdot m^2$) in the range of 0.5 to 10, preferably 5. In addition, a maximum solids loading ($Q_s = Q \cdot S_a/A_s$) of 5 kg TSS/h · m^2 should be applied. The flotation chamber, independent of its geometric form, is designed for a mean retention time of between 10 and 20 minutes.

The design parameters for the air contact tank are: (i) pressure, and (ii) net area (a, m^2/m^3) of the inside elements. Pressure must be maintained at a value of between 250 and 500 kPa, and the retention time in the contact tank is less than five minutes. The recirculation flow can vary from 5–15% (and up to 100%) of the total water flow to be clarified. The quality of raw water and the pretreatment conditions influence the retention time in the flotation chamber and the SOR. For this reason it is recommended to run pilot tests before developing the final design.

The design parameters for *diffused* air flotation are provided by the manufacturer of the air diffuser. Prefearbly they should be in the low range of values given for *dissolved* air flotation. An outline of a typical installation of a diffused air DAF system is shown in Figure 10.8.

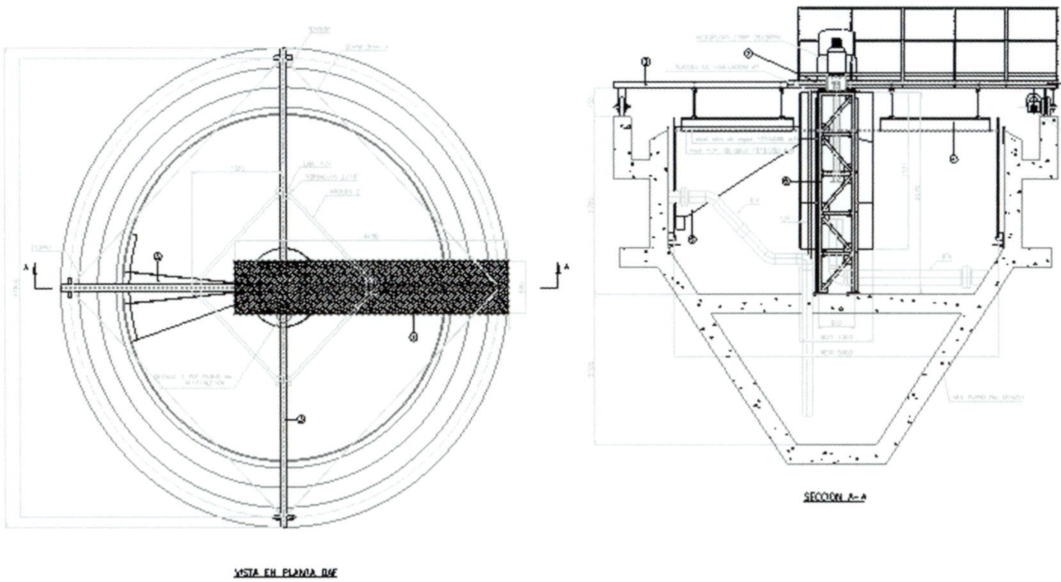

Figure 10.8 Plan and cross section of a typical diffused air flotation installation

10.3.3 Basic design

The following procedure is used to calculate the dimensions of a DAF system, applied to the case of a population of 20,000.

The main variables selected by the designer are: (i) the flow, which is calculated using Equation 3.4, (ii) The filtration rate (q_F) which is determined from Tables 10.1 and 10.2, (iii) the backwash flow rate (Q_{BW}), which is selected as at least $3q_F$, (iv) the number of filters (n_F) which should be at least two. The design variables applied to a population of 20,000 are presented in Table 10A.

Variables selected by the designer

The main variables selected by the designer are: (i) the flow rate (Q_{HD}), which is calculated using Equation 3.4, (ii) S_a which is determined from the SST concentration in the outflow of preceding process, (iii) The SOR which should be between 0.5 and 10 $m^3/m^2 \cdot h$, (iv) the solids load which should be between 0.5 and 10.0 $kg/m^2 \cdot h$, (v) the ratio (A/S) which should be in the range 0.006 to 0.2, (vi) the recirculation ratio (R) which should be between 0.05 and 1.0; (vii) the fraction of dissolved air pressure (P), preferably from pilot test with a TSS removal of 95% or more; and (viii) the air solubility (s_a) which is calculated by the following equation constructed from data in Table 10.6:

$$s_a = 28.638e^{-0,021T} \tag{10.14}$$

The design variables applied to a population of 20,000 are presented in Table 10C.

Basic design

Having defined the design variables, the sizing of the DAF unit is done as follows:

- From the design variables, the pressure is calculated using Equation 10.13.

- An estimated P is calculated from definitions in Equation 10.13:

$$P = \frac{p + 101.35}{101.35} \tag{10.15}$$

- Calculate A′ from the equation,

$$SOR = (Q + Qr)/A' \tag{10.16}$$

- Calculate A″ from the equation,

$$Q_s = (Q + Qr) \cdot X_T/A''z \tag{10.17}$$

- Select design area (A_D) of the highest between A′ and A″.
- Calculate diameter of the DAF unit with

$$D_{DAF} = \sqrt{(4A_D/\pi)} \tag{10.18}$$

Table 10C

Designer variables					
Variable	**Value**	**Unit**	**Value**	**Unit**	**Source**
DAF					
Q_{HD}	7760.44	m^3/d			Equation 3.4
TSS, S_a	60.78	mg/L			Influent quality
$SOR = (Q + Q_r)/A$	10	$m^3/h \cdot m^2$			0.5 a 10.0
Q_s	5	$kg/m^2 \cdot h$			0.5 a 10.0
A/S	0.01				0.006 a 0.2
R	0.5				Q_r/Q_D
f	0.5				0.5–0.6
$Q_x = Q_{HD} \cdot S_a$	471.68	kg/h			
E, % TSS	95	%			Pilot test
Temperature	25.00	°C			Influent quality
s_a	17.11	mL/L			Equation 10.15

A photo of a typical DAF installation is shown in Figure 10.9 and also in Figure 1.71. For the example with a population of 20,000, the results of the DAF unit sizing (when it is located after a UASB and an Anaerobic Filter, as is explained in Chapter 11, and calculated by the Excel program COMB 3-UASB-SAND FILTER-UV.xls) is presented in Table 10D. It is possible to construct several smaller DAFs instead of one large unit, but this is more expensive.

Figure 10.9 Photo of a typical DAF installation. [*Source:* http://www.sereco.it/share/img_prodotti/32img1.jpg]

Table 10D

Design			
Parameter	**Value**	**Unit**	**Observation**
DAF			
Pressure, P	2.1	atm	Equation 10.14
Manometric pressure,	112.4	kPa	Equation 10.16
Area fro SOR, A′	776.0	m^2	Equation 10.17
Area from Q_s, A″	141.5	m^2	Equation 10.18
Design area, A_D	776.0	m^2	Largest from A′ and A″
Diameter, DAF	31.4	m	Equation 10.19
TSS_e	3.0	mg/L	According to efficiency

10.4 UV DISINFECTION (BY ULTRAVIOLET RAYS)

10.4.1 Introduction

Ultraviolet (UV) radiation is one of the methods of disinfection of water and wastewater. It destroys bacteria, fungal spores, viruses and other microorganisms. It is a more effective when the water or effluent to be disinfected is clear. This method does not have a residual effect on the disinfected liquid and this is one of its advantages. It has been shown that when organisms are exposed to UV radiation, the germicidal effect of a high intensity UV radiation over a short period of time provides the same disinfection as does a lower intensity over a longer time period of time. The product of radiation intensity (I) by the contact time (t) is called the radiation dose (D). Most commercial disinfection equipment operates with UV lamps of medium and low power and with a wavelength of between 200 and 300 nm (10^{-9} m) (Russell, 2006).

One ways to produce ultraviolet light for disinfection is with the low pressure mercury arc lamp, because 85% of the light output is monochromatic with a wavelength of 253.7 nm. This wavelength is within the range of ultraviolet radiation UV-B, the optimum range for germicidal effect as shown in Figure 10.10. UV-B rays range are of wavelength lower than or equal to 320 nm and this is a dangerous range because it may cause skin cancer, so it is important that plant operators are not exposed to the radiation. Figure 10.11 shows the germicidal effect as function of the wavelength of UV rays (Sincero & Sincero, 2003).

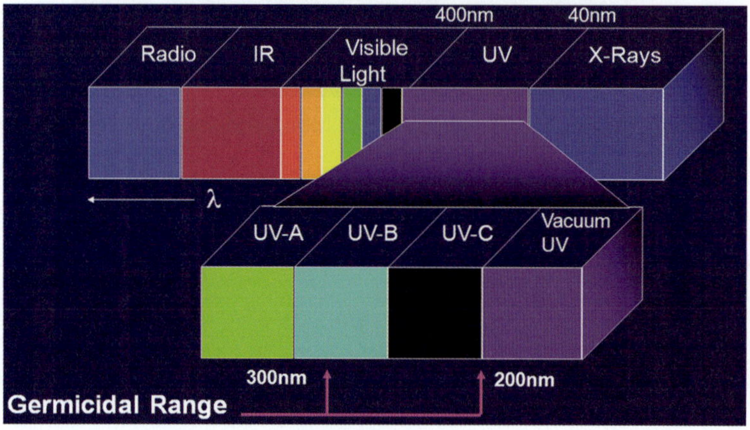

Figure 10.10 The Electromagnetic spectrum and Germicidal range of UV rays [*Source:* Park (2006)]

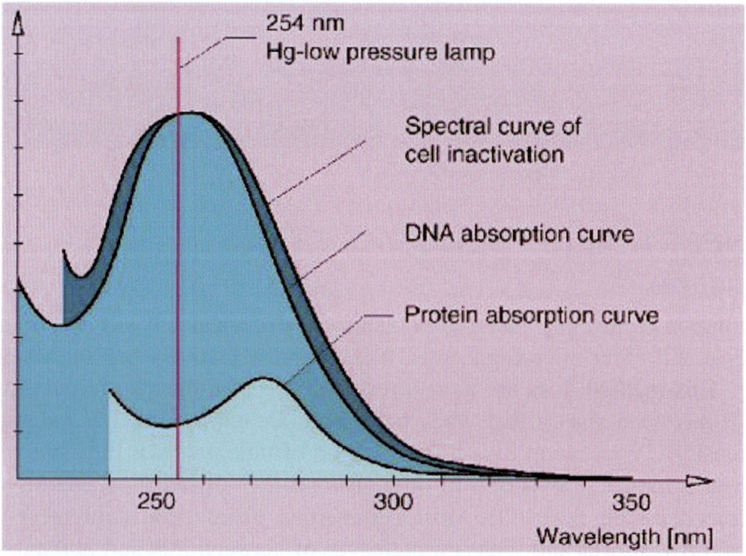

Figure 10.11 Curve of germicide response [*Source:* Park (2006)]

10.4.2 Basics of the process

The UV radiation dose (D) is calculated by the following equation:

$$D = I \cdot t \qquad (10.19)$$

where:

D = UV dose in $W \cdot s/m^2$ or J/m^2.
I = UV ray intensity in W/m^2.
t = contact time in s.

The lethal dose for different pathogens is presented in Table 10.7. However, pilot plant testing with the specific wastewater is recommended to determine the real efficiency of a UV disinfection equipment. The turbidity or water clarity has a strong influence on the required radiation dose, as shown in Figure 10.12.

Table 10.7 UV dose for four $\log_{10}$ cycles (99.99%) removal of various microorganisms.

Microorganism	Dose (J/m^2)
E. coli	80
Hepatitis C	110
Total coliforms	150
V. Cholera	30
S. Typha	70
L. Pneumophila	90

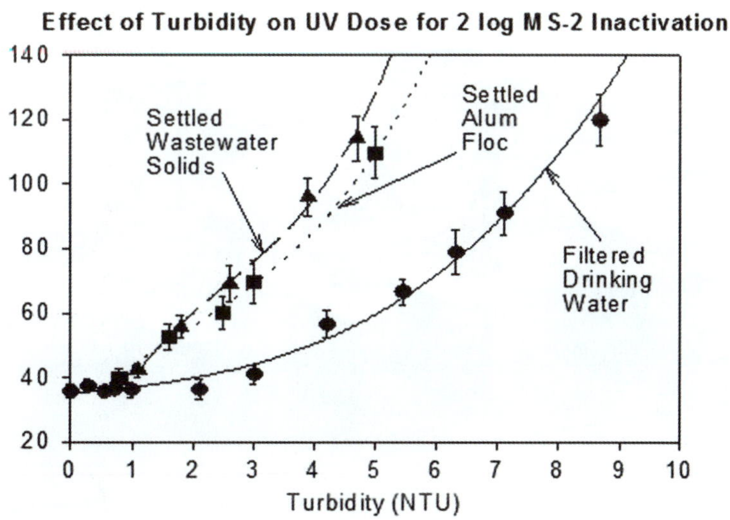

Figure 10.12 Effect of Turbidity on UV dose for the inactivation of MS-2 (virus, *enterica coliphages*) by two $\log_{10}$ cycles [*Source:* Park (2006)]

UV reactor designs (UV) are classified as: (i) open channel systems, (ii) closed channel systems without contact, and (iii) closed-pipe system with contact. Open channel reactors are the most widely used in the municipal wastewater treatment industry and are composed of UV lamps groups oriented in any required direction, crossed by the flow of wastewater. The flow in a channel is gravity flow and the channel is open to the atmosphere. Closed channel non-contact systems are used for water and wastewater flowing through a UV-transmitting tube, usually of Teflon. The lamps are located out of the pipes and the flow type may be pressure or gravity. Closed pipe or channel contact systems consist of UV lamps placed inside of covered UV-transmitting quartz, submerged in a flow of water or wastewater (Figures 10.13 and 10.14).

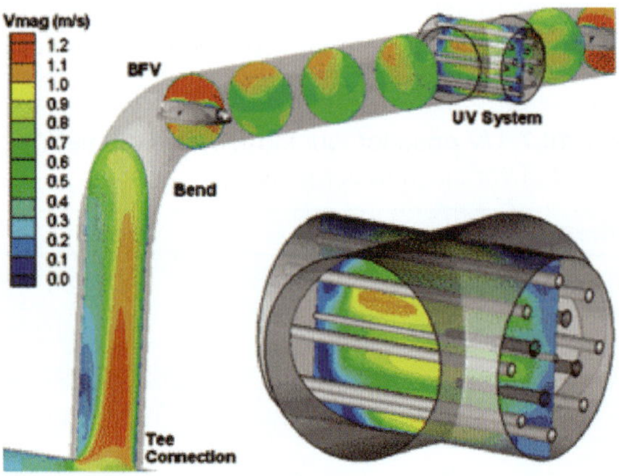

Figure 10.13 A Closed pipe UV reactor with contact the reactor and the liquid [*Source:* http://ww.pennnet.com/articles/article_display.cfm?article_id=249723]

Figure 10.14 Layout of a UV system in a wastewater treatment plant [*Source:* http://konchewater.com/product/device/disinfection_system/ultraviolet_dis/2011/0714/77.html (2011)]

The hydraulics for ideal UV disinfection is designed as a longitudinal plug flow with complete transversal mixing. The reactor inlet conditions can be designed with baffles to minimize dead spots and short circuits. High flows can be maintained through the reactor to promote turbulence and transversal mixing. The relationship between the length and hydraulic radius of the reactor must be high to minimize longitudinal dispersion. The UV dose generated by the reactor at a flow rate Q can be calculated using Equation 10.19:

$$D = I \cdot t = I \cdot V/Q$$

Where t is the contact time in a reactor with an effective volume (V) and a water flow (Q). The effective volume of a UV reactor is the volume of water exposed to UV light, and the average radiation intensity is calculated for the effective volume. The UV dose calculated using the average intensity is often higher than the dose calculated using the intensity measured at the reactor wall. The UV intensity inside the reactor depends on the energy at the UV lamp output, reflectance, refraction and absorption of UV light as it passes through the quartz cover, and the UV light absorption by organic and inorganic chemicals. The UV intensity at a point in the water is the sum of the contributions of light from each point along the arc of each UV lamp immersed in water. The death rate of microorganisms is calculated by the equation:

$$N = N_0 e^{kIt} = N_0 e^{kD} \qquad (10.20)$$

Where N_0 is the initial microorganism concentration in MPN[2]100 ml and N the survivors concentration after a time t. From equation 10.20 it is clear that $\text{Log}(N/N_0) = kD$. The layout of a UV system in a wastewater treatment plant and the lamps placement arrangement in the treatment plant is shown in Figure 10.14.

10.4.3 Basic design

The following procedure is used to design the UV disinfection system, applied to the case of a population of 20,000.

Variables selected by the designer

After choosing the reactor type, the following designer variables are selected: (i) the flow rate (Q_{HD}) which is calculated using Equation 3.4, (ii) the dose of UV (D) is determined, and it must be between 16 and 38 $mW \cdot s/m^2$, (iii) the velocity (v) in the channel or pipe is determined, and it must be between 0.6 and 2.0 m/s, (iv) the channel depth (h_c) is selected and it should be between 0 3 and 1.0 m, and (v) the ration (L_c/w_c) is selected and it should be between 1 to 4. The variables selected by the designer for the example of 20,000 inhabitants are shown in Table 10E.

Basic design

Once the design variables are defined, the sizing is as follows:

- Calculate the channel width (w_c) by:

$$w_c = Q_{HD}/v \cdot h_c \qquad (10.21)$$

[2]MPN: most probable number of microorganisms.

Table 10E

			Designer variables		
Variable	**Value**	**Unit**	**Value**	**Unit**	**Observation**
UV					
Q_{HD}	89.8	L/s	0.090	m^3/s	
Dose, *D*	16.0	mW · s/cm	160	Ws/m^2	16–38
v, channel velocity	0.60	m/s			0.6–2.00
h_c, channel depth	0.30	m			0.3–1.00
(L_c/w_c) ratio	1.50				1 to 4

- Calculate the length of the channel (L_c) with a selected (L_c/w_c) ratio:

$$L_c = (L_c/w_c) \cdot h_v \tag{10.22}$$

- Calculate the channel volume (V),

$$V = L_c \cdot h_c \cdot w_c \tag{10.23}$$

- Calculate the contact time (t) with,

$$t = V/Q_{HD} \tag{10.24}$$

- If a tube reactor with closed contact is used (see Figure 10.13), calculate the cross-sectional area $A_T = Q_{HD}/v$ and the diameter (φ) by:

$$\varphi = \sqrt{(4A_T/\pi)} \tag{10.25}$$

- Calculate the UV dose (D) using Equation 10.19.

The results for the example for a "closed-pipe UV reactor with contact" are shown in the Table 10F.

Table 10F

			Design		
Variable	**Value**	**Unit**	**Value**	**Unit**	**Observation**
UV					
Channel width, w_c	0.499	m			Equation 10.22
Channel length, L_c	0.450	m^2			Equation 10.23
Disinfection volume, V	0.067	m^3			Equation 10.24
Contact time, t	0.750	s			Equation 10.25
Tube reactor diameter, φ	0.262	m	10.0	inch	Equation 10.26
UV intensity, I	213	$W \cdot s/m^2$			Equation 10.20

In this case it is necessary to apply UV rays with an intensity $I = 213\ W \cdot s/m^2$ and a contact time $t = 1$ s (0.750 s), in an installation similar to that shown in Figure 10.13.

10.5 MEMBRANES

10.5.1 Introduction

The membrane separation technologies are used both for treatment of potable water and of wastewater. The most commonly used are Micro Filtration (MF), Ultra Filtration (UF), Nano Filtration (NF) and Reverse Osmosis (RO). Figure 10.15 shows the types of membranes, their pore size openings in microns and types of particles and dissolved matter removed by the different types of membranes.

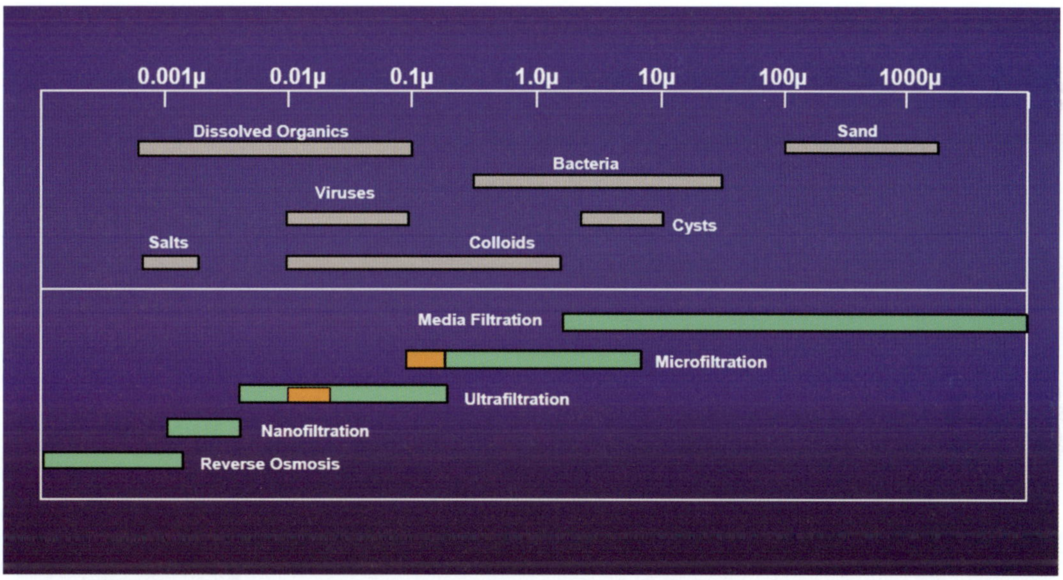

Figure 10.15 Types of membranes, pore size openings and classification of particles and dissolved matter removed by each membrane type [*Source:* Gravel (2002)]

In membrane processes the stream of water (solvent) that crosses the membrane is defined as *permeate* (or *filtrate*), while the stream that does not flow through the membrane is the *reject* (or *concentrate*) and it contains concentrated solutes. In the membrane filtration process one pump feeds tangentially the membranes, while a valve in the concentrate (reject) pipe controls the necessary pressure so that the permeated water exits without pressure. See Figure 10.16.

The type of material that is removed depends on the membrane pore size and applied pressure, as shown in Table 10.8. The membranes come in different shapes, as shown by Figures 10.17 and 10.18.

A facility with membranes as one of its processes consists of:

- Pre-treatment, which eliminate or reduce substances that could impair the proper functioning of the membrane.
- Membrane modules, which is where the membrane is housed.

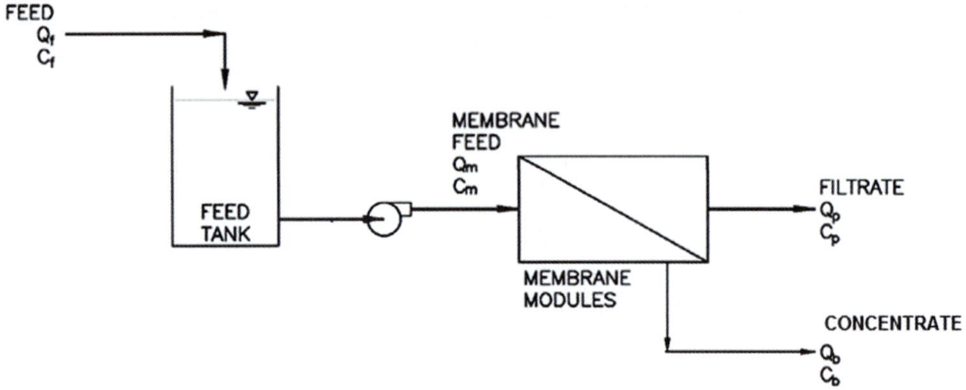

Figure 10.16 Diagram of a membrane plant [*Source:* EPA (2005)]

Table 10.8 Operation parameters of membranes.

Technology	Operation range μm	Pressure kPa	Flow rate $l/m^2 \cdot d$	Membrane material
Micro filtration	0.08–2.0	7–100	400–1600	Polypropylene
Ultra filtration	0.005–0.2	70–700	400–815	Cellulose acetate
Nano filtration	0.001–0.01	500–1000	200–815	Cellulose acetate
Reverse osmosis	0.0001–0.001	850–7000	320–490	Cellulose acetate

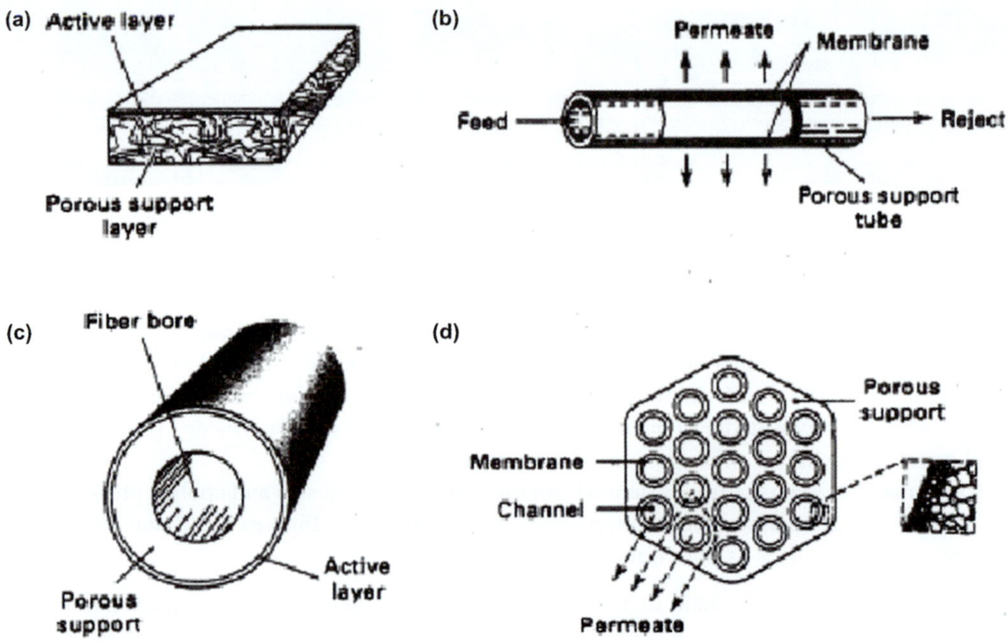

Figure 10.17 Membrane shapes: (a) flat; (b) tubular; (c) hollow fiber; (d) monolithic

- Circuit cleaning as necessary to periodically remove the membrane substances that cause reduction of the permeate flux and/or quality of the permeate stream. Usually an acid wash and another alkaline wash are done for cleaning.

The principle of functioning of a package of pressurized hollow fiber membranes is presented in Figure 10.19. In this figure the feed solution is pressurized. Another way of operating such a system is with an external vaccum, but this can be done only with microfiltration membranes and to some extent with ultrafiltration membranes, where the required pressure is less than 1 atmosphere. A module of hollow fiber membranes is presented in Figure 10.20.

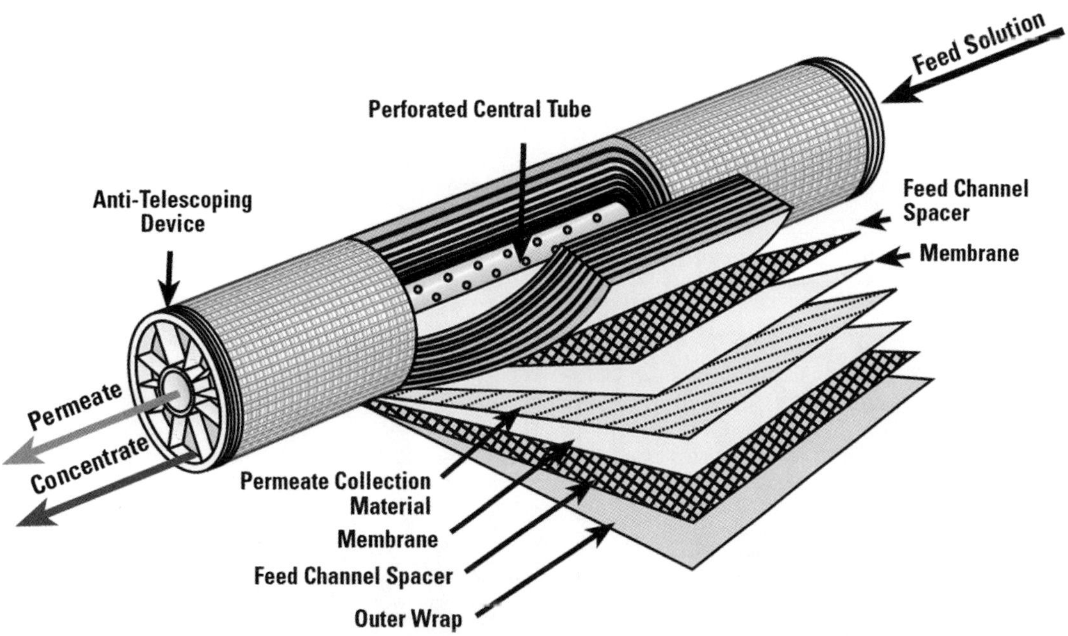

Figure 10.18 Details of a module or package of hollow fiber membrane [*Source:* Wagner, Jorgen (2001)]

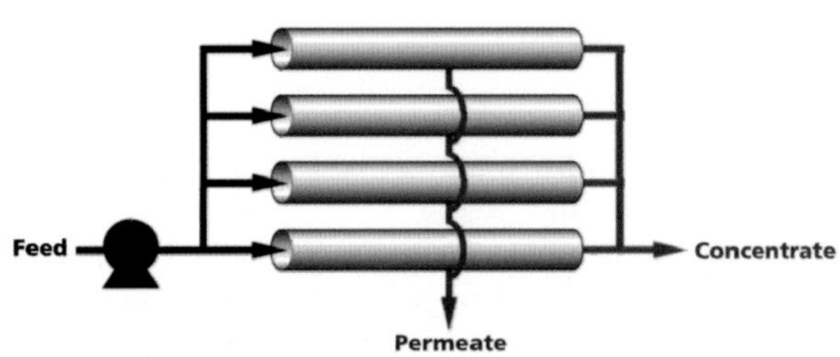

Figure 10.19 The principle of functioning of a package of hollow fiber membranes

Membrane filtration is usually combined with conventional activated sludge to a process denominated Membrane BioReactor (MBR). There are three options for locating the membrane modules in an activated sludge treatment plant: (i) leave the activated sludge process as it is and pass the effluent through a membrane treatment as a tertiary treatment process in a separate membrane tank; (ii) immerse the membrane modules in the aeration tank, and (iii) immerse the membrane modules in a separate membrane tank which replaces the secondary sedimentation tank. All three options work well; however, the third option is in most widespread use. The first option is the most expensive and therefore scarcely used. In the case of using membrane filtration as polishing of the effluent of another appropriate technology processes, the membrane filtration would be done by immersing the membrane modules in a separate membrane tank. A section of a membrane tank with several membrane modules immersed in it is presented in Figure 10.21. It illustrates the form of operation of the membrane tank.

Figure 10.20 A hollow fiber membrane module

Figure 10.22 shows a general view of a membranes plant which consists of several membrane tanks in which the membrane modules are immersed.

Membranes need to be cleaned periodically to maintain their solids separation capacity. The cleaning is being done automatically by the control system through backflushing with chemical solutions, usually Citric Acid and Sodium Hypochlorite. Cleaning cycles vary by application and raw water quality and there are various types of cleaning. Typical cleaning cycles are the following: Backpulse of Permeate every 15 minutes for 30 seconds, maintenance cleaning occurs about once every 40 hours and recovery cleaning about once every 20 days minimum. The membranes also need to be replaced periodically, once every 5 to 9 years, depending on the composition of the raw sewage. For typical municipal wastewater, replacement is required about once every 8 years. Membrane replacement increases O&M

costs, however, membrane systems cost has been steadily declining over the past 10 years, and if this trend continues, replacement will become less of a burden.

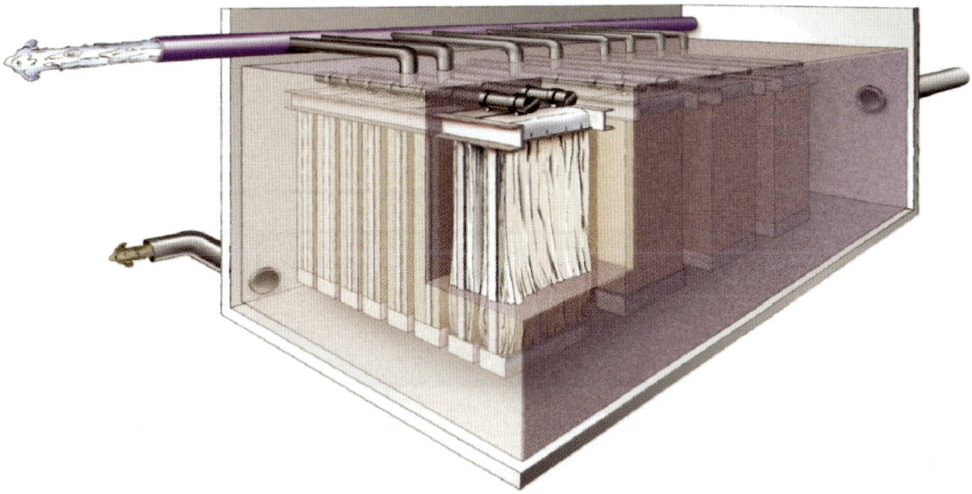

Figure 10.21 A Section of a membrane tank with several membrane modules immersed in it

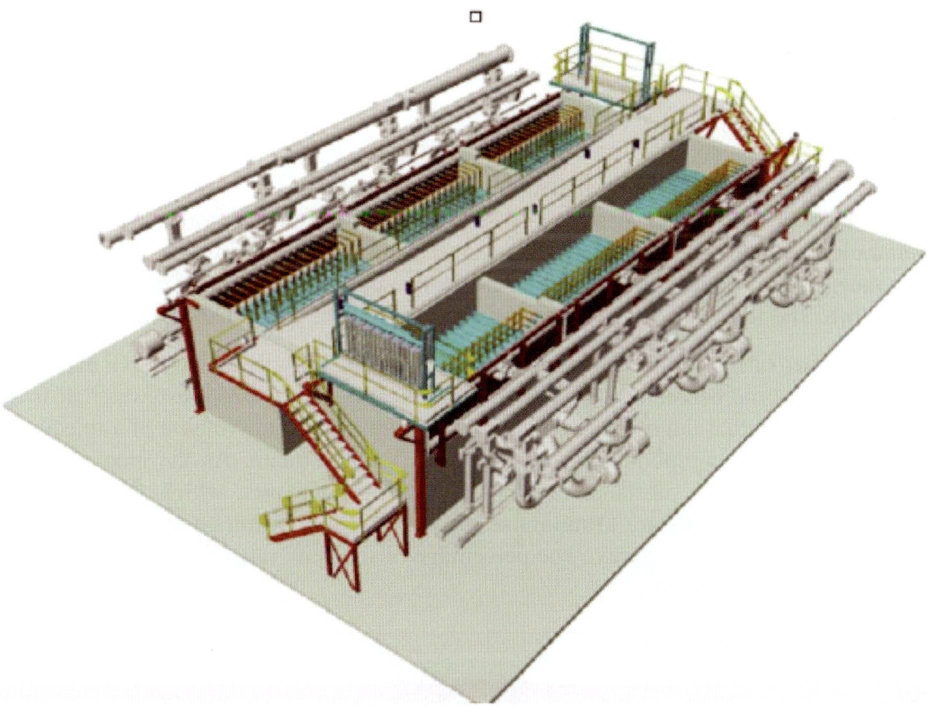

Figure 10.22 General view of a typical membrane plant

10.5.2 Basics of the process

The basic principle[3] of membrane filtration is the transport of matter through a "barrier" called semi-permeable membrane with the help of a driving force. Some substances in the stream of municipal wastewater can pass through a membrane, while others are retained and do not pass. The preparation of mass balances, global and for the substance to be separated, are an essential approach in design and operation of membrane filtration. The basic diagram for a mass balance is presented in Figure 10.23.

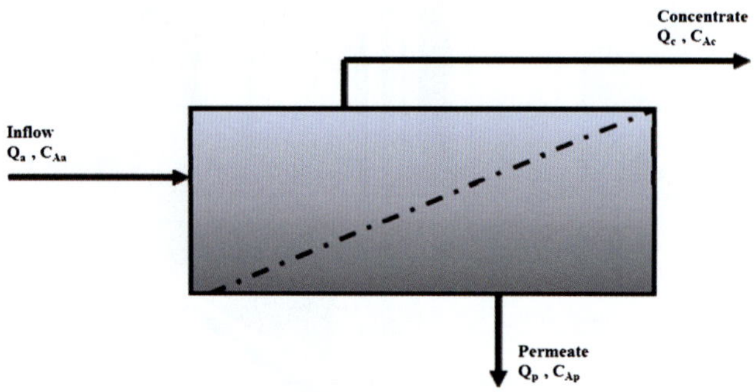

Figure 10.23 Mass balance diagram of a membrane system [*Source:* Llosá (2005)]

Referring to Figure 10.23 we obtain:

$$Q_a = Q_p + Q_c \qquad (10.26)$$

$$Q_a.C_{Aa} = Q_p \cdot C_{Ap} + Q_c \cdot C_{Ac} \qquad (10.27)$$

where:

Q_a = wastewater inflow
C_{Aa} = substance A concentration in the wastewater to be treated
Q_p = Permeate flow (treated wastewater) that is determined by the installed membrane surface
C_{Ap} = concentration of substance A in the permeated, which is determined by the type of membrane
C_{Ac} = is the concentration of substance A established for the rejection stream (the concentrate), hopefully as high as possible. The limit is determined by the solubility of the separated (rejected) substances.

There are four types of filtration membranes, known by the acronyms MF, UF, NF and OR as defined above. The Recovery is the percentage of flow that is processed as treated wastewater. It is desirable to maximize the recovery so that the rejected compounds remain in a small volume of a reject stream. In an ideal system, the separated compounds go into the concentrate stream. As the recovery increases, the concentration of the separated substances in the concentrate increases, which in turn leads to lower permeate purity. The ideal results are to minimize the volume of concentrate and get at the same time a permeate with the desired level of purity. The percentage of the flow rate of the reject Stream (R) can be calculated by the following equation:

$$R = (C_a - C_p)/(C_c - C_a) \cdot 100\% \qquad (10.28)$$

[3]Based on Llosá (2005)

where:

C_a = is the initial concentration of a contaminant in the wastewater to be separated
C_p = is the final concentration of the contaminant in the permeate
C_c = is the concentration of the contaminant in the concentrate (or reject) stream

Usually membrane processes use tangential filtration, in which the flow is tangential to the membrane. Pumping needs to be provided so as to achieve a sufficiently high speed, parallel to the membrane surface (see Figure 10.24). The particles are carried by the shear action of the fluid, so the only resistance to filter the pressurized stream is caused by the membrane itself. Therefore, one of the most important control parameters in processes such as the NF and UF is the tangential velocity.

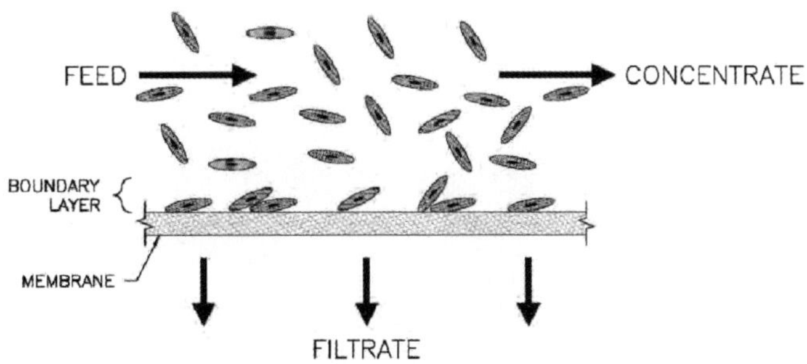

Figure 10.24 Diagram of tangential filtration by membranes Llasó (2005)

The advantages of using membrane processes are: (i) mainly, the extremely high effluent quality which they produce, so they have to be used when the required effluent quality is very high; (ii) membrane units do not require addition of chemicals; (iii) the performance of membranes is not much effected by the variation of temperature so membrane process can be used in any climate zone, and (iv) they are easy to install.

The disadvantages of membranes processes are: (i) high cost, mainly due to the cost of the membranes, although with progress of research their cost decreases with time, (ii) rapid aging, clogging and deterioration of the membranes, although with progress of research they become more resistant; (iii) the majority of membranes are sensitive to oxidants such as chlorine; and (iv) the membrane processes generate a concentrate stream which presents problems for safe disposal.

Membranes should be first tested in a pilot plant before making the final design on the process and on the specific membranes to be selected. This is required to find: (i) the material and membrane type, depending on the required effluent, (ii) the total area of the membrane, the number of membranes to be use and the general arrangement of the modules, and (iii) the required pretreatment.

The cost of a membrane unit decreases with the increase of wastewater flow as seen in Figure 10.25.

Design of membrane systems should be prepared for the maximum flow (Q_{HD}) because membranes do not stand up well to hydraulic shocks. The diurnal variation of municipal wastewater flow is significant. A daily peak factor of 2.5 or more is common, depending on the size of the city and on infiltration. A membrane unit designed for the average flow would have difficulties to cope with such loads variations.

There is no information on the effluent quality of membrane units used to polish effluents of appropriate technology processes, but effluent quality of the MBR process is a good indication of the quality which can be achieved with membrane polishing of effluents of appropriate technology processes. The effluent quality of the MBR process is exceptionally high, superior to any other wastewater treatment process including tertiary treatment. This effluent is suitable for industrial reuse applications or as a high quality feed water source for reverse osmosis treatment. Indicative effluent quality of an MBR system is BOD < 2 mg/l, SS < 1 mg/l, Turbidity <1 NTU, practically complete void of bacteria and with up to 4 log removal of virus (depending on the membrane nominal pore size). In addition, it provides a barrier to certain chlorine resistant pathogens such as Cryptosporidium and Giardia. As to Nitrogen and Phosphorous removal, the membrane itself does not remove these elements and if their elimination is needed, specific removal process can be incorporated into the MBR process.

Figure 10.25 Samples of effluent and raw sewage of the ben gurion MBR treatment plant

The effluent quality of MBR is stable, while that of other treatment processes fluctuates with time in response to fluctuations of sewage flow rate and variations in climatic conditions. The high and stable quality of the MBR effluent enables it to be reused in industrial plants, which require a stable quality. Industrial plants need sometimes desalinated water, and the MBR effluent can be a good feed to the reverse osmosis process, because the ultra filtration prevents fouling of RO membranes.

The MBR effluent can certainly be reused in agriculture and also in landscape irrigation within urban areas. This is very important in water scarce regions. The MBR effluent is less offensive to the environment when discharged to receiving water bodies. It is practically free of pathogens since membranes are effective means of bacteria and virus removal, and therefore presents a lower public health risk.

A photo of a sample of effluent of an MBR treatment plant at the Ben Gurion International Airport in Israel aside a sample of the raw sewage is presented in Figure 10.25. The effluent looks as pure water, and indeed in terms of turbidity it is better than natural surface water and even treated water, as its turbidity is 0.075 NTU according to the reading of the continuous monitoring effluent tubidimeter. In spite of the low turbidity content of the effluent, it of course does not reach the quality of potable water since it contains dissolved organic compounds whose origin is the wastewater, which are not present in natural fresh water. The Ben Gurion MBR plant effluent is reused for irrigating all the green areas in the airport, for feeding cooling towers in the airport area, and for cleaning the runways, when necessary.

The main problem of membrane treatment is the disposal of the concentrate, which in many cases is considered a hazardous waste. The current methods of handling this concentrate consist in: (i) concentration by evaporation, and (ii) converting the concentrated solutions to inert solutions. These techniques, in addition to having low efficiency and to generating different types of waste, are complex and require more than one unit operation to treat a single waste. Recently spray-drying technique appeared as another promising technique for concentrate treatment. This technology has higher energy efficiency and lower emissions than the currently used processes (Reina, 2005).

10.5.3 Basic design

The following procedure is used to size a membrane filtration system, applied to the case of a population of 20,000.

Variables selected by the designer

The following variables are selected by the designer: (i) the flow rate (Q_{HD}) which is calculated using Equation 3.4, (ii) determine the values for BOD_5, TSS and any other parameter of importance in the influent, in the effluent (according to the environmental regulation) and in the concentrate (tolerable limit in concentrate), (iii) estimate the percentage of the flow rate of the reject Stream (R) using Equation 10.28 for BOD_5, TSS, and so on. Select the highest R, (iv) calculate the reject flow (RQ_{HD}) and permeate flow or effluent flow $(1-R)\ Q_{HD}$, and (v) determine in a pilot test the types of membrane required to achieved the effluent concentrations of BOD_5, TSS and other quality parameters under the prevailing regulation; and the filtration rate (q_F) for the type of membrane (MF, UF, NF or RO), with the help of Figure 10.15 and Table 10.8. The designer variables for the 20,000 sample are presented in the Table 10G.

Table 10G

Designer variables					
Variable	**Value**	**Unit**	**Value**	**Unit**	**Source**
Membrane					
BOD_{5i}	48.05	mg/L			Influent
BOD_{5e}	5.00	mg/L			Norm
BOD_{5R}	100.00	mg/L			Reject max limit
TSS_i	10.13	mg/L			Influent
TSS_e	1.00	mg/L			Norm
TSS_R	100.00	mg/L			Reject max limit
Q_{HD}	323.35	m^3/h			Equation 3.4
R(BOD)	0.45		45.32	%	Equation 10.29
R(TSS)	0.09		9.22	%	Equation 10.29
R	0.45		45.32	%	Highest R
Q_{DR}	146.54	m^3/h			RQ_{HD}
Q_{DP}	176.82	m^3/h			$(1-R)Q_{HD}$

(Continued)

Table 10G Samples of effluent and raw sewage of the ben gurion MBR treatment plant (*Continued*).

Designer variables					
Variable	**Value**	**Unit**	**Value**	**Unit**	**Source**
Define Membrane	MF				Figure 10.16
q_{MF}	1000.00	$L/m^2 \cdot d$			Table 10.7
Define Membrane	UF				Figure 10.16
q_{UF}	0.00				Table 10.7
Define Membrane	NF	$L/m^2 \cdot d$			Figure 10.16
q_{NF}	500.00				Table 10.7
Define Membrane	RO				Figure 10.16
q_{OR}	0.00				Table 10.7

Note that the BOD concentration in the inflow is 48 mg/l and the TSS is 10 mg/l. These values are quite low, because the inflow to the membrane tank is an effluent of preceding appropriate technology unit processes. The required effluent quality is 5 mg/l BOD and 1 mg/l TSS, which are values that a membrane process can achieve.

Basic design

Having defined the design variables, the dimensioning of the membranes is as follows:

- Calculate the filtration area for each type of filtration required using:

$$A_F = q_F \cdot Q_{DP} \tag{10.29}$$

For the example chose MF and NF. The filtration areas and the working pressure for MF and NF are shown in Table 10H.

Table 10H

Design			
Parameter	**Value**	**Unit**	**Observation**
Membrane			
A_{MF}	176815.8	m^2	Equation 10.30
A_{UF}	0.0	m^2	Equation 10.30
A_{NF}	88407.9	m^2	Equation 10.30
A_{OR}	0.0	m^2	Equation 10.30
Pressure MF	50.0	kPa	Table 10.7
Pressure UF	0.0	kPa	Table 10.7
Pressure NF	750.0	kPa	Table 10.7
Pressure OR	0.0	kPa	Table 10.7

Chapter 11

Combinations of unit processes of appropriate technology

11.1 INTRODUCTION

A series of unit processes of appropriate technology for wastewater treatment was presented in the previous chapters, each yielding a different effluent quality. The presented series of processes might not be the complete set of existing appropriate technology unit processes and perhaps there are others which can be considered as part of this group. Sometimes an effluent quality higher than what a single unit process of appropriate technology can produce is required. In such cases a treatment plant consisting of a series of appropriate technology unit process can be used (2, 3 or more), in which the effluent of the first unit process is fed into the second unit process, the effluent of the second is fed to the third and so on. Another way to look at it is that to improve the quality of the effluent of a certain unit process, this effluent can be subjected to one or more polishing steps, each of which consisting of an additional unit process. This approach can produce practically any final effluent quality required. As presented in Chapter 1, *the main message of this book is the idea that it is possible to combine unit processes to create various treatment plants, each based on a series of appropriate technology processes, and that it is possible to combine unit process in such a manner that jointly they can generate any required effluent quality*. A plant based on a combination in series of appropriate technology unit processes is still easy to operate and is usually of lower costs than conventional processes in terms of investments and certainly in operation and maintenance. However, the use of a complex process as a polishing step of an effluent of an appropriate technology process should be avoided, because it transforms the entire treatment plant to a complex and costly treatment plant. There are cases of plants consisting of a UASB unit followed by an activated sludge unit or a trickling filter unit. Such plants are not appropriate technology plants.

A large number of combined processes are presented and discussed in Chapter 1, part of them already in use in full scale plants and others may be used in the future. It is impossible to provide design procedures for all the combined processes mentioned in the book. In this chapter we present the design procedures of five combined processes. We consider that a sample of five combined processes which consist of a total of nine appropriate technology unit processes constitute a sufficiently large sample. This sample can serve for the readers as the basis for developing the design of other combined processes in which they might be interested. The processes dealt with in this chapter are defined in Table 11.1. The unit processes that compose each of the five combined processes have been described in previous chapters, so for any technical considerations on fundamentals, design equations and design procedures the reader is referred to the corresponding chapters.

Table 11.1 The five selected combined appropriate technology processes for wastewater treatment whose design procedures are discussed in this chapter.

Combined process number	Combined process structure	Combined process performance (% removal of BODT, SST and fecal coliforms)
1	UASB followed by Facultative Lagoons	BODT (80–90%) SST (70–80%) FC in effluent (30–10^5) MPN/100 ml, depending on the design of the lagoons. If necessary, effluent can be disinfected
2	UASB followed by Anaerobic Filter	BODT (80–90%) SST (80–90%) FC removal very low, if necessary, effluent can be disinfected
3	UASB followed by Sand Filter followed by UV Disinfection	BODT (80–90%) SST (80–90%) FC Practically 100%
4	Chemically Enhance Primary Treatment (CEPT) followed by Sand Filtration followed by UV Disinfection	BODT (80–90%) SST (80–90%) FC Practically 100%
5	UASB followed by Anaerobic Filter followed by Dissolved Air Flotation followed by Membrane Filtration	BODT > 98% SST > 99% FC Practically 100%

Each process in the table includes a preliminary treatment unit which is not shown in the description of the processes and which precedes the treatment units mentioned.

In all the combined processes, Rotating Micros Screens (RMS) is proposed as the first unit of the preliminary treatment section of any treatment plant. If grit removal is required, a Vortex Grit Chamber is recommended as the preferred unit process of the preliminary treatment, complementing the RMS. If the raw wastewater contains large amounts of oil and grease which need to be removed, then an aerated grit chamber should be used as part of the preliminary treatment installation, complementing the RMS. In the schematic flow diagrams of the treatment processes which are presented in the book, the scheme of a Rotating Micro Screen is used to represent the entire preliminary treatment unit. Also, when discussing combined treatment processes, the mention of preliminary treatment is sometimes omitted, but it is clarified that preliminary treatment forms part of every treatment process.

The suspended solids content in the effluents of all the combined processes mentioned in the table is sufficiently low to allow for performing effective disinfection of these effluents by any existing disinfection method. Since the final unit process in processes 1 in the table above is a lagoons unit it can be designed to achieve an effluent of low pathogens content even without disinfection. The effluent produced by process 5 is void of bacteria as a result of the nature of the process, so there is no need to disinfect this effluent. The effluents of processes 2, 3 and 4 can be effectively disinfected. The fact that all these processes can be managed to yield effluents free of pathogens is important from the standpoint of public health, especially for developing countries.

Paragraph 1.19 of Chapter 1 present Table 1.11 with all the programs used in the book, including the Combinations, and a short explanation of how to use them. Note that that combination 2 is calculated with program CHAP 6-ANAEROBIC-UASB-AF.xls, and each of the other combinations has its own program working analogous to the process programs. Also note that in combination programs the effluent of a preceding unit is the influent to the following unit. For all other details the programs work exactly as presented in each process program, of course with the specific influent for each case.

11.2 COMBINATION 1: ROTATING MICRO SCREENS FOLLOWED BY UASB FOLLOWED BY FACULTATIVE LAGOONS

11.2.1 Introduction

In this combined processes the effluent of a UASB plant is polished by facultative lagoons. This is a very logical scheme and an exemplary appropriate technology process. Preliminary treatment is achieved by rotating screens and, if needed, a vortex grit chamber, which is the best available preliminary treatment scheme. The screen opening should not be less than of 6 mm in order to avoid removing organic matter which is necessary for the proper functioning of the UASB reactor. The removal of the main portion of organic matter is achieved in this combination by the USAB process, which is one of the main unit processes of appropriate technology and is discussed in detail in Chapter 6. It is a simple, effective low cost process. The facultative or maturation lagoons system used for polishing the UASB effluent is a simple reliable system discussed in Chapter 5. Removal of pathogenic organisms can be achieved in this process naturally (without disinfection), if sufficient detention time is provided in the lagoons system. The emphasis of the design of the lagoon must be on the removal of pathogens, since the bulk of organic matter is removed in the UASB Reactor. In several plants visited the lagoon are designed with a very short detention time, practically operating as clarifier where settled sludge is decomposed in the bottom. Such an approach should be avoided and the lagoons need to be designed for the removal of pathogens. The design basis for disinfection in lagoons is discussed by Arthur (1983).

This combined process generates solid waste material in the rotating screen and excess sludge in the UASB reactor. This process includes therefore the disposal of the screened material, usually in a sanitary landfill, and the handling of the UASB excess sludge, mainly by drying beds, without mechanical equipment. The main disadvantage of this combined process is that the lagoons occupy large extensions of land. If area is scarce, deeper lagoons with mixers can be used. An ultrasound device can be used to decrease algae content in the final effluent. Because of the use of lagoons as part of this process, it is suitable for small and medium size cities, but not for large cities (Libhaber, 2007). The biogas generated in the UASB reactor can be collected and put to use, if economically feasible. As a whole this is a good and recommended process. A schematic flow diagram of this process is presented in Figure 11.1.

This process is being applied in several plants in Brazil and in Colombia, using conventional preliminary treatment, not RMS. The largest plant of this type in Colombia is the Rio Frio plant in Bucaramanga. A photo of the plant is presented in Figure 11.2. This plant, commissioned in 1991, is one of the oldest in the world large scale UASB-based facilities treating municipal wastewater. The plant's original design flow was 740 l/s serving a population of 240.000. The UASB reactors are located at the bottom of Figure 1.2 and next to them are the sludge drying beds (not seen in the figure). The UASB reactors are equipped with aluminium covers and the collected biogas is flared. Pretreatment for the UASB reactors is provided by 6 mm screens followed by grit chambers. The reactors are followed by two facultative lagoons. Removal efficiency of BOD is about 80% in the USAB plant and 33% in the facultative lagoons, achieving a total BOD removal in the range of 85–90%.

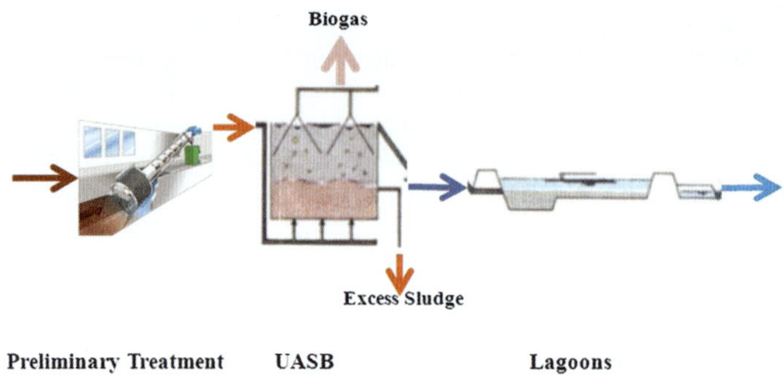

Figure 11.1 A schematic flow diagram of the combined process of preliminary treatment followed by UASB followed by facultative lagoons

Figure 11.2 A photo of the Rio Frio Plant, Bucaramanga, Colombia

The plant has been able to achieve essentially secondary treatment levels compatible with that of conventional processes at a very low capital and O&M costs, thereby demonstrating the potential of application of this process in developing countries. The Rio Frio plant has two problems: (i) the detention time in the facultative lagoons is way too low (about 2 days) and this limits their performance; and (ii) the potable water in Bucaramanga has elevated sulphate levels, and most of the sulphates present in the raw wastewater are effectively reduced to H_2S in the anaerobic reactors, generating odour around the plant and promoting corrosion of plant equipment. Addition of aluminium covers and the collection and flaring of the biogas has helped to reduce the intensity of the odours and the corrosion of the equipment.

With time the population served by the Rio Frio plant increased and its capacity needed to be augmented. The plant is now being upgraded and its flow diagram is being modified. The UASB reactors section is being expanded but the facultative lagoons have been eliminated since there was no area available to increase them. The lagoons are being replaced by an activate sludge unit. As previously mentioned, this transforms the entire plant to a conventional treatment plant, and it ceases to be an appropriate technology treatment plant. However, the UASB unit performed well during a period of 20 years and has not been eliminated, so it will continue to function in the foreseeable future, and the configuration of UASB followed by lagoons proved to be a successful configuration which performed well and provided during 20 years a high quality effluent, comparable to that of conventional treatment, at a fraction of the cost.

Several plants of UASB followed by lagoons are in operation in Brazil. SANEPAR, the water utility of the state of Parana in Brazil operates several plants of this type. Figure 1.62 shows a photo of a UASB plant followed by a facultative lagoon in Ronda, a small town, part of the city of Ponta Grossa in the state of Parana. Another plant in Parana based on this process is the Padilha Sul Plant located in the city of Curitiba, the capital of Parana. A photo of this plant is presented in Figure 1.63. The first stage of this plant was designed for an average wastewater flow of 440 l/s and is serving an equivalent population of 318,000. It will be expanded to handle an average flow of 600 l/s. The UASB reactors occupy the small area at the left hand side of the bottom of the figure. As can be seen, the UASB reactors unit occupies only a small portion of the total plant area. Also seen in the photo is that the UASB reactors are located at a close vicinity to residential areas. That means that the UASB units do not present an environmental nuisance to the nearby residential areas.

11.2.2 Performance

The advantages of combined process of UASB followed by lagoons are: (i) low capital and operation cost, (ii) easy maintenance, it does not use mechanical equipment, (iii) small quantities of sludge, and (iv) good effluent quality. The main disadvantage of this process is that the lagoons occupy a large land area. The approach of deeper lagoons equipped with mixer can reduce the area of the lagoons).

The plant of Rio Frio, Colombia, obtained BOD removal of about 80% in the UASB reactor and 33% in the maturation ponds, for a total removal of BOD of 85–90%. In general, this process achieves 80–90% BOD removal and 70–80% SST removal. The removal of FC depends on the design of the lagoons. If a short detention time is provided, the FC removal is low. However, properly designed lagoons can achieve a high FC removal.

Construction costs for the existing facilities of the Rio Frio treatment plant in Bucaramanga were approximately 15 US$/Capita. Operating costs are 0.79 US$/Yr/Capita or 0.015 US$/m^3 treated (Osorio, 1996). However, this plant was constructed in the early 1990s. It is estimated that today The investment cost in a UASB reactor followed by a lagoons system is in the range 30–50 US$/Capita, depending on the design of the lagoons, and the O&M cost is in the range 1.0–1.5 US$/Yr/Capita.

11.2.3 Design

(The model program for this Example is available online at http://www.iwawaterwiki.org/xwiki/bin/view/Articles/Software+Developed+for+Sustainable+Treatment+and+Reuse+of+Municipal+Wastewater)

The design procedures and a detailed example of the design of a UASB reactor and a lagoons system which make up this combination are presented in Chapter 6 and 5 respectively. The reader is referred to the design examples in these chapters. The results of application in series of a UASB reactor and a lagoons system for the example of 20,000 people, with an influent Fecal Coliform of 5×10^6 MPN/100 ml are presented below. The Excel program COMB 1-USAB-MATURATION LAGOON-01.xls is the program used to calculate example of this combined process.

Variables selected by the designer

The quality parameters of the raw wastewater are the same of all other examples and presented in Table 11A. The specifics of each process design are explained in the corresponding chapter (in this case, Chapters 5 and 6). Please note that the following Tables identified by alphabetic order contain computer calculated results.

Table 11A

WW quality				
Variable	**Value**	**Unit**	**Value**	**Unit**
BOD_5	277.8	mg/L	0.28	kg/m^3
COD	596.1	mg/L	0.60	kg/m^3
COD/BOD_5	2.1			
TKN	40.0	mg/L	0.04	kg/m^3
N-Nitrate	2.0	mg/L	0.00	kg/m^3
Total Phosphorus	5.8	mg/L	0.01	kg/m^3
pH	7.1	UN		
Alkalinity	100.0	mg/L	0.10	kg/m^3
TSS	202.6	mg/L	0.20	kg/m^3
VSS	173.6	mg/L	0.17	kg/m^3
O&G	100.0	mg/L	0.10	kg/m^3
Fecal Coli	5000000.0	MPN/100 mL	50000000000.00	MPN/m^3

UASB

The designer variables selected for the UASB are presented in Table 11B. For design details see Chapter 6, Section 6.3, Paragraph 4. See also the important note in the Box in Section 11.1.

Table 11B

Designer variables					
Variable	**Value**	**Unit**	**Value**	**Unit**	**Observation**
UASB standard					
Settling basin width, w_s	4.0	m			Figure 6.10: UASB-SM
GSLS heigth, H_G	2.5	m			Figure 6.10: UASB-SM
Baffle width, w_b	0.15	m			Figure 6.10: Baffle width
Reactor velocity, v_r	0.75	m/h			<1.00 m/h; $v_r < v_s$; Table 6.3
Gas velocity, v_g	1.0	$m^3/h \cdot m^2$			>1.00 m/h; Tabla 6.3
Yield coefficient, Y	0.08				Table 3.2: 0.08
Fraction of CH_4, η	0.65				Typical
Maximum Efficiency, E_{max}	90.00	%			80 to 90%
Volumetric load, L_v at 15°C	2.0	$kg/m^3 \cdot d$			2.0–4.0 $kg/m^3 \cdot d$ at 15°C, applied to liquid volume
Efficiency, E	89.31	% DBO_5 or sCOD			Equation 6.3
Actual volumetric load, L_v (T)	5.44	$kg/m^3 \cdot d$			Equation 6.4

(Continued)

Table 11B (*Continued*).

Designer variables					
Variable	**Value**	**Unit**	**Value**	**Unit**	**Observation**
Water temperature	25.00	°C			Coldest month of year
$sCOD_e$ or BOD_5, S	29.69	mg/L	0.03	kg/m^3	Equation 6.5
Methane production, V_{CH_4}	124.34	m^3/d			Equation 6.6
Biogas production, V_{biogas}	191.29	m^3/d	7.97	m^3/h	Ecuación 6.7
Inlet structures					
Branch manifold flow	0.020	m^3/s			$Q_D/2$
Manifold velocity, v	0.03	m/s			Goal seek: see note in cell N38
orifices per m^2, ε_O	2.00				1 a 2 per m^2
# orfices, preliminary, n_O	114.00				Equation 6.27
# distribuitors, n_d	11.00				Equation 6.27a
# orifices per distributor, n_{od}	5.00				Equation 6.28
Outlet structures					
Module flow, $Q_M = Q_D/n_M$	0.01	m^3/s			
Channel widthl b	0.30	m			Select
Draying bed					
Applying rate, q_x	150	kg/m^2 · yr			Between 120–150

Maturation lagoon

The designer variables selected for the maturation lagoons are presented in Table 11C. For design details see Chapter 5, Section 5.3, Paragraph 6.

Table 11C

Designer variables					
Variable	**Value**	**Unit**	**Value**	**Unit**	**Observation**
Maturation lagoon					
Depth, H_M	1.5	m			2–5 m
T_{min} air	18.0	°C			Coldest month
T_i water	25.0				
# Lagoons M, n_M	2				Parallel
L/w (ratio length to width)	2				Between 1 and 3
Pipe velocity, v	1	m/s			0.6–2.0 m/s
$L_S = 350\ (1.107 - 0.002T)^{(T-25)}$ or 400	216.5	kg BOD_5/ha · d			Equation 5.14 or 400

(*Continued*)

Table 11C (*Continued*).

Designer variables					
Variable	**Value**	**Unit**	**Value**	**Unit**	**Observation**
K_L (20°C)	0.35	d^{-1}			Tipical
θ (BOD_5)	1.06				1.04–1.09
Influent BOD_{5i}, $S_{OM} = S_F$	29.7	mg/L			UASB Effluent
K_B (20°C)	2.7	d^{-1}			Tipical FC
θ (CF)	1.19				Tipical FC
Influent FC: B_{OM}	5000000	MPN/100 mL			UASB does not remove
Influent helminthes	10	Ova/L			Between 300–800
Target FC	20000	MPN/100 mL			Between 1000 and 20000
Target helminthes eggs	1	Ova/L			<1 Ova/L

Design

UASB

The results of the design calculations of the UASB unit are presented in Table 11D. See Chapter 1, Section 1.19 for explanation of how the programs work, and Chapter 5 for design details. Note that the UASB main objective is to remove BOD (to about 30 mg/L BOD_5) and the maturation lagoons objectives are to remove Fecal Coliforms and helminths so as to be close to the targets of FC < 20,000 MPN/100 mL and helminths of no more than 1 Ova/L.

Table 11D

Design					
Parameter	**Value**	**Unit**	**Value**	**Unit**	**Observation**
UASB standard					
Reactor area, Ar	192.4	m^2			Equation 6.8
Net height, H_N	0.9	m			Equation 6.9
Effective liquid height, H_L	2.5	m			Equation 6.10 (minimo $2 \cdot 5$ m)
UASB heigth, H_T	5.0	m			Equation 6.11
Detention time, t_d	6.7	h			Equation 6.12
UASB total volume, V_{UASB}	962.2	m^3			Equation 6.13
Sludge blanket volume, V_L	481.1	m^3			Equation 6.14
Total gas extraction area, A_g	8.0	m^2			Equation 6.15
Design UASB length, L_{UASB}	17.0	m			Equation 6.16
Net UASB width, W_{UASB}	11.3	m			Equation 6.17
Total gas width, $W_g = n_{MP} \cdot w_g$	0.47	m			Equation 6.18

(*Continued*)

Table 11D (*Continued*).

	Design				
Parameter	**Value**	**Unit**	**Value**	**Unit**	**Observation**
# preliminar modules, n_{MP}	2.52				Equation 6.19
# modules, n_M	3.00				Integer rounded up
Gas throat width, w_g	0.16	m			Equation 6.20
Design UASB total width, W_{UASBD}	13.4	m			Equation 6.21
Settling basisn total area, A_s	204.0	m^2			Equation 6.22
Design SOR or $v_{sD} = Q_D/A_s$	0.71	m/h			$<$1,00 and$>v_{rD}$. If otherwise change change v_r or/and v_g
Design UASB area, $A_{rD} = W_{UASBD} \times L_{UASB}$	227.27	m^2			For n_M modules: Equation 6.22a
Design v_r, $v_{rD} = Qb/A_{rD}$	0.64	m/h			Equation 6.23
Design UASB volume, $V_{UASBD} = A_{rD} \cdot H_T$	1136.35	m^3			6.24a
Design detention time, $t_{dD} = V_{UASB}/Q_D$	7.87	h			
Sludge production					
$Q_x = V_{UASBD}\ Y(S_0 - S)/t_{dD}$	68.76	kg/d			Equation 6.25
Biogas production					
Volume per day, V_g	191.3	m^3/d			Equations 6.6 and 6.7
Inlet structure					
Manifold area a_m	0.696	m^2			Equation 6.26, con $Q_D/2$
Manifold diameter, $\phi = (4\ a_m/\pi)\wedge(1/2)$	0.942	m	37.0	pulg	
Distribuitor area, $a_d = 0.4\ a_m/n_D$	0.025	m^2			From Equations 6.29 and 6.27a
Distributor diameter, $= (4\ a_d/\pi)\wedge(1/2)$	0.180	m	7.0	pulg	
Orifice area, $a_o = 0.4\ a_d/n_o$	0.002	m_2			From Equations 6.30 and 6.28
Orifice diameter, $\phi_0 = (4\ a_o/\pi)\wedge(1/2)$	0.051	m	2.00	pulg	Goal seek: set cell L38 to 2 by changing C14
Outlet structures					
Channel water depth, h	0.102	m			Equation 6.31
Total channel depth, h_c	0.202	m			$h + 0.10$
Drying bed					
$A_{bed} = Q_x/q_x$	167.3	m^2			Equation 6.33

Maturation lagoons

The results of the design calculations of the lagoons system are presented in Table 11E. For details see Chapter 5. Note that the effluent contains 20,000 MPN/100 mL of Fecal Coliforms and 0.2 Ova/L of helminths, meeting the proposed targets.

Table 11E

Parameter	Value	Unit	Value	Unit	Observation
Maturation Lagoon					
L_S	216.5	kg BOD_5/ha · d			Selected in input
t_d	38.6	d			Equation 5.11; $t_{dmin} = 3$ d
Volume, $V'_F = Q \cdot t_d$	133868.6	m^3			Equation 5.21
Final Volume, V_F	133868.6	m^3			Greater of t_d and $t_{dmin} = 3$ d
Area, $A_s = V_F/h$	89245.7	m^2			Equation 5.20
Lagoon area, $a_s = (A_s/n_F)$	44622.9	m^2			
Lagoon width, $w = [a_s/(L/w)]^{0,5}$	149.4	m			
Lagoon length, $L = (L/w)\ w$	298.7				
$T_L = (A_s f T_a + QT_i)/(A_s f + Q)$	18.5	°C			Equation 5.8
$K_L = K_L(20°C)\ \theta^{(T\text{-}20)}$	0.32	d^{-1}			Equation 5.7
Effluent BOD_5 $S_M = S_O/(1 + K_L \cdot t_d)$	2.22	mg/L			Equation 5.4, parallel
Effluent BOD_5 $S_M = S_O/(1 + K_L \cdot t_d/n)^n$	0.57	mg/L			Equation 5.5, series
$v_F = 0{,}00091 V_F (S_O - S)/(1 + 0.05 t_d)$	1141.39	kg/yr	1141.39	m^3/yr	Equation 5.24
Filling years $= 0.5\ V_F/v_F$	58.6	years			1/2 volume
Maturation lagoon: fecal coliform and helminthes removal and efficiencies					
$K_B = K_B(20°C)\ \theta^{(T\text{-}20)}$	6.4	d^{-1}			Equation 5.26
Effluent FC: $B = B_O/(1 + K_B \cdot t_d)$	20000.0	MPN/100 mL			Equation 5.25; Goal seek Target effluent FC
Effluent Helminthes	0.2	Ova/L			
Efficiency FC $= 100 \cdot (B_O - B)/B_O$	99.6	%			From Equation 5.25
Efficiency Helminthes $= 100\ [1 - 0.41 \exp(-0.49\ t_d + 0.0085 t_d^2)]$	98.2	%			Equation 5.15
Efficiency BOD_5 removal maturation	92.5	%			
Total efficiency BOD_5	99.2	%			Combination system
Total efficiency fecal coliform	99.6	%			Combination system
Total efficiency helminthes	98.2	%			Combination system
Inlet and outlet structures					
Total diameter, $\phi = (4 \cdot PI \cdot v \cdot Q_D/1000)^{0,5}$	0.226	m	9.00	In	With Q_D
Lagoon diameter, $\phi_L = \phi = [4 \cdot PI \cdot v \cdot (Q_D/4)/1000]^{0.5}$	0.160	m	6.00	in	With $Q_D/4$

A drawing of a typical plant based on the combined process of UASB followed by lagoons is presented in Figure 11.3.

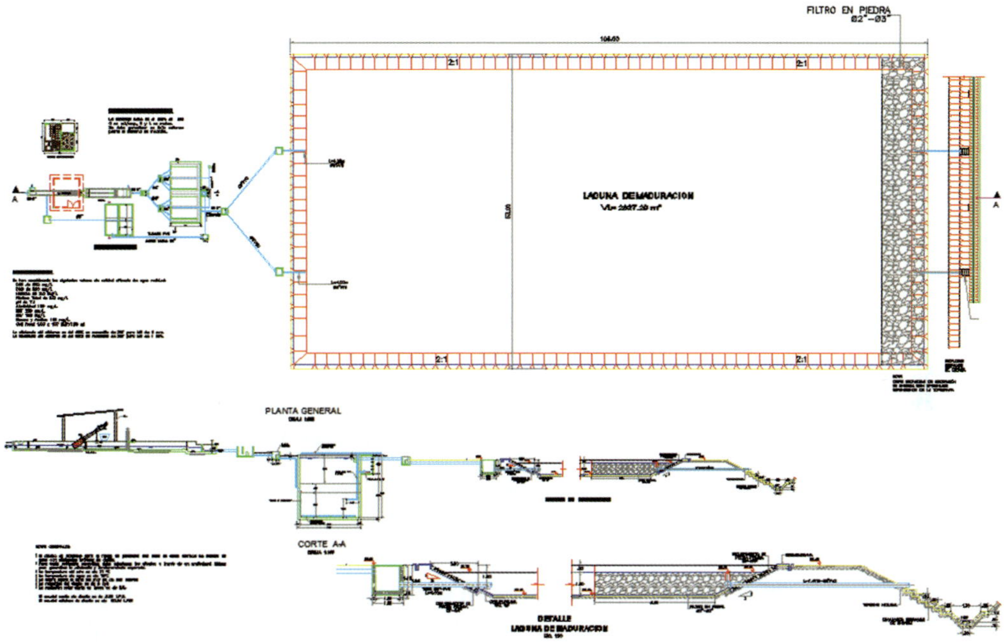

Figure 11.3 Plan and typical cross section of the combined process of preliminary treatment followed by UASB followed by facultative lagoons

11.3 COMBINATION 2: ROTATING MICRO SCREENS FOLLOWED BY UASB FOLLOWED BY ANAEROBIC FILTER

11.3.1 Introduction

In addition to the use of anaerobic filters as the main treatment unit, they can be used as polishing units for improvement of the quality of the effluent of a preceding unit. They are in fact more suitable to perform as polishing units since when fed with treated effluent of a preceding unit they are less bound to be clogged by inorganic suspended solids.

A combined process composed of Rotating Micro Screens as preliminary treatment, followed by UASB as the unit for removal of the main portion of organic matter, then followed by an anaerobic filter for polishing of the UASB effluent, and finally, if necessary, followed by a disinfection unit (chlorination or UV disinfection) for removal of pathogens. is a reasonable schemes and an effective appropriate technology process. The screen opening in the preliminary treatment unit should not be less than of 6 mm in order to avoid removing organic matter which is necessary for the proper functioning of the UASB reactor. This combined process is similar to the process described in the previous section and is basically the replacement of the lagoons by an anaerobic filter followed by a disinfection unit. The advantages of this process are that it yields a high quality effluent and is compact, i.e. does not occupy a large land area. This process can be applied in cases where area for locating a treatment plant is scarce and insufficient for installation of lagoons. The process includes also the disposal of the screened material of the primary treatment and the handling of the UASB and anaerobic filter excess sludge, mainly by drying beds, without mechanical equipment. The process is adequate for application in small, medium and large cities. The biogas generated in the UASB reactor and in the anaerobic filter can be collected and put to use, if economically feasible. A schematic flow diagram of the process of UAB followed by anaerobic filter is presented in Figure 11.4.

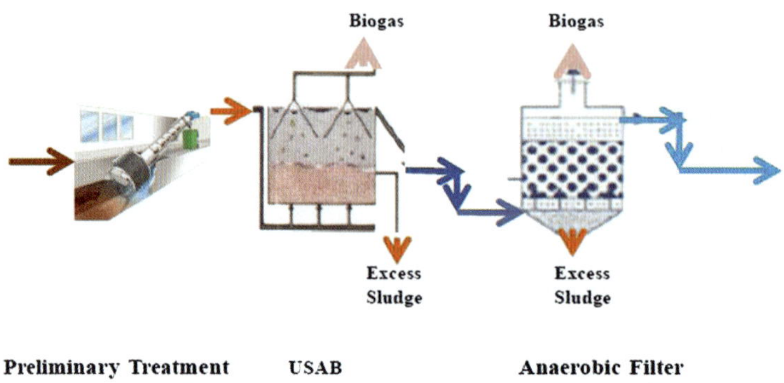

Figure 11.4 A schematic flow diagram of the combined process of preliminary treatment followed by UASB followed by an anaerobic filter

This combination can achieve effluent quality levels of secondary treatment with low capital investment and low operating and maintenance costs, making use of limited extension areas. Table 11.2 presents design considerations of an Anaerobic Filter as a polishing unit. Note that the total COD reduction in the Anaerobic Filter used for polishing is around 20%.

Table 11.2 Design parameters of anaerobic filter used as a polishing unit.

Parameter	Value
t_d minimum	2 h
T_d average	4 h
False bottom depth	0.40–0.60 m
Anaerobic bacteria support bed (gravel) depth	0.40–0.60 m
Freeboard	0.40–0.60 m
Total height	1.50–1.80 m

This process has been applied in several pilot plants and full scale plants in Brazil, using conventional preliminary treatment or an Imhoff tank as the preliminary treatment stage. In pilot plant experiments reported by Chernicharo (2000) a final effluent containing an average of 22 mg/l total BOD (soluble and suspended) and 15 mg/l of TSS was achieved. Such an effluent quality is similar to that of an activated sludge process effluent. Total BOD and total COD removal efficiencies achieved were 90%, soluble COD removal achieved was 81% and TSS removal achieved was 95%. Additional investigations reported by Chernicharo (2007) demonstrated that plants based on UASB followed by anaerobic filters are capable to maintain in the final effluents concentrations of COD, BOD and SST lower than 120 mg/l, 60 mg/l and 30 mg/l respectively.

Andrade Neto (2006) reports that several plants based on the UASB followed by an anaerobic filter process are under operation since 1996 in the state of Parana, Brazil, serving populations in the range of 1,500 to 50,000 persons. Effluent quality achieved in these plants is total BOD lower than 60 mg/l and TSS lower than 20 mg/l. The same type of plants is under operation since 1997 in the state of Minas Gerais, Brazil, serving populations in the range 2,000 to 15,000 persons. One plant in Minas Gerais, at Ipatinga, serves a population of about 200,000.

A photo of a plant based on this combined process in the small town of Tibagi, in the State of Parana, Brazil, serving a population of about 20,000 is presented in Figure 1.65. This plant, operated by SANEPAR, the water and sanitation utility of Parana, achieves an effluent quality of BOD 5–25 mg/l, COD 40–100 mg/l and TSS 4–10 mg/l. A photo of the effluent of this plant is presented in Figure 1.67.

COPASA, the water and sanitation utility of the State of Minas Gerais in Brazil operates several plants based on the process of UASB followed by anaerobic filters. The treatment plants of Arenal, Bela Vista, Horto, and Ipanema located in the municipality of Ipatinga are some of these plants. The performance of the Ipanema plant was evaluated during the period July 2005–September 2006 (Greco, 2007). The wastewater flow to this plant was 180 l/s and it serves a population of about 200,000. The plant achieved an average BOD removal efficiency of 84%, an average COD removal efficiency of 87% and an average SST removal efficiency of 80%. A photo of the wastewater treatment plant in Ipatinga, Minas Gerais, which is a UASB followed by anaerobic filters plant serving a population of about 200,000 is presented in Figure 1.68. The UASB units and the Anaerobic Filters units in this plant are rectangular and that enables to achieve a compact configuration of the plants. As seen in the figure, the plant also includes drying bed of the excess sludge, located in the upper part of the photo. The plant contains 10 UASB units located in two rows (at the bottom of the photo) and two anaerobic reactors located in a raw between the UASB reactors and the drying beds.

11.3.2 Performance

This combination has been applied in pilot plants and large-scale plants in Brazil, using conventional preliminary treatment as the pre-treatment stage of the process (not Rotating screens). The operation results of these plants show that in this combined process the total BOD removal is in the range 80–90% and total SS removal is also 80–90% and sometimes higher, up to 95%. The FC removal very low, if necessary, the effluent can be disinfected.

The advantages of this process are: (i) it is compact (does not occupy large land area), (ii) it generates a high effluent quality, comparable to that of an activates sludge process, (iii) the management of excess sludge generated by the UASB and anaerobic filter can be done by the use of drying beds, without the need of mechanical equipment, and (iv) the process is suitable for use in small medium and large towns. A disadvantages assigned to this process is that it produces an anaerobic (oxygen depleted) effluent. This disadvantage is disputable.

This process is of special importance since it is able to achieve essentially a secondary treatment level effluent at low capital and O&M costs, occupying a small area similar and even smaller than that of activated sludge, thereby demonstrating the potential of applying this process in developing countries as a favourable alternative to activated sludge.

Andrade Neto (2006) reports that the investments costs of the UASB followed by anaerobic filter systems in Parana were in the range of 5 to 30 US$/Capita. It is estimates that currently the investment cost in this combined process is in the range 20–40 US$/Capita and the O&M cost is in the range 1.0–1.5 US$/Yr/Capita.

11.3.3 Design

The development of a detailed example for UASB and Anaerobic Filter processes which jointly make up this combined process is presented in Chapter 6, to which the reader is referred (see Excel program CHAP 6-Anaerobic-UASB-AF.xls). The results of application in series of a UASB reactor and Anaerobic Filter system for the example of 20,000 people are shown in Chapter 6 and their presentation is not repeated here. See Chapter 6 for details. It is noted that in chapter 6 the Anaerobic Filter process was not considered as the main treatment process but rather as the polishing process of a UASB effluent, which is in fact combined process 2.

A Drawing of a typical plant based on the combined process of UASB followed by Anaerobic Filter is presented in Figure 11.5.

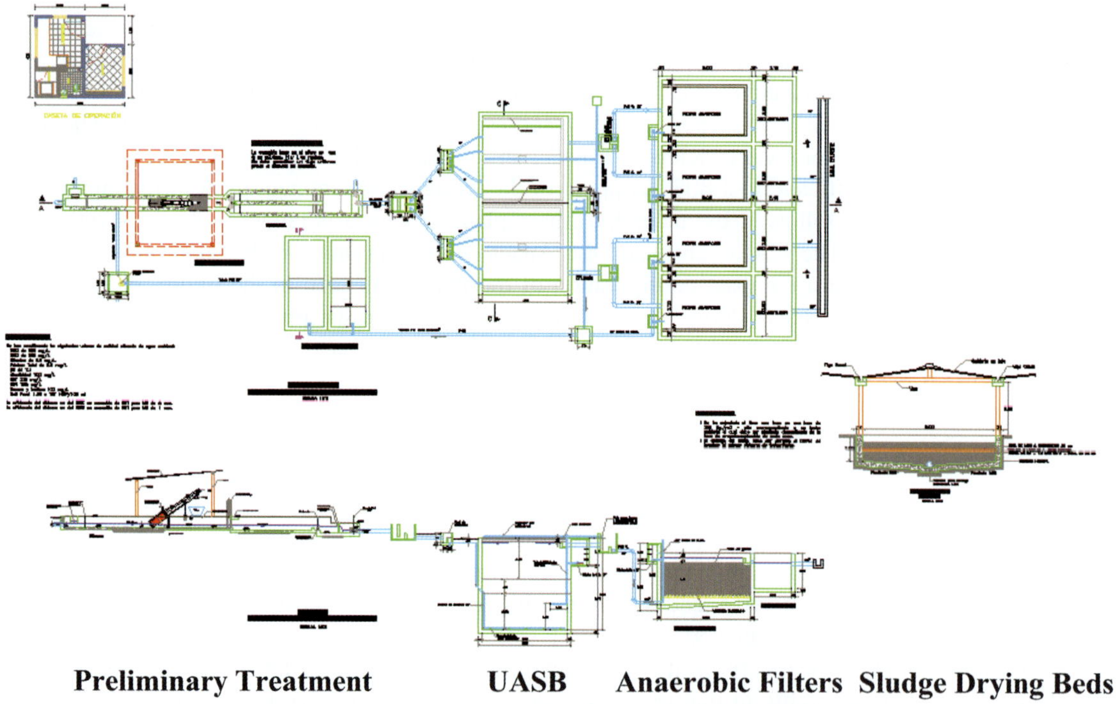

Figure 11.5 Plan and typical cross section of the combined process of preliminary treatment followed by UASB followed by anaerobic filter

11.4 COMBINATION 3: ROTATING MICRO SCREENS FOLLOWED BY UASB FOLLOWED BY SAND FILTRATION FOLLOWED BY UV DISINFECTION

11.4.1 Introduction

Preliminary Treatment followed by UASB followed by Sand Filtration is a simple, reasonable process which occupies a very small land area. A UASB effluent contains a significant amount of suspended solids which can be removed to improve the quality of the effluent. Their removal by sand filtration can reduce the suspended solids content of the effluent and also its BOD content, since the suspended solids are mostly organic matter. The concept of this process is similar to the concept of UASB followed by DAF, but the suspended solids separation from the UASB effluent is achieved by different polishing processes. Sand filtration seems to be a more reliable process than DAF. The operation of a sand filtration unit is usually well known to utilities since they operate such plants for treatment of potable water. The process consumes very little energy and its operation and maintenance costs are low. We do not know about full scale treatment plants based on the UASB followed by Sand Filtration combined process (while we know about several full scale plants based on USAB followed by DAF). The effluent quality is expected to be similar in both. This system does not significantly remove fecal coliforms but the sand filtration removes helminth eggs. Disinfection is required in order to obtain a pathogens free effluent. The sand filter will capture and remove most of the solids contained in the UASB effluent, so the final effluent will be clear and can be effectively disinfected by the UV unit which is proposed as part of this combined process. Coagulants and/or Flocculants need to be injected to the UASB effluent to achieve an efficient filtration. The biogas generated in the UASB reactor can be collected and put to use, if economically feasible. A schematic flow diagram of this process is presented in Figure 11.6.

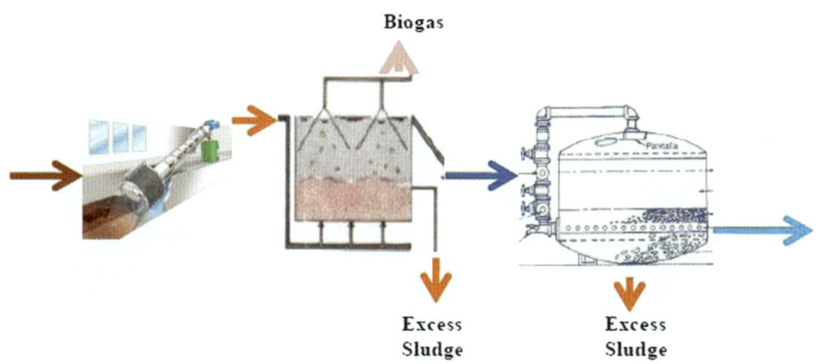

Figure 11.6 A schematic flow diagram of the combined process of preliminary treatment followed by UASB followed by a sand filter

11.4.2 Performance

There are no effluent quality data from existing plants. It is estimated that the effluent quality will be similar to that of the process of UASB followed by DAF, that is, total BOD removal in the range 80–90%, and TSS removal 80–90%. With UV disinfection, the effluent will be free of pathogens.

The advantages of this combined process are: (i) it occupies a very small land area, (ii) it produces a good effluent quality at a low investment cost; (iii) it has small energy consumption, (iv) it is easy to operate and utilities are familiar with the operation of sand filtration, (v) it is suitable for cities of all sizes.

There is no cost information from existing plants, but it is estimated that the investment costs for this combined process are in the range 30–50 US$/Capita and the operation and maintenance costs are in the range 1–1.5 US$/Yr/Capita.

11.4.3 Design

The development of a detailed example for UASB is presented in Chapter 6 and in Combination 1. The development of the Sand Filter and UV disinfection examples are presented in Chapter 10, to which the reader is referred. The Excel program which performs the design calculation of combined process 3 is COMB3-UASB-SAND FITER-UV-02.xls. The results of application in series of a UASB reactor followed by a Sand Filter followed by UV disinfection for the example of 20,000 people are shown below.

Variables selected by the designer

UASB

The design is exactly as presented in Chapter 6 and in Paragraph 11.2 of this chapter and is not repeated here.

Sand filter

The designer variables selected for the Sand Filter are presented in Table 11F. See Chapter 10, Paragraph 10.2, and the OBSERVATION COLUMN of the table below for details. Note that the flow is the same as in the example of a treatment plant for 20,000 people and that (q_{BW}) is at least three times the filtration rate (q_F) of 120 $L/min \cdot m^2$. In this case the backwash flow was selected to be four times the filtration rate, that is 480 $L/min \cdot m^2$.

Table 11F

Designer variable					
Variable	**Value**	**Unit**	**Value**	**Unit**	**Observation**
Filtro De Arena					
Q_D	89.8	L/s			Equation 3.4
Pressure filter?	Yes				Yes for P < 100,000 inhab
q_F	120.00	$L/min \cdot m^2$	2.94	gpm/ft^2	Tables 10.1 and 10.2
q_{BW} (select)	480.00	$L/min \cdot m^2$	11.76	gpm/ft^2	Minimum 3_{qF}
n_F	3.00				>2

UV disinfection

The designer variables selected for the UV disinfection system are presented in Table 11G. See Chapter 10, Section 10.4 for design details. The OBSERVATION column of the table below given ranges for the selecting parameters.

Table 11G

Designer variables					
Variable	**Value**	**Unit**	**Value**	**Unit**	**Observation**
UV					
Q_{HD}	89.8	L/s	0.090	m^3/s	
Dose, D	16.0	$mW \cdot s/cm$	160	Ws/m^2	16–38
v, channel velocity	0.60	m/s			0.6–2.00
h_c, channel depth	0.30	m			0.3–1.00
(L_c/w_c) ratio	1.50				1 to 4

Design

UASB

The calculations of the AUSB unit are the same presented in Chapter 6 and they are not repeated here. See Chapter 6 and Paragraph 11.2 of this chapter for more details.

Sand filter

See Chapter 10, Section 10.2 for design details. The applications of Equations 10.10 to 10.13 is straightforward. The results are presented in Table 11H.

Table 11H

Design			
Parameter	**Value**	**Unit**	**Observation**
Filtro De Arena			
Q_F	29.9	L/s	Equation 10.10
A_F	15.0	m^2	Equation 10.11
D_F	4.4	m	Equation 10.12
Q_{BW}	119.8	L/s	Equation 10.13

UV disinfection

The results of the design calculations of the UV disinfection unit are presented in Table 11I. The results of the design calculations of the Sand Filter unit are presented in Table 11N. These results are calculated by the Excel program COMB 3-UASB-Sand Filter-UV.xls. The design details are presented in Chapter 10 Section 10.4. The Equations applied in each case are specified in the OBSERVATION column of the table below.

Table 11I

Design					
Variable	**Value**	**Unit**	**Value**	**Unit**	**Observation**
UV					
Channel width, w_c	0.499	m			Equation 10.22
Channel length, L_c	0.450	m^2			Equation 10.23
Disinfection volume, V	0.067	m^3			Equation 10.24
Contact time, t	0.750	s			Equation 10.25
Tube reactor diameter, ϕ	0.262	m	10.0	inch	Equation 10.26
UV intensity, I	213	$W \cdot s/m^2$			Equation 10.20

A Drawing of a typical plant based on the combined process of UASB followed by Sand Filter followed by UV Disinfection is presented in Figure 11.7.

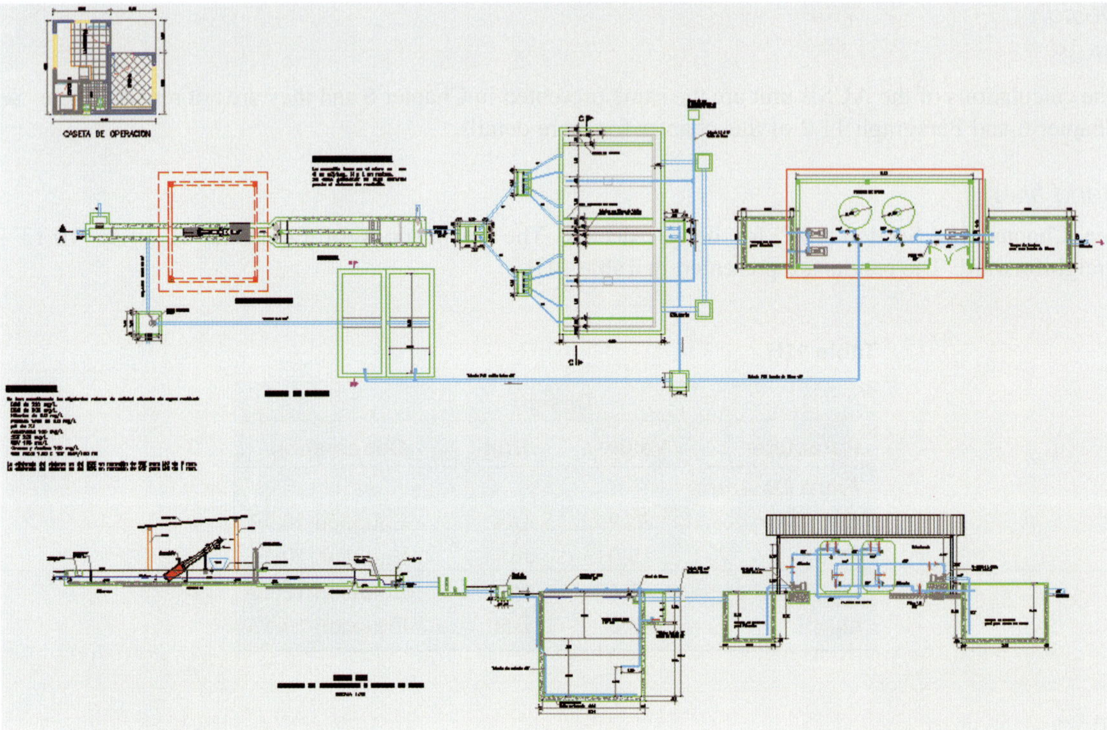

Preliminary Treatment **UASB** **Sand Filters**

Figure 11.7 Plan and typical cross section of the combined process of preliminary treatment followed by UASB followed by sand filter followed by UV disinfection

11.5 COMBINATION 4: ROTATING MICRO SCREENS FOLLOWED BY CEPT FOLLOWED BY SAND FILTRATION FOLLOWED BY UV DISINFECTION

11.5.1 Introduction

CEPT is an important appropriate technology unit process because of its capacity to function at the entire range of temperatures, including very low temperatures. If a effluent quality higher than that which can be produced by the CEPT unit process is required, sand filtration can be used to polish the CEPT effluent, so a combination of CEPT followed by sand filtration and, if required, UV disinfection can provide a good quality effluent, similar to that of conventional treatment. The filtration can be a single or multimedia sand filtration. The schematic process flow diagram of this combined process is presented in Figure 11.8. The CEPT followed by Sand Filtration process is based on physicochemical and physical processes, with no biological processes involved, so it is not very sensitive to temperature and will function well at very low temperatures, and as long as the wastewater does not freeze the process will perform even if the air temperature in the plant site is much lower than zero. The sand filtration produces a clear effluent which can be effectively disinfected. It is recommended to perform pilot plant test to determine the appropriate filtration rate and the effectiveness of the UV disinfection. The design of UV disinfection unit should be done according to the objectives presented in Table 10.6. This combined process is adequate for medium and large cities; but is not recommended for small towns because a certain level of technical capacity is required to operate it.

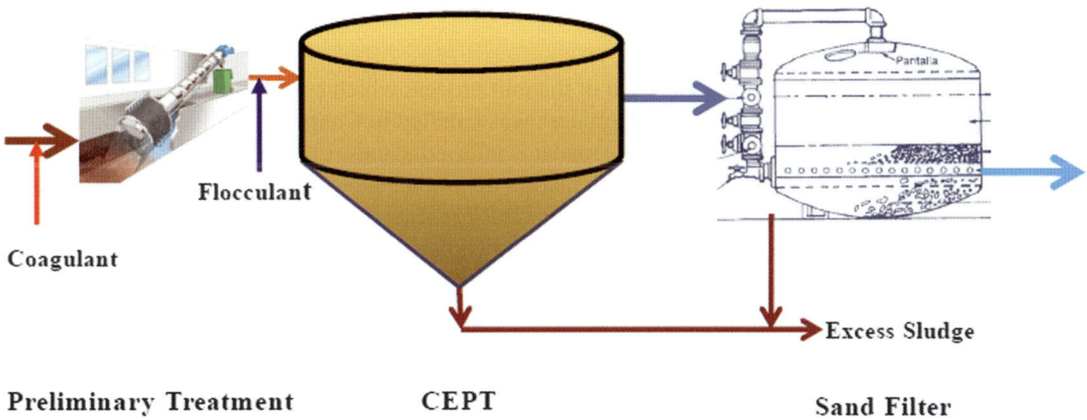

Figure 11.8 A schematic flow diagram of the combined process of preliminary treatment followed by CEPT followed by a sand filter

According to Cooper-Smith (2001) an extensive pilot scale investigation showed that the combination of CEPT followed by sand filtration and UV disinfection proved capable of producing a high quality effluent (averaging 20 mg/1 BOD, 15 mg/1 TSS) using a Tetra Deep Bed Filter at a filtration rate of 5 m/h and a coagulant (Kemira PAX XL60, a polyaluminium silicate) dose of 40 mg/l. UV disinfection of the filtered effluent complied with the required Bathing Water Directive standard of the EU and Net Present Value calculations showed this option to be considerably cheaper than a secondary biological treatment option. Sludge treatment is identical to that of CEPT sludge.

CEPT followed sand filtration is an appropriate technology process since it is a relatively simple technology, which provides low cost, easy to implement and effective treatment. The CEPT unit process is an easy to operate process since it is basically consisting of a settling unit and a sand filter. Sand filtration is a process that most water utilities know how to operate since they operate filtration plants in the water supply system, and it is basically an easy process to operate. The only complication might be the handling of the sludge, but as a whole, it is an appropriate technology process.

11.5.2 Performance

Based on pilot plant tests it is considered that the combined process of CEPT followed by sand filtration can achieve a total BOD removal in the range 80–90%, and TSS removal 80–90%. With UV disinfection, the effluent will be free of pathogens.

The advantages of this combined process are: (i) it is capable of functioning at the entire range of temperatures, including very low temperatures, (ii) it produces an effluent quality comparable to that of secondary treatment at lower costs, (iii) It occupies a very low land area, smaller than that of conventional processes, (iv) it has a low energy consumption, and (v) it is easy to operate.

The disadvantages of the combined process are: (i) it produces large quantities of sludge which needs to be properly managed, and (ii) it is not adequate for small towns.

The estimated investment cost of the combined system CEPT followed by a Sand Filter is in the range 40–50 US$/Capita and the O&M cost is in the range 1.5–2.0 US$/Yr/Capita.

11.5.3 Design

The development of a detailed example for each of the unit process which constitute this combined process are presented in previous chapters. The CEPT process was handled in Chapter 9 while the Sand Filter and UV disinfection are presented in Chapter 10. The reader is referred to these chapters for detailed information. The Excel program which performs the design calculation of combined process 4 is COMB 4-CEPT-SAND FITER-UV.xls. The results of application in series of a CEPT reactor followed by a Sand Filter followed by UV disinfection for the example of 20,000 people are shown below.

Variables selected by the designer

CEPT

The designer variables selection and the design procedure in this case is exactly the same as presented in Chapter 9, Section 9.3, which can be consulted for details.

Sand filter

The designer variables selection and the design of the sand filter is same as the one presented in paragraph 11.4 of this chapter, which can be consulted for details.

UV disinfection

The designer variables selection is the same as in Paragraph 11.4 of this chapter, which can be consulted for details.

Design

CEPT

The design procedure in this case is exactly the same as presented in Chapter 9, Section 9.3, which can be consulted for details.

Sand filter

The design of the sand filter is same as the one presented in Paragraph 11.4 of this chapter, which can be consulted for details.

UV disinfection

The procedure in this case is the same as in Paragraph 11.4 of this chapter, which can be consulted for details.

A Drawing of a typical plant based on the combined process of CEPT followed by Sand Filter followed by UV Disinfection is presented in Figure 11.9.

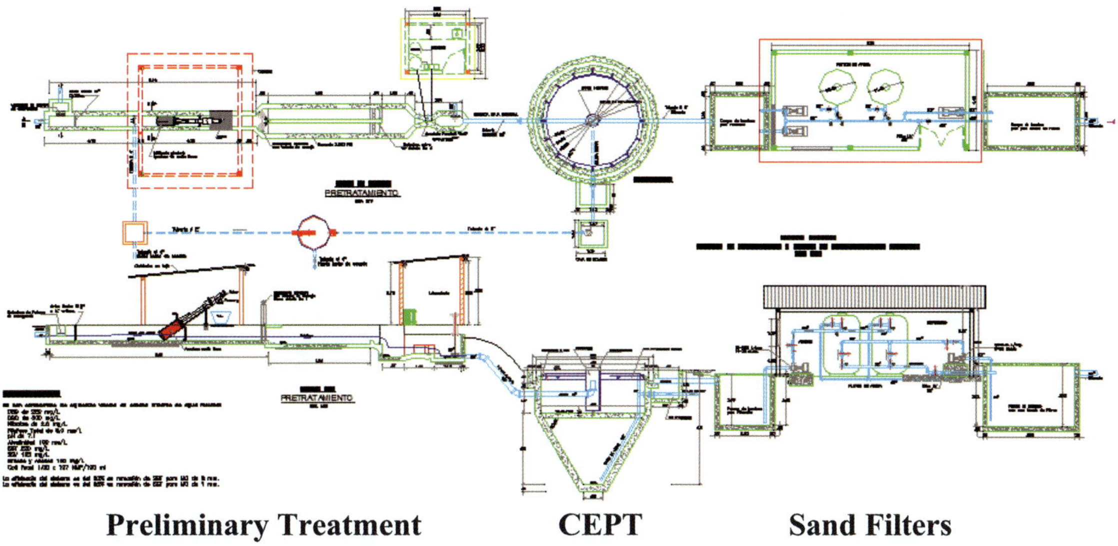

Figure 11.9 Plan and typical cross section of the combined process of preliminary treatment followed by CEPT followed by sand filtration followed by UV disinfection

11.6 COMBINATION 5: ROTATING MICRO SCREENS FOLLOWED BY UASB FOLLOWED BY ANAEROBIC FILTER FOLLOWED BY DAF FOLLOWED BY MEMBRANE FILTRATION (MICRO FILTRATION AND IF NECESSARY ULTRA FILTRATION)

11.6.1 Introduction

Most of the combined processes which were presented above are processes used in full scale plants or in pilot scale plants. The combined process discussed in this section, preliminary treatment followed by UASB followed by Anaerobic Filter followed by Dissolved Air Flotation followed by Membrane Filtration, has most probably not yet been applied in a full scale plant. It cannot be considered an appropriate technology process because the membrane filtration units are not based on an appropriate technology process; however the entire section of the plant preceding the membrane filtration is an appropriate technology section. This process is presented to show that it is possible to combine appropriate technology units with the cutting edge existing technology to obtain an effluent of the highest possible quality. By using combined appropriate technology units it is possible to produce an effluent adequate to be fed to a membrane filtration unit. A schematic process flow diagram is presented in Figure 11.10. The effluent of this process would be of very low content of BOD and SST and practically void of Fecal Coliforms. An effluent of such quality can be reused in industrial plants.

The section preceding the membrane filtration unit would not be costly in investment and in O&M, nor would it be difficult to operate. The membrane filtration unit would add cost and its setup will depend on the required effluent quality, which will determine if one or a combination of Micro and Ultra filtration units would need to be used. The membrane filtration units are more complex to operate, but not too complex. The process is not adequate for small cities but utilities of medium and large cities would have no problem to operate it. Information regarding cost data and operation problems is unavailable because this process has not yet been in use.

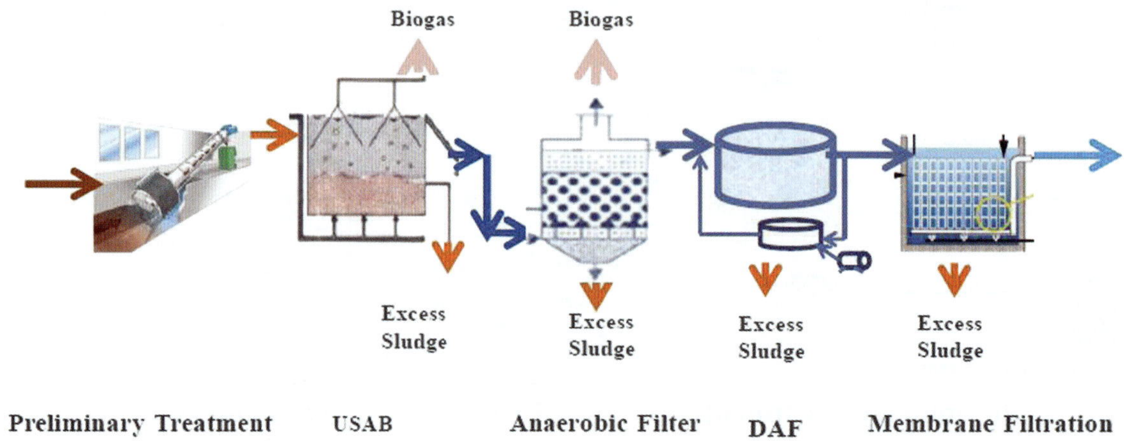

Figure 11.10 A schematic process flow diagram of the combined process of preliminary treatment followed by UASB followed by anaerobic filter followed by DAF followed by membrane filtration

The combination of UASB followed by Anaerobic Filter was described above and also in Chapter 6. The effluent from the anaerobic filter is treated in a DAF, requiring the use of flocculants to improve the removal of TSS (see Chapter 10). The quality of effluent in this combined process is very high, it achieves almost complete removal of BOD and TSS and complete removal of fecal coliforms. Despite the number of units in this combined process it does not consume large amounts of energy, and does not reproduce large amounts of excess sludge. This process would also be more expensive than the other combined processes mentioned, although without ultrafiltration the cost is more reasonable.

11.6.2 Performance

The combined process discussed in this section, preliminary treatment followed by UASB followed by Anaerobic Filter followed by Dissolved Air Flotation followed by Membrane Filtration, has most probably not yet been applied in a full scale plant. Nevertheless, we consider that it can function well and yield an extremely high quality effluent. The effluent of the DAF system, which is inflow to the membrane unit, is of a quality comparable or even higher than an effluent of an activated sludge plant, is there is no reason that the membrane unit will have problems processing this influent. The expected overall BOD removal in this combined process is over 98% SST removal of 99%a and the effluent is expected to be void of pathogens. Partial phosphorous removal may take place because of phosphorous precipitation as a result of the coagulant added before the DAF unit. However, Nitrogen removal will not be achieved by this process. The advantage of this combined process is that it can yield in effluent of very high quality. Its disadvantages are: (i) due to the large number of units processes in series it becomes relatively complex, and (ii) it can be costly. There is no knowledge on costs and operational problems of this combined process.

11.6.3 Design

The development of a detailed example for each of the unit process which constitute this combined process are presented in previous chapters. The UASB and Anaerobic Filter processes are handled in Chapter 6. The DAF and membrane processes are handled in Chapter 10. The reader is referred to these chapters for detailed information. The Excel program which performs the design calculation of combined process 5 is COMB 5-UASB-AF-DAF-MEMBRANES-02.xls. The results of application in series of a UASB reactor followed by a Sand Filter followed by a DAF unit followed by a Membrane Filtration unit for the example of 20,000 people are shown below.

Variables selected by the designer

UASB

The calculations are the same as in Chapter 6 and in Section 11.2 of this chapter, and are not repeated here. Consult there for details.

Anaerobic filter

The calculations are the same as in Chapter 6 and in Section 11.2 of this chapter, and are not repeated here. Consult there for details.

DAF

The designer variables selected for the DAF are presented in Table 11J. For design details see Chapter 10, Section 10.3.

Table 11J

Designer variables					
Variable	**Value**	**Unit**	**Value**	**Unit**	**Source**
DAF					
Q_{HD}	7760.44	m^3/d			Equation 3.4
TSS, S_a	60.78	mg/L			Influent quality
SOR = $(Q+Q_r)/A$	10	$m^3/h \cdot m^2$			0.5 a 10.0
Q_s	5	$Kg/m^2 \cdot h$			0.5 a 10.0
A/S	0.01				0.006 a 0.2
R	0.5				Q_r/Q_D
f	0.5				0.5–0.6
$Q_x = Q_{HD} \cdot S_a$	471.68	kg/h			
E, % TSS	95	%			Pilot test
Temperature	25.00	°C			Influent quality
S_a	17.11	mL/L			Equation 10.15

Membrane filtration

The designer variables selected for the Membrane Filtration unit are presented in Table 11K. For design details see Chapter 10, Section 10.5.

Table 11K

Designer variables					
Variable	**Value**	**Unit**	**Value**	**Unit**	**Source**
Membrane					
BOD_{5i}	48.05	mg/L			Influent
BOD_{5e}	5.00	mg/L			Norm
BOD_{5R}	100.00	mg/L			Reject max limit
TSS_i	10.13	mg/L			Influent
TSS_e	1.00	mg/L			Norm
TSS_R	100.00	mg/L			Reject max limit
Q_{HD}	323.35	m^3/h			Equation 3.4
R(BOD)	0.45		45.32	%	Equation 10.29
R(TSS)	0.09	m^3/h	9.22	%	Equation 10.29
R	0.45		45.32	%	Highest R
Q_{DR}	146.54	m^3/h			RQ_{HD}
Q_{DP}	176.82	m^3/h			$(1-R)Q_{HD}$
Define Membrane	MF				Figure 10.16
q_{MF}	1000.00	$L/m^2 \cdot d$			Table 10.7
Define Membrane	UF				Figure 10.16
q_{UF}	0.00				Table 10.7
Define Membrane	NF				Figure 10.16
q_{NF}	500.00	$L/m^2 \cdot d$			Table 10.7
Define Membrane	RO				Figure 10.16
q_{OR}	0.00				Table 10.7

Design

UASB

The calculations are the same as in Chapter 6 and in Section 11.2 of this chapter, and are not repeated here. Consult there for details.

Table 11L

Design			
Parameter	**Value**	**Unit**	**Observation**
DAF			
Pressure, P	2.1	atm	Equation 10.14
Manometric pressure, p	112.4	kPa	Equation 10.16
Area fro SOR, A'	776.0	m^2	Equation 10.17
Area from Q_s, A''	141.5	m^2	Equation 10.18
Design area, A_D	776.0	m^2	Largest from A' and A''
Diameter, $_{DAF}$	31.4	m	Equation 10.19
TSS_e	3.0	mg/L	According to efficiency

Table 11M

Design			
Parameter	**Value**	**Unit**	**Observation**
Membrane			
A_{MF}	176815.8	m^2	Equation 10.30
A_{UF}	0.0	m^2	Equation 10.30
A_{NF}	88407.9	m^2	Equation 10.30
A_{OR}	0.0	m^2	Equation 10.30
Pressure MF	50.0	kPa	Table 10.7
Pressure UF	0.0	kPa	Table 10.7
Pressure NF	750.0	kPa	Table 10.7
Pressure OR	0.0	kPa	Table 10.7

Anaerobic filter

The calculations are the same as in Chapter 6 and in Section 11.2 of this chapter, and are not repeated here. Consult there for details.

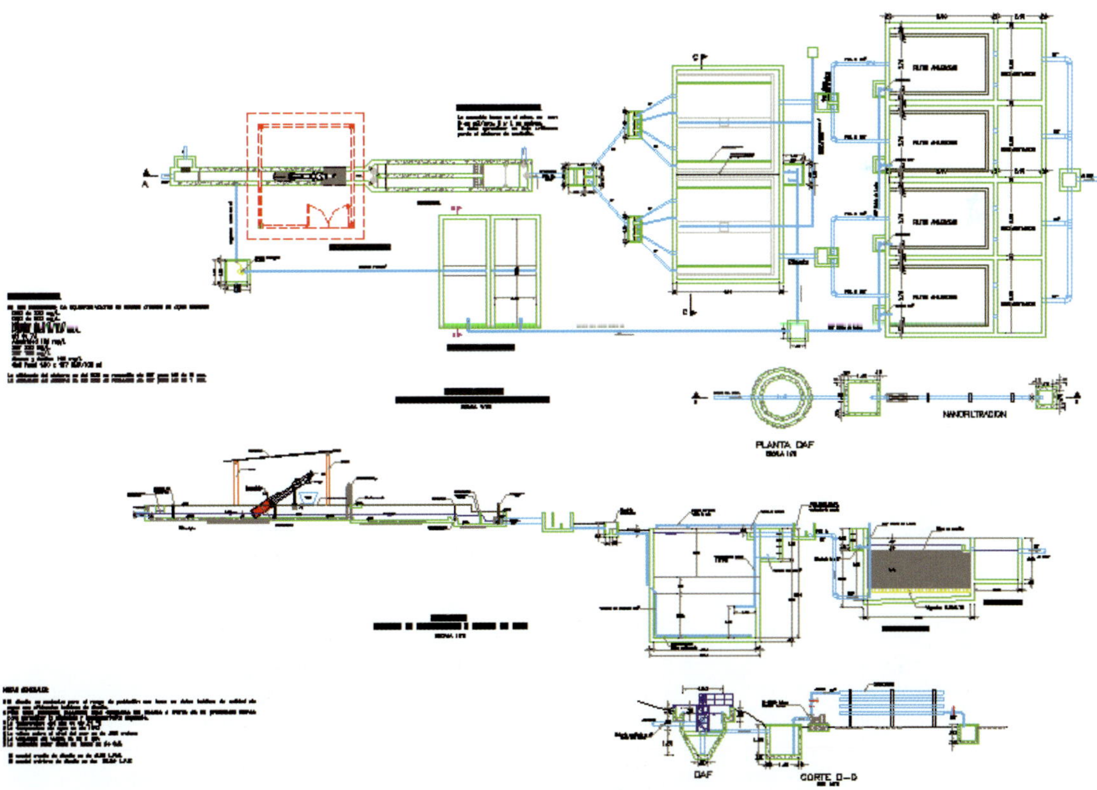

Figure 11.11 Plan and typical cross section of the combined process of preliminary treatment followed by UASB followed by anaerobic filter followed by DAF followed by membrane filtration

DAF

The results of the design calculations of the DAF unit are presented in Table 11L. For design details see Chapter 10, Section 10.3.

Membrane filtration

The results of the design calculations of the Membrane Filtration unit are presented in Table 11M. For design details see Chapter 10, Section 10.5.

A Drawing of a typical plant based on the combined process of UASB followed by Anaerobic Filter followed by DAF followed by Membrane Filtration is presented in Figure 11.11.

Chapter 12

Global warming and wastewater treatment impact on climate

12.1 GLOBAL WARMING[1]

12.1.1 Introduction

Global warming is the increase in average temperature of the earth's surface and oceans, which is occurring since the mid-twentieth century, and the projection of its continuation. Global warming is probably a result of the observed increase in the concentration of Greenhouse Gases (GHG) emissions of anthropogenic origin (mainly CO_2), which are largely generated by the burning of fossil fuels (coal, oil, natural gas, etc.). Other GHG emissions also contribute to global warming (CH_4, N_2O, SF_6). Water vapour is also a greenhouse gas but its concentration depends on the temperature of the Earth, and is considered in global climate models (GCM)[2] as the feedback of water vapour. The main non-greenhouse gas contributors to greenhouse effect are the clouds which also absorb and emit infrared radiation and have a similar effect to that of the radiative properties of greenhouse gases. The most abundant greenhouse gases are: (i) water vapour, which contributes about 36–70%, (ii) carbon dioxide, which contributes 9–26%, (iii) methane, which contributes 4–9%; and (iv) ozone, which contributes about 3–7%.

The Kyoto Protocol on climate change is an international agreement that aims to reduce emissions of six greenhouse gases that contribute to global warming: carbon dioxide (CO_2), methane (CH_4) and nitrous oxide (N_2O), plus three fluorinated industrial gases: hydro fluorocarbons (HFC), perfluorocarbons (PFC) and sulphur hexafluoride (SF_6). This instrument is within the United Nations Framework on Climate Change Convention (UNFCCC), signed in 1992 during what is known as the Earth Summit in Rio de Janeiro.

On 11 December 1997, industrialized countries agreed in the city of Kyoto to execute a set of measures to reduce greenhouse gases. Signatory governments agreed to reduce during 2008–2012 the average contaminating emissions by 5%. with reference to 1990 levels, known as the base year. The agreement came into force on February 16, 2005, after ratification by Russia on November 18, 2004. The protocol came to give binding force to what the UNFCCC could not do at the time.

[1]This section is based on presentations of A. Orozco: (i) "La lucha contra el calentamiento global: una visión diferente", CORPORACIÓN OTRAPARTE, Medellín-Colombia, 2007" and (ii) "The impact of global warming on wastewater treatment, Water Week 2009, The World Bank, Washington DC, USA, 2009.

[2]*Global Climate Models* include the *General Circulation Models* are also known as GCM. Both meanings are valid and relate basically to models for the projection of global climate.

On the other hand, other signatories to the Kyoto Protocol which are large producers of CO_2, such as China and India (see Table 12.1 and Figure 12.1), did not acquire reduction commitments because they are classified as developing countries, whose treatment is different from that of developed countries. However, if we analyze the ranking from the standpoint of per capita emissions, the order is quite different as China and India rank among the lower CO_2 emitters as shown in Figure 12.2.

Table 12.1 World ranking producer of CO_2 (2005).

Rank	Country	CO_2 MTM
1	United States of America	1576537
2	China (Mainland)	1514126
3	Russian Federation	410290
4	India	382740
5	Japan	335706
6	Germany	213969
7	United Kingdom	149131
8	Canada	146704
9	Republic of Korea	123422
10	Italy (Including San Marino)	123392
11	Islamic Republic of Iran	123246
12	Mexico	115043
13	Indonesia	114518
14	South Africa	111570
15	Saudi Arabia	104015
16	France (Including Monaco)	103082
17	Australia	100671
18	Spain	93805
19	Ukraine	89282
20	Brazil	88834

The main objective of the Kyoto Protocol is to lower anthropogenic greenhouse gas concentrations which cause greenhouse effect. According to UN Global Climate Model (GCM) projections, it is expected that the average temperature of the planet's surface will rise by between 1.4 and 5.8°C by the year 2100.

The Intergovernmental Panel on Climate Change, known by the acronym IPCC, was established in 1988 by the World Meteorological Organization (WMO) and the United Nations Environmental Programme (UNEP). The objective of the IPCC is to assess the risk of climate change caused by human activities and its reports are based on publications in scientific and peer-reviewed journals. In 2007 it was awarded the Nobel Peace Prize, shared with Al Gore. The IPCC is a multinational body charged with conducting negotiations on global climate change and lead the scientific debate on global warming, the emission of carbon particles, the greenhouse effect, and others. Different scenarios of global climate change are included among its lines of action, as required in the context of the Kyoto Protocol. In 2007 the Fourth IPCC Assessment Report (AR4) was presented, showings the measured changes in temperature, sea level and snow cover in the Northern Hemisphere during the past 150 years (Figure 12.3a). Note that in the decade 1998–2008 there have been no increases in temperature (Figure 12.3b).

Figure 12.1 Map of CO_2equivalent emissions (2002) (*Source*: Worldmapper: http://www.worldmapper.org/display.php?selected=295)

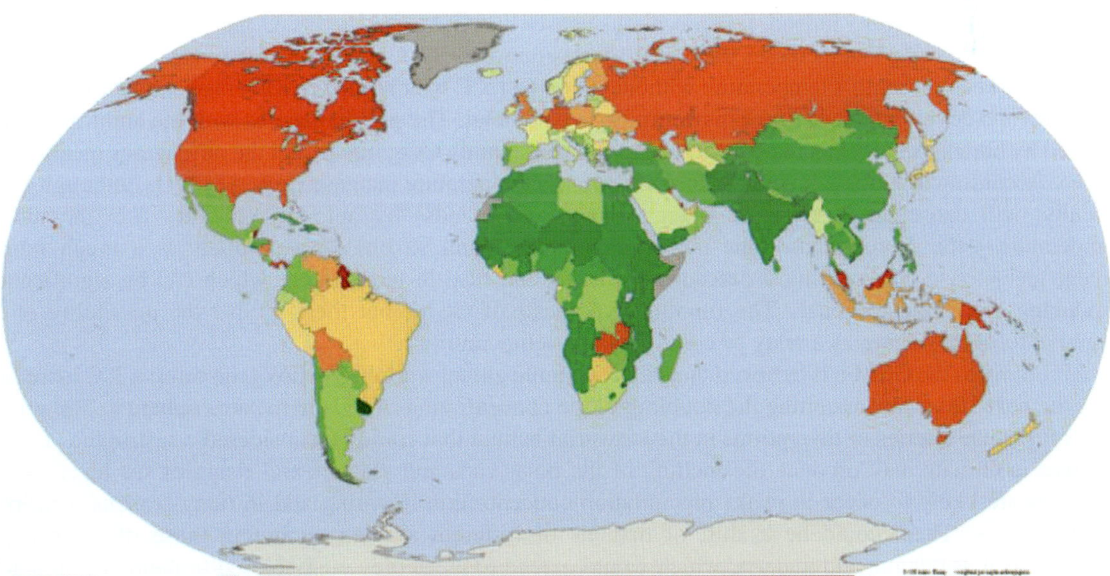

Figure 12.2 Ranking of per capita emission of CO_2 equivalent (*Source*: http://upload.wikimedia.org/wikipedia/commons/8/84/CO2_responsibility_1950–2000.svg)

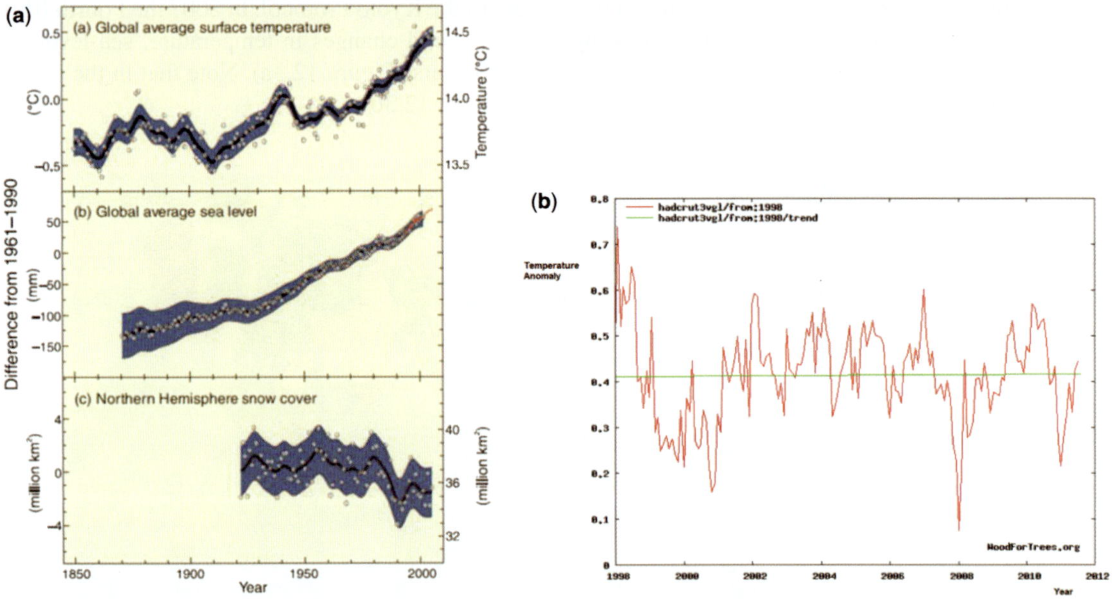

Figure 12.3 a) Measured changes in earth's average temperature, sea level and snow cover in the northern hemisphere from IPCC; b) temperature anomaly since 1998 from hardcrut (*Source*: IPCC's AR4 (2007), http://www.scalgon.co.uk/page10.html)

It is important to understand that GHGs have existed on earth for millions of years and without them life here would be impossible. Indeed, it is greenhouse gases like CO_2, CH_4 and N_2O (naturally produced) that cause radiative forcing that increases the planet's temperature by 33°C (i.e. from an average of −18°C which would have been the earth's temperature without GHGs to 15°C which is the real average earth temperature) and allow the development of life. This warming is beneficial. The problem arises with the anthropogenic (caused by human activity) increase in greenhouse gases, mainly CO_2, but also of the other gases mentioned above. According to global climate models (GCM), the temperature increase caused by CO_2, although low initially, will cause an increase of water vapour (and clouds) by auto-sustainability (i.e. the more temperature → the more steam → the more temperature → and so on). Water vapour is a much more potent greenhouse gas, causing a cascade of a larger increase in temperature which can be significant, producing a positive feedback. The running hypothesis of GCM, but there is also the possibility of a negative feedback acting as a relay to control earth's temperature.

According to the GCM it is believed that anthropogenic global warming (AGW) can cause a 3°C increase over the next 100 years, assuming the doubling of the concentration of CO_2 in the atmosphere to 560 ppm. This may cause a series of disruptions in the terrestrial habitat that could be considered catastrophic. These changes in climate may produce the melting of the polar caps and glaciers and rising of sea level. Also increases are likely to occur in peaks precipitation concentration in spring and in rainy seasons in warm climate zones. There could be a shift of habitats of different species, including those of vectors of diseases such as dengue and malaria, which would extend to areas previously closed to them. The impact on agriculture and arable areas would be enormous with a transfer of fertility to more northern areas. Even though these events are presented as "likely" by the IPCC, there is a lot of uncertainty due to the typical chaotic behaviour of climatic phenomena. Moreover, there is not unanimity among scientists regarding the assumption that anthropogenic CO_2 can cause such a temperature change, and some suggest that the GCM are inaccurate. But invoking the precautionary principle over the magnitude of the global consequences that this phenomenon may cause, serious actions are immediately required by the IPCC.

Moreover, it is important to note that in this case enormous economic and political interests are involved. Supporters of the Kyoto Protocol, which promotes the reduction of CO_2 emissions, benefit in several ways. The Kyoto Protocol is very favourable for developing countries that can make big dividends from their forests, as CO_2 sequestration agents. So the Kyoto Protocol becomes a source of great economic value, without the obligation to comply with limits on greenhouse gas emissions produced by their industries. They are also direct beneficiaries of the Clean Development Mechanism (CDM). On the other hand, those who are sceptics that CO_2 is responsible for global warming are opposed to the Kyoto Protocol because they do not want to spend money on controlling emissions from industries, arguing that it is going to be wasted money. Each side takes positions on the subject according to its interests.

There is a group of scientists as important as Gregory Benford (professor at the University of California at Irvine, an advisor to NASA and also a science fiction novelist), Paul Crutzen (Nobel Prize laurite in Chemistry shared with Molina and Rowland for his research on the ozone hole) and Edward Teller (inventor of the Hydrogen Bomb, now deceased) who while accepting the possibility of CO_2 as an agent of global warming, argue that it is not economically feasible to sufficiently control its emissions to stop the warming. Rather, they proposed global environmental management solutions, which are known as geoengineering solutions. For example, under one solution it is proposed to place a mylar lens 1000 km in diameter in a geosynchronous orbit to reduce solar radiation by 1%, which would control the effects of global CO_2 emission in the next 50 years. A most viable solution proposed by Crutzen is to reflect the 1% of solar radiation with aerosols artificially injected into the atmosphere in sufficient quantity. Lately, a proposed geoengineering solution known as Budyko's blanket would cost only US$ 250 million per year in contraposition to the US$ 1.2 trillion per year required by the IPCC's proposed solution of fossil fuels substitution by clean energy sources. However, the negative global effects of these solutions are unknown (Orozco, 2011).

12.1.2 Earth's temperature and warming

In this section the main factors involved in calculating the earth's temperature and the greenhouse effect are explained in a simple form. To begin, it is confirmed that the main energy source of the planet is the sun. Although there are other sources of geological origin, they are so insignificant compared to solar radiation that the latter can be considered for all practical purposes the sole energy source.

The planet's temperature is the result of: (i) the heat coming from solar radiation (which radiates mainly in the electromagnetic spectrum of visible light) less the radiation that bounces by reflection (like a mirror) of the polar ice caps and clouds, and (ii) the heat emitted by the earth as a black body, which itself depends on temperature. The GHG influence warming because they are transparent to solar radiation (visible light), but opaque to infrared radiation, which is the wavelength at which the planet gives off heat, given its low temperature.

How much radiation comes from the sun?

With reference to the Figure 12.4, it is clear that radiation from the Sun that is intercepted by Earth corresponds to the shaded area projected onto a plane placed behind the planet. On average the radiation is 1,370 W/m^2. However, this radiation is distributed to the entire area of the globe which is four times greater, that is, $1{,}370/4 = 342\ W/m^2$ reaches theearth's surface. Moreover, from the solar radiation that reaches the earth about 30% is reflected by the polar ice caps, clouds, etc. The percentage reflected, written in decimal form, is called albedo, α.

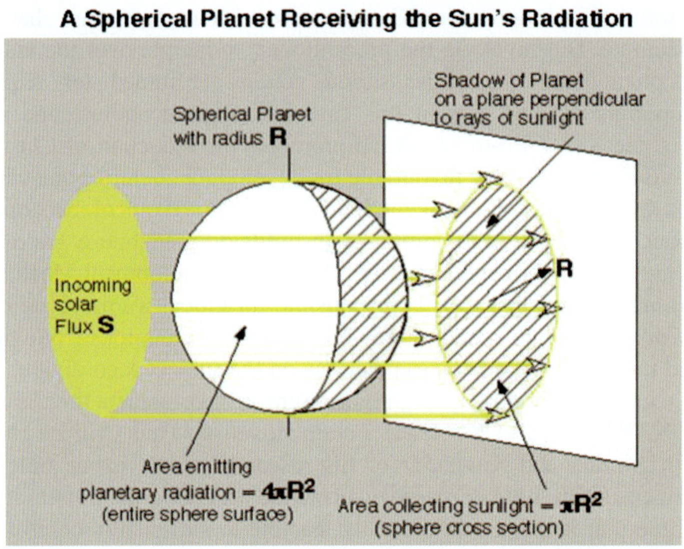

Figure 12.4 The terrestrial sphere receiving solar radiation (*Source*: Yochanan Kushnir, "The Greenhouse effect and the two dimensional pattern of earth's radiation budget", greehouse pdf.)

What amount of heat is emitted by the Earth?

Assuming that the Earth is in thermal equilibrium, then the radiation or heat it receives from the Sun must be equal to the heat radiation emitted by the Earth as a black body, which follows the Stefan-Boltzmann's law:

$$E = \sigma Tg^4 \tag{12.1}$$

Where σ is the Boltzmann's constant (5.67×10^{-8} W/m^{2}K^4) and T_g the temperature of the Earth. The balance between what is received from the sun, taking into account the albedo, α = 0.3, and what is emitted by the entire surface of the Earth, implies the equality:

$$1370 \cdot (1 - \alpha) \cdot (\pi R^2) = \sigma T_g{}^4 \cdot (4\pi R^2).$$

By solving this equation we find that $T_g = 255$°K (degrees Kelvin) or −18°C, which is incorrect. At such temperature, life on the planet would be impossible. What happens then?: the Earth's atmosphere contains a certain class of gases called greenhouse gases, which reflect the heat emitted by the planet. Indeed, the sun emits its radiation at a wavelength at which GHGs are transparent (corresponding to the solar surface temperature of 6000°K, which covers a wide spectrum, mainly visible light). But the planet in turn emits its radiation or heat as a black body in much larger wavelengths (corresponding to the temperature of the Earth) which are in the infrared spectrum. These waves are reflected by greenhouse gases. So part of the energy emitted by the planet is brought back to the ground by this reflection. This phenomenon is known as the *greenhouse effect*. The thermal balance of energy coming from the sun (E) between the atmosphere and Earth's surface is shown in Figure 12.5. According to Kirchhoff's law the *absorbance* (a) or fraction of radiation absorbed in the atmosphere is equal to the *emissivity* (e), or fraction emitted by the atmosphere. For sunlight a = e = 0, and for infrared radiation (IR) e = a varying between 0.7 and 0.99. With that in mind, from the heat balance in the figure the following equilibrium equations emerge: Energy absorbed = Energy Issued.

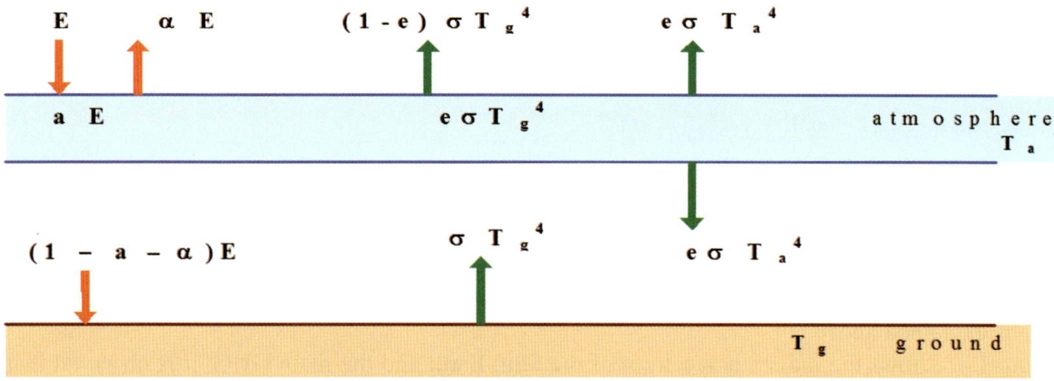

Figure 12.5 Thermal balance between solar radiation (E) and radiation emission of the atmosphere and surface. The fraction of E which absorbs the atmosphere is a = 0, and the fraction of IR light it emits is e = a_a. T_g is the ground temperature (°K) and T_a is atmosphere temperature

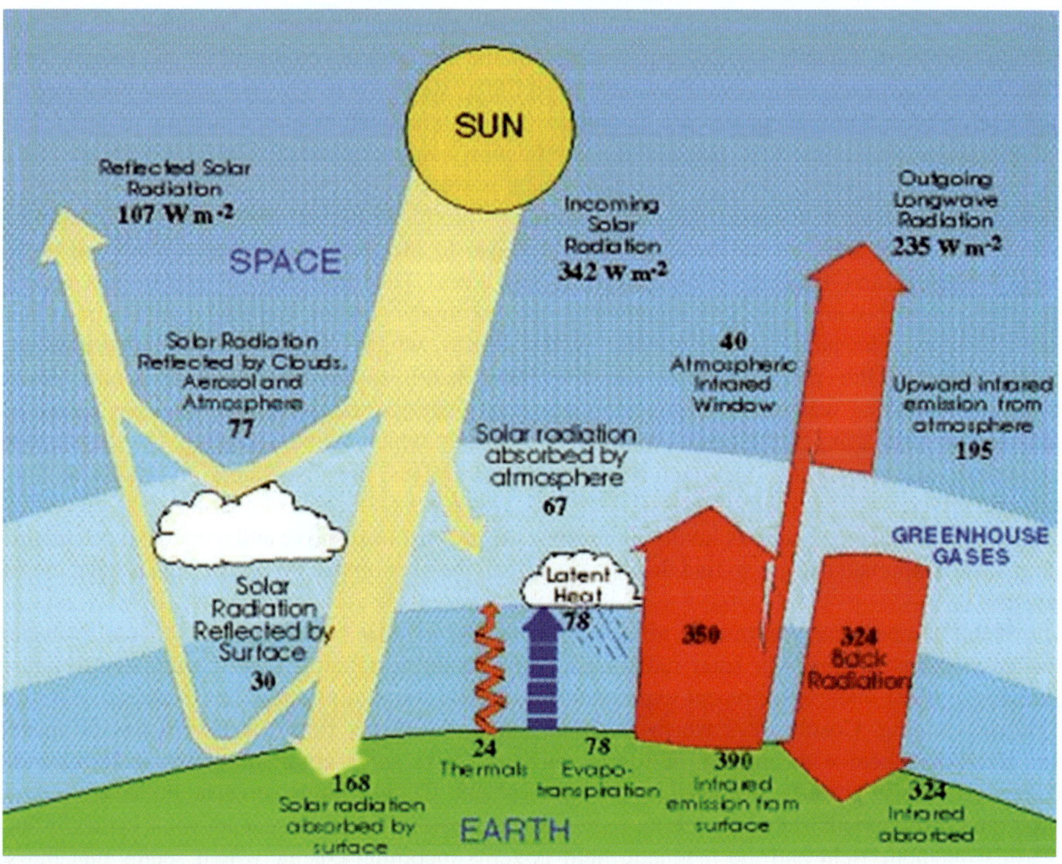

Figure 12.6 The greenhouse effect (*Source*: Yochanan Kushnir, "The Greenhouse effect and the two dimensional pattern of earth's radiation budget, pdf Greehouse.)

For the atmosphere: $e\sigma T_g{}^4 = 2e\sigma T_a{}^4$;

from which: $T_g = 2^{1/4}T_a$ (12.2)

For the surface: $(1 - \alpha)E + e\sigma T_a{}^4 = \sigma T_g{}^4$ (12.3)

By solving these equations for $\alpha = 0.3$ and $e = 0.76$ we receive that $T_g = 15°C$, which corresponds to reality. This means that in the heat balance of GHGs reflection is 155 W/m^2.

Figure 12.6 shows the radiative balance between the Sun and Earth, taking into account the earth's albedo and the reflection of GHGs. Note that the average figure that reaches the earth's surface is $342 = 1370/4$ W/m^2, as explained above.

Figure 12.7 shows the radiation spectrum of the Sun, Earth and the main GHG: it is observed that the sun's radiation spectrum is broad but focuses on the window of visible light (picture top) while the spectrum of the Earth focuses on the infrared wavelength of 5 micrometers (figure bottom). The other graphs correspond to the absorbance of specific greenhouse gases and the overall atmosphere (penultimate graph from top to bottom). The blue part corresponds to the absorption of given wavelengths. Note that CO_2 absorption is centered at 15 micrometers.

From the above discussion, it is clear then that the greenhouse effect is necessary to maintain a temperature on earth suitable for the requirements of the life and development of the human species. Without this phenomenon there would not have been life as we know it. This is a natural phenomenon.

However, if greenhouse gases increase disproportionately, produced by human activity, which is known as anthropogenic emissions, then the additional reflection caused by these new GHGs called radiative forcing can lead to a global warming above that which is suitable for the complex system that makes up the biosphere. This can be caused by the CO_2 emissions produced by burning fossil fuels. Heating may lead to defrosting of some of the polar ice cap, and also to increasing sea levels because of seawater thermal expansion and flooding of populated inland areas. The defrosting of polar ice cap do not increase sea level because the unfreezed water only occupies the ice cap buoyant volume thawed. In turn, it reduces the polar cap size, lowering the albedo, which causes more sunlight to reach the earth by reducing the reflective area. On the other hand, increasing the temperature promotes evaporation, increasing water vapour (the most powerful GHG) which in turn favours the formation of clouds, thereby increasing the albedo caused by the clouds which favours cooling. At the same time water vapour as greenhouse gas and the clouds themselves are conducive to warming by the greenhouse effect.

In short, the problem is complex involving feedback from various positive and negative mechanisms for heating, so the important thing is to find the net effect of all the phenomena involved. This is what the GCM models do.

The global temperature is primarily affected by changes in solar radiation, which occur in cycles of eleven years. These changes in solar radiation have a strong correlation with the number of sunspots (Figure 12.8). Most interesting, recent experiments at CERN found a possible link between cosmic rays and global warming. Cosmic rays which are bombarding the planet from the far reaches of space are pelting the atmosphere with protons, and those protons ionize some compounds that in turn condense into aerosols which produce droplets in the atmosphere. Clouds might in turn build around those droplets, and those clouds shield the Earth, reducing temperatures. And as sunspot solar cycles influence the earth's magnetic field and the magnetic field governs incoming cosmic ray, it seems that the solar cycle variations may have an important influence on climate change. Additional information on this matter can be found in http://www.nature.com/news/2011/110824/full/news.2011.504.html

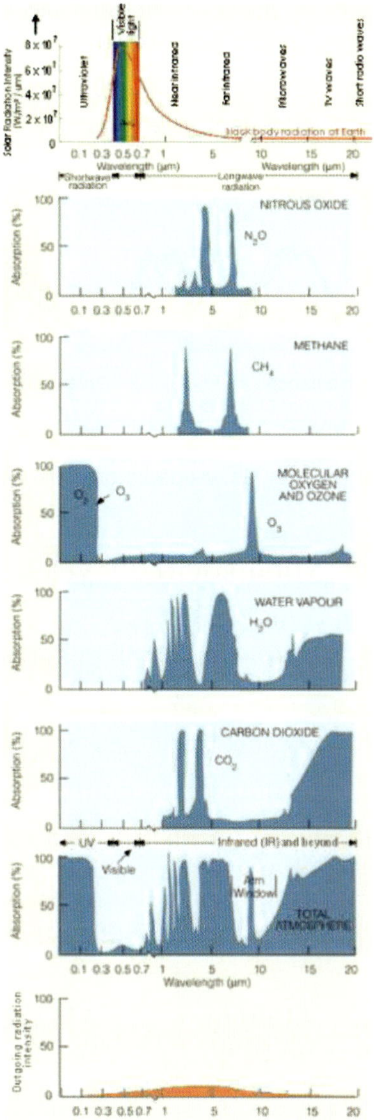

Figure 12.7 (i) Intensity of solar radiation (top) and terrestrial (bottom) according to wavelength, (ii) wavelengths at which main GHG and the atmospheric reflect radiation (note that CO_2 absorbs at wavelengths of 15 micrometers) (*Source*: http://www.abelard.org/briefings/global_warming_atmospheric_chemistry_physics.php)

In addition, solar radiation is affected by Milankovitch cycles[3], which are related to changes in the position of the planet, caused by the eccentricity, inclination and precession of the orbit of the Earth,

[3]Milutin Milankovitch: Serbian civil engineer and mathematician (1879–1958).

with a period 100,000 years. When all these changes occur simultaneously they cause a significant warming or cooling, that is, changes in the temperature of the Earth, as shown in Figure 12.9.

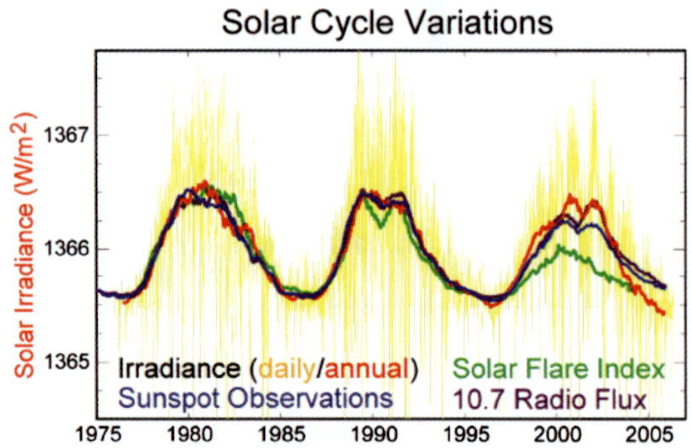

Figure 12.8 Cycle of sunspots (*Source*: http://es.wikipedia.org/wiki/Variaci%C3%B3n_solar)

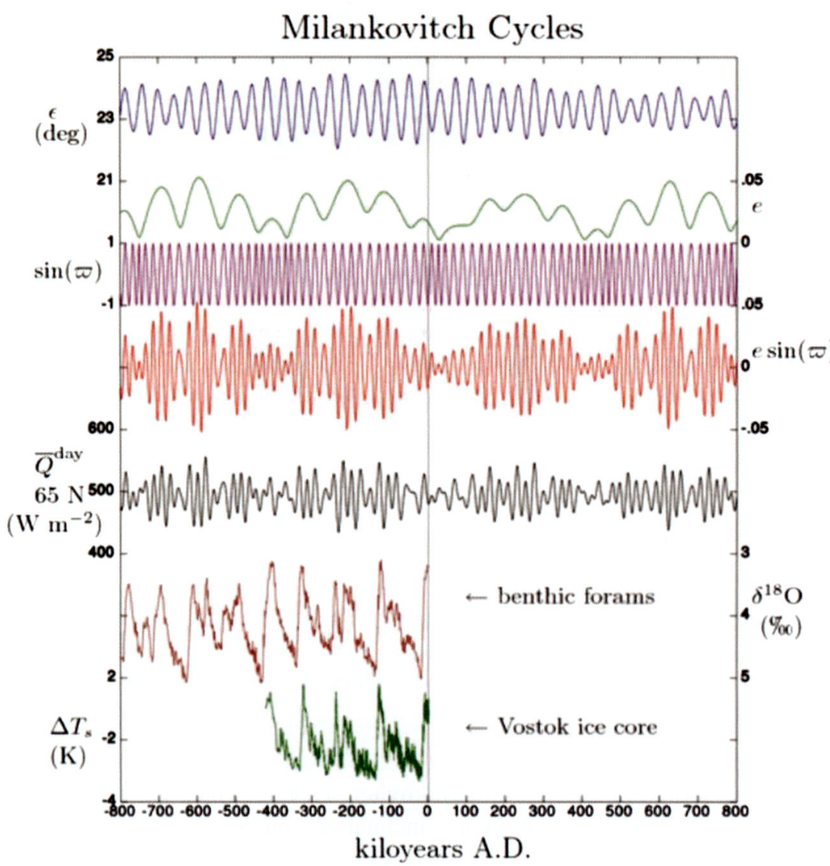

Figure 12.9 Milankovicht cycles (*Source*: http://en.wikipedia.org/wiki/Milankovitch_cycles)

Another important point is that the radiative forcings caused by carbon dioxide are not the most important. The most important greenhouse gas is in fact the water vapour, but in the GCM it is considered as a *feedback* that depends on temperature and is not of anthropogenic origin. Indeed, the basic proposed mechanism of global warming in the GCM is that by increasing (slightly) the temperature because of the CO_2, the increase caused by water vapour and clouds will generate a positive feedback increasing temperature in an uncontrolled manner (up to 3°C over the next 100 years)[4]. Warming may also be caused by other GHGs, but aerosols cause cooling. The most important sources of aerosols are the particles emitted by volcanic eruptions and, surprisingly, the fall of meteorites. Indeed, it was a meteorite that caused the cooling of the Earth that led to the extinction of dinosaurs. To project future warming trends, it is necessary to model all these effects in the Global Climate Models, and additionally, to incorporate the changes in CO_2 and other greenhouse gases caused by anthropogenic emissions.

12.1.3 CO_2 emission

The first alarm about the possibility of global warming caused by anthropogenic emission was given by atmospheric CO_2 measurements on the island of Mauna Loa, made by Keeling. These measurements showed that regardless of variations caused by the change of seasons, there was a significant trend of increasing CO_2. Over the years, the trend continued to increase until shaping the hockey stick curve which shows CO_2 levels above the natural variations of the gas in the pre-industrial era (before 1840), as shown in Figures 12.10 and 12.11. The hockey stick curve somehow overlooked the Medieval Warm Period between 1000 and 1200 and the Little Ice Age which occurred around 1600. For these reasons there is a controversy about its exactitude.

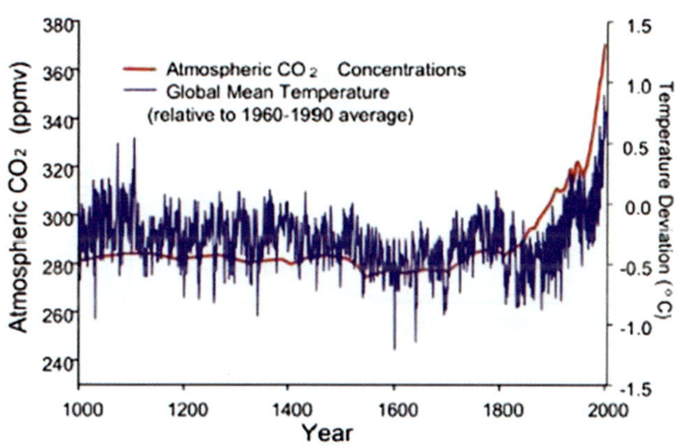

Figure 12.10 CO_2 increase of CO_2 in the industrial age (blue) correlated with the increase in the earth's temperature in respect to the average temperature during 1960–1990 (red) (*Source*: http://imageshack.us/photo/my-images/244/1000yearsco2smallej4.jpg/sr=1)

Figure 12.11 shows that this increase in CO_2 (top graph) is correlated with the increase in emissions from burning fossil fuels which occurred during the industrial era (bottom graph), from the year 1840 on. It seems that all the pieces fit in perfectly and that if the atmospheric discharge of CO_2 and other greenhouse gases

[4]However, it is not fully demonstrated that there is positive feedback. Some authors claim that water vapor decreases with temperature causing a negative feedback, that is, rather tends to reduce the temperature to the original temperature of the planet (see Bill Gray, 2009).

(which show similar increasing trends) continues, it is apparent that the fate of the planet is in question, unless necessary precautionary measures are taken. These measures include decrease in the current trends of fuel use, by application of new technologies that promote energy saving and substitution of fossil fuels with other energy sources such as solar, wind, and in spite of the 2011 explosion at Fukuyama's (Japan) atomic plant, the nuclear energy. At present (2011) global emissions of CO_2 amount to more than 9 billion tons, and the tendency is that this figure will double in the next 50 years. However it is important to note that the each KWh of energy produced by solar panel or wind is four and two times, respectively, more expensive than conventional energy produced from fossil fuels. Further developments in the clean energy field are necessary to lower prices before they can obtain a generalized use in times of economic crisis.

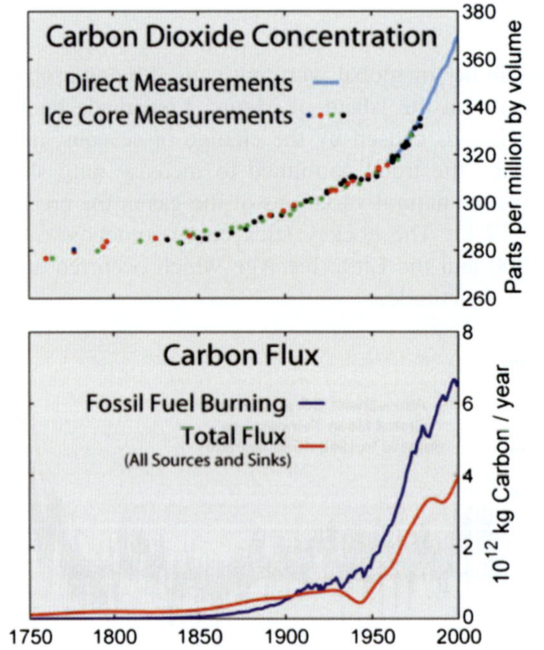

Figure 12.11 Growth in CO_2 in the industrial era correlated with fossil fuels emissions (*Source*: http://en.wikipedia.org/wiki/Greenhouse_gas)

12.1.4 GCM: global climate models

The GCM have became the widespread tools to determine how much CO_2 can be tolerated in the atmosphere without causing catastrophic global warming and how much fossil fuel emissions may be allowed to obtain the desired level of CO_2. The final temperature of the earth depends on solar radiation, albedo, the concentration of greenhouse gases, the CO_2 sequestered in biomass and dissolved in the ocean, and so on. These variables in turn depend on temperature, thus constituting a feedback phenomenon. In addition GHG concentrations also depend on the anthropogenic emissions and aerosols depend on volcanic events. And if we consider that each of these variables has a particular value for each of the different geographical locations in the planet's surface since the temperature, humidity, cloudiness and other variables vary from the Poles to the Equator and from east to west, one can imagine the complexity of the climate models. Predicting the climate differs from the predicting the weather. The latter applies to the knowledge of weather in one day and given place, while climate predictions refer to monthly averages at fairly large geographic areas. Climate is less exigent than weather forecasting which requires a degree of precision that GMC do not posses.

To achieve its goal, in this type of program the entire volume of the atmosphere is divided in cubic cells whose bases are on the surface of the Earth divided like a chessboard, and whose heights are the depth of the atmosphere divided into several layers from the surface to the stratosphere. In each cell, starting from given initial conditions, the fundamental equations of physics and thermodynamics are applied. The interrelationships between each cell and its neighbouring cells are applied every hour, simulating the evolution of Earth's climate. Figures 12.12 and 12.13 show the distribution of cells on the surface of the Earth taken in one of the most important models, and the formation of the cubic cells with unit volumes. It is possible to build models with smaller cells and estimates of shorter time intervals (e.g. one minute intervals) but these computations are increasingly complicated, demanding a greater number of operations, that is, greater computing power. But the chaotic nature of the Navier-Stoke's Equation at the core of the program model will always shown growing uncertainties. Until recently the implementation of the GCM were undertaken in powerful supercomputers located in a selected number of institutions which could afford it. But with the increase in computing capacity of personal computers (PC) model versions such as EdGMC of NASA have been produced, which can run on a PC to generate acceptable results as research tools. Many scientists prefer in fact using simpler versions for their research.

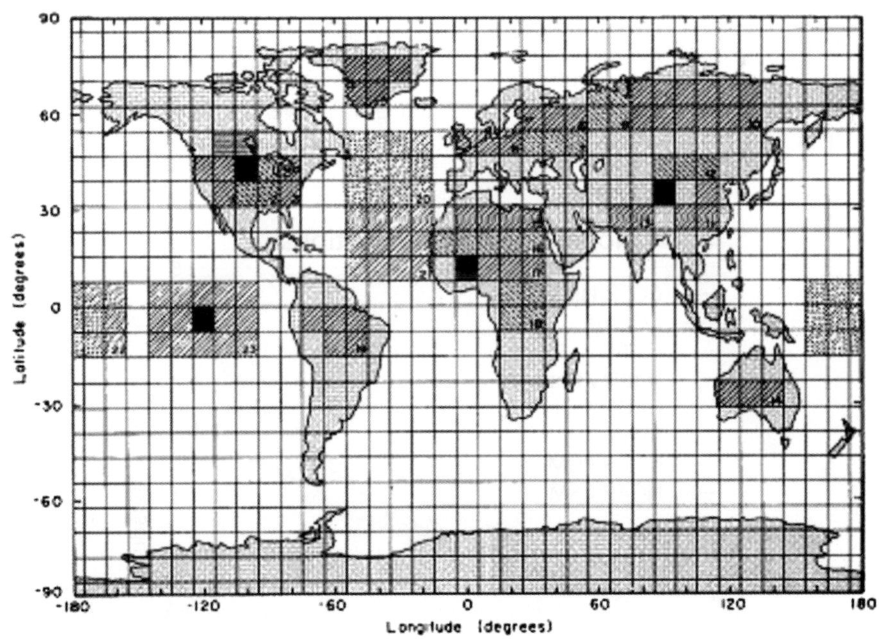

Figure 12.12 Distribution of the grid that forms the cells on the surface of the earth. the third dimension of the cubic cells is given by the height of the atmosphere (*Source*: Hansen *et al.* (1983) "efficient three-dimensional global models for climate studies: Models I and II", Monthly Weather Review, New York, USA)

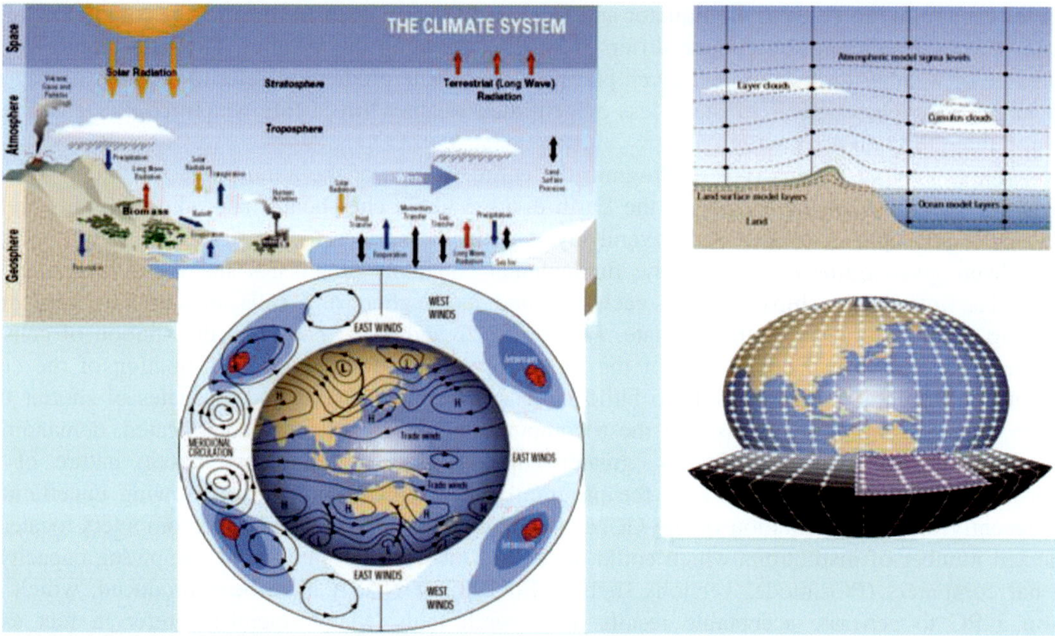

Figure 12.13 Determination of unit volumes (cubic cells) and other elements necessary for the GMC (*Source*: Gail Box (2000), "The physics of climate and climate change", ppt)

There are different types of models which apply to different scenarios of behaviour (i.e. business as usual-. nothing is done; or CO_2 doubling, etc.) and then the results can be compared with each other. Figure 12.14 shown the results of the predicted global average temperature by eight different GCMs for the IPCC *SRES A2* scenario which refers to a "very heterogeneous world, with increased population."

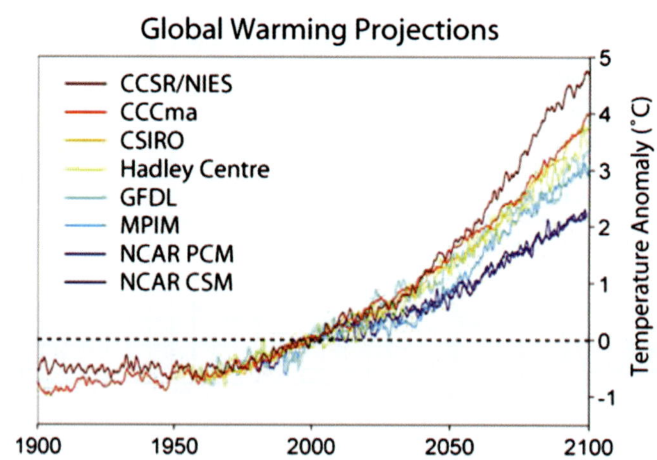

Figure 12.14 Temperature increase after 2000 predicted by eight different GCMs for the IPCC SRES A2 scenario (*Source*: http://www.theoildrum.com/story/2006/1/27/44052/9337)

As seen in Figure 12.14, all the models predict that by the year 2100, the temperature will increase. The range of the predicted temperature increase is between 2.3 and 4.8°C, concentrating around 3.5°C. The difference between the models is not big, but not small either. The IPCC offers various scenarios of action to anticipate probable warming for different cases[5] simulated by different models, which yield a probable value and range. The GCM is the tool available to anticipate the effects of the conduct of human society on the climate and on this tool scientists base their proposals to remedy the impact of the forecasted trends.

On the other hand, a new theory has recently been proposed[6] with a constraint in the initial conditions of the modelling that, if correct, could convert the GCM results obtained until now into irrelevant data. According to this theory, the planet has a regulatory system that maintains a constant temperature, regardless of internal phenomena, and it can only change with incidence of changes in solar radiation.

12.1.5 The data of *vostok* and other analyses

One of the most surprising modern science accomplishments has been undertaken by a group of Russian, French and North-American scientists in 1998 to measure the earth's climate in the past, including temperature and concentrations of GHGs, from analysis of the contents of air bubbles trapped in ice in the Russian station at Vostok, Antarctica. For this research 3623 meters were drilled in order to obtain ice cores. The conclusions of the analysis of data obtained in this study are surprising and have been confirmed by additional ice drilling in different locations. Given the importance of these data to validate the results obtained by the IPCC, it is important to give a brief discussion to understand their meaning. In the drilling of ice cores, age can be inferred from seasonal layers that are clearly visible, GHGs can be measured directly from the bubbles, and temperature can be estimated quite accurately from the variation in the bubble's frozen oxygen 18 isotope ($\delta^{18}O$).

Analysis of data from Vostok produced the results shown in Figure 12.15. At a first glance the results are surprising and at a time of their generation they were considered as the definite proof of the importance of the impact of CO_2 on global warming. A version of these curves is presented by Al Gore in his famous documentary.

However, these same data analyzed in detail showed phenomena that have become a Gordian knot for the community of the IPCC. Indeed, a careful analysis of the curves shows two things: (i) increase of the temperature always occurs before the increase of CO_2 concentration, about 800 years earlier, and (ii) the glaciations (variation of the glacial-interglacial temperature change of approximately 6°C) occur every 100,000 years, the same period of one of Milankovitch's cycles which are related to changes in solar radiation or sunlight that reaches Earth. This means that the temperature triggers climate changes and the CO_2 concentration increases in response to this change, and is not the reason for the change. Interestingly, it also triggered changes in CH_4 and N_2O, which follow the same patterns of temperature change. And temperature changes coincide with the extreme heat, as shown in Figure 12.16, which also shows the correlation of the glaciations to the Milankovitch's cycles. Although there are explanations of these graphs in the IPCC scenarios, they have not generated consensus within the scientific community itself.

---

[5]See the IPCC report "Climate Change 2007: The Physical Science Basis-Summary for Policymakers", for a clarification of scenarios and outcomes. The article can be downloaded from http://www.ipcc.ch/SPM2feb07.pdf.

[6]See F. Miskolczi (2007), "Greenhouse Effect in Semi-Transparent Planetary Atmospheres", Idojaras, 111, No. 1, Hungary.

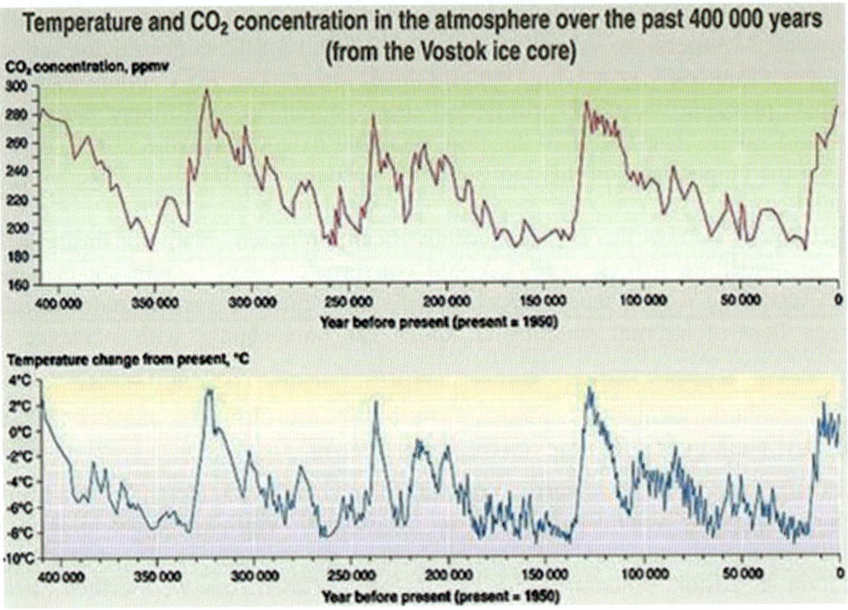

Figure 12.15 Variations in CO_2 concentration and temperature in the atmosphere over the past 400,000 years according to data from Vostok ice core (*Source*: http://www.worldviewofglobalwarming.org/images/vostok.jpg)

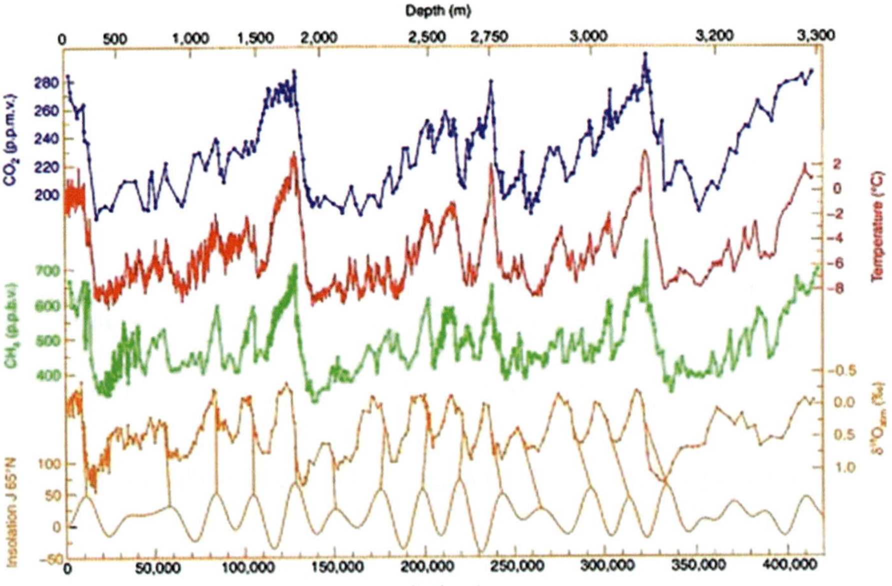

Figure 12.16 Variation of CO_2, CH_4, and temperature ($\delta^{18}O$) with solar radiance (*Source*: http://www.usgcrp.gov/usgcrp/images/Vostok.jpg)

The most common explanation is that: (i) radiation triggers temperature change but then, by feedback, the CO_2 takes command of climate change; and (ii) $\delta^{18}O$ (the *proxy* by which temperature id calculated) does not correspond to the same period of the gases, because of diffusion within the ice. This is not a satisfactory explanation to the community of climate scientists.

12.1.6 The Kyoto Protocol

The Kyoto Protocol is an agreement made within the framework of the United Nations Convention on Climate Change. This international instrument aimed at reducing within the period from 2008 to 2012 approximately 5% emissions, compared to 1990 emissions, of six greenhouse gases which are probable promoters of warming Global: (i) carbon dioxide (CO_2), (ii) methane (CH_4), and (iii) nitrous oxide (N_2O), plus three fluorinated industrial gases: (iv) hydrofluorocarbons (HFCs), (v) perfluorocarbons (PFCs), and (vi) sulphur hexafluoride (SF_6). The protocol required that it be ratified by countries with a combined CO_2 production greater than 55% of global emissions. Currently 160 countries have ratified the Kyoto protocol. The protocol follows, in general terms, the following principles[7]:

(a) Governments are separated into different categories: developed countries, referred to as Annex I countries (which have the obligation of reducing GHG emissions and must undergo an annual inventory of greenhouse gas) and developing countries, designated as non-Annexed (which have no obligation to reduce GHG emissions, but rather, can participate in the Clean Development Mechanism). There is a special category within the annexed countries, which is that of "transition countries" (towards a market economy), typically the countries of the former "iron curtain" that can benefit through the mechanisms of Joint Investment (JI).

(b) Any Annex I country which cannot meet its Kyoto obligation will be penalized 1.3 times its GHG emission permits and still have to meet them in the second quarter of the commission for each ton of GHG emissions that exceed their top in the first period of commission (i.e. 2008–2012).

(c) By 2008–2012, Annex I countries must reduce their GHG emissions by an average 5% below their 1990 levels. For many countries, such as the European Union (EU) Member States, this percentage corresponds in some cases to 15% below their expected GHG emissions in 2008. While the average emissions reduction is 5%, national limitations range from 8% reductions for the EU to a 10% emissions increase for Iceland, but since the EU intends to meet its target by distributing different rates among its member states, much larger increases (up to 27%) are allowed for some of the least developed countries of the EU. The limitations of the reduction will expire in 2013.

(d) The Kyoto Protocol includes "flexible mechanisms" that do not prevent economies of Annex I to obtain the limits by purchasing GHG emission reductions elsewhere. These can be purchased or obtained in financial exchanges (such as the new emissions trading system in the EU, which has no relation to Kyoto) or projects that reduce emissions in non-Annexed countries under the Clean Development Mechanism (CDM), or in other Annex I countries in transition, under the mechanism of JI (Joint Implementation).

(e) Only the certified emission reduction (CER) accredited by the board of the CDM can be bought and sold in this way. Under the aegis of the UN, Kyoto established the board of the clean development mechanism to determine and approve "CDM projects" in economies that are not annexed before granting CER. A similar scheme called "Joint Implementation (JI) is applied in transitional economies mainly covering the former Soviet Union and Eastern Europe before.

[7]http://en.wikipedia.org/wiki/Kyoto_Protocol.

This means that non-Annexed countries do not need to comply the GHG emission limits, but if they develop a GHG reduction project, it can provide them carbon credits to be sold to Annex I countries. This also creates a global system of carbon trade credits through the CER. In this way, it helps to highly developed countries to meet targets with cheaper projects in developing countries, and at the same time stimulates the reduction of greenhouse gases in these countries. This protocol favours developing countries. It also favours the developed nations that have no indigenous sources of fossil fuels, since these restrictions are tremendously expensive thus forcing the search for less expensive systems in energy use and research and development of alternative fuels. However, Germany made an exception with its coal industry, which does not need to comply with the protocol.

An interesting case is that of Russia. President Putin ratified Kyoto in November 2004 against the opinion of the Russian Academy of Sciences, the Ministry of Industry and Energy and the president's economic adviser, Andrey Illaroniov, in return for the EU support for Russia's admission to WTO (World Trade Organization). With this ratification, the Kyoto Protocol could enter into force after completing more than 55% of global CO_2 emissions. It is noteworthy that in the base year for emissions of 1990 Russia had a terrible environmental control system, so it is relatively easy for Russia to meet its goals.

The Kyoto Protocol allows the possibility of obtaining carbon credits through sequestration of CO_2 permanently in the jungles and forests, with sustained techniques of forestation and reforestation. They are being implemented gradually while obtaining agreements on verification methods. There are other methods of carbon sequestration in oceans, soil, and so on, not eligible for CER. The CO_2 can also be stored geologically (in the subsoil)[8].

The main countries that initially opposed the Kyoto Protocol were USA (most recently demanding that China and India be included among countries requested to provide legally binding commitments on curbing greenhouse gas emissions) and Australia (which signed in 2009 and now is one of the first countries to tax oil).

As of today (November 2011) Russia, Japan and Canada told the G8 that they would not join a second round of carbon cuts under the Kyoto Protocol at United Nations talks this year, and the US reiterated it would remain outside the treaty. The future of the Kyoto Protocol has become central to the efforts to negotiate reductions of carbon emissions under the UN's Framework Convention on Climate Change, whose annual meeting will take place in Durban, South Africa, from November 28 to December 9, 2011.

12.1.7 IPPC proposals

The IPCC is a multinational body in charge of conducting negotiations on global climate change and direct the scientific debate on global warming, the emission of carbon particles, the greenhouse effect, and others. The different scenarios of global climate change (those arising in the context of the Kyoto Protocol) are included among its lines of action.

There is a range of proposals to meet the proposed limits of GHGs. A summary of those proposals is presented in the September 2006 issue of *Scientific American*, "*Energy's Future-Beyond Carbon*". Basically, these proposals aim at the following goals: (i) avoiding the use of fossil fuels, (ii) implementation of nuclear energy, (iii) use of hydrogen as a fuel, (iv) use of biofuels and renewable energy (solar, wind, biomass, etc.), and (v) development of new methods of energy generation such as nuclear fusion, the capture of wind at high altitude, the use of solar panels in space to obtain energy from the sun, the production solar cells by nanotechnology, the implementation of a global electric

[8]http://en.wikipedia.org/wiki/Carbon_sequestration

interconnection by superconductors, utilization of the energy of waves and tides, the genetic creation of microbes to produce energy. All these methods are still far from immediate implementation and may be in use in the future. Curiously, one of the most promising clean energy sources, Low Energy Nuclear Reactions (LENR)[9] is not mentioned in this description. On the other hand, big hopes are assigned to nuclear energy, now banned in several countries because of the Fukuyama atomic plant explosion during the aftermath of 2011 earthquake in Japan of 8.9 degrees on Richter scale. Biofuels are still considered a promising path to meet the limits, in spite of the fact the breakthroughs are still needed to be able to replace oil with plant-based fuels, since efforts are proving that this is difficult to achieve[10].

In any case, the intended effect is to reduce atmospheric CO_2 emissions, for which action have been proposed in different fields, namely: (i) generation of energy, (ii) efficiency and energy conservation by the end-user, (iii) carbon sequestration, (iv) use of alternative energies, and (v) best practices in agriculture and reforestation. Each measure has several proposals to cut a wedge on the CO_2 production equivalent to the reduction of 25 Giga-tons of carbon in 50 years. Figure 12.17 shows the trend of growth of CO_2 emissions in the "business as usual" scenario – nothing is done-, and how, with seven wedges it can be maintained constant at 7 billion tons of emissions per year.

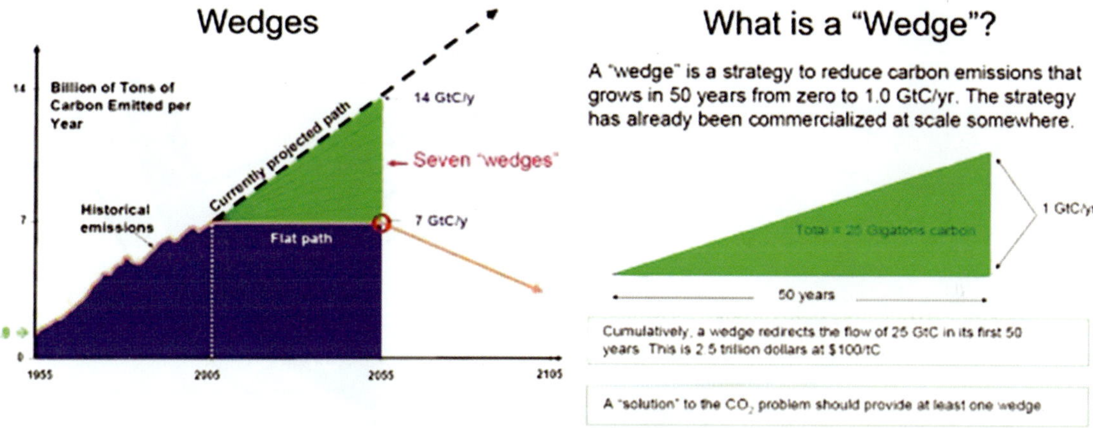

Figure 12.17 With the reduction of seven wedges it is possible to maintain the CO_2 emission levels of 2005 (*Source*: http://climatechangeaction.blogspot.com/2007/01/guest-post-by-almuth-ernsting-global.html)

The proposal of the wedges was made by Socolow and Pacala in the article "A *Plan to Keep Carbon in Check*" in the September 2006 issue of *Scientific American* (pp 28–35). According to the authors, the reduction of each wedge starts at zero and ends at 1 GtC/yr in 50 years (2055 in Figure 12.17). In the article the authors propose fifteen feasible methods of lowering a wedge, each method defined within one of the five fields mentioned above. For example, increasing the fuel economy from 30 to 60 miles per gallon in two billion cars produce a wedge of savings (25 GtC) in the field: "(ii) energy efficiency and conservation by the end user"; raising the efficiency of 1600 coal-based power plants from 40 to

[9]http://landshape.org/enm/rossis-e-cat-a-black-swan-event/

[10]Biello (2011).

60%, and making another wedge in the area: "(i) generation of energy". The fifteen proposed methods are shown in Figure 12.18 of which only seven are needed to curb global warming. Table 12.2 lists the 15 wedges in a different way.

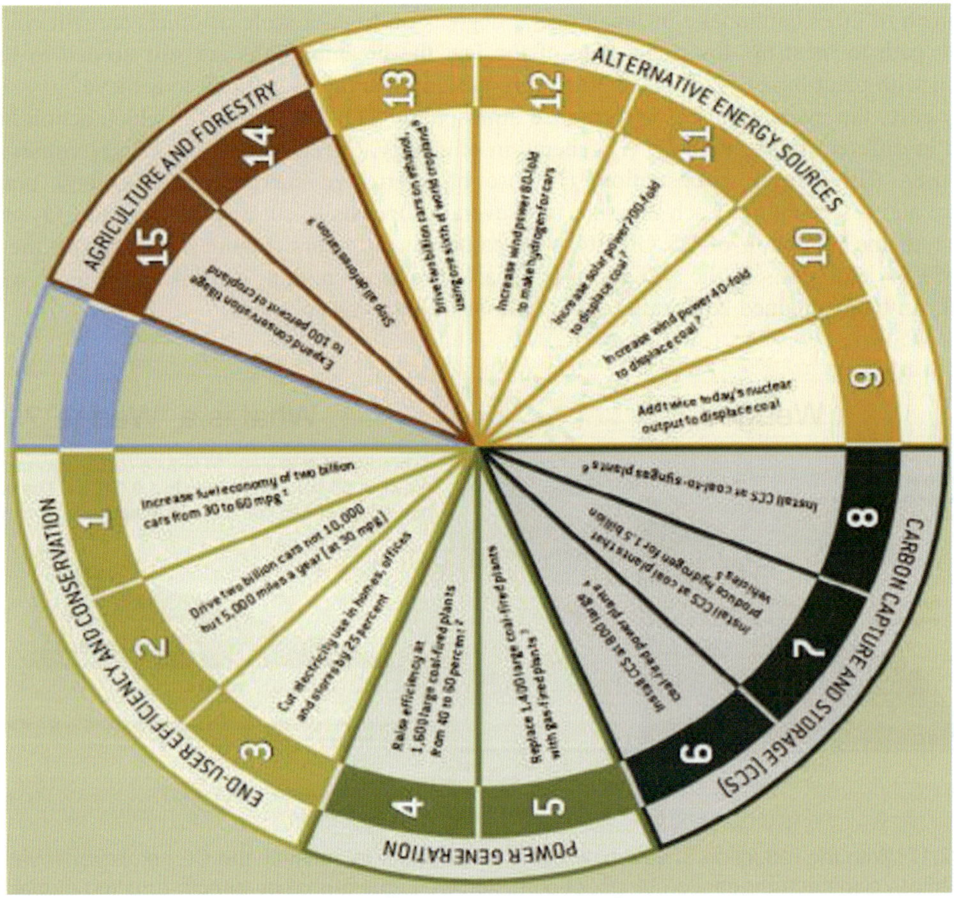

Figure 12.18 Fifteen ways to Obtain a wedge (*Source*: Socolow and Pacala, Scientific American, september 2006, pp. 50–57)

It is important to note that since the publication of Sokolow and Pacala paper in 2006, several developments has changed the wedge's perspective, such as: (i) the recent crash of Fukuyama atomic plant in Japan after the earthquake of 2011 which cooled the enthusiasm for nuclear reactor as a clean energy source, (ii) the failed results of biofuels in GHG reduction capabilities and cost (Biello, 2011), and (iii) the unsuccessful experiences to achieve low cost of wind and solar energy production in comparison to those of fossil fuel. However many of the proposed fifteen way to reduce a wedge are still feasible.

This shows that it is difficult to obtain the necessary reductions of CO_2 emissions to prevent global warming, without a strong political will. However, many people doubt that governments are willing to take actions that go against their interests in the short term, and it is therefore necessary to evaluate other alternatives to control the phenomenon of global warming, by cutting CO_2 global emissions.

Table 12.2 Fifteen ways to reduce a wedge of 25 GtC in 50 years.

1.	Fuel Economy: increase mileage from 30 to 60 mpg in 2 billion cars.
2.	Reduce annual mileage of 2 billion cars from 10,000 to 5000 miles (30 mpg).
3.	Efficient use of electricity in homes, offices, warehouses with cuts of 25% (a 50% cut would yield two wedges)
4.	Efficient use of oil and gas in space heating and industrial processes.
5.	Biofuels: drive two billion cars on ethanol (*shown ineffective and unfeasible*, see Biello, 2011)
6.	Increase wind power 80 times to make hydrogen for cars (*wind power twice as costly as coal power)*
7.	Increase 700-fold solar energy to displace coal (*solar power four times as costly as coal power)*
8.	Increase wind power 40-fold to displace coal (*wind power twice as costly as coal power*)
9.	Increase twice today's nuclear output to displace coal (*banned in many countries today*)
10.	Install CCS (Carbon Capture and Storage) in 800 large coal-fired power plants.
11.	Install CCS at coal plants to produce hydrogen for 1.5 billion vehicles
12.	Install CCS at coal plants-syngas
13.	Raise efficiency at 1,600 large coal-fired power plants from 40 to 60%
14.	Stop logging
15.	Expand conservation tillage methods to the entire arable area.

Source: https://www.commentmgr.com/projects/1148/docs/IAF_5Feb07_Part_One_B.ppt

12.1.8 Geoengineering proposals

Geoengineering is defined as "options that would involve large-scale engineering of our environment in order to combat or counteract the effects of changes in atmospheric chemistry". Many believe geoengineering is a politically more realistic method of controlling climate change. In fact, when analyzing the complexity of actions required meeting the objectives of the Kyoto Protocol and the multiplicity of political interests of different states, one can conclude that it is required to seek mechanisms to achieve more direct control of global climate. When one of the authors of this book (Orozco) became interested in the details of global warming, it seemed to him that the obvious thing was to control solar radiation, the main source of energy that heats the Earth. In fact the relevant calculations were made with the help of 1D (one dimension) programs written with *symbolic algebra* software *Mathematica 5.2* and concluded that the a deviation of 1% of solar radiation would control global warming the next 100 years. The result is presented in Figure 12.19, which shows that with a 1% deviation of solar radiation global average temperatures will vary by less than 1°C in the next century.

As the average temperature of the planet is composed of extreme temperatures that vary from the poles to the equator, and has multiple causes that the GCM can assess more accurately, then the EdGMC program from NASA was ran with 1% reduction in solar radiation during the next 20 years and it was found that the variation of the average surface temperature is minimal (see Figure 12.20), and the average temperature distribution in the air from the surface of the Earth in September for the years 2023–2027 according to the results of the program is as shown in Figure 12.21.

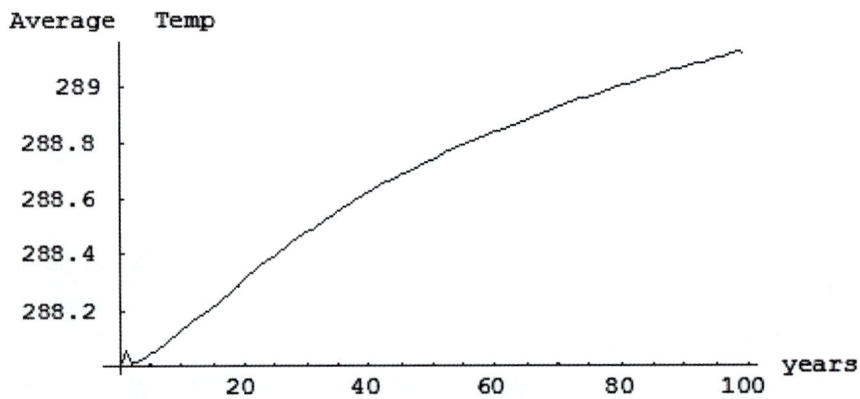

Figure 12.19 Temperatures in °K in the next 100 years in response to a 1% deviation of solar radiation [*Source*: A. Orozco, "umbrellaproject.nb", programa 1D en *Mathematica 5.2* (2007)]

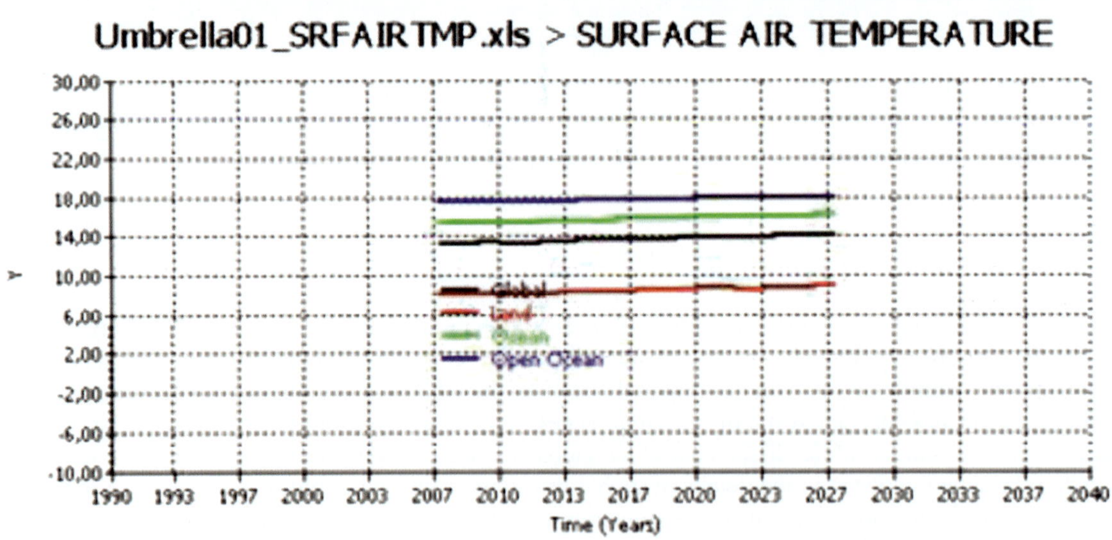

Figure 12.20 Temperatures in the next 20 years in response to a 1% deviation of solar radiation, according to EdGMC program of NASA [*Source*: A. Orozco, "Umbrella01" EdGMC run in 3D (2007)]

The question now is: How to divert 1% of solar radiation? One way is to place one (or thousands) umbrella(s) or screen(s) with an area of 1.3 million km^2 to intercept sunlight before entering the Earth's atmosphere. This will cost about 500 billion dollars, spread over 50 years, which means 10 US

$billion/year[11], not really much compared with the costs of Kyoto (USD1.2 trillion per year during 50 year).

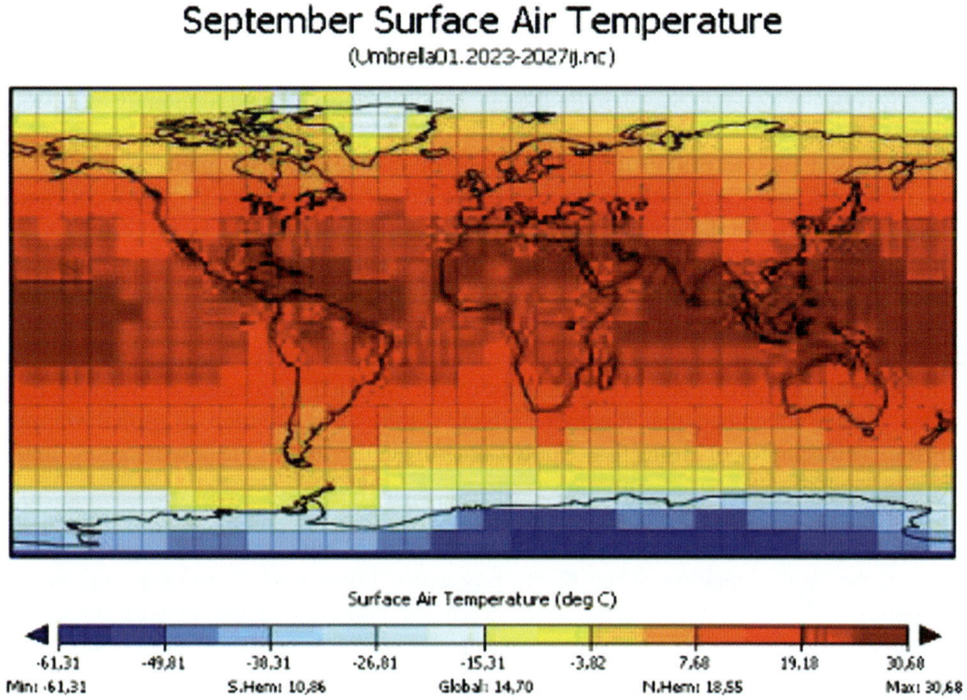

Figure 12.21 Average surface air temperatures in september during 2023–2027 in response to a 1% deviation of solar radiation, according to EdGMC program of NASA [*Source*: A. Orozco, "Umbrella01" EdGMC run in 3D (2007)]

However, there have been several proposals in this regard since years ago. In fact in 1991 the patent 5'003.186 was granted in the U.S. to Chang and collaborators, to control global warming with the launch into the stratosphere of Welsbach particles (metal oxides, e.g. aluminium) which reflect sunlight of certain wavelengths[12]. In 1998 Edward Teller, father of the hydrogen bomb, proposed the reduction of 1% of solar radiation reaching the Earth at an estimated cost of US$ one billion per year[13]. In 2006 the Nobel laureate Paul Crutzen proposed altering Earth's albedo deliberately adding sulphate aerosols to mimic the volcanic eruptions that have cooled the planet in the past, obviously making it in a controlled manner[14].

[11]A. Orozco, "EconomicsUmbrella.nb" (2007) en *Mathematica 5.2.*

[12]See http://www.airapparent.ca/library/abstract/welsbach_seeding.htm

[13]See http://www.hoover.org/publications/digest/3522851.html

[14]See http://www.realclimate.org/index.php/archives/2006/06/geo-engineering-in-vogue/

And even before that, Gregory Benford science fiction novelist and professor at the University of California at Irvine, had proposed to launch a Fresnel lens made of Mylar of 1,000 km in diameter to a geosynchronous orbit at the Lagrangian point L-1 (located at a distance of 1.5 million kilometres from earth, where the gravity of the Sun and of the Earth cancel each other). In fact others had proposed similar ideas before. But in June 2006 in a presentation at the *Skeptics Society*[15] Benford improved the idea proposing to locate a curtain of particles of diatomaceous earth of specific size in the stratosphere (of 25 km height) to reflect the income of 1% of solar radiation. Interestingly, we can select the wavelength that is rejected and the first option is very seductively, ultraviolet rays.

The estimated cost of one billion dollars per year of this solution is hundreds of times less than the proposed by Kyoto Protocol. Benford proposed the idea along with colleagues from Stanford University and the Lawrence Livermore National Laboratory. He projected to conduct a small scale test in the Arctic, where the released particles can be controlled by circulation patterns present in the area. NASA has shown interest in the project and there are private investors willing to fund the research.

In fact since 1992 the U.S. Congress had funded a study that was published in 2000 by the National Academy Press, with the participation of a member of the National Academies of Science, Engineering and Medicine, which concluded that the more effective mitigation solution to global warming is the launch in the atmosphere of reflective spray aerosol compounds via commercial military and private aircrafts.

It is clear then that the U.S., as the main opponent of Kyoto, has in its sleeve studied options for the case that global warming is a reality and not, as some think, an implausible idea. According to the apologists of geoengineering, the implementation of the Kyoto protocol is not feasible because trying to regulate the climate by policy actions is simply impossible, given the diversity of interests of states and due to human nature.

Projects to control solar radiation have the advantage of the speed of implementation. Other techniques that fall into the geoengineering category include:

- *Cool Roof*: using light-coloured materials to cover floors and ceilings to increase the albedo of the earth.
- Increasing the reflectivity of clouds, using fine aerosols of sea water to whiten the clouds with salt crystals. A very interesting proposal in this regard is the Roto-Nave Buckau. Stephen Salter proposed robots versions of these ships that could spray seawater into the air to create clouds, shielding the earth from the sun (see Figure 12.22).

There are other proposals for geoengineering in radically different directions. The main other, perhaps, is to add iron into the ocean to promote the growth of phytoplankton, which in turn sequester CO_2 from the atmosphere, known as the Geritol[16] solution. In a trial in which 500 kg of iron were added into a 100 km^2 area of the Pacific Ocean the stimulation to the formation of plankton served to sequester about 350,000 kg of CO_2. Performing this procedure on a global scale could capture billions of tons. However, many fear the side effects of such a planktonic soup spreading through all the oceans of the world, which would certainly be unpredictable.

[15]See http://www.heartland.org/Article.cfm?artId=19484

[16]See http://www.airapparent.ca/library/full_text/iron_vs_greenhouse.htm

Figure 12.22 The ship roto-nave buckau that can spray sea water into the air

12.1.9 Final reflections

Thanks to the documentary "An Inconvenient Truth" by Al Gore, the IPCC reports (with the support of the UN) and the Virgin Earth Challenge Prize of US$ 25 million for finding a technically and economically feasible solution to the problem of global warming, it was possible to convey to the public the notion that anthropogenic global warming is a reality and that the effects of this phenomenon has implications which may be disastrous for mankind if not prevented in a timely manner.

Consequently, strategies have been adopted as the Kyoto protocol, with the support of the IPCC, leading to controling the emissions of CO_2 and other greenhouse gases. The protocol is supported almost by all non-Annex countries, including two of the largest global contributors of GHG: China and India.

On the other hand, the geoengineering community has made proposals, some a little dangerous as the Geritol solution, others to be implemented somewhat in the future as a *Mylar* lens to be placed 1,500,000 km away from the Earth and other plausible as the inflation of aerosol of Diatomaceous earth of 1 micron in diameter to reject 1% of solar radiation and with it a great deal of ultraviolet radiation. The most practical up to day is the solution known as the Budyko' Blanket which proposes to pump to the stratosphere SO_2 with a cost of USD 250 million per year estimated by *Intellectual Ventures* (Orozco, 2011). The world needs to keep working on these type of solutions which do not require monumental political effort as that required for implementing Kyoto Protocol.

What then needs to be done? the solution probably is in the middle, a combination of several initiatives. To resolve this impasse it will be necessary to reduce some emissions wedges of 25,000 million tons of carbon, reflect away from the earth part of the solar radiations at a level of some digits per thousand spraying aerosols in the stratosphere, and replace some fossil fuels with alternative forms of energy. One recently emerging alternative form of energy is the known as Low Energy Nuclear Reactions (LENR), which is in an experimental stages in Bologna, Italy[17] and other places (including at NASA), which if successful would represent a completely new avenue to confront global warming.

[17]http://landshape.org/enm/academics-eat-rossis-dust/

Global warming will probably not be the predicted warming, so it will be necessary to refine the technical tools, especially the GMC, and above all to improve the knowledge of the laws that govern the planet's climate. With these and other actions, the human species will manage to reign on Earth for another thousand years. However, major environmental organizations such as Friends of the Earth and Greenpeace are reluctant to endorse geoengineering. Some have argued that any public support for this type of solution could undermine the fragile political consensus to reduce GHG emissions.

12.2 WASTEWATER TREATMENT IMPACT ON CLIMATE

12.2.1 Emission factors (EF) of green house gases in wastewater treatment systems[18]

The wastewater originating from a variety of domestic, commercial and industrial sources can be treated *in situ*, or can be conveyed by sewerage systems to a centralized plant, or be discharged through sub-aquatic outfalls. Domestic wastewater is generated by household activity, while industrial wastewater is the waste of industrial processes. The treatment and discharge systems can be differentiated for rural and urban users and for high income and low income users. The wastewater treated in a wastewater treatment system (WWTS) includes both domestic and industrial wastewater, as well as some rainwater, all of which jointly are known as municipal wastewater.

Most WWTS consist of several combined processes including mechanical treatment, biological and chemical processes. In the mechanical treatment suspended particles, sand and oil are separated from the wastewater and dehydrated sludge is generated and further stabilized by different processes. In general, the stabilization of organic sludge can be divided into biological and chemical processes. Biological processes include anaerobic stabilization where sludge is stabilized in anaerobic digesters, and aerobic stabilization is effected by extended aeration of the sludge.

The wastewater and sludge produced can be a source of methane (CH_4) when treated anaerobically or disposed of without treatment. They can also be a source of nitrous oxide (N_2O) if they are nitrification and then denitrification. *Emissions of carbon dioxide (CO_2) from wastewater are not considered in the IPCC guidelines because they are of biogenic origin and should not be included in the counting national total emissions.* "Biogenic origin" means that the carbonaceous organic matter of the wastewater comes from the atmospheric CO_2: which is used by CO_2-fixing vegetation to form organic biomass, which in turn serves as food for animals, and so on. This organic matter is consumed by man, and part of it, due to its metabolic transformation, is discharged into the wastewater. When the organic matter is decomposed in the wastewater treatment plants, the CO_2 originally captured in food is recycled again (liberated to the athmosphere), so there is no new addition of carbon dioxide to the atmosphere. The carbon dioxide came from the atmosphere and is returned to the atmosphere in the same amount.

This is the reason that GHGs considered in wastewater treatment plants are CH_4 (generated by anaerobic treatment processes) and N_2O (generated by nitrification and denitrification). The methane gas produced in anaerobic treatment has an equivalence of 1ton $CH_4 = 23$ ton-equivalent CO_2 (CO_{2e}) from the standpoint of the Global Warming Potential (GWP) of 100 years. This locates the methane as the most important GHG in wastewater treatment industry. Nitrous oxide, produced in the process of nitrification-denitrification, has an equivalence of 1ton $N_2O = 320$ ton CO_{2e}, that is, it is a stronger GHG than methane, but it is produced only

[18]Based On "2006 IPCC guidelines for national greenhouse gas inventories, Volume 5: WASTE-Chapter 6: wastewater treatment and discharge", IPCC, WMO, UNEP (2006); In "Guidelines for College-Level Greenhouse Gas Emissions Inventories" by Julian Dautremont-Smith (2002); "Approved baseline and monitoring methodology AM0039: Methane emissions reduction from organic wastewater and bioorganic solid waste using composting" de UNFCCC/CCNUCC (1996).

in small quantities so its impact is not large and it is less important than the methane. If nitrification-denitrification does not occur in the WWTS it might occur in the effluent receiving water body and the generation of N_2O may occur there.

Emission factor of N_2O

According to FAO, the consumption of protein in the USA in 1999 was 114.9 g protein/capita · d, while in Colombia it was 60 g (0.060 kg) in 2000[19]. The default fraction of nitrogen in the proteins provided by the IPCC is 0.16 kg-N/kg-protein. Default Emission Factor (EF) in sewage given by the IPCC is 0.005 kg N_2O-N/kg-N (range 0.0005 to 0.25). Using the default values as described above and taking into account that the fraction of N in N_2O is (28/44) the equation for the production of N_2O based on the nitrogen discharged in the wastewater directly into a water body, is thus:

$$\text{kg}\,N_2O/\text{yr} = N_{AR}EF_{AR}(44/28) \tag{12.4}$$

where:

N_2O = nitrous oxide, kg N_2O /year
N_{AR} = nitrogen discharged to the aquatic environment, kg N/year
EF_{AR} = emission factor of N_2O discharged in wastewater, N_2O-N kg/kg N
The (44/28) is the conversion factor from kg N_2O-N to kg N_2O

On the other hand, the N_2O produced during nitrification and denitrification in wastewater treatment, as measured in the USA plants, is 3.2 g N_2O/capita · yr[20] and if we have industrial and commercial discharges, it is necessary to apply a default factor ($F_{ind\text{-}com}$) of 1.25, so another equation for N_2O production in WWTS is:

$$(\text{kg}\,N_2O/\text{yr})_{WTSP} = P \times 3.2\,\text{kg}\,N_2O/\text{capita} \cdot \text{year} \times F_{ind-com} = P \times 4\,\text{kg}\,N_2O/\text{capita} \cdot \text{yr} \tag{12.5}$$

where P is the population.

Emission factors of CH_4

The IPCC provides the following equation to calculate CH_4 emissions from treated wastewater:

$$\text{kg}\,CH_4 = P \cdot (\text{DOM}) \cdot [1 - (\%\text{DOM})\,\text{eliminated as sludge}](CH_4)_{max}\text{MCF} - \text{kg}(CH_4)_{burned} \tag{12.6}$$

where DOM is the per capita degradable organic matter and MCF is the methane correction factor. Similarly, to calculate the emissions from sludge treatment, the IPCC suggests the following equation:

$$\text{kg}\,CH_4 = P \cdot (\text{DOM}) \cdot [(\%\text{DOM})\text{eliminated as sludge}] \cdot (CH_4)_{max} \cdot \text{MCF} - \text{kg}(CH_4)_{burned} \tag{12.7}$$

The per capita degradable organic matter can be measured as BOD. The IPCC suggests a default value (DOM) = 0.05 kg DBO_5/capita · d, unless we know the real value for the country. The latest IPCC guidelines on best practices suggest a maximum output of 0.6 kg CH_4/kg DBO_5. The methane correction factor (MCF) depends on the type of treatment (see Table 12.3 below) and the proposed

[19]Food Security Statistics – Colombia, FAO (2006)

[20]"2006 IPCC guidelines for national greenhouse gas inventories, Volume 5: WASTE-Chapter 6:wastewater treatment and discharge", IPCC, WMO, UNEP (2006)

production should be affected by (0.6) to obtain the actual maximum methane production. According to the IPCC, the MCF varies between 0.0 for a fully aerobic system to 1.0 for an anaerobic system. If no data are available, the default is to use 0 for aerobic systems and 1.0 for anaerobic systems. The operator of the WWTS must be able to provide information on the proportion of (DOM) removed as sludge and total CH_4 captured and burned, in both wastewater and sludge treatment. If there are no data on the amount of CH_4 recovered and burned, the IPCC suggests a default of 0. In its inventory of GHG emissions the US EIA (US Energy Information Administration) supposes that "the methane recovery in facilities for municipal wastewater treatment is negligible."

Table 12.3 Default MCF values for domestic wastewater.

Type of treatment and discharge pathway or system	Comments	MCF[1]	Range
Untreated system			
Sea, river and lake discharge	Rivers with high organics loadings can turn anaerobic.	0.1	0–0.2
Stagnant sewer	Open and warm	0.5	0.4–0.8
Flowing sewer (open or closed)	Fast moving, clean. (Insignificant amounts of CH_4 from pump stations, and so on)	0	0
Treated system			
Centralized, aerobic treatment plant	Must be well managed. Some CH_4 can be emitted from settling basins and other pockets.	0	0–0.1
Centralized, aerobic treatment plant	Not well managed. Overloaded.	0.3	0.2–0.4
Anaerobic digester for sludge	CH_4 recovery is not considered here.	0.8	0.8–1.0
Anaerobic reactor	CH_4 recovery is not considered here.	0.8	0.8–1.0
Anaerobic shallow lagoon	Depth less than 2 metres, use expert judgment.	0.2	0–0.3
Anaerobic deep lagoon	Depth more than 2 metres	0.8	0.8–1.0
Septic system	Half of BOD setdes in anaerobic tank.	0.5	0.5
Latrine	Dry climate, ground water table lower than latrine, small family (3–5 persons)	0.1	0.05–0.15
Latrine	Dry climate, ground water table lower than latrine, communal (many users)	0.5	0.4–0.6
Latrine	Wet climate/flush water use, ground water table higher than latrine	0.7	0.7–1.0
Latrine	Regular sediment removal for fertilizer	0.1	0.1

[1]Based on expert judgment by lead authors of this section.
Source: IPCC, WMO, UNEP (2006), "2006 IPCC guidelines for national Greenhouse gas inventories, Volume 5".

Indirect emissions

For fully aerobic WWTP, up to 1.4 kilograms of CO_2/kg COD (removed) originate from the production of energy for aeration. This means that more CO_2 is produced in generating energy for the aeration than in the actual treatment process of the wastewater, and this energy generation is typically from fossil fuels, while the carbon of wastewater comes mostly from sources of renewable energy (i.e. food, which is of biogenic origin).

With this consideration, a change from anaerobic to aerobic processes for sludge treatment or for the liquid wastewater treatment has a massive impact on the production of fossil fuel: 0.8 kg extra of CO_2 produced per kg COD removed because of the shift to aerobic sludge digestion. In terms of GHG production, the total output (in CO_2 equivalent) may be reduced from 2.4 kg of CO_2/kg COD (removed) in fully aerobic treatment to 1.0 kg CO_2/kg COD (removed) in anaerobic processes.

In this respect, significant advantages are associated with the use of anaerobic processes because it is possible to largely eliminate the net energy input to the aerobic processes when using anaerobic processes, and therefore reduce the production of greenhouse gases from fossil fuels.

12.2.2 Methodologies of quantification of green house gases in wastewater treatment systems

Based on the foregoing discussion we suggest the following methodology for calculating GHG in the WWTP:

Quantification of N_2O

Emissions of nitrous oxide (N_2O) can take place as direct emissions at treatment plants, or indirectly as emissions after the disposal of wastewater or effluents into watercourses, lakes or the sea. Direct emissions from nitrification and denitrification processes in wastewater treatment plants can be considered small sources. Typically, these emissions are much lower than those of the effluent. and can only be of interest to advanced countries whose wastewater treatment plants predominantly include nitrification and denitrification. The calculation of N_2O emission from a WWTS is as follows:

(1) Determine the contributing population (P) to the WWTP
(2) Determine if there are industrial and commercial contributions. If there are, assume $F_{ind\text{-}com} = 1.25$.
(3) Apply Equation 12.5:

$$(\text{kg N}_2\text{O/yr})_{\text{STAP}} = \text{P} \times 3.2\ \text{kg N}_2\text{O/capita} \cdot \text{yr} \times \text{Find-com}$$

(4) Determine equivalent CO_2 (CO_{2e}) with:

$$\text{CO}_{2e} = 320\ \ (\text{kg N}_2\text{O/yr})_{\text{STAP}}/1000\,\text{ton}$$

Quantification of CH_4

Methane production of a WWTP depends on the watewater treatment process and on the sludge production and treatment. If treatment is aerobic in both cases (wastewater and sludge), then there is no production of CH_4 and the CO_2 produced is not taken into account because it is biogenic, that is, CO_2 eventually will be trapped again in the wastewater through the biological mechanisms of organic matter formation.

If the wastewater treatment is aerobic but there is anaerobic sludge treatment, then, for the calculation of CH_4 produced, use the equation:

$$(\text{kg CH}_4)_{\text{WWTP}} = \text{P} \times \text{DOM} \times (\%\text{DOM})\ \ \text{eliminated as sludge} \times (\text{CH}_4)_{\max} \times \text{MCF}$$

If part or all of the CH_4 is recovered or burned, subtract it from the total amount produced, and it eventually may be subject to obtain Carbon certificates in the case of a non-annexed countries under the Kyoto protocol. The calculation is as follows:

(1) Determine the contributing population (P) to the WWTP (use the concept of equivalent population if there are industrial wastewaters).
(2) Determine country's DOM or set a DOM $= 0.05$ kg DBO_5/capita · d.
(3) Determine the fraction of the DOM trapped as primary or secondary sludge. If the exact figure is not known for the WWTP, use 0.30.
(4) Define $(CH_4)_{max} = 0.05$ kg DBO_5/capita · d
(5) From Table 12.3 determine the MCF = 0.8.
(6) For aerobic wastewater treatment apply the equation:

$$(\text{kg}\,CH_4)_{WWTP} = P \times DOM \times (\%DOM)\ \text{eliminated as sludge} \times (CH_4)_{max} \times MCF$$

(7) Determine the CO_{2e} with:

$$CO_{2e} = 23\ (\text{kg}\,CH_4/\text{yr})_{WWTP}/1000\ \text{ton}$$

(8) If the wastewater treatment is anaerobic, then use the equation:

$$(\text{kg}\,CH_4)_{WWTP} = P \times DOM \times (1 - \%DOM)\ \text{eliminated as sludge} \times (CH_4)_{max} \times MCF$$

and follow the previous steps.

12.2.3 The impact of wastewater on global warming

To have some measure of the participation of wastewater in the production of GHG, it is important to obtain the proportions by sector from the country's GHG inventories (see the different sectors in Figure 12.23) and the GHG for different countries (see Figures 12.24 and 12.25). From these figures it is observed that the proportions for different countries are reasonably constant, so if you get data from a country that has good inventories of GHGs which were estimated with sufficient resources, it is possible to extend the findings to the rest of the world. As a basis of inventory for these proportions the inventories of the USA since its GHGs inventories were carefully estimated.

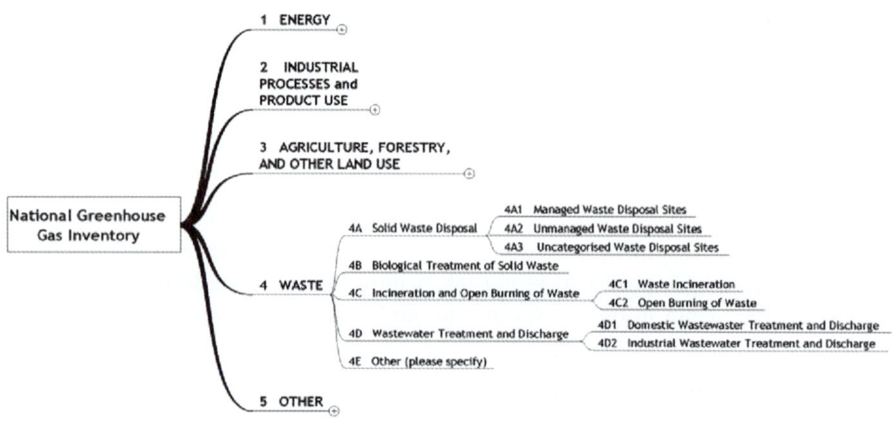

Figure 12.23 Structure of GHG-producing sectors [*Source*: IPCC, WMO, UNEP (2006)]

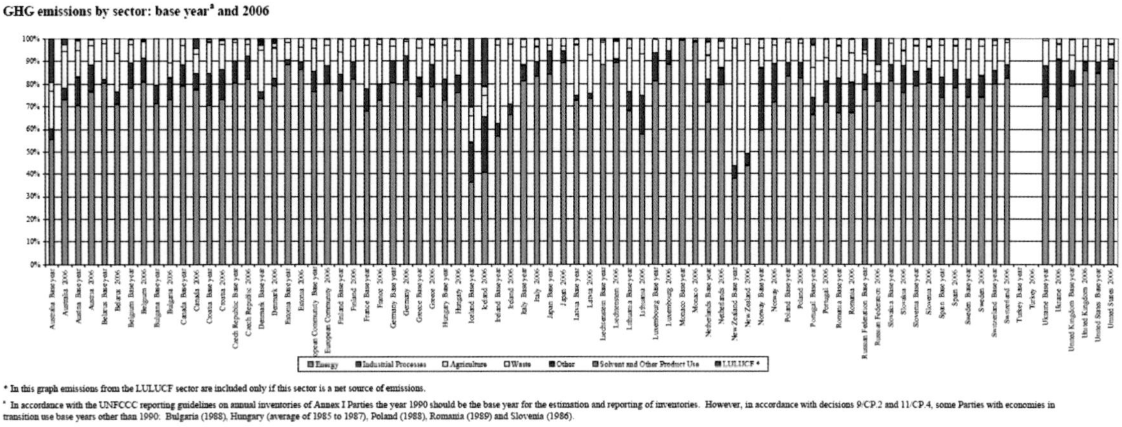

Figure 12.24 Share of GHG by sector and country [*Source*: EPA (2008), "Inventory of U.S. Greenhouse gas emissions and sinks: 1990–2006"]

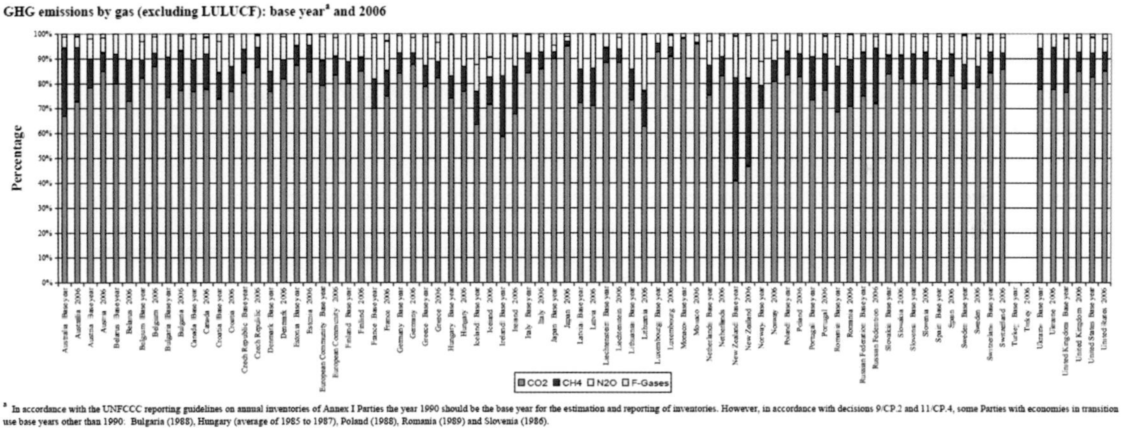

Figure 12.25 Share of GHG by GHG and country [*Source*: EPA (2008), "Inventory of U.S. Greenhouse gas emissions and sinks: 1990–2006"]

Table 12.4 taken from the EPA (2008) - "Inventory of U.S. Greenhouse Gas Emissions and Sinks: 1990–2006", USA, shows that in terms of CO_{2e}, the waste sector contributes 2.3% of the total GHG. In addition, Table 12.5 shows the different proportions of contributions within the waste sector and indicates that in terms of CO_{2e} the percentage of GHG contribution by the wastewater's sector to the waste sector is 20%. Figure 12.26 summarizes the above conclusions and provides that wastewater contributes 0.46% of the total GHG emissions. Consequently, it is concluded that the wastewater sector is not a significant sector in terms of generating GHGs, but it still has an influence on the final results of GHG emissions, so it should not be overlooked.

Table 12.4 Recent trends in US Greenhouse gas emissions and sinks by chapter/IPCC sector.

Chapter/IPCC sector	1990	1995	2000	2001	2002	2003	2004	2005	2006
Energy	5,203.9	5,529.6	6,067.8	5,982.8	6,036.3	6,078.3	6,150.9	6,174.4	6,076.9
Industrial processes	299.9	315.7	326.5	297.9	308.6	301.2	315.9	315.5	320.9
Solvent and other Product use	4.4	4.6	4.9	4.9	4.4	4.4	4.4	4.4	4.4
Agriculture	447.5	453.8	447.9	463.7	449.0	434.3	432.1	453.6	454.1
Land use, land-use change, and forestry (Emissions)	13.1	13.6	30.0	20.0	28.4	19.7	17.1	23.2	36.9
Waste	179.6	176.8	155.6	152.1	154.5	160.3	157.7	158.7	161.0
Total Emissions	**6,148.3**	**6,494.0**	**7,032.6**	**6,921.3**	**6,981.2**	**6,998.2**	**7,078.0**	**7,129.9**	**7,054.2**
Net CO_2 Flux from land use, land-use change, and forestry (Sinks)*	(737.7)	(775.3)	(673.6)	(750.2)	(826.8)	(860.9)	(873.7)	(878.6)	(883.7)
Net emissions (sources and sinks)	**5,410.6**	**5,718.7**	**6,359.0**	**6,171.1**	**6,154.4**	**6,137.3**	**6,204.3**	**6,251.3**	**6,170.5**

* The net CO_2 flux total includes both emissions and sequestration, and constitutes a sink in the United States. Sinks are only included in net emissions total.

Note: Totals may not sum due to independent rounding. Parentheses indicate negative values or sequestration.

Source: EPA (2008), "Inventory of U.S. Greenhouse Gas Emissions and Sinks: 1990–2006".

Note that Waste Sector Emissions are 2.3% of Total Emissions.

Table 12.5 Emissions from wastes.

Gas/Source	1990	1995	2000	2001	2002	2003	2004	2005	2006
CH_4	**172.9**	**169.1**	**146.7**	**143.0**	**145.5**	**151.0**	**148.1**	**149.0**	**151.1**
Landfills	149.6	144.0	120.8	117.6	120.1	125.6	122.6	123.7	125.7
Wastewater Treatment	23.0	24.3	24.6	24.2	24.1	23.9	24.0	23.8	23.9
Composting	0.3	0.7	1.3	1.3	1.3	1.5	1.6	1.6	1.6
N_2O	**6.6**	**7.7**	**8.9**	**9.2**	**9.0**	**9.3**	**9.6**	**9.7**	**9.9**
Domestic Wastewater Treatment	6.3	6.9	7.6	7.8	7.6	7.7	7.8	8.0	8.1
Composting	0.4	0.8	1.4	1.4	1.4	1.6	1.7	1.7	1.8
Total	**179.6**	**176.8**	**155.6**	**152.1**	**154.5**	**160.3**	**157.7**	**158.7**	**161.0**

Note: Totals may not sum due to independent rounding.

Note that Emissions of Domestic Wastewater Treatment are 20% of the Emissions of the Waste Sector.

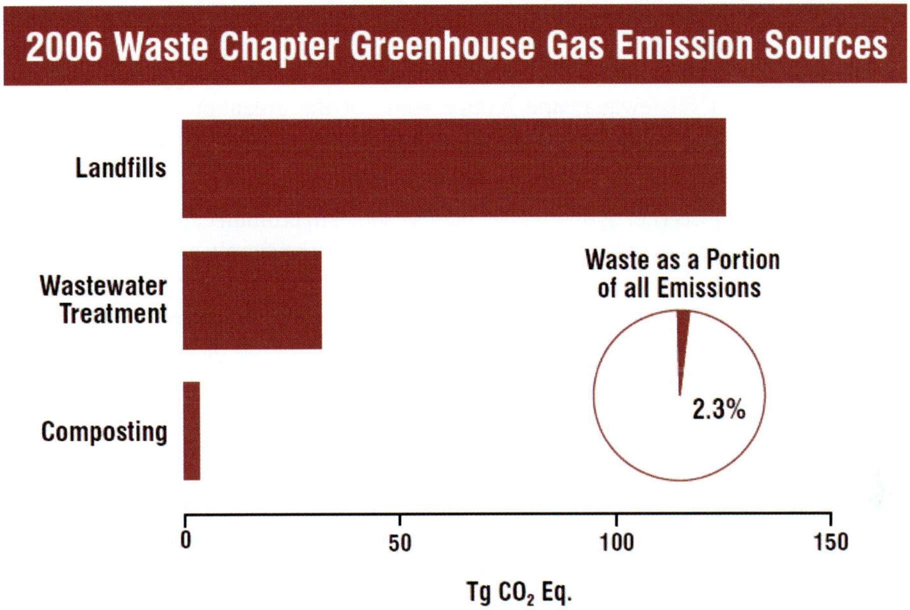

Figure 12.26 Emissions from wastewater treatment are 0.46% (0.20 × 0.023) of total GHG emissions [Source: EPA (2008), "Inventory of U.S. Greenhouse gas emissions and sinks: 1990–2006"]

12.3 CLEAN DEVELOPMENT MECHANISMS (CDM)

12.3.1 The kyoto protocol and the CDM

According to Wikipedia[21], the Clean Development Mechanism (CDM) is an agreement on the Kyoto Protocol, established in Article 12, which allows governments of industrialized countries (also called developed countries or Annex 1 countries of the Kyoto Protocol) and businesses (natural or legal persons, public or private entities) enter into agreements to meet targets for reducing greenhouse gases (GHGs) in the first commitment period between the years 2008–2012, investing in emissions reduction projects in developing countries (also referred to countries not included in Annex 1 of the Kyoto Protocol) as an alternative to purchase certified emission reductions (CERs) at a lower costs than in their markets.

The CDM can provide a reduction in costs for the countries annexed, while allowing the possibility of transferring clean technologies to developing countries. By investing in these CDM projects governments or companies receive "certified emission reductions" (CERs, one of three types of carbon bonds) at lower cost than can be purchased in the carbon market and simultaneously manage to comply with the reduction targets to which they are committed.

The CDM is supervised by the CDM Executive Board (CDM-EB) and is under the guidance of the Conference of the Parties (COP) of the United Nations Framework Convention on Climate Change (UNFCCC). It is also necessary to obtain an approval of voluntary participation and certification by the Designated National Authority (DNA) of contribution to the sustainable development country in which the project is located. The DNA is usually the Ministry for the Environment, which in turn may establish internal procedures for approval. To obtain the certification of emissions, interested parties

[21]http://en.wikipedia.org/wiki/Clean_Development_Mechanism

(industrialized country and developing country recipient of the project) should demonstrate a long-term and measurable real emission reduction.

CDM projects can be carried out in various sectors and activities such as energy; industry; transportation; agriculture; solids waste and wastewater; and reduce some of the greenhouse gases or capture CO_2 in forestry projects (reforestation). They can be small or large scale project. Projects that are classified as small-scale are:

- Renewable energy projects (i.e. hydroelectric plants) with a maximum capacity equivalent to 15 MW,
- Energy efficiency projects to reduce energy consumption by the supply and/or demand side, up to 60 GWh per year;
- Other projects that reduce anthropogenic emissions and directly emit less than 60 kt CO_{2e};
- Reforestation activities with a capture capacity of less than 16 kton of CO_{2e} annually.

Some useful definitions are:

COP: The Conference of the Parties, the supreme body of the Kyoto Protocol which is responsible for the implementation of the protocol.

Executive Board (EB): is responsible for overseeing the operation of the CDM, to review and prepare the detailed decisions of the mechanism and ensure satisfactory performance. It is also the entity that approves the list of projects and issues CERs.

Designated Operational Entity (DOE): Institutions responsible for validating CDM projects and verify and certify the CERs.

Designated National Authority (DNA): In the host countries, is responsible for evaluating the sustainable development criteria country on a CDM project. After assessing the project, the DNA issuesd a letter approving the project. For example, in the case of Colombia the DNA is the Ministry of Environment, Housing and Territorial Development.

CER: Certified Emission Reductions.

PDD: Project Design Document.

The format and guidelines for preparing a PDD is available on the website of the UNFCCC http://cdm.unfccc.int/Reference/Documents. The CDM project cycle is presented in Figure 12.27. Table 12.6 presents some applicable projects to CDM by sector.

12.3.2 Requirements of the CDM

To develop a practical methodology for preparation of a CDM project, use the PDD form which is available in pdf format on the web:

http://cd4cdm.org/Publications/cdm%20guideline%202nd%20edition.pdf (as of 2011). This brochure shows the guidelines and procedures, step by step, to fill in the paperwork.

In order to be eligible for CDM, take into account the following:

- Projects should be developed on a voluntary basis.
- GHG emissions reductions must be real and measurable.
- GHG emissions reductions must be additional. Some definitions:

  Additionality: A project activity is additional if GHG emissions of anthropogenic origin are reduced below those that occurred in the absence of the proposed project activity.

  Baseline or reference scenario for a CDM project activity is the scenario that reasonably represents the GHG emissions of anthropogenic origin that occurred in the absence of the proposed project activity.

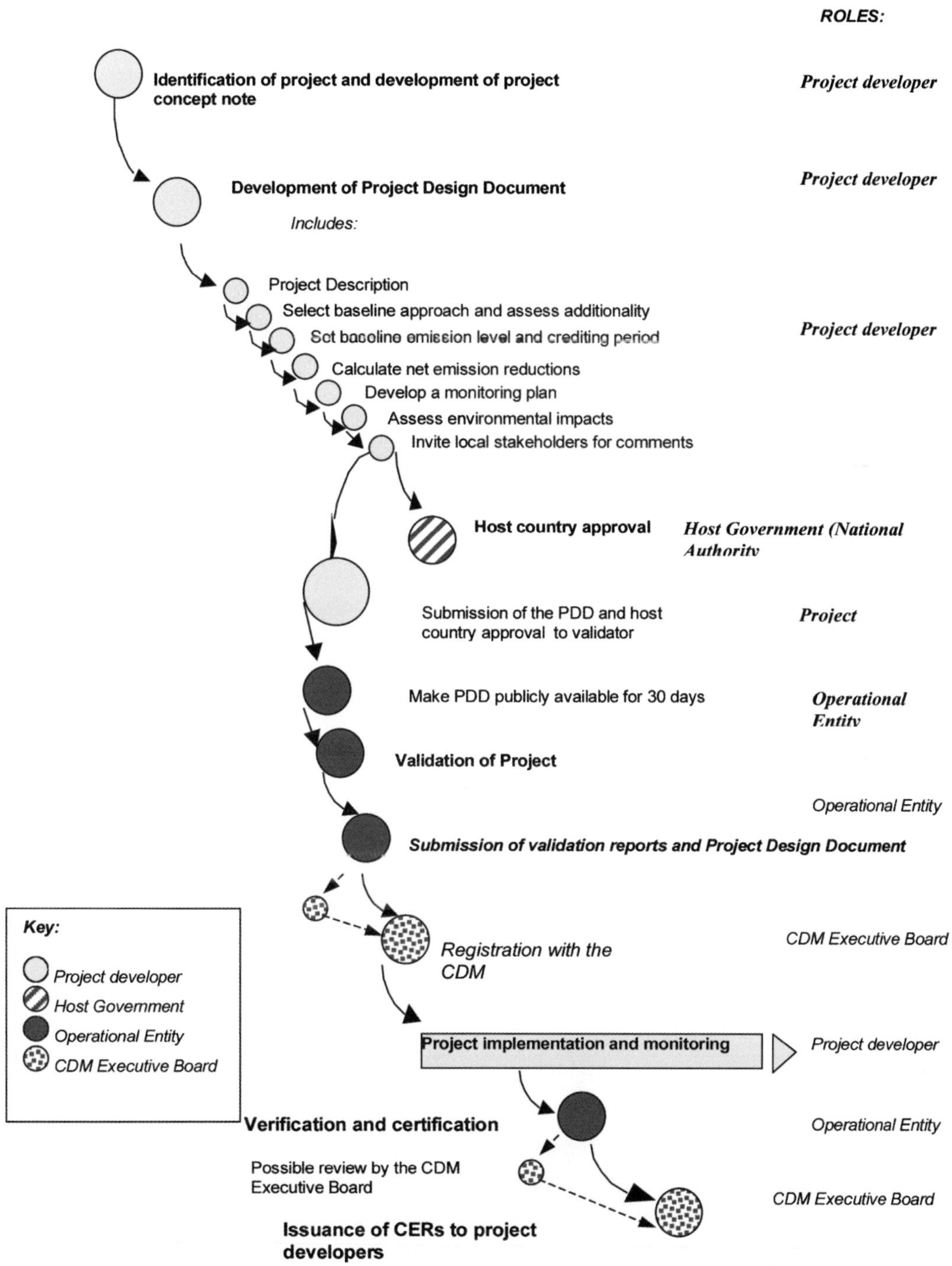

Figure 12.27 Cycle of a proposed clean development mechanism (CDM) project [*Source*: http://www.undp.org/energy/docs/cdmchapter2.pdf]

Table 12.6 Projects applicable to CDM.

Project types	Small-scale CDM project activity categories
Type 1: Renewable energy projects	A. Electricity generation by the user B. Mechanicaf energy for the user C. Thermal energy for the user D. Renewable electricity generation for a grid
Type II: Energy efficiency improvement projects	A. Supply side energy efficiency improvements – transmission and distribution B. Supply side energy efficiency improvements – generation C. Demand-side energy efficiency programmes for specific technologies D. Energy efficiency and fuel switching measures for industrial facilities E. Energy efficiency and fuel switching measures for buildings
Type III: Other project activities	A. Agriculture B. Switching fossil fuels C. Emission reductions by low-greenhouse gas emission vehicles D. Methane recovery E. Methane avoidance
Types 1 – III:	Other small-scale project

Source: http://cd4cdm.org/Publications/cdm%20guideline%202nd%20edition.pdf (2011)

Methodologies for baselines are the methodologies used to establish the baseline and monitoring plan. They are of two types: (i) methodologies approved by the UNFCCC and (ii) new development methodology.

- The project should have contributions to Sustainable Development of the host country, according to:
  Social criteria.
  Economic criteria.
  Environmental criteria.
- The project should have a limited crediting period. There are two types:
  Renewable: 3 periods of 7 years.
  Mixed: 10 years, not renewable.

The format for the PDD contains the following sections:

A.	General description of project activity
B.	Baseline methodology
C.	Duration of the project activity/crediting period
D.	Monitoring methodology and plan
E.	Calculation of GHG emission by sources
F.	Environmental impacts
G.	Stakeholder comments

(*Continued*)

Annex 1.	Contact information on project participants
Annex 2.	Information regarding public funding
Annex 3.	New baseline methodology
Annex 4.	New monitoring methodology
Annex 5.	Table of baseline data

Table 12.7 defines the conditions which the investing country and the host country have to comply (Rozado, 2006).

Table 12.7 Conditions for investing and host countries for CDM.

Investing country	Host country
• Ratification of Kyoto Protocol. • To be the country in Annex I. • Voluntary participation. • Designate a national authority. • Establish national working guides and procedures to approve CDM projects. • Establish a National Registration. • Allocate and control the rights of emission in the country. • Submit the yearly national inventory most recently required.	• Ratification of Kyoto Protocol. • To be country not Annex I. • Designate a national authority for the activity of the CDM project. • Participation in an activity of CDM projects must be voluntary.

The methodology for calculating emissions follows the procedures presented in paragraph 12.2.2. The nomenclature of the PDD is different from the one presented in this chapter, However, the example presented in the following section will follow the official PDD nomenclature so as to avoid confusion.

12.3.3 A CDM case study: Santa Cruz, Bolivia[22]

The total population of Santa Cruz, Bolivia, in 2007 (see city's location in Figure 12.28) was about 1.3 million and it keeps growing at a high annual rate of about 6%. Forecasts indicate that by 2025 the city's population will be about 2.76 million and in 2039 it will reach about 3.7 million. In Santa Cruz, sanitation services are provided by 10 cooperatives of which SAGUAPAC is the largest, serving 65% of the city's area. Sewerage coverage in SAGUAPAC's service area is about 50%, giving Santa Cruz an overall level of sewerage coverage of 32% that represents approximately 450,000 inhabitants. The collected wastewater is conveyed to three existing municipal wastewater treatment plants: Plant North 1 (denominated plant N1), Plant North 2 (denominated N2) and Plant East (denominated Plant E). The

[22]Based on SAGUAPAC (2006), "Project design document form-Clean Development Mechanisms". Santa Cruz de la Sierra, Bolivia; and Libhaber, Menahem (2006), "*Annex 15: Wastewater Treatment Technical Aspects: Increasing the Treatment Capacity of the Existing Wastewater Treatment Plants of Santa Cruz*", in The World Bank (2006) Project Appraisal Document on the Bolivia Urban Infrastructure Project, Report No. 36796-BO, October 13, 2006, Washington D.C, USA.

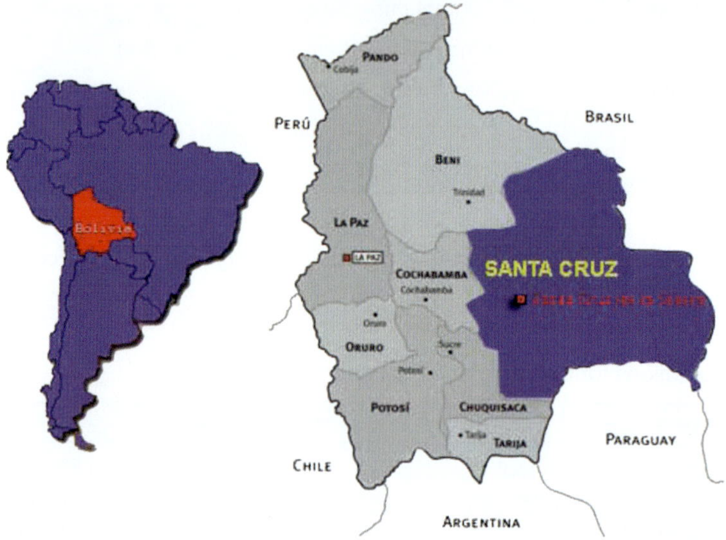

Figure 12.28 The location of the city of Santa Cruz

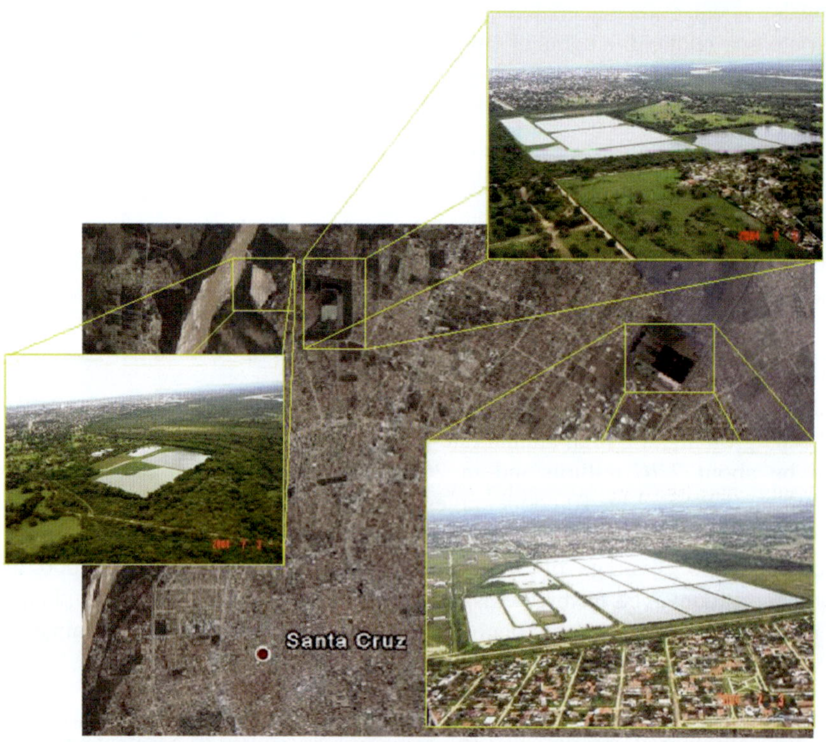

Figure 12.29 Locations of the lagoon systems in Santa Cruz [*Source*: SAGUAPAC (2007), personal communication]

location of these plants is presented in Figure 12.29. When the three plants were constructed they were located outside the city limits. As a result of the city's growth, they are currently located within the city and are surrounded by residential neighbourhoods.

Treatment plants have proved to be very efficient for the removal of organic matter and fecal coliforms; however, by including in each plant anaerobic lagoons, biogas is being generated by them as a by-product and the main component in the biogas is methane. Figure 12.30 shows the layout of Plants N1 and N2, indicating the location of the anaerobic lagoon.

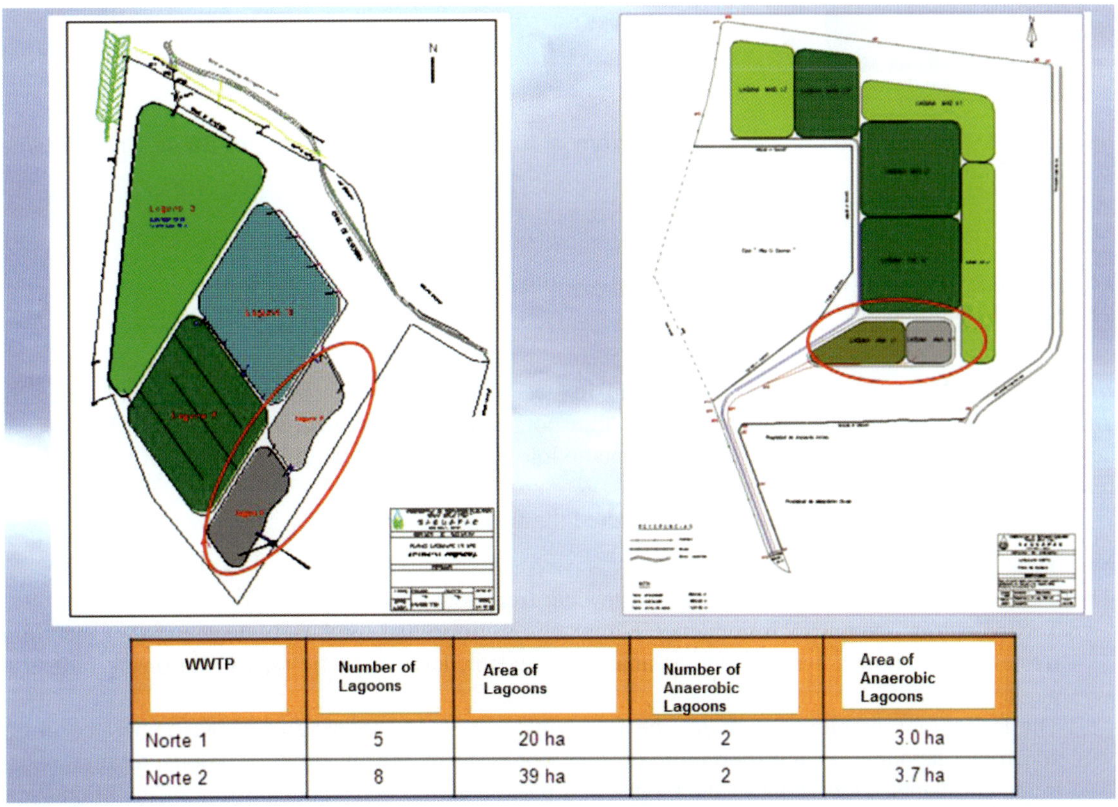

WWTP	Number of Lagoons	Area of Lagoons	Number of Anaerobic Lagoons	Area of Anaerobic Lagoons
Norte 1	5	20 ha	2	3.0 ha
Norte 2	8	39 ha	2	3.7 ha

Figure 12.30 Wastewater treatment plants N1 and N2 in Santa Cruz [*Source*: SAGUAPAC (2007)]

SAGUAPAC, fulfilling its quality policy of continuous improvement, conducted a study to define the works of upgrading its treatment plants, seeking to increase their capacity and to reduce the emission of greenhouse gases that contribute to global warming. As a result of the study the treatment process was modified. The new treatment process flow diagram is presented in Figure 12.31. The treatment plants were upgraded in accordance with the new flow diagram. The project involved the capture of the biogas generated in the anaerobic lagoons, washing and flaring it. The main facilities included in the project of capturing and flaring the biogas are:

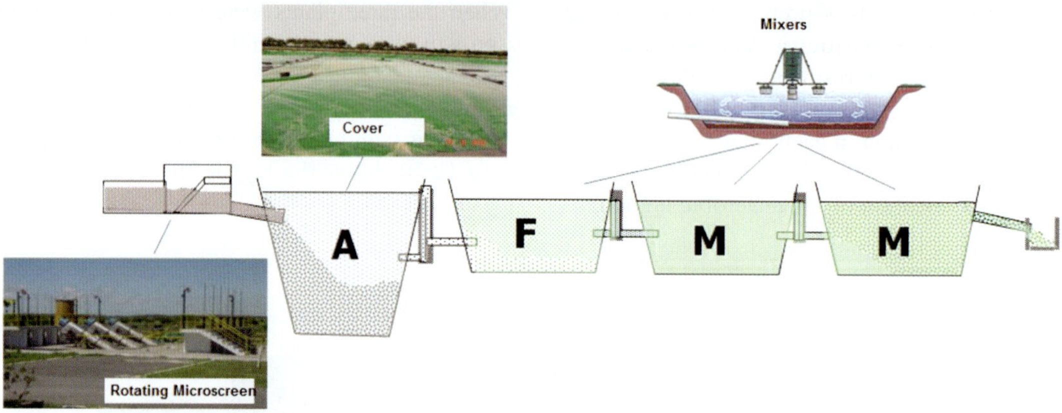

Figure 12.31 The upgraded process flow diagram of the wastewater treatment in Santa Cruz [*Source*: SAGUAPAC (2007)]

- Floating cover on the anaerobic lagoons, including a system of biogas collection.
- Gas extraction and conveyance system.
- Gas flaring unit.

One of the objectives of the project was to reduce biogas emissions from the treatment plants of SAGUAPAC by installing a system to capture, transport and flare it, framed within the guidelines of the Clean Development Mechanism (CDM). The methodology for the capture and transport of the biogas is presented in the Chapter 5.

To apply for the CDM, SAGUAPAC considered the following aspects:

- Type III: Other project activities;
- Category III.H: Methane recovery in wastewater treatment.
- Technology/measure (iv): Introduction of methane recovery and combustion to an existing anaerobic wastewater treatment system such as anaerobic reactor, lagoon, septic tank or an onsite industrial plant.
- The technology employed includes the following stages: 1) capturing of biogas generated at the anaerobic lagoons, 2) transportation of the biogas from the anaerobic lagoons to the biogas treatment plants, 3) washing of the biogas in a packed tower, and 4) combustion of the biogas in a closed flare.

The main components of the biogas system and their main technical specifications are:

(a) geomembrane to cover the lagoons: HDPE, 1.5 mm, UV resistant, 0.94 g/cm^3
(b) floaters, counterweights and inspection openings
(c) collection pipes: HDPE, 6″ diameter, ½″ holes every 0.20 m
(d) transportation pipes: HDPE, 4″ diameter, 3.5 m/s, 3.5 kg/cm^2
(e) suction pool: 18″ diameter, 6 m long
(f) regenerative turbines: multistage, 2 bars
(g) scrubbing tower (for the industrial wastewater treatment plant only): absorption of H_2S, 24″ diameter tower, 3 m long, relief valve, manometer, safety valve

(h) closed flare: 1400°C resistant lining, exhaust gas temperature sensor, flame arrestor, UV flame detector, automatic air/gas ratio regulation, PLC, pilot flame for start up.
(i) electrical main board
(j) control system: flow meters, temperature sensors, pressure gauges, continuous gas analyzer, PC.

The dimensions of the covered anaerobic lagoons are shown in Figure 12.30. The methane mass balance in the treatment plants was prepared according to the methodology presented in Figure 12.32.

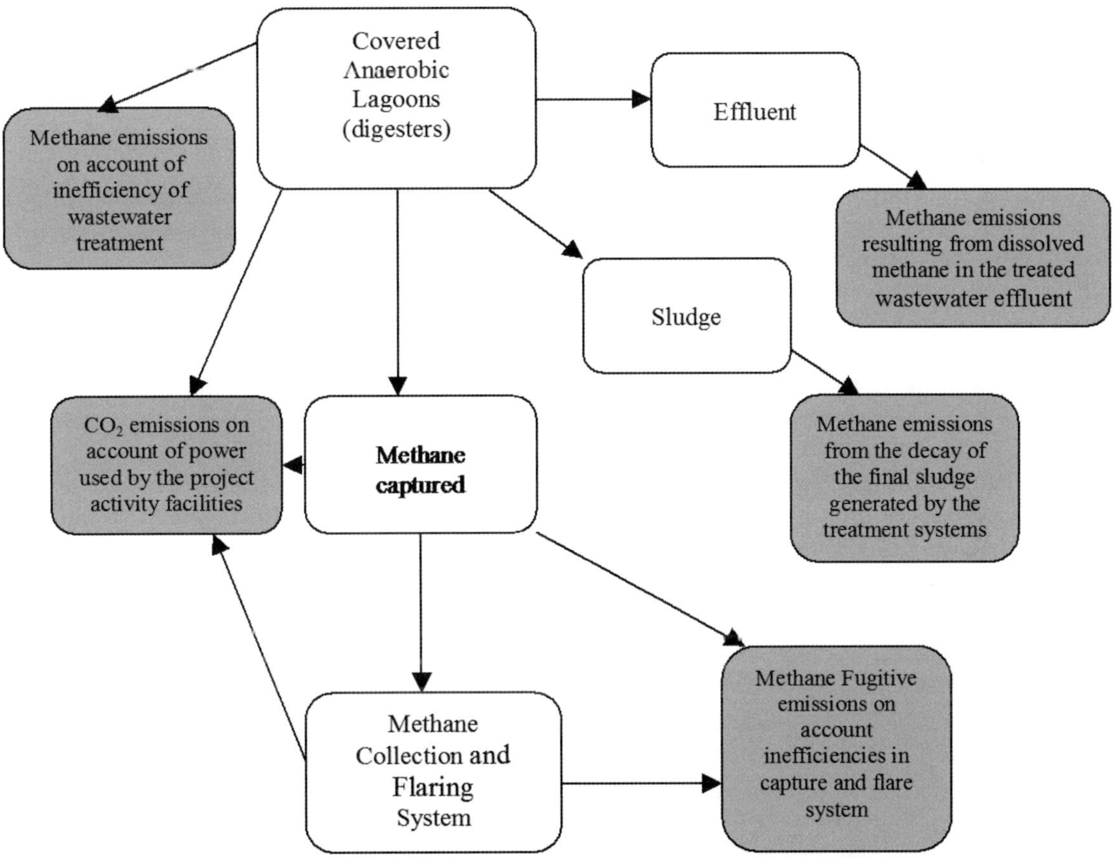

Figure 12.32 CH_4 mass balance in the wastewater treatment plants of Santa Cruz

The figure shows that only the methane produced in the anaerobic lagoons, captured and burned can be effectively applied to the CDM. The inefficiencies of the capture system, i.e. methane coming out with the sludge disposed or dissolved in the effluent are not counted unless it is also captured by autonomous systems. The methane burning system is not 100% efficient. The CO_2 produced by power generation must also take into account, however, in the plants of SAGUAPAC there is very little power consumption by the treatment processes.

Baseline emissions[23]

Emissions that would happen in the absence of the project activity and project emissions were estimated according to baseline methodology AMS III.H Version 04. Baseline emissions were estimated *ex-ante* as follows:

$$BE_y = MEP_{y,ww,treatment} * GWP_{CH4} + MEP_{y,s,treatment} * GWP_{CH4} \quad (12.8)$$

where: $MEP_{y,ww,treatment}$ is methane emission potential of wastewater treatment plant in the year "y" (tonnes) calculated from:

$$MEP_{y,ww,treatment} = Q_{y,ww} * COD_{y,ww,untreated} * B_{o,ww} * MCF_{sww,treatment} \quad (12.9)$$

$MEP_{y,s,treatment}$ is methane emission potential of the sludge treatment system in the year "y" (tonnes) not applicable for the proposed project activity, because the treatment plant does not generate sludge.

$Q_{y,ww}$ is the volume of wastewater treated in the year "y" (m^3) obtained from operative data sheet from SAGUAPAC for each anaerobic open lagoon.

$COD_{y,ww,untreated}$ is the chemical oxygen demand of the wastewater entering the anaerobic treatment reactor/system with methane capture in the year "y" (tonnes/m^3). These values were provided by SAGUAPAC from operative data sheet.

$B_{o,ww}$ is the methane producing capacity of the wastewater. The IPCC default value for domestic wastewater of 0.21 kg CH_4/kg.COD) was used.

$MCF_{ww,treatment}$ is the methane correction factor for the wastewater treatment system that will be equipped with methane recovery and combustion. MCF lower value of 0.8 for *Anaerobic deep lagoon (depth more than 2 m)* according to type of wastewater treatment and discharge pathway or system was used.

GWP_{CH4} is the Global Warming Potential for methane. The value of 21 was used.

Taking into account this situation and from the above equations, the *ex-ante* estimated baseline emissions was calculated varying from 106,606 ton CO_2e in 2008 to 170,155 ton CO_2e in 2012. Table 12.8 shows the baseline emissions for each wastewater treatment plant.

Table 12.8 Baseline emissions for Santa Cruz wastewater methane capture project.

Anaerobic lagoons plant	Year	Baseline emissions (ton CO_{2e})
North 1	2008	9,459
	2012–2028	12,842
North 2	2008	33,394
	2012–2028	48,175
East	2008	28,217
	2012–2028	49,987
Industrial	2008	35,536
	2012–2028	59,151

Source: "Project Design Form for CDM", SAGUAPAC (2006).

[23]Based on "Project Design Form for CDM", SAGUAPAC (2006)

The emission reduction

The emission reduction achieved by the project activity was estimated as the difference between the baseline emission BE_y and the sum of the project emission PE_y and leakage. Leakage of the project activity is assumed zero due to that the technology employed.

$$ER_y = BE_y - (PE_y + Leakage_y) \tag{12.10}$$

The Project emissions PE_y consist of:

(i) CO_2 emissions on account of power used by the project activity facilities $PE_{y,power}$. Emission factors for grid electricity or diesel fuel use was calculated as described in category AMS I.D, yielding a emission factor of 0.436 $tonCO_2e/MWh$;

(ii) Methane emissions on account of inefficiency of the wastewater treatment and presence of degradable organic carbon (Degradable Organic Matter, DOM) in the treated wastewater, $PE_{y,ww,\ treated}$ calculated from:

$$PE_{y,ww,treated} = Q_{y,ww^*} COD_{y,ww,treated^*} B_{o,ww^*} MCF_{ww,final^*} GWP_{CH_4} \tag{12.11}$$

where: $COD_{y,ww,treated}$ is the chemical oxygen demand of the treated wastewater in the year "y" (tonnes/m^3) obtained from operative data sheets of SAGUAPAC.

$MCF_{ww,final}$ is the methane correction factor based on the type of treatment and discharge pathway of the wastewater. MCF higher value of 0.1 was used for *Aerobic treatment well managed*.

(iii) Methane emissions from the decay of the final sludge generated by the treatment systems $PE_{s,y,final}$ calculated as:

$$PE_{y,s,final} = S_{y,final^*} DOC_{y,s,final^*} MCF_{s,final^*} DOC_{F^*} F^* 16/12^* GWP_{CH_4} \tag{12.12}$$

where: $S_{y,final}$ is the amount of final sludge generated by the wastewater treatment in the year y (tonnes). These values were obtained from operational data sheet from SAGUAPAC.

$DOC_{y,s,final}$ is the degradable organic content of the final sludge generated by the wastewater treatment in the year y. For ex-ante estimates using the IPCC default values of 0.05 for domestic sludge and 0.09 for industrial sludge were used.

$MCF_{s,final}$ is the methane correction factor of the landfill that receives the final sludge, estimated as described in category AMS III.G. The value of 0.4 was used according to the *IPCC 2006 Guidelines for National Greenhouse Gases Inventories* for unmanaged shallow landfill (deep below 5 m). DOC_F is the fraction of DOC dissimilated to biogas. The IPCC default value of 0.5 was used.

F is the fraction of CH_4 in landfill gas. The IPCC default of 0.5 was used.

(iv) Methane fugitive emissions on account of inefficiencies in capture and flare systems $PE_{y,fugitive}$ calculated from:

$$PE_{y,fugitive} = PE_{y,fugitive,ww} + PE_{y,fugitive,s} \tag{12.13}$$

where: $PE_{y,fugitive,ww}$ are fugitive emissions through capture and flare inefficiencies in the anaerobic wastewater treatment in the year "y" (ton CO_{2e}) calculated as:

$$PE_{y,fugitive,ww} = (1 - CFE_{ww})^{*} MEP_{y,ww,treatment*} GWP_{CH_4} \tag{12.14}$$

where: CFE_{ww} is the capture and flare efficiency of the methane recovery and combustion equipment in the wastewater treatment plant. The default value of 0.9 was used.

$MEP_{y,ww,treatment}$ is the methane emission potential of wastewater treatment plant in the year "y" (tonnes) calculated as:

$$MEP_{y,ww,treatment} = Q_{y,ww*} COD_{y,ww,untreated*} B_{o,ww*} MCF_{ww,treatment} \tag{12.15}$$

where: $MCF_{ww,treatment}$ is the methane correction factor for the wastewater treatment system that will be equipped with methane recovery and combustion. MCF higher value of 1.0 for *Anaerobic deep lagoon (depth more than 2 m)* from table III.H.1 was used.

$PE_{y,fugitive,s}$ are fugitive emissions through capture and flare inefficiencies in the anaerobic sludge treatment in the year "y" (ton CO_{2e}) that are not taken into account since the proposed project activity does not include anaerobic sludge treatment, nor does the treatment process generate sludge.

(v) Methane emissions resulting from dissolved methane in the treated wastewater effluent $PE_{y,dissolved}$ is calculated as:

$$PE_{y,dissolved} = Q_{y,ww*} [CH4]_{y,ww,treated*} GWP_{CH_4} \tag{12.16}$$

where: $[CH_4]_{y,ww,treated}$ is the dissolved methane content in the treated wastewater (ton/m^3). The default value of 10e-4 tonnes/m^3 was used.

Finally, the project emissions are the sum of all sources of emissions detailed above:

$$PE_y = PE_{y,power} + PE_{y,ww,treated} + PE_{y,s,final} + PE_{y,fugitive} + PE_{y,dissolved} \tag{12.17}$$

From the *ex-ante* estimate, the project emissions during the 21-year crediting period vary from 80,308 ton CO_{2e} in 2008 to 126,142 tonCO_2e from 2012. Table 12.9 shows the project emissions for each wastewater treatment plant.

Table 12.9 Project emissions for Santa Cruz wastewater methane capture project.

Plant	Year	$PE_{y,ww,treated}$ (ton CO_{2e}/yr)	$PE_{y,Dower}$ (ton CO_{2e}/yr)	$PE_{y,s,final}$ (ton CO_{2e}/yr)	$PE_{y,dissolved}$ (ton CO_{2e}/yr)	PE_y (ton CO_{2e}/yr)
North 1	2008	398	2.5	746	6,775	8,918
	2012–2028	541	2.5	1,013	9,198	12,106
North 2	2008	1,378	2.5	2,914	25,519	33,260
	2012–2028	1,988	2.5	4,203	36,814	47,980
East	2008	1,197	2.5	2,137	17,881	24,211
	2012–2028	2,121	2.5	3,787	31,676	42,888
Industrial	2008	1,613	2.5	1,643	6,631	13,920
	2012–2028	2,684	2.5	2,734	11,038	23,169

Table 12.10 shows the estimated project and baseline emissions and emission reductions for each year of the crediting period while Figure 12.33 shows the contribution of each anaerobic lagoon to the total emission reductions of THE Santa Cruz Wastewater Methane Capture Project.

Table 12.10 Project and baseline emissions and emission reductions.

Year	Estimation of project activity emissions (ton CO_{2e})	Estimation of baseline emissions (ton CO_{2e})	Estimation of leakage (ton CO_{2e})	Estimation of overall emission reductions (ton CO_{2e})
2008	80,308	106,606	Not applicable	26,298
2009	91,767	122,493	Not applicable	30,727
2010	103,225	138,381	Not applicable	35,155
2011	114,684	154,268	Not applicable	39,584
2012	126,142	170,155	Not applicable	44,013
2013	126,142	170,155	Not applicable	44,013
2114	126,142	170,155	Not applicable	44,013
2015	126,142	170,155	Not applicable	44,013
2016	126,142	170,155	Not applicable	44,013
2017	126,142	170,155	Not applicable	44,013
2018	126,142	170,155	Not applicable	44,013
2019	126,142	170,155	Not applicable	44,013
2020	126,142	170,155	Not applicable	44,013
2021	126,142	170,155	Not applicable	44,013
2022	126,142	170,155	Not applicable	44,013
2023	126,142	170,155	Not applicable	44,013
2024	126,142	170,155	Not applicable	44,013
2025	126,142	170,155	Not applicable	44,013
2026	126,142	170,155	Not applicable	44,013
2027	126,142	170,155	Not applicable	44,013
2028	126,142	170,155	Not applicable	44,013
Total (ton of CO_{2e})	**2,534,405**	**3,414,384**		**879,979**
Average 21 year (2008–2028) crediting period (ton of CO_{2e})	**120,686**	**162,590**		**41,904**

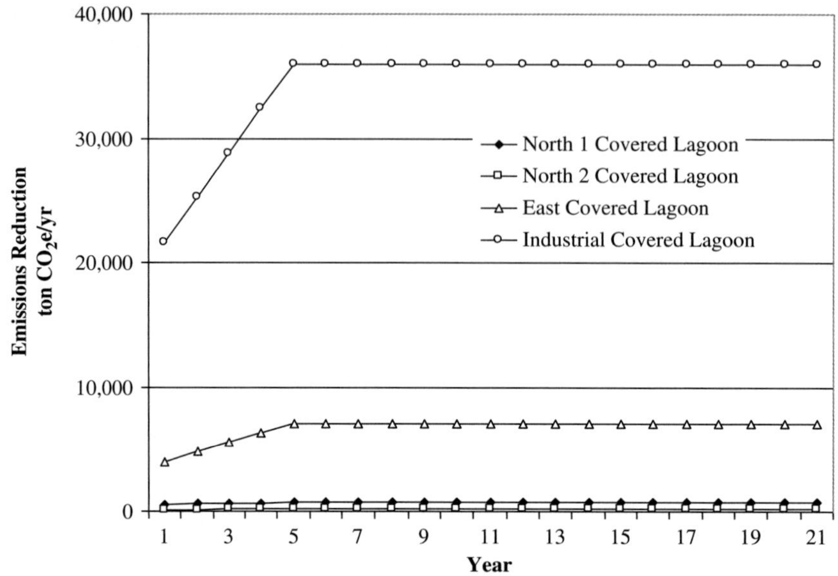

Figure 12.33 Contribution of each anaerobic lagoon to the emission reductions of the Santa Cruz wastewater methane capture project

References

Aguas de Cartagena. (2007). Environmental Management Plan for Handling, Transportation and Final Disposal of Solid Wastes of the Wastewater Treatment in of Punta Canoas, Cartagena, Colombia (Internal Document, In Spansih).

Aiyuk S., Amoako J., Raskin L., van Haandel A. and Verstraete W. (2004). Removal of carbon and nutrients from domestic wastewater using a low investment, integrated treatment concept. *Water Research*, **38**, 3031–3042.

Aktug S. (2002). Dynamic Modelling of Facultative Stabilization Ponds. Izmir, Turkey. In (http://www.veribaz.com/viewdoc.html?dynamic-modelling-of-facultative-stabilization-ponds-355648.html)

Alley E. R. (2007). Water Quality Control-Handbook. Mc Graw-Hill, New York, USA.

Andoh R. Y. G., Smith B. P. and Saul A. J. (1999). The screen efficiency of a novel self-cleansing cso, 8th International Conference on Urban Storm Drainage, Sydney, Australia.

Andrade Neto C. O. (1997). Sistemas Simples para Tratamento de Esgotos Sanitarios, Experiencia Brasilera (Simple Systems for Treatment of Sanitary Wastewater, Brazilian Experience), ABES, Rio de Janeiro, Brazil (in Portuguese), pp. 35–74.

Andrade Neto C. O., Guimarães P., Pereira M. G. and de Souza Melo H. N. (2000). Decanto-Digestor e Filtros Anaeróbios (Decantor-Digester and Anaerobic Filters), Centro de Tecnologia, Campus Universitário, 59072–970, Natal, RN, Brasil (in Portuguese).

Andrade Neto C. O. (2006). O uso do Filtro Anaerobio para Tratamento de Esgoto Sanitario (The use of Anaerobic Filter for Treatment of Sanitary Wastewater). *Medio Filtrante Journal*, Marco/Abril, Brazil (in Portuguese). pp. 12–16.

Andrade Neto C. O., Campos J. R., Sobrinho P. A., Chernicharo C. A. d. L., Nour E. A. (2009). Filtros Anaerobios, Chapter 6 in: Campos J. R. (2009). Tratamento de Esgotos Sanitarios por Procerso Anaerobio e Disposicao Contolada no Solo (Treatment of Sanitary Wastewater by Anaerobic Process and Controlled Disposal on Soil) PROSAB. ABES, Brazil, (in Portuguese), pp. 139–154.

Andreoli V. C., Von Sperling M. and Fernandes F. (2001). Lodos de Esgotos: Tratamento e Dispsicao Final (Wastewater Sludge: Treatment and Final Disposal). DESA-UFMG, SANEPAR, Capa Editoracao, Brazil, (in Portuguese).

Arboleda J. (1992). Teoría y Práctica de la Purificación del Agua (Theory and Practice in Water Purification). ACODAL, Bogotá, Colombia (in Spanish).

Arthur J. P. (1983). Notes on the design and development of waste stabilization ponds. In: Warm Climates of Developing Countries, World Bank Technical Paper Number 7, The World Bank, Washington DC, USA, **18**, pp. 50–53.

Asano T., Burton F., Leverenz H., Ryvjiro T. and Tchobanoglous G. (2007). Water Reuse, Issues, Technologies and Application. Metcalf & Eddy Inc.

Biello D. (2011). The false promise of biofuels, *Scientific American*, August 2011, USA, **302 N2**, pp. 38–45.

Blanco G. and Wagner Santallana E. L. (2007). Environmental Impact Study: Expanding the Plant Capacity of Wastewater Treatment of Saguapac-Santa Cruz de la Sierra. The World Bank, Bolivia.

Botiva S. C. (1977). Evaluación de parámetros de calidad de aguas residuales domésticas para cálculo de población equivalente (Evaluation of Domestic Wastewater Quality Parameters for Calculating Equivalent Population), College of Engineering National University (in Spanish).

Brock T. (1974). Biology of Microorganisms. Prentice-Hall, New Jersey, USA.

Butts A. T. (1988). Development of Design Criteria for Sidestream Elevated Pool aeration Stations, Illinois State Water Survey Division, Water Quality Section.

CAESB (2006). Estacoes de Tratamento de Esgotos (Wastewater Treatment Plants). The Water Utility of Brasilia, Brasil, October 2006 (in Portuguese).

Campos J. R. (2009). Tratamento de Esgotos Sanitarios por Procerso Anaerobio e Disposicao Contolada no Solo (Treatment of Sanitary Wastewater by Anaerobic Process and Controlled Disposal on Soil) PROSAB. ABES, Brazil, (in Portuguese), pp. 73–99, 101–115, 144–145, 155–198, 321–407.

Catalog HUBER Rotamat (2007). High Performance Screen for Efficient and Economical Purification of Waste Water, Germany.

Cavalcanti P. F. F. (2009). Aplicacao de Reatores UASB e Lagoas de Polimento no Tratamento de Esgoto Domestico (Application of UASB Reactores and Polishing Lagoons in Domestic Wastewater Treatment), Federal University of Campina Grande – UFCG, Brazil (in Portuguese), pp. 48–50.

CEPIS (2004). Diseño de plantas de tecnología apropiada – Capítulo cinco: Batería de filtros de tasa declinante y lavado mutuo (Design of appropriate Technology Plants), Chapter 5, A battery of Filters of Declining Rate and Mutual Washing, CEPIS-MANUAL II: in http://www.cepis.ops-oms.org/bvsatr/fulltext/tratamiento/manualII/cinco.pdf (in Spanish).

Chernicharo C. A. d. L. (2000). Reatores Anaerobios (Anaerobic Reactors). DESA-UFMG, Brazil, (in Portuguese), **150**, pp. 218–224.

Chernicharo C. A. d. L. (2007). Reatores Anaerobios (Anaerobic Reactors). DESA-UFMG, Capa Editoracao, Brazil (in Portuguese), pp. 127–152, 360–361, 363–364, 365.

Chernicharo C. and Augusto de Lemos (2007). Reatores Anaerobios (Anaerobic Reactors). DESA-UFMG, Capa Editoracao, Brazil (in Portuguese).

COD/BOD5 (2005). COD/BOD5 Reduction with ROTAMAT® Fine and Micro Screens: Huber Technology, Catalog of Huber Technology, Germany.

Colorado Department of Public Health and Environment (2007). Design Criteria Considered in the Review of Wastewater Treatment Facilities-Policy 96-1, Colorado, USA.

Cooper-Smith G. (2001). Chemically Enhanced Primary Treatment: The UK Experience, Water 21, p. 10, August 2001.

Crites R. W., Middlebrooks E. J. and Reed S. C. (2006). Natural Wastewater Treatment Systems. Taylor & Francis Group, LLC, FL, USA.

Crutzen P. J., Mosier A. R., Smith K. A. and Winiwarter W. (2008). N_2O release from agro-biofuel production negates global warming reduction by replacing fossil fuels, *Atmos. Chem. Phys.*, **8**, pp. 389–395.

Daiger G., Peterson R., Whiterspoon J. and Allen E. (2000). Impact of Global Warming Concerns on Wastewater Treatment Plant Design and Operation. ASCE, USA.

Dautremont-Smith J. (2002). Guidelines for College-Level Greenhouse Gas Emissions Inventories.

de Vargas L. (2007). Flotación (Filtration), CEPIS, MANUAL I, http://www.cepis.ops-oms.org/bvsatr/fulltext/tratamiento/manualI/tomoII/ocho.pdf (in Spanish).

DeGarie C. J., Crapper T., Howe B. M., Burke B. F. and McCarthy P. J. (2005). Floating geomembrane covers for odour control and biogas collection and utilization in municipal lagoons. *Water Science & Technology*, **52**(10), pp. 291–298, © IWA Publishing 2000.

EPA (1999). Wastewater Technology Fact Sheet: Screening and Grit Removal, USA.

EPA (2005). Membrane Filtration Guidance Manual. Cincinnati, USA.

EPA (2008). Inventory of U.S. Greenhouse Gas Emissions and Sinks: 1990–2006, USA.

FAO (2006). Food Security Statistics – Colombia.

Fernades C. (2007). Digestào Anaeróbia (Anaerobic Digestion). http://www.dec.ufcg.edu.br/saneamento/DigeAnae.html

Fernandez C. (2006). "El Protocolo de Kioto y los mecanismos flexibles: MDL", PPT, España, The Kyoto Protocol and the Flexible Mechanisms: CMD, PPT, PPT, Spain (in Spanish).

FORET (2007). http://www.fmcforet.com/productes/OHP.htm

Foxon K. M., Pillay S., Lalbahadur T., Rodda N., Holder F. and Buckley C. A. (2003). The Anaerobic Baffled Reactor (ABR): An Appropriate Technology for On-Site Sanitation, Available on website http://www.wrc.org.za

Frassinetti P. C. and van Haandel A. C. (1996). Improved performance and increased applicability of waste stabilization ponds by pretreatment in a UASB reactor. *Water Science and Technology*, **33**(7), 147–156.

Freni G., Maglionico M. and Di Federico V. (2003). State of the Art Urban Drainage Modeling. University of Bologna, University of Palermo, SINTEF, NTNU and Wrc, Italy.

Garcia Agudo E. and Russell G. L. (1992). Militamices como sistema de pretratamiento, (Micro Screens as Pretraetment Systems), CEPIS, OPE, WHO (in Spanish).

Gloyna E. F. (1971). Waste Stabilization Ponds. WHO, Monograph Series N° 60, Geneva.

Gravel M. (2002). Membrane Filtration-Application from the Inside to the Outside. Newfoundland Design Associates Ltd, Canada.

Gray B. (2009). Appendix of the paper: On the Hijacking of the American Meteorological Society (AMS).

Greco N. H. C. (2007). Desempenho de um Sistema Reator UASB – Filtro Anaerobio en Escala Real: ETE Ipanema (Performance of a System of UASB Reactor – Anaerobic Filter in a Full Scale Plant: The Ipanema Wastewater Treatment Plant), 24ª Congresso Brasileiro de Engenharia Sanitaria e Ambiental, 2–7 de Setembro de 2007, Belo Horizonte, Minas Gerias, Brazil (in Portuguese).

Gurney Environmental, www.gurneyenvironmental.com

Gurovich L. and Ramos R. (1990). Tasa de Riego Mensual para una Hectárea Tipo en la Zona de Maipú (Monthly Irrigation Rate of a Typical Hectar in the Maipu Zone, Santiago, Chile), Unpublished Internal Report by Ifarle-Tahal Consulting Engineers prepared for the Santiago Poniente Project, Chile (in Spanish).

Hackman B. (2000). Membrane Filtration. Strand Ass, USA.

Hans Huber AG Catalogue (2007). Innovative Equipment and Processes for Water and Wastewater Treatment.

Harleman D. R. F. (2004). Personal Communication.

Harleman D. R. F. and Chagnon F. (2004). Personal Communication.

Harleman D. R. F. and Murcott S. (1998). Low Cost Nutrient Removal Demonstration Study, Tech. Rep., World Bank/UNDP, BRA/90/010.

Harleman D. R. F. and Murcott S. (2001). An Innovative Approach to Urban Wastewater Treatment in the Developing World, Water 21, June 2001, pp. 44–48.

Hermann L. F. and Schittini M. (2011). Estudio de las Opciones de Aprovechamiento del Biogas de las PTARs de SAGUAPAC y Analisis Finnaciero en el Mercado de Santa Cruz (Study of the Options for Utilizing the Biogas Generated in the Wastewater Treatment Plants of SAGUAPAC and Financial Analysis in the Santa Cruz Market), SAGUAPAC Unpublished Internal Report Prepared by the Consulting Firm Incisa Bioenregia (in Spanish).

http://en.wikipedia.org/wiki/Milankovich_Effect

http://icecap.us/images/uploads/On_The_Hijacking_of_the_American_Meteorological_Society.pdf

http://themes.eea.europa.eu/Sectors_and_activities/energy/indicators/EN01, 2007.04/Fig1.gif/view

http://unfccc.int/files/meetings/cop_13/application/pdf/cp_bali_action.pdf

http://www.abelard.org/briefings/global_warming_atmospheric_chemistry_physics.php

http://www.iwawaterwiki.org/xwiki/bin/view/Articles/Helmintheggs (May 12, 2011).

Huber Technology Group (2003). Personal Communication.

Huber-Bombas y Riegos (2007). Personal Comunication. Medellin, Colombia.

Hulshoff Pol L. W., de Castro Lopes S. I., Lettinga G. and Lens P. N. L. (2004). Anaerobic Sludge Granulation, *Water Research* **38**, UK.

Idelovitch E. and Michail M. (1981). Nitrogen removal by free Ammonia stripping from high pH ponds. *Journal of Water Pollution Control Federation*, **53**(9), September 1981, pp. 1391–1401.

Idelovitch E. and Michael M. (1984). Soil aquifer treatment: a new approach to an old method of wastewater reuse. *Journal of Water Pollution Control Federation*, **56**(8), pp. 963–943.

Indian & Northern Affairs Canada Gatineau, Quebec, K1A 0H4 (2006). Design Guidelines for First Nations Water Works, Canada.

Instituto de Investigación e Información Geocientífica, Minero Ambiental y Nuclear Ingeominas (2000). Estudio Básico para la Conformación de un Parque Minero Industrial para el Ordenamiento y Desarrollo de la Actividad Extractiva y Transformadora de Minerales Arcillosos con Fines a Desmarginalización de Barrios en el Distrito Capital. Informe Final. Proyecto PNUD/DAMA/COL/96-023. Bogotá, Colombia, p. 202.

IPAC (2007). http://www.llnl.gov/ipac/technology/profile/environment/UV-H2O2TreatmentOfAqueousMixedWastes/index.php

IPCC (2008). Climate Change 2007: Synthesis Report-Summary for Policymakers, Switzerland.

IPCC, WMO, UNEP (2006). 2006 IPCC Guidelines for National Greenhouse Gas Inventories, Vol. **5**.

Johansen J. O. (2009). Sewage biogas fuels public transport buses in Oslo. *World Water and Environmental Engineering*, March/April, 47–48.

Juanico M. (1993). Alternative schemes for municipal sewage treatment and disposal in industrialized countries: Israel as a case study. *Ecological Engineering*, **2**, 101–118.

Juanico M. (1995). The Effect of the Operational Regime on the Performance of Wastewater Storage Reservoirs, IAWQ Waste Stabilization Ponds Conference, Joao Pessoa, Brazil, March 1995.

Juanico M. (2005). Wastewater reservoirs. In: Pond Treatment Technology, A. Shilton (ed.). IWA Publishers, pp. 357–380.

Juanico M. and Dor I. (1999a). Reservoirs for Wastewater Storage and Reuse, Ecology, Performance and Engineering Design. Springer.

Juanicó M. and Dor I. (eds.) (1999b). Hypertrophic Reservoirs for Wastewater Storage and Reus. Springer, Germany, pp. 87–90.

Juanicó M. and Milsten A. (2004). Semi-intensive treatment plants for wastewater reuse in irrigation. *Water Science and Technology*, **50**(2), 62–103.

Juanico M. and Shelef G. (1994). Design, operation and performance of stabilization reservoirs for wastewater irrigation in Israel, *Water Research*, **28**(1), 175–186.

Juanico M. and Weinberg H. (1995). Diseno de Procesos de Lagunas de Estabilizacion a Diferentes Altitudes para las Condiciones de Bolivia (Design of Stabilization Lagoons Process at Different Altitudes for the Conditions of Bolivia), PRORPAL (in Spanish).

Keller J. and Hartley K. (2002). "Greenhouse Gas Production in Wastewater Treatment: Process Selection is the Major Factor", *Water Science and Technology*, ISSN 0273-1223 CODEN WSTED4, London, UK.

Kemmer and Frank M. (ed.) (1988). Nalco, "The Nalco Water Handbook", McGraw-Hill Co., New York, USA.

Koehler R. (2003). Innovative and Economic Processes for Water and Wastewater Treatment, Presentation of Huber Technology in the World Bank.

Koppl S. and Formmann C. (2004). Eco-efficient Wastewater Treatment of River and Sea Outfalls, MWWD 3rd International Conference on Marine Wastewater Disposal and Marine Environment, Catania, Italy, September 27–October 2.

Lara J. A. (1999). Depuración de Aguas Residuales Municipales con Humedales Artificiales, (Municipal Wastewater Purification by Constructede Wetlamds), Trabajo Final de Maestria, Universidad Politécnica de Catalumña, España (MSc thesis, Politechnic University of Catalumña, Spain), in Spanish.

Leppe A. Z. and Padilla L. B. (1999). Emisarios Submarinos de Penco y Tome, 5 Años de Vigilancia y una Evaluacion Global de sus Efectos Ambientales (The Submarine Outfalls of Penco and Tome, 5 Years of Monitoring and a Global Evaluation of their Environmental Impact), XIII Congreso Chileno de Ingenieria Sanitaria y Ambiental, AIDIS Chile, Antofagasta, October 1999 (in Spanish).

LG Sonic (2003). A Microcosm Trial of Novel Controls for Phytoplankton – The Effects of Ultrasound on Algae, http://www.lgsonic.com/pond-algae-control-detail/

Libhaber M. (1987). Integrated Planning and On-Farm Economics of Reclaimed Wastewater Irrigation, Workshop on Treatment and Use of Sewage Effluents for Irrigation and other Purposes, UNEP, Mediterranean Action Plan, Split, Yugoslavia, November 25–27.

Libhaber M. (1996). Reuso de Aguas Servidas Tratadas y Almacenadas en Embalses de Estabilizacion (Reuse of Treated Wastewater Stored in Stabilization Reservoirs), International Seminar on Management of Urban Wastewater, Organized by the World Bank in Medellin, Colombia, December 2–4 (in Spanish).

Libhaber M. (2003). Wastewater Reuse for Irrigation, the Stabilization Reservoirs Concept, Presentation in the World Bank Water Week, Washington DC, USA.

Libhaber M. (2004). Wastewater Treatment in Developing Countries: the Use of Physico-Chemical Processes for Achieving Affordable Disposal Schemes, Paper presented in the Seminar "Wastewater Treatment without Biology – May Advanced Physico-Chemical Treatment Methods Replace Biological Treatment or Supplement It?", Technical University of Munich, Munich, July 15, 2004.

Libhaber M. (2006). Annex 15: Wastewater Treatment Technical Aspects: Increasing the Treatment Capacity of the Existing Wastewater Treatment Plants of Santa Cruz, in The World Bank (2006) Project Appraisal Document on the Bolivia Urban Infrastructure Project, Report No. 36796-BO, October 13, 2006, Washington D.C, USA.

Libhaber M. (2007). Appropriate Technology for Wastewater Treatment and Reuse in Developing Countries, Paper presented in the Conference "Water Supply and Sanitation for All" Huber Technology, Berching, Germany, September 27–28, 2007.

Libhaber M. and Michail M. (1987). Innovative Methods in Wastewater Treatment for Agricultural Reuse in Israel, Water Reuse Symposium IV, AWWA Research Foundation, Denver, Colorado, August 2–7.

Libhaber M. and Roberts J. W. P. (2002). Social Aspects of Wastewater Submarine Outfalls in Developing Countries, the Case of Cartagena, Colombia, MWWD Conference, Istanbul.

Llosá O. M. (2005). Generalidades de las tecnologías de membranas (Generality of Membrane Technology), Seminario Internacional de Aguas Residuales, Valencia, España.

Lyle J. T. (1994). Regenerative Design for Sustainable Development. John Wiley & Sons, Inc, New York, USA, pp. 245–246.

Maldonado-Yactayo V. (2007). Fitración (Filtration), http://www.cepis.ops-oms.org/bvsatr/fulltext/tratamiento/manualI/tomoII/nueve.pdf

Mara D. (1987). Waste stabilization ponds: problems and controversies. *Water Quality International*, 1, 20–22.

Mara D. (2004). Domestic Wastewater Treatment in Developing Countries. Earthscan, UK, USA, pp. 91, 105–113, 165, 189–191, 194–199.

Mara D. (2006). "Waste Stabilization Ponds, WSP", http://www.personal.leeds.ac.uk/~cen6ddm/WSP.html

Mara D. and Pearson H. W. (1999). A hybrid waste stabilization pond and wastewater storage and treatment reservoir system for wastewater reuse for both restricted and unrestricted crop irrigation. *Water Research*, **33**(2), 591–594, Great Britain.

Martin J. E. and Martin T. E. (1991). Technologies for Small Water and Wastewater Systems, Van Nostrand Reinhold, pp. 320–322.

McGauhey P. H. (1968). Engineering Management of Water Quality. McGraw-Hill, NY.

McKinney R. T. (1962). Microbiology for Sanitary Engineers. McGraw-Hill Book Co, New York.

Medina Hoyos I. R. (1993). Process Pilot Plant Anaerobic Wastewater Treatment. Fondo Nacional Para el Medio Ambiente, Bolivia.

Melo J. C. (2008). Sistema Condominial: Uma Resposta ao Desafio de Universializaco do Saneamento (Condominial System: A Response to the Challenge of Universal Sanitation), Program of Modernization of the Sanitation Sector, Ministry of Cities, Government of Brazil (in Portuguese).

Merz S. K. (2000). Guidelines for Using Free Water Surface Constructed Wetlands to Treat Municipal Sewage. Queensland Department of Natural Resources, Queensland State Government, Brisbane, Queensland, Australia.

Metcalf and Eddy (2003). Wastewater Engineering: Treatment and Reuse. 4th, and 5th edn, McGraw-Hill Book Co, New York, p. 280.

Ministry of Environment, Housing and Regional Development of Colombia (MAVDT) (2008). Studies Leading to the Determination of Technologies for Municipal Wastewater Treatment Applicable to Populations of Less than 300,000 inhabitants (unpublished Report).

Moreira-Oliveira C., Martins Godinho V. and Von Sperling M. (2006). Influência de diferentes Granulometrias de Filtros de Pedra na Composição e Concentração de Algas em Efluentes de Lagoas de Polimento (Influence of Different Granulometries of Rock Filters on the Composition and Concentration of Algae in Polishing Lagoons Effluents), 25° Congresso Brasileiro de Engenharia Sanitária e Ambiental, Brasil (25th Brazilian Congress of Sanitary Engineering), in Portuguese.

Morrisey S. P. and Harleman D. R. F. (1992). Retrofitting Conventional Primary Treatment for Chemically Enhanced Primary Treatment in the USA, in Chenical Water and Wastewater Treatment II, Proceedings of the Fifth Gothenburg Simposium, Nice, France, Springer-Verlag, 1992.

Mota F. S. B. and von Sperling M. (2009). Nutrientes de Esgotos Sanitarios: Utilizacao e Remocao (Nutrients in Sanitary Wastewater: Utilization and Removal), PROSAB, Editora ABES (in Portuguese), pp. 317.

Murcott S., Dunn A. and Harleman D. R. F. (1996). Chemically Enhanced Wastewater Treatment for Agricultural Reuse in Mexico, International Association of Water Quality, Biennal Conference, Singapore.

OPS (2005). Guías para el Diseño de Tanques Sépticos, Tanques Imhoff y Lagunas de Estabilización (Guidelines for Design of Septic Tanks, Inhoff Tanks and Stabilization Lagoons) OPS-CEPIS, Lima, Peru (in Spanish).

Orozco Á. (1976). Comparison of Complete Mixing Activated Sludge Models Using a Glicerol Feed. The Pennsylvania State University, State College.

Orozco Á. (1977). Teoría de la Remoción de Sustrato Soluble en Plantas de Lodos Activados (Theory of Soluble Subtsrate Removal in Activated Sludge Plants). Revista ACODAL, Bogotá, Colombia (in Spanish).

Orozco Á. (1988). Anaerobic wastewater treatment using an open flow baffled reactor at low temperature, Fifth International Symposium on Anaerobic Digestion, Poster paper book, Bologna, Italy.

Orozco Á. (1997). Pilot and full-scale anaerobic treatment of low-strength wastewater at sub-optimal temperature (15 °C). with a hybrid plug flow reactor, Proceedings of the 8th International Conference on Anaerobic Digestion, Vol. **2**, Oral presentation, Sendai, Japan, May 25–29.

Orozco Á. (1999). Prediseño de Alternativas de Tratamiento de Agua Residual (Pre-design of Alternatives for Wastewater Treatment), Final Report for SERAQA, Tunja, Colombia (in Spanish).

Orozco A. (2005). Bioingeniería de aguas residuales: teoría y diseño, (Biengineering of Wastewater: Theory and Design). ACODAL, Bogotá, Colombia (in Spanish).

Orozco Á. (2007). La lucha contra el calentamiento global: una visión diferente, (The against Global Warming: A Different Vison) Corporación Otraparte, Medellín, Colombia (in Spanish).

Orozco Á. (2008). Los gases de efecto invernadero (GEI) y su relación con los sistemas de tratamiento de aguas residuales (STAR), Manejo de Vertimientos Líquidos para contribuir a la Mitigación del Cambio Climático, Seminario Internacional, ACODAL-Seccional occidente, Cali, Colombia.

Orozco Á. (2009). The Impact of Wastewater Treatment on Global Warming, Presentation in Water Week 2009, The World Bank, Washington DC-USA.

Orozco Á and Salazar A. (1988). Tratamiento Biológico de Aguas Residuales (Biological Treatment of Wastewater). CESET, U de A, Medellín, Colombia.

Orozco-Jaramillo Á. (2005). Bioingenieria de Aguas Residuales (Wastewater Bioengineering). ACODAL, Colombia, (in Spanish).

Osorio E. S. (1996). Sistema de Tratamiento de Aguas Residuales Tipo UASB y Lagunas Facultativas, en Bucaramanga, Colombia (Wastewater Treatment System by UASB followed by Facultative Lagoons), International Seminar on Management of Urban Wastewater, Organized by the World Bank in Medellin, Colombia, December 2–4 (in Spanish).

Paradowska M. (2004). Tailored Chemical Oxidation Techniques for the Abatement of Bio-Toxic Organic Wastewater Pollutants. Doctoral thesis, Universitat Rovira I Virgili, España.

Park J. (2006). Application of UV Light for Drinking Water Disinfection – Is It The Magic Bullet?, PPT presentation, University of Wisconsin-Madison, USA.

Pavlostathis G. and Giraldo-Gómez E. (1991). Kinetics of anaerobic treatment. *Water Science and Technology*, **24**(8), IAWPRC, Oxford, UK, pp. 35–59.

Peña-Varón M. and Mara D. (2005). Waste Stabilization Ponds, Universidad del Valle, Instituto Cinara. Cali, Colombia and School of Civil Engineering, University of Leeds, Leeds, UK.

Potabilización de Aguas Naturales (Potabilization of Natural Water). http://www.puc.cl/quimica/agua/potabiliz.htm

Prado P. L. F. (2010). Analise da Potencialidade do Reuso Indireto Potavel: Estudo de Caso da ETE Atuba Sul, Regiao Metropolitana de Curitiba (Analysis of the Potential Indirect Potable Reuse: Case Study of the Atuba Sul Treatment Plant, Metropolitan Region of Curitiba), MSc dissertation, Federal University of Parana in Cooperation with the University of Stuttgart (in Portuguese), p. 36.

RAS (2000). Reglamento Técnico del Sector de Agua Potable y Saneamiento Básico, (Technical Regulation for the Water and Sanitation Sector) MINDESARROLLO (Ministry of Economic Development), Colombia (in Spanish).

Rebhum M., Ronen D. and Eren J. (1987). Monitoring and study program of an interregional wastewater reclamation system for agriculture. *Journal Water Pollution Control Federation*, Washington DC, USA, **59**(5), pp. 242–248.

Reina J. (2005). El secado por Aspersión. Una Alternativa para el Tratamiento de los Concentrados Procedentes de Planta de Osmosis Inversa (Drying by Spraying. An Alternative for Treatment of Concentrates of Reverse Osmosis Plants). Barcelona, España (in Spanish).

Roberts J. W. P., Salas J. H., Reiff M. F., Libhaber M., Labbe A. and Thomson C. J. (2010). Marine Wastewater Outfall and Treatment Systems. IWA Publishing, London, New York, pp. 1–13, 188–190.

Rodríguez J. P. (2007). Tratamiento de aguas residuales con lagunas de oxidación (Treatment of Wastewater in Oxidation Ponds). ACODAL Diplome in Wastewater and Industrial Wastes Treatment, Barranquilla, Colombia (in Spanish).

Rolim S. M. (2000). Sistemas de Lagunas de Estabilizacion (Stabilization Lagoons Systems). McGraw-Hill Publishing, (in Spanish), pp. 341–347.

Romem M. (1991). Manual for Design of Stabilization Reservoirs. Israeli Association of Engineers and Architects (in Hebrew).

Russell D. L. (2006). Practical Wastewater Treatment. John Wiley & Sons, USA.

SAGUAPAC (2006). Project Design Document form-Clean Development Mechanisms", Santa Cruz, Bolivia.

SAGUAPAC (2007a). "Proyecto de captura y quemado de biogas en PTAR" (The Project of Capture and Flaring of the Biogas of the Wastewater Treatment Plants), PPT in ACODAL Conference, Santa Marta, Colombia.

SAGUAPAC (2007b). Proyecto de captura y quemado de biogás en las plantas de tratamiento de aguas residuales-Estudios Técnicos (Collection and Burning of the Biogas Produced in the wastewater Treatment Plants – Technical studies), Cooperativa de Servicios Públicos, Santa Cruz Ltda., Bolivia (in Spanish).

Scheierling M. S., Bartone C., Mara D. D. and Drechsel P. (2010). Improving Wastewater Use in Agriculture: An Emerging Priority, World Bank Policy Research Paper 5412, Available at http://siteresources.worldbank.org/INTWAT/Resources/ESWWastewaterAg.pdf

Schirg P. G. (2001). Introducción a la Teoría y Práctica de la Técnica de Membranas (Introduction to the Theory and Practice of Membrane Techniques). PS Prozesstechnik, Switzerland.

Shammas N. K., Wang L. K. and Wu Z. (2009). Waste stabilization ponds and lagoons. In: Biological Treatment Processes, L. K. Wang, N. C. Pereira and Y. T. Hung (eds.), Humana Press, New York, USA.

Shuler L. and Kargi F. (1992). Bioprocess Engineering. Prentice Hall International Series, New Jersey.

Shuval H., Adin A., Fattal B., Rawitz E. and Yekutiel P. (1986). Wastewater Irrigation in Developing Countries, Health effects and Technical Solutions, World Bank Technical Paper 51, UNDP Project Management Report Number 6, The World Bank, Washington DC, p. 276.

Silva Martínez A. E., Vega González E. and Martínez Pereda P. (2000). Criterios de Selección de Procesos de Separación por Membranas, (Criteria for Selection of Separation Processes by Memebranes). Univ. Nal., Autónoma, Mexico (in Spanish).

Sincero A. P. and Sincero G. (2003). Physical-Chemical Treatment of Water and Wastewater. IWA Publishing, USA.

Soler A., Moreno M. D., Saez J. and Moreno J. (2000). Kinetic model for deep waste stabilization ponds operating in batch mode. *Water Science and Technology*, **42**(10), pp. 315–325. © IWA Publishing 2000.

Tahal Consulting Engineers and The Hebrew University (1981) Storage of Effluents in the Naan Reservoir, Tahal Report 04/81/87 (in Hebrew), PP 6, 9.

Tchobanoglous G. (1981). Wastewater Engineering: Collection and Pumping of Wastewater. Metcalf Eddy Inc., McGraw-Hill Publishing.

Tchobanoglous G., Burton L. F. and Stensel H. D. (2003). Wastewater Engineering: Treatment and Reuse. 4th edn, Metcalf Eddy Inc., McGraw-Hill Publishing.

The World Bank (2003). World Development Report (WDR), p. xiii.

The World Bank (2007). Project Appraisal Document on a Proposed Purchase of Emission Reduction by the Community Development Carbon Fund in the Amount of US$ 2.080,000 from the Water Supply and Sanitation

Cooperative of Santa Cruz for the Urban Wastewater Methane Gas Capture Project, Report No. 39377-BO, May 17, 2007, Washington DC, USA.
Thomsen M. and Lyck E. (2005). Emission of CH_4 and N_2O from Wastewater Treatment Plants (6B). National Environmental Research Institute, Denmark.
Tsukamoto R. (2005). "Tratamiento Físico-Químico de Aguas Residuales Municipales: Estrategia Eficiente, Rápida y de Bajo Costo para Latinoamérica" (Physico-Chemical Treatment of Municipal Wastewater: An Efficient, Rapid and Low Cost Strategy for Latin America). Seminario Internacional de Aguas Residuales, Armenia, Colombia.
Tsukamoto R. Y. (2002). Tratamento Primário Avançado: O Paradigma Moderno do Tratamento de Esgotos, Paper 1 of 3, (Chemically Enhanced Primary Treatment: The Modern Paradigm of Wastewater Treatment), Agua Latino América, São Paulo, Vol. March–April 2002 (in Portuguese and Spanish) http://www.agualatinoamerica.com/docs/PDF/3-4-02basico.pdf (Acessado em 20/03/2008).
Types of filters (2011). http://water.me.vccs.edu/concepts/filters.html
UNFCCC/CCNUCC (1996). Approved Baseline and Monitoring Methodology AM0039: Methane Emissions Reduction from Organic Waste Water and Bioorganic Solid Waste Using Composting.
United Nations (2008a). National greenhouse gas inventory data for the period 1990–2006, FCCC/SBI/2008/12.
United Nations (2008b). Synthesis and Assessment Report on the Greenhouse Gas Inventories submitted in 2008, FCCC/WEB/SAI/2008 21 July 2008.
Uribe J. H. Caracterización de los residuos líquidos domésticos de la ciudad de Medellín (Characterization of the liquid domestic wastewater of the city of Medellin). Empresas Publicas de Medellin, Medellin, Colombia.
USEPA (1993). Constructed Wetlands for Wastewater Treatment and Wildlife Habitat: 17 Case Studies, EPA832-R-93-005.
Vall P., Michel T. and Rituerto J. M. (2007). "A Radical Approach to Industrial Wastewater", Water 21.
Van Handel A. C. and Lettinga G. (1994). Anaerobic Treatment of Wastewater. Campina Grande, Brazil.
Vargas-Ramirez M. and Lampoglia T. C. (2006). Scaling-up access to sewerage services using condominial technology. *Waterlines*, **25**(2), 11–13.
Von Sperling M. (2002). Lagoas de Estabilizacao (Stabilization Lagoons). DESA-UFMG, Capa Editora UFMG, Brazil, (in Portuguese), pp. 61–69, 145.
Von Sperling M. and Oliveira S. C. (2007). Challenges in the Implementation and Operation of Secondary Wastewater Treatment. A comparison between actual and expected performance in Brazil, Department of Sanitary and Environmental Engineering – Federal University of Minas Gerais; Belo Horizonte, Brazil.
Wagner J. (2001). Membrane Filtration Handbook – Practical Tips and Hints. Osmonics Inc. p. 16.
Wahl T. (1995). Hydraulic Testing of Static Self-Cleaning Inclined Screens, The First International Conference on Water Resources Engineering, ASCE, USA.
Wahl T. and Einhellig R. F. (2000). Laboratory testing and numerical modeling of coanda effect screens. Joint Conference on Water Resources Engineering and Water Resources Planning & Management, July 30–August 2, 2000—Minneapolis, Minnesota, USA: http://www.usbr.gov/pmts/hydraulics_lab/pubs/PAP/PAP-0841.pdf.
Wallace S. (2007). The Wetland Wastewater Alternative, Water 21 February 2007, pp. 38–40.
Wehner J. F. and Wilhelm R. H. (1958). Boundary Conditions of Flow Reactor. Chemical Engineering Science, New York, Volume **6**, p. 89.
WHO (2003). Guidelines for Safe Recreational Water Environments, Vol. **12**, Coastal and Fresh Waters, World Health Organization, Geneva, Switzerland, p. 80 (http://www.who.int/water_sanitation_health/bathing/srwe1/en/).
Wikibooks (2007). "Ingeniería de Aguas Residuales/Tratamiento Físico-químico" (Wastewater Engineering/ Physico-Chemical Treatment). http://es.wikibooks.org/wiki/IngenierC3ADa_de_aguas_residuales/ Tratamiento_fC3ADsico-quC3ADmico
Wikilibros (2007). Wastewater Engineering/Design of anaerobic digestion processes. http://es.wikibooks.org/wiki/Ingenier%C3%ADa_de_aguas_residuales/Dise%C3%B1o_de_procesos_en_digesti%C3%B3n_anaerobia#6._REACTOR_DE_FLUJO_SUSPENDIDO_.28UASB.29
Wikipedia (2007). http://es.wikipedia.org/wiki/Efecto_Coanda
Wright H. B. and Caims W. L. (2006). Rayos ultravioleta (Ultraviolet Rays). Trojan Technologies Inc., London, Ontario, Canada.

WSP – Water and Sanitation Program of the World Bank (2002). Condominial Water and Sewerage Systems – Costs of Implementing of the Model – Economic and Financial Evaluation, Washington DC, USA.

WSP – Water and Sanitation Program of the World Bank (2008). Constructed Wetlands: A Promising Wastewater Treatment System for Small Localities, Prepared by Martin Gauss, April 2008.

Yanez F. C. (1993). Lagunas de Estabilizaci6n (Stabilization Lagoons). ETAPA, Cuenca, Ecuador, (in Spanish), pp. 252–276.

Yochanan K., The Green House Effect and the Two Dimensional Pattern of Earth's Radiation Budget, Greehouse, pdf. Columbia University, Palisades, NY, USA.

Zonglian S., Xilai Z., Bairen Y., Chunji J. and Mengchun G. (2006). Granule development and performance in sucrose fed anaerobic baffled reactors. *Journal of Biotechnology*, **122**, 198–208.

Index

Note: t = Table, f = Figure

B